ABALONE OF THE WORLD

Biology, Fisheries and Culture

En vano hemos llegado
hemos brotado en la tierra
¿solo así he de irme?
¿como las flores que perecieron?
¿Nada quedará de mi nombre?
¿Nada de mi fama aquí en la tierra?
!Al menos flores, al menos cantos!

Ayocuan Cuetzpaltzin
Cantares Mexicanos
16th Century

To no purpose are we here,
We have sprung up from the earth,
Like the flowers that perish,
And like them we fade.
Neither name
Nor memory
Shall remain.
And yet,
And yet the flowers shall return,
And yet the songs.

Translation: Alan Cooper and Dinora Escobar

ABALONE OF THE WORLD

Biology, Fisheries and Culture

PROCEEDINGS OF THE 1ST INTERNATIONAL
SYMPOSIUM ON ABALONE

EDITED BY

S.A. SHEPHERD
Department of Fisheries, Adelaide, South Australia

MIA J. TEGNER
Scripps Institution of Oceanography, La Jolla, California, United States

S.A. GUZMAN DEL PROO
Instituto Politecnico Nacional, México City, México
Organiser of the Symposium held in La Paz, México

Fishing News Books

Fishing News Books
A division of Blackwell Science Ltd
Editorial Offices:
Osney Mead, Oxford OX2 0EL
25 John Street, London WC1N 2BL
23 Ainslie Place, Edinburgh EH3 6AJ
238 Main Street, Cambridge
Massachusetts 02142, USA
54 University Street, Carlton
Victoria 3053, Australia

Other Editorial Offices:
Arnette Blackwell SA
1, rue de Lille, 75007 Paris
France

Blackwell Wissenschafts-Verlag GmbH
Kurfürstendamm 57
10707 Berlin
Germany

Feldgasse 13, A-1238 Wien
Austria

First published 1992
Reprinted 1996

DISTRIBUTORS

Marston Book Services Ltd
PO Box 87
Oxford OX2 0DT
(*Orders*: Tel: 01865 206206
Fax: 01865 721205
Telex: 83355 MEDBOK G)

USA
Blackwell Science, Inc.
238 Main Street
Cambridge, MA 02142
(*Orders*: Tel: 800 759-6102
617 876-7000)

Canada
Oxford University Press
70 Wynford Drive
Don Mills
Ontario M3C 1J9
(*Orders*: Tel: 416 441-2941)

Australia
Blackwell Science Pty Ltd
54 University Street
Carlton, Victoria 3053
(*Orders*: Tel: 03 347-5552)

British Library
Cataloguing in Publication Data

International Symposium on Abalone (1st, 1989, La Paz, Mexico)
Abalone of the world: biology, fisheries and culture.
1. Gastropods
I. Title II. Shepherd, S.A. III. Tegner, Mia J. IV. Guzmán del Próo, S.A.
639.4832

ISBN 978-0-85238-181-6

Contents

PART IV FEEDING AND GROWTH

PART V DISEASES

PART VI FISHERY BIOLOGY

PART VII CASE STUDIES OF ABALONE FISHERIES

PART VIII STOCK ENHANCEMENT: INTRODUCTIONS

PART IX CULTURE

Preface

Abalone live on the coasts of every continent of the world, and have been known as a food delicacy since historic times. The first fisheries were in China and Japan over 1500 years ago, but it is only in the last 30 years that fisheries for abalone have burgeoned world-wide and become economically important in many countries. During this brief period research on abalone has likewise grown exponentially and, not surprisingly, much of our newly-acquired knowledge relates to management of the fisheries and lately to culture.

Yet increase in knowledge has not always been followed by good management. Some abalone stocks have remained stable, while others have declined seriously through overfishing, and now apparently disease. Too often, management has limped along, sadly inadequate to meet the challenges posed by increasing fishing pressure, environmental problems and the economic needs of fishermen. Past failures, however, do not mean that fisheries science has exhausted its potential for creativity or its capacity for change. On the contrary, we believe that a revolution is taking place in, for example, knowledge of population processes and modelling those processes leading to better approaches, and providing better tools, for management.

The culture of abalone has also made great advances, most remarkably in Japan, and now techniques, often with local adaptations, are developing in many countries. We predict that the culture of abalone will expand to production levels that will eventually rival or exceed production from wild stocks.

In this climate of rapid change and expansion the time is ripe to synthesize existing knowledge of abalone and to consider where we should go from here. This book originated from the First International Symposium on Abalone Biology, Fisheries and Culture, which was held in La Paz, Mexico, in November 1989. In addition to papers submitted at the symposium we solicited reviews of current topics of importance in fisheries biology and accounts of the history of management of nearly every major abalone fishery in the world. We believe that such a synthesis is necessary for progress, and that much can be learned from these histories. We hope that this will be an ongoing process, and to this end a second abalone symposium is tentatively scheduled for 1993 in Australia.

We would have liked to present a coherent theory for managing every abalone fishery, but that is unattainable because each fishery exists in its own unique matrix of economic and social conditions. The best we can hope to do is, like John the Baptist, to point researchers and managers in the right direction.

This volume should appeal to a diverse audience: ecologists will be interested in the life history papers and reviews, fishery biologists in population processes and modelling, managers in the accounts of fisheries, abalone divers in the complex problems that they are a part of, malacologists in the wealth of accumulated knowledge about a single genus, *Haliotis*, and finally aquaculturists in the developments in nutrition and artificial diets.

Above all, we hope the book will prove useful to all those concerned with the conservation and management of the natural resources of coastal waters. Such concern needs to be matched by an understanding of the component species and their population processes. If this book promotes this understanding its editors and contributors will be well pleased.

The Editors

Acknowledgements

The First International Symposium on Abalone Biology, Fisheries and Culture, held in La Paz, Mexico, on 21–25 November 1989, was supported financially by the governments of Mexico and Australia under their bilateral agreement on Science and Technology.

Sergio Guzmán del Próo assumed the onerous responsibility for organizing the symposium in La Paz; the astonishing success of the symposium attests to the enthusiasm and energy which he devoted to the task.

Many organizations contributed financially and many individuals worked to ensure the success of the meetings. The principal supporting institutions and organizers were: The Escuela Nacional de Ciencias Biologicas, the Centro Interdisciplinario de Ciencias Marinas del Instituto Politécnico Nacional, the Instituto Nacional de Pesca, the Centro Regional Investigación Pesquera de La Paz de la Secretaria de Pesca and the Fisheries Department of South Australia.

These organizations gave financial or logistic support: Consejo del Sistema Nacional de Educación Tecnológica (COSNET, SEP), Consejo Nacional de Ciencia y Tecnología (CONACYT), Secretaria de Pesca, Instituto Politecnico Nacional, Gobierno del Estado de Baja California Sur, Universidad Autónoma de Baja California Sur, CET del Mar, Centro de Investigaciones Biologicas and Federación Regional de Cooperativas Pesqueras de Baja California, Delegación de Pesca. Their appreciation of the significance of the event was crucial to its success.

Many diverse and onerous duties relating to the symposium were carried out by Cruz del Carmen Juarez, Rogelio Gonzalez, Marco Sanchez H., Rene Torres V., Hilario Perez F., Gilberto Leon, M.A. Reinecke, Filiberto Pena, Ma Estela Parra and Patricia Troyo. We are also indebted to Guadalupe Rochin, Gloria Romero, Lorena Arenas, R. Isela Ruiz, Pila Leon, Gilda Michel and Alma Ley who typed the abstracts and other symposium works with efficiency and enthusiasm.

The following persons reviewed papers in English submitted for publication: N. Andrew, F. Azam, P.A. Breen, S.M. Clarke, B. Campbell, F. Conte, G.E. Davis, R.W. Day, J.D. DeMartini, M. Denny, E.E. Ebert, J. Estes, S. Farlinger, W. Fenical, C.S. Friedman, J. Geibel, J.L. Greenier, P.L. Haaker, M. Hadfield, K.O. Hahn, W.S. Hearn, G.D. Hamer, P. Hone, R. Hilborn, K.A. Karpov, M. Keough, G. Kirby, D. Leighton, C.L. Lewis, D.R. Lindberg, S. McBride, T. McCormick, D.T. Manahan, W.J. Nash, K. Norman-Boudreau, P.E. McShane, N. Murray, D.O. Parker, G. Rohan, R. Sainz, D.R. Schiel, A. Smith, A. Staniford, P. Sluczanowski, B.N. Tissot, V. Vacquier, G. Van Blaricom and F. Winter.

The Mexican papers were reviewed by: Alfonso Maeda, Rodolfo Ramirez, Ma Luisa Sevilla, Ana Maria Escofet, Hector Fernandez, Fernando Martinez, Ernesto A. Chavez O. and Martin Ortiz Q.

Abstracts were translated into Spanish by: Benjamin Anguas, Enrique Gonzalez, Sara de la Campa de G., Andres Levy P., Marco Sanchez Hidalgo, Sergio Guzmán del Próo and Dinora Escobar.

The California Sea Grant College Program and the Australian Department of Industry, Technology and Commerce generously provided funds for editing and retyping some English language contributions. Lastly, South Australian Department of Fisheries staff Amanda Edwards, Colleen Doley, Colleen Lower and Robyn Matthews bore the brunt of the typing; Dr Fernando Arbolida and Estrella Atong (both of Sultan Qaboos University, Oman) proof-read the Spanish papers and typed the index respectively. We especially thank Dinora Escobar for carefully checking the manuscripts.

Keynote address

This symposium was the culmination of a four-year cooperative research programme between Australia and Mexico.

By 1984 the Mexican abalone fishery had collapsed from an annual production of 5000 t to a low of 400 t. In contrast, the Australian abalone fishery, despite increasing fishing pressure, has remained stable, although biologists studying the fishery believe that, if warning signs are ignored, it too might follow the same path as the Mexican fishery. The conviction that biologists and managers in Australia and Mexico might profitably participate in a common forum to exchange experiences on research and management prompted my Australian colleague Dr Shepherd and me to develop first a research programme and then this symposium, both funded by the Australia–Mexico agreement on Science and Technology.

Biological studies of abalone began in the nineteenth century with the works of the French anatomists Lacaze-Duthiers in 1859 and Wegmann in 1884. The next important work was that of Doris Crofts in 1929 in the notable series of *Memoirs of the Liverpool Marine Biology Committee*. From Europe the scene of research shifted to Japan, with the classic work of Dr Takashi Ino in 1952. In 1962 Keith Cox produced his well known study on California abalone. The remaining history of abalone research is contemporary, and nearly all papers published since then have been written by authors still living today. Many of them are now here in La Paz. I am especially glad to welcome Dr D.L. Leighton, who pioneered much research on the larval stage of Californian species and is present with us.

It is a happy circumstance that biologists from all the main abalone-producing countries of the world are present at this meeting. Here we have the first opportunity ever to compare the biology of our different species and to consider the unique problems each country faces in management. We have a splendid opportunity to analyse the reasons for the collapse of some stocks, and to seek ways of bringing about their recovery. The environment is favourable for the discussion of, for example, new approaches to management and questions about the population dynamics and ecology of different species.

Mexico desperately needs to develop new ideas and a new focus for our research so that we can restore the natural productivity of our abalone resource in Baja California. In spite of its severe decline, abalone are still important to the economy of coastal fishing villages and their inhabitants.

A thorough review of all the work done to the present, the application of new modelling techniques to evaluate the stocks, a fresh assessment of the optimum yield, an examination of the use of quotas as an alternative management strategy to avoid overfishing, an evaluation of juvenile seeding for stock enhancement – these are some of the matters that will be keenly probed in this symposium.

There are some population processes that are still poorly known for abalone. They include: larval settlement and juvenile recruitment and their relation with the abundance of the parent stock. Dr J.D. Prince has proposed new hypotheses that challenge current paradigms about what constitutes a unit stock and the extent of larval transport in the sea. Discussion of these issues is fundamentally important because they are initial inputs for new models of the population dynamics of abalone.

The undoubted advances in Japan and the United States in artificial culture seem to many to offer a panacea to the problem of overfishing in Mexico. The technology has stimulated the construction of many hatcheries in Mexico, but unfortunately we lack the scientific infrastructure and funding so necessary for their success. Early optimism has been shown to be illusory. Nevertheless, analysis of the successes and problems of abalone culture in Japan and the United States and assessment of the potential for stock enhancement, or even for production of small abalone for the cocktail market – these are all matters of great interest in Mexico. But even some of the basic biological questions, such as the natural diet and diseases of Mexican species, are still unanswered and provide a rich field for discussion. Japanese and North American experience is certain to be valuable both for Mexico and for those countries which are commencing aquaculture.

This symposium concludes with two days of workshops and I invite everyone to participate in them, especially in the discussions with my Mexican colleagues about the Mexican abalone fishery and aquaculture in Baja California. We are determined to critically assess research and management of our abalone fishery. I do not imagine that solutions and formulae for success will magically appear. Yet the exposure to other experiences and constructive criticism will shake any complacency and compel us to look at the problems in different ways.

Finally I must express my grateful appreciation to my colleagues of CICIMAR and my own institution, The National School of Biological Sciences of the National Polytechnical Institute; to the members of the organizing committee of the symposium, especially the National Institute of Fisheries, who have enthusiastically contributed to preparing this symposium; to Drs S.A. Shepherd, M.J. Tegner and N. Uki, who have co-ordinated the organization in their own countries; to the various institutions in Mexico and Australia that have supported me financially; and lastly to all participants who are, after all, the principal actors in this great event.

S.A. Guzmán del Próo
La Paz, 21 November 1989

PART 1
EVOLUTION, DISTRIBUTION AND GENETICS

Chapter 1
Evolution, distribution and systematics of Haliotidae

DAVID R. LINDBERG *Museum of Paleontology, University of California, Berkeley, California 94720, USA*

ABSTRACT: Members of the Haliotidae are globally distributed in the modern ocean. Especially specious faunas are found in the tropical western Pacific, Australia, Japan, South Africa and along the coast of the northeastern Pacific margin. Bathymetrically, they occur from the intertidal to depths in excess of 400 m. In Japan, North America and South Africa, abalone species are commonly associated with large, brown algae.

Specimens referable to *Haliotis* first appear in the Cretaceous of California (about 70 Ma). Dubious taxa have also been described from the Cretaceous of Europe. Neogene fossil occurrences are common and taxa are known from the Miocene of Australia, New Zealand, Japan and North America. Abalone are placed in the taxon Pleurotomariacea. Members of this group have a dextrally coiled shell with marginal opening(s). The interior of the shell is typically nacreous, and the animal possesses a pair of bipectinate ctenidia, well-developed epipodial tentacles and a rhipidoglossate radula. Putative ancestors within the Pleurotomariacea are unrecognized among Paleozoic and Mesozoic taxa. However, some abalone-like characters are found in other taxa. For example, the Triassic Temnotropidae have auriform shells while members of the early Mesozoic Trochotomidae have single trema that appear analogous to the tremata found in abalone shells.

The genus *Haliotis* has been divided into about 15 subgenera. The characters on which these taxa are based include the ratio of shell to body size, shell sculpture, epipodial structures and the morphology of the tremata. The probability that these taxonomic groupings reflect ancestry is low. Reduction of shell size is known in many other gastropod groups and can clearly be a convergent character. Shell sculpture is notoriously labile in gastropods and the morphology of the tremata is undoubtedly subject to habitat conditions rather than phylogenetically constrained.

RESUMEN: Los Haliotidae están globalmente distribuidos en los oceános modernos. Faunas especialmente ricas en especies se encuentran en trópico Pacífico Oeste, Australia, Japón, Africa del Sur, y a lo largo de las costas del Pacífico Noreste. Batimétricamente se localizan desde la zona entre mareas hasta profundidades mayores de 400 m. En Japón, Norteamérica, y Africa del Sur, las especies de abulón están comunmente asociadas con algas gigantes. Especímenes atribuíbles a *Haliotis* aparecen en el Cretáceo de California, aproximadamente hace 70 milliones de años. También se han descrito taxas dudosos del Cretáceo en Europa. Los fósiles del Neoceno son comunes, y se conocen taxas del Mioceno de Australia. Nueva Zelanda, Japón y Norteamérica. Los abulones están situados en los Arqueogasterópoda taxon Pleurotomariacea. Los miembros de este grupo tienen una concha dextrógira con aberturas marginales. El interior de la concha es típicamente nacarado y el animal posee un par de ctenidios bipectinados, tentáculos epipodiales bien desarrollados y una rádula ripidoglosada. Los ancestros putativos dentro de los Pleurotomariacea no se reconocen entre los taxa del Paleozoico y Mesozoico. Sin embargo, algunos caracteres semejantes a aquellos de los abulones se encuentran otros taxa. Por ejemplo los Tremnotropidae del Triásico, tienen conchas auriformes mientras que los miembros los Trochotomidae del Mesozoico temprano tienen tremata sencillos que parecen análogos a los que se encuentran en las conchas de abulón. El Género *Haliotis* está dividido en mas de 15 subgéneros. Los caracteres sobre los cuales están basados estos taxa incluyen la proporción entre la concha y el tamaño del cuerpo, esculturaciones de la concha, estructuras apipodiales y la morfología del tremata. La probabilidad de que estos agrupamientos taxonómicos reflejen un ancestro es baja. La reducción del tamaño de la concha es bien conocida en otros grupos de gasterópodos y puede claramente ser un carácter convergente. La estructura de la concha es notoriamente lábil en gasterópodos y la morfológia de los tremata está indudablemente sujeta más a las condiciones del habitat que costreñida filogenéticamente.

INTRODUCTION

The last world-wide systematic treatment of *Haliotis* was undertaken 100 years ago by H.A. Pilsbry (1890). Since that time there have been numerous regional treatments in both the scientific and popular literature (e.g., Cotton, 1959; Yaron, 1983; Muller, 1984; Ubaldi, 1986) and several global checklists (e.g., Wagner & Abbott, 1967; Pickry, 1980). But for the most part, our knowledge of the evolution and phylogeny of the genus *Haliotis* remains sketchy and poor, and the classification system wholly phenetic and of little use in recognizing relationships between the species. This is surprising given the commercial importance of these species in so many parts of the world.

A modern systematic revision is beyond the scope of this paper. Anatomical character states for most of the species are not known, and few molecular studies have been done. The following review has been prepared exclusively from the literature. Systematic arrangements and synonymies reflect the most recent regional reviews and checklists. The recent distribution of *Haliotis* species, especially in the temperate regions, and its geological history present intriguing patterns that demand further investigation. A phylogenetic study and its resulting classification would provide workers with a new tool for understanding the biology and history of this important group of marine molluscs.

Abbreviations used in text are as follows: CAS, California Academy of Sciences, San Francisco, California; LACM, Natural History Museum of Los Angeles County, Los Angeles, California; UCMP, Museum of Paleontology, University of California, Berkeley, California.

EVOLUTION

The ancestors of the Haliotidae are unknown. In most modern systematic treatments haliotids are grouped with the Pleurotomaridae and Scissurellidae in the taxon Pleurotomariacea. Members of this group are diagnosed by the presence of an excurrent opening along the margin of their shells, paired bipectinate ctenidia and the presence of a well-developed epipodium. Moreover, the gill and internal organs on the right side of the body are typically reduced in size (McLean, 1966). Haliotids are secondarily flattened species, that is their ancestors are assumed to have been tightly coiled. Secondary flattening is a common morphological expression in gastropod molluscs, and has convergently evolved in numerous lineages including the patellogastropods, vetigastropods, caenogastropods and opisthobranchs. In some taxa secondary flattening has little effect on internal anatomy and shared derived anatomical characters can still be used to argue affinities between coiled and limpet-like taxa. However, the haliotids, like other pleurotomariaceans, are highly modified and their derivation and relationships are not known (Hickman, 1984). The characters that the pleurotomariaceans do share are all primitive (plesiomorphic), and therefore are of little use in reconstructing phylogeny.

There are some extinct taxa that have abalone-like characters. For example, the Triassic Temnotropidae have auriform shells while members of the early Mesozoic Trochotomidae have single trema that appear analogous to the tremata found in abalone shells. However, presently there is no way to determine whether these similarities result from convergence or common ancestry.

Although the fossil record of the pleurotomariaceans dates from the Cambrian (Hickman, 1984), haliotids first appear in the Cretaceous of California (Anderson, 1902). This fossil species, *H. lomaensis* Anderson, 1902 (Fig. 1.1(a)), has all the shell characters seen in Recent haliotid species and certainly does not represent an intermediate or an early morphological stage in the evolution of the genus. Although over 66 Ma old, the specimen is so similar to Recent taxa that Durham (1979a) was able to assign this species to the Recent subgenus *Paua*. Cretaceous abalones have also been reported from Europe (e.g., *H. antiqua* Binkhorst, 1861; *H. cretacea* Lundgren, 1894), but these specimens are suspect and several authors have suggested that these specimens are actually portions of the outer whorls of large coiled gastropods (Kaunhowen, 1898; Woodring 1931). The Cretaceous California haliotid occurs in sediments that formed offshore and the species was probably transported from a shallower, neashore habitat; other associated fauna also suggest a mix of nearshore and offshore taxa (Durham, 1979b). Included among these nearshore taxa, and stratigraphically above *H. lomaensis*, is the rudistid bivalve *Coralliochama orcutti*. During the Cretaceous, coral–algal communities virtually disappeared from shallow-water, high-energy reef sites in the Tethys, and during the Upper Cretaceous rudistid bivalves formed most of the reef frameworks (Kauffman & Johnson, 1988). Thus, *Haliotis* may have evolved as a component of rudistid-dominated communities during the Cretaceous and subsequently became associated with coral reef communities as they became re-established following the extinction of rudistids at the end of the Mesozoic.

Fossil abalone are poorly represented in the Paleogene (Paleocene to Oligocene; 66.4–23.7 Ma). There is a single report of a *Haliotis* species from the Eocene-Oligocene of Australia (Hertlein, 1937), but it has not been repeated in subsequent palaeontological treatments (Cox, 1960). However, beginning in the Miocene, haliotids are found throughout the world including Europe, North America, Asia, Asia Minor, New Zealand, Australia and the Caribbean (Fig. 1.1(b)) (Hertlein, 1937; Cox, 1960). An especially rich fossil record is found around the Pacific Rim (Table 1.1). The absence of abalone in the Paleogene fossil record is difficult to evaluate. Many abalone live in shallow, high-energy environments where taphonomic processes tend to destroy shells. The shells are nacreous and highly susceptible to disintegration.

Nakamura (1986), studying chromosome number and karyotypes of Japanese haliotids, made several observations regarding abalone evolution. Nakamura pointed out that although chromosome number is often thought to be conservative within a particular taxonomic group (Burch, 1961), members of the genus *Haliotis* have been reported to have 14, 16 or 18 pairs of chromosomes. Moreover, species with identical chromosome numbers could also have different karyotypes (e.g., *H. varia* and *H. japonica*). In contrast, Nakamura found that the Japanese species

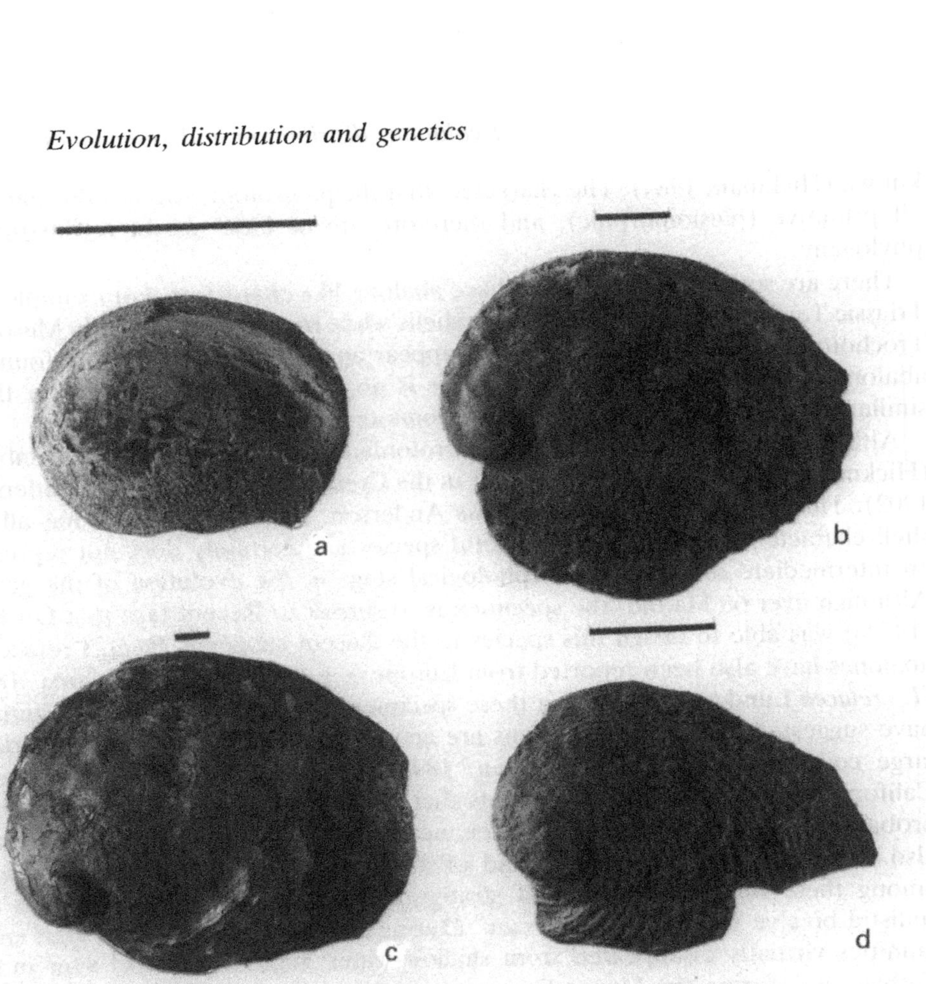

Fig. 1.1 California fossil abalone. Scale bars = 10 mm. (a) *Haliotis lomaensis* Anderson, 1902. Cretaceous, Pt Loma, San Diego County, California. CAS type No. 61772.01. (b) *Haliotis palaea* Woodring, 1931. Miocene, Santa Monica Mountains, Los Angeles County, California. LACM type No. 4156. (c) *Haliotis assimilis* Dall, 1878. Pliocene, Arroyo Drive Quarry, San Diego County, California. LACM Loc. No. 107. (d) *Haliotis pourtalesi* Dall, 1881. Pliocene, San Gorgonio Pass, Riverside County, California. UCMP type No. 16074.

H. discus and the North American species *H. cracherodii* had identical chromosome numbers and karyotypes. The lowest number of chromosomes (14 pairs) was reported in the Atlantic/Mediterranean species *H. tuberculata* and its synonyms (Colombera & Tagliaferri, 1983). Based on the patterns he observed, Nakamura (1986) suggested that haliotids evolved quickly because of chromosomal alteration (saltational chromosome speciation, Lewis, 1962; White, 1973) and that karyotypic transformation took place at the time of origin and thus the current karyotypes do not reflect the ancestral condition.

Although taxonomists have been able to name between 60 and 70 species of *Haliotis* (Table 1.2), the speciation events that supposedly produced this relatively large group have left some tenuous boundaries between the 'species'. This is reflected in the taxonomy of the group. For example, although the subspecies category (morphologically or electrophoretically distinct, allopatric forms) is seldom invoked for nearshore marine molluscs with external fertilization and planktonic

Table 1.1 Fossil Pacific Rim *Haliotis* species. Fossil *Haliotis* have also been reported from Europe, Cyprus, Asia Minor (Hertlein, 1937) and the Caribbean (Jung, 1968).

Taxon	Period or epoch	Locality
**H. lomaensis*	Cretaceous	California, USA
H. kurosakiensis	Miocene	Aomori Pref, Japan
**H. kochibei* (= *H. gigantea*)	Miocene	Ibaragi Pref, Japan
	Pliocene	Ishikawa Pref, Japan
H. kamtschatkana glabrosa	Miocene	Shizuoka Pref, Japan
H. kamtschatkana koyamai	Miocene	Nagano Pref, Japan
H. notoensis	Miocene	Ishikawa Pref, Japan
H. moniwaensis	Miocene	Miyagi Pref, Japan
H. fujioka	Miocene	Miyagi Pref, Japan
**H. japonica*	Miocene	Miyagi Pref, Japan
H. lasia	Miocene	California, USA
H. palaea	Miocene	California, USA
H. flemingi	Miocene	Kawau Island, New Zealand
	Pleistocene	Whakatane, New Zealand
H. waitemataensis	Miocene	Kawau Island, New Zealand
H. naevosoides	Miocene	Flemington & Victoria, Australia
	Pliocene	Adelaide, Australia
H. mooraboolensis	Miocene	Moorabool River, Australia
H. ovinoides	Miocene	Victoria, Australia
H. gigantoides	Pliocene	Okinawa Pref, Japan
H. kokei	Pliocene	Kanagawa Pref, Japan
**H. diversicolor*	Pliocene and Pleistocene	Okinawa Pref, Japan
H. santacruzensis	Pliocene	Galápagos Islands
H. elsmerensis	Pliocene	California, USA
H. pourtalesi[1]	Pliocene	California, USA
**H. fulgens*	Pliocene	California, USA
**H. cracherodii*	Pliocene	California, USA
**H. rufescens*	Pliocene	California, USA
**H. corrugata*	Pliocene	California, USA
**H. assimilis*	Pliocene	California, USA
H. powelli	Pleistocene	Bay of Plenty, New Zealand
**H. cyclobates*	Pleistocene	Fleurieu Peninsula, Australia
**H. emmae*	Pleistocene	Victoria, Australia
**H. laevigata*	Pleistocene	Victoria, Australia
**H. rubra*	Pleistocene	Victoria, Australia

[1] Also Pleistocene Barbados (Jung, 1968).
* Extant species.

Table 1.2 Extant abalone species; this list was compiled from available regional checklists (see references); synonyms are listed after each species when applicable.

Haliotidae Rafinesque, 1815
Haliotis Linnaeus, 1758

H. asinina Linnaeus, 1758

H. australis Gmelin, 1791
  H. aleata Röding, 1798

H. barbouri Foster, 1946

H. brazieri Angas, 1869

H. coccinea Reeve, 1846
  H. janus Reeve, 1846

H. coccoradiata Reeve, 1846

H. corrugata Wood, 1828
  H. nodosa Philippi, 1845

Table 1.2 *Continued*

Haliotidae Rafinesque, 1815
Haliotis Linnaeus, 1758

- *H. cracherodii cracherodii* Leach, 1814
    - *H. interrupta* Valenciennes, 1832
- *H. cracherodii californiensis* Swainson, 1821
- *H. crebrisculpta* Sowerby, 1914
- *H. cyclobates* Péron & Lesueur, 1816
    - *H. excavata* Lamarck, 1822
- *H. dalli* Henderson, 1915
- *H. discus discus* Reeve, 1846
- *H. discus hannai* Ino, 1953
- *H. dissona* Iredale, 1929
- *H. diversicolor* Reeve, 1846
    - *H. tayloriana* Reeve, 1846
    - *H. gruneri* Philippi, 1848
    - *H. supertexta* Lischke, 1870
- *H. dohrniana* Dunker, 1882
- *H. elegans* Philippi, 1848
- *H. exigua* Dunker, 1877
- *H. fulgens fulgens* Philippi, 1845
    - *H. splendens* Reeve, 1846
    - *H. planilirata* Reeve, 1846
- *H. fulgens turvei* Bartsch, 1942
- *H. fulgens guadalupensis* Talmadge, 1964
- *H. gigantea* Gmelin, 1791
    - *H. tubifera* Lamark, 1822
    - *H. gigas* Röding, 1798
- *H. glabra* Gmelin 1791
    - *H. ziczac* Reeve, 1846
    - *H. picta* Röding, 1798
- *H. guineensis* Gmelin, 1791
    - *H. rosacea* Reeve, 1846
    - *H. decussata* Philippi, 1850
    - *H. virginea* Reeve, 1846
- *H. hanlevi* Ancey, 1881
- *H. hargravesi* Cox, 1869
- *H. howensis* Iredale, 1929
- *H. iris* Gmelin, 1791
- *H. jacnensis* Reeve, 1846
    - *H. echinata* Sowerby, 1883
- *H. japonica* Reeve, 1846
    - *H. aquatilis* Reeve, 1846
    - *H. incisa* Reeve, 1846
- *H. kamtschatkana kamtschatkana* Jonas, 1845
- *H. kamtschatkana assimilis* Dall, 1878
    - *H. aulaea* Bartsch, 1940
    - *H. smithsoni* Bartsch, 1940
- *H. laevigata* Donovan, 1808
    - *H. albicans* Quoy & Gaimard, 1834
    - *H. excisa* Gray, 1856
- *H. mariae* Wood, 1928
- *H. melculus* Iredale, 1927
- *H. midae* Linnaeus, 1758
    - *H. capensis* Dunker, 1844
    - *H. elatior* Pilsbry, 1890
- *H. ovina* Gmelin, 1791
    - *H. caelata* Röding, 1798
    - *H. latilabris* Philippi, 1851
- *H. parva* Linnaeus, 1758
    - *H. canaliculata* Lamarck, 1822
    - *H. carinata* Swainson, 1822
    - *H. cingulata* Röding, 1798
    - *H. rubicunda* Röding, 1798
- *H. planata* Sowerby, 1833
- *H. pourtalesi* Dall, 1881
- *H. pustulata* Reeve, 1846
    - *H. alternata* Sowerby, 1882
    - *H. ancile* Reeve, 1846
    - *H. nebulata* Reeve, 1846
    - *H. pertusa* Reeve, 1846
    - *H. scutulum* Reeve, 1846
    - *H. revelata* Reeve, 1846
    - *H. zealandica* Reeve, 1846
    - *H. strigata* Weinkauff, 1883
- *H. queketti* Smith, 1910
- *H. roberti* McLean, 1970
- *H. roei* Gray, 1827
- *H. rubra rubra* Leach, 1814
    - *H. conicopora* Péron, 1816
    - *H. cunninghami* Gray, 1826
    - *H. granti* Pritchard & Gatliff, 1903
    - *H. improbulum* Iredale, 1924
    - *H. naevosa* Martyn, 1786
    - *H. vixlirata* Cotton, 1943
- *H. rubra clathrata* Reeve, 1815
- *H. rufescens* Swainson, 1822
    - *H. californiana* Valenciennes, 1832
    - *H. ponderosa* C.B. Adams 1848
- *H. scalaris* Leach, 1814
    - *H. emmae* Reeve, 1846
    - *H. rubicunda* Gray, 1846
    - *H. tricostalis* Lamark, 1822
    - *H. tricostata* Wood, 1828

H. semiplicata Menke, 1843
  H. lauta Reeve, 1846

H. sieboldii Reeve, 1846

H. sorenseni Bartsch, 1940

H. spadicea Donovan, 1808
  H. ficiformis Menke, 1844
  H. sanguinea Hanley, 1841

H. speciosa Reeve, 1846
  H. alfredensis Bartsch, 1915

H. squamata Reeve, 1846
  H. elevata Sowerby, 1883

H. tuberculata Linnaeus, 1758
  H. incisa Reeve, 1846
  H. bistriata Gmelin, 1791
  H. lamellosa Lamark, 1822
  H. lucida Requien, 1848
  H. reticulata Reeve, 1846
  H. rugosa Lamark, 1822
  H. vulgaris da Costa, 1778

H. unilateralis Lamark, 1822

H. varia varia Linnaeus, 1929
  H. concinna Reeve, 1846
  H. semistriata Reeve, 1846
  H. viridis Reeve, 1846
  H. varia pustulifera Pilsbry, 1890
  H. astricta Reeve, 1846
  H. granulata Röding, 1798
  H. papulata Reeve, 1846
  H. rubiginosa Reeve, 1846
H. varia stomatiaeformis Reeve, 1846
H. varia aliena Iredale, 1928

H. virginea virginea Gmelin, 1791
  H. subvirginea Weinkauff, 1833
H. virginea crispata Gould, 1847
H. virginea huttoni Filhol, 1880
H. virginea morioria Powell, 1938

H. walallensis Stearns, 1899

---

larvae, it is commonly proposed for members of the genus *Haliotis*, and in most cases, subspecies are associated with different temperature regimes (see below). From the tropical west Pacific, Talmadge (1963) described three subspecies of *H. varia*. In the northeastern part of the region (Okinawa–Fiji–Philippines), the subspecies *varia* was reported. In the central part of the region (Indonesia) *stomatiaeformis* was present, while in northern Australia the subspecies *aliena* occurred. Talmadge could find no sharp line demarcating these subspecies and proposed that they were interbreeding and intergrading where they came into juxtaposition. Electrophoretically distinct subspecies have been reported from Japan (*H. discus discus* (southern) and *H. discus hannai* (northern)) and like the tropical example, interbreeding at boundaries is known to occur (Chapter 37). Well-documented cases of subspecific taxa have also been described from Mexico, New Zealand and Australia.

Barriers to interbreeding or hybridization between *Haliotis* spp. within a region can also be tenuous. Owen *et al.* (1971) reported hybridization between all of the species present in southern California; of 15 potential combinations, only three could not be documented. The evidence for hybridization included shell and anatomical characters, morphometrics, biochemical evidence and cross-breeding experiments. Most of the hybrids were from areas of habitat disturbance.

Speciation events within haliotid lineages have produced regional radiations in both tropical and temperate faunas. The largest of the temperate radiations appears to have occurred in southern Australia, North America and Japan. The North American radiation has produced the largest abalone in the world; individuals of *H. rufescens* may exceed 290 mm in shell length (Chapter 26). In the tropical western Pacific there are over 10 species, all typically small (about 50 mm in shell length). Some tropical species are restricted to small island groups, such as *H. pulcherrima* in the Tuamotus, *H. roberti* in the Cocos Islands and *H. dalli* in the Galápagos Islands. Because of the lack of morphological characters to determine relationships within *Haliotis*, it is not possible at this time to determine if the tropical fauna represents a single radiation or several radiations of a single lineage, or an aggregation of different lineages which have experienced different speciation and extinction rates in the tropics.

Evidence from the temperate regions suggests that both processes have occurred. Probable aggregate faunas include Japan and Australia, whereas the New Zealand and North American faunas probably represent endemic radiations. Eleven species of abalone are known from Japan, however they are divisible into three distinct groups (Fig. 1.2). In the southernmost islands the abalone are members of the tropical group which reach their northern limit along the coast of Japan. In the central part of Japan there are two short-range endemics in the transitional zone between the warm and cold water faunas (see Newman (1979) for a discussion of transition zones and short-range endemics), and in the northern portion of the islands are cool temperate species that extend down from the north and reach the southern limit of their distributions along the Japanese coasts. The Australian and South African faunas also appear to be composed of provincial aggregates.

In contrast, in North America and New Zealand the faunas appear to be composed of only single-lineage radiations. Six of the seven North American species have ranges restricted to a 3000 km section of the warm temperate coast (Fig. 1.3); the only cool temperate species, *H. kamtschatkana*, also occurs in the cool temperate, western Pacific (see below). The North American species have well partitioned the nearshore environment and there tends to be little habitat overlap (Table 1.3) (McLean, 1966); similar habitat partitioning also occurs in the temperate Australian species (Shepherd, 1973, 1975). Although the North American species hybridize readily (Owen *et al.*, 1971), the fossil record suggests that these species (or their ancestors) have been extant and distinct in the region for over 5 million years (Fig. 1.1(c)). Hertlein (1937) pointed out that many of the Californian Miocene species were remarkably similar to the Recent ones. For example, *H. koticki* and *H. kamtschatkana*, *H. lasia* and *H. fulgens*, *H. palaea* and *H. corrugata* and *H. elsmerensis* and *H. rufescens*.

DISTRIBUTION

Recent members of the Haliotidae are globally distributed in the modern ocean (Table 1.4); there are no extralimital fossil occurrences of the genus. Talmadge

Table 1.3 Bathymetric distribution of North American *Haliotis* species; data from McLean, 1966.

Species	Depth (m)	Habitat
H. corrugata	5–20	Kelp forests
H. cracherodii	0–5	Intertidal
H. fulgens	0–10	*Eisenia* beds
H. kamtschatkana[1]	0–10	Coralline algae and kelp forest
H. rufescens[1]	0–20	Coralline algae and kelp forest
H. sorenseni	20–60	Below kelp beds
H. walallensis[1]	5–20	Kelp forests

[1] Species that submerge in the southern portion of their range.

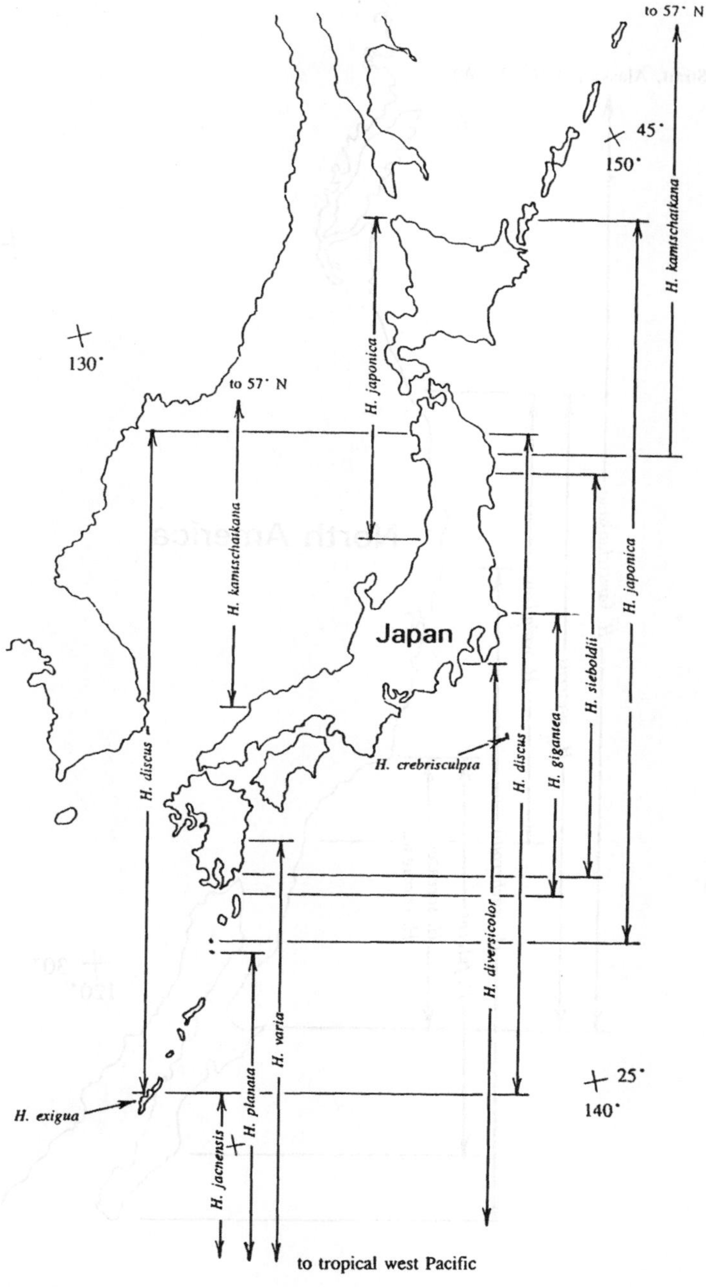

Fig. 1.2 Distribution of *Haliotis* species in Japan. Data from Kuroda & Habe (1952).

(1963) noted that the distributional patterns of Indo-Pacific species indicated four distinct Recent faunal areas: (1) East Africa and the Indian Ocean, (2) Australia–New Zealand, (3) Japan–China coasts and (4) the Pacific insular arc; he considered the Philippines–Indonesian archipelagos to be the source area of many of the species found in the Indo-Pacific region. Especially specious faunas are also found

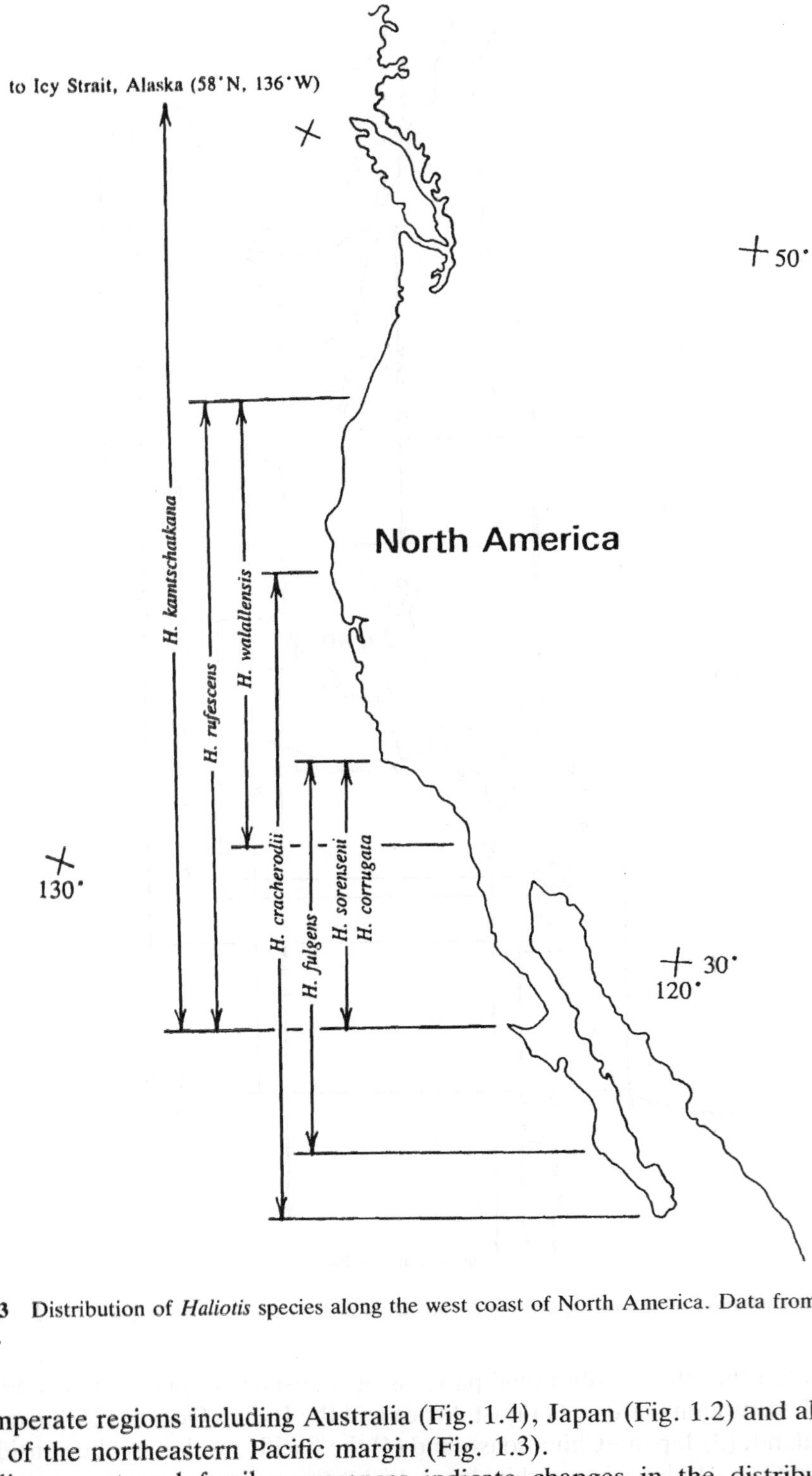

Fig. 1.3 Distribution of *Haliotis* species along the west coast of North America. Data from McLean (1966).

in temperate regions including Australia (Fig. 1.4), Japan (Fig. 1.2) and along the coast of the northeastern Pacific margin (Fig. 1.3).

Relict recent and fossil occurrences indicate changes in the distribution of several tropical species. For example, the Recent *H. crebrisculpta* has a markedly

Table 1.4 Diversity and mean size of members of the genus *Haliotis*. For regions that include several distinct climatic zones, the total number of species is followed by their distribution in each respective climatic zone.

Region	No of species	Mean size (mm)	Climatic zone
Europe/Africa	3	57	Warm temperate/subtropical
Southern Africa	5		
	3	103	Warm temperate
	2	51	Subtropical
Indian Ocean	3	50	Subtropical/tropical
Australia	15–17		
	5	58	Tropical
	7	124	Warm temperate
	4	84	Temperate
New Zealand	6	65	Warm temperate/temperate
West Pacific	10–13	47	Tropical
Japan	11		
	6	43	Warm temperate/subtropical
	5	125	Temperate
North America	7	142	Temperate
Panamic Caribbean	3	21	Tropical

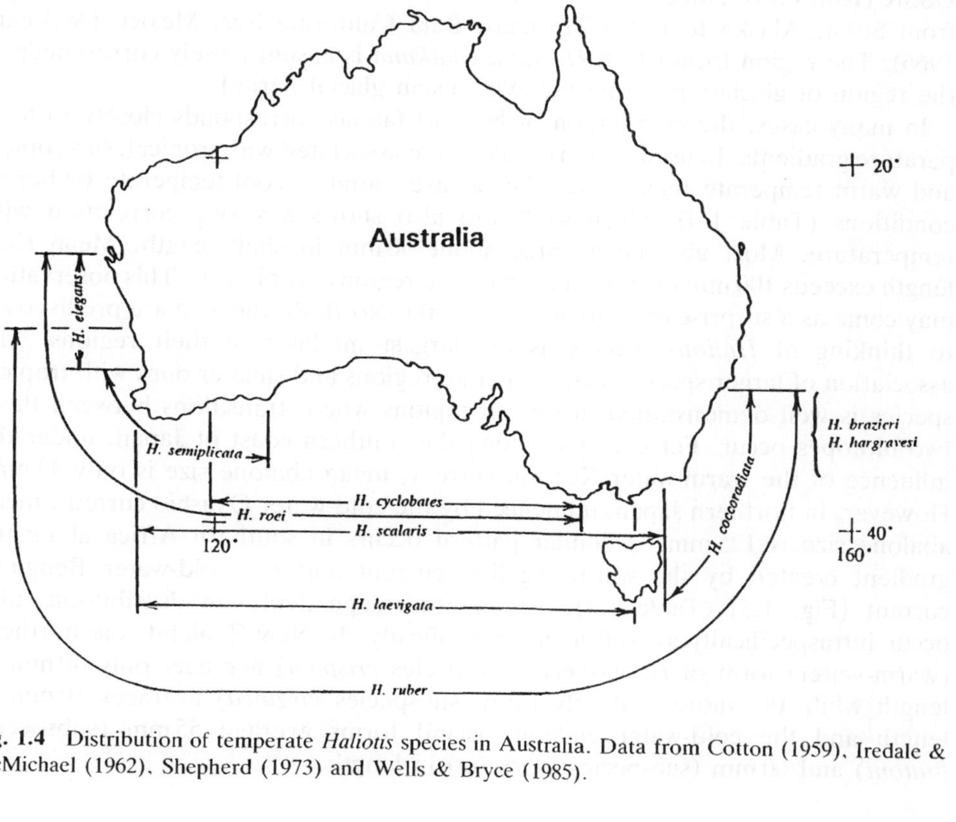

Fig. 1.4 Distribution of temperate *Haliotis* species in Australia. Data from Cotton (1959), Iredale & McMichael (1962), Shepherd (1973) and Wells & Bryce (1985).

disjunct distribution in the tropical west Pacific. In the southern region of the tropical west Pacific, the species is found in New Caledonia and along the coast of tropical Australia. To the north *H. crebrisculpta* occurs in the Hachijo Jima Group off the coast of southern Japan; there are no intermediate records (Talmadge, 1963).

In the tropical eastern Pacific and Caribbean there occurs a complex of three species which appear closely related: *H. pourtalesi* of the Florida Keys, *H. roberti* at the Cocos Islands and *H. dalli* at the Galápagos Islands. A fourth fossil species, *H. santacruzensis*, is known from the Pliocene of the Galápagos. Although insular and divided by Central America today, the group was presented in the early Pliocene of California. Schremp (1981) has reported two specimens of *H. pourtalesi* from the warm-water Imperial formation in southern California (Fig. 1.1(d)). Thus, it appears that this group of small (<20 mm) tropical abalone was probably distributed along the mainland of North America prior to the closing of the Panamic portal about 5 Ma ago. Moreover, because the Galápagos Islands are also less than 4 Ma old (Hickman & Lipps, 1985) and have never been connected to the mainland, the arrival of an abalone species in the archipelago also suggests a larger distribution of this group during the late Neogene. Local extinctions of this group on the North American mainland combined with changes in the regional tectonic setting have left a patchy distribution in the tropical eastern Pacific and Caribbean.

In the North Pacific *H. kamtschatkana* shows a prominent disjunct distribution. In the western Pacific *H. kamtschatkana* extends from Honshu, Japan to Kamchatka, USSR (Kuroda & Habe, 1952). In the eastern Pacific, *H. kamtschatkana* ranges from Sitka, Alaska to Bahía Tortugas, Baja California Sur, Mexico (McLean, 1966). The region from which *H. kamtschatkana* is absent closely corresponds to the region of glaciation during the Wisconsian glacial period.

In many cases, the distribution of haliotid faunas corresponds closely to temperature gradients. In most regions abalone are associated with tropical, subtropical and warm temperate regions; few faunas are found in cool-temperate or boreal conditions (Table 1.4). Mean shell size also shows a strong correlation with temperature. Most abalone average about 50 mm in shell length. Mean shell length exceeds 100 mm in only four temperate regions (Table 1.4). This observation may come as a surprise to many workers in the North Pacific who are predisposed to thinking of *Haliotis* species as the largest molluscs in their regions. The association of larger species with temperate regions and smaller ones with tropical species is well demonstrated in several regions where transitions between these two biotopes occur. For example, along the southern coast of Japan, under the influence of the warm-water Kuroshi current, mean abalone size is only 43 mm. However, in northern Japan, influenced by the cold-water Oyashio current, mean abalone size is 125 mm. A similar pattern occurs in southern Africa along the gradient created by the warm Agulhas current and the cold-water Benguela current (Fig. 1.5) (Table 1.4). Temperature-dependent size distributions also occur intraspecifically as well as interspecifically. In New Zealand, the northern (warm-water) form of *H. virginea* (subspecies *crispata*) averages only 40 mm in length while the more southerly form (subspecies *virginea*) averages 50 mm in length and the cold-water endemic island forms average 55 mm (subspecies *huttoni*) and 60 mm (subspecies *morioria*) in length.

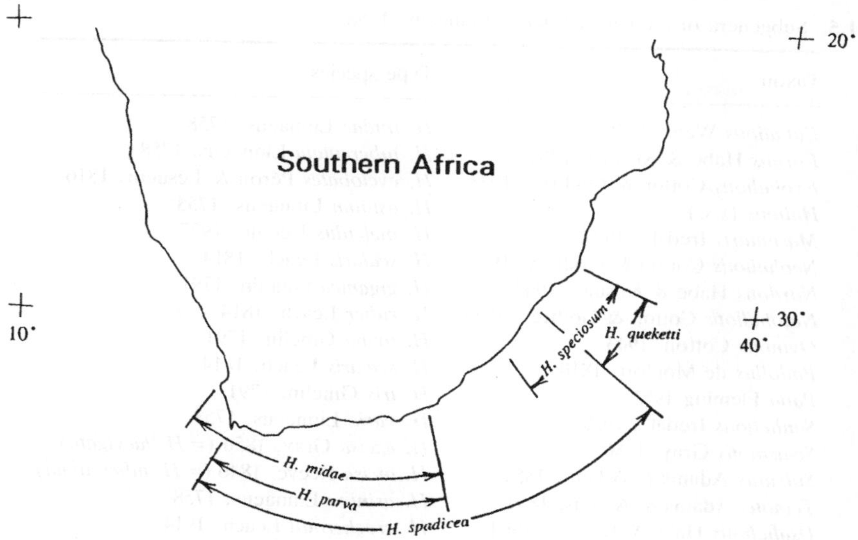

Fig. 1.5 Distribution of *Haliotis* species in southern Africa. Data from Kensley (1973).

Bathymetrically, abalone are distributed from the intertidal to depths in excess of 400 m (Foster, 1946). In southern Australia most abalone occur from the intertidal to about 20 m; *H. scalaris* occurs to depths of 50 m; similar bathymetric distributions occur in southern Africa and in New Zealand. Most of the North American species also occur in less than 20 m of water (Table 1.3).The bathymetric distributions of temperate species often submerge when their geographical distributions extend into warmer environs. These patterns suggest that despite the likely Tethyan origin of the genus, groups that have subsequently invaded temperate regions are restricted by warmer water temperatures.

Although tropical species typically occur in shallow reefal communities, the tropical New World species are exceptional in that they are usually found at depths between 50 and 200 m. Tropical species are generally associated with coral and stone substrates (Talmadge, 1974). This habitat has probably changed little since the origin of the group, except perhaps for the replacement of the rudist bivalves by hermatypic corals in the reef community (Kauffman & Johnson, 1988).

In temperate environs abalone are associated with rock and boulder substrates, often in high-energy environments; crevices and caves are common microhabitats. In Japan, North America and southern Africa, many abalone species are associated with large, brown algae such as *Ecklonia*, *Nereocystis*, *Macrocystis*, *Laminaria* and *Eisenia*. In southern Australia *H. cyclobates* is found in seagrass habitats.

SYSTEMATICS

There has not been a comprehensive, world-wide, systematic treatment of the genus *Haliotis* since Pilsbry (1890). Although there have been numerous regional treatments, a classification of the systematic relationships below the level of the

Table 1.5 Subgenera of the genus *Haliotis* Linnaeus, 1758.

Taxon	Type species
Euhaliotis Wenz, 1838	*H. midae* Linnaeus, 1758
Eurotis Habe & Kosuge, 1964	*H. tuberculata* Linnaeus, 1758
Exohaliotis Cotton & Godfrey, 1933	*H. cyclobates* Péron & Lesueur, 1816
Haliotis (*s.s.*)	*H. asinina* Linnaeus, 1758
Marinauris Iredale, 1927	*H. melculus* Iredale, 1927
Neohaliotis Cotton & Godfrey, 1933	*H. scalaris* Leach, 1814
Nordotis Habe & Kosuge, 1964	*H. gigantea* Gmelin, 1791
Notohaliotis Cotton & Godfrey, 1933	*H. ruber* Leach, 1814
Ovinotis Cotton, 1943	*H. ovina* Gmelin, 1791
Padollus de Monfort, 1810	*H. scalaris* Leach, 1814
Paua Fleming 1952	*H. iris* Gmelin, 1791
Sanhaliotis Iredale, 1929	*H. varia* Linnaeus, 1758
Schismotis Gray, 1856	*H. excisa* Gray, 1856 (= *H. laevigata*)
Sulculus Adams & Adams, 1854	*H. incisa* Reeve, 1846 (= *H. tuberculata*)
Teinotis Adams & Adams, 1854	*H. asinina* Linnaeus, 1758
Usahaliotis Habe & Kosuge, 1964	*H. cracherodii* Leach, 1914

genus is unavailable at this time. There is little doubt that members of the Haliotidae are monophyletic (all share a common ancestor), but relationships below this taxonomic level are unknown.

The genus *Haliotis* contains approximately 70 species (Table 1.2) and has been divided into over 15 subgenera (Table 1.5). The characters on which these taxa are based include the ratio of shell to body size, shell sculpture, epipodial structures and the morphology of the tremata. The probability that these taxonomic groupings reflect ancestry is low. Reduction of shell size is known in many other gastropod groups and can clearly be a convergent character. Shell sculpture is notoriously labile in gastropods and the morphology of the tremata is undoubtedly subject to habitat conditions rather than phylogenetically constrained. Until new characters are studied and their distributions among species determined, the division of the genus *Haliotis* into lower taxa is fallacious.

Many of the subgenera have been elevated to generic rank by some Australian and Japanese workers. Given the questionable validity and usefulness of these taxa, this action seems premature.

However, the haliotid radula has not been sufficiently studied for characters that may prove useful in diagnosing taxa within the family.

Cotton & Godfrey (1933) and more recently Owen *et al.* (1971) have used the structure of the epipodium to diagnose species. This structure is so conservative that hybrids have epipodial characters intermediate between those of the parent species (Owen *et al.*, 1971).

ACKNOWLEDGEMENTS

The author thanks G.E. Davis, S.A. Shepherd and M.J. Tegner for reviewing the manuscript. L. Kool (CAS) and G.L. Kennedy (LACM) provided specimens of *Haliotis* from the collections under their care for which the author is most grateful. This is contribution No 1530 from the Museum of Paleontology, University of California at Berkeley.

REFERENCES

Anderson F.M. (1902) Cretaceous deposits of the Pacific coast. *Proc. Calif. Acad. Sci.*, **2**, 1–154.

Burch J.B. (1961) Chromosomes of *Planorbis corneus*, with a discussion of chromosome numbers in snail systematics. *Basteria*, **25**(4/5), 45–52.

Colombera D. & Tagliaferri F. (1983) Chromosomes from male gonads of *Haliotis tubercolata* and *Haliotis lamellosa* (Haliotidae, Archaeogastropoda, Mollusca). *Caryologia*, **36**, 231–4.

Cotton B.C. (1959) *South Australian Mollusca Archaeogastropoda*. W.L. Hawes, Government Printer, Adelaide.

Cotton B.C. & Godfrey K. (1933) South Australian shells. Part 9. *S. Aust. Natural.*, **51**, 14–24.

Cox L.R. (1960) Systematic descriptions, family Haliotidae. In *Treatise on Invertebrate Paleontology, Part I, Mollusca 1* (Ed. by R.C. Moore), pp. I221–I223. Geological Society of America University of Kansas Press, Lawrence.

Durham J.W. (1979a) California's Cretaceous *Haliotis*. *Veliger*, **21**(3), 373–4.

Durham J.W. (1979b) A fossil *Haliotis* from the Galápagos Islands. *Veliger*, **21**(3) 369–72.

Foster R.W. (1946) The family Haliotidae in the western Atlantic. *Johnsonia*, **2**(21) 37–40.

Hertlein L.G. (1937) *Haliotis koticki*, a new species from the lower Miocene of California. *Bull. S. Calif. Acad. Sci.*, **36**(3) 93–7.

Hickman C.S. (1984) *Pleurotomaria*: pedigreed perseverance. In *Living Fossils* (Ed. by N. Eldredge & S.M. Stanley), pp. 225–31. Springer-Verlag, New York.

Hickman C.S. & Lipps J.H. (1985) Geologic youth of Galápagos Islands confirmed by marine stratigraphy and paleontology. *Science*, **227**, 1578–80.

Iredale T. & McMichael D.F. (1962) A reference list of the Marine Mollusca of New South Wales. *Aust. Mus. Sydney, Memoir*, **9**, 1–109.

Jung P. (1968) *Pleurotomaria* and *Haliotis* from Barbados and Carriacou, West Indies. *Ecol. Geol. Helv.*, **61**, 593–601.

Kauffman E.G. & Johnson C.C. (1988) The morphological and ecological evolution of Middle and Upper Cretaceous reef-building rudistids. *Palaios*, **3**, 194–216.

Kaunhowen F. (1898) Die Gastropoden der Maestrichter Kreide. *Paläont. Abh.*, **8**, 1–132.

Kensley B. (1973) *Sea Shells of Southern Africa. Gastropods*. M. Miller, Cape Town.

Kuroda T. & Habe T. (1952) *Check List and Bibliography of the Recent Marine Mollusca of Japan*. Hosokawa Printing, Tokyo.

Lewis H. (1962) Catastrophic selection as a factor in speciation. *Evolution*, **16**, 257–71.

McLean J.H. (1966) *West American prosobranch Gastropoda: superfamilies Patellacea, Pleurotomariacea, Fissurellacea*. PhD dissertation, Stanford University.

Muller S. (1984) Taxonomy of the genus *Haliotis* in South Africa. *Trans. R. Soc. S. Africa*, **44**(1) 69–77.

Nakamura H.K. (1986) Chromosomes of Archaeogastropoda (Mollusca: Prosobranchia), with some remarks on their cytotaxonomy and phylogeny. *Publ. Seto Mar. Biol. Lab.*, **31**(3/6) 191–267.

Newman W.A. (1979) California transition zone: significance of short range endemics. In *Historical Biogeography, Plate Tectonics, and the Changing Environment* (Ed. by J. Gray & A.J. Boucot), pp. 399–416. Oregon State University Press, Oregon.

Owen B., McLean J.H. & Meyer R.J. (1971) Hybridization in the eastern Pacific abalones. *Bull. Los Angeles County Mus. Nat. Hist. Sci.*, **9**, 1–37.

Pickry R. (1980) Checklist of the Haliotidae of the World. *Gloria Maris*, **19**(2), 23–32.

Pilsbry H.A. (1890) *Manual of Conchology*, Vol. XII. Academy of Natural Sciences, Philadelphia.

Schremp L.A. (1981) Archaeogastropoda from the Pliocene Imperial Formation of California. *Journal of Paleontology*, **55**(5), 1123–36.

Shepherd S.A. (1973) Studies on southern Australian abalone (genus *Haliotis*) I. Ecology of five sympatric species. *Aust. J. Mar. Freshwat. Res.*, **24**, 217–57.

Shepherd S.A. (1975) Distribution, habitat and feeding habits of abalone. *Aust. Fish.*, **34**, 12–15.

Talmadge R.R. (1963) Insular haliotids in the western Pacific (Mollusca: Gastropoda). *Veliger*, **5**(4), 129–39.

Talmadge R.R. (1974) Some notes on the distribution and taxonomy of haliotids. *Veliger*, **16**(4), 411–12.

Ubaldi R. (1986) Le *Haliotis* della Nuova Zelanda. *Argonauta*, **2**(1–2), 115–34.

Wagner R.J.L. & Abbott R.T. (1967) *Van Nostrand's Standard Catalog of Shells*, 2nd edn. Van Nostrand, Princeton.

Wells F.E. & Bryce C.W. (1985) *Seashells of Western Australia*. Western Australia Museum, Perth.

White M.J.D. (1973) *Animal Cytology and Evolution*, 3rd edn. Cambridge University Press, Cambridge.

Woodring W.P. (1931) A Miocene *Haliotis* from southern California. *Journal of Paleontology*, **5**(1), 34–9.

Yaron I. (1983) The Haliotidae of the Red Sea. *Levantina*, **42**, 486–93.

Chapter 2
Genetic relationships within the genus *Haliotis*

L.D BROWN and N.D. MURRAY *Department of Genetics and Human Variation, La Trobe University, Bundoora, Victoria 3083, Australia*

ABSTRACT: The relationships of twelve species of *Haliotis* have been investigated by protein gel electrophoresis. Genetic distance measures, based on 10 gene loci, indicate that, amongst Australian abalone, *H. conicopora* is probably conspecific with *H. rubra*, whilst *H. rubra*, *H. roei*, *H. laevigata* and *H. scalaris* form a related cluster of species. All other species examined (*H. cyclobates* and *H. coccoradiata* (Australia), *H. iris*, *H. australis* and *H. virginea* (New Zealand), *H. midae* (South Africa) and *H. discus hannai* (Japan)) are only distantly related to this cluster and to each other.

Hybridization between *H. rubra* and *H. laevigata* has been recorded, and appears to lead to introgression between these species.

The observed genetic differences between some *Haliotis* species are larger than are normally associated with species in the same genus. We conclude that revision of the genus is desirable, but that this will need to be based on genetic information and will be difficult if introgression is widespread. Such a revision would allow both local and global biogeographic problems to be addressed.

RESUMEN: Se han investigado las relaciones entre doce especies de *Haliotis* utilizando electroforesis en gel de proteinas. Las medidas de las distancias genéticas, basadas en los loci de diez genes, indican que entre los abalones Australianos, *H. conicopora* es probablemente conespecifica con *H. rubra*, mientras que *H. rubra*, *H. roei*, *H. laevigata* y *H. scalaris* forman un grupo de especies relacionadas. Todas las otras especies examinadas (*H. cyclobates* y *H. coccoradiata* (Australia), *H. iris*, *H. australis* y *H. virginea* (Nueva Zelandia), *H. midae* (Sudafrica) y *H. discus hannai* (Japón)) están relacionadas a dicho grupo y entre ellas únicamente en forma distante.

La hibridacion entre *H. rubra* y *H. laevigata* ha sido registrada y parece llevar a la introgresion entre estas especies.

Las diferencias geneticas observados entre algunas de las especies de *Haliotis* son mayores que las diferencias que se encuentran normalmente entre especies del mismo género. Concluímos en que es aconsejable una revisión del género, pero es necesario que ésta se base en información genética. Concluimos también en que la tarea será difícil si la introgresión se extiende. Dicha revisión permitiría tratar los problemas biogeográficos locales y globales.

INTRODUCTION

All abalone are currently treated as species of the genus *Haliotis* in the monogeneric family Haliotidae. This practice dates from Cox (1960), who considered that the morphological characters previously used to erect other genera (shell form, configuration and sculpture) were inadequate to provide conclusive evidence of generic status and led to much taxonomic confusion. As well, Cox (1962) suggested that some of the taxonomic confusion within the genus could arise from hybridization.

While this pragmatic approach to classification gives a more useful result than an unreliable and over-split taxonomy, it is nevertheless likely that the single genus name conceals much evolutionary diversity, especially since the genus dates from at least the Cretaceous (Cox, 1962). It is important that any genuine phylogenetic relationships amongst abalone be detected. Only in this way can the composition of the fauna of a particular region, or the biogeography of the group as a whole be understood.

Genetic studies allow the recognition of cases of convergent morphological similarity or rapid, and perhaps environmentally-caused, morphological divergence. They can also be used to identify hybrids. Here we report the results of an electrophoretic survey of genetically determined protein variation in seven Australian, three New Zealand, one South African and one Japanese species of *Haliotis*. This study reveals very large genetic differences among some species of *Haliotis* and strongly suggests the need for taxonomic review accompanied by further genetic comparisons.

MATERIALS AND METHODS

The species studied and their collection localities are listed in Table 2.1. Whole specimens were frozen and transported to the laboratory on dry ice (frozen CO_2). Samples of hepatopancreas and muscle were removed and stored at −80°C, and the radula was removed from some individuals of all species for subsequent electron microscopy (Brown, 1991).

Electrophoretic methods follow those of Richardson *et al.* (1986). Ten enzyme loci have been studied: *AAT-1*, *AAT-2*, *GPD*, *GPI*, *LDH*, *MDH-1*, *MDH-2*, *MPI*, *PGM* and *TPI*, as described in Chapter 3.

Allele frequencies were used to calculate Nei's genetic identity I and distance D (Nei, 1972) between all pairs of populations. A dendrogram based on D was constructed using the unweighted pair group method of analysis (UPGMA) (Sneath

Table 2.1 Sample sizes of *Haliotis* species and populations with collection localities and abbreviations.

Species	Locality	Abbreviation	Sample size
H. rubra	NSW, Australia	RUBNED	20
	Tasmania, Australia	RUBTBH	20
	South Australia	RUBSPI	20
H. conicopora	Western Australia	CONIC	20
H. laevigata	Furneaux Islands, Australia	LAETFI	20
	South Australia	LAESCJ	20
	South Australia	LAESPI	20
H. scalaris	South Australia	SCAL	20
H. roei	South Australia	ROESWI	15
	Western Australia	ROEWCA	15
H. cyclobates	South Australia	CYCLO	10
H. coccoradiata	Victoria, Australia	COCCO	10
H. midae	South Africa	MIDAE	8
H. iris	Wellington, New Zealand	IRIS	10
H. australis	Wellington, New Zealand	AUST	10
H. virginea	Auckland Islands, New Zealand	VIRGIN	10
H. discus hannai	Hokkaido, Japan	DISCUS	6

& Sokal, 1973). In addition, 17 individuals that appeared, on the basis of shell characteristics, to be hybrids between *H. rubra* and *H. laevigata* were examined electrophoretically.

RESULTS

Figure 2.1 shows the dendrogram based on genetic distances amongst the species and populations studied. (Details of allele frequencies – Brown, 1991.)

Haliotis conicopora clusters with the three populations of *H. rubra*, with a maximum D of 0.021 (CONIC and RUBNED). Similar small genetic distances are found between all conspecific populations studied (0.007 in *H. roei* and 0.025 within *H. laevigata*).

At a D region of 0.41–0.52 two clusters of Australian species are recognizable: *H. rubra/conicopora* with *H. roei* and *H. laevigata* with *H. scalaris*. These two clusters in turn form a major Australian cluster and are more similar to each other

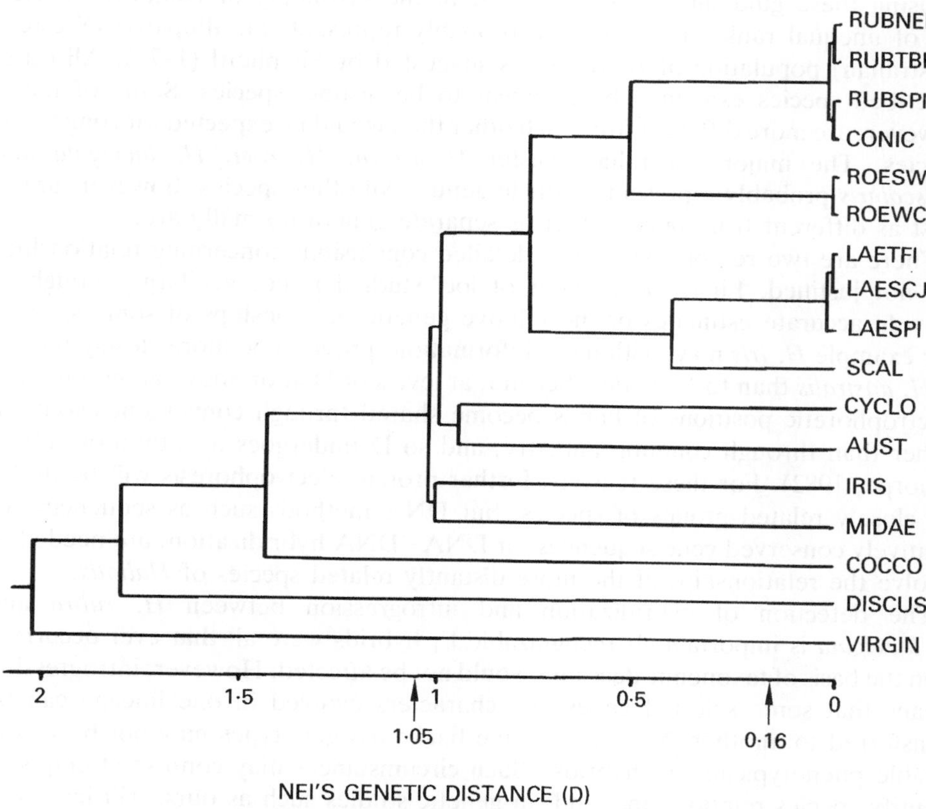

Fig. 2.1 Unweighted pair group method of analysis dendrogram for 17 operational taxonomic units (12 species) of *Haliotis* using Nei's (1972) genetic distance based on ten genetic loci. See Table 2.1 for explanation of abbreviated species names. For explanation of arrows see discussion.

(D = 0.77) than any other pairs of species are. All other pairs of species have genetic distances of approximately one or greater than one.

The existence of fixed allelic differences between *H. rubra* and *H. laevigata* at seven loci (L.D. Brown 1991) allows the identification of 'hybrids' as first-generation (F_1) crosses or as the products of backcrossing to either parent species. Of the 17 individuals examined, 11 appeared to be F_1 hybrids, i.e. they were heterozygous at all discriminating loci. Three individuals had *H. rubra* genotypes at at least one of these loci, and must have resulted from backcrosses of F_1 hybrids to *H. rubra*. The other three, for similar reasons, were identified as backcrosses to *H. laevigata*.

DISCUSSION

Thorpe (1982) reviewed studies of genetic distances between populations and species of invertebrates, vertebrates and plants. He related distances to levels of systematic divergence and found that it was possible to identify critical levels for D values that distinguish between conspecific populations and species (D = 0.16) and between species and genera (D = 1.05).

Using these guidelines it is clear that, in the taxonomy of *Haliotis*, 'species' are of unequal rank. *H. conicopora* probably represents an allopatric (Western Australian) population of *H. rubra*, as suggested by Shepherd (1975). All other described species examined here appear to be distinct species. Some of them, however, are more different from each other than would be expected for congeneric species. The major Australian cluster *H. rubra*, *H. roei*, *H. laevigata* and *H. scalaris* probably represents a single genus. All other species, however, are at least as different from one another as separate genera normally are.

There are two reasons why more detailed conclusions concerning relationships are not justified. First, the number of loci studied is not yet large enough to provide accurate estimates of the relative genetic relationships of some species. For example *H. iris* may, with more information, prove to be more closely related to *H. australis* than to *H. midae*. Second, at levels of D at or above about one, the electrophoretic positions of bands become shared through convergent evolution rather than through common ancestry, and so D undergoes a 'saturation' effect (Thorpe, 1982). For these reasons, further protein electrophoresis will be useful for closely related groups of species, but DNA methods such as sequencing of relatively conserved gene sequences, or DNA–DNA hybridization, are needed to resolve the relationships of the more distantly related species of *Haliotis*.

The detection of hybridization and introgression between *H. rubra* and *H. laevigata* is important. If recognizable F_1 hybrids were all that ever occurred, then the basis of taxonomic decisions would not be affected. However, introgression means that some selected genes and characters evolved in one lineage can be transferred to another. Moreover, some backcross genotypes may not be recognizable phenotypically as 'hybrids'. Such circumstances may confuse attempts to identify species relationships, and, in genetic studies such as ours, will lead to a smaller estimate of D than would be the case without introgression.

Other possible hybrids have been reported between *Haliotis* species. *Haliotis roei* and *H. laevigata* are reported to hybridize in Western Australia (S. Slack-

Smith, pers. comm.); both Cox (1962) and Owen *et al.* (1970) report several Californian hybrids. The taxonomic consequences of introgression are therefore likely to be widespread.

Our results also have implications for understanding regional abalone faunas. The Australian species studied fall into three distinct lineages: the major cluster of *H. rubra/conicopora*, *H. roei*, *H. laevigata* and *H. scalaris*, one lineage represented only by *H. cyclobates* and one represented only by *H. coccoradiata*. These are all temperate species. It can be asked whether these lineages have diverged from each other *in situ*, or whether each lineage has closer relatives amongst the tropical species of *Haliotis*. The answers would give important insights into Australia's marine biogeography.

Similar questions arise with the three New Zealand species, which are all very different from one another.

On a global scale, the question of the relationships of southern temperate, northern temperate and tropical species must also be raised. The large amount of genetic diversity revealed makes the genus *Haliotis* a potentially valuable group of species for distinguishing between competing theories of marine biogeography (Nelson, 1985; Briggs, 1987), provided that genetic affinities can be properly measured.

ACKNOWLEDGEMENTS

This work has been financially supported by the Australian Fishing Industry Research and Development Council and La Trobe University. The authors are grateful to S.A. Shepherd, L. Joll, R.J.Q. Tarr, L.J. Tong, G.A. Moss, H. Momma and P.E. McShane for providing samples, and to Dr G.C. Kirby for criticism of the manuscript.

REFERENCES

Briggs J.C. (1987) Antitropical distribution and evolution in the indo-west Pacific Ocean. *Syst. Zool.*, **36**(3), 237–47.

Brown L.D. (1991) *Evolutionary genetics and population structure in abalone (genus* Haliotis*)*. PhD thesis, Trobe University, Melbourne.

Cox K.W. (1962) California abalones, family Haliotidae. *Fisheries Bulletin No. 118*, California Department of Fish and Game.

Cox L.R. (1960) Family Haliotidae Rafinesque 1915. In *Treatise on Invertebrate Paleontology*, (Ed. by E. Moore), (1) Mollusca 1, pp. 221–33. University of Kansas Press, Lawrence.

Nei N. (1972) Genetic distance between populations. *American Naturalist*, **106**, 283–92.

Nelson G. (1985) A decade of challenge: the future of biogeography. In *Plate Tectonics and Biogeography* (Ed. by A.E. Leviton & M.L. Aldrich). Earth Sciences History (*J. Hist. Earth Sci. Soc.*): **4**(2): pp. 187–96.

Owen R.S., McLean J.H. & Meyer R.J. (1970) Hybridization in the eastern Pacific abalones (*Haliotis*). *Bull. Los Angeles C. Mus. Nat. His. Sci.*, **9**, 1–37.

Richardson B.J., Baverstock P.R. and Adams M. (1986) *Allozyme Electrophoresis. A Handbook for Animal Systematics and Population Studies*. Academic Press, Australia.

Shepherd S.A. (1975) Distribution, habitat and feeding habits of abalone. *Aust. Fish.*, **34**(1), 12–15.

Sneath P.H.A. & Sokal R.R (1973) *Numerical Taxonomy: The Principles and Practice of Numerical Classification*. W.H. Freeman, San Francisco.

Thorpe J.P. (1982) The molecular clock hypothesis: biochemical evolution, genetic differentiation and systematics. *Ann. Rev. Ecol. Syst.*, **13**, 139–68.

Chapter 3
Population genetics, gene flow, and stock structure in *Haliotis rubra* and *Haliotis laevigata*

LINDSAY D. BROWN and NEIL D. MURRAY *Department of Genetics and Human Variation, La Trobe University, Bundoora, Victoria 3083, Australia*

ABSTRACT: Stock structures of *Haliotis rubra* and *H. laevigata* have been investigated using protein gel electrophoresis.

Breeding populations of both species appear to be genetically large. Genetic differences between populations of *H. rubra* are small and geographically cumulative but small-scale genetic heterogeneity suggests that local populations may be predominantly recruited from local stock.

Haliotis laevigata populations are genetically distinct from one another even over small distances.

It is concluded that management of both species needs to be focused on local populations.

RESUMEN: La estructura de poblaciones de *Haliotis rubra* y *H. laevigata* ha sido investigada utilizando electroforesis en gel de proteinas.

Las poblaciones de cría de ambas especies parecen ser genéticamente grandes. Las diferencias genéticas entre poblaciones de *H. rubra* son pequeñas y geográficamente cumulativas pero una heterogeneidad genética a pequeña escala sugiere que las poblaciones locales son reclutadas en forma predominante de las mismas poblaciones.

Las poblaciones de *H. laevigata* son genéticamente distintas unas de otras aun mediando cortas distancias entre ellas.

Se concluye que el manejo de ambas especies necesita enfocarse sobre las poblaciones locales.

INTRODUCTION

Sustainable harvest of natural populations requires management based on sound biological principles such as the recognition of local breeding populations and conservation of genetic variability (Berst & Simon, 1981). Information on the geographic distribution of allele frequencies reveals local populations and the extent and patterns of dispersal (Allendorf & Phelps, 1981), and so helps define the units for management.

Haliotis rubra and *H. laevigata* (blacklip and greenlip abalone respectively) are two widely distributed, commercially fished species in southern Australia. Samples from throughout the geographic ranges of both species in southeastern Australia have been analysed for genetic variation using protein gel electrophoresis. The broad-scale and localized population structure of both species, their gene flow patterns and the implications for management are discussed.

MATERIALS AND METHODS

Sampling

Tissue samples (hepatopancreas) were collected as byproducts of commercial processing. As well, personal collections were made with the aid of government fisheries departments. In all, 1354 adult *H. rubra* from 15 locations and 579 *H. laevigata* from 8 locations were collected between September 1985 and April 1988. Collection sites are shown in Fig. 3.1. The tissue samples were transported in dry ice (frozen CO_2) and stored at −80°C.

Electrophoresis

A sample of hepatopancreas was taken from each individual for electrophoresis. Protein gel electrophoresis on cellulose acetate gels and histochemical staining of enzymes followed the methods of Richardson *et al.* (1986). The genetic basis of electrophoretic variation was inferred by close agreement of phenotype patterns with likely subunit structure of the enzyme and by approximate fit of phenotype frequencies to Hardy-Weinberg proportions. The enzymes examined were aspartate aminotransferase (AAT, two loci), glycerol-3-phosphate dehydrogenase (αGPD),

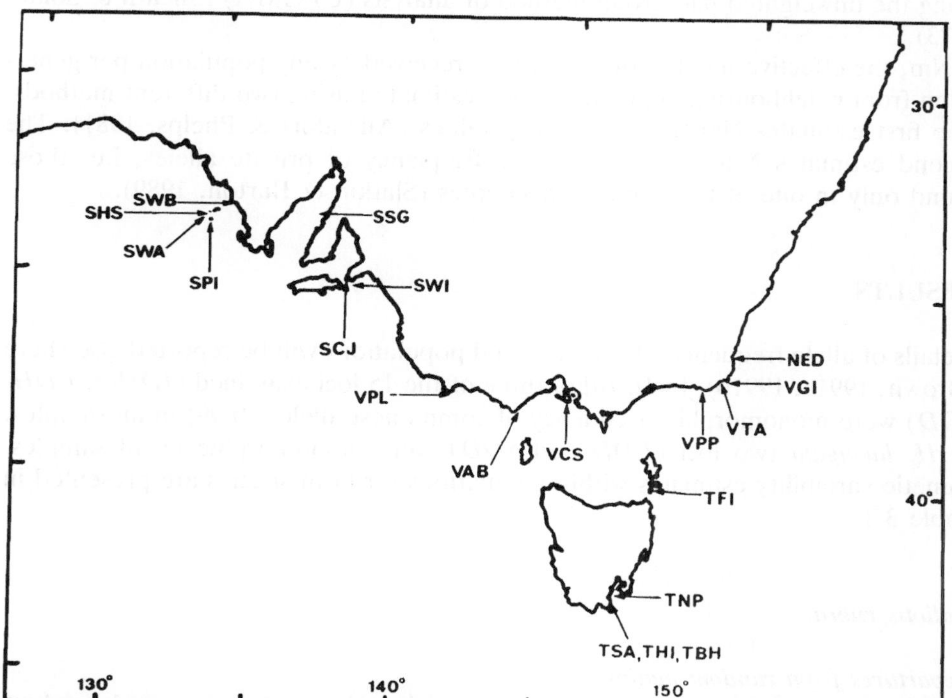

Fig. 3.1 Collection sites for *Haliotis rubra* and *H. laevigata* in southeastern Australia. *H. rubra* (NED, VGI, VTA, VPP, VCS, VAB, VPL, TNP, TBH, THI, TSA, SWI, SCJ, SWB, SPI). *H. laevigata* (TFI, SWI, SCJ, SSG, SWB, SHS, SWA, SPI).

glucose phosphate isomerase (GPI), β-hydroxybutyrate dehydrogenase (HBDH), isocitrate dehydrogenase (IDH, two loci), lactate dehydrogenase (LDH), malate dehydrogenase (MDH, two loci), mannose phosphate isomerase (MPI), phosphoglucomutase (PGM), superoxide dismutase (SOD), sorbitol dehydrogenase (SORDH) and triosephosphate isomerase (TPI). All running buffers were buffer L of Richardson *et al.* (1986), except for IDH where the 0.05 M Tris-citrate pH 7.8 buffer was used.

Statistical analysis

Gene frequency variation was analysed using the 'Genestats' program of Black & Krafsur (1985), which calculates F statistics (Wright, 1951) to describe non-random mating within and among subpopulations. The coefficient F_{IS} describes the departure from random mating within subpopulations. Significance of departure from Hardy-Weinberg equilibrium was tested using χ^2 analysis of observed and expected numbers of heterozygotes (Weir & Cockerham, 1984).

The amount of variation between populations was quantified by the standardized genetic variance F_{ST}, as modified by Weir & Cockerham (1984). The significance of differences in allele frequency among samples was tested using χ^2 analysis.

Rogers genetic distances (R) were calculated for pairwise combinations of samples (Rogers, 1972). Dendrograms based on genetic distance were constructed using the unweighted pair group method of analysis (UPGMA) (Sneath & Sokal, 1973).

Nm, the effective number of immigrants received by any population per generation from neighbouring populations, was estimated using two different methods. The first estimates Nm from mean F_{ST} values (Allendorf & Phelps, 1981). The second estimates Nm from the average frequency of private alleles, i.e. those found only in one of the population samples (Slatkin & Barton, 1989).

RESULTS

Details of allele frequencies for all loci and populations will be reported elsewhere (Brown, 1991a; 1991b). In *H. rubra* three of the 15 loci examined (*IDH-2*, *LDH*, *SOD*) were monomorphic (frequency of commonest allele >0.99) in all samples. In *H. laevigata* two loci (*LDH* and *SOD*) were monomorphic in all samples. Genetic variability estimates within populations for both species are presented in Table 3.1.

Haliotis rubra

Departures from random mating

A total of 156 χ^2 tests were performed comparing observed to expected numbers of heterozygotes, for a null hypothesis of a population in Hardy-Weinberg equilibrium, for each locus in each population sample. Seventeen tests proved significant ($P < 0.05$), but nine of these were dismissed because of low expected

Table 3.1 Genetic variability estimates in *Haliotis rubra* and *H. laevigata* (mean ± sd).

	Average number of alleles per locus ($\bar{A}$)	Proportion of polymorphic loci (P)	Average heterozygosity	
			observed ($\bar{H}_o$)	expected ($\bar{H}_s$)
*H. rubra**				
Mean	2.58	0.631	0.140	0.147
sd	0.1782	0.0611	0.0013	0.0110
H. laevigata†				
Mean	2.67	0.733	0.195	0.207
sd	0.1127	0.0873	0.0186	0.0122

* 15 populations, 15 loci, average sample size = 90.267
† 8 populations, 15 loci, average sample size = 72.375

numbers in genotype classes. The remaining eight all indicated heterozygote deficiencies. These were *AAT-2* in NED, *IDH-1* in SWI and *HBDH* in SWB, SCJ, SWI, VCS, VGI and VPP. Although 7.8 significant results from 156 χ^2 tests ($P = 0.05$) would be expected by chance alone, the pattern of heterozygote deficiency for *HBDH* across six population samples is notable.

Summarized F_{IS} values for each locus appear in Table 3.2. Two loci, *IDH-1* and *HBDH*, show consistent heterozygote deficiencies. The small negative F_{IS} values result from the large numbers of rare alleles in *H. rubra*.

Population structure

The mean standardized genetic variances, F_{ST}, for each polymorphic locus over the broad geographic scale are summarized in Table 3.2. Allele frequencies differed significantly ($P < 0.05$) for 8 of the 12 polymorphic loci. 'Genestats' calculates a heterogeneity χ^2 for each allele and shows the relative contribution of each population sample to significant ($P < 0.05$) results. The relative contributions of population samples to heterogeneity for each polymorphic locus show that no population stands out as genetically discrete. The major contributors to heterogeneity are predominantly found at the edges of the sampling area (NED, SPI, TBH) while those at the centre (VCS, VAB, VPL) make little contribution.

Analyses on a smaller geographic scale were carried out with three sample populations (TBH, THI, TSA) found within Port Esperance on the east coast of Tasmania. TBH is located on the northern headland, TSA on the southern headland and THI on a small island inside the bay. The approximate distance between TBH and THI is 1.5 km, between THI and TSA 1 km and between TBH and TSA 2.25 km. Significant differences in allele frequencies were found for two loci, *TPI* and *MPI*, and in both cases the island population, THI, contributed most to the heterogeneity.

Rogers genetic distances (R) have been calculated and used to construct a UPGMA dendrogram (Fig. 3.2). The relationship between pairwise genetic distances between populations and geographic (shortest across water) distance was also analysed (Fig. 3.3). The correlation was positive and significant ($r = 0.787$, $P < 0.01$, df = 14). From this graph neighbourhood size (the region within which

Table 3.2 Mean F-statistics (Weir & Cockerham, 1984) for 12 polymorphic loci in *Haliotis rubra* and 13 polymorphic loci in *H. laevigata*. X^2 test is for departure of F_{ST} from zero.

Locus	F_{IS}	F_{IT}	F_{ST}	χ^2
H. rubra				
AAT-1	−0.010	−0.012	−0.002	66.67
AAT-2	0.019	0.025	0.007	132.40**
αGPD	0.041	0.047	0.006	101.59**
GPI	0.013	0.021	0.009	206.26***
HBDH	0.160	0.184	0.028	215.27***
IDH-1	0.109	0.162	0.060	196.45***
MDH-1	−0.018	−0.017	0.001	45.60
MDH-2	−0.002	−0.001	0.001	26.12
MPI	−0.006	−0.002	0.003	181.60***
PGM	0.029	0.038	0.009	190.67***
SORDH	−0.007	−0.003	0.004	23.57
TPI	−0.030	−0.024	0.006	96.31***
Mean	0.056	0.071	0.016	
sd	0.0101	0.0115	0.0018	
H. laevigata				
AAT-1	−0.006	−0.009	−0.003	4.15
AAT-2	0.083	0.098	0.016	67.23***
αGPD	0.010	0.019	0.009	51.12*
GPI	0.005	0.008	0.003	87.62
HBDH	0.138	0.180	0.048	118.35***
IDH-1	0.377	0.381	0.006	32.72
IDH-2	−0.033	−0.012	0.020	26.52***
MDH-1	−0.039	−0.036	0.003	22.79
MDH-2	0.124	0.125	0.001	13.62
MPI	0.067	0.078	0.012	69.91*
PGM	0.048	0.049	0.001	57.21
SORDH	0.015	0.108	0.004	25.36*
TPI	0.012	0.025	0.013	20.17**
Mean	0.062	0.076	0.014	
sd	0.00664	0.00216	0.00800	

* $P<0.05$, ** $P<0.01$, *** $P<0.001$

mating is effectively random) can be estimated (Richardson *et al.*, 1986). The average geographical size of a neighbourhood is the X coordinate of the regression of genetic distance (R) on geographic distance for the average 'R' value found between replicate sample sets. For *H. rubra* this is approximately 600 km.

The gene flow estimate Nm for *H. rubra* is 15.38 based on the mean F_{ST} and 19.67 based on the average frequency of private alleles ($\bar{p}(1) = 0.00715$, $N_{sam} = 90.267$).

Haliotis laevigata

Departures from random mating

A total of 95 χ^2 tests were performed comparing observed and expected numbers of heterozygotes for a null hypothesis of a population in Hardy-Weinberg equilib-

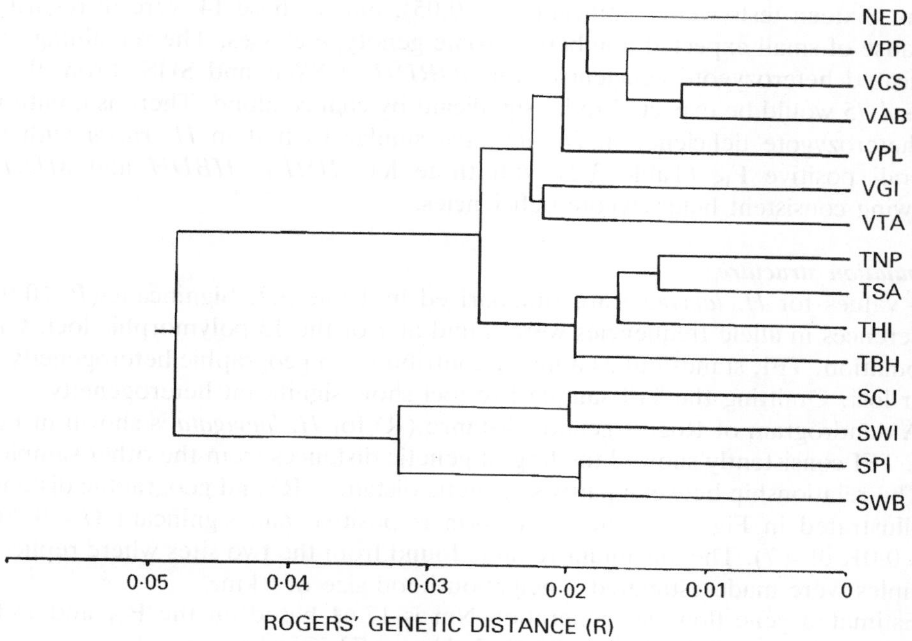

Fig. 3.2 Unweighted pair group method of analysis dendrogram for 15 populations of *H. rubra*. Genetic distance based on 15 loci.

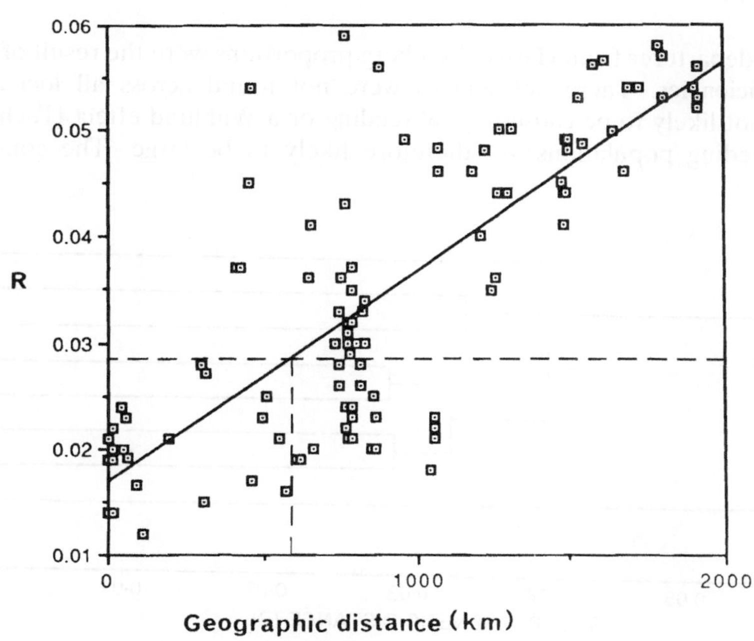

Fig. 3.3 Regression of Rogers' genetic distance (R) on geographic distance for *H. rubra*. The solid line is the line of best fit. The dashed line indicates the average 'R' value found between six pairs of replicate samples.

rium. Sixteen tests were significant ($P<0.05$), but of these 14 were disregarded because of small expected numbers in some genotype classes. The remaining two indicated heterozygote deficiencies for *SORDH* at SWB and SHS. From 95 χ^2 tests 4.75 would be expected to be significant by chance alone. There is a pattern of heterozygote deficiency in *H. laevigata* similar to that in *H. rubra* with an overall positive F_{IS} (Table 3.2) with three loci *IDH-1*, *HBDH* and *MDH-1* showing consistent heterozygote deficiencies.

Population structure

F_{ST} values for *H. laevigata* are summarized in Table 3.2. Significant ($P<0.05$) differences in allele frequencies were found at 7 of the 13 polymorphic loci. One population, TFI, stands out as a major contributor to geographic heterogeneity at four loci. Omitting the TFI sample five loci show significant heterogeneity.

A dendrogram of Rogers genetic distance (R) for *H. laevigata* is shown in Fig. 3.4. TFI consistently showed the largest genetic distances from the other samples.

The relationship between pairwise genetic distance (R) and geographic distance is illustrated in Fig. 3.5. The correlation is positive and significant ($r=0.805$, $P<0.01$, df$=7$). The maximum R value found from the two sites where replicate samples were made estimated a neighbourhood size of 0 km.

Estimated gene flow per generation Nm is 17.61 based on the F_{ST} and 14.01 based on private alleles ($\bar{p}(1)=0.01063$, $N_{sam}=72.375$).

DISCUSSION

Haliotis rubra

Significant departures from Hardy-Weinberg proportions were the result of heterozygote deficiencies. These deficiencies were not found across all loci and are therefore not likely to be caused by inbreeding or a Wahlund effect (Richardson, 1982). Breeding populations are therefore likely to be large. The consistently

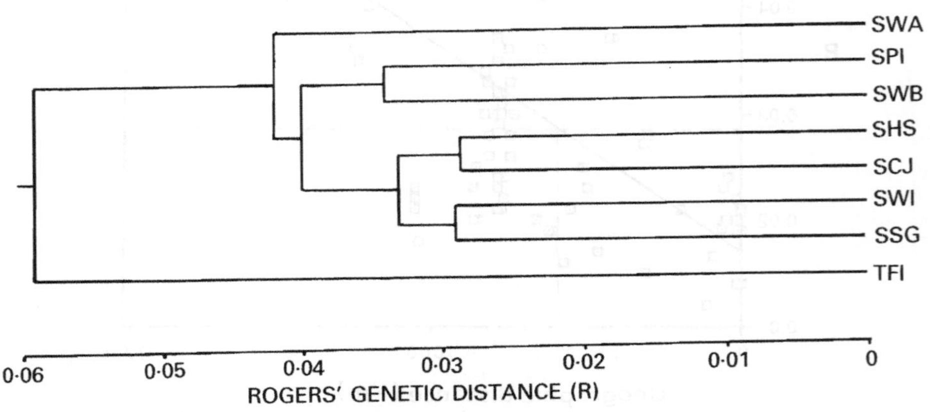

Fig. 3.4 Unweighted pair group method of analysis dendrogram for eight populations of *H. laevigata*. Genetic distance based on 15 loci.

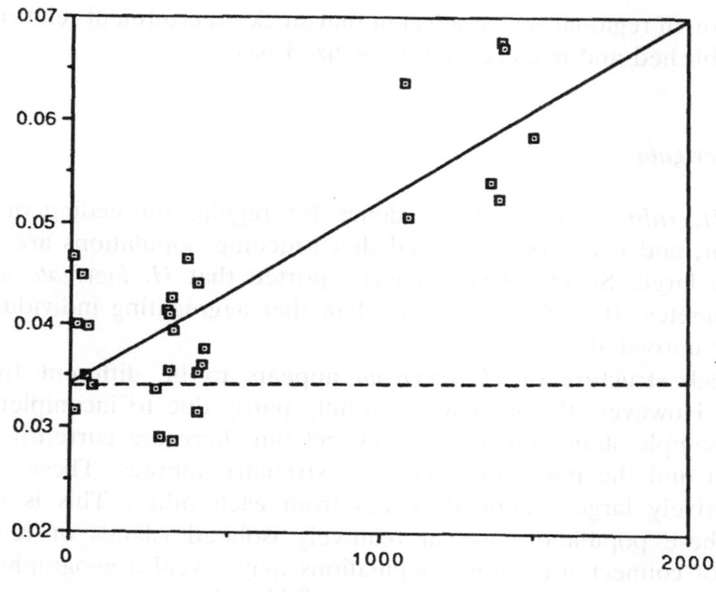

Fig. 3.5 Regression of Rogers' genetic distance (R) on geographic distance for *H. laevigata*. The solid line is the line of best fit. The dashed line indicates the maximum 'R' value found between two pairs of replicate samples.

large F_{IS} values for *HBDH* and *IDH* could be explained by locus-specific effects such as gel-scoring bias or errors, null alleles, selection or assortative mating (Richardson *et al.*, 1986). It is not clear which is the most likely explanation, but heterozygote deficiency and homozygote excess were found in all classes of genotypes and there was no evidence of null alleles, i.e. no null homozygotes were found. Moreover, all individuals were scorable, so that scoring bias is unlikely. The possibilities of assortative mating and selection need to be investigated further.

The broad-scale gene pool of *H. rubra* is not geographically homogeneous as shown by the significant F_{ST} values. However there are no very distinct populations. Rather, differences are small and geographically cumulative, as shown by their relative positions in Fig. 3.2 and by the greater contribution of edge populations to the significant F_{ST} values. The gene flow estimates indicate this as well. The estimates obtained here (15.38 and 19.67) are large, although not as large as the estimate of 42.0 for the mussel *Mytilus edulis* (Slatkin, 1985).

Even though populations have been large and interconnected during the species' history, local populations may still be predominantly recruited from local stock. The significant F_{ST} values for the three Port Esperance populations suggest that this is likely to be true for physically isolated reef and island populations such as THI. This observation supports the local recruitment hypothesis of Prince *et al.* (1987) for Tasmanian *H. rubra*.

From a management viewpoint it seems reasonable that zones of about 600 km of coastline (corresponding to neighbourhood size) could be recognized for the conservation of regional gene pools, but that stock–recruitment relationships need to be established and managed on a localized basis.

Haliotis laevigata

As with *H. rubra*, there is no evidence for regular inbreeding or population subdivision, and it can be concluded that breeding populations are likely to be genetically large. Shepherd (1986) has reported that *H. laevigata* aggregate to release gametes. It is clear from our data that aggregating individuals must be essentially unrelated.

The stock structure of *H. laevigata* appears to be different from that of *H. rubra*. However, this is almost certainly partly due to incomplete sampling. The TFI sample stands out as very distinct but there are currently no samples between it and the remaining (South Australian) samples. These samples also show relatively large genetic distances from each other. This is likely to be because these populations are at relatively isolated islands or reefs. Further sampling of connected coastline populations may reveal a geographically cumulative population structure similar to that of *H. rubra*.

For the South Australian populations, the large genetic distances between nearby populations (Fig. 3.5) and the corresponding neighbourhood estimate of 0 km essentially means that these populations need to be regarded as separate gene pools and management units.

ACKNOWLEDGEMENTS

This research has been supported by a Fishing Industry Research and Development Council grant (No. 85/8). The authors thank the Mallacoota abalone fishermen's co-operative, the Department of Conservation, Forests and Lands of Victoria, the South Australian Department of Fisheries, the Department of Sea Fisheries of Tasmania, the New South Wales Department of Agriculture and Fisheries and Peter Frost (Flinders Island) for help in collection of abalone. Dr G.C. Kirby criticized the manuscript.

REFERENCES

Allendorf F.W. & Phelps S.R. (1981) Use of allelic frequencies to describe population structure. *Can. J. Fish. Aquat. Sci.*, **38**, 1507–14.

Berst A.H. & Simon R.C. (Eds) (1981) Proceedings of the Stock Concept Symposium. *Can. J. Fish. Aquat. Sci.*, **38**(1), 1458–921.

Black IV W.C. & Krafsur E.S. (1985) A Fortran program for the analysis of genotype frequencies and description of the breeding structure of populations. *Theor. Appl. Genet.*, **70**, 484–90.

Brown L.D. (1991a) Genetic variation and population structure in the blacklip abalone, *Haliotis rubra*. *Aust. J. Mar. Freshwat. Res.*, **42**, 47–90.

Brown L.D. (1991b) Evolutionary genetics and population structure in abalone (genus Haliotis). PhD thesis, La Trobe University, Melbourne.

Prince J.D., Sellers T.L., Ford W.B. & Talbot S.R. (1987) Experimental evidence for limited dispersal of Haliotid larvae (genus *Haliotis*; Mollusca: Gastropoda). *J. Exp. Mar. Biol. Ecol.*,

106, 243–63.

Richardson B.J. (1982) Geographical distribution of electrophoretically detected protein variation in Australian commercial fishes. 1. Jack mackerel, *Trachurus declivus*. Jenyns. *Aust. J. Mar. Freshwat. Res.*, **33**, 917–26.

Richardson B.J., Baverstock P.R. & Adams M. (1986) *Allozyme Electrophoresis. A Handbook for Animal Systematics and Population Studies*. Academic Press, Australia.

Rogers J.S. (1972) Measures of genetic similarity and genetic distance. *Studies in Genetics VII*, pp. 145–53. University of Texas Publication 7213.

Shepherd S.A. (1986) Studies on Southern Australian abalone (genus *Haliotis*) VII. Aggregative behaviour of *H. laevigata* in relation to spawning. *Marine Biology*, **90**, 231–6.

Slatkin M. (1985) Rare alleles as indicators of gene flow. *Evolution*, **39**(1), 53–65.

Slatkin M. & Barton N.H. (1989) A comparison of three indirect methods for estimating average levels of gene flow. *Evolution*, **43**(7), 1349–68.

Sneath P.H.A. & Sokal R.R. (1973) *Numerical Taxonomy: The Principles and Practice of Numerical Classification*. W.H. Freeman, San Francisco.

Weir B.S. & Cockerham C.C. (1984) Estimating F-statistics for the analysis of population structure. *Evolution*, **38**, 1358–70.

Wright S. (1951) The genetical structure of populations. *Ann. Eugenics*, **15**, 323–54.

Chapter 4
Water movement and the ecology and evolution of the Haliotidae

BRIAN N. TISSOT *Department of Zoology, Oregon State University, Corvallis, OR 97331, USA*

ABSTRACT: This study focuses on water movement as an important selective factor influencing the morphology and vertical distribution of abalone. Measurements of drag in flow tanks indicated that species with strong shell sculpture experienced higher drag forces than smooth shelled species at velocities which occur in nature. Thus, shell sculpture, which may serve as armour against shell-crushing predators in the biologically-intense subtidal, may limit the distribution of species due to their relatively high drag shell.

Large, elevated tremata (respiratory pores), which predominate in subtidal species, may serve to enhance respiratory exchange passively in areas that experience low water movement. Induced mantle cavity flow in live, dead and morphologically-manipulated abalone showed striking differences among species. Large, elevated tremata were more efficient at promoting passive mantle cavity circulation than small, unelevated tremata. Induced flow is hypothesized to represent a substantial energetic saving, and is likely to be an important selective factor in the ecology of modern abalone as well as in the evolution of the Haliotidae.

RESUMEN: El presente estudio se centra en el movimiento de agua como un importante factor de selección que influencia la morfología y distribución vertical de abalones. Medidas de arrastre tomadas en tanques con agua circulante, a velocidades que ocurren en la naturaleza, indican que las especies con concha fuertemente esculpida experimentan mayores fuerzas de arrastre que las especies con concha con suave escultura. Así, la escultura de la concha, la que puede servir como defensa contra depredadores 'quiebra-conchas' en el biológicamente intenso submareal, puede también limitar la distribución de las especies debido a las relativamente altas fuerzas de arrastre que ellas presentan.

Poros respiratorios (tremata) grandes y elevados, los que predominan en especies submareales, pueden servir para aumentar pasivamente el intercambio respiratorio en áreas con bajo movimiento de agua. Flujo inducido en la cavidad del manto de abalones vivos, muertos y manipulados morfológicamente muestran importantes diferencias entre las especies. Tremata grandes y elevados son más eficientes en términos de promover la circulación pasiva en la cavidad del manto que tremata pequeños y bajos. Se propone que el flujo inducido representa un ahorro energético substancial y es probable que sea un importante factor de selección en la ecología de abalones modernos, asi como en la evolución de los Haliotidae.

INTRODUCTION

Water movement is known to be an important factor influencing the size and shape of marine organisms (Riedl, 1971; Denny *et al.*, 1985). This study focuses on water movement as a factor influencing the shell morphology of abalone (Mollusca: Haliotidae). It is proposed that much of the morphological diversity within the Haliotidae has evolved in response to two results of water moving over the shell surface:

(1) drag forces resulting from shell sculpture and
(2) pressure gradients between respiratory pores (tremata) which promote passive mantle cavity circulation.

Abalone are characterized by a flattened shell, paired gills and a row of tremata. Among species there is striking diversity in shell sculpture, the number, size and shape of tremata, and in the epipodium, an elaborate sensory extension of the muscular foot (Reeve, 1846; Owen *et al.*, 1971) (Table 4.1). These diverse characters covary and form three distinct morphological groups within the family (Tissot, 1985). Smooth-shelled species, such as the black abalone, *Haliotis cracherodii*, frequently possess numerous (5–16), small tremata that are flush with the dorsal shell surface, and a thin, simple epipodium. In contrast, species with conspicuous shell sculpture (spiral or oblique ribs), such as the pink abalone,

Table 4.1 Description of morphological groups within the Haliotidae (based on Ino, 1952; Cox, 1962; Forster, 1962; Sinclair, 1963; Newman, 1966; Guice, 1969; Poore, 1972; Shepherd, 1973; Tutschulte, 1976 and personal observations).

Shell shape	oval, arched	elongate, flat	oval, variable
Shell sculpture	obscure spiral ribs	obscure to strong spiral ribs, imbricate growth lines	elevated, prominent spiral, oblique and axial ribs
Tremata	5–16, flush	4–8, slightly elevated	2–7, highly elevated
Epipodium	thin, simple plates	thick, simple, papillate	medium thick, branched plates and tubercles
Habitat	open, shallow (to 10 m)	semi-protected, and open, shallow to moderate (to 20 m)	protected, shallow to deep (to 600 m)
Representative species (partial list)	*H. asinina* *H. australis* *H. cracherodii* *H. cyclobates* *H. glabra* *H. iris* *H. laevigata* *H. midae* *H. planata* *H. virginea*	*H. coccinea* *H. diversicolor* *H. elegans* *H. lamellosa* *H. mariae* *H. pustulata* *H. squamata* *H. tuberculata* *H. walallensis*	*H. brazieri* *H. corrugata* *H. fulgens* *H. dalli* *H. discus* *H. gigantea* *H. kamtschatkana* *H. ovina* *H. parva* *H. pourtalesii* *H. pulcherrima* *H. semiplicata* *H. scalaris* *H. sieboldii* *H. sorenseni* *H. varia*

H. corrugata, usually possess fewer (3–5), large tremata elevated on tubular projections, and a thicker, more complex epipodium. An intermediate group, represented by *H. walallensis*, has a mixed morphology: fine to moderate spiral sculpture, 6 to 8 small, slightly elevated tremata, and a thick, papillate epipodium.

Consistent associations between morphology and habitat suggest the hypothesis that shell sculpture and tremata shape are of adaptive significance. Smooth-shelled species (e.g. *H. cracherodii*, *H. laevigata*, *H. iris*) generally occur in open intertidal and shallow subtidal habitats (Cox, 1962; Sinclair, 1963; Shepherd, 1973). Species intermediate in sculpture (e.g. *H. tuberculata*, *H. walallensis*, *H. lamellosa*) occupy habitats ranging from shallow to deep waters (Cox, 1962; Forster, 1962; B. Owen, pers. comm.). In contrast, elaborately sculptured forms (e.g. *H. corrugata*, *H. scalaris*, *H. sorenseni*) occupy deeper and relatively more protected areas in a wide variety of subtidal habitats (Shepherd, 1973; Tutschulte, 1976). Thus, diversity in shell and trematal structure is associated with habitats that vary in depth and microhabitat.

Water movement varies markedly with depth and generates forces due to drag, lift and accelerations (Denny *et al.*, 1985). Drag, however, is the predominant force on shallow-water marine organisms (Denny, 1988). In general, drag is maximal in the mid-intertidal zone and decreases with increasing depth (Denny, 1988). Because a rough shell surface generates large drag forces when exposed to high velocity currents (Branch & Marsh, 1978), abalone with pronounced shell sculpture may be restricted to deeper and/or more protected habitats. In contrast, smooth-shelled species may inhabit open shallow water areas as drag is significantly reduced on their streamlined shells.

Large, elevated tremata openings, which predominate in subtidal species, may serve to enhance respiratory exchange passively in areas that experience low water movement. Until recently, the tremata of abalone were assumed to be exclusively exhalant structures for mantle cavity water propelled by ctenidial cilia and laden with digestive wastes and reproductive products (Crofts, 1929). Voltzow (1983), however, revealed that elevated tremata in the pinto abalone, *H. kamtschatkana*, induced a passive circulation of mantle cavity water in response to the dynamics of water flowing over the shell surface. Diversity among abalone in tremata structure, therefore, may be of functional significance in a manner similar to the apical pore of keyhole limpets (Murdock & Vogel, 1978) and the raised ostia of marine sponges (Vogel, 1974).

Based on these morphological trends, the following hypotheses and their corollaries are proposed concerning distributional patterns within the Haliotidae:

H_1: Shell sculpture promotes drag; the distribution of abalone with sculpture is restricted by forces generated in high velocity, shallow habitats.

H_2: Diversity in tremata structure is associated with variation in water movement among habitats; abalone are energetically dependent on induced mantle cavity flow.

Preliminary tests of these hypotheses are provided in the form of

(1) measurements of maximal water velocities in the field near intertidal black abalone, *H. cracherodii*;

(2) measurements of drag forces on the shells of 13 species;
(3) induced mantle cavity flow rates in three morphologically dissimilar species, *H. cracherodii*, *H. rufescens* and *H. kamtschatkana*;
(4) the response of induced flow to morphological manipulations in abalone models.

MATERIALS AND METHODS

Maximum water velocities were measured in the intertidal zone near black abalone using dynamometers similar to those described by Palumbi (1984). Measurements were made at two sites in California, Santa Cruz Island (34°00′N, 114°50′W) and Año Nuevo Island (37°6′N, 122°20′W), during quarterly surveys in 1988–90. Measurements were taken at 24-hour intervals using three to six dynamometers along four transects at each site (see Tissot, 1990 for more information).

Shells for 13 species were obtained from a variety of sources: each species was represented by a single specimen of average adult size and morphology as described by Cox (1962), Poore (1972) and Shepherd (1973). A subjective index was derived to describe the extent of shell surface complexity:

1 = no sculpture
2 = faint sculpture
3 = moderate sculpture
4 = strong sculpture.

Drag was measured by suspending clay-filled shells inside flow tanks similar to those described by Vogel & LaBarbara (1978). Shells were attached to a 10 cm piece of brass tubing and mounted upside-down within 2 to 5 mm of the surface of a plexiglas mount. The brass rod passed through a slit in the mount and was attached to an aluminum beam bearing two copper-foil strain gauges (W.T. Bean, BAE-13-250BB-SE). Free stream velocities attained in the flow tank ranged from 0.13 to 1.25 m/s.

Drag coefficients (C_D), dimensionless indices of drag forces, were calculated from the equation:

$$C_D = 2D/p\ S\ U^2 \qquad (4.1)$$

where D = drag, p = density of sea water, S = reference area of the object and U = velocity (Vogel, 1981). Reference areas (projected area perpendicular to flow) were measured by tightly wrapping a stiff piece of copper wire around the widest point of the shell. The outline made by the wire was traced onto paper and its area determined.

Reynolds numbers (Re), dimensionless indices of velocity, were calculated using the equation:

$$Re = l\ U/v \qquad (4.2)$$

where l = characteristic length of the object, U = velocity and v = kinematic viscosity of seawater (Vogel, 1981). Maximum shell length was used as the characteristic length of abalone. For each species, drag coefficients were determined over a range of Reynolds numbers.

Induced flow was measured on three live individuals each of *H. cracherodii*, *H. rufescens*, and *H. kamtschatkana* in a flow tank maintained at ambient temperatures (12–15°C) by a continuous input of fresh seawater. The velocity of water moving through the mantle cavity of live, dead and model abalone was determined by timing the passage of fluorescein dye, injected with a micro dye injector mounted on a micromanipulator, between anterior and posterior tremata. This procedure was repeated three to six times on each individual at each free stream velocity. To measure induced flow in dead abalone in natural postures, abalone were anaesthetized with an injection of 1 ml per 100 g body weight of succinyl choline chloride solution (5 mg/ml in seawater) and quickly frozen in the bottom of a cryostat maintained at −20°C. Before measurement of induced flow, abalone were thawed in seawater to ambient temperature.

The effects of individual tremata characteristics on induced flow were tested using clay abalone models. Models were constructed of clay-filled shells equipped with a 'mantle cavity' of dimensions and volume similar to live animals. Analysis of covariance (ANCOVA) was used to test for differences among species in the rate of induced flow as a function of external velocity.

RESULTS

A total of 930 dynamometer readings were taken over the course of 252 observation days in 1988–90 (Table 4.2). Maximum water velocities adjacent to black abalone ranged from 0.7 to 3.4 m/s. Average daily maximum velocities were 2.18 m/s at Año Nuevo Island and 2.15 m/s at Santa Cruz Island.

Drag coefficients for clay-filled shells of all species ranged between 0.21 and 0.61 at flow tank velocities of 0.13–1.25 m/s (Fig. 4.1). Drag coefficients for all species at the highest overlapping Re tested (32 000 or 0.7–1.25 m/s) ranged between 0.30 and 0.53 (Table 4.3). There was a significant product-moment correlation between drag coefficients at the highest overlapping Re and the shell

Table 4.2 Maximum daily water velocities (m/s) measured near intertidal black abalone along four transects at two study sites in California; water velocities were measured using dynamometers similar to those described in Palumbi (1984).

	Transect			
Locality	A	B	C	D
Año Nuevo Island				
Mean	2.06	2.02	2.48	2.34
Standard deviation	0.47	0.41	0.42	0.50
Number of days	21	21	19	20
Total *N*	69	84	45	44
Santa Cruz Island				
Mean	2.01	2.10	2.32	2.21
Standard deviation	0.40	0.29	0.59	0.43
Number of days	47	42	42	40
Total *N*	190	170	162	166

N = total number of measurements (sum of dynamometer readings).

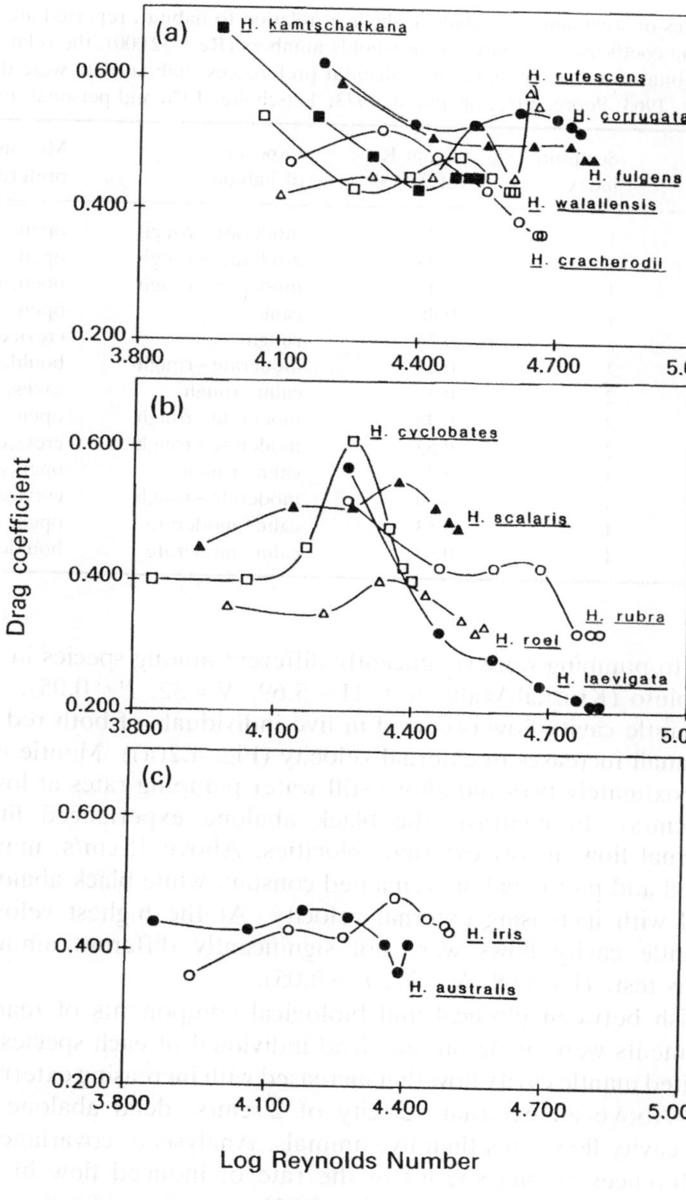

Fig. 4.1 Relationships between drag coefficients (C_D) and Reynolds numbers (Re) for shells of 13 species of abalone: (a) eastern Pacific species, (b) South Australia species, (c) New Zealand species.

sculpture index ($r = 0.58$, $N = 13$, $P < 0.05$). Thus, at the highest common Re tested, species with raised shell sculpture, such as *H. corrugata*, *H. fulgens* and *H. scalaris*, experienced greater drag than smooth-shelled species, such as *H. laevigata*, *H. iris* and *H. cracherodii*.

Mantle cavity flow measurements for live individuals of *H. cracherodii*, *H. kamtschatkana* and *H. rufescens*, hereafter referred to as black, pinto and red abalone, respectively, are presented in Fig. 4.2(a). In still water, mantle cavity

Table 4.3 Measures of streamlining in abalone shells in relation to habitat; reported are an index of shell sculpture, drag coefficients at equivalent Reynolds numbers (Re = 32 000), the relative exposure of each species' habitat to waves and adult microhabitat preferences (habitat data were derived from Cox, 1962; Sinclair, 1963; Poore, 1972; Shepherd, 1973; Tutschulte, 1976 and personal observations).

Species	Sculpture index	C_D at Re of 32 000	Exposure of habitat	Microhabitat preference
H. laevigata	1	0.30	moderate–rough	open
H. iris	1	0.43	moderate–rough	open, crevices
H. cracherodii	1	0.43	moderate–rough	open, crevices
H. cyclobates	1	0.48	calm	open
H. roei	2	0.34	rough	crevices
H. australis	2	0.41	moderate–rough	boulders, crevices
H. ruber	2	0.42	calm–rough	caves, crevices
H. walallensis	2	0.48	moderate–rough	open, crevices
H. rufescens	2	0.53	moderate–rough	crevices
H. kamtschatkana	3	0.45	calm–rough	open, crevices
H. fulgens	3	0.50	moderate–rough	crevices
H. corrugata	4	0.53	calm–moderate	open, crevices
H. scalaris	4	0.53	calm–moderate	boulders, crevices

flow rates due to pumping were significantly different among species in the order: red > black > pinto (Kruskal-Wallis test, $H = 5.69$, $N = 32$, $P < 0.05$).

Increased mantle cavity flow occurred in live individuals of both red and pinto abalone with small increases in external velocity (Fig. 4.2(a)). Mantle cavity flow increased approximately two-fold above still-water pumping rates at low external velocities (10 cm/s). In contrast, the black abalone experienced little to no increased internal flow at low external velocities. Above 15 cm/s, mantle cavity flow of both red and pinto abalone remained constant while black abalone mantle flow increased with increasing external velocity. At the highest velocity tested (45 cm/s), mantle cavity flows were not significantly different among species (Kruskal-Wallis test, $H = 0.68$, $N = 37$, $P > 0.05$).

To distinguish between physical and biological components of mantle cavity flow, measurements were made on one dead individual of each species. All dead abalone exhibited mantle cavity flow that increased with increasing external velocity (Fig. 4.2(b)). Above an external velocity of 20 cm/s, dead abalone possessed higher mantle cavity flow rates than live animals. Analysis of covariance revealed significant differences among species in the rate of induced flow in the order: pinto > black > red ($F = 6.22$, $df = 150$, $P < 0.01$).

Clay-filled shells allowed a measure of induced flow rates as a function of shell morphology independent of the animals' internal morphology (gills, tentacles, etc.). Induced flow in clay-filled shells was significantly different among species in the order: pinto > red > black (ANCOVA, $F = 6.22$, $df = 156$, $P < 0.05$) (Fig. 4.3(a)). Induced flow rates in a pinto abalone in which tremata elevations were removed by the addition of clay to the dorsal surface were significantly lower compared to those before manipulation (ANCOVA, $F = 57.2$, $df = 63$, $P < 0.01$). In contrast, induced flow rates were not significantly different between the normal black abalone model and a model with clay elevations added to the tremata (ANCOVA, $F = 2.60$, $df = 62$, $P > 0.05$).

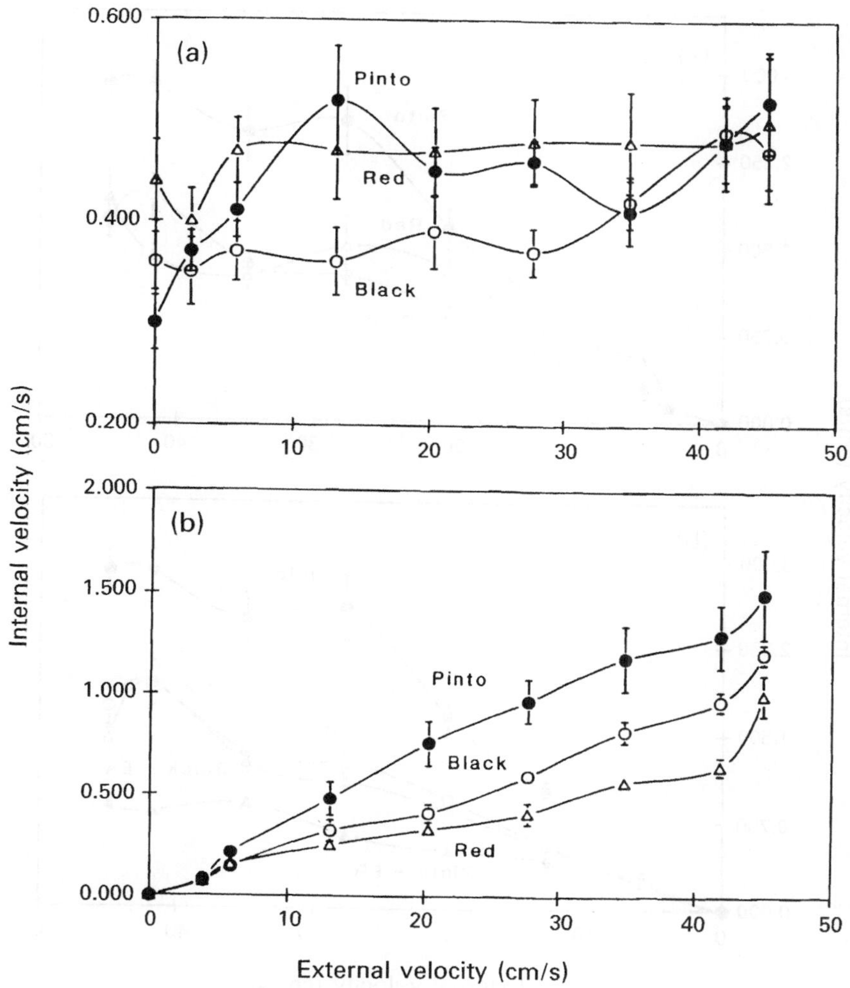

Fig. 4.2 Mean values of mantle cavity flow rates as a function of external velocity in black (*H. cracherodii*), red (*H. rufescens*) and pinto abalone (*H. kamtschatkana*). (a) Mean responses (±1 se) of live animals; (b) Mean induced flow rates (±1 se) in dead animals.

DISCUSSION

Measurements of drag on 13 species of abalone demonstrated that shell sculpture increases drag forces on the external shell surface. Species with elaborate shell sculpture, such as *H. corrugata*, *H. fulgens* and *H. scalaris*, had higher drag coefficients than smoother-shelled abalone, such as *H. cracherodii*, *H. iris* and *H. laevigata*. The former group of species generally occupy subtidal habitats or protected crevices, while the latter group occur more frequently on open substrates in shallow water habitats (Tables 4.1 and 4.3).

Several studies on marine gastropods have shown similar relationships between shell morphology, drag and vertical distribution. South African limpets (*Patella*) with shell sculpture were shown to have higher drag coefficients than species

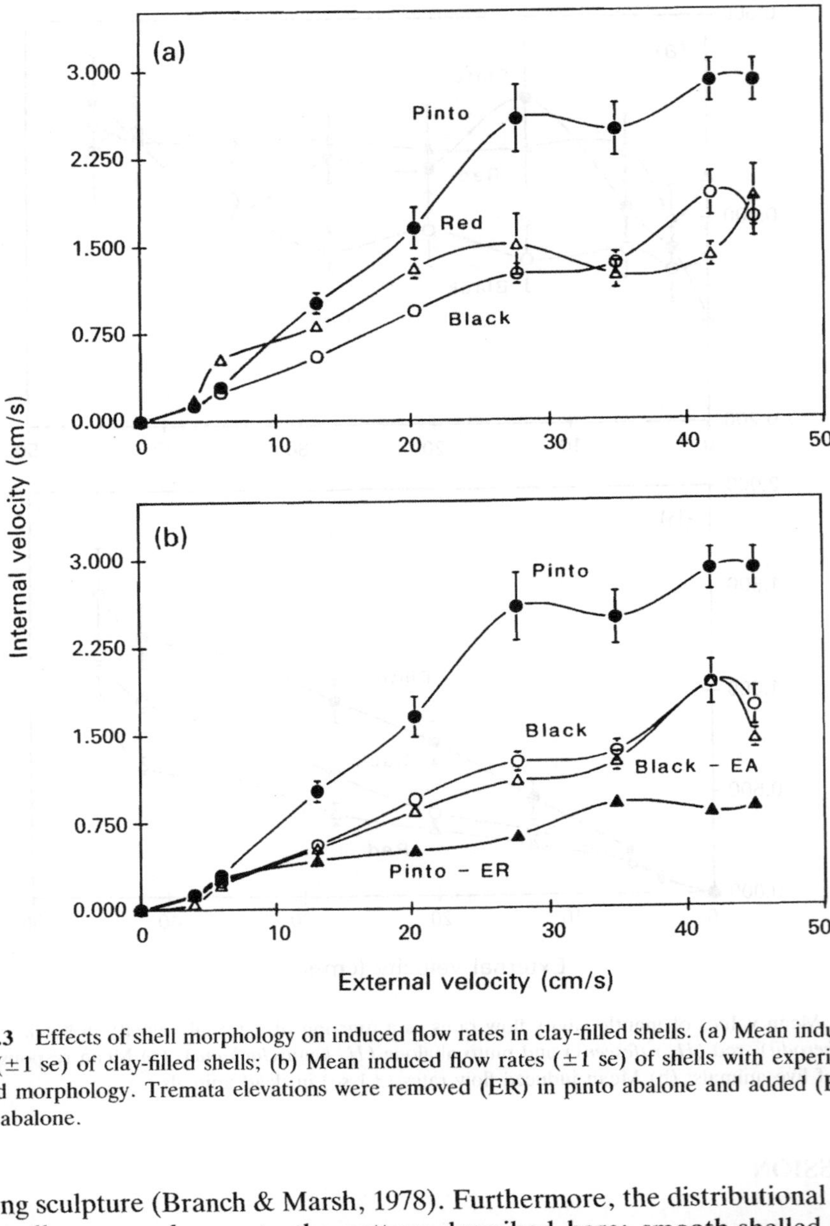

Fig. 4.3 Effects of shell morphology on induced flow rates in clay-filled shells. (a) Mean induced flow rates (±1 se) of clay-filled shells; (b) Mean induced flow rates (±1 se) of shells with experimentally altered morphology. Tremata elevations were removed (ER) in pinto abalone and added (EA) to black abalone.

lacking sculpture (Branch & Marsh, 1978). Furthermore, the distributional pattern in *Patella* was analogous to the pattern described here: smooth-shelled limpets occurred in more exposed, high intertidal water habitats than sculptured limpets, which occurred in more protected, low intertidal habitats. Based on similar distributional–morphological patterns, several authors have proposed hydrodynamic forces as a major selective factor influencing the distribution of gastropods (e.g. Warburton, 1976; Purchon, 1977).

Of critical importance to the role of drag forces as limits to distribution is the relationship between maximum forces in nature and abalone tenacity (attachment to the substrate). The observation that individuals are dislodged during strong storms (Cox, 1962; Tissot, 1990) indicates that hydrodynamic forces may be an important factor, although dislodgment could occur for a variety of reasons (e.g. moving boulders). If the tenacity of abalone is similar to that of the owl limpet (*Lottia gigantea*), which is 1.38×10^5 Pascals (Denny, 1988), then for an exposed 10 cm black abalone ($C_D = 0.43$) water velocities of over 20 m/s would be required for dislodgment (assuming all forces are due to drag). Water velocity measurements taken next to intertidal black abalone ranged between 1 and 3.43 m/s (Table 4.2). Thus, these animals are not typically threatened with dislodgment under 'normal' conditions. However, velocities of over 15 m/s are not uncommon in the marine environment (Vogel, 1981; Denny *et al.*, 1985). If the forces due to lift and acceleration are added to those calculated for drag above, then large storms may generate forces great enough to dislodge abalone. This chapter has demonstrated that species with strong shell sculpture will experience greater drag forces relative to smooth-shelled species at the same water velocity. However, interspecific differences in tenacity remain an equally important factor which is critical in the evaluation of the role of drag forces as limits to distribution in the Haliotidae.

Although interactions between velocity gradients and shell sculpture may limit the shallow-water distribution of some species, sculpture may play a greater role in expanding their deep-water distribution by serving as armour against shell-crushing predators, as has been documented in numerous other gastropods (Vermeij, 1978; Palmer, 1979). Major abalone predators such as crabs, octopuses, fishes, stingrays and sea otters (reviewed in Chapter 21, this volume) are generally more abundant in subtidal relative to intertidal habitats. To evaluate this hypothesis, however, the role of abalone sculpture as a deterrent against these predators needs to be examined.

Induced water flow through the mantle cavity in response to changing external velocity showed striking differences among species. A passive, two-fold increase in mantle cavity flow occurred in red and pinto abalone at low external velocities (5–15 cm/s). In contrast, black abalone experienced little induced flow at low velocities but displayed elevated flow rates at velocities which occur in its habitat (>50 cm/s). Thus, red and pinto abalone, which have large, elevated tremata, are more effective at inducing mantle circulation than the small, unelevated tremata of black abalone. However, due to the high velocity environment typical of black abalone, strong mantle cavity circulation undoubtedly occurs in the field.

Changes in induced flow rates in externally manipulated shells indicated morphological features that may have evolved to exploit velocity gradients on the shell surface. Interactions between moving water and shell traits suggest that covariation of several features has occurred which results in utilization of induced flow. Covariation involving a reduction in hole number, increase in size and elevation of tremata contributes to increased flow rates. Alteration of single traits (for example, the addition of elevations to small tremata) did not increase the effectiveness of induced flow at low external velocities. These observations are consistent with existing patterns of morphological covariation among abalone (Table 4.1), and thus support the hypothesis that trematal diversity results from induced flow.

In general, induced flow may represent an energetic saving for organisms, as work is done without the expenditure of metabolic energy (Vogel, 1976). In keyhole limpets, and similarly in abalone, the advantage of utilizing passive mantle cavity flow does not appear to be directly related to increased pumping rates *per se*, as only a small percentage of the total energy budget is devoted to ciliary pumping (Hughes, 1971). Rather, increased flow probably decreases the volume fraction of oxygen which must be removed from mantle cavity water for aerobic metabolism and allows a decrease in respiratory surface area (Murdock & Vogel, 1978). In still water, abalone extract 50% of available oxygen from the incurrent water (Hazelhoff, 1938). The existence of cardiac supplementation to ciliary mantle cavity pumping (Bourne, 1984) and the avoidance response of abalone to water low in oxygen (Jan & Chang, 1983) support the hypothesis that induced mantle cavity flow may be important. Abalone with large, elevated tremata are more efficient at promoting induced flow at low external velocities, and therefore are capable of maintaining a constant mantle cavity flow rate in a wider array of habitats than species with small, unelevated tremata.

The functional interpretation of variation in tremata shape and shell sculpture suggests that water movement is an important selective factor in the ecology of the Haliotidae. Therefore, it is likely that these functional relationships have also been an important factor during evolution. Unfortunately, the fossil record is fragmentary and incomplete, owing in large part to their rocky, wave-swept habitats (Boucot, 1981). As a result, an evaluation of the role of water movement as an evolutionary force is limited to phylogenetic analysis of modern species based on a combination of shell, epipodial and respiratory features. This approach could determine if the three morphological groups described here represent distinct evolutionary lineages or the result of convergent evolution in the face of similar selective pressures.

ACKNOWLEDGEMENTS

The author thanks the 1983 Biomechanics class taught at Friday Harbor Marine Laboratories by M. LaBarbara and S. Vogel. The manuscript has benefited from reviews by M. Hixon, J. McLean, M. LaBarbara, P. Lebednik, S. Vogel, J. Voltzow, G. Brenchley, K. Justice, R. MacMillen, B. Shaffer, S.A. Shepherd, M.J. Tegner, S. Walker and two anonymous reviewers. The author thanks J. Steinbeck for sending live abalone to Friday Harbor. Funds were provided by the University of Washington and the University of California, Irvine. The author would especially like to extend gratitude to J. Voltzow, who freely communicated her ideas on induced flow, and to G. Brenchley, who initially suggested that induced flow may differ among species.

REFERENCES

Boucot A.J. (1981) *Principles of Benthic Marine Paleoecology*. Academic Press, New York.

Bourne G.B. (1984) Muscular enhancement of ventilation in the pinto abalone, *Haliotis kamtschatkana*. *Am. Zool.*, **24**(3), 19A.

Branch G.M. & Marsh A.C. (1978) Tenacity and shell shape in six *Patella* species: adaptive features. *J. Exp. Mar. Biol. Ecol.*, **34**, 111–30.

Cox K.W. (1962) California abalones, family Haliotidae. *Cal. Fish. Game Bull.*, **118**.

Crofts D.R. (1929) *Haliotis*, Memoir 29. University Press, Liverpool.

Denny M.W. (1988) *Biology and the Mechanics of the Wave-Swept Environment*. Princeton University Press, Princeton.

Denny M.W., Daniels T.L. & Koehl M.A.R. (1985) Mechanical limits to size in wave-swept organisms. *Ecol. Monogr.*, **55**(1), 69–102.

Forster G.R. (1962) Observations in the ormer population of Guernsey. *J. Mar. Biol. Ass. UK*, **42**, 493–8.

Guice C.J. (1969) *Haliotis pourtalesii* Dall, 1881 from Florida Waters. *Veliger*, **11**(2), 140.

Hazelhoff E.H. (1938) Uber die ausnutzung des sauerstoffs bie verschiedenen wassertieren. *Z. Vergl. Physiol.*, **26**, 306–27.

Hughes R.N. (1971) Ecological energetics of the keyhole limpet *Fissurella barbadensis* Gmelin. *J. Exp. Mar. Biol. Ecol.*, **6**, 167–78.

Ino T. (1952) Biological studies on the propagation of the Japanese abalone (genus *Haliotis*). *Tokai Reg. Fish. Res. Lab. Bull.*, **5**.

Jan R.Q. & Chang K.H. (1983) The oxygen consumption by Formosan abalone, *Haliotis diversicolor supertexta* Lischke, during decline of ambient oxygen. *Bull. Inst. Zool. Academia Sinica*, **22**(1), 43–8.

Murdock G.R. & Vogel S. (1978) Hydrodynamic induction of water flow in a keyhole limpet (Gastropoda, Fissurellidae). *Comp. Biochem. Physiol.*, **61A**, 227–31.

Newman G.G. (1966) Movements of the South African abalone *Haliotis midae*. *Invest. Rep. Div. Sea Fish. S. Africa*, **56**, 1–19.

Owen B., McLean J.H. & Meyer R.J. (1971) Hybridization in the eastern Pacific abalones (*Haliotis*). *Bull. LA County Mus. Nat. Hist.*, **9**.

Palmer A.R. (1979) Fish predation and the evolution of gastropod shell sculpture: experimental and geographical evidence. *Evolution*, **33**(2), 697–713.

Palumbi S.R. (1984) Measuring intertidal wave forces. *J. Exp. Mar. Biol. Ecol.*, **81**, 171–9.

Poore G.C.B. (1972) Ecology of New Zealand abalones, *Haliotis* species (Mollusca: Gastropoda). 2. Seasonal and diurnal movement. *NZ J. Mar. Freshwat. Res.*, **6**(3), 246–58.

Purchon R.D. (1977) *The Biology of the Mollusca*, 2nd edn. Pergamon Press, New York.

Reeve L. (1846) *Monograph of the genus* Haliotis, **3**. Conchologia Iconica, London.

Riedl R.J. (1971) Water movement, animals. In *Marine Ecology*, (Ed. by G. Kinne), pp. 1124–49. Wiley-Interscience, London.

Shepherd S.A. (1973) Studies on southern Australian abalone (genus *Haliotis*): I. Ecology of five sympatric species. *Aust. J. Mar. Freshwat. Res.*, **24**(3), 217–57.

Sinclair M. (1963) Studies on the paua, *Haliotis iris* Martyn, in the Wellington district, 1945–46. *Zool. Publ. Victoria Univ. Wellington*, **35**, 1–16.

Tissot B.N. (1985) *Mechanisms and selective factors associated with morphological variation and diversity in two families of marine gastropods*. Master's thesis, University of California, Irvine.

Tissot B.N. (1990) *Geographic variation in the black abalone: the roles of development, ecological factors, and mass mortality*. PhD thesis, Oregon State University.

Tutschulte T.C. (1976) *The comparative ecology of three sympatric abalones*. PhD thesis, University of California, San Diego.

Vermeij G.J. (1978) *Biogeography and Adaptation, Patterns in Marine Life*. Harvard University Press, Cambridge, Mass.

Vogel S. (1974) Current-induced flow through the sponge, *Halichondria*. *Biol. Bull.*, **147**, 443–56.

Vogel S. (1976) Flows in organisms induced by movements in external medium. In *Scale Effects in Animal Locomotion* (Ed. by T.J. Pedley), pp. 285–97. Academic Press, London.

Vogel S. (1981) *Life in Moving Fluids, the Physical Biology of Flow*. Willard Grant Press, Boston.

Vogel S. & LaBarbara M. (1978) Simple flow tanks for research and teaching. *BioScience*, **28**, 638–43.

Voltzow J. (1983) Flow through and around the abalone *Haliotis kamtschatkana*. *Veliger*, **26**(1), 18–21.

Warburton K. (1976) Shell form, behavior, and tolerance to water movement in the limpet, *Patina pellucida* (L.) (Gastropoda: Prosobranchia). *J. Exp. Mar. Biol. Ecol.*, **23**, 307–28.

PART II
PHYSIOLOGY AND REPRODUCTION

Chapter 5
Review of endocrine regulation of reproduction in abalone *Haliotis* spp.

KIRK O. HAHN *Pacific Aquaculture Consulting, 14431 Holt Ave, Santa Ana, California 92705, USA*

ABSTRACT: Neurosecretion in the cerebral, pleural-pedal and visceral ganglia of abalone has been investigated to determine the role of hormones in the regulation of reproduction. The amount of neurosecretory material in specific cells in the ganglia has been found to correlate with different physiological cycles during reproduction. Neurosecretion in the cerebral ganglia has been found to correlate with gametogenesis and vitellogenesis, and neurosecretion in the pleural-pedal ganglion has been found to correlate with induction of spawning. Also, a structure has been found which stores neurosecretory material with a proposed role in spawning. A juxtaganglionic organ of glandular tissue has been described in the connective tissue surrounding each cerebral ganglion and has been proposed to have a function regulating reproduction. A prostaglandin-like chemical has been indicated to have a possible role in the induction of spawning. Induction of spawning by hydrogen peroxide or UV irradiation of seawater is probably caused by the production of a prostaglandin, although the exact mechanism or target tissue has not been determined. Continued research on the endocrine regulation of reproduction in abalone will help the development of culture methods to condition and spawn novel culture species.

RESUMEN: Se investigó la neurosecreción de los ganglios cerebral, pleuropedal y visceral del abulón para determinar el papel de las hormonas en la regulación de la reproducción. Se encontró que la cantidad de material neurosecretado por células específicas en los ganglios, se correlaciona con diferentes ciclos fisiológicos durante la reproducción. Se encontró que la neurosecreción en el ganglio cerebral se correlaciona con la gametogenesis y la vitellogenesis, mientras que la secreción en el ganglio pleuropedal se correlaciona con la inducción al desove. Se encontró también una estructura que almacena material neurosecretor que se piensa tiene un papel en el desove. Se describe un órgano juxtaganglionic de tejido glandular en el tejido conectivo que rodea a cada ganglio cerebral; se propone que dicho órgano, tiene una función reguladora de la reproducción. Se indica que una sustancia quimica aprecida a la prostaglandina podria tener una participación en la inducción al desove. La inducción al desove por peróxido de hidrógeno o por agua de mar irradiada con UV probablemente sea causada por la producción de una prostaglandina, aunque el meccanismo exacto o tejido-blanco, aun no ha sido determinado. La continuidad de la investigación sobre la regulación endocrina de la reproducción del abulón, ayudará al desarrollo de métodos de cultivo para acondicionar y desovar otras especies susceptibles de ser cultivadas.

INTRODUCTION

This chapter will review the information presently known about the brain ganglia in abalone and the role neurosecretion in these ganglia plays in the regulation of reproduction. Understanding the role the endocrine system plays in regulating reproduction offers an opportunity to increase the fecundity and spawning rate of presently cultured abalone and the expansion of the industry into novel abalone species. A brief discussion on the endocrine and nervous systems in abalone is presented to give a background on the current information in this field.

Hormones are produced by both neurosecretory cells and non-nervous cells usually grouped together into endocrine organs. However, very few non-nervous organs which have a proven endocrine function have been described in molluscs. Neurosecretory cells are found in large numbers and variety in the molluscan ganglia, and are the principal source of hormones (Joosse, 1979). Neurohormones are the phylogenetically oldest blood-borne chemical messengers, and this type of physiological regulation is a major function of the neuronal activity in lower invertebrates (Scharrer, 1978). Neurosecretory systems are fundamental to the endocrine regulation of important physiological functions, including growth and reproduction (Bern & Hagadorn, 1965).

The typical animal neurosecretory system is composed of neurosecretory cells and axons that terminate in close contact with the circulatory system to form a diffuse neurohaemal area or a neurohaemal organ for storage and release of the secretory product. A neurohaemal organ provides a site for the storage of pre-formed hormone with release subject to regulation by nervous impulses transmitted to the axon terminus.

The size and shape of molluscan neurosecretory cells do not differ from conventional neurons. The cells are often unipolar and located at the periphery of the ganglion. Cells of the same type are often located together in a specific area of the ganglion with the smaller cell bodies near the neuropil (concentration of axons in the medulla of the ganglion) and the larger ones at the periphery (Joosse, 1979; Joosse & Geraerts, 1983).

The perineurium (tissue sheath surrounding a nerve or ganglion) of molluscs is permeated with lacunae which are bathed by haemolymph and in many places it is penetrated by fine axonal processes, containing neurosecretory material, which terminate in the connective tissue (Dorsett, 1986). Release of neurosecretory material occurs inside the perineurium (Joosse & Geraerts, 1983). The hormones must diffuse through the perineurium and the connective tissue before entering the haemolymph (Roubos *et al.*, 1981; Joosse & Geraerts, 1983). Despite the general permeability of the perineurium, the axons of some neurosecretory groups travel through the neuropil to concentrated release sites or neurohaemal areas on particular nerves or connectives (nervous connection between different ganglia) (Joosse & Geraerts, 1983).

The location and configuration of release sites of molluscan neurohormones are unusual. In contrast to annelids, crustaceans, insects and vertebrates, molluscs usually lack special neurohaemal organs for the release of neurohormones into the haemolymph, but have evolved neurohaemal regions of large surface area. The regions are not concentrated and are widely distributed along the periphery of nerves, connectives and commissures (nervous connection between paired ganglia) (Joosse, 1979).

The central nervous system of *Haliotis* spp. is streptoneurous (twisted due to torsion) with a noticeable lack of concentration of ganglia into complex organs. The ganglia are elongated and flattened, and a sheath of ganglion cells usually extends to the flat commissures and connectives (Crofts, 1929). The ganglia are surrounded by a connective tissue sheath, and axon processes from the ganglion cells, muscle fibres and blood vessels are found within the sheath. There is no continuous outer epithelium separating the interior of the ganglion from the haemocoel which is equivalent to a blood-brain barrier (Dorsett, 1986).

The cerebral ganglia are the most anterior ganglia in the head. They are paired, lie above the oesophagus or buccal mass, and are connected by a long commissure (Dorsett, 1986). The long cerebro-pleural and cerebro-pedal connectives pass on either side of the odontophore, and connect to the pleural-pedal ganglion. The pleural-pedal ganglion is composed of paired pleural and pedal ganglia, with all four ganglia fused together into one complex ganglion. The pleural and pedal ganglia are intimately fused together to form a single rectangular mass and cannot be considered separately. Paired pleuro-visceral connectives pass dorsally from the pleural-pedal ganglion and form an extensive visceral loop on the dorsal surface of the abdominal viscera. The visceral ganglion lies slightly anterior to the posterior end of the right ctenidium, is diffuse and unpaired and innervates the gonad (Crofts, 1929; Bullock, 1965).

ENDOCRINE REGULATION OF REPRODUCTION

Most work on the endocrine regulation of the reproductive cycle in prosobranch gastropods has concentrated on the release of gametes (Joosse, 1988). However, the mechanisms are still poorly understood because the necessary experimental work is difficult to conduct on these animals and the neurosecretory cells are small (Fretter, 1984). Endocrine regulation in primitive prosobranchs is probably achieved by first-order neurosecretory mechanisms (neurohormones have a direct effect on the target tissues) (Choquet, 1971; LeGall & Streiff, 1975; Joosse, 1979; Fretter, 1984).

Cerebral ganglion

Four cell types (A–D) are observed in the cerebral ganglia, however, only cell type A has been indicated to have a role in regulating reproduction (Yahata, 1971; Hahn, 1990). A-cells are concentrated along the ganglion edge of the middle region with most along the inside surface and nerve junctions. These cells occur either singly or in groups of several cells. The edges of the cell body have many neurosecretory granules and the granules sometime fill the entire cell body (Yahata, 1971; Hahn, 1990).

Cell type A shows the characteristics of a neurosecretory cell. The cell body is full of intensely paraldehyde fuchsin (paraldehyde fuchsin, light green, orange G, chromotrope 2R) staining granules, and the amount and stain intensity of the neurosecretory material vary with the reproductive cycle. The cell bodies are concentrated along the ganglion periphery which would facilitate the release of 'neurohormones' into the haemolymph (Yahata, 1971; Hahn, 1990).

The stain intensity of cell type A shows a correlation with vitellogenesis in the ovary (Fig. 5.1). The stain intensity of A-cells shows some variation during the initial phase of gonad maturation. The number of oocytes increases during this time but the size of the oocytes remains very small. The stain intensity and amount of granules increase in the A-cells as the oocytes begin growing. The

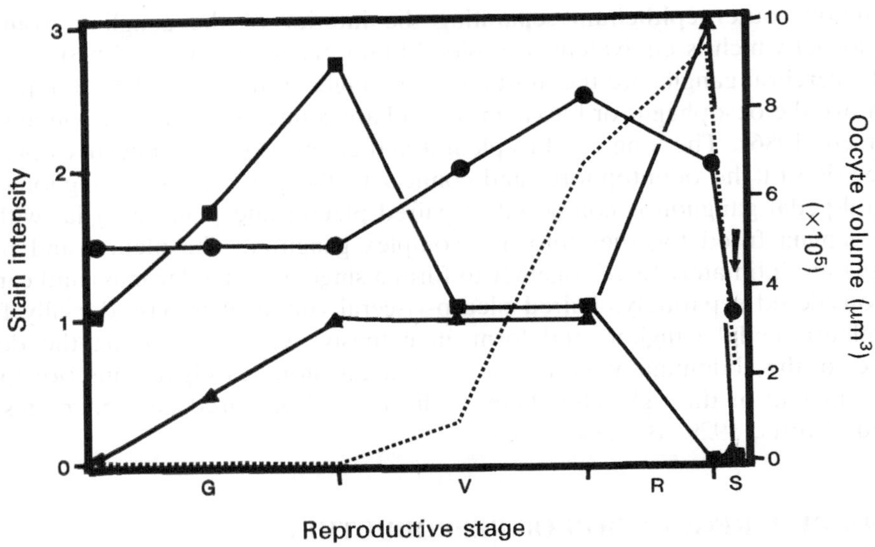

Fig. 5.1 Paraldehyde fuchsin stain intensity in the brain ganglia during gonad maturation. (●) Cell type A in cerebral ganglion. (■) Cell type #1 in pleural-pedal ganglion. (▲) Cell type #7 in pleural-pedal ganglion. (....) Mean oocyte volume during gonad development. G, gametogenesis; V, vitellogenesis; R, ripe; S, spent; arrow, induction of spawning.

neurosecretory material in A-cells remains at this level while the oocyte volume increases dramatically during the period of vitellogenesis. Although the material inside the A-cells shows a decrease after spawning, it probably does not regulate spawning (Hahn, 1990).

A study by Yahata (1971) found a similar relationship in cell type A, although he did not collect the abalone in the same manner. Yahata collected animals from a natural population and the reproductive cycle of the abalone used during the experiment was never verified histologically, only inferred by the date of sample collection. The Japanese abalone, *H. discus hannai*, begins gonad development in early spring and is fully ripe by September and October (Uki & Kikuchi, 1984). Paraldehyde fuchsin positive granules became visible in A-cells in June and the neurosecretory staining continued to increase until it reached a maximum in September. The stain intensity decreased after this point.

Injections of crude homogenates of the cerebral ganglion into ripe females do not induce spawning in ripe *H. discus hannai* females (Yahata, 1973). Cerebral ganglion crude homogenate does not cause any discharge of normal eggs and the ovary does not change histologically from unspawned controls. However, there is a considerable gain in the mean body weight from an increase in water uptake without any noticeable change in the ovary. Therefore, it was concluded that the cerebral ganglion does not contain a factor involved in the induction of spawning, although this ganglion may contain a factor which regulates the body water uptake (Yahata, 1973).

The secretory material in cell type A probably regulates vitellogenesis in the females, since the cycle of neurosecretory material closely follows the increase in the oocyte volume (Hahn, 1990). The production of a vitellogenic factor in the cerebral ganglia may be a general feature of prosobranch gastropods. A vitellogenic factor has been found in the cerebral ganglia of *Calyptraea sinensis, Crepidula fornicata* and *Patella vulgata* (Streiff, 1967; Choquet, 1969; Choquet, 1971; Joosse, 1979; de Jong-Brink *et al.*, 1983; Joosse, 1988).

Pleural-pedal ganglion

There are several cell types found in the pleural-pedal ganglion, but only two cell types, #1- and #7-cells, have neurosecretory activity which correlates to a physiological event in the reproductive cycle (Hahn, 1990).

Cell type #1

Most cells in the pleural-pedal ganglion are cell type #1. These large, oblong cells are located along the ganglion edge. The cell body is 8–20 μm in diameter and contains a large nucleus (5–10 μm in diameter). When stained with Herlant's (aniline blue, orange G, erythrosin B, acid fuchsin), the cell body is light to dark blue with a tinge of yellow (Hahn, 1990).

There is evidence of neurosecretory material in the cell bodies of #1-cells. The stain intensity of the neurosecretory material is very low at the beginning of gonad development, with a steady increase in the amount and stain intensity up to the point in the gonad development that gametogenesis ends and vitellogenesis begins. After this point, the neurosecretory material decreases dramatically. The neurosecretory material in #1-cells remains at this low level until the animals spawn, after which the cells appear completely devoid of neurosecretory material (Hahn, 1990).

The variation of neurosecretory material in cell type #1 clearly correlates with gametogenesis (Fig. 5.1). The secretion is maximum when the number of gametes is rapidly increasing in the gonads. Once vitellogenesis begins in the females, the neurosecretory material decreases since it is no longer needed. The #1-cells are completely empty and inactive when the animal is fully mature, and the cells are the same as before and after spawning (Hahn, 1990).

Cell type #7

The #7-cells in the pleural-pedal ganglion are found primarily at the periphery of the tissue surrounding the statocyst vesicle, with most of the cells toward the anterior. The cell body is large (16 μm) with a small nucleus (4 μm). The cytoplasm is light blue and granules are present in the cytoplasm. The granules stain dark purple with paraldehyde fuchsin (Hahn, 1990).

Cell type #7 shows a strong correlation to the induction of spawning (Fig. 5.1). Cells of this type with neurosecretory material are very rare at the beginning of gonad development. Those cells which do have neurosecretory material have very few granules and the granules have a low stain affinity. The number of granules in the cells and the stain intensity of the neurosecretory granules increase as gonad maturation continues. In addition, #7-cells with neurosecretory material increase in number. The amount of material and stain intensity reach a maximum level at full gonad maturation. After the animals spawn, #7-cells are completely empty and lack paraldehyde fuchsin staining. The difference in the paraldehyde fuchsin staining before and after spawning is very striking. The cells change from being completely full of darkly staining granules to being complete empty with no paraldehyde fuchsin staining. These observations indicate that cell type #7 produces a neurohormone which is released to induce spawning (Hahn, 1990).

Additional evidence for this conclusion comes from the detection of a possible neurohaemal organ in close proximity to the #7-cells in the statocyst tissue. This structure is the first neurohaemal organ reported in a prosobranch gastropod. The material inside the neurohaemal organ has the same stain affinity as the material inside the #7-cell body. The granular material stains blue-green with Herlant's and dark purple with paraldehyde fuchsin, indicating that it is probably neurosecretory in nature (Hahn, 1990). A storage area for a neurohormone to induce spawning is to be expected, although one has never been found before.

Spawning in abalone occurs rapidly, within 3 to 4 hours from induction to spawning, and all gametes are released during a short period (15–30 minutes) (Kikuchi & Uki, 1974; Morse *et al.*, 1978; Uki & Kikuchi, 1982). This type of spawning requires a rapid release of hormone at a concentration sufficient to induce the physiological responses which occur during spawning in the gonad. The time period appears too short to rely upon a steady production and release of neurohormone from the cells during the spawning event. The release of hormone in sufficient quantity could only be accomplished by either having the neurosecretory cells in large numbers along the ganglion periphery or having the hormone stored in a neurohaemal organ for rapid release.

The role of the pleural-pedal ganglion in the induction of spawning was demonstrated by Yahata (1973). Injections of pleural-pedal ganglion crude homogenate into female *H. discus hannai* cause a large swelling and softening of the ovary, which is followed by a slight increase in body weight caused by an increase in water content of the ovary. Most significantly, these animals spawn a large number of fertilizable eggs after one or two injections of the homogenate. The increase in the body weight, which occurs before the induced spawning, is due to an increased uptake of water in the ovary. This indicates that the uptake of water by the gonad may be essential to bring about the spawning and the water uptake might be directly regulated by a secretion in the pleural-pedal ganglion. These results indicate that the pleural-pedal ganglion of *H. discus hannai* may produce and release a factor(s) which could induce spawning, either directly or indirectly (Yahata, 1973).

Visceral ganglion

When female *H. discus hannai* are injected with crude homogenates of visceral ganglion or a combination of pleural-pedal and visceral ganglia, the ovary exhibits the same characteristics seen with injections of pleural-pedal ganglion alone. Spawning of a large quantity of fertilizable eggs follows the injections. However, the quantity of eggs spawned is significantly larger with injections of visceral ganglion or a combination of pleural-pedal and visceral ganglia as compared to injections of pleural-pedal ganglion alone. This finding suggests that the visceral ganglion plays a more central role in the induction of spawning than the pleural-pedal ganglion. The latter may initiate a cascade of physiological events which lead to the induction of spawning by a substance produced or stored in the visceral ganglion or the neurosecretory material from both ganglia may work jointly to induce spawning (Yahata, 1973).

Connective tissue

The connective tissue surrounding the cerebral ganglia of *H. discus hannai* shows no signs of secretory activity or function in the regulation of reproduction (Yahata, 1971; Hahn, 1990). Although there are some variations in the histology of the connective tissue during the reproductive cycle, these changes do not correlate with any physiological function in reproduction.

Miller *et al.* (1973), however, have proposed that the connective tissue and an associated juxtaganglionic tissue have a function in the reproduction of *H. rufescens*. A juxtaganglionic organ has been found associated with the cerebral ganglia of some prosobranch gastropods and has been suggested as a possible neurohaemal area (Martoja, 1968). The juxtaganglionic organ is a well-vascularized, poorly innervated tissue of glandular cells within the connective tissue surrounding the dorsal and posterior regions of the cerebral ganglia (Switzer-Dunlap, 1987). Cyclic changes in the appearance of the connective tissue are observed in opisthobranchs, i.e. advanced development at the time of gametocyte maturation and atrophy at the time of gamete release. The juxtaganglionic organ has been suggested to have a role in reproduction but there are no experimental results to indicate a regulatory function (Joosse & Geraerts, 1983).

The juxtaganglionic tissue in *H. rufescens* is characterized by large eosinophilic cells with a granular cytoplasm and large nucleoli. The juxtaganglionic tissue occurs dorsally and laterally, as well as medially on the inner surface of the ganglion. The juxtaganglionic tissue is generally distributed diffusely within the connective tissue surrounding the ganglion, but it is especially concentrated above the root of the cerebro-pleural connective and at the bases of the optic and tentacular nerves. The juxtaganglionic organ is well developed at the time of gametocyte maturation but begins to atrophy at spawning. Prior to spawning, the juxtaganglionic cells of mature abalone are numerous, very large and contain an acidophilic, finely granular cytoplasm. Shrinkage of the cytoplasm of the juxtaganglionic cells occurs immediately after induction of spawning, which suggests a correlation between the juxtaganglionic tissue and the reproductive cycle, although the experimental methods lack the precision to prove this relationship (Miller *et al.*, 1973).

Prostaglandin

Although the endocrine regulation of reproduction in abalone is not completely understood, it is suspected that a prostaglandin-like chemical may be involved in the induction of spawning (Morse *et al.*, 1977). Prostaglandin-dependent egg production and spawning have been demonstrated in some molluscan species (e.g. oysters, scallops, mussels and clams) (Joosse, 1988). Mammalian prostaglandins (PGE and PGF) added to seawater in very low concentrations (about 0.1 ppm) can specifically induce spawning in gravid male and female *H. rufescens* (Morse *et al.*, 1977, 1978) In *Helisoma* spp. (a freshwater pulmonate), introduction of nanogram amounts of prostaglandin E_2 (PGE_2) into the female genital tract of mated individuals increases both the number of egg masses produced and the number of eggs per mass. The action of PGE_2 is localized in the female genital tract, since injection of PGE_2 (1000 fold increase) into the haemocoel has no effect on reproduction (Kunigelis & Saleuddin, 1986).

A model for the effect of prostaglandin on reproduction in *Helisoma* spp. has been proposed. Prostaglandin precursors, such as arachidonic acid, are transferred with semen during copulation. Prostaglandin synthetase present in the bursa copulatrix converts these precursors to active prostaglandin. Prostaglandin stimulates the production of a bursal matedness factor which is released into the haemocoel. This factor acts upon the dorsal bodies to stimulate the production and release of dorsal body hormone. This hormone induces vitellogenesis in the ovotestis to initiate egg production in virgins or to regulate egg production in regularly-mating animals. Prostaglandin produced within the ovotestis may induce ovulation, and oviposition may sequentially follow ovulation or may be under the regulation of a second endocrine centre (Kunigelis & Saleuddin, 1986).

The induction of spawning in abalone by UV irradiation of seawater or addition of H_2O_2 to seawater suggests that prostaglandins have a role in the induction of spawning. Enzymological and physiological studies indicate that spawning may result from a peroxide-induced stimulation of the endogenous enzymic synthesis of potent hormone-like prostaglandin molecules (Morse *et al.*, 1978). The biosynthesis of prostaglandins is known to depend on the prior synthesis of prostaglandin endoperoxides. The short-lived intermediates contain both endoperoxides and hydroperoxy moieties which are sequentially introduced from molecular oxygen by the catalytic activity of the enzyme fatty acid cyclooxygenase (prostaglandin endoperoxide synthetase) (Morse *et al.*, 1977).

Both UV irradiation and addition of H_2O_2 to seawater are believed to produce hydroperoxy free radicals, $HOO^{\bullet}$, or peroxy diradicals, $^{\bullet}OO^{\bullet}$ (Morse *et al.*, 1977; Uki & Kikuchi, 1984). These radicals probably act as donors of electronically-activated oxygen suitable for the cyclooxygenase-catalyzed addition of oxygen to precursors to form prostaglandins.

It is not known where the arachidonic acid, which is the precursor needed for conversion to prostaglandin, is stored in the abalone, but the ovary is a possible source. There is a high level of arachidonic acid present in the ovary and the fatty acid composition in the ovary is very similar in the different abalone species. Furthermore, the presence of arachidonic acid in the ovary is unusual for a marine species (Olley & Thrower, 1977).

Spawning induction can be blocked by the addition of aspirin (an inhibitor of the fatty acid cyclooxygenase-catalyzed first step in the conversion of arachidonic acid to prostaglandin endoperoxide) or catalase (an enzyme which specifically destroys H_2O_2) to the water before adding hydrogen peroxide. The inhibition of the H_2O_2 induction of spawning by aspirin suggests that cyclooxygenase activity and prostaglandin endoperoxide synthesis may be required in the animal during the exposure to H_2O_2 for the induction of spawning. This function is probably normally regulated by endocrine factors, since the fatty acid cyclooxygenase reaction is rate limiting in the physiological sequence of events leading to spawning in the gravid abalone (Morse *et al.*, 1977).

Table 5.1 Structures which have a possible role in the endocrine regulation of reproduction.

Structure	Possible role
Cerebral ganglion	
A-cells	vitellogenesis
Pleural-pedal ganglion	
#1-cells	gametogenesis
#7-cells	induction of spawning
neurohaemal organ	induction of spawning
Visceral ganglion	induction of spawning
Gonad	induction of spawning

CONCLUSIONS

The use of ganglion homogenates and hormonal therapies offer an opportunity to the culturist to improve control over the brood stock (Table 5.1). Adult abalone could be conditioned during normally quiescent periods; alternatively the time needed for conditioning could be greatly reduced. Additionally, spawning of all species of abalone could become relatively easy and the need for expensive equipment or chemicals eliminated. Research in these areas is continuing and new techniques should help the culturist to be more efficient.

REFERENCES

Bern H.A. & Hagadorn I.R. (1965) Neurosecretion. In *Structure and Function in the Nervous System of Invertebrates* (Ed. by T.H. Bullock & G.A. Horridge), pp. 353–429. W.H. Freeman, San Francisco.

Bullock T.H. (1965) Mollusca: Gastropoda. In *Structure and Function in the Nervous System of Invertebrates* (Ed. by T.H. Bullock & G.A. Horridge), pp. 1283–386. W.H. Freeman, San Francisco.

Choquet M. (1969) *Contribution a l'étude du cycle biologique et de l'inversion du sexe chez* Patella vulgata *L. (Mollusque Gastropode Prosobranch)*. These de Science, No 185, Lille.

Choquet M. (1971) Study of the life cycle and sexual inversion in *Patella vulgata* L. (Prosobranch, Gastropod, Mollusk). *Gen. Comp. Endocrinol.*, **16**(1), 59–73.

Crofts D.R. (1929) *Haliotis. Liverpool Marine Biology Committee Memoirs on Typical British Marine Plants and Animals*, *29*, 1–174.

de Jong-Brink M., Boer H.H. & Joosse J. (1983) Mollusca. In *Reproductive Biology of Invertebrates* Vol. 1 (Ed. by K.G. Adiyodi & R.G. Adiyodi), pp. 297–355. John Wiley, New York.

Dorsett D.A. (1986) Brains to cells: The neuroanatomy of selected gastropod species. In *The Mollusca*, Vol. 9 (Ed. by A.O.D. Willows), pp. 101–87. Academic Press, New York.

Fretter V. (1984) Prosobranchs. In *The Mollusca*, Vol. 7 (Ed. by A.S. Tompa, N.H. Verdonk & J.A.M. van den Biggelaar), pp. 1–45. Academic Press, New York.

Hahn K.O. (1990) *Neurosecretory staining in the cerebral and pleural-pedal ganglia of* Haliotis discus hannai *and* Trochus niloticus, *and its relationship to reproduction*. PhD thesis, University of California, Davis.

Joosse J. (1979) Evolutionary aspects of the endocrine system and of the hormonal control of reproduction of molluscs. In *Hormones and Evolution*, Vol. 1 (Ed. by E.J.W. Barrington), pp. 119–57. Academic Press, New York.

Joosse J. (1988) The hormones of molluscs. In *Endocrinology of Selected Invertebrate Types*, (Ed. by H. Laufer & R.G.H. Downer), pp. 89–140. Alan R. Liss, New York.

Joosse J. & Geraerts W.P.M. (1983) Endocrinology. In *The Mollusca*, Vol. 4 (Ed. by A.S.M.

Saleuddin & K.M. Wilbur), pp. 317–406. Academic Press, New York.

Kikuchi S. & Uki N. (1974) Technical study on artificial spawning of abalone, genus *Haliotis* II. Effect of irradiated sea water with ultraviolet rays on inducing to spawn. *Bull. Tohoku Reg. Fish. Res. Lab.*, *33*, 79–86.

Kunigelis S.C. & Saleuddin A.S.M. (1986) Reproduction in the freshwater gastropod, *Helisoma*: involvement of prostaglandin in egg production. *Int. J. Invert. Reprod. Develop.*, **10**, 159–67.

LeGall S. & Streiff W. (1975) Protandric hermaphroditism in prosobranch gastropods. In *Intersexuality in the Animal Kingdom* (Ed. by R. Reinboth), pp. 170–8. Springer-Verlag, New York.

Martoja M. (1968) Existence d'une organe juxtaganglionaire chez *Aplysia punctata* Cuv. (Gastropoda Opistobranchia). *C. R. Acad. Sci. Paris*, **260**, 2907–9.

Miller W., Nishioka R.S. & Bern H.A. (1973) The 'Juxtaganglionic' tissue and the brain of the abalone, *Haliotis rufescens* Swainson. *Veliger*, **16**(2), 125–9.

Morse D.E., Duncan H., Hooker N. & Morse A. (1977) Hydrogen peroxide induces spawning in mollusks, with activation of prostaglandin endoperoxide synthetase. *Science*, **196**, 298–300.

Morse D.E., Hooker N. & Morse A. (1978) Chemical control of reproduction in bivalve and gastropod molluscs, III: An inexpensive technique for mariculture of many species. *Proc. World Maricul. Soc.*, **9**, 543–7.

Olley J. & Thrower S.J. (1977) Abalone – an esoteric food. In *Advances in Food Research*, Vol. 23 (Ed. by C.O. Chichester, I.M. Mark & G.F. Stewart), pp. 143–86. Academic Press, New York.

Roubos E.W., Keijzer A.N. & Buma P. (1981) Adenylate cyclase activity in axon terminal of ovulation hormone producing neuroendocrine cells in *Lymnaea stagnalis. Cell Tissue Res.*, **220**, 665–8.

Scharrer B. (1978) Current concepts on the evolution of the neurosecretory neuron. In *Neurosecretion and Neuroendocrine Activity. Evolution, Structure and Function. VII. International Symposium on Neurosecretion*, Vol. 7 (Ed. by W. Bargmann, A. Oksche, A.L. Polenov & B. Scharrer), pp. 9–14. Springer-Verlag, New York.

Streiff W. (1967) Etude endocrinologique du determinisme du cycle sexual chez un mollusque hermaphrodite protandre, *Calyptra sinensis. Ann. Endocrinol.*, **28**, 461–72.

Switzer-Dunlap M. (1987) Ultrastructure of the juxtaganglionar organ, a putative endocrine gland associated with the cerebral ganglia of *Aplysia juliana. Int. J. Invert. Reprod. Develop.*, **11**, 295–304.

Uki N. & Kikuchi S. (1982) Technical study on artificial spawning of abalone, genus *Haliotis* VIII. Characteristics of spawning behavior of *H. discus hannai* induced by ultraviolet irradiation stimulus. *Bull. Tohoku Reg. Fish Res. Lab.*, **44**: 83–90.

Uki N. & Kikuchi S. (1984) Regulation of maturation and spawning of an abalone, *Haliotis* (Gastropoda) by external environmental factors. *Aquaculture*, **39**, 247–61.

Yahata T. (1971) Demonstration of neurosecretory cells in the cerebral ganglion of the abalone, *Nordotis discus* Reeve. *Bull. Fac. Fish. Hokkaido Univ.*, **22**(3), 207–14.

Yahata T. (1973) Induced spawning of abalone (*Nordotis discus* Reeve) injected with ganglional suspensions. *Bull. Jpn. Soc. Sci. Fish.*, **39**(11), 1117–22.

Chapter 6
Abalone egg vitelline layers and hybrid gamete interactions

CYNTHIA L. LEWIS, ROBIN EICKHOFF and CHARLES STRINGHAM
Biology Department, Life Sciences 135, San Diego State University, San Diego, California 92182, USA
KIRK MILHOAN *School of Medicine, Physiology & Pharmacology Program, University of California, San Diego, La Jolla, California 92093, USA*
KEVIN TUBBS *University of New England, College of Osteopathic Medicine, Hills Beach Road, Middeford, Maine 04005, USA*

ABSTRACT: For fertilization to be successful in abalone, the sperm's acrosome granule must release a lysin which solubilizes the egg vitelline layer, one of the physical blocks to fertilization. The abalone egg vitelline layer is about 64% carbohydrate and 36% protein by weight (Lewis *et al.*, 1982). The major components separated by polyacrylamide gel electrophoresis (PAGE) are glycoproteins. We attempted to understand how these components fit together in the intact vitelline layer by adding enzymes to disrupt its structure, then staining it with ruthenium red, a dye that binds preferentially to acid glycosaminoglycans, and viewing the eggs by electron microscopy. Various ruthenium red-stained components are removed by hyaluronidase, trypsin, pronase and ethylene diamine tetraacetic acid. Fourier transform infrared spectroscopic (FT-IR) analysis of vitelline layer-lysin mixtures suggests that lysin induces conformational change in the egg vitelline layer which is specifically indicated by changes in the amide III vibration of the vitelline layer, reflecting a loss of β-sheet protein structure. A turbidometric assay of heterospecific lysin-vitelline layer mixtures shows that heterospecific lysin is incapable of completely solubilizing the egg vitelline layers. We speculate that normally lysin binds to the vitelline layer in a stoichiometric fashion, possibly at a carboxylate region, based on changes in the carboxylate wave intensity of the vitelline layers measured by FT-IR spectroscopy after solubilization by lysin, and that lysins of different species have different active sites for this binding.

RESUMEN: A fin de que la fecundación del abulón pueda efectuarse, el gránulo acrosómico del espermatozoide debe liberar una lisina que solubilice la capa vitelina del óvulo, que es uno de los obstáculos físicos para la fecundación. La capa vitelina del óvulo de abulón tiene cerca de 64% de carbohidratos y 36% de proteinas por peso. Los principales componentes separadoes por PAGE son glicoproteinas. Nosotros pretendemos entender como estos componentes interaccionan en la capa vitelina intacta, agregando enzimas para destruir su estructura, después ésta se tiñe con rojo de rutenina, colorante que se une preferentemente a ácidos glicosaminoglicanos, y se observa el óvulo al microscopio electrónico. Varios componentes teñidos con el rojo de rutenio se eliminan con hialuronidasa, tripsina, pronasa y EDTA. El análisis de nuestro material por espectroscopía FT-IR llevado a cabo en los laboratorios Batelle, sugiere que la lisina del espermatozoide provoca cambios en la estructura molecular de la capa vitelina del óvulo. Un análisis de la lisina de espermatozoides de *Haliotis corrugata* con las capas vitelinas del óvulo de *H. rufescens* usando un ensayo turbidométrico, mostró que la lisina de espermatozoides de *H. corrugata* fue incapaz de solubilizar completamente las capas vitelinas del óvulo de *H. rufescens*. Especulamos que la lisina se une a la capa vitelina de una manera estequiométrica y que las lisinas de diferentes especies pueden tener diversos sitios activos para esta unión.

INTRODUCTION

About 0.2% of all abalone collected off California are interspecies hybrids (Owen *et al.*, 1971; Talmadge, 1977; Hahn, 1989). Hybrids have been produced among four commercially important abalone species native to California (Leighton & Lewis, 1982): the red, *Haliotis rufescens*, the pink, *H. corrugata*, the green, *H. fulgens* and the white, *H. sorenseni*. However, hybrid production is difficult and success may reflect genetic affinities between species (Meyer, 1967). Leighton & Lewis (1982) determined that sperm concentration must be 10 times greater than for homologous crosses and eggs must be freshly spawned for optimal hybrid fertilization to occur. Juveniles of the *H. rufescens* × *H. fulgens* and *H. rufescens* × *H. sorenseni* crosses displayed superior growth and juveniles from the *H. rufescens* × *H. corrugata* cross had higher survival rates than those of *H. corrugata* (Leighton & Lewis, 1982). Thus, it has been our goal to promote desirable traits in cultured abalone by facilitating production of these crosses.

Abalone sperm are attractive models for the investigation of fertilization events due to the large size of their acrosome granule (Lewis *et al.*, 1980). One of the proteins contained in the acrosome granule of abalone is a powerful egg vitelline layer lysin (reviewed by Dan, 1967; Hoshi, 1985). Lysin apparently acts non-enzymically to break apart the vitelline layer that surrounds the egg into monomers (Lewis *et al.*, 1982), and may also promote fusion between sperm and egg plasma membranes (Hong & Vacquier, 1986). A less soluble 20 kD molecular weight protein also located in the acrosome granule facilitates lysin's activity in an unknown manner in the Japanese abalone (Haino-Fukushima & Usui, 1986). Evidence supporting a stoichiometric, non-enzymic mechanism of dissolution of the molluscan egg vitelline layer is compelling (Haino-Fukushima, 1974; Lewis *et al.*, 1982; Ogawa & Haino-Fukushima, 1984). This hypothesis has stimulated considerable interest in the abalone as a model in which to study sperm–egg interaction (Sakai *et al.*, 1982, 1985; Ogawa & Haino-Fukushima, 1984; Shiroya & Sakai, 1984; Fridberger *et al.*, 1985; Shiroya *et al.*, 1986; Usui, 1987).

Our ultimate goal has been to facilitate entry of non-homologous sperm through the egg vitelline layer, promoting its fusion with the egg plasma membrane and subsequent initiation of hybrid development. We measured vitelline layer–lysin interaction by visualizing the vitelline layer with a stain for glycosaminoglycans, standard biochemical characterization of lysin and the vitelline layers of several species, an assay of lysin–vitelline layer interactions and by observing FT-IR spectral changes that reflect the molecular configuration of the vitelline layer when lysin is present.

METHODS AND MATERIALS

Gamete collection

Abalone were collected near La Jolla, or from Santa Catalina Island (*Haliotis sorenseni*), California, and were induced to spawn by exposure to Tris-buffered hydrogen peroxide (Morse *et al.*, 1977). The gametes were collected in two to four litres of filtered seawater.

Preparation of gametes for transmission electron microscopy

The sperm and eggs of *H. rufescens* were collected separately in artificial seawater (ASW). Sperm were concentrated to 10^7 sperm ml^{-1}, added to 1 ml packed eggs at that concentration to induce polyspermy, and the gametes were fixed by adding 5 ml aldehyde fixative for 20 minutes followed by postfixation in osmium containing ruthenium red (Longo, 1973). Eggs were made polyspermic to improve the likelihood of observing sperm–egg interactions. The inorganic dye ruthenium red (RR) preferentially binds to acid glycosaminoglycans (GAGs, Luft, 1971). Some fertilized eggs were treated with 250 μg ml^{-1} trypsin, 100 IU ml^{-1} testicular hyaluronidase, 250 μg ml^{-1} pronase, or 500 μg ml^{-1} ethylene diamine tetra-acetic acid (EDTA) prior to fixation (all chemicals from Sigma). These treatments were selected to disrupt but not solubilize the vitelline layers (Kelley, 1975). The gametes were dehydrated in a graded ethanol series, propylene oxide, and infiltrated and embedded in Spurr's resin. The blocks were sectioned using a Reichert ultracut ultramicrotome and viewed with a Philips 200 TEM at 40 kV using a 25 or 40 micrometer objective aperture. Some thin sections were stained with lead citrate and uranyl acetate (Lewis *et al.*, 1980). As controls, gametes were viewed without RR in the fixative and with grids unstained.

Isolation of the lysin and vitelline layers

Vitelline layers were prepared from spawned eggs of *H. rufescens*, *H. corrugata* and *H. fulgens* using the method of Lewis *et al.* (1982). Sperm lysin was isolated from *H. rufescens* and *H. fulgens* as in Lewis *et al.* (1982), and from *H. corrugata* by reducing the ionic strength of the dialysing solution in the final step to 25 mM Tris-HCl, pH 7.4, containing 5 mM NaCl and 0.02% sodium azide.

Physicochemical methods

For the subunit molecular weight determination of lysin and vitelline layers, PAGE was performed in a 12.5% gel containing 0.1% sodium dodecyl sulphate (SDS) (Laemmli, 1970). The samples were boiled, reduced and hydrolysed with 2% mercaptoethanol and 2% SDS, then the gel was stained with silver (Morrissey, 1981) or Coomassie blue (Weber & Osborne, 1969). Subunit molecular weights were determined (Weber & Osborne, 1969) using breast muscle crude supernatant (BMCS) as a molecular weight standard (Petell *et al.*, 1981) or β galactosidase (116 kD), phosphorylase (97.4 kD), bovine serum albumin (66 KD), egg albumin (45 kD), carbonic anhydrase (29 kD) and lysozyme (14.6 kD).

Infrared spectroscopy

Atoms and groups of atoms within a molecule vibrate with characteristic frequencies and will absorb light at those frequencies which are in the infrared region. If light containing infrared frequencies is focused on a molecule and the amount of light absorbed is measured as a function of frequency, then the frequency and intensity of light absorption can be used to identify compounds or to determine features of molecular structure. This plot of intensity versus frequency is the infrared spectrum (Barnwell, 1972). Mixtures of purified *H. rufescens* lysin and vitelline layers were sent to the National Center For Biomedical Infrared Spectroscopy, Battelle Laboratories, Columbus, Ohio. The infrared spectra were acquired using a liquid attenuated total reflectance cell requiring highly concentrated *H. rufescens* lysin and vitelline layers (Harrick, 1979); 3.5 mg ml^{-1}, 4.75 mg ml^{-1} and 6.0 mg ml^{-1} vitelline layer protein was combined with 5.0 mg ml^{-1} lysin protein in 50 mM Tris-HCl and 10 mM NaCl, pH 7.4 (Lewis *et al.*, 1982) in 0.5 ml aliquots. Above 5.0 mg ml^{-1}, lysin will not stay in solution in the isolation medium. The 3.5 mg ml^{-1} vitelline layer mixture was the only one completely solubilized by the lysin. The 4.75 and 6.0 mg ml^{-1} vitelline layer mixtures included unsolubilized material in the form of a sticky gel that was pelleted in the samples.

Turbidometric assay of sperm lysin activity

The dissolution of isolated vitelline layers by lysin from homospecific and heterospecific crosses was determined by measuring the decrease in turbidity of a vitelline layer suspension at 340 nm following the addition of lysin (Lewis *et al.*, 1982). For this assay the vitelline layers were fragmented by mild sonication in seawater containing 20 mM Tris, pH 8.0.

RESULTS

Visualization of acid glycosaminoglycans in the vitelline layer

Ruthenium red binds to microvilli on the egg plasma membrane, material in the perivitelline space, the inner vitelline layer surface, ovoid structures inside the vitelline layer, the outer vitelline layer surface and egg jelly. The electron-dense RR granules are enhanced by standard staining with heavy metals (Fig. 6.1 (A)–(C)). Loss of RR labelling where sperm have released lysin onto the vitelline layers is most obvious in the vitelline layer ovoid structures and along the outer vitelline layer surface (Fig. 6.1(B), (D), (E)). Specific RR-binding sites are apparent at regular intervals where the vitelline layer subunits have been dispersed by lysin. These sites give the appearance of round heads (encircled labelled sites) attached to fibrous tails (unlabelled sites, Fig. 6.1(D), (E)).

Incubation of polyspermic eggs in testicular hyaluronidase prior to RR staining removes most of the jelly coat, all of the previously-bound sperm (Fig. 6.2(A)) and some, but not all, of the vitelline layer ovoid material (Fig. 6.2(D)). Ruthenium red-stained GAGs at the vitelline layer surface are more dense (Fig. 6.2(D)) than in the seawater control (Fig. 6.2(A)); 250 μg ml^{-1} trypsin removes much of the RR-stained vitelline layer ovoid material, but leaves their dense outlines intact (Fig. 6.2(B)). When 500 μg ml^{-1} trypsin was used to treat the eggs, all RR-stained ovoid material was removed (not shown). Trypsin, which clips carboxyl ends of arginine and lysine residues, did not remove RR-stained GAGs from the

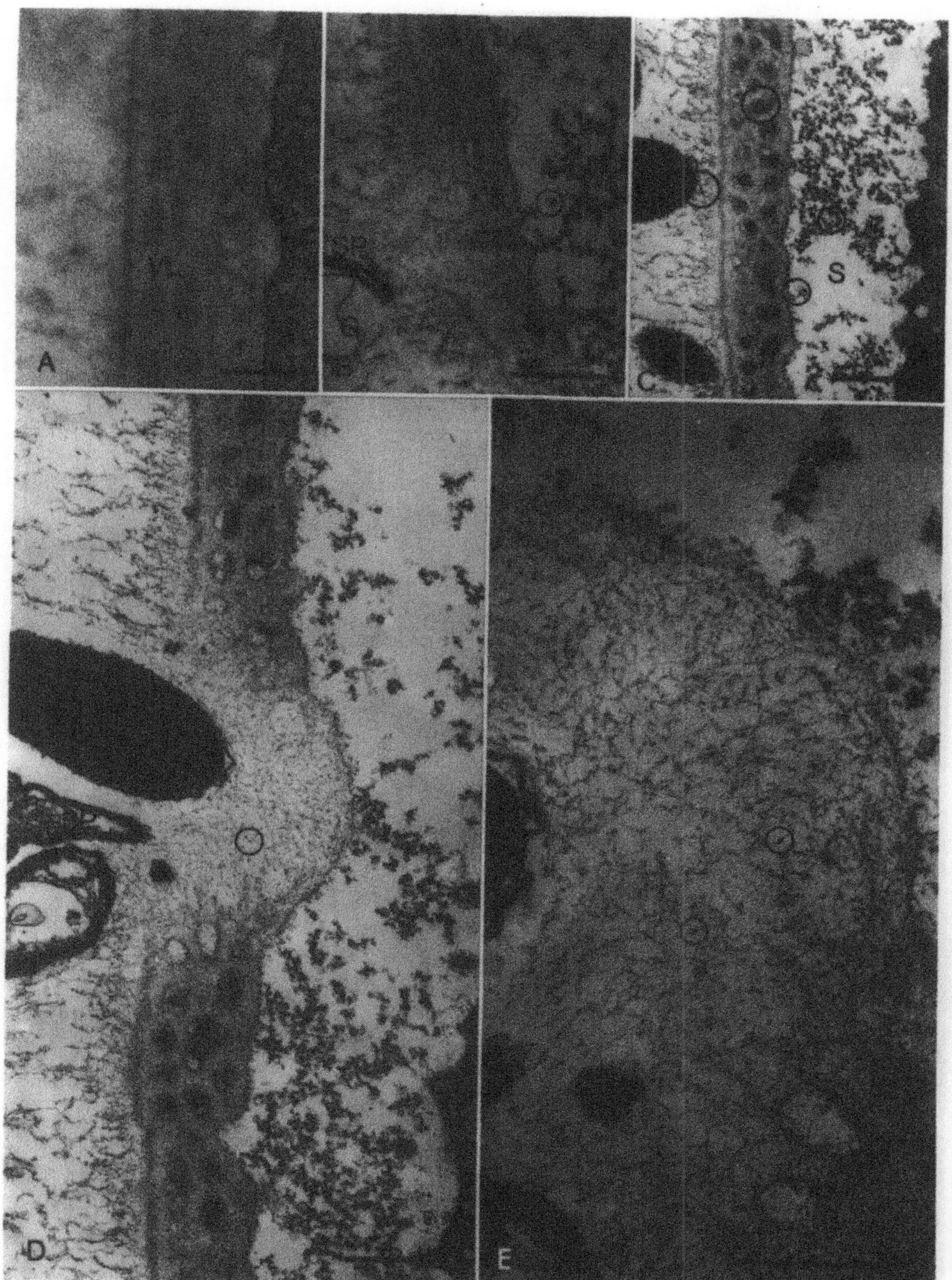

Fig. 6.1 Polyspermic *Haliotis rufescens* eggs fixed with or without ruthenium red (RR) in the fixative. The egg is to the right in all micrographs. (A) Control, no RR in fixative, no stain. Scale is 0.5 μm. (B) Control, RR in fixative, no stain. Circles are around RR-stained material in the perivitelline space. RR binds to the same areas as in (C). Note the acrosomal granule ghost (G) and acrosome filament from an entering sperm (SP). Scale is 1.0 μm (C) RR in fixative and section stained with aqueous uranyl acetate and lead citrate. RR binds to the plasma membrane (circled), material in the perivitelline space (S, circled), the vitelline layer surfaces (circled) and vitelline layer ovoid structures (circled). Scale is 1.0 μm. (D) and (E) Fixed with RR and stained. RR-bound dye is located on round heads (circled) of the vitelline layer components dispersed by lysin from acrosome-reacted sperm (SP). Note the loss of dye attached to the vitelline layer ovoid structures where the sperm has penetrated the vitelline layer. Scales are 0.5 μm.

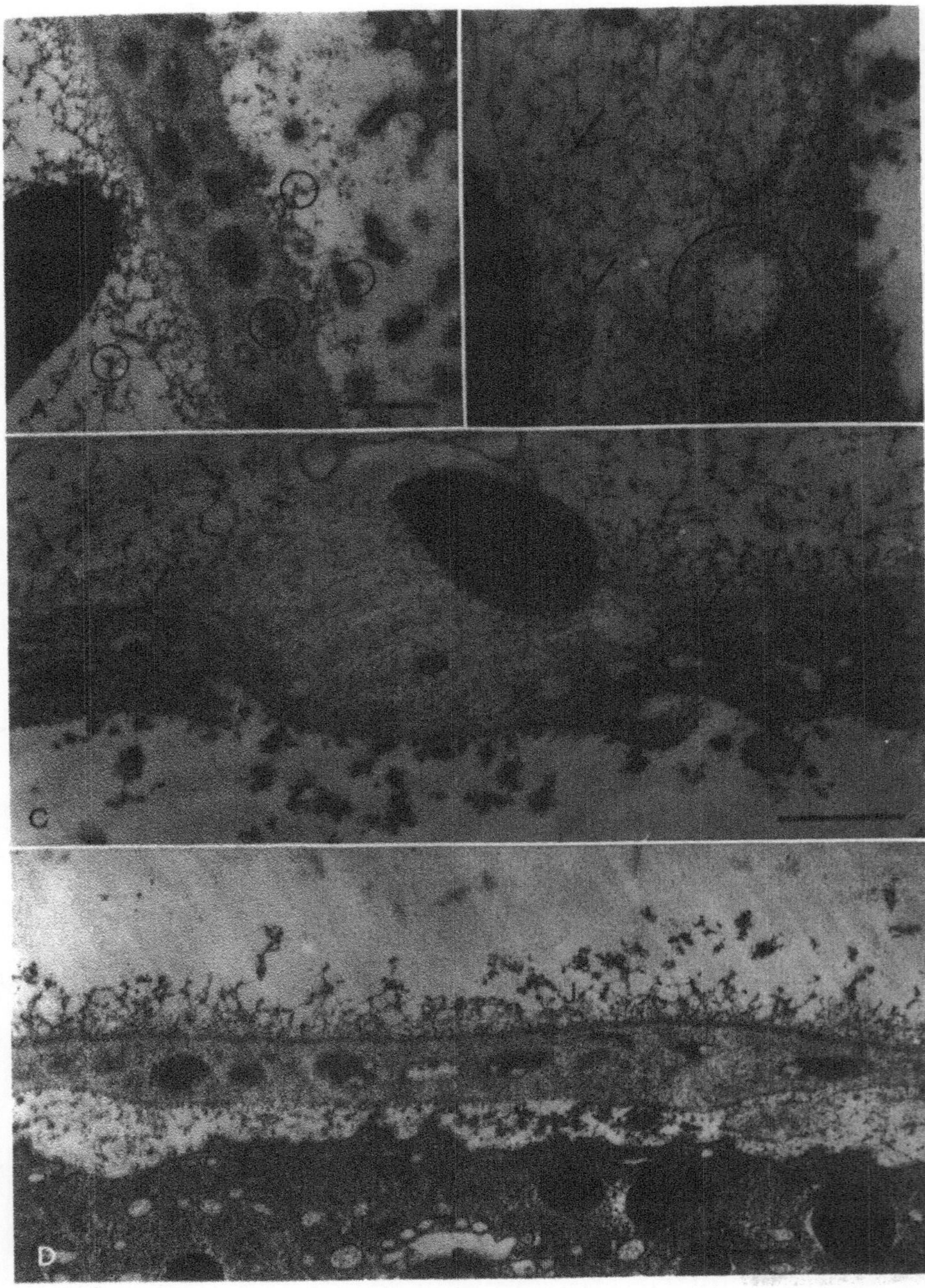

Fig. 6.2 Polyspermic eggs with the sperm coming from the left side in (A) and (B), and from the top in (C) and (D). All eggs are fixed with RR and stained. (A) Control egg incubated in artificial seawater for 30 minutes at 15°C. RR dye is bound to the same areas (circles) as in the egg in Fig. 6.1(C). Scale is 0.5 μm. (B) Egg incubated in 250 μg/ml trypsin for 30 minutes at 15°C, then fixed. Ovoids (one is circled) in the vitelline layer no longer stain with RR, although RR-stained material is in the perivitelline space, and on dispersed vitelline layer heads. Two stained vitelline layer heads are indicated by arrows. Scale is 0.5 mm. (C) Egg treated with trypsin. Ovoids (two are circled and indicated by arrows) in the vitelline layer no longer stain with RR. Scale is 1.0 μm. (D) Egg incubated in 100 IU ml^{-1} testicular hyaluronidase for 30 minutes at 15°C. Note the reduction of egg jelly (J). Scale is 1.0 μm.

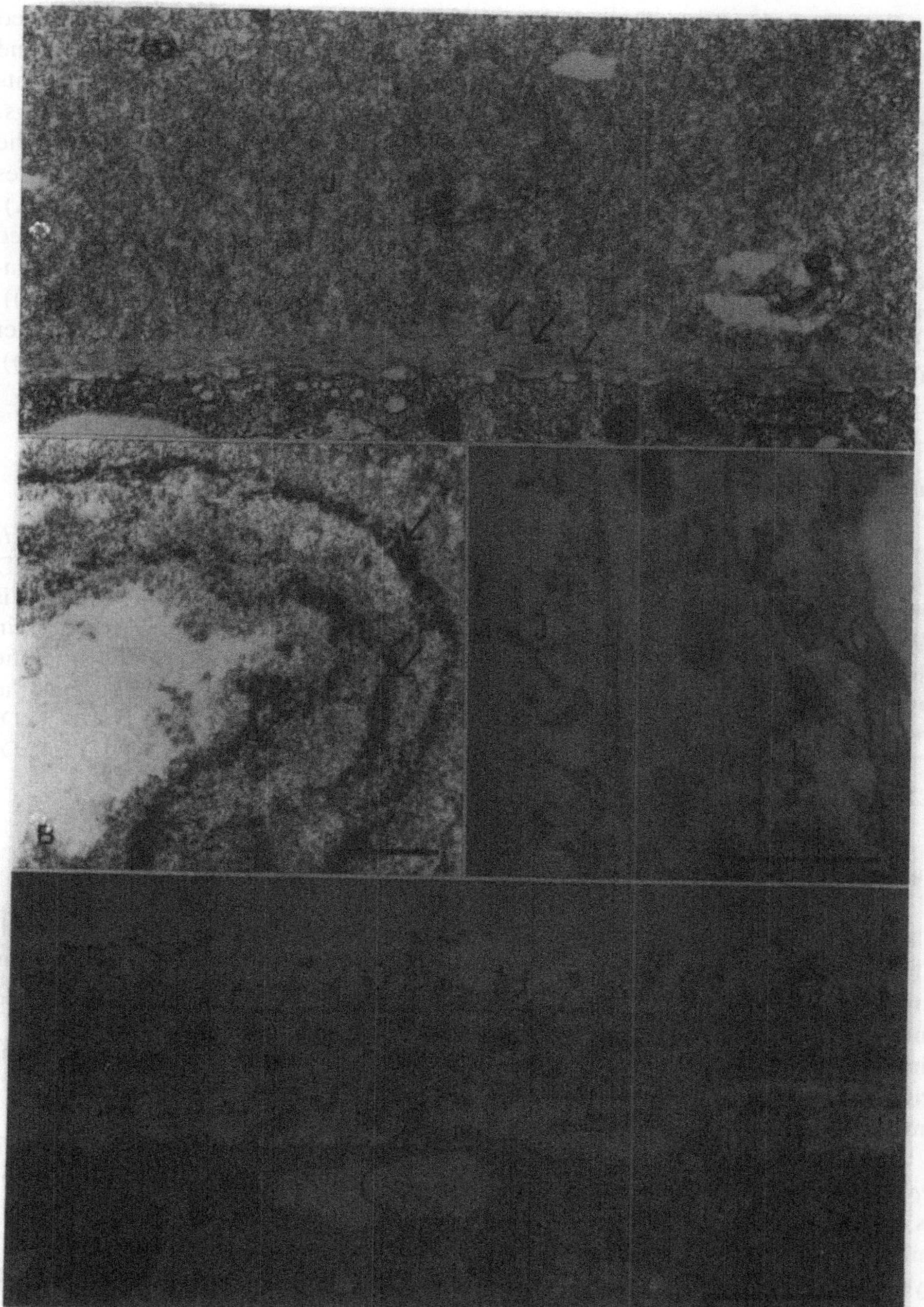

Fig. 6.3 Polyspermic eggs fixed with RR. (A) and (B) are stained; (C) and (D) are not stained. All eggs are to the right or to the bottom in the micrographs. (A) and (B) Eggs incubated in 250 $\mu m\ ml^{-1}$ pronase for 30 minutes at 15°C. Note the increased density of the egg jelly (J) and presence of approximately three new RR-dense bands (three arrows) in the vitelline layer. (A) Scale is 1.0 μm. (B) Close up of the vitelline layer with three dense bands. Scale is 0.1 μm. (C) and (D) Eggs incubated in 500 $\mu g\ ml^{-1}$ EDTA for 30 minutes at 15°C. Note the reduction of egg jelly (J) and sperm. Scales are 1.0 μm.

jelly or from the outer vitelline layer surface (Fig. 6.2(C)), but few sperm remained bound to the vitelline layer; thus, sperm-binding receptors on the egg vitelline layer must be glycoproteinaceous. Egg vitelline layer head and tail components separated by interaction with sperm lysin appear the same as in untreated eggs. Pronase, a broader spectrum protease, on the other hand, drastically alters the egg's constitution. The egg jelly appears denser, the perivitelline space collapses and the vitelline layer no longer has RR-stained ovoid structures (Fig. 6.3(A), (B)). The vitelline layer has three lines of RR-stained material. These stained areas may outline the positions of the original ovoid areas or may reflect reorganization of GAGs upon removal of more of the protein component (Fig. 6.3(B)). EDTA removes all of the sperm, most of the jelly coat and alters the outer vitelline layer surface so GAGs appear clumped together (Fig. 6.3(C), (D)). EDTA-treated eggs appear similar to hyaluronidase-treated eggs.

Subunit molecular weights

Lysin (the lowest molecular weight band in purified sperm extract) from *H. fulgens* is 22.0 kD (insoluble material is 24.0 kD), from *H. corrugata* is 18.4 kD (insoluble material 22.6 kD), from *H. rufescens* is 16.6 kD (insoluble material is 19.6 kD) and from *H. sorenseni* is 19.0 kD (Fig. 6.4 (B)–(E)). No differences in electrophoretic mobility are apparent in the nine major bands of the vitelline layers of *H. rufescens*, *H. corrugata* and *H. fulgens* (31 kD to 50 kD), but the lower molecular weight subunits are different: *H. fulgens* (16.0 kD, 14.5 kD, 13.5kD), *H. corrugata* (16.2 kD, 14.7 kD, 14.5 kD) and *H. rufescens* (15.5 kD, 12.0 kD) (Fig. 6.5 (B)–(D)).

Fourier-transform infrared spectroscopy

The spectrum of sperm lysin alone shows the amide I (1650 cm^{-1}) and amide II (1550 cm^{-1}) vibrations typical of all proteins, but has only three frequencies that are near to (1110, 1250 cm^{-1}) or coincide with (1430 cm^{-1}) the 10 frequencies of intact vitelline layers (compare Fig. 6.6(A) with (B) or (C)) in the 1000 to 1500 cm^{-1} fingerprint region. Thus, changes in the fingerprint region indicate changes in the vitelline layer structure rather than changes in the lysin structure when they interact.

Lysin apparently interacts with vitelline layers in a stoichiometric fashion (Lewis *et al.*, 1982). The concentration of solubilized vitelline layers that can be obtained is limited by the solubility of lysin in isolation medium (5.0 mg lysin protein ml^{-1}). When FT-IR spectra of intact vitelline layers are compared to those of lysin-solubilized ones, changes can be determined in the conformation of the vitelline layers. Ten of the major vitelline layer wave frequencies with or without lysin align (Fig. 6.6(B)), although their intensities differ. These differences in intensity are enhanced by developing a 'difference' spectrum (Fig. 6.6(C)) where a spectrum of partially solubilized vitelline layers is subtracted from a spectrum of intact ones. Nine of the 10 'difference' wave frequencies align with those of intact vitelline layers. Graphs were drawn to show wave frequency alignment, although the Y-axes are on different scales.

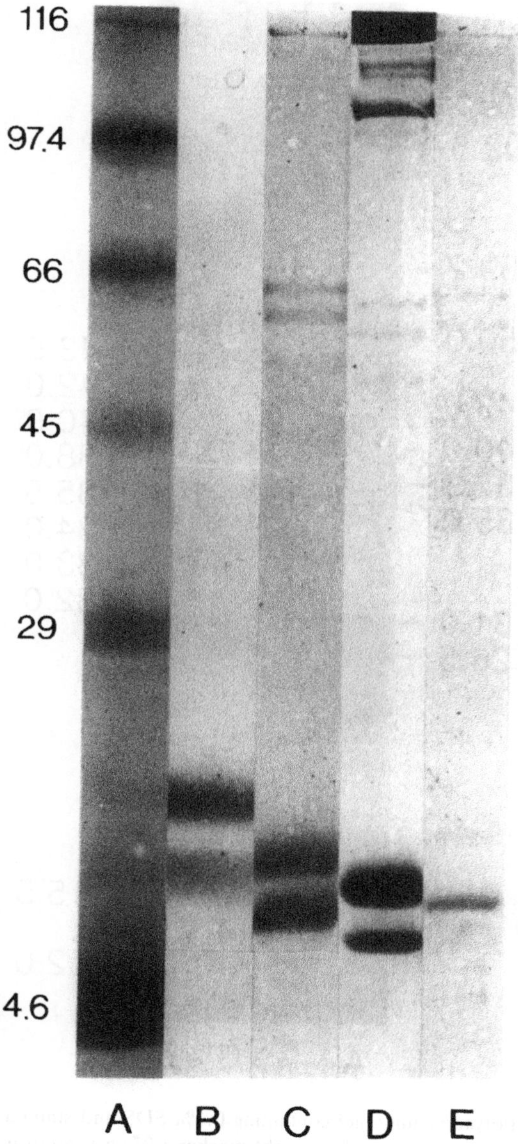

Fig. 6.4 A–E 12.5% polyacrylamide gel containing 0.1% SDS and stained with Coomassie blue; 25 μg protein applied to each lane. A Molecular weight markers. B Partially purified *H. fulgens* lysin 22.0 kD protein and an insoluble 24.0 kD protein. C Partially purified *H. corrugata* lysin, 18.4 kD, and a 22.6 kD band. D Partially purified *H. rufescens* lysin, 16.6 kD, and an insoluble 19.6 kD band. E Purified *H. sorenseni* lysin, 19.0 kD. The insoluble, higher molecular weight bands in B, C and D may be comparable to Haino-Fukushima & Usui's (1986) insoluble lysin-facilitating protein.

The most noticeable differences between intact and solubilized vitelline layer spectra are in the intensities of the 1120 cm^{-1} and 1040 cm^{-1} bands. Intact vitelline layers have a less intense 1120 cm^{-1} (Fig. 6.6(B), (C)) and a more intense 1040 cm^{-1} band than in the solubilized ones (Fig. 6.6(C), see arrows on the difference spectrum). Thus, the 1120 cm^{-1} is characteristic of the solubilized

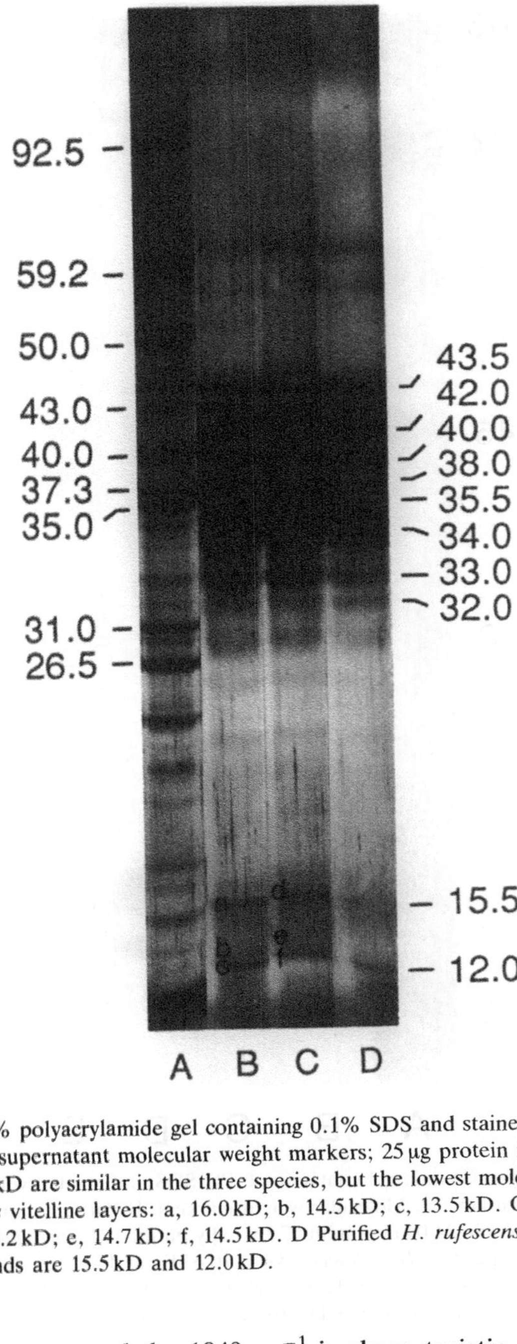

Fig. 6.5 A–D 12.5% polyacrylamide gel containing 0.1% SDS and stained by the silver method. A Breast muscle crude supernatant molecular weight markers; 25 μg protein in each channel. B–D Bands from 32 to 43 kD are similar in the three species, but the lowest molecular weight bands differ. B Purified *H. fulgens* vitelline layers: a, 16.0 kD; b, 14.5 kD; c, 13.5 kD. C Purified *H. corrugata* vitelline layers: d, 16.2 kD; e, 14.7 kD; f, 14.5 kD. D Purified *H. rufescens* vitelline layers, low molecular weight bands are 15.5 kD and 12.0 kD.

vitelline layer structure and the 1040 cm^{-1} is characteristic of the intact vitelline layer structure. These two bands may be related to one another, that is, as one increases during lysin–vitelline layer interaction, the other decreases.

Another difference between intact and solubilized vitelline layers is in the amide III vibration at 1240 cm^{-1}. Intact vitelline layers have a greater intensity of the 1240 cm^{-1} (Fig. 6.6(C)) than solubilized ones, and a new band at 1260 cm^{-1} is seen in the solubilized vitelline layers (Fig. 6.6(B)) which was buried beneath the

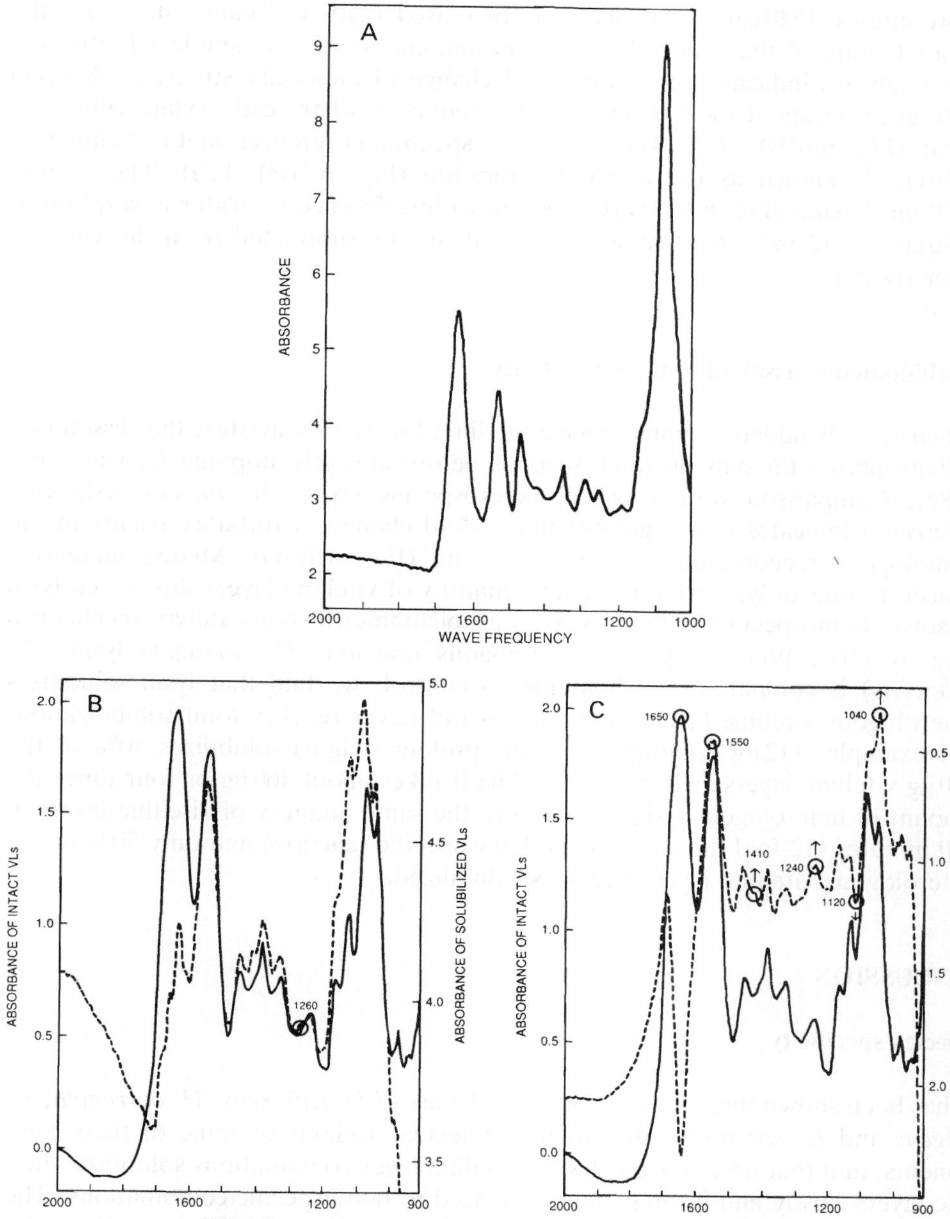

Fig. 6.6 Transmission FT-IR spectra of *H. rufescens* sperm lysin and intact and solubilized vitelline layers, egg treated in different ways. (A) 5 mg lysin protein ml^{-1} in vitelline layer isolation medium. (B) 10 mg intact vitelline layer protein ml^{-1} (solid line) compared to 5.0 mg protein ml^{-1} partially solubilized by 5.0 mg lysin protein ml^{-1} (broken line). Note that the scale for the intact vitelline layers is on the left and the scale for the solubilized vitelline layer spectrum is on the right. (C) 10 mg intact vitelline layer protein ml^{-1} (solid line) compared to a difference spectrum where the spectrum of 4.75 mg vitelline layer protein ml^{-1} solubilized by 5.0 mg lysin protein ml^{-1} is subtracted from the spectrum of 10 mg intact protein ml^{-1} (broken line). Note that the scale for the intact vitelline layers is on the left and the scale for the solubilized ones is on the right. Wave frequencies of interest are circled and arrows indicate an increase or decrease in wave intensity of intact compared to solubilized vitelline layers.

more intense 1240 cm^{-1} in intact ones. Increased 1240 cm^{-1} band intensity is the major feature of the amide III vibration, and changes in the amide III vibration nearly always indicate a conformational change of molecular structure. A small difference occurs at the 1410 cm^{-1} wave frequency where carboxylate vibrations occur (Fig. 6.6(B), (C)). There are also spectral differences in the band near 1550 cm^{-1}, known as the amide II vibration (Fig. 6.6(B), (C)). The intense 1650 cm^{-1} band (Fig. 6.6(B), (C)) is due to interference by water absorption at 1640 cm^{-1} and indicates that water was not totally subtracted from the vitelline layer spectrum.

Turbidometric assay of biological activity

When lysin is added to homospecific vitelline layers in seawater, the dissolution activity occurs for only 30 to 60 seconds before abruptly stopping (Lewis *et al.*, 1982). Comparison with a typical heterospecific cross (*H. fulgens* lysin × *H. rufescens*) indicates that a greater than 5-fold change in turbidity occurs in the homologous reaction during the same time (Fig. 6.7(A)). Mixing increasing concentrations of lysin with a constant quantity of vitelline layers shows that lysin dissolves homospecific vitelline layers by a stoichiometric, non-catalytic mechanism (Fig. 6.7(B)). When a typical heterologous reaction (*H. corrugata* lysin × *H. rufescens*) is compared to a homologous control, we find that lysin solubilizes heterologous vitelline layers more slowly and never reaches total solubilization. For example, 112 μg homospecific lysin (protein weight) solubilizes 50% of the 100 μg vitelline layers in 10 minutes, while it takes about 460 μg or four times the amount of heterologous lysin to solubilize the same quantity of vitelline layers in 120 minutes (12-fold the time for the homospecific reaction) and only 50% of the heterologous vitelline layers becomes solubilized.

DISCUSSION

Species specificity

It has been shown that lysins and vitelline layers of *H. rufescens*, *H. corrugata*, *H. fulgens* and *H. sorenseni* differ in the molecular weights of some of their components, and that heterologous lysin–vitelline layer combinations solubilize vitelline layers slowly and incompletely compared to homospecific combinations. The PAGE determination of the molecular weight of *H. rufescens* lysin (16.6 kD) agrees closely with the calculation of Fridberger *et al.* (1985) by amino acid sequencing (16.07 kD), and the *H. corrugata* lysin ran at a slightly higher molecular weight (18.4 kD) than on Fridberger's (1987) PAGE (16.4 kD). Fridberger and colleagues found *H. corrugata* lysin to contain three more amino acids than *H. rufescens* lysin, although V.D. Vacquier & C.L. Carner (pers. comm.) found only one additional amino acid in *H. corrugata* lysin. V.D. Vacquier (pers. comm.) also found a significant difference in molecular weights of lysin from *H. rufescens* and *H. fulgens*. Sperm have difficulty binding to and acrosome reacting with heterologous egg vitelline layers (C.L. Lewis, pers. obs.). As the

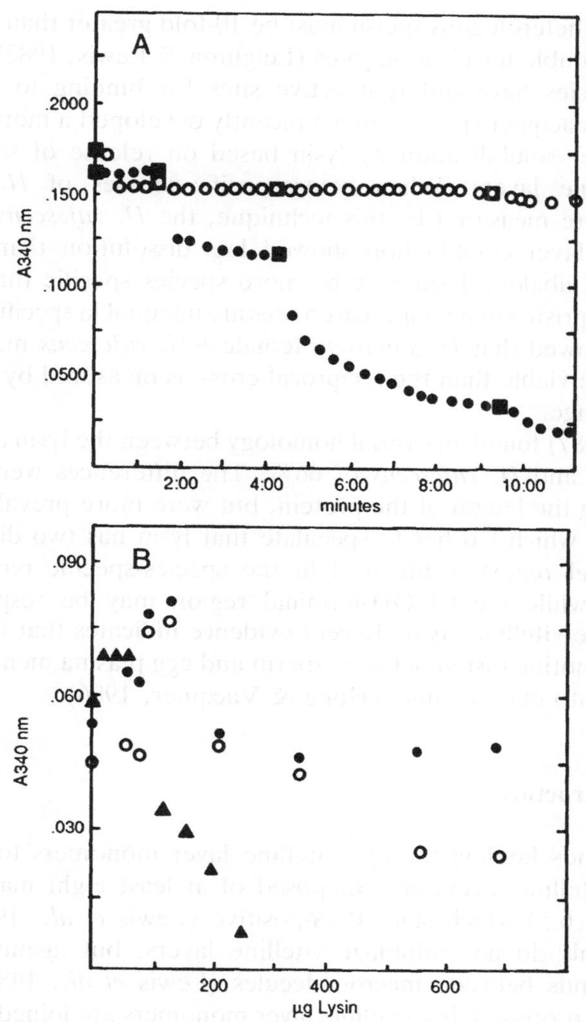

Fig. 6.7 (A) and (B) Each point represents the average of four readings of the same sample made at 15 s intervals after the suspension was stirred in a cuvette. Each sample contained 100 µl BSA at 20 mg ml^{-1} to decrease adsorption of lysin to the cuvette, 100 µg (200 µl in A and 400 µl in B) isolated *H. rufescens* vitelline layer protein and variable quantities of lysin, with the total volume brought to 1.0 ml with calcium-free artificial seawater (ASW), pH 8.0. When complete clarity occurred, the A_{340} averaged 0.015 against a calcium-free ASW blank. This background was subtracted from each sample. (A) Time-course turbidometric assay of homospecific and heterospecific lysin–vitelline layer combinations. 75 µl aliquots of 0.5 mg ml^{-1} lysin (37.5 µg protein) were added to vitelline layers at intervals. Boxes around data points indicate the last reading taken before an addition of lysin. Filled circles: readings from *H. rufescens* pure lysin and vitelline layers; open circles: readings from partially pure *H. fulgens* lysin and *H. rufescens* vitelline layers. (B) Turbidometric assay of increasing lysin concentrations of homospecific and heterospecific lysin–vitelline layer combinations; 400 µl aliquots of 40 to 700 µg protein of pure *H. rufescens* or partially pure *H. corrugata* lysin. Filled triangles: readings made 10 minutes after the addition of *H. rufescens* lysin; filled circles: readings made 10 minutes after the addition of *H. corrugata* lysin; open circles: readings made 120 minutes after the addition of *H. corrugata* lysin.

concentration of heterologous sperm must be 10-fold greater than for homologous sperm for comparable fertilization rates (Leighton & Lewis, 1982), perhaps lysins of different species have different active sites for binding to their respective vitelline layers. Vacquier (pers. comm.) recently developed a more sensitive assay for vitelline layer solubilization by lysin based on release of soluble ^{125}I from iodinated vitelline layers. When heterospecific crosses of *H. rufescens* and *H. corrugata* were measured by this technique, the *H. rufescens* lysin + *H. corrugata* vitelline layer combination showed less dissolution than the reciprocal cross. Thus, red abalone lysin may be more species specific than pink abalone lysin. It is not surprising that reciprocal crosses are unequal in specificity. Leighton & Lewis (1982) showed that *H. sorenseni* female + *H. rufescens* male crosses were three times more viable than the reciprocal cross as measured by development to the 4–16 cell stage.

Fridberger (1987) found an overall homology between the lysin amino sequences of *H. corrugata* and *H. rufescens* of 66%. The differences were not randomly distributed along the length of the protein, but were more prevalent in the NH_2-terminal region, which led her to speculate that lysin has two distinct functions: the NH_2-terminal region is involved in the species-specific recognition of the vitelline layer, while the COOH-terminal region may be responsible for the dissolution of the vitelline layer. Recent evidence indicates that lysin may have a second role promoting fusion between sperm and egg plasma membranes after the gametes come into close contact (Hong & Vacquier, 1986).

Vitelline layer structure

The type of bonds holding the egg vitelline layer monomers together was also investigated. Vitelline layers are composed of at least eight major glycoprotein monomers (Fig. 6.5) which stain PAS-positive (Lewis *et al.*, 1982). High ionic strength materials do not solubilize vitelline layers, but agents which destroy hydrophobic bonds between macromolecules (Lewis *et al.*, 1982) do solubilize them. Thus, it is proposed that vitelline layer monomers are joined by hydrophobic bonds and that lysin has a detergent-like activity. In *H. rufescens*, most of the vitelline layer carbohydrate is fucose (49%) with about 15% hexosamine; the most abundant amino acid is threonine (Lewis *et al.*, 1982).

How are the sugar and protein components arranged and how do they change when lysin interacts with them? That lysin binds to the vitelline layers rather than just solubilizing them seems likely as a material similar to slightly-set gelatin was produced when highly concentrated vitelline layers and lysin were added together. Ruthenium red-stained glycosaminoglycan (GAG) components that appear like small heads are distributed evenly throughout vitelline layers which are solubilized by lysin. Carbohydrate components are also concentrated in ovoid structures within the intact vitelline layer and along both surfaces. Vitelline layer ovoid structures must contain both carbohydrate and protein since they are trypsin-, pronase- and hyaluronidase-sensitive and stain with RR. Hyaluronidase-sensitive material along the outer vitelline layer may be sperm-binding receptors since sperm are lost after trypsin, hyaluronidase, and EDTA treatment.

Lysin activity on the vitelline layer

The FT-IR data suggest that conformational changes occur in the vitelline layer structure upon contact with lysin. If vitelline layers contain phosphorus, changes in band frequencies at 1120 and 1040 cm^{-1} are likely due to lysin binding on to vitelline layer phosphate linkages. One could speculate that calcium or other ions attached to carboxylate moieties of vitelline layer monomers hold the β-sheet structure together. This would explain an observation than 20 mM ethylene glycol bis (β-aminoethyl ether): *N,N′*-tetraacetic acid (EGTA) disrupts the sheet structure. The decreased intensity of the 1240 cm^{-1} of lysin-solubilized vitelline layers indicates a smaller amount of β-sheet protein structure. This may be due to intermolecular disruption of β-sheet protein conformation, although we have no evidence of the nature of the β-sheet conformation, only that the total amount of β-sheet decreases after interaction with homospecific lysin. One could also speculate that the binding of lysin to the vitelline layer may involve a carboxylate linkage due to the shift in solubilized vitelline layers at the 1410 cm^{-1} wave frequency.

These results support our previous suspicion that a block to abalone hybridization lies in the species specificity of the lysin–vitelline layer interaction (Leighton & Lewis, 1982). The turbidometric assay data clearly show incomplete solubilization of vitelline layers by heterospecific lysin. The spectroscopic data indicate direct binding of lysin onto the vitelline layer which changes the latter from β-sheet protein conformation to some unknown structure. Unfortunately, the addition of homospecific lysin to break the vitelline layer apart before adding heterospecific sperm has not led to high rates of hybrid fertilization (Leighton & Lewis, 1982). Thus, there may be some requirement for sperm to bind physically to the egg vitelline layer for sperm to acrosome react or for some other necessary step to occur (Lewis *et al.*, 1980, 1982). We predict that a greater understanding of these requirements will facilitate precise control of hybrid fertilization in the future.

ACKNOWLEDGEMENTS

The authors sincerely appreciate the long-term efforts of R. McConnaughey, collector of marine invertebrates at the Scripps Institution of Oceanography, for supplying the abalone. The authors thank V.D. Vacquier for many helpful discussions. K. Smith and R. D'Luhy, National Center for Biomedical Infrared Spectroscopy, Battelle Laboratories, Columbus, Ohio, performed the FT-IR spectroscopic data collection and analysis and suggested formats for data presentation. Drs Smith and D'Luhy are supported by NIH, Division of Research Resources, Grant No. RR-01367.

REFERENCES

Barnwell C. (1972) *Fundamentals of Molecular Spectroscopy*. McGraw-Hill, London.

Dan J.C. (1967) Acrosome reaction and lysins. In *Fertilization: Comparative Morphology, Biochemistry, and Immunology* (Ed. by C.B. Metz and A. Monroy), pp. 237–93. Academic Press, New York.

Fridberger A. (1987) Amino acid sequence differences between species-specific egg lysin proteins from sperm of two closely related abalones. PhD dissertation, University of Uppsala.

Fridberger A., Sundin J., Vacquier V.D. & Peterson P.A. (1985) Amino acid sequence of an egg-lysin protein from abalone spermatozoa that solubilizes the vitelline layer. *J. Biol. Chem.*, **260**, 9092–9.

Hahn K.O. (1989) *Handbook of Culture of Abalone and other Marine Gastropods*. CRC Press, Boca Raton.

Haino-Fukushima K. (1974) Studies on the egg-membrane lysin of *Tegula pfeifferi*: Purification and properties of the egg-membrane lysin. *Biochim. Biophys. Acta*, **229**, 459–70.

Haino-Fukushima K. & Usui N. (1986) Purification and immunocytochemical localization of the vitelline coat-lysin of abalone spermatozoa. *Dev. Biol.*, **115**, 27–34.

Harrick N.J. (1979) *Internal Reflection Spectrocopy*. Harrick Scientific, Ossining, NY.

Hong K. & Vacquier V.D. (1986) Fusion of liposomes induced by a cationic protein from the acrosome granule of abalone spermatozoa. *Biochemistry*, **25**, 543–9.

Hoshi M. (1985) Lysins. In *Biology of Fertilization*, Vol 2 (Ed. by C.B. Metz & A. Monroy), pp. 431–62. Academic Press, New York.

Kelley R.D. (1975) Ultrastructural identification of extracellular matrix and cell surface components during limb morphogenesis in man. *J. Embryol. Exp. Morphol.*, **34**, 1–18.

Laemmli U.K. (1970) Cleavage of structural proteins during the assembly of the head of bacteriophage T4. *Nature*, **227**, 680–5.

Leighton D.L. & Lewis C.A. (1982) Experimental hybridization in abalones. *Int. J. Invert. Reprod.*, **5**, 273–82.

Lewis C.A., Leighton D.L. & Vacquier V.D. (1980) Morphology of abalone spermatozoa before and after the acrosome reaction. *J. Ultrastruct. Res.*, **71**, 39–46.

Lewis C.A., Talbot C.F. & Vacquier V.D. (1982) A protein from abalone sperm dissolves the egg vitelline layer by a nonenzymatic mechanism. *Dev. Biol.*, **92**, 227–39.

Longo F.J. (1973) An ultrastructural analysis of polyspermy in the surf clam *Spisula solidissima*. *J. Exp. Zool.*, **183**, 153–80.

Luft J.H. (1971) Ruthenium red and violet. II. Fine structural localization in animal tissues. *Anat. Rec.*, **171**, 369–416.

Meyer R. (1967) *Hemocyanins and the systematics of California* Haliotis. PhD dissertation, Stanford University.

Morrissey J.H. (1981) Silver stain for proteins in polyacrylamide gels: a modified procedure with enhanced uniform sensitivity. *Anal. Biochem.*, **117**, 307–10.

Morse D.E., Duncan H., Hooker N. & Morse A. (1977) Hydrogen peroxide induces spawning in mollusks, with activation of prostaglandin endoperoxide synthetase. *Science*, **196**, 298–300.

Ogawa A. & Haino-Fukushima K. (1984) Isolation and purification of vitelline-coat lysin from testis of *Turbo cornatas* (Mollusca). *Dev. Growth Diff.*, **26**, 345–60.

Owen B., McLean J. & Meyer R. (1971) Hybridization in the Eastern Pacific abalone. *Bull. Los Angeles Co. Mus. Nat. Hist. Sci.*, **9**, 1–37.

Petell J.K., Sardo M.J. & Lebherz H.G. (1981) A simple procedure for the isolation of seven abundant muscle enzymes. *Prep. Biochem.*, **11**, 69–89.

Sakai Y.T., Shiroya Y. & Haino-Fukushima K. (1982) Fine structural changes in the acrosome reaction of the Japanese abalone, *Haliotis discus*. *Dev. Growth Diff.*, **24**, 531–42.

Sakai Y.T., Suzuku F. & Shiroya Y. (1985) Distribution of intramembrane particles and its changes during the acrosome reaction in spermatozoa of the Japanese abalone. *Dev. Growth Diff.*, **27**, 787–802.

Shiroya Y. & Sakai Y.T. (1984) Ultrastructural changes of the 'Truncated cone' during the acrosome reaction in the Japanese abalone spermatozoa. *Dev. Growth Diff.*, **26**, 25–32.

Shiroya Y., Hosoya H., Mabuchi I. & Sakai Y.T. (1986) Actin filament bundles in the acrosome of abalone spermatozoa. *J. Exp. Zool.*, **239**, 105–15.

Talmadge R.R. (1977) Notes on a California hybrid *Haliotis*. *Veliger*, **20**, 37–9.

Usui N. (1987) Formation of the cylindrical structure during the acrosome reaction of abalone spermatozoa. *Gam. Res.*, **16**, 37–45.

Weber K. & Osborne M. (1969) The reliability of molecular weight determinations by dodecyl sulfate-polyacrylamide gel electrophoresis. *J. Biol. Chem.*, **244**, 4406–12.

Chapter 7

The effects of gas supersaturation on the behaviour, growth and mortality of red abalone, *Haliotis rufescens* (Swainson)

AMY LEITMAN *PO Box 353, Port Townsend, Washington 98368, USA*

ABSTRACT: Invertebrate aquaculture has undergone rapid expansion over the past few decades in the United States. This industry has often been accompanied by a series of problems inherent to a laboratory system. One problem is gas supersaturation, a common occurrence in hatcheries that pump natural waters into an intensive culture system. This can cause serious gas bubble disease problems. Previous studies have identified the effects of lethal and sublethal gas supersaturation levels on fin fish, bivalves and crustaceans; however, this problem has not been analysed for commercially reared red abalone (*Haliotis rufescens* Swainson).

This experiment examined the effects of several levels of gas supersaturation (from 105 to 143%) on the behaviour, growth and mortality of red abalone. Treatment levels of gas supersaturation of 100–110% induced significantly higher growth than those of 120–143%; intermediate growth was observed in the 115% gas supersaturation level. Mortality and qualitative behavioural results from all gas level treatments supported the trends that were established by the length and weight data. This suggests that increased gas supersaturation levels were more lethal to the red abalone. Data from this experiment should help aquaculturists assess the health of their abalone stock during a gas supersaturation episode.

RESUMEN: En las últimas décadas, el cultivo de invertebrados ha experimentado una rápida expansión en los Estados Unidos. Esta industria a menudo ha sido acompañada por una serie de problemas inherentes a los sistemas de laboratorio. Uno de esos problemas es la supersaturación gaseosa, que se presenta frecuentemente en los criaderos que bombean agua natural a un sistema de cultivo intensivo. Esto puede causar serios problemas de enfermedad por burbujas de gas. Estudios previos han identificado los efectos de los niveles letal y subletal de la supersaturación de gas en peces, bivalvos y crustáceos. Sin embargo este problema no ha sido analizado para la cría comercial del abulón rojo (*Haliotis rufescens* Swainson).

Este experimento examinó los efectos de varios niveles de supersaturación gaseosa (de 105–143%) en el comportamiento, crecimiento y mortalidad del abulón rojo. Niveles de 100–110% de supersaturación gaseosa produjeron un crecimiento significativamente mayor que los niveles de tratamiento de 120–143% de gas; se observó un crecimiento intermedio en el nivel de 115% de supersaturación. Los resultados de mortalidad y comportamiento cualitativo de todos los niveles de tratamiento de gas apoyaron las tendencias que fueron establecidas con datos de longitud y peso. Esto sugiere que los incrementos de los niveles de supersaturación gaseosa fueron mas letales para el abulón rojo. Los datos de este experimento deberían ayudar a los acuicultores a evaluar la salud de las existencias de abulón durante el período de supersaturación gaseosa.

INTRODUCTION

In the last decade the aquaculture industry has expanded globally to meet the growing nutritional needs of an increasing human population. Among these industries are marine invertebrate hatcheries that, in addition to growing animals for food, often also intend to restock decimated natural populations. As most targeted species have suffered severe losses caused by 'gas bubble disease' (Marsh & Gorham, 1905; Malouf *et al.*, 1972; Lightner *et al.*, 1974) it is suspected that the abalone industry may not be meeting its potential due to these problems. Effects of exposure to seawater supersaturated with gas have been well documented in fin fish (Marsh & Gorham, 1905; Weitkamp & Katz, 1980; Marking, 1987). Marsh & Gorham (1905) were the first to categorize the pathological effects as gas bubble disease. Colt (1984) later attributed the symptoms to uncompensated hyperbaric pressure of dissolved gases that form bubbles in the tissues of aquatic organisms. However, this disease has only recently been recognized in aquatic invertebrates (Malouf *et al.*, 1972; Goldberg, 1978; Elston & Lockwood, 1983; Bisker & Castagna, 1986, 1987), and only Goldberg and Bisker & Castagna attempted to define the levels of gas that are lethal to selected commercially important molluscs. With the surging interest in invertebrate aquaculture, the implications and expense of hatchery-induced gas bubble disease on the aquaculture industry may be significant. The primary objective of this study was to understand the effects of gas bubble disease on red abalone (*Haliotis rufescens* Swainson).

Supersaturated seawater has been a source of many gas bubble disease episodes since the turn of the century. For example, at Woods Hole in 1900 a small leak in the intake pipe forced air into solution which caused the water to become supersaturated (Marsh, 1903). More recently, E.E. Ebert (pers. comm.) and Weitkamp & Katz (1980) noticed that supersaturation problems have occurred as a result of leaks on the suction side of pumps. Rapid temperature increases can also cause gas supersaturation since gases are less soluble at higher temperatures. When abalone and other molluscs are cultured, water temperature is often increased by between 1 and 10°C to optimize growth (Ebert, 1974). Malouf *et al.* (1972) hypothesized that this process was responsible for supersaturation of water within a bivalve hatchery.

Three epizootic outbreaks occurred at the California Department of Fish and Game (CDFG) Laboratory at Granite Canyon in the last 15 years (E.E. Ebert, pers. comm.). All of these outbreaks resulted in high mortality of juvenile and adult red abalone and all occurred in the spring, when a 2–5°C decrease in temperature commonly occurs along the California coast. Since cold water can hold more gases than warm water, when the cold water is heated more gases come out of solution; the lag time to equilibrate can cause gas supersaturation. In all cases the abalone had numerous lesions on the sole of the foot and associated necrotic tissue that had elevated numbers of bacterial pathogens (*Vibrio* spp.). Elston (1983) reported that juvenile abalone exposed to short-term supersaturated oxygen conditions (12 hours) showed systemic infection with the bacterium *Vibrio alginolyticus*. At the Granite Canyon laboratory, efforts were made to combat mortality by the use of sulfa drugs, fungicides and bactericides. When these methods failed, it was suspected that bacteria played a secondary role and that stress, possibly due to gas supersaturation, was the precursor to the high mortality of abalone. In this study, the critical limits of gas supersaturation on the behaviour, growth and mortality of red abalone were examined.

MATERIALS AND METHODS

The experiments were performed at CDFG's Marine Resource Laboratory, Granite Canyon, which is located on the central California coast, 20 km south of Monterey. The experiments began 1 May 1987 and ran continuously to 31 July 1987.

Experimental seawater

Gas supersaturation was accomplished by pumping water with a 0.14 J/s (0.01 horsepower), 26.5 l/min (7 gal/min) centrifugal pump (Cole-Parmer Instrument Corporation) which was connected to a 'pressure chamber' made of PVC pipe 5 m in length and 0.25 m in diameter. The chamber was plumbed with outlet, inlet, compressed air injection and overflow valves (Fig. 7.1). The combined pressure capacity of the pump, the compressed air and the head pressure provided approximately 2.1 atm (30 psi) of water pressure. The laboratory-supplied compressed air was maintained at 2.2 ± 0.06 atm (32 psi ± 1 psi), just slightly higher than the pressurized water. The higher pressure forced the air to dissolve in the seawater inside the pressure chamber.

All laboratory seawater in the experiment was filtered to 10 μm by means of a cartridge filtration system (Filtrex Inc); half of this water was first diverted through the pressure chamber where it was supersaturated to approximately 143% ± 3%. A gas tensionometer was used (Common Sensing Inc total dissolved gas meter, model FTR/TGO) to measure total gas (mmHg TG), nitrogen (mmHg N_2), oxygen (mmHg O_2) and per cent total gas supersaturation levels (%TG) every other day throughout the experiment to ensure that the levels remained constant. The supersaturated and ambient water supply lines were otherwise treated identically.

Seven different supersaturation concentrations were produced by mixing the appropriate amount of water from the pressure chamber (143% ± 3% supersaturation) with the ambient laboratory seawater (100% ± 0.45%). To eliminate significant variation, each treatment was mixed in a 15 l header tank before being introduced into the individual test containers (Fig. 7.2). Intermediate levels were maintained at supersaturation levels of 105, 110, 115, 120 and 130%. The gas supersaturation level of 115% was initiated 30 days into the experiment. The six treatments and the control were run simultaneously. Initial tests indicated that supersaturation levels in the individual test containers were identical to the header tank supersaturation levels. A modified Winkler analysis (Parsons *et al.*, 1984) was performed once a month to calibrate the oxygen probe. Temperature was maintained at 10–14°C.

There were three replicate test container troughs for each gas level treatment. Water was fed by gravity into these test troughs from the corresponding header tanks. Each container held twenty 12 mm (±1 mm) *H. rufescens* chosen from healthy laboratory stock.

One large blade (approximately 44 cm long) of *Macrocystis pyrifera* was placed in each test container once a week; this was more than could be consumed by the abalone during this period. The test containers were decanted and rinsed once a week to prevent accumulation of detritus from both the abalone and the kelp.

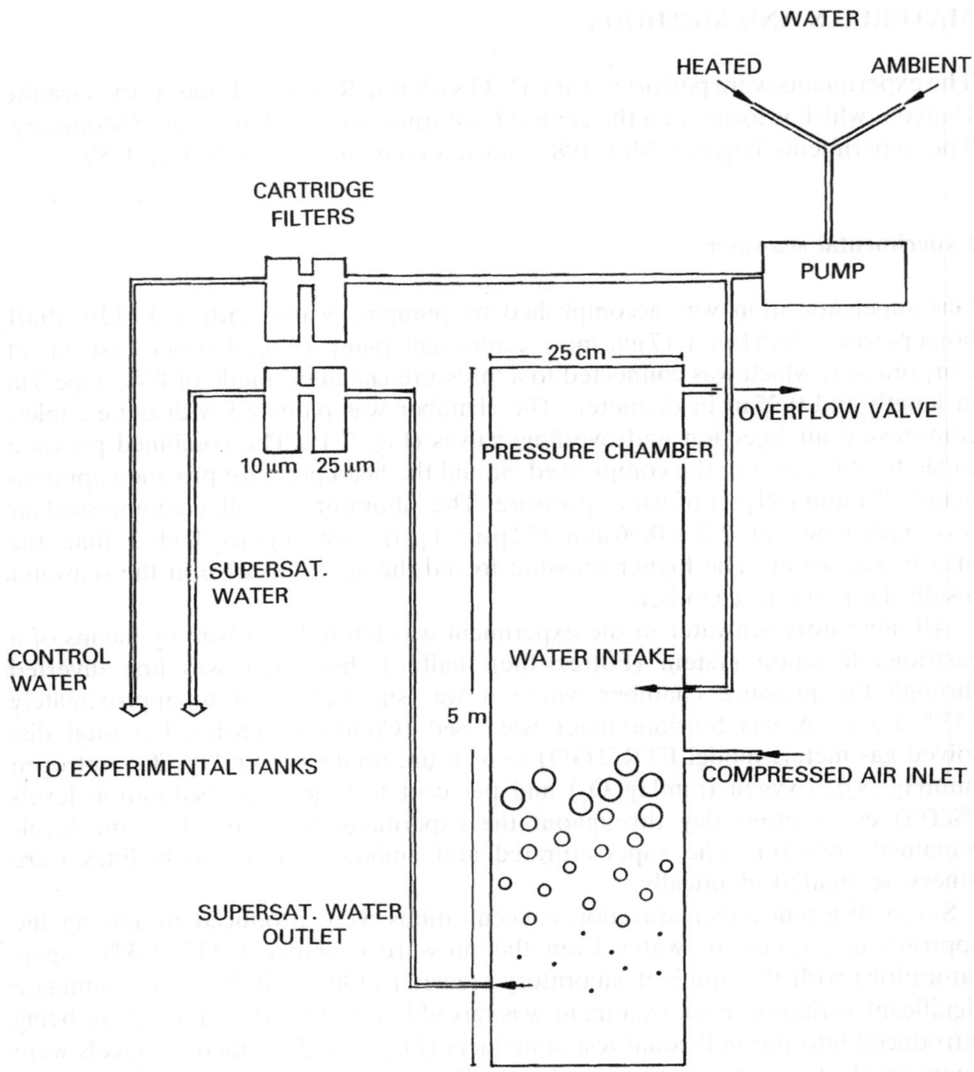

Fig. 7.1 Schematic of experimental design.

Bacterial analysis

A simple membrane filtration method was chosen to estimate the abundance of bacteria (Patrick, 1978) because of its low risk to the juvenile abalone. Every two to three days three replicate samples were taken from the effluent water line in each gas level. The three test containers per treatment were sampled on a rotating schedule that allowed the sampling to cover all test containers every 10 days. *Vibrio* spp. were identified and recorded throughout the experiment.

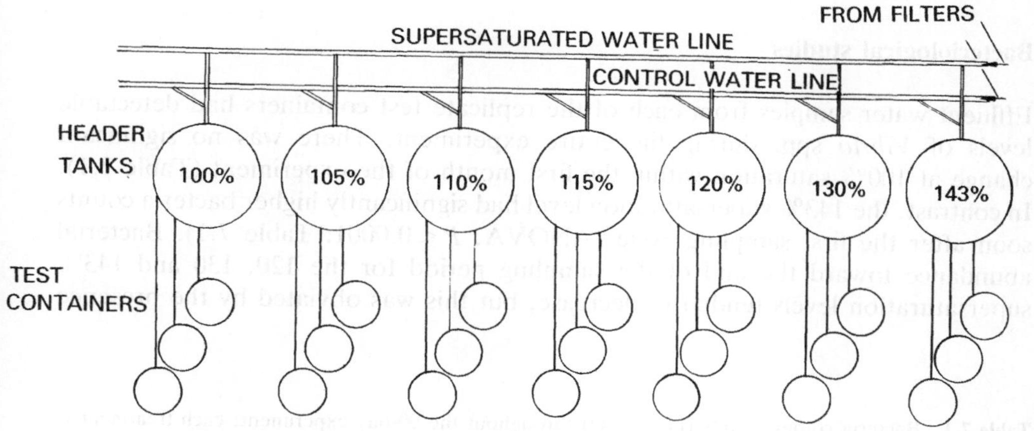

Fig. 7.2 Schematic of experimental tank set-up.

Growth measurements

On the first day of each month during the experiment, abalone length and weight measurements were taken for each of the experimental groups and the control. Total weight, individual weight and the range of weights were recorded to the nearest 1.0 mg for each treatment. Calipers were used to determine each animal's maximum shell length to the nearest 0.01 mm.

Mortality rates were recorded weekly for each test container.

Behavioural measurements

Several qualitative parameters were monitored three times a week throughout the experiment to detect any behavioural and/or physical changes in the experimental animals. Three times a week the amount of food consumed per container was given an index of 0–5 (0 = least eaten, 5 = most eaten) to estimate feeding rates.

One characteristic of a healthy abalone is the ability to 'right' itself quickly after being turned over. The length of time each individual required to right itself was measured to quantify its 'health'. If an animal had not righted itself within 15 seconds, it was considered a failed attempt. Another characteristic of a healthy abalone is the animal's ability to maintain a strong hold on the substrate (tenacity). A comparable force was applied to five randomly chosen animals from each test container; an index of 0–5 (0 = weakest hold, 5 = strongest hold) was used to describe the animal's strength.

Other physical characteristics that were easily observed, such as air bubbles in the digestive gland and emaciation, were recorded using the same index of 0–5; the more severe the trait, the higher the number employed.

RESULTS

Bacteriological studies

Effluent water samples from each of the replicate test containers had detectable levels of *Vibrio* spp. during the entire experiment. There was no significant change at 100% saturation within the first month of the experiment (Table 7.1). In contrast, the 143% supersaturation level had significantly higher bacteria counts soon after the first sampling cycle (ANOVA, $P<0.0001$; Table 7.1). Bacterial abundance toward the end of the sampling period for the 120, 130 and 143% supersaturation levels tended to decrease, but this was obviated by the presence

Table 7.1 Bacteria colony counts (mean ± sd) throughout the 90-day experiment; each treatment is sampled once during each 10-day cycle; sample size (n) changes according to survivorship of each replicate treatment.

	Gas treatment levels						
Cycle	*100%	*105%	*110%	*115%	120%	130%	143%
1	17.0 (±9.0)	57.7 (±71.9)	11.7 (±11.7)	–	12.3 (±6.8) n = 59	83.0 (±92.7) *	32.3 (±29.1) *
2	10.3 (±14.6)	30.3 (±44.0)	8.0 (±6.1)	–	14.0 (±15.9) n = 59	35.7 (±12.5) n = 59	134.3 (±105.2) *
3	22.3 (±13.1)	110.3 (±78.7)	35.7 (±24.1)	–	51.0 (±32.6) n = 59	73.7 (±58.0) n = 58	80.3 (±100.8) n = 56
4	76.7 (±56.9)	68.3 (±14.6)	85.0 (±24.1)	64.7 (±58.2)	137.7 (±44.8) n = 58	115.7 (±31.1) n = 53	179.0 (±36.4) n = 46
5	89.0 (±100.9)	101.7 (±51.8)	125.3 (±64.7)	73.7 (±27.2)	138.7 (±24.2) n = 52	178.7 (±35.2) n = 39	139.0 (±41.8) n = 31
6	111.7 (±78.5)	96.7 (±37.5)	92.3 (±68.1)	118.0 (±72.5)	99.3 (±26.0) n = 38	88.3 (±32.0) n = 31	8.7 (±15.0) n = 13
7	90.3 (±33.9)	97.3 (±28.7)	110.3 (±18.2)	84.3 (±47.1)	125.0 (±38.9) n = 25	–	–
8	122.3 (±31.1)	141.7 (±26.8)	89.3 (±30.0)	73.3 (±53.5)	–	–	–
9	95.0 (±22.9)	94.3 (±78.8)	86.7 (±60.1)	60.0 (±16.1)	–	–	–

* These replicate samples have a total of 60 animals/treatment.
– Refers to samples not taken: 115% treatment level started one month later; 120%, 130% and 143% treatment levels did not contain sufficient numbers of animals for meaningful data analysis by the seventh cycle.

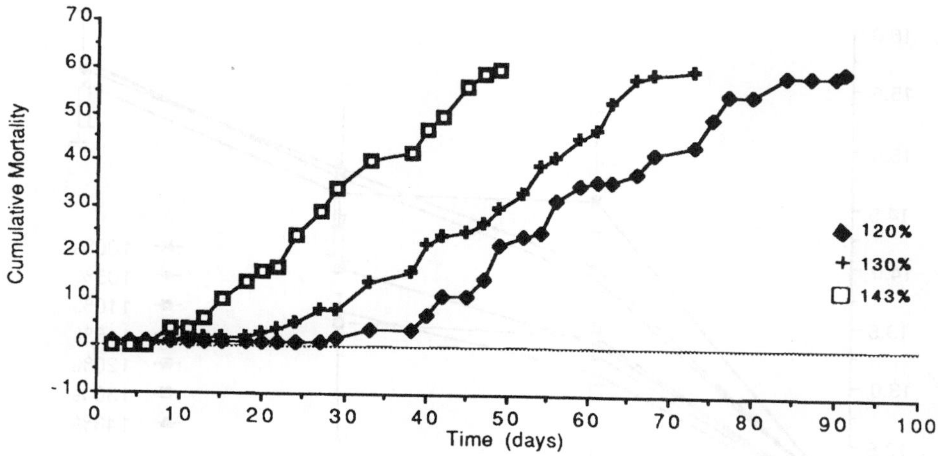

Fig. 7.3 Cumulative mortality at three gas supersaturation levels. Each data point represents the mean mortality of three replicate tanks. All gas level treatments were initiated with 60 animals each.

of fewer abalone (Table 7.1, Fig. 7.3). Other than the discrete separation of bacteria counts in the lower and higher gas levels at the initial phase of the experiment, the treatments all fell within a similar range and showed no other significant trends in bacteria (Table 7.1).

Sublethal effects

Scheffe's multiple comparison test (Zar, 1984) indicated that there were no significant differences in final shell lengths within the upper gas level group of 120 to 143% and the lower gas level group of 100 to 115%; however, a significant difference was found between the two groups (Fig. 7.4). The lower gas level group had a much larger shell length increase.

Weight change over the course of the experiment was significantly different among the treatments (ANOVA, $P < 0.0001$) and *a posteriori* tests indicated that two distinct groups existed. The abalone in the 100, 105, 110 and 115% levels all gained weight, while the abalone in the 120, 130 and 143% levels lost weight (Fig. 7.5). Although the 115% gas level abalone were statistically grouped with the heavier animals, they maintained a more intermediate level of weight change (Fig. 7.5).

Lethal effects

There were no mortalities in gas supersaturation treatments below 120% during the experiment, with the exception of one animal at the 115% level. The trend of increased mortality rate with increased gas supersaturation levels is clearly shown by the cumulative mean mortality data (Fig. 7.3).

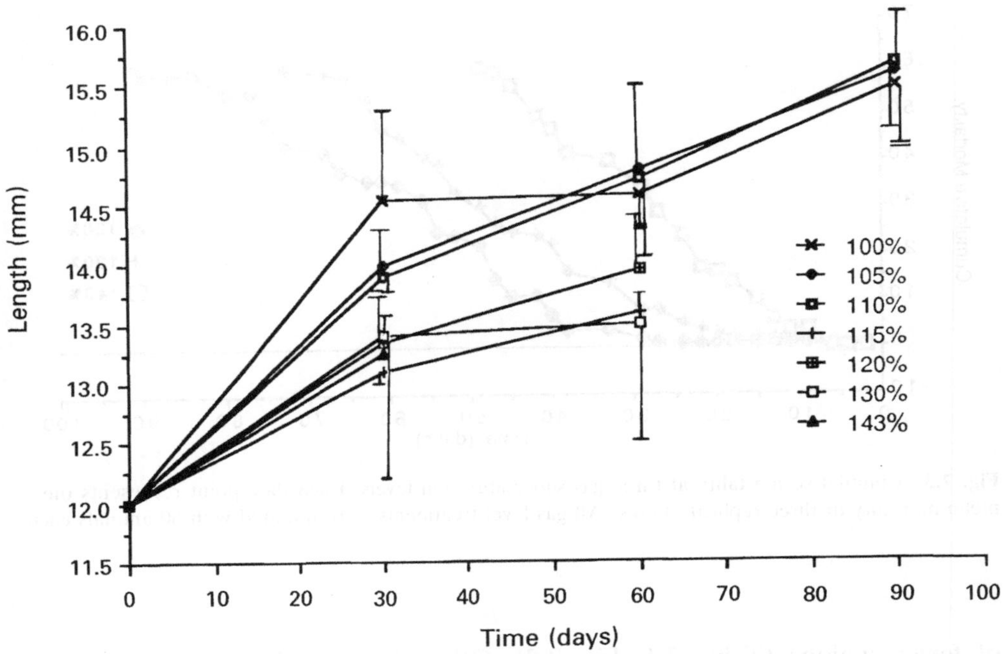

Fig. 7.4 Shell length increase of individuals for all gas level treatments throughout the 90-day experiment or until mortality occurred. Mean ± sd, $N = 3$.

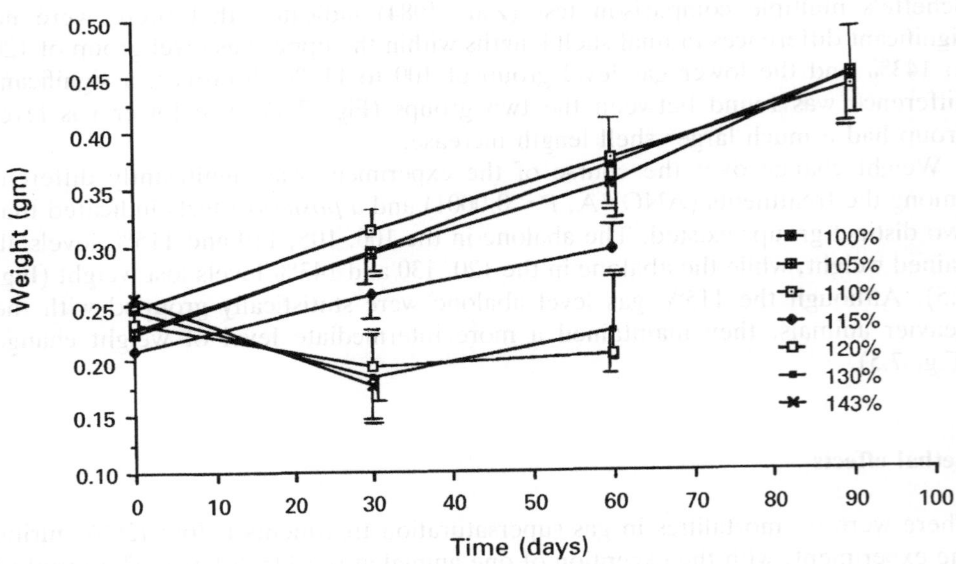

Fig. 7.5 Weight increase of individuals in all gas treatments throughout the 90-day experiment or until mortality occurred. Mean ± sd, $N = 3$.

Table 7.2 Qualitative values for food consumption and tenacity of abalone in all gas treatment levels throughout the experiment (means of three experimental troughs per gas level).

Gas level (%)	Food consumption mean (±sd)	Tenacity mean (±sd)	N
100	4.9 (±0.3)	4.8 (±0.3)	99
105	4.6 (±0.6)	4.5 (±0.5)	99
110	4.3 (±0.6)	4.0 (±0.5)	99
115	3.2 (±0.4)	3.1 (±0.4)	63
120	1.0 (±0.8)	1.5 (±1.1)	76
130	0.1 (±0.1)	0.5 (±0.8)	69
143	0.0 (±0.0)	0.1 (±0.0)	48

N = sampling days.
0–5 scale used in all qualitative tests: 0 = least eaten or weakly attached, 5 = most eaten or strongly attached.

Behaviour

Macrocystis pyrifera consumption data show the same two conspicuous groupings previously found in the quantitative tests (Table 7.2). The control, 105% and 110% levels had the highest consumption rates, the 115% gas supersaturation level was intermediate and the abalone in the more supersaturated treatment levels (120, 130 and 143%) fed at substantially lower rates. Animals in the highest gas supersaturation levels were also less tenacious (Table 7.2). Moreover, abalone in the control, 105 and 110% treatment levels were usually able to right themselves in a few seconds (average time was 1–4 seconds), while approximately 50% of the abalone in the 120, 130 and 143% treatment levels often took over 15 seconds to right themselves during the first 6 weeks of the experiment. As exposure time increased beyond 6 weeks, 100% of the abalone in the three highest gas treatment levels ultimately lost the ability to right themselves. A small percentage of these animals were often found detached and floating on the surface.

DISCUSSION

These experiments showed clear trends of the effects of gas supersaturation on the growth, behaviour and mortality of red abalone. Stress induced in the 120% and higher gas supersaturation treatments caused mortality.

The control animals in this experiment gained weight at an average rate of 2.4 mg/day. Each higher gas level had an incrementally decreasing growth rate. At the intermediate gas level of 115%, the abalone showed a slow and steady growth rate of 0.8 mg/day (Fig. 7.5), which at best represents a maintenance rate. In the 120, 130 and 143% supersaturation levels, the abalone actually lost weight. The weight loss increased as the gas levels increased, with a maximum weight loss of 0.08 mg/day for the 143% treatment level. These growth data suggest that there is a gas supersaturation threshold beyond which the abalone will lose biomass. Apparently the stress from supersaturation directly or indirectly causes loss of biomass which ultimately results in mortality.

Similar trends are apparent in the length data. The control abalone grew at a rate of 15.8 μm/day. The highest mean growth rate of 22.1 μm/day occurred in the 105% level, suggesting that there is a potential for increasing red abalone growth rate by culturing them with low level gas supersaturated seawater. Separating the two distinct groups, the 115% level gained a maximum of 11.8 μm/day, an intermediate shell growth rate. In the higher gas supersaturation treatment groups the lengths did not increase much beyond the initial 12 mm ± 1 mm range (Fig. 7.4); this is probably the result of stress-induced starvation and is in accordance with the weight data. Alternating phases of length (shell) and weight (tissue) growth (punctuated inverse growth) are thought to exist for abalone (E.E. Ebert, pers. comm.). This is probably represented by the 105% gas level, the 115% gas level and, to a more severe degree, by all of the higher gas level groups (120, 130 and 143%) that continued to increase in length although they did not gain weight (Figs 7.4, 7.5).

Significant mortality occurred in the three highest gas supersaturation treatments. In the 143% treatment level, 50% of the abalone died by the twenty-seventh day; 50% mortality occurred on the forty-ninth day in the 130% gas level and the fifty-sixth day in the 120% level. Stress-induced emaciation was probably the main cause of mortality, although some mortalities were a result of animals that exploded (either the digestive gland or mantle cavity membrane) due to the pressure exerted on the tissues by gas emboli. These gas bubbles seemed to influence all the other behavioural parameters. For example, if an abalone had a large gas embolus in the mantle cavity it would then float; the floating abalone would not be able to attach to any substrate to eat and would appear to lose weight and eventually starve. Once emaciated, the shell would often impede the abalone's ability to flip over because of the grotesquely exaggerated shell to body ratio. The 120, 130 and 143% gas levels are too high for successful abalone cultivation. However, the slower mortality rates in the 120 and 130% levels suggest that once high gas supersaturation levels are detected, additional mortalities may be averted by degassing the water.

Behavioural observations in this study supported the quantitative data. Feeding rates, ability to hold on to a substrate and motor skills decreased incrementally with increasing gas supersaturation values. This pattern was conspicuous throughout the experiment and helped define the lethal and sublethal levels of gas supersaturated seawater for *H. rufescens*.

Elston's (1983) experiments suggest that the exposure to hyperoxygenation reduced the abalone's immunity to disease. In the experiment described here, the 143% treatment showed a positive correlation between bacterial abundance and mortality. However, the high numbers of bacteria found in the 143% gas treatment level seemed to be a result of opportunistic *Vibrio* spp. infecting already moribund abalone.

To elucidate further the role bacteria play in abalone mortalities, we need to understand the effects of varying gas supersaturation levels on bacterial abundance in the absence of abalone.

With some of the lethal and sublethal gas supersaturation levels and their effects on the red abalone identified, the aquaculturist will have better water quality guidelines and health indicators to follow if and when gas supersaturation is suspected to be the cause of a problem. However, additional experiments need to be undertaken to define more clearly the effects and recovery of abalone exposed to intermediate gas supersaturation levels between 110 and 120%. If abalone can recover from gas bubble disease symptoms after exposure to a known gas supersaturation level, a culturist will be better equipped to decide whether or not to employ the time and cost expenditures necessary to degas the water. This should ultimately result in higher abalone survival rates and lower costs to the aquaculturist.

ACKNOWLEDGEMENTS

This work was supported in part by a Packard Foundation Research Grant and a joint grant with the California Department of Fish and Game and the San Jose State University Foundation. Randy Hamilton at the Monterey Bay Aquarium helpfully offered the use of necessary analytical equipment. The author is indebted to Dave Rose, Earl Ebert and Bridget McAfee for their endless hours of scientific support.

REFERENCES

Bisker R. & Castagna M. (1986) The effect of various levels of air-saturated seawater on *Mercenaria mercenaria* (Linne), with reference to Gas Bubble Disease. *J. Shellfish Res.*, **5**, 97–102.

Bisker R. & Castagna M. (1987) Effect of air supersaturated seawater on *Argopectin irradians concentricus* (Say) and *Crassostrea virginica* (Gmelin). *J. Shellfish Res.*, **6**, 79–83.

Colt J. (1984) Computation of dissolved gas concentrations in water as functions of temperature, salinity and pressure, *Am. Fish. Soc. Spec. Publ.*, 14.

Ebert E.E. (1974) Elevated temperature tolerance of the red abalone, *Haliotis rufescens*. Oral presentation at the joint annual meeting of the California–Nevada chapter of the American Fisheries Society and the Western Section of the Wildlife Society in Monterey, CA.

Elston R. (1983) Histopathology of oxygen intoxication in the juvenile red abalone, *Haliotis rufescens* Swainson. *J. Fish Dis.*, **6**(2), 101–10.

Elston R. & Lockwood G.S. (1983) Pathogenesis of Vibriosis in cultured juvenile red abalone, *Haliotis rufescens* Swainson. *J. Fish Dis.*, **6**, 111–28.

Goldberg R. (1978) Some effects of gas supersaturated seawater on *Spisula solidissima* and *Argopectin irradians*. *Aquaculture*, **4**(14), 281–7.

Lightner D.V., Salser B.R. & Wheeler R.S. (1974) Gas Bubble Disease in brown shrimp *Pendalus aztecus*. *Aquaculture*, **4**, 81–4.

Malouf R., Keck R., Maurer D. & Epifanio C. (1972) Occurrence of Gas Bubble Disease in three species of bivalve molluscs. *J. Fish. Res. Bd Can.*, **29**, 588–9.

Marking L.L. (1987) Evaluation of gas supersaturation treatment equipment at fish hatcheries in Michigan and Wisconsin. *Prog. Fish Cult.*, **49**, 208–12.

Marsh M.C. (1903) A fatality among fishes in water containing an excess of dissolved air. *Trans. Am. Fish. Soc.*, **32**, 192–3.

Marsh M.C. & Gorham F.P. (1905) The gas disease in fishes. *Rep. Bur. Fish.*, (1904), 343–76.

Parsons R.R., Maita Y. & Lalli C.M. (1984) *A Manual of Chemical and Biological Methods for Seawater Analysis*. Pergamon Press, Oxford, New York.

Patrick F.M. (1978) The use of membrane filtration and marine agar 2216E to enumerate marine heterotrophic bacterial. *Aquaculture*, **13**, 369–72.

Weitkamp D.E. & Katz M. (1980) A review of dissolved gas supersaturation literature. *Trans. Am. Fish. Soc.*, **109**, 659–702.

Zar J.H. (1984) *Biostatistical Analysis*, 2nd edn. Prentice-Hall, Englewood Cliffs.

Chapter 8
Fecundity and optimal sperm density for fertilization in the ormer (*Haliotis tuberculata* L.)

J. CLAVIER *Orstom, BP A5, Noumea, New Caledonia*

ABSTRACT: The potential and instantaneous fecundity of *Haliotis tuberculata* L. was studied in the vicinity of St Malo (northern Britanny). A linear relationship is found between potential fecundity of males or females and weight of individuals but the total quantity of sperm and ova shed during laboratory-induced spawning (instantaneous fecundity) is generally lower than expected from this correlation. Incubations of ova with increasing concentrations of sperm reveal zero or very low fertilization rates for sperm densities less than 10^3 cells ml^{-1}, and a maximal rate at between 10^5 and 10^6 cells ml^{-1}. During spawning, 2×10^9 to 3×10^9 sperm are released at 30- to 45-second intervals. In natural grounds, this gamete concentration may be diluted by water movement and sperm concentration may be lower than the threshold of 10^3 cells ml^{-1}. *Haliotis tuberculata* is sedentary even during the reproductive period and good reproductive success may be ineffective on beds with low ormer densities.

RESUMEN: Se estudia la fecundación instantánea y absoluta de *H. tuberculata* L. en la vencindad de St Malo (norte de la Bretaña). La relación entre la fecundidad absoluta de los machos o de las hembras con el peso es lineal, pero la cantidad de esperma y de huevos emitidos durante el desove provocado en el laboratorio es generalmente menor que tal correlación deja suponer. La incubación de los huevos con concentración creciente de esperma revela tasas de fertilización nulas o muy bajas para densidades de espermatozoides inferiores a $10^3 ml^{-1}$ y tasa máximas para densidades de 10^5 a $10^6 ml^{-1}$. Durante el desove, 2×10^9 a 3×10^9 espermatozoides fueron emitidos cada 30 a 40 segundos. En el medio natural de vida, esa 'unidad de fecundación' puede sufrir una dilución por los movimientos del agua y la concentración de espermatozoides puede diminuir debajo de $10^3 ml^{-1}$. *Haliotis tuberculata* estando sedentario incluso durante el período de reproducción, se supone que la fecundación podría ser ineficaz cuando la densidad de abulones queda baja.

INTRODUCTION

The ormer (*Haliotis tuberculata* L.) lives along the Atlantic coast of Europe from the English Channel to Africa. It occurs in some abundance off the north coast of Brittany and the Channel Islands where it is regarded as a desirable shellfish delicacy. Nowadays, diving for ormers is not authorized in Europe and shore gathering is the only legal fishing method (Chapter 33, this volume).

The first accounts of ormer reproduction were published by Stephenson (1924) and Crofts (1929). Since then, an abundant literature has been devoted to the reproductive biology of *H. tuberculata* (Girard, 1972; Cochard, 1980; Hayashi, 1980). *Haliotis tuberculata* is dioecious with an equal sex ratio. Maturation of the gonad is mainly related to both photoperiod and temperature; spawning occurs during summer (July–September). The early development of *H. tuberculata* has been described (Koike, 1978) and the technology for mass production of juveniles is controlled (Flassch & Aveline, 1984).

Fecundity, representing the total number of oocytes in the ovary, is usually involved in models of population dynamics. However, for sedentary species like

H. tuberculata, other factors such as spawning behaviour, the sperm density required and fertilization are also important.

In the present study, measurements of the number of gametes shed during induced spawning, observations on spawning behaviour and fertilization experiments for *H. tuberculata* were undertaken. The implications for fishery management are discussed.

MATERIALS AND METHODS

Haliotis tuberculata were collected from the northern coast of Britanny, in the vicinity of Saint-Malo (48°40′N–2°00′W) during June 1984, just prior to the spawning season. The total number of mature oocytes contained in the ovaries of *H. tuberculata* was estimated. After formalin fixation, each gonad was carefully dissected, drained on filter paper and weighed. Subsamples were taken from near the base, mid and apical parts of the ovary and weighed to 0.0001 g. Oocytes were loosened from their follicles by tearing the tissues with glass needles and their number was counted under a dissecting microscope. Estimates of the number of oocytes per whole ovary were then calculated. Because some oocytes may not be released during spawning and may be resorbed, the total number of oocytes is referred to as 'potential fecundity'.

The total number of gametes shed during spawning was estimated by the use of induced-spawning techniques. Mature *H. tuberculata* were selected according to the swelling of their gonad and induced to spawn. They were first left out of water for one hour (Genade *et al.*, 1988) and then isolated in containers filled with exactly 5 l filtered seawater irradiated with UV light (Kikuchi & Uki, 1974a) and heated 3° above ambient temperature to give them a temperature shock (Imai, 1967). Following spawning, ormers were removed and the number of eggs in each container was calculated by counting 10 samples of 1 ml. Sperm concentration was determined by haemocytometer readings on 10 successive samples. The number of ova shed during induced spawning is referred to as 'instantaneous fecundity'.

Fertilizations trials were conducted by combining gamete stocks collected according to the above method. A range of sperm densities was obtained by progressive dilution of a sample collected after spawning. Sperm density in each dilution was measured with a haemocytometer; 10 ml of each sperm suspension was added to an equal volume of water containing 100 to 200 ova spawned synchronously with sperm shedding or less than 30 minutes after sperm. Percentage fertilization was estimated by direct counts of cleaving eggs after 3 hours at ambient temperature (19°C). Spawning behaviour including frequency of pulses of gametes through shell perforation was noted during these experiments.

RESULTS

The number of ripe oocytes in the ovaries (potential fecundity) ranged from 4×10^5 for 20 g *H. tuberculata* to 3.8×10^6 for 160 g (Fig. 8.1). Potential fecundity (NO) of *H. tuberculata* is given by:

$$NO = 2.35\ 10^4\ W - 1.855\ 10^5$$

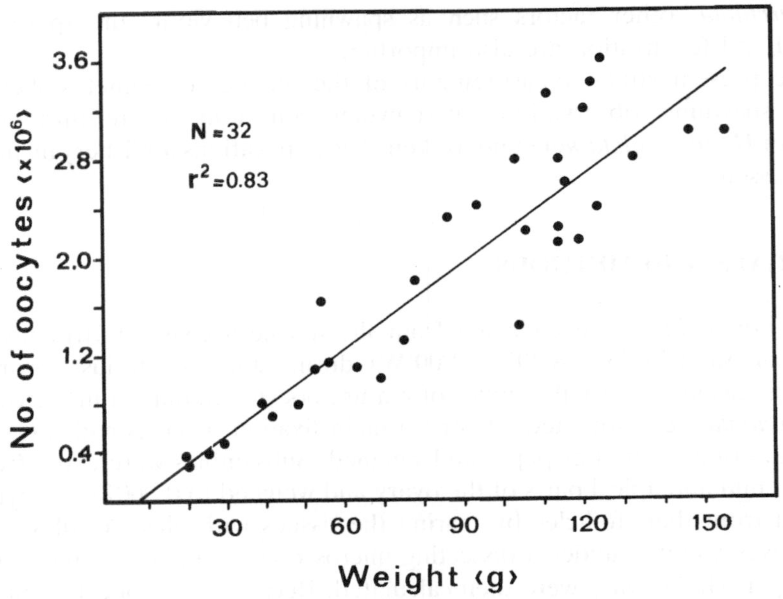

Fig. 8.1 Relationship between the number of oocytes in the ovary and total weight. *N* is the number of observations and r is the correlation coefficient.

where *W* is the total weight of *H. tuberculata* in g. From this relationship the number of eggs is nil for an individual weight of 8 g corresponding to a shell length of about 40 mm and an age of 3 years (Clavier & Richard, 1986).

The number of eggs shed during induced spawning (instantaneous fecundity) ranged from 2×10^5 to 1.6×10^6 (Fig. 8.2(A)). The values are generally lower than would have been expected from a full ovary. As pointed out by Hayashi (1980), this may be a result of a partial release of the ripe eggs in the ovary. In fact, the potential fecundity line drawn on Fig. 8.2(A) is the upper limit of induced spawning figures.

The relationship between the number of sperm released during induced spawning and the weight of *H. tuberculata* is given in Fig. 8.2(B). The general disposition of experimental points is similar to Fig. 8.2(A) and the number of sperm varies widely for a given body weight. By analogy with females, we should assume that potential fecundity of males is linear with individual weight and corresponds to the upper limit of the induced-spawning values. The total number of sperm in the gonad (NS) may be estimated by:

$$NS = 2.56\ 10^9\ W - 2.3\ 10^{10}$$

where *W* is the total weight of *H. tuberculata* in g.

During the experiments, the total spawning time varied from 40 to 80 minutes. The frequency of sperm pulses shows a fairly constant pattern for observed *H. tuberculata* (Fig. 8.3). After a short initial phase, sperm were released every 30 to 45 seconds. This frequency decreases progressively at the end of the process and spawning was completed after 30 to 70 pulses according to shell length. As a

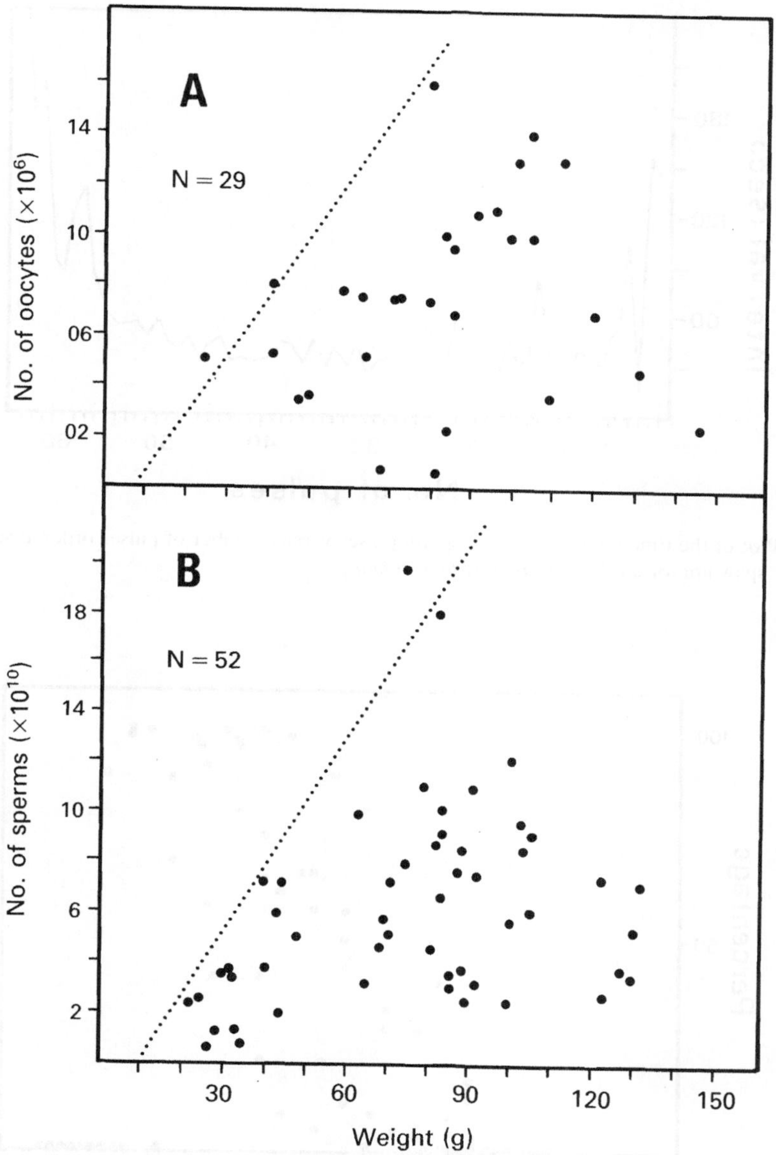

Fig. 8.2 Relationship between numbers of ova (A) and sperm (B) induced to spawn and total weight. Dotted lines represent relationships between the total number of gametes in reproductive glands calculated for females (A) in Fig. 8.1 and assumed for males (B), and the total weight. *N* is the number of observations.

general rule, male *H. tuberculata* release 2 to 3×10^9 spermatozoa every 30 to 45 seconds during a period of about one hour.

The effect of sperm concentration on fertilization was studied on six different male–female pairs (Fig. 8.4). No cleaving eggs were observed at sperm concentrations lower than 1000 cells ml^{-1}. Fertilization rates increased progressively with increasing sperm concentration, but replicate test runs disclosed wide variations.

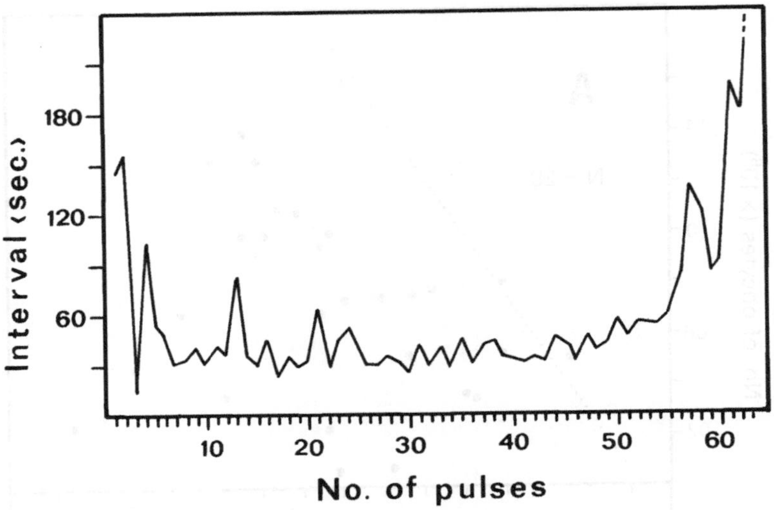

Fig. 8.3 Plot of the time interval between sperm pulses versus number of pulses ordered sequentially during the spawning of an 87 mm *Haliotis tuberculata*.

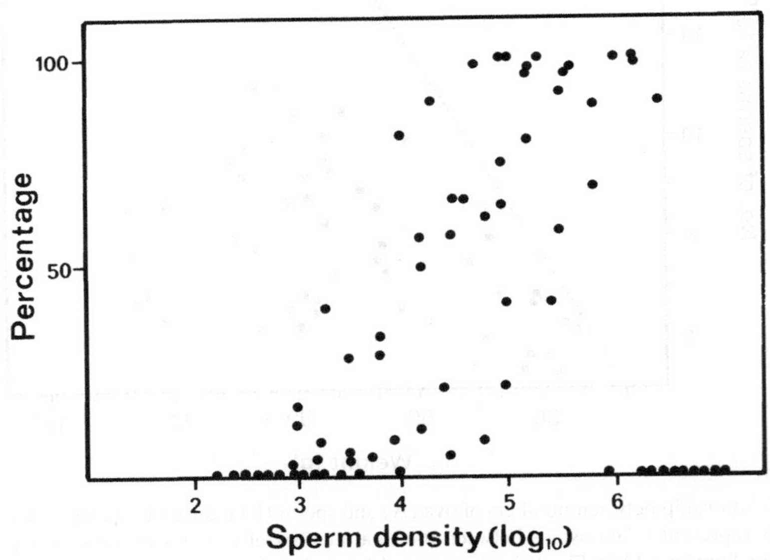

Fig. 8.4 Relation between sperm density and percentage fertilization rate in six male–female pairs.

Fertilization rates of 100% were obtained with concentrations of 10^5 sperm ml^{-1}, although results were more reliable at 10^6 sperm ml^{-1}. Concentrations greater than these levels caused lysis of the vitelline layer with destruction of the ova or abnormal development of embryos.

DISCUSSION

First maturity of *H. tuberculata* at a length of 40 mm is consistent with the observations of Hayashi (1980) on Guernsey populations. A linear relationship between the number of mature eggs and body weight has been reported for other *Haliotis* species (Newman, 1967; Poore, 1973; Kikuchi & Uki, 1975). It corresponds to a power relationship between the number of ripe eggs and the length of the shell (Girard, 1972; Hayashi, 1980). This result may have practical significance in stock management. For example, when potential fecundity is linearly related to biomass, a similar reproductive output may be provided by a low number of old, slow-growing individuals or by a greater number of relatively young specimens with high production. In areas with high recruitment, preservation of old *H. tuberculata* to maintain the propagation of the species (Girard, 1972) is therefore useless from the theoretical point of view.

Fertilization is not effective at all sperm concentrations. In the course of laboratory experiments, concentrations of 10^3 to 10^6 sperm ml^{-1} were required for fertilization and the sperm densities yielding maximum fertilization were between 10^5 and 10^6 sperm ml^{-1}. Similar results have been found on the Japanese abalone *H. discus hannai* Ino by Kikuchi & Uki (1974b). Moreover, a sperm density of 10^6 cells ml^{-1} yielded maximal fertilization among several species of *Haliotis* from the California coast (Leighton & Lewis, 1982).

Like most other abalone, *H. tuberculata* is sedentary and movement does not increase during the spawning season (Clavier & Richard, 1984). This contrasts with the studies of Breen & Atkins (1980) on *H. kamtschatkana* Jonas and of Shepherd (1986) on *H. laevigata* Donovan, who recorded aggregative behaviour of the two species for spawning. If the author's laboratory observations could be transposed to natural grounds, male ormer would release 2 or 3×10^9 sperm every 30 to 45 seconds. Tidal currents are typically strong along the northern coast of Britanny and dilution of this 'fertilization unit' below 10^3 sperm ml^{-1} would occur quickly and at a relatively short distance from the spawning male. Hence, fertilization may be ineffective on beds with low *H. tuberculata* densities. In contrast, on dense *H. tuberculata* grounds gametes from several males could raise the local concentration of sperm and increase the chances of fertilization.

The results of the present study suggest that, in the case of sedentary species like *H. tuberculata*, fecundity is related more to the sperm concentration at spawning than the number of eggs in the ovary. This has implications for stock management of abalone populations. It is generally assumed that the highest fishing mortality rates occur on *Haliotis* aggregations (Newman, 1969; Breen & Atkins, 1980; Shepherd, 1986). The removal of *H. tuberculata* aggregations is likely to have a drastic effect on fertilization and therefore to reduce recruitment (Shepherd, 1986). Management of the fishery of *H. tuberculata* should provide for maintenance of aggregations such that recruitment is not endangered.

ACKNOWLEDGEMENTS

The author would like to thank O. Richard for his valuable assistance. The author is also grateful to P.E. McShane, S.A. Shepherd and M. Kulbicki for reading and criticizing the manuscript. Research facilities were kindly provided by the Laboratoire maritime du Muséum national d'Histoire naturelle, Dinard.

REFERENCES

Breen P.A. & Atkins B.E. (1980) Spawning in a British Columbia population of northern abalone, *Haliotis kamtschatkana*. *Veliger*, **23**, 177–9.

Clavier J. & Richard O. (1984) Etude expérimentale du déplacement de l'ormeau (*Haliotis tuberculata*) dans le milieu naturel. *Rev. Trav. Inst. Pêches marit.*, **46**, 315–26.

Clavier J. & Richard O. (1986) Growth of juvenile *Haliotis tuberculata* (Mollusca: Gastropoda) in their natural environment. *J. Mar. Biol. Assoc. UK*, **66**, 497–503.

Cochard J.C. (1980) *Recherches sur les facteurs déterminant la sexualité et la reproduction chez* Haliotis tuberculata *L. en rade de Brest.* 3e Cycle thesis, Université de Bretagne Occidentale, Brest.

Crofts D.R. (1929) *Haliotis. LMBC Mem. Typ. British Plants Anim.*, **29**.

Flassch J.P. & Aveline C. (1984) Production de jeunes ormeaux à la station expérimentale d'Argenton. *Rap. Scient. Tech. CNEXO*, **50**.

Genade A.B., Hirst A.L. & Smit C.J. (1988) Observations on the spawning, development and rearing of the South African abalone *Haliotis midae* Linn. *S. Afr. J. Mar. Sci.*, **6**, 3–13.

Girard A. (1972) La reproduction de l'ormeau *Haliotis tuberculata* L. *Rev. Trav. Inst. Pêches marit.*, **36**, 163–84.

Hayashi I. (1980) The reproductive biology of the ormer, *Haliotis tuberculata*. *J. Mar. Biol. Assoc. UK*, **60**, 415–31.

Imai T. (1967) Mass production of molluscs by means of rearing the larvae in tanks. *Venus*, **25**, 159–67.

Kikuchi S. & Uki N. (1974a) Technical study on artificial spawning of abalone (genus *Haliotis*). 2–Effect of irradiated sea water with U.V. rays on inducing spawning. *Bull. Tohoku Reg. Fish. Res. Lab.*, **33**, 79–86.

Kikuchi S. & Uki N. (1974b) Technical study on artificial spawning of abalone (genus *Haliotis*). 3–Reasonable sperm density for fertilization. *Bull. Tohoku Reg. Fish. Res. Lab.*, **34**, 67–71.

Kikuchi S. & Uki N. (1975) Technical study on artificial spawning of abalone (genus *Haliotis*). 6–On sexual maturation of *Haliotis gigantea* (Gmelin) under artificial conditions. *Bull. Tohoku Reg. Fish. Res. Lab.*, **35**, 85–90.

Koike Y. (1978) Biological and ecological studies on the propagation of the ormer, *Haliotis tuberculata Linnaeus*. 1–Larval development and growth of juveniles. *La Mer*, **16**, 124–36.

Leighton D.L. & Lewis C.A. (1982) Experimental hybridization in abalones. *Int. J. Invert. Reprod.*, **5**, 273–82.

Newman G.G. (1967) Reproduction of the South African abalone *Haliotis midae*. *Invest. Rep. Div. Sea Fish., Rep. S. Afr.*, **64**, 1–24.

Newman G.G. (1969) Distribution of the abalone (*Haliotis midae*) and the effect of temperature on productivity. *Invest. Rep. Div. Sea Fish., Rep. S. Afr.*, **74**, 1–8.

Poore G.C.B. (1973) Ecology of New Zealand abalones, *Haliotis* species (Mollusca, Gastropoda). 4–Reproduction. *NZ J. Mar. Freshwat. Res.*, **7**, 67–84.

Shepherd S.A. (1986) Studies on southern Australian abalone (genus *Haliotis*). VII. Aggregative behaviour of *H. laevigata* in relation to spawning. *Mar. Biol.*, **90**, 231–6.

Stephenson T.A. (1924) Note on *Haliotis tuberculata*. *J. Mar. Biol. Assoc. UK*, **13**, 480–95.

PART III
LARVAL ECOLOGY AND SETTLEMENT

Chapter 9

Implications of dissolved organic material in seawater for the energetics of abalone larvae *Haliotis rufescens:* a review

DONAL T. MANAHAN and WILLIAM B. JAECKLE *Department of Biological Sciences, University of Southern California, Los Angeles, California, 90089–0371, USA*

ABSTRACT: Larvae of the abalone *Haliotis rufescens* lack a digestive system and have been considered to be 'non-feeding' (lecithotrophic) throughout their lifespan. In this review, we summarize some of the recent research in our laboratory to illustrate that larval stages of abalone can 'feed' by utilizing organic carbon that is in solution in natural seawater. When abalone larvae were cultured in chemically-unamended natural seawater, the organic biomass either stayed the same or increased from the egg to the larva (2-day old). Most of the biomass increase was in the form of lipid. This increase occurred when larvae were reared in particle-free natural seawater (here defined as seawater passed through a 0.2 μm pore size filter). However, sibling larvae, cultured in parallel, always showed a decrease in dry organic weight when cultured in seawater which had first been passed through a sand filter, thereby altering the chemistry. Measurements of the energy requirements (oxygen consumption) for larval development were compared to the energy made available from the utilization of endogenous reserves (carbohydrate, lipid and protein). For the larval cultures studied, the rates of utilization of biochemical reserves did not supply the energy needs of the larvae. Thus, it would seem that these 'non-feeding' larvae require an exogenous source of organic carbon. Under the experimental conditions used, the only organic material available to the larvae was in the form of dissolved organic material (DOM) in seawater. Estimates of the contribution that DOM in seawater could provide to larvae showed that this pool of exogenous material could supply the missing energy not supplied by the egg. These findings suggest that alterations to the organic chemistry of seawater can affect the metabolism of larvae. If such alterations are not controlled for in aquaculture facilities, the presence or absence of dissolved organic compounds in seawater can alter the growth and development of abalone.

RESUMEN: Las larvas de abulón *Haliotis rufescens* carecen de un sistema digestivo y son consideradas 'ayunantes' (lecitotroficas) en esta etapa de su vida. Mas aun, cuando se cultivan larvas de abulón en agua de mar natural, (no alterada químicamente), la biomasa orgánica siempre se incrementó (5 cultivos), desde el huevo hasta la larva (de 2 dias de edad). La mayor parte del incremento en biomasa se dió en forma de lípidos. Esto siempre ocurrió cuando las larvas avivaron en agua de mar libre de particulas. En contraste las larvas testigo cultivadas en paralelo, mostraron siempre un decremento en el peso orgánico seco cuando se cultivaron en agua de mar que había pasado primero por un filtro de arena, alterando por lo tanto su composición orgánica. Se hicieron mediciones de los requerimientos de energía (consumo de oxígeno) para el desarrollo de larvas y se compararon con las de la energía disponible a partir de la utilización de reservas endógenas (carbohidratos, proteinas y lípidos). Para los 5 cultivos de larvas estudiados, las tasas de utilización de las reservas bioquímicas no proveyeron las necesidades energéticas de las larvas. Las estimaciones variaron de menos del 25% a mas del 71% de las necesidades metabólicas de las larvas. Por lo tanto parecería ser que estas larvas 'ayunantes' requieren de una fuente exógena de carbón orgánico. Bajo las condiciones experimentales utilizadas, el único material orgánico disponible para las larvas estuvo presente en forma de material orgánico disuelto (DOM). Nuestras mediciones de la contribución que el DOM del agua de mar podría proporcionar a las larvas de abulón mostraron que este suministro de material exógeno podría proveer la energía que no aporta el huevo. Por ejemplo, la tasa de incorporación sólo de aminoacidos por las larvas veliger (aproximádamente 1% del total del DOM) podría en promedio contribuir a cubrir el 55% de las necesidades metabólicas de estas larvas. Nuestros resultados sugieren que la alteración de la química orgánica del agua de mar puede afectar el metabolismo de las larvas de abulón. Si tales alteraciones no son controladas en los laboratorios de cultivo, la presencia o ausencia de compuestos orgánicos disueltos en el agua de mar, pueden alterar el crecimiento y desarrollo de los abulones.

INTRODUCTION

The decline of abalone fisheries has stimulated interest in the mariculture of these organisms. As with most marine organisms, the larval and early post-settlement stages are probably the most vulnerable phases of the abalone's life cycle. Abalone have been successfully reared from the egg to the adult under laboratory conditions, although the process is not trouble free (Ebert & Houk, 1984). There is a large body of data on factors affecting larval settlement and metamorphosis (Morse *et al.*, 1979; Morse & Morse, 1984), but the relationship between larval energy reserves and post-settlement survival is unknown.

Larvae of marine invertebrates are traditionally grouped into those that have a functional digestive system and can feed on particulate foods ('feeding larvae'), and those that lack a digestive system during larval life ('non-feeding larvae') (Thorson, 1950). It is essential to understand the energetic and nutritional needs of any organism to grow it under controlled conditions. For feeding larvae, this theme has stimulated numerous studies on the algal foods and feeding preferences of the larval stages of commercially important molluscs, such as oysters (see reviews by Loosanoff & Davis, 1963; Walne, 1974).

In contrast, abalone larvae (*Haliotis* spp.) are non-feeding (Crofts, 1937) and only begin feeding several days after settlement. Considerations of the energy needs of abalone larvae have started from the premise that these animals meet their metabolic requirements by utilizing the maternal reserves provided to the egg without any input of energy from the environment. This suggests that larval biomass should decrease during development (i.e., the larvae are 'starving'). In this chapter, we summarize some of our recent data to show that abalone can increase in biomass (dry organic weight) during development and that the energy reserves in the egg do not supply the cost of larval development. We also show that larvae of *H. rufescens* obtain energy from their environment by taking up dissolved amino acids and sugars from seawater. We conclude from these data that there are environmental factors affecting the energetics of larval development beyond those of the initial energy reserves in the egg.

MATERIALS AND METHODS

Animals

All eggs and larvae of the red abalone *H. rufescens* were supplied by the 'Ab Lab' located in Port Hueneme, California. Larvae were reared at a temperature of 15 to 17°C in seawater that was passed through a 0.2 μm (pore size, Nuclepore) filter.

Biomass changes

The changes in biomass were measured each day (for veliger larvae) until settlement (see Jaeckle & Manahan, 1989a, for details). Biomass is given as dry organic weight per larva. Dry organic weight is defined as the difference between total dry weight and ash dry weight. The ash weight was determined after burning the organic material in a muffle furnace. All weighings were performed on a Cahn Model 29 electrobalance (accuracy of 0.1 μg).

Uptake by larvae of sugars and amino acids from seawater

High-performance liquid chromatography (HPLC): substrate concentrations
High-performance liquid chromatography was used to measure the rates of net uptake by larvae of amino acids and sugars from seawater and the concentrations of dissolved amino acids in the seawater used to culture the larvae. A known number of larvae was placed in a volume of seawater (50 to 100 ml, depending on the experiment). A mixture of amino acids or sugars (separate experiments) was then added to the flask to give a known concentration. Where mixtures were used (e.g. as in Fig. 9.2, lower graph), each substrate was present in equimolar concentrations. The change in the concentration of each compound with time was measured by HPLC. Individual compounds were separated using HPLC and the amount of each chemical was measured with the appropriate detector. Following reaction with *o*-phthaldialdehyde, picomole amounts of individual amino acids were measured as their fluorescence products by means of a fluorometer (details in Manahan, 1989). Picomole amounts of sugars in seawater were measured using a new procedure (Welborn & Manahan, 1990). Individual sugars were detected with a post-column reaction technique involving an electrochemical mediator, copper (II) bis(phenanthroline) (CBP). Individual sugars were separated on the column and the CBP reagent was added by a mixing tee (post-column). The sugars in the eluent reduced the copper in the CBP complex from Cu^{+2} to Cu^{+1}. The resulting zones of reduced CBP in the eluent stream were then reoxidized as they passed across the working electrode (glassy carbon) in the cell of the electrochemical detector.

Isotope influx: kinetics of amino acid transport
A known number of larvae was transferred into each of several (9−11) 20 ml vials and sterile seawater was added until the final volume was 10 ml; 1 μCi ^{14}C-alanine was added to each vial. In addition, a known amount of cold carrier (^{12}C-alanine, Sigma) was added to give the required substrate concentration in each vial. The concentrations used ranged from 0.5 to 100 μM. A time course experiment of 6 to 8 minutes duration was then carried out with each vial to measure the rate of alanine transport into larvae at each different substrate concentration. At approximately 1 minute intervals, a sample of the larval suspension was taken, excess

isotope removed (see methods in Manahan *et al.*, 1989), and the radioactivity in the larvae was determined with a liquid scintillation counter.

Biochemical reserves

The energy reserves of eggs (carbohydrate, lipid and protein), and their subsequent changes during development, were measured using biochemical fractionation of the homogenates of eggs and larvae (basic procedures of Holland & Gabbott, 1971; modifications given by Jaeckle & Manahan, 1989a).

Respiration rates

Rates of oxygen consumption of larvae were measured using polarographic oxygen sensors (Jaeckle & Manahan, 1989a).

RESULTS AND DISCUSSION

Abalone larvae lack a digestive system and cannot, therefore, feed on particulate foods during early development. Yet they can increase in biomass (dry organic

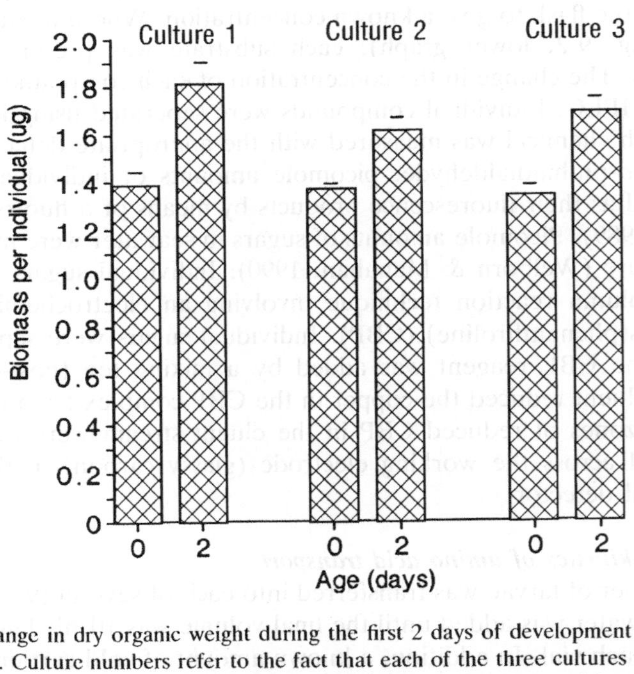

Fig. 9.1 The change in dry organic weight during the first 2 days of development of the abalone *Haliotis rufescens*. Culture numbers refer to the fact that each of the three cultures were reared using gametes obtained from three independent spawnings. (Day 0 = dry organic weight of egg; day 2 = dry organic weight of newly-formed veliger larva.) (Figure redrawn from data given by Jaeckle & Manahan, 1989a). Error bars represent 1 standard error of the mean. The differences between the means for day 0 and day 2 for each of cultures 1, 2 and 3 are statistically significant ($P \leq 0.05$) based on single classification ANOVA (Sokal & Rohlf, 1969) with two groups of equal sample size: $N = 5$ per sample.

weight) during early development (Fig. 9.1). The data given in Fig. 9.1 were obtained using gametes that were from independent spawnings of three different pairs of females and males (labelled as cultures 1, 2 and 3 in Fig. 9.1). In all three cases, there was an increase in biomass (statistics given in legend to figure) from the egg (day 0) to the veliger-stage larvae (day 2). Thus, growth occurred even in the absence of a digestive system.

Given the lack of a digestive system during these stages of development, dissolved organic material (DOM) in seawater was the most likely source of nutrients being utilized by the larvae. The data shown in Fig. 9.2 support this suggestion. The upper graph shows the net depletion of five amino acids by 2-day-old veliger larvae. Note that the Y-axis is on a log scale (i.e., the rate of removal of substrate is exponential with time). The rates of uptake of multiple substrates can be calculated from these data using first-order depletion constants for each individual substrate (see Jaeckle & Manahan, 1989b, for equations). These rates are given in the lower graph for 16 amino acids. The data show that veliger larvae have transport systems in their integument capable of net uptake from seawater of several classes of amino acids (acidic, basic and neutral). Veliger larvae are also capable of net uptake of sugars. Figure 9.3 shows that larvae (5-day-old) can take up the monosaccharide glucose and the disaccharide maltose. The transport systems for monosaccharides seem to be specific for certain substrates. For

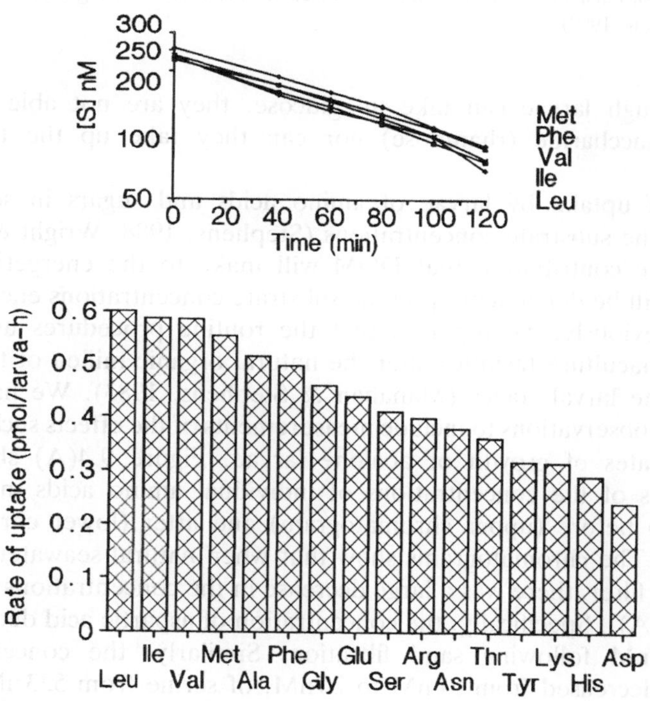

Fig. 9.2 Upper graph: the net uptake of amino acids from seawater by veliger larvae of *Haliotis rufescens* (2-day-old). The concentration of each substrate was measured at each time interval using HPLC. Lower graph: the relative rates of uptake of 16 amino acids by veliger larvae of *H. rufescens*. Each of the 16 amino acids was present in a mixture at 250 nM for each amino acid (data from Jaeckle & Manahan, 1989b).

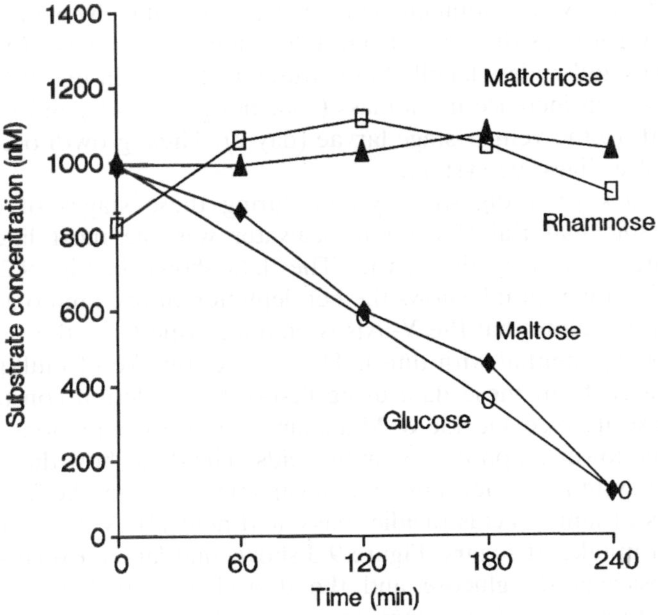

Fig. 9.3 The net uptake of sugars from seawater by veliger larvae of *Haliotis rufescens* (5-day-old). The concentration of each substrate was measured at each time interval using HPLC (data from Welborn & Manahan, 1990).

example, although larvae can take up glucose, they are not able to take up another monosaccharide (rhamnose) nor can they take up the trisaccharide maltotriose.

The rates of uptake by larvae of amino acids and sugars in seawater are dependent on the substrate concentrations (Stephens, 1988; Wright & Manahan, 1989). Thus the contribution that DOM will make to the energetics of larval development will be dependent upon the substrate concentrations encountered by the larvae. Previously, we reported that the routine procedures used to treat seawater in aquaculture facilities alter the natural organic milieu of the seawater used to rear the larval stages (Manahan & Stephens, 1983). We have recently extended these observations to include measurements of the effects such alterations have on the rates of growth of abalone larvae. Figure 9.4(A) shows HPLC chromatographs of the concentrations of individual amino acids in samples of seawater taken before (upper chromatogram) and after (lower chromatogram) sand filtration. The chromatograms show that when natural seawater was passed through a sand filter, there was a large decrease in the concentrations of all amino acids (Fig. 9.4(A)). Specifically, the concentration of aspartic acid decreased from 167 nM to 16 nM following sand filtration. Similarly, the concentrations of glutamic acid decreased from 51 nM to 20 nM, of serine from 533 nM to 69 nM, of alanine from 216 nM to 31 nM and of tyrosine from 57 nM to 10 nM. In two batches of larvae, reared from gametes obtained from two independent spawnings, there was either a significant increase ($P \leq 0.05$, details in legend to Fig. 9.4) in dry organic weight of a veliger larva relative to the egg at day 0 (Fig. 9.4(B)) or no net change ($P \geq 0.05$, Fig. 9.4(C)) when the larvae were cultured in seawater which had not been passed through a sand filter (Fig. 9.4(A), upper chromatogram).

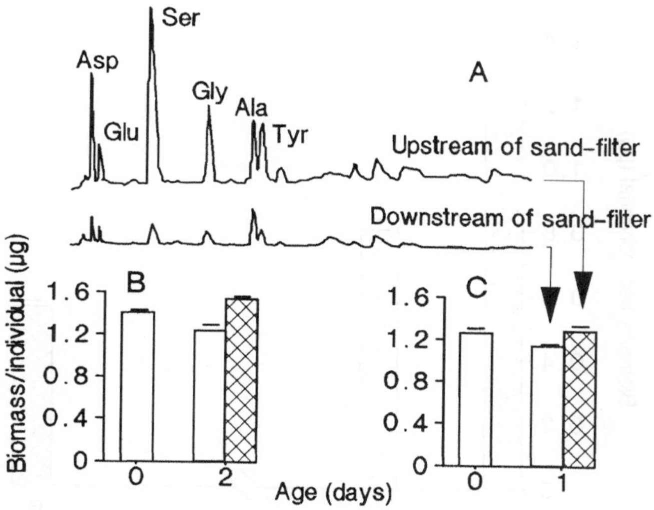

Fig. 9.4 The upper portion (A) shows HPLC chromatograms depicting the change in total free amino acid concentrations in seawater which was sampled upstream and downstream of a sand filter. Peak areas are proportional to substrate concentrations. The bar graphs (B and C) show the changes in the dry organic weight during development of *Haliotis rufescens* larvae reared in the differently-treated seawater. Parts (B) and (C) represent the results for larvae obtained from two independent spawnings. Open bars represent the dry organic weight of larvae reared in seawater which had been passed through a sand filter, and then a 0.2 µm (pore size) filter. Cross-hatched bars depict larvae reared in filtered seawater (0.2 µm pore size) which was not passed through a sand filter. Day 0 = egg; day 1 = trochophore-stage larva; day 2 = veliger-stage larva. Error bars represent one standard error of the mean (data from Jaeckle & Manahan, 1990). (B) The difference between the means for day 0 ($N = 6$) and day 2 (open histogram, $N = 5$) is statistically significant ($P \leq 0.05$) based on single classification ANOVA (Sokal & Rohlf, 1969) with two groups of unequal sample size. The difference between day 0 and day 2 (hatched histogram) is statistically significant ($P \leq 0.05$), based on single classification ANOVA (Sokal & Rohlf, 1969) with two groups of equal sample size, $N = 6$ per sample. (C) The difference between the means for day 0 and day 1 (open histogram) is statistically significant ($P \leq 0.05$, single classification ANOVA, Sokal & Rohlf, 1969) with two groups of equal sample size, $N = 6$ per sample. The difference between the means for day 0 and day 1 (hatched histogram) is not statistically significant ($P \geq 0.05$, single classification ANOVA with two groups of equal sample size, $N = 6$ per sample.

In contrast, the corresponding sibling larvae reared in sand-filtered seawater (Fig. 9.4(A), lower chromatogram) showed a significant decrease ($P \leq 0.05$) in dry organic weight during the same time period (Fig. 9.4(B),(C)).

An additional batch of larvae (different from those used in Fig. 9.4(B),(C)) was reared through to settlement in sand-filtered seawater. Again these larvae showed a continuous decrease in dry organic weight relative to the egg (Fig. 9.5). However, when larvae were cultured in seawater that had not been sand filtered (Fig. 9.6), a net increase in biomass was observed during the first 2 days of development (cf Fig. 9.1), and subsequently only a small loss of biomass (relative to Fig. 9.5) was observed. Samples of the larvae used for the determinations of biomass in Fig. 9.6 were collected for biochemical fractionation. The amounts of carbohydrate, lipid and protein were measured at each sampling point (Fig. 9.7). The biomass at each point sampled is now presented as lipid (solid area of histogram) and protein (hatched area). No values are presented for the amount of carbohydrate in eggs and larvae because this fraction was insignificant, being

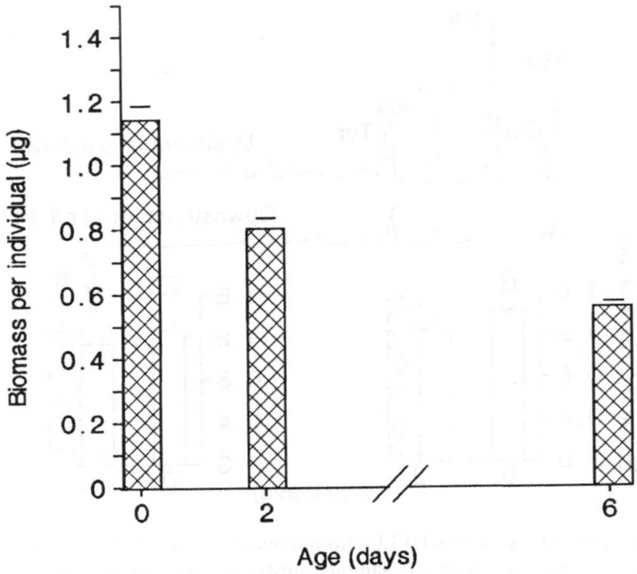

Fig. 9.5 The decrease in dry organic weight of larvae of *Haliotis rufescens* reared from eggs (day 0) in sand-filtered seawater (data from Jaeckle & Manahan, 1992).

below the limit of detection of the analytical procedures used. The area given as the open part of the histogram is that fraction of the dry organic weight not accounted for by the amount of lipid and protein. We have termed this fraction the 'remainder' fraction. A comparison of the data given in Figs 9.6 and 9.7 shows that when an increase in dry organic weight was observed (e.g. Fig. 9.6, day 0 to day 2), a corresponding increase in lipid and protein was also observed (in addition to an increase in the remainder fraction).

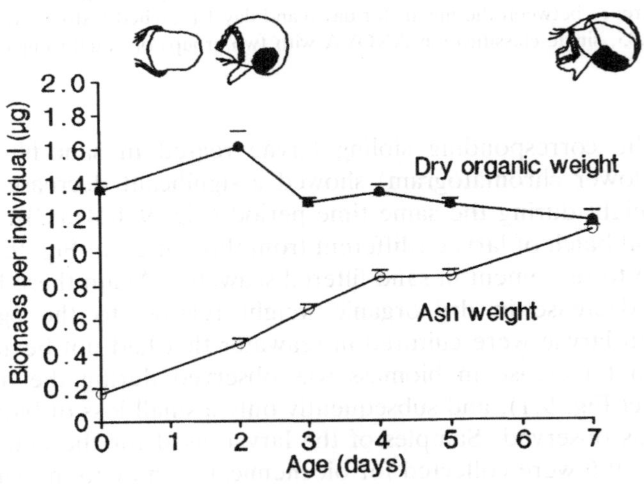

Fig. 9.6 Changes in dry organic weight and ash dry weight during the complete larval development of *Haliotis rufescens* reared in seawater which had not been passed through a sand filter (data from Jaeckle & Manahan, 1989a).

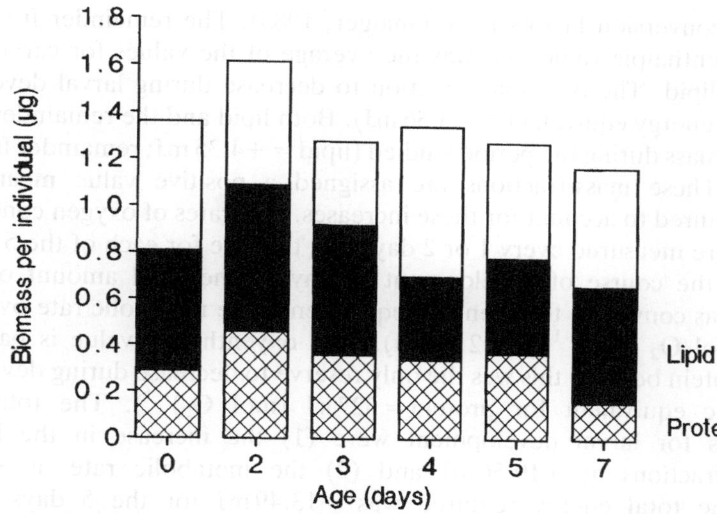

Fig. 9.7 Changes in the amount of protein and lipid in larvae during the development of *Haliotis rufescens*. Hatched portions of histograms represent that fraction of the dry organic weight accounted for by protein. Solid portions represent that fraction accounted for by lipid. Open portions represent the difference between the dry organic weight (see Fig. 9.6) and the combined amounts of lipid and protein (data from Jaeckle & Manahan, 1989a).

Do the rates of utilization of energy reserves in the egg provide all the energy required for the cost of larval development? The data presented in Table 9.1 address this point. The major biochemical reserves in the egg (day 0) were determined and measured again in the larval stage just prior to metamorphosis (day 5). The larvae used for these measurements were reared using gametes obtained from a separate spawning to that used for Figs 9.6 and 9.7. Net changes in the amounts of each biochemical component were assigned an enthalpic

Table 9.1 An energy budget for the complete larval development of the gastropod *Haliotis rufescens* (data summarized from Jaeckle & Manahan, 1989a; errors and statistical analyses given therein).

Available energy (changes in biochemical content)	
Protein	−3.36 mJ
Lipid	+4.35 mJ
Remainder	+6.21 mJ
Sum of available energy	−3.36 mJ
Required energy (metabolic rate and growth)	
Metabolic rate over 5 days:	
5.56 nmol O_2 larva^{-1} =	+2.93 mJ
Increases in biochemical content:	
Lipid and remainder	+10.56 mJ
Sum of required energy	+13.49 mJ
Energy budget	
Ratio of available to required energy = 3.36/13.49 = 25%	

+ Energy required.
− Energy available.

equivalent (conversion factors from Gnaiger, 1983). The remainder fraction was assigned an enthalpic value that was the average of the values for carbohydrate, protein and lipid. The only mass-fraction to decrease during larval development was protein (energy equivalent = −3.36 mJ). Both lipid and the remainder fraction increased in mass during the period studied (lipid = +4.35 mJ; remainder fraction = +6.21 mJ). These mass-fractions are assigned a positive value meaning that energy is required to account for these increases. The rates of oxygen consumption by larvae were measured every 1 or 2 days and the rate for each of the 5 days was added over the course of development (5 days). The total amount of oxygen consumed was converted to an energy equivalent. The metabolic rate over 5 days was 5.56 nmol O_2 larva^{-1} (= +2.93 mJ). This oxyenthalpic value is based on a value for protein because this was the only reserve to decrease during development (oxyenthalpic equivalent for protein = 527 kJ (mol O_2)$^{-1}$). The total energy requirements for larval development were (1) the increase in the lipid and remainder fractions at +10.56 mJ and (2) the metabolic rate at +2.93 mJ. Summed, the total energy required was +13.49 mJ for the 5 days of larval development. The only energy available from endogenous reserves is that from protein at −3.36 mJ. Thus, the amount of energy contributed from the energy reserves of the egg to the cost of larval development is given as the ratio of available energy to required energy = 3.36/13.49 = 25%. This calculation reveals that 75% of the energy demands of development are not provided for by the reserves initially present in the egg of *H. rufescens*.

To balance the energy needs of development, larvae would require 10.13 mJ (i.e., the difference between the available energy at 3.36 mJ and the required energy at 13.49 mJ). Could the rate of DOM uptake account for this missing energy needed for larval development? The kinetics of alanine transport from seawater by trochophore (24-h-old) and veliger (48-h-old) larvae of *H. rufescens* are given in Fig. 9.8. The value for K_t, the substrate concentration (of alanine) at half-maximum velocity (J_{max}), was 8.0 µM for a trochophore (1-day-old) and J_{max} was 22.8 pmol alanine trochophore^{-1} h^{-1}. Later in development (2-day-old

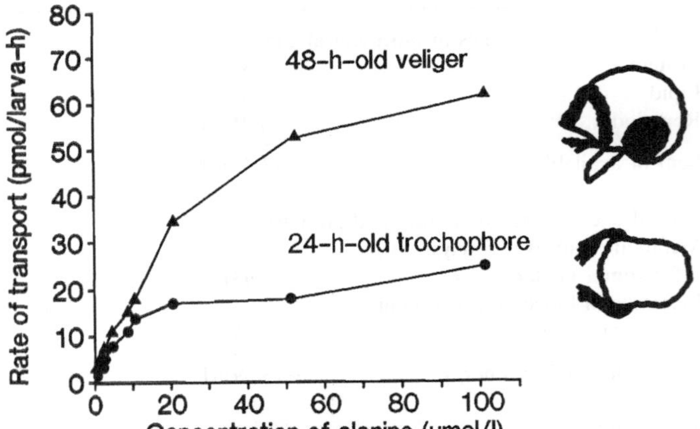

Fig. 9.8 Ontogenic changes in the rate of transport of alanine by trochophore (24-hours-old) and veliger (48-hours-old) larvae of *Haliotis rufescens* as a function of increasing substrate concentration (data from Manahan *et al.*, 1989).

veliger larvae), the K_t increased to 22.7 μM and the J_{max} increased to 71.1 pmol alanine veliger^{-1} h^{-1} (Manahan *et al.*, 1989). Using the Michaelis-Menten equation, a rate of amino acid transport by embryos can be calculated from a concentration of 1 μmol/l. The resulting rates of alanine transport would range from 2.53 pmol to 3.00 pmol alanine h^{-1} for a trochophore and a veliger larva, respectively. Taking the average of 2.53 and 3.00, and multiplying by the molecular weight of alanine (89), a larva would obtain 246 pg alanine per hour, or 30 ng per 5-day period. The developing abalone larva would need 10.13 mJ to meet its energy needs over the 5 days of development (Table 9.1). The components of the DOM pool that are likely to be biologically available are primarily carbohydrates and protein-like compounds. These components can be assigned an enthalpic equivalent of 20.75 kJ g^{-1}, which is the average of the value for carbohydrate (17.5 kJ g^{-1}) and protein (24.0 kJ g^{-1}) (enthalpic equivalents from Gnaiger, 1983). Thus 10.13 mJ would be equivalent to 488 ng organic material (as DOM). The proportion of this amount that could be supplied by the transport of amino acids alone at a rate of 30 ng per 5 days would be 6% (i.e., 30 divided by 488). An estimate of the contribution of DOM to the energy demand of larval development can then be calculated as follows. Amino acids make up approximately 1% of the total pool of dissolved organic carbon in seawater (Williams, 1975); larvae of *H. rufescens* transport amino acids at a rate that could account for 6% of the calculated energy needs over 5 days; thus, if 17% of the total DOM pool was transported at rates similar to those measured for amino acids, then 100% (i.e., 6 × 17) of the energy needs of the development of abalone larvae could be met by the uptake of DOM.

CONCLUSIONS

The increases in both dry organic weight (Figs 9.1 and 9.6) and specific biochemical components (Fig. 9.7) during the development of *H. rufescens* suggest that these larval stages are feeding, but the source of exogenous food is in a dissolved form. Our findings with lecithotrophic larvae of abalone presented here are supported by studies of the energy cost for direct development in other species (echinoderms: Turner & Rutherford, 1976; Lawrence *et al.*, 1984; McClintock & Pearse, 1986). These authors reported little change in total biomass and energy content between the egg and the juvenile. The assumption that 'non-feeding' larvae of abalone are totally dependent on energy reserves in the egg, and independent of energy in the environment, seems incorrect. The interaction between developing abalone larvae and the organic chemistry of their environment should be kept in mind in the design of aquaculture facilities used to treat seawater. Differences in the quality and quantity of DOM in seawater used to culture abalone may help to explain variations in the successful rearing of larvae.

ACKNOWLEDGEMENTS

The authors thank John McMullen (Ab Lab) for generously providing facilities and abalone gametes and larvae. This research was supported by a grant from NOAA, Office of Sea Grant (NA 86 AA-D-SG 199).

REFERENCES

Crofts D.R. (1937) The development of *Haliotis tuberculata*, with special reference to organogenesis during torsion. *Philos. Trans. R. Soc. Ser. B.*, **228**, 219–68.

Ebert E.E. & Houk J.L. (1984) Elements and innovations in the cultivation of red abalone *Haliotis rufescens*. *Aquaculture*, **39**, 375–92.

Gnaiger E. (1983) Calculation of energetic and biochemical equivalents of respiratory oxygen consumption. In: *Polarographic Oxygen Sensors* (Ed. by E. Gnaiger & H. Forstner), pp. 337–45. Springer-Verlag, Berlin.

Holland D.L. & Gabbott P.A. (1971) A microanalytical scheme for the determination of protein, carbohydrate, lipid and RNA levels in marine invertebrate larvae. *J. Mar. Biol. Assoc. UK*, **51**, 659–68.

Jaeckle W.B. & Manahan D.T. (1989a) Growth and energy imbalance during the development of a lecithotrophic molluscan larva (*Haliotis rufescens*). *Biol. Bull.*, **177**, 237–46.

Jaeckle W.B. & Manahan D.T. (1989b) Feeding by a 'nonfeeding' larva: uptake of dissolved amino acids from seawater by lecithotrophic larvae of the gastropod *Haliotis rufescens*. *Mar. Biol.*, **103**, 87–94.

Jaeckle W.B. & Manahan D.T. (1992) Experimental manipulations of the organic chemistry of seawater: implications for studies of energy budgets in marine invertebrate larvae (Echinodermata, Mollusca). *J. Exp. Mar. Biol. Ecol.*

Lawrence J.M., McClintock J.B. & Guille A. (1984) Organic level and caloric content of eggs of brooding asteroids and an echinoid (Echinodermata) from Kerguelen (South Indian Ocean). *Int. J. Invert. Reprod. Devel.*, **7**, 249–57.

Loosanoff V.L. & Davis H.C. (1963) Rearing of bivalve molluscs. *Adv. Mar. Biol.*, **1**, 1–136.

McClintock J.B. & Pearse J.S. (1986) Organic and energetic content of eggs and juveniles of Antarctic echinoids and asteroids with lecithotrophic development. *Comp. Biochem. Physiol.*, **85A**, 341–5.

Manahan D.T. (1989) Amino acid fluxes to and from seawater in axenic veliger larvae of a bivalve (*Crassostrea gigas*). *Mar. Ecol. Prog. Ser.*, **53**, 247–55.

Manahan D.T., Jaeckle W.B. & Nourizadeh S. (1989) Ontogenic changes in the rates of amino acid transport from seawater by marine invertebrate larvae (Echinodermata, Echiura, Mollusca). *Biol. Bull.*, **176**, 161–8.

Manahan D.T. & Stephens G.C. (1983) The use of high performance liquid chromatography to measure dissolved organic compounds in bivalve aquaculture systems. *Aquaculture*, **32**, 339–46.

Morse A.N.C. & Morse D.E. (1984) Recruitment and metamorphosis of *Haliotis* larvae induced by molecules uniquely available at the surfaces of crustose red algae. *J. Exp. Mar. Biol. Ecol.*, **75**, 191–215.

Morse D.E., Hooker N., Duncan H. & Jensen L. (1979) Gamma-aminobutyric acid, a neurotransmitter, induces planktonic abalone to settle and begin metamorphosis. *Science*, **204**, 407–10.

Sokal R.R. & Rohlf S.J. (1969) *Principles and Practice of Statistics in Biological Research*. W.H. Freeman, San Fransisco.

Stephens G.C. (1988) Epidermal amino acid transport by marine invertebrates. *Biochim. Biophys. Acta*, **947**, 113–38.

Thorson G. (1950) Reproductive and larval ecology of marine bottom invertebrates. *Biol. Rev. Camb. Philos. Soc.*, **25**, 1–45.

Turner R.L. & Rutherford J.C. (1976) Organic, inorganic and caloric composition of eggs, pentaculae, and adults of the brooding sea cucumber *Cucumaria curata* Cowles (Echinodermata: Holothuroidea). *J. Exp. Mar. Biol. Ecol.*, **24**, 49–60.

Walne P.R. (1974) *Culture of Bivalve Molluscs: 50 Years Experience at Conway*. Fishing News Books, Oxford.

Welborn J.R. & Manahan D.T. (1990) Direct measurements of sugar uptake from seawater into molluscan larvae. *Mar. Ecol. Prog. Ser.*, **65**, 233–9.

Williams P.J. LeB. (1975) Biological and chemical aspects of dissolved organic material in seawater. In *Chemical Oceanography*, Vol 2 (Ed. by P.J. Riley & G. Skirrow), pp. 307–64. Academic Press, London.

Wright S.H. & Manahan D.T. (1989) Integumental nutrient uptake by aquatic organisms. *Ann. Rev. Physiol.*, **51**, 585–600.

Chapter 10
Molecular mechanisms controlling metamorphosis and recruitment in abalone larvae

DANIEL E. MORSE *Marine Biotechnology Center and Department of Biological Sciences, University of California, Santa Barbara, California 93106, USA*

ABSTRACT: Abalone larvae have proved especially useful for experimentally resolving the molecular mechanisms controlling the settlement and metamorphosis of marine invertebrate larvae in response to chemical signals from the environment. Recent studies *in vitro* have confirmed the existence of two separate families of chemosensory receptors and signal transduction pathways that process such chemical information from the environment to control metamorphosis in abalone larvae. A variety of chemical stimuli have been found that act at the chemosensory receptors controlling these two pathways, while other chemical stimuli have been found to induce larval settlement, or enhance the sensitivity of the larvae to settlement inducers, by interacting directly with the signal transducers. The receptors and signal transducers of the regulatory (amplifier) pathway have been characterized *in vitro*, on cilia purified from the larval epithelium; results of these analyses confirm conclusions from earlier experiments *in vivo*. Properties of the receptors of the regulatory and morphogenetic pathways suggest how these may interact to control settlement in response to two different classes of chemical signals (in the water column and on surfaces) in the natural environment. At last count, 13 *Haliotis* species (including *H. discus hannai*) have been found to respond similarly to chemical inducers of settlement and metamorphosis. We have begun to clone genes expressed in the abalone early after metamorphosis, to facilitate analysis of the mechanisms controlling their developmental activation and their contribution to early growth.

RESUMEN: Las larvas de abulón son especialmente útiles para explicar experimentalmente los mecanismos moleculares que controlan el asentamiento y metomorfosis de las larvas marinas invertebradas como respuesta a señales químicas del medio ambiente. Recientes estudios *in vitro* han confirmado la existencia de dos familias distintas de receptores quimicosensores y senderos de transducción de señales que procesan dicha información química del medio ambiente para controlar las metamorfosis en las larvas de abulón. Se ha encontrado una variedad de estímulos químicos que actúan en los receptores quimicosensores que controlan estos dos senderos; asimismo, se ha visto que otros estímulos químicos inducen al asentamiento de las larvas, o aumentan la sensibilidad de las larvas a los inductores de asentamiento, a través de la interacción directa con los transductores de señales. Los receptores y los transductores de señales del sendero (amplificador) regulador han sido caracterizado *in vitro*, sobre cilios purificados del epitelio larval; los resultados de estos análisis confirman conclusiones de anteriores experimentos *in vivo*. Las propiedades de los receptores de los senderos reguladores y morfogenéticos indican cómo estos pueden actuar recíprocamente para controlar el asentamiento respondiendo a dos tipos diferentes de señales químicas (en la columna de agua y en las superficies) en el medio ambiente natural. En un recuento final, se han encontrado 13 especies de *Haliotis* (incluyendo la *H. discus hannai*) que responden en forma similar a los inductores químicos de asentamiento y metamorfosis. Hemos empezado a hacer clones de genes que se manifiestan en el abulón justo después de la metamorfosis, para facilitar el análisis de los mecanismos que controlan su activación durante el desarrollo y su contribución al crecimiento temprano.

INTRODUCTION

Larval metamorphosis in species of *Haliotis*, the abalone, provides a uniquely tractable system for the analysis of chemosensory and molecular mechanisms that control development and recruitment in marine organisms. The advantages of *Haliotis* for such studies include the large numbers of larvae that can be produced, with complete control of spawning; the relatively rapid and synchronous development of the larvae; the fact that the larvae are lecithotrophic; and the nearly absolute requirement of the larvae (when cultured under defined and controlled conditions) for an exogenous inducer of metamorphosis (Morse, 1990). This chapter presents new findings not previously published, reviews earlier work on the control of metamorphosis and recruitment in abalone larvae and discusses recent progress in the application of these findings.

INDUCERS OF SETTLEMENT AND THE MORPHOGENETIC PATHWAY

A wide variety of biotic and chemical substances have been found or have been proposed capable of inducing the settlement and metamorphosis of abalone larvae. The biotic materials include crustose coralline red algae (Crofts, 1929; Morse D. *et al.*, 1979a,b, 1980c; Morse A. *et al.*, 1984; Morse & Morse, 1984a; Morse D. 1988), various microalgae and bacteria (Morse D. *et al.*, 1979b, Morse A. *et al.*, 1984; Akashige *et al.*, 1981; Leighton, 1985) and abalone mucus (Seki & Kanono, 1981; Akashige *et al.*, 1981; Slattery, 1987). While each of these findings has led to the development of useful improvements in the technology for cultivation of abalone, only the crustose red algae thus far have been demonstrated, in quantitative and controlled analyses, to enhance substratum-specific recruitment of *Haliotis* spp. in the natural environment. Such results have been obtained for *H. discus hannai* (Saito, 1981), *H. laevigata* and *H. scalaris* (Shepherd & Turner, 1985); *H. cracherodii* (Douros, 1985) and *H. rubra* (Prince *et al.*, 1987). The results of Douros and Prince *et al.* also appear to exclude any significant contribution to recruitment from conspecific (e.g. mucus) cues in the natural environment, as both studies found that settlement preference for microhabitats was independent of conspecific density.

It has not been determined whether the settlement-inducing molecules found at the surfaces of crustose red algae are encoded in the red algal genome or in the genomes of associated microorganisms. Recently, however, Aileen Morse (unpub. obs) found that 33 different strains of bacteria isolated from samples of crustose red algae all failed to induce settlement or metamorphosis of abalone larvae.

The precise identity of the inductive chemical from the crustose red algae still eludes us. We do know that this inducer is a molecule first found associated with a colourless protein from the algae; biochemical fractionation resolves this activity into a smaller fraction that is associated with a peptide which contains several unusual amino acid residues (Morse A. *et al.*, 1984; Morse & Morse, 1984a; Morse A., 1988). While other (i.e. foliose) red algae (and the biochemically and evolutionarily related cyanobacteria) also contain closely related inducing molecules intracellularly, these algae and bacteria, while intact, fail to induce the settlement and metamorphosis of abalone larvae, in contrast to the induction caused by intact crustose red algae (Morse A. *et al.*, 1984; Morse & Morse 1984a,b). This difference was shown to be due to the unique availability of the inducing molecules at the surfaces of the crustose red algae (Morse & Morse, 1984a). The inducers may be transported to the surfaces of the crustose coralline red algae by the high secretory and exfoliative activities of these algae.

The algal inducer, in both its protein-associated and peptide-associated forms, has proved to be GABA-mimetic. That is, the molecule (purified 20 000-fold from the recruiting alga) mimics the action of the simple amino acid neurotransmitter, gamma-aminobutyric acid (GABA), in binding strongly, specifically and saturably to GABA receptors purified from mammalian brain (Morse A., 1985; Morse D., 1985; Morse D., 1990; A.N.C. Morse & H.G. Trapido-Rosenthal, unpub. obs.). Apparently, similar receptors on the abalone larvae mediate the recognition of the algal inducer, causing settlement and metamorphosis (Trapido-Rosenthal and Morse D., 1986a; A.N.C. Morse & H.G. Trapido-Rosenthal, unpub. obs.).

Several lines of evidence suggest that the externally accessible receptors on the abalone larvae that can be labelled with radioactive GABA analogues are those which mediate the induction of metamorphosis (Trapido-Rosenthal & Morse, 1986a; Morse D., 1990). The fact that the natural inducer of larval settlement and metamorphosis purified from a naturally recruiting crustose coralline algal host, *Lithothamnium californicum*, competes with radioactive GABA analogues for binding to these receptors is the strongest evidence suggesting that the receptors that recognize the natural algal inducer are the same receptors that also recognize GABA and its analogues in the induction of settlement and metamorphosis by these compounds (Trapido-Rosenthal & Morse, 1986a; Morse D. 1990; A.N.C. Morse & H.G. Trapido-Rosenthal, unpub. obs.). In addition:

(1) the affinity of these larval receptors for various GABA analogues predicts the effectiveness of the analogues as inducers of settlement and metamorphosis;
(2) the receptors appear on the larvae as a function of time of development and their appearance precedes (and apparently is required for) the development of larval competence to respond to both algal and GABA inducers; and
(3) precocious exposure of larvae to GABA 'down regulates' or habituates the larvae, blocking the subsequent developmental increase in the number of externally accessible receptors and rendering the larvae 'blind' to induction by GABA (although they are fully inducible by compounds that act at the level of signal-transduction downstream from the receptor; see below).

At the present, more is known about the nature of these receptors than about their location. This is because it has not yet proved possible to label these receptors with sufficient specificity to ensure that only the receptors controlling metamorphosis have been tagged, as would be required for their unequivocal direct visualization, whereas biochemical methods do permit resolution and analysis of the receptor-binding properties under these conditions (Trapido-Rosenthal & Morse, 1986a). A number of observations indirectly suggest that the receptors controlling metamorphosis may be external, although GABA also is known to be taken up by *Haliotis* larvae (Trapido-Rosenthal & Morse, 1986a; Jaeckle & Manahan, 1989). Thus, the following are all consistent with the suggestion that the receptors controlling metamorphosis are peripheral and possibly external:

(1) the finding that the most efficient inducer is the macromolecular GABA-mimetic (Morse A., 1985);
(2) the fact that baclofen (a GABA analogue) is a better inducer than GABA itself, yet is taken up (internalized) by *Haliotis* larvae less efficiently than GABA (Trapido-Rosenthal & Morse, 1986a); and
(3) the observation that the receptors specifically labelled with baclofen are shed from the larvae at the time during metamorphosis when cilia and other external larval structures are shed (Trapido-Rosenthal & Morse, 1986a).

Experimental evidence reviewed elsewhere (Morse D., 1990) has led to the suggestion, as a simplified working hypothesis, of a pathway of signal transduction events that may follow recognition of the exogenous morphogenetic stimulus by abalone larvae. This evidence suggests that binding of the exogenous inducing molecule by specialized chemosensory receptors results in the sequential activation of a receptor- and membrane-associated adenyl cyclase, synthesis of cyclic AMP, activation of a cyclic AMP- and calcium-regulated protein kinase (likely to be protein kinase A), phosphorylation of an endogenous protein, opening of chloride (or other anion) channels in the chemosensory cell membrane and an efflux of chloride (or other anion), believed to result in an excitatory depolarization, or firing, of the chemosensory cell. In this way, the morphogenetic chemical signal from the environment may be transduced to an electrochemical signal that can be propagated by the larval nervous system. Evidence leading to this simplified working hypothesis has been presented and discussed in detail previously (Morse *et al.*, 1980a; Baloun & Morse, 1984; Morse D., 1985, 1990).

Consistent with the above, a large number of chemicals known to affect intracellular cyclic AMP and calcium levels, protein phosphorylation, cell membranes, membrane ion channels, ion translocation and ionic depolarization of cell membranes have been found also to induce settlement and metamorphosis of abalone larvae (Table 10.1). These include cyclic AMP analogues, forskolin, substituted xanthines such as isobutylmethylxanthine and theophylline, calcium ionophores and chelators, an elevated external concentration of potassium ion (which conveniently depolarizes externally accessible, electrically excitable membranes), a low external chloride concentration, specific and non-specific ion-channel openers (such as ivermectin and organic solvents, respectively) and free arachidonic and palmitoleic acids (Morse *et al.*, 1980a; Baloun & Morse, 1984; Baxter & Morse, 1987; Morse D., 1990; Jensen *et al.* 1990).

The case of the free fatty acids, arachidonic and palmitoleic acids, is particularly interesting. These compounds, first erroneously suggested to be the natural and species-specific inducers of settlement and metamorphosis of the tube-building polychaete, *Phragmatopoma californica*, (Pawlik, 1986) are widely known in other systems as potent disrupters of membrane structure (Orly & Schramm, 1975; Chan *et al.*, 1983; Baba *et al.*, 1984; Kim & Clapham, 1989; Ordway *et al.*, 1989), with activities including the resultant activation of membrane-associated adenyl cyclase (causing an increase in cyclic AMP synthesis) and potassium

Table 10.1 Chemical inducers of settlement and metamorphosis for *Haliotis* spp. and the cellular levels at which they apparently act, see text for explanation.

Probable sensory receptor stimuli
Algal, microbial GABA-mimetics
GABA, GABA analogues
Mucus
Cyclic AMP effectors
db-cAMP
Forskolin
Isobutylmethylxanthine, theophylline
Fatty acids
Ca^{++} *Effectors*
A23187 ionophore
Chelators
Membrane depolarizers
High external K^+
Low external Cl^-
Ivermectin
Solvents

db-cAMP = dibutyryl-3′, 5′-cyclic AMP.
Forskolin = activator of enzyme that synthesizes cAMP.
Isobutylmethylxanthine and theophylline = inhibitors of enzyme that degrades cAMP.
Ivermectin = chloride ion channel opener.

channel-mediated transmembrane ion flux. Apparently as a consequence of one or more of these activities, arachidonic and palmitoleic acids induce settlement (plantigrade attachment) and metamorphosis of abalone larvae, at concentrations even lower than those required to induce these changes in *Phragmatopoma* (Jensen *et al.*, 1990). Recent evidence shows that these fatty acids are in fact not taxon-specific or substratum-specific natural inducers, but instead are widely active artifacts of the method of extraction (Jensen *et al.*, 1990). Like forskolin, isobutylmethylxanthine and excess potassium ion (Yool *et al.*, 1986; Jensen & Morse, 1990; Morse D., 1990; see Table 10.1 and below), their activity on both abalone and polychaete larvae apparently reflects action(s) at sites physiologically downstream from (or parallel to) the actions of the natural, substratum-specific inducing molecules (Jensen *et al.*, 1990).

GENERALITY OF RESULTS

As found by a number of researchers in several different countries, the results discussed above are applicable to several abalone species. At the last count, 13 different species of temperate and tropical *Haliotis* have been found in which the larvae are induced to settle (attach to substratum), metamorphose and commence normal post-metamorphic growth in response to crustose coralline red algae and in response to GABA (reviewed in Morse D., 1990). Akashige *et al.* (1981)

reported that *H. discus hannai* failed to show such results in response to GABA and that after induction, evidence of toxicity was seen. However, examination of the experimental conditions employed in the research laboratories of Akashige and colleagues and of the author revealed several significant differences which apparently account for the different results observed. The principal differences appear to include the degree and manner of bacterial prophylaxis. Indeed, in the results of Morse *et al.* (1979b, Table 10.1), when GABA was used with larval cultures containing bacteria without antibiotics, high mortality and failure of metamorphosis similar to those seen by Akashige *et al.* (1981) were observed; in these same experiments, the inclusion of antibiotics prevented the mortality and ensured a high success rate of GABA-induced metamorphosis (Table 10.1). Uncontrolled bacterial growth in the larval cultures can thus alter the results obtained, perhaps as a consequence of bacterial metabolism and its products (Morşe 1984, 1990).

That crustose coralline red algae and GABA are each capable of efficiently inducing normal settlement, attachment and metamorphosis of *H. discus hannai* larvae under conditions similar to those originally described by Morse *et al.* (1979a,b, 1980a,b) is shown in Figs 10.1 and 10.2. It was observed that larvae of *H. discus hannai* cultivated and analysed at 22°C in filtered seawater containing antibiotic became fully competent to attach and metamorphose in response to 1 mM GABA at approximately 70 hours post fertilization (Fig. 10.1). This was

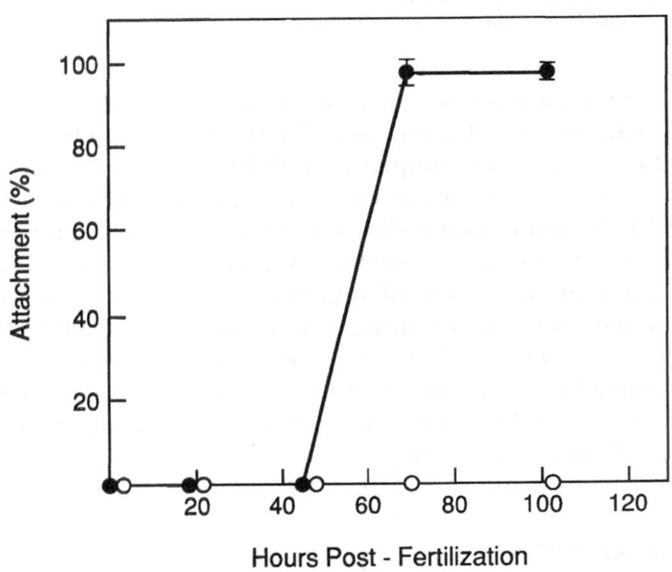

Fig. 10.1 Development of competence of *Haliotis discus hannai* to metamorphose in response to stimuli. Adult Ezo abalone were induced to spawn with hydrogen peroxide (Morse *et al.*, 1977); the gametes were fertilized and larvae reared at low density (≤1/ml) in seawater (1 μm-filtered, containing 100 ppm penicillin G) changed every 12 hours. Larvae were tested for competence as a function of time following fertilization, by exposure of duplicate samples of 100 ± 25 larvae incubated in 15 ml fresh seawater (1 μm-filtered, 100 ppm penicillin G; 2.4 cm depth), either with no additions (open circles) or with 10^{-3} M GABA (filled circles). Results (mean ± range) were scored 30 mins after the start of incubation. Attached refers to plantigrade attachment. All operations were at 22°C.

observed to be the same time required for the same population to become competent to respond to induction of attachment and metamorphosis by crustose coralline red algae (data not shown). As in the case of *H. rufescens* and other species tested, although the high concentration (1 mM) of GABA used in Fig. 10.1 is useful for rapidly (i.e. in 30 minutes) detecting the development of competence in *H. discus hannai*, this high concentration is toxic on prolonged incubation (cf Morse 1979a, 1980a,b); a lower concentration is needed to induce metamorphosis with minimal mortality. Both the concentration dependence and the concentration optimum for GABA-induced metamorphosis of competent *H. discus hannai* (Fig. 10.2) are similar to those observed for *H. rufescens* (Morse *et al.*, 1980b) and most other *Haliotis* species tested (cf Morse, 1984). At the optimal concentration (ca 10^{-6} mol/l) of GABA, 86% ± 2% (range) of the larvae were induced to undergo normal metamorphosis, with negligible mortality. Metamorphosis induced by parallel samples of crustose coralline algae was 79% ± 4% (range), while zero-addition and rock controls yielded 0% ± 0% metamorphosis. Preliminary results of these experiments with *H. discus hannai*, conducted in collaboration with Professor Zhang Fusui at the Institute of Oceanology in Qingdao, were reported elsewhere (Morse, 1984, 1990).

At least two commercially successful abalone cultivation companies in California (Ab Lab, Inc, Port Hueneme, CA and The Cultured Abalone, Inc, Santa Barbara) are now using GABA to induce larval settlement and metamorphosis, verifying

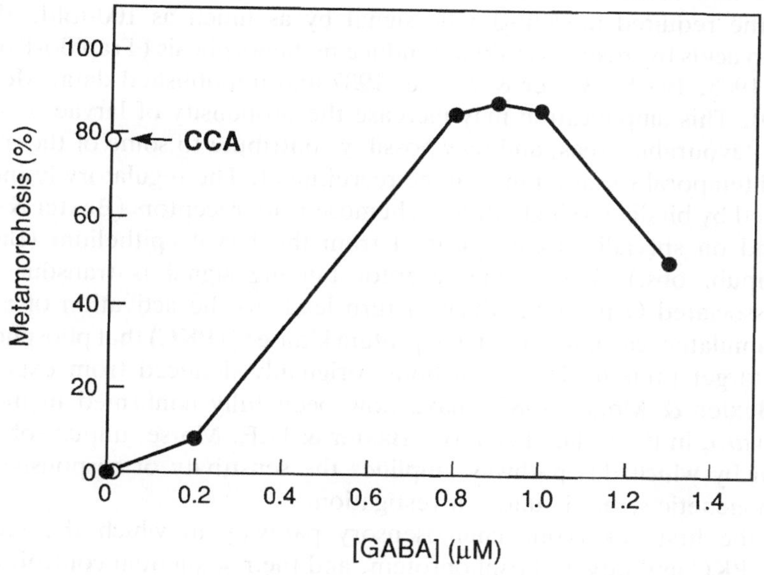

Fig. 10.2 Induction of metamorphosis of *Haliotis discus hannai*. Experiment as in Fig. 10.1, except that larvae were 80 hours post fertilization, the concentration of GABA varied from 0 to 1.3×10^{-6} M, and results were scored after 48 hours. Metamorphosis refers to abscission of velum and growth of flared, ribbed shell. Mortality in all treatments ≤ 4%. All operations were at 22°C. CCA = crustose coralline red algae on rock, collected with parallel samples of rock without CCA (yielding zero metamorphosis; lower open circle) from same intertidal. CCA and rock samples were incubated in 250 ml volumes containing same numbers of larvae and same concentration of penicillin as in smaller incubations.

the usefulness of the method for production purposes. The author's laboratory has been using the method, with high survival of the resulting progeny, for more than 10 years. Other production facilities are also finding other sources of inducers, particularly abalone mucus, especially useful for this purpose (Chapter 44; cf Seki & Kan-no, 1981; Uki, 1989). It is likely that each type of inducer will be found to have its own specific set of costs and benefits associated with its use (Morse 1984; Uki, 1989).

Elements involved in the post-receptor pathway of signal transduction suggested to control settlement and metamorphosis in the red abalone, *Haliotis rufescens*, (Table 10.1) have also been implicated in similar processes in the larvae of several other species and phyla (Baloun & Morse, 1984; Yool *et al.*, 1986; Jensen & Morse, 1990; Morse D., 1990). While specific differences exist between the pathways implicated in *H. rufescens* and those in the non-haliotid species, the ability to induce metamorphosis efficiently by direct depolarization with excess potassium is proving widely useful for studies of morphogenesis in a number of marine invertebrates (Yool *et al.*, 1986; Morse D., 1990).

FACILITATION AND THE REGULATORY PATHWAY

A second pathway of chemical signal recognition and transduction that regulates the metamorphic response of abalone larvae has recently been found. Lysine and related diamino acids dissolved in seawater can amplify the sensitivity of the larvae to the required morphogenetic signal by as much as 100-fold, although these amino acids by themselves do not induce metamorphosis (Trapido-Rosenthal & Morse, 1985, 1986b; Baxter & Morse, 1987 and unpublished data; Morse D., 1988, 1990). This amplification may increase the propensity of larvae to settle in potentially favourable areas, and may possibly contribute to some of the observed spatial and temporal variation in abalone recruitment. The regulatory lysine signal is recognized by binding to high-affinity chemosensory receptors (Baxter & Morse, 1987) found on specialized cilia purified from the larval epithelium (Baxter & Morse, unpub. obs.). This lysine receptor binding signal is transduced by a receptor-associated G protein, which in turn leads to the activation of a diacylglycerol-stimulated, calcium-stimulated protein kinase C (PKC) that phosphorylates a specific target protein. These reactions, originally deduced from experiments *in vivo* (Baxter & Morse, 1987), have now been fully confirmed in molecular studies *in vitro*, in the isolated cilia (G. Baxter & D.E. Morse, unpub. obs.). The mechanism by which this pathway amplifies the sensitivity or responsiveness to the morphogenetic signal is under investigation.

This is the first eukaryotic chemosensory pathway in which the receptors, G protein, PKC and target phosphoprotein, and their sequential control, have all been functionally demonstrated and resolved *in vitro*. The regulatory receptor, G protein and phosphorylated target protein have been labelled and the latter two have been purified. Recently, Lisa Wodicka has purified, amplified and characterized the G protein mRNA and cDNA sequences from the isolated cilia (Wodicka & Morse, unpublished data). From these molecular studies, we are learning that the receptors and signal transducers controlling metamorphosis in abalone larvae are closely related to those controlling sensory, developmental and neuronal processes in other organisms.

INTERACTION BETWEEN THE MORPHOGENETIC AND REGULATORY PATHWAYS

From the studies discussed above, it is becoming clear that larval settlement, attachment and metamorphosis are regulated by the interaction between chemosensory pathways (and other biological and non-biological factors) with a degree of complexity far greater than previously anticipated. Table 10.2 summarizes the principal features of the two pathways controlling metamorphosis in response to external chemical stimuli. The interaction of these pathways appears to provide the larvae with a capacity for fine tuning the sensitivity of recognition of, and response to, chemical signals from the environment that is greater than was anticipated for marine invertebrate larvae just a few years ago.

Analyses of the binding properties of the chemosensory receptors controlling the two pathways are providing some clues about the way this complex process may work (Table 10.3). The morphogenetic responses (attachment and metamorphosis of the larvae) are normally dependent on the recognition of the algal GABA-mimetic molecules that are associated with the recruiting crustose red algal surface or other functionally similar signals (e.g. in mucus). This recognition is apparently mediated by highly specific GABA-type receptors (R_G) that have a very low affinity or weak binding capacity (indicated by the relatively high K_D value of 10^{-6} M) (Trapido-Rosenthal & Morse, 1986a). There is a large number

Table 10.2 Principal features of the two chemosensory pathways controlling attachment and metamorphosis of *Haliotis rufescens* larvae in response to chemical signals from the environment.

	Morphogenetic	Regulatory
Signal	Algal GABA-mimetic	DOM diamino acid
Receptor	GABA-type	Lysine-type
Transducers	Adenyl cyclase Cyclic AMP Ca^{++} Protein kinase Chloride channel	G Protein Phospholipase Diacylglycerol Ca^{++} Protein kinase
Output	Excitatory depolarization	Amplification of Morphogenetic Signal

Table 10.3 Interaction between chemosensory pathways controls attachment and metamorphosis of *Haliotis rufescens* larvae; values for affinity and number/larva of the GABA-type (R_G) and lysine-type (R_L) externally-accessible chemosensory receptors are approximate and taken from the data of Trapido-Rosenthal & Morse (1986) and Baxter & Morse (1987 and unpub. obs.), respectively, see text for explanation.

	R_G	R_L
Sensing medium	Surface	Water
Ligand	GABA-mimetic	Lysine
Affinity	Low (K_D 10^{-6} M)	Higher (K_D 10^{-7} M)
Number/larva	High (10^{10})	Low (10^8)
Controls	Attachment	Sensitivity

of these receptors on the larvae, suggesting a kind of fail-safe function, in which a high concentration of the required inducing signal may normally be required to induce attachment.

In contrast, the regulatory (or 'amplifier') pathway is apparently adapted to recognize relatively low concentrations of amino acid signals (specifically lysine and its analogues) dissolved in seawater. These receptors (R_L) have a higher affinity (lower K_D, as measured by Scatchard analyses), tuned to the concentration range over which lysine varies in dissolved organic material (Trapido-Rosenthal & Morse, 1985, 1986b), and there is a lower number of these receptors on the larvae (G. Baxter & D.E. Morse, 1987, and unpub. obs.; see Table 10.3). The output of this lysine-sensing pathway increases the threshold sensitivity of the abalone larvae to low concentrations of the required surface-associated inducer, thus potentially at least enhancing attachment in response to the detection of specific dissolved nutrients. Whether these two types of receptors and the two pathways they control operate within a single cell or in two chemosensory cells whose outputs are distally convergent is the object of our present investigations. It also remains to be demonstrated whether the operation of these two pathways may be of ecological or adaptive significance in the natural environment.

GENE ACTIVATION AND THE CONTROL OF POST-METAMORPHIC DEVELOPMENT AND GROWTH

Metamorphosis in abalone is the culmination of events following activation of differential gene expression and resulting cellular differentiation in the developmentally arrested larvae (Morse D., 1990). In *Haliotis rufescens*, metamorphosis is relatively slow, involving new shell synthesis (which requires mantle cell proliferation and conchiolin protein synthesis (Cariolou & Morse, 1988)) and conversion of the non-feeding larva to a feeding juvenile expressing newly-synthesized digestive enzymes (Spaulding, 1989; D. Spaulding & D.E. Morse, unpub. obs.). In this respect, abalone metamorphosis may differ somewhat from the more rapid transition observed in such species as the nudibranch *Phestilla sibogae*, which can quickly jettison its larval shell and transform other larval structures apparently without extensive new gene expression (Hadfield, 1978). To determine how the recognition of chemical signals from the environment and the chemosensory signal transduction processes described above can regulate the activation of differential gene expression in abalone larvae we have begun to characterize genes coding for cell-specific protein products that must be synthesized in unique cell lines early in the metamorphic process. Our objective is to trace the mechanisms of activation from the environmental signal receptor and transduction cascade to the promoter and enhancer sequences in the DNA controlling the stage-specific and cell-specific transcription of these genes.

Thus far, we have characterized three specific protein families that appear to be developmentally regulated in *Haliotis rufescens*. These include the calcium-binding conchiolin shell peptides (Cariolou & Morse, 1988), the aryl sulphatases (Spaulding, 1989; Spaulding & Morse, 1991), and a unique family of serine proteases (Groppe & Morse, 1989 and unpub. data). The latter two are digestive enzymes produced in specific tissues of the digestive tract, whereas conchiolin peptides are synthesized specifically in the proliferating cells of the mantle epithelium following metamorphosis. In all three cases, we have found evidence for a multi-gene family; in the case of conchiolin and aryl sulphatase, strong evidence for gene switching at or following metamorphosis has been obtained.

Recently, Jay Groppe has developed techniques which have permitted the molecular cloning and sequence determination of the unique serine protease cDNAs and messengers (Groppe & Morse, 1989). Molecular hybridization studies detecting the protease messenger RNAs demonstrate that the genes for these proteases are very highly expressed, in a very tissue-specific pattern. In addition to providing unexpected information on the apparent importance of protein digestion in nutrition of the growing abalone, these genes and their enzyme products, as well as the genes for the other specialized proteins mentioned above, will serve as the focus for our studies of the mechanisms that control the developmental activation of differential gene expression in a cell-specific manner.

Several years ago, we found that mammalian insulin and growth hormone at low concentrations (10^{-7} M and 10^{-9} M, respectively) specifically and significantly accelerated the rate of early growth in post-metamorphic abalone (Morse, 1984). These observations, and the exciting recent observations of Kawauchi and his colleagues (unpublished data) in Japan, suggest that cell differentiation and early growth in *Haliotis* spp. may be regulated by endogenous neuroendocrine mechanisms homologous at the molecular level to those emerging from contemporary studies in mammals. The potential applications of these findings for accelerating the growth of abalone in cultivation have been discussed (Morse, 1984).

ACKNOWLEDGEMENTS

Research reported here was conducted with support from the National Science Foundation (grant #DCB87–18224), National Institute of Health Grant #1–R01–RR06640, the US Navy Office of Naval Research (contracts #N00014–87–K–0762 and N00014–88–K–0288) and the US Department of Commerce (NOAA)-California Sea Grant College Program.

REFERENCES

Akashige S., Seki T., Kan-no H. & Nomura T. (1981) Effects of γ-aminobutyric acid and certain neurotransmitters on the settlement of the larvae of *Haliotis discus hannai* Ino (Gastropoda). *Bull. Tohoku Reg. Fish. Res. Lab.*, **43**, 37–45.

Baba A., Tatsuno T. & Iwata H. (1984) Modulation by unsaturated fatty acids of norepinephrine- and adenosine-induced formation of cyclic AMP in brain slices. *J. Neurochem.*, **42**, 192–7.

Baloun A.J. & Morse D.E. (1984) Ionic control of settlement and metamorphosis in larval *Haliotis rufescens* (Gastropoda). *Biol. Bull.*, **167**, 124–38.

Baxter G. & Morse D.E. (1987) G protein and diacylglycerol regulate metamorphosis of planktonic molluscan larvae. *Proc. Natl Acad. Sci. USA*, **84**, 1867–70.

Cariolou M.A. & Morse D.E. (1988) Purification and characterization of calcium-binding conchiolin shell peptides from the mollusc, *Haliotis rufescens*, as a function of development. *J. Comp. Physiol. B.*, **157**, 717–29.

Chan P.H., Fishman R.A., Chen S. & Chew S. (1983) Effects of temperature on arachidonic acid-induced cellular edema and membrane perturbation in rat brain cortical slices. *J. Neurochem.*, **41**, 1550–7.

Crofts D.R. (1929) *Haliotis. Liverpool Mar. Biol. Comm. Mem. Typical Br. Mar. Plants Anim.*, **29**, 1–174.

Douros W.J. (1985) *Density, growth, reproduction and recruitment in an intertidal abalone: effects of intraspecific competition and prehistoric predation.* MA dissertation, University of California, Santa Barbara.

Groppe J.C. & Morse D.E. (1989) Molecular cloning of novel serine protease cDNAs from abalone. In *Current Topics in Marine Biotechnology* (Ed. by S. Miyachi, I. Karube & Y. Ishida), pp. 285–8. Japan Society of Marine Biotechnology, Tokyo.

Hadfield M.G. (1978) Metamorphosis in marine molluscan larvae: an analysis of stimulus and response. In *Settlement and Metamorphosis of Marine Invertebrate Larvae* (Ed. by F.-S. Chia & M.E. Rice), pp. 165–75. Elsevier, New York.

Jaeckle W.R. & Manahan D.T. (1989) Feeding by a 'nonfeeding' larva: Uptake of dissolved amino acids from seawater by lecithrotrophic larvae of the gastropod *Haliotis rufescens. Mar. Biol.*, **103**, 87–94.

Jensen R.A. & Morse D.E. (1990) Chemically induced metamorphosis of polychaete larvae in both the laboratory and the ocean environment. *J. Chem. Ecol.*, **16**, 911–30.

Jensen R.A., Morse D.E., Petty R.L. & Hooker N. (1990) Artificial induction of larval metamorphosis by free fatty acids. *Mar. Ecol. Prog. Ser.*, **67**, 55–71.

Kim D. & Clapham D.E. (1989) Potassium channels in cardiac cells activated by arachidonic acid and phospholipids. *Science*, **244**, 1174–6.

Leighton D.L. (1985) Early growth of green abalone in hatchery and field. In *Diving for Science ... 85, Proceedings of Joint International Scientific Diving Symposium* (Ed. by C.T. Mitchell), pp. 235–46. American Academy of Underwater Sciences, La Jolla.

Morse A.N.C. (1985) Characterization of inducers of molluscan larval metamorphosis. Doctoral dissertation, University of California, Santa Barbara.

Morse, A.N.C. (1988) The role of algal metabolites in the recruitment process. In *Marine Biodeterioration* (Ed. by M.-F. Thompson, R. Sarojini & R. Nagabhushanam), pp. 463–773. Oxford Press, New Delhi.

Morse A.N.C., Froyd C. & Morse D.E. (1984) Molecules from cyanobacteria and red algae that induce larval settlement and metamorphosis in the mollusc *Haliotis rufescens. Mar. Biol.*, **81**, 293–8.

Morse A.N.C. & Morse D.E. (1984a) Recruitment and metamorphosis of *Haliotis* larvae are induced by molecules uniquely available at the surfaces of crustose red algae. *J. Exp. Mar. Biol. Ecol.*, **75**, 191–215.

Morse A.N.C. & Morse D.E. (1984b) GABA-mimetic molecules from *Porphyra* (Rhodophyta) induce metamorphosis of *Haliotis* (Gastropoda) larvae. *Hydrobiology*, **116**, 155–8.

Morse D.E. (1984) Biochemical and genetic engineering for improved production of abalones and other valuable molluscs. *Aquaculture*, **39**, 263–82.

Morse D.E. (1985) Neurotransmitter-mimetic inducers of settlement and metamorphosis of marine planktonic larvae. *Bull. Mar. Sci.*, **37**, 697–706.

Morse D.E. (1988) Trigger and amplifier pathways: Sensory receptors, transducers and molecular mechanisms controlling larval settlement, adhesion and metamorphosis in response to environmental chemical signals. In *Marine Biodeterioration* (Ed. by M.-F. Thompson, R. Sarojini & R. Nagabhushanam), pp. 453–72. Oxford Press, New Delhi.

Morse D.E. (1990) Recent progress in larval settlement and metamorphosis: Closing the gaps between molecular biology and ecology. *Bull. Mar. Sci.*, **46**, 465–83.

Morse D.E., Duncan H., Hooker N. & Morse A. (1977) Hydrogen peroxide induces spawning in molluscs, with activation of prostaglandin endoperoxide synthetase. *Science*, **196**, 298–300.

Morse D. E., Hooker N., Duncan H. & Jensen L. (1979a) γ-aminobutyric acid, a neurotransmitter, induces planktonic abalone larvae to settle and begin metamorphosis. *Science*, **204**, 407–10.

Morse D.E., Hooker N., Jensen L. & Duncan H. (1979b) Induction of larval abalone settlement and metamorphosis by γ-aminobutyric acid and its congeners from crustose red algae, II: Applications to cultivation, seed-production and bioassays; principal causes of mortality and interference. *Proc. World Maricul. Soc.*, **10**, 81–91.

Morse D.E., Duncan H., Hooker N., Baloun A. & Young G. (1980a) GABA induces behavioral and developmental metamorphosis in planktonic molluscan larvae. *Fed. Proc.*, **39**, 3237–41.

Morse D.E., Hooker N. & Duncan H. (1980b) GABA induces metamorphosis in *Haliotis*, V: Stereochemical specificity. *Brain Res. Bull.*, **5** (Suppl. 2), 381–7.

Morse D.E., Tegner M., Duncan H., Hooker N., Trevelyan G. & Cameron A. (1980c) Induction of settling and metamorphosis of planktonic molluscan (*Haliotis*) larvae. III: Signaling by metabolites of intact algae is dependent on contact. In *Chemical Signaling in Vertebrate and Aquatic Animals* (Ed. by D. Muller-Schwarze & R.M. Silverstein), pp. 67–86. Plenum Press, New York.

Ordway R.W., Walsh J.V., & Singer J.J. (1989) Arachidonic acid and other fatty acids directly activate potassium channels in smooth muscle cells. *Science*, **244**, 1176–9.

Orly J. & Schramm M. (1975) Fatty acids as modulators of membrane functions: catecholamine-activated adenylate cyclase of the turkey erythrocyte. *Proc. Natl. Acad. Sci. USA*, **72**, 3433–7.

Pawlik J.R. (1986) Chemical induction of larval settlement and metamorphosis in the reef-building tube worm *Phragmatopoma californica* (Sabellardiidae: Polychaeta). *Mar. Biol.*, **91**, 59–68.

Prince J.D., Sellers T.L., Ford W.B. & Talbot S.R. (1987) Experimental evidence for limited dispersal of haliotid larvae (genus *Haliotis*; Mollusca: Gastropoda). *J. Exp. Mar. Biol. Ecol.*, **106**, 243–63.

Saito K. (1981) The appearance and growth of 0-year-old Ezo abalone. *Bull. Japan. Soc. Sci. Fish.*, **47**, 1393–400.

Seki T. & Kan-no H. (1981) Induced settlement of the Japanese abalone, *Haliotis discus hannai*, veliger by the mucus of the juvenile and adult abalones. *Bull. Tohoku Reg. Fish. Res. Lab.*, **43**, 29–36.

Shepherd S.A. & Turner J.A. (1985) Studies on Southern Australian abalone (genus *Haliotis*). VI. Habitat preference, abundance and predators of juveniles. *J. Exp. Mar. Biol. Ecol.*, **93**, 285–98.

Slattery M. (1987) *Settlement and metamorphosis of red abalone* (Haliotis rufescens) *larvae: A critical examination of mucus, diatoms and γ-aminobutyric acid (GABA) as inductive substrates*. MA dissertation, San Jose State University.

Spaulding D. (1989) *Purification, characterization, and developmental patterns of arylsulfatases from the abalone*, Haliotis rufescens. Doctoral dissertation, University of California, Santa Barbara.

Spaulding D. & Morse D.E. (1991) Purification and characterization of sulfatases from *Haliotis rufescens*: evidence for changes in synthesis and heterogeneity during development. *J. Comp. Physiol. B.*, **161**, 498–515.

Trapido-Rosenthal H.G. & Morse D.E. (1985) L-α, δ-Diamino acids facilitate induction of settlement and metamorphosis in larvae of a gastropod mollusc (*Haliotis rufescens*). *J. Comp. Physiol. B.*, **155**, 403–14.

Trapido-Rosenthal H.G. & Morse D.E. (1986a) Availability of chemosensory receptors is down-regulated by habituation of larvae to a morphogenetic signal. *Proc. Natl Acad. Sci. USA*, **83**, 7658–62.

Trapido-Rosenthal H.G. & Morse D.E. (1986b) Regulation of receptor-mediated settlement and metamorphosis in larvae of a gastropod mollusc (*Haliotis rufescens*). *Bull. Mar. Sci.*, **39**, 383–92.

Uki N. (1989) Abalone seedling production and its theory. *Int. J. Aqu. Fish. Technol.*, **1**, 3–15.

Yool A.J., Grau S.M., Hadfield M.G., Jensen R.A., Markell D.A. & Morse D.E. (1986) Excess potassium induces larval metamorphosis in four marine invertebrate species. *Biol. Bull.*, **170**, 255–66.

Chapter 11
Early life history of abalone: a review

PAUL E. McSHANE *Fisheries Research Division, Ministry of Agriculture and Fisheries, PO Box 297, Wellington, New Zealand*

ABSTRACT: Salient studies of dispersal, settlement and recruitment of abalone larvae are reviewed. Little is known of the life history of abalone larvae mainly because larvae and post-settlement juveniles have eluded detection by scientists. Computer simulations accent the difficulty of finding a small mobile aggregation of larvae in the open sea, particularly because abalone larvae are episodically and unpredictably released. Recent observations on recruitment of *Haliotis rubra* are consistent with swarms of larvae remaining near the reef surface where gametes were first released. Dispersal of abalone larvae is primarily influenced by regional hydrodynamics. Localized dispersal of passively transported abalone larvae is possible only if water movement is attenuated or if larvae are entrained in current eddies. The short duration of larval life and the abalone larva's selectivity for substrata are factors which act against successful long-distance dispersal of larvae. Studies of recruitment near the time of settlement reveal high settlement in sheltered rather than exposed habitat. These studies also show that even neighbouring populations of abalone can be isolated and show little evidence of recruits intermixing.

RESUMEN: Se revisan los estudios mas relevantes sobre dispersión larval, asentamiento y reclutamiento de larvas de abulón. Muy poco se conoce de las etapas de desarrollo temprano del abulón, principalmente porque las larvas y los estadios juveniles despues de la fijación, han eludido la detección de los científicos.

Las simulaciones con comportadoras destacan la dificultad de encontrar una pequeña agregación móvil de larvas en mar abierto, particularmente porque las larvas del abulón son liberadas episódica e impredeciblemente. Observaciones recientes sobre el reclutamiento de *Haliotis rubra* son consistentes con 'manchas' de larvas que permanecen cercanas a la superficie de arrecifes donde de liberaron gametos previamente.

La dispersión de las larvas está influida principalmente por las condiciones hidrodinámicas regionales. La localización de la dispersión de larvas de abulón transportadas pasivamente, sólo es posible si el movimiento de las aguas es moderado o si las larvas son acarrreadas dentro de corrientes de remolino.

La corta duración del período larval del abulón y su selectividad por el sustrato son factores que actúan en contra del éxito de una dispersión larval en distancias largas. Los estudios de reclutamiento cercanos al tiempo de asentamiento, revelan un alto asentamiento en un habitat protegido mas bien que en un habitat desprotegido. Estos estudios demuestran tambien que aun las poblaciones vecinas de abulón pueden estar aisladas y muestran poca evidencia de entremezcla de los reclutas.

INTRODUCTION

Abalone are commercially valuable prosobranch gastropods and the continued exploitation of some species has threatened the viability of many populations (Mottet, 1978). The great commercial value of abalone has prompted much research on various aspects of their population biology (Mottet, 1978). In this review I have concentrated on studies concerning the larval ecology and recruitment of abalone. In particular, I have focused on recent evidence for localised dispersal of larval *Haliotis rubra*. My interpretation of the available evidence is that localized dispersal of the larvae is the primary determinant of variations in recruitment of the species and abalone in general. Recruitment is defined here as the stage at which an individual that has settled can first be observed (Keough & Downes, 1982; Connell 1985).

It is fashionable for fisheries biologists to relate recruitment to stock (population) abundance (see Sluczanowski, 1984; Prince *et al.*, 1988 for abalone). While such a relationship is keenly sought by managers of exploitable populations, there is little evidence for any such relationship especially for invertebrates (Hancock, 1973; Yoshioka, 1986).

The hypothesis that recruitment processes have importance in determining the distribution and abundance of adults for species of marine benthic invertebrates has recently gained theoretical and empirical support (Denley & Underwood, 1979; Underwood & Denley, 1984; Connell, 1985; Gaines & Roughgarden, 1987; Caffey, 1985; Lewin, 1986). Factors governing recruitment, such as the availability and survival of larvae, will determine a pattern which may be modified by competition (Connell, 1961, 1972), predation (Paine, 1969, 1974, 1980) or environmental disturbance (e.g. Dayton 1971, 1975). Yet ecological factors such as availability, dispersal and survival of larvae determine a pattern of settlement, whereas recruitment processes merely modify these patterns (Keough & Downes, 1982; Connell, 1985; McGuinness & Davis, 1989).

Most studies of recruitment processes have until recently been focused on intertidal taxa; the literature is dominated by examples from a few species of barnacles (Connell, 1985). Recently, recruitment patterns in subtidal species have been investigated: sea urchins (Rowley, 1989), holothurians (Young & Chia, 1982), ascidians (Davis, 1987, 1988; Young, 1989), corals (Gerrodette, 1981), polychaetes (Hannan, 1984; Butman, 1987, 1989), bryozoans (Yoshioka, 1982, 1986; Keough, 1983, 1986) and gastropods (Fletcher, 1987; McShane *et al.*, 1988).

These more recent studies provide a better insight into recruitment within a more general phylogenetic framework, but for many species of marine invertebrate events in the life history prior to recruitment remain undescribed. For example, the dispersal capacity of those species with pelagic larvae is often considered to be related to the duration of planktonic life (Harrigan, 1972; Heck & McCoy 1978; Olson, 1985; Keough, 1989). Yet dispersal of pelagic larvae from complex benthic habitat, in particular coral or temperate reef, could well depend primarily on regional hydrodynamics (McShane *et al.*, 1988; Butman, 1987, 1989).

Considering the complexity of the abalone's natural habitat it is not surprising that little is known of the ecology of abalone larvae. Studies of abalone in their natural habitat are more often anecdotal accounts rather than detailed surveys. There are no accounts of settlement of abalone larvae and the extent of their dispersal is equivocal (Tegner & Butler, 1985; Prince *et al.*, 1987; McShane *et al.*, 1988). Yet such information is important for elucidating the stock structure in abalone populations.

Before reviewing information on larval ecology it is worthwhile summarizing features of the reproductive biology of abalone relevant to recruitment processes. Comprehensive reviews of the reproductive biology of abalone have been presented by others (Shepherd & Laws, 1974; Mottet 1978; Tutschulte & Connell, 1981; Hahn, 1989a).

REPRODUCTIVE BIOLOGY

Abalone are dioecious broadcast spawners and usually have a seasonal reproductive cycle consisting of an annual spawning preceded by gametogenesis. The periodicity and duration of spawning can vary both intra- and interspecifically (Shepherd & Laws, 1974; McShane *et al.*, 1986, 1988; Tutschulte & Connell 1981). *Haliotis rubra* releases gametes over several weeks (McShane & Smith, 1991). The eggs of abalone are negatively buoyant and generally hatch within 24 hours of fertilization (Mottet, 1978).

The fecundity of the various species of abalone differs, but a female can produce as many as 10 million eggs (Newman, 1967; Poore, 1972; Tutschulte & Connell, 1981; McShane *et al.*, 1986, 1988). Thus abalone are fecund despite the larvae being lecithotrophic (feeding on stored nutrients, Strathmann, 1985). However, not all eggs are necessarily released at spawning (Chapter 33) and an individual abalone might well release eggs over an extended period.

Behavioural changes in adult abalone before and during spawning, for example aggregation (Breen & Adkins, 1980; Shepherd, 1986) and differential movement (Quayle, 1971; Breen & Adkins, 1980) have been invoked to explain recruitment patterns. Aggregation is obviously advantageous to broadcast spawners in that synchrony of spawning is promoted and fertilization is enhanced. It has been suggested that the removal of aggregations of abalone by fishermen could affect reproductive success and therefore recruitment to exploitable populations (Sluczanowski, 1984; Shepherd, 1986; Prince *et al.*, 1987; McShane & Smith, 1989a).

LARVAL ECOLOGY

Abalone larvae are pelagic and lecithotrophic (Ino, 1952; Mottet, 1978). Lecithotrophy is not common among marine benthic invertebrates (Mileikovsky, 1971). Lecithotrophic larvae may derive some nutrients from seawater (Jaeckle & Manahan, 1989; see also Olson & Olson, 1989), but abalone larvae are non-feeding and have a relatively brief pelagic phase compared with feeding (planktotrophic) larvae.

In aquaria the behaviour of adult abalone of different species differs but the behaviour of abalone larvae of all species is similar (Ino, 1952; Yano & Ogawa, 1977; Seki & Kan-no, 1981a,b; Ebert & Houk, 1984) and typical of that of marine invertebrates which produce pelagic larvae (Thorson, 1950). The larval life cycle of abalone is generalized and illustrated in Fig. 11.1. Initially an upward swimming larva (trochophore) is produced, probably as an adaptation to avoid predation by benthic filter feeders (Thorson, 1964; Mileikovsky, 1971; Crisp, 1974). Veliger larvae swim downwards (Thorson, 1964). The direction in which invertebrate larvae move is dictated by photo or pressure receptors (Yano & Ogawa, 1977; Forward *et al.*, 1989). The rate of development of the lecithotrophic larva and, therefore, the duration of larval life is controlled primarily by temperature and varies from 4 to 15 days (Ino, 1952; Leighton, 1972, 1974).

Results from these studies and those of Tanaka (1978), Lee *et al.* (1978) and Seki & Kan-no (1981a,b) are only clues to the abalone larva's behaviour in nature because so few larvae have been found in the sea (Tomita *et al.*, 1977; Tanaka

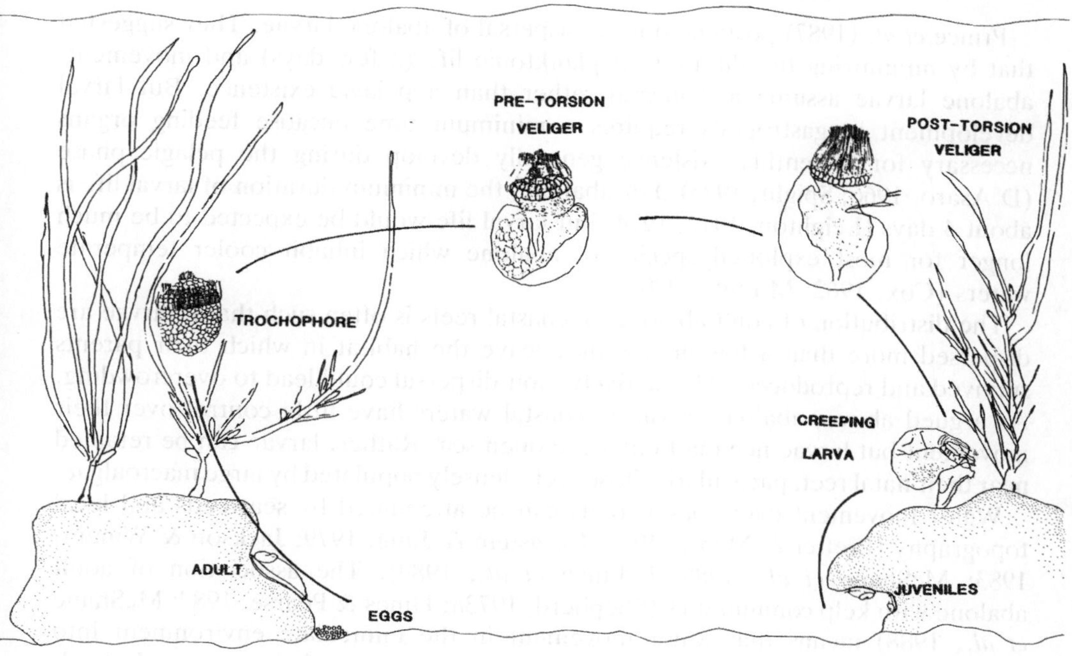

Fig. 11.1 The larval life cycle of abalone. Larvae are entrained near their natal reef by complex reef topography and stands of kelp, as illustrated.

et al., 1986; McShane *et al.*, 1988). For instance, abalone larvae swim, as do most benthic invertebrates with a pelagic larval phase, by cilial beating – a mode of swimming that confers little directional movement. Furthermore, abalone larvae are roughly spherical and, unlike the long thin cyprid larvae of barnacles (Crisp, 1955; Gaines *et al.*, 1985; Bushek, 1988), have difficulty orientating in the water. Consequently, swimming behaviour might be expected to confer little directional movement on pelagic larvae in the sea; not surprisingly their directional movement is impaired by the slightest water movement (Mileikovsky, 1973; Chia *et al.*, 1984). Because the coastal environment inhabited by abalone is often subject to violent water movement and waves near shore act to mix the water column (Pond & Pickard, 1983; McShane *et al.*, 1988), abalone larvae may be assumed to behave as if lacking directional movement and their dispersing as passively transported particles (cf Hannan, 1984; Butman, 1989). In water masses where vertical mixing is minimal (e.g. oceanic or estuarine waters), vertical movement by larvae, for example the crab *Neopanope sayi*, may confer directional movement by the larvae utilizing or avoiding surface currents (Forward *et al.*, 1989).

Invertebrates with long-lived pelagic larvae might well be expected to disperse widely (Scheltema, 1986). Abalone larvae whose duration of planktonic life is estimated to vary from 3 to 10 days might not cross oceans but they may be expected to disperse over considerable distances (Forster *et al.*, 1982; Tegner & Butler, 1985). However, the hydrodynamics in nearshore environments typical of those of many reef-dwelling species are often complex and simple analytical solutions to dispersal can be misleading (Strathmann *et al.*, 1981; Eckman, 1983; Shanks & Wright, 1987; McShane *et al.*, 1988).

Prince *et al.* (1987) postulated local dispersal of abalone larvae. They suggested that by minimizing the duration of planktonic life (a few days) and movement, abalone larvae assume a demersal rather than a pelagic existence. But larval development in gastropods requires a minimum time because feeding organs necessary for a benthic existence generally develop during the pelagic phase (D'Asaro, 1966; Spight, 1975). For abalone, the minimum duration of larval life is about 4 days (Leighton, 1972, 1974) but larval life would be expected to be much longer for most exploited species of abalone which inhabit cooler temperate waters (Cox, 1962; Mottet, 1978).

The distribution of adult abalone on coastal reefs is often such that if larvae are dispersed more than a few metres they leave the habitat in which their parents survived and reproduced. Alternatively, non-dispersal could lead to overcrowding. As argued above, abalone larvae in coastal waters have little control over their movement but larvae need not enter the open sea. Rather, larvae can be retained near their natal reef, particularly those reefs densely populated by large macroalgae.

Water movement over coastal reefs can be attenuated by seaweeds and local topography (Fretter & Manly, 1977; Bernstein & Jung, 1979; Jackson & Winant, 1983; McShane *et al.*, 1988; Eckman *et al.*, 1989). The association of adult abalone with kelp communities (Shepherd, 1973a; Hines & Pearse, 1982; McShane *et al.*, 1988) means that water movement in the immediate environment into which larvae are first released may be much less than that of open water. Indeed, studies of the hydrodynamics at the surface of one abalone reef have shown that local reef topography and kelp attenuate near-bed currents sufficiently for larvae to stay on their natal reef for the normal duration of larval life (McShane *et al.*, 1988; see also Eckman *et al.*, 1989).

Of the surveys conducted in coastal waters, few have found abalone larvae (Quayle, 1971; Breen & Adkins, 1980; McShane *et al.*, 1988). Prince *et al.* (1987) interpret this lack of success as evidence of a demersal larval behaviour. However, studies of coastal hydrodynamics have accented the difficulty of finding a small, mobile larval patch by sampling at random (McShane *et al.*, 1988). This difficulty, and the fact that larvae are generally released episodically and unpredictably, mean that the probability of finding abalone larvae in the open sea is low. Indeed, surveys for abalone larvae soon after the start of the spawning season by Breen & Adkins (1980) and McShane *et al.* (1988) yielded no larvae; a similar survey by Tomita *et al.* (1977) yielded only a few veliger larvae. A more detailed survey by Tanaka *et al.* (1986) yielded more than 1000 larvae m^{-3} (both trochophore and veliger), but on most occasions none were found. However, the larvae Tanaka *et al.* did find were uniformly distributed in surface waters. They concluded that eddies in the coastal waters where the survey was conducted concentrated larvae, an observation consistent with passive transport of larvae.

Larvae concentrated in discrete patches by eddies are more readily detected than widely dispersed or diffused larvae. The concentration of larvae is especially important for the detection of short-lived larvae, such as those of abalone, whose occurrence in the sea is more ephemeral than longer-lived, planktotrophic larvae (Lobel & Robinson, 1986; Yoshioka, 1986). Eddies are a common feature of continental shelf flows (Black & Gay, 1987). Black (1989) found that circulation patterns around coral reefs on the Great Barrier Reef caused free-drifting larvae to remain longer in some regions than others. Eddies can therefore act to limit dispersal of larvae and concentrate larvae in discrete areas, often near habitat favourable for settlement. For example, the formation of eddies off coastal headlands favours the retention of abalone larvae near the sublittoral reefs associated with the headlands (Fig. 11.2; see also McShane *et al.*, 1988).

Information on the survival of larvae of marine invertebrates is scant, but the available evidence suggests that the survival rate of planktonic larvae is low (Mileikovsky, 1971; Strathmann, 1985; Keough, 1989). Estimates of mortality for lecithotrophic larvae are lacking, but presumably their overall larval mortality is lower because the duration of planktonic life is shorter than that of planktotrophic larvae (Strathmann, 1985). Larvae die because of predation (Hines, 1979, 1986; Pennington *et al.*, 1986; Young, 1989), starvation (reviewed by Olson & Olson 1989) and transport away from suitable habitat (Thorson, 1950). For reef-dwelling species such as abalone, transport of larvae away from reef habitat would cause high mortality, particularly because abalone larvae have only a few days during which they are competent to settle (Ino, 1952; Leighton, 1972; 1974). Larvae lost from natal sites by advection and diffusion decrease the concentration of colonizers at a site (Strathmann, 1985), Thus dispersal of abalone larvae can decrease survival rates. Survival of abalone larvae would be enhanced if they could restrict

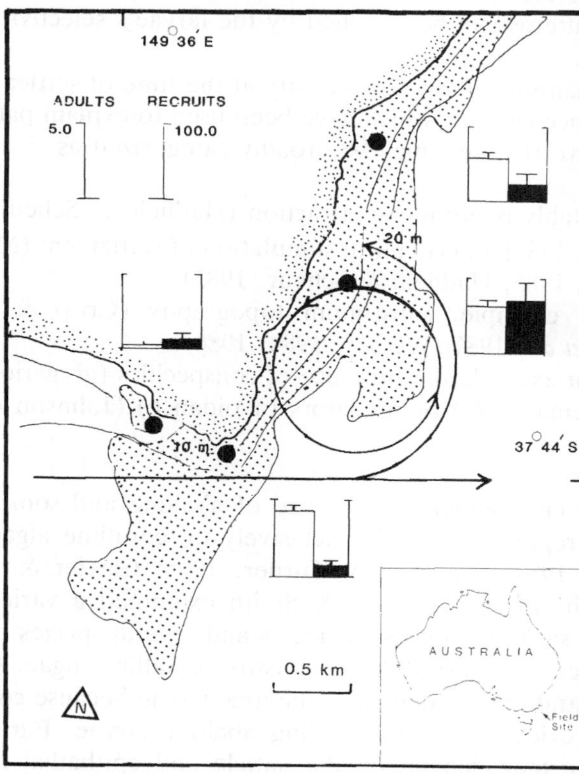

Fig. 11.2 Distribution of recruits relative to adults of *Haliotis rubra* off Sandpatch. The direction of the prevailing current and the presence of an eddy (suggested from the results of hydrodynamic studies) are indicated. Data are relative densities of recruits (mean n.m^{-2}, black) and adults (mean n.m^{-2}, white) of *H. rubra*. The reef complex offshore is shown, as is bathymetry.

or control their dispersal so that they remained near favourable settlement sites rather than being transported in the sea where the probability of encountering a suitable site for settlement is low. Successful colonization by larvae would be further favoured by a brief larval life during which predation of planktonic larvae could be minimized (Strathmann, 1985).

SETTLEMENT

Settlement of benthic marine invertebrates is defined here as contact with substrata leading to metamorphosis from the pelagic to the benthic form (Keough & Downes, 1982). For lecithotrophic species such as abalone, successful settlement therefore depends on competent larvae making contact with suitable substrata before the larvae starve to death.

In the absence of suitable substrata, many benthic invertebrates can postpone settlement (Strathmann *et al.*, 1981). However, settlement of abalone larvae can be delayed only a few days before the larvae's yolk reserves are exhausted (Ino, 1952; Leighton, 1972).

Passive dispersal of larvae can explain many of the settlement patterns observed for marine invertebrates (Eckman, 1983; Hannan, 1984; Bushek, 1988; Butman, 1989); but such patterns can be modified by the larvae's selectivity for substrata on which to settle.

Numerous mechanisms of larval selectivity at the time of settlement have been proposed. These mechanisms, which have been used to explain patterns of settlement in marine invertebrates, may be broadly categorized as

(a) chemical, notably pheromonic attraction (Hadfield & Scheur, 1985; Banse, 1986; Pawlik, 1986) and chemical stimulation of settlement (Morse & Morse, 1984; Morse, 1985; Hadfield & Scheur, 1985);
(b) physical, for example, substratum topography (Crisp & Barnes, 1954; Strathmann *et al.*, 1981; Wethey, 1984, 1986);
(c) biological, for example the presence of conspecifics (gregariousness) (Crisp, 1974; Scheltema, 1974) or predators (avoidance) (Johnson & Strathmann, 1989).

In their natural environment, the larvae of abalone and some other benthic invertebrates are reported to settle exclusively on coralline algae (Saito, 1981; Morse & Morse, 1984; Shepherd & Turner, 1985; Clavier & Richard, 1986; McShane & Smith, 1988). McShane & Smith examined a variety of substrata including epibiota such as sponges, tunicates and several species of seaweed, but found newly settled *H. rubra* only on crustose coralline algae. Coralline algae provide good substrata for settlement of benthic larvae because concavities in the coralline algae provide shelter for settling abalone larvae. Furthermore, post-settlement abalone may depend on the cuticle and epithallial contents grazed from coralline algae (Garland *et al.*, 1985). But in aquaria, abalone larvae can settle on diatom films or artificial substrata (Ebert & Houk, 1984). Settlement of larvae in aquaria is enhanced if the diatom films have been previously grazed by conspecifics (Seki & Kan-no, 1981a,b; Ebert & Houk, 1984; Slattery, 1987).

Yet Shepherd & Turner (1985), after examining wild populations, rejected the hypothesis that abalone were attracted to conspecific adults. In the wild, the grazing of epiphytes by adult abalone and other herbivores such as sea urchins helps maintain coralline algae which would otherwise become overgrown (Paine & Vadas, 1969; Ayling, 1981; Andrew & Choat, 1982; Breitburg, 1984; Clarkson & Shepherd, 1985; Steneck, 1982, 1986; Fletcher, 1987). In the absence of grazers, potential settlement sites on coralline algae could be overgrown to the detriment of the settlement of abalone larvae in natural communities (Saito, 1981; Shepherd & Turner, 1985).

Grazers might also release settlement-inducing chemicals from coralline algae (Morse & Morse, 1984). Substances unique to crustose coralline algae promote settlement and metamorphosis of abalone larvae; the biochemical basis of this process is well established (Hadfield, 1984; Morse *et al.*, 1979a,b, 1980; Morse & Morse, 1984). The results of the studies by Morse and his colleagues demonstrate the promotion of settlement of abalone with the neurotransmitter gamma-amino-butyric acid (GABA). GABA is not a 'true' inducer of settlement (Hahn, 1989b) but it is claimed to mimic effectively natural substances, present in coralline algae, that stimulate settlement and metamorphosis in abalone (Morse & Morse, 1984). Yet studies by Akashige *et al.* (1982) and Slattery (1987) suggest that GABA encourages 'settlement' by depression of ciliary beating in abalone larvae and that rates of metamorphosis and survival of GABA-induced abalone are poor. However, Morse (Chapter 10, this volume) points out that secondary conditions involved in laboratory trials of GABA induction, for example the presence or absence of harmful bacteria, have a significant influence on the results. Indeed, high concentrations of GABA are toxic to abalone larvae (Morse *et al.*, 1979b; Morse, 1984).

It is tempting to invoke chemotaxis as the mechanism by which larvae are attracted to substrata suitable for their subsequent development (Wilson 1968; Meadows & Campbell, 1972; Hadfield & Scheur, 1985; Morse, 1985; Pawlik, 1986). However, for many species including *Haliotis*, settlement in response to chemical cues depends on direct contact with the substratum (Crisp, 1974, 1976; Morse *et al.*, 1980; Burke, 1986). Water turbulence could therefore keep larvae away from suitable sites, whereas those larvae in relatively calm waters are more likely to make contact with surfaces and respond to specific cues in the substrata (McShane *et al.*, 1988). The feeble swimming ability of abalone larvae may be more important as a means of moving away from unsuitable sites or avoiding predation (Bingham & Walters, 1989; Young, 1989) rather than a means of actively colonizing suitable sites (Hannan, 1984).

RECRUITMENT

Traditionally, abalone recruits have been surveyed by direct observation by divers (e.g. Sainsbury, 1982). Such surveys are obviously limited to those abalone large enough to be detected by divers. While underwater magnification can enhance detection of tiny, recently-settled abalone (Kennelly & Underwood, 1984; Shepherd & Turner, 1985), observations based on such methods are qualitative rather than quantitative because they underestimate the numbers of post-settlement abalone (McShane & Smith, 1988). Methods based on the use of anaesthetic to remove juvenile abalone from the substrata (Tomita *et al.*, 1977; Prince & Ford, 1985; Tanaka *et al.*, 1986) provide more realistic estimates of recruitment in abalone, but the application of such methods is limited to removable substrata such as small boulders. A suction sampler developed by McShane & Smith (1988) has overcome the limitations of the other methods and allows collection of abalone recruits soon after settlement on a variety of substrata. A disadvantage with the suction method is that the separation of tiny abalone from other epibiota and sediment typically contained in an epibenthic sample is laborious and identification of tiny abalone to species level is difficult (McShane & Smith, 1988).

Most observations of recruitment in abalone were taken some months after settlement (Sainsbury, 1982; Shepherd & Turner, 1985; Prince *et al.*, 1987). The timing of observations is important in minimizing effects of differential mortality and in allowing realistic estimates of annual recruitment. The size distribution of recruits can provide an indication of post-settlement age and the frequency of settlement (McShane *et al.*, 1988; McShane & Smith, 1991). Several surveys during the annual spawning season may be necessary to quantify recruitment in populations of abalone because recruitment can occur over an extended period during a spawning season (McShane & Smith, 1991). Studies of *H. rubra* show that recruitment can occur throughout a year, albeit at a lower rate than that of the 'main' spawning season (McShane & Smith, 1991). Although estimation of recruitment in abalone populations is problematic, surveys can yield valuable information on recruitment processes. In particular, patterns of recruitment can provide clues as to dispersal mechanisms, settlement periodicity, site selectivity and sources of variation in recruitment.

Surveys in the natural habitat of abalone reveal that recruitment of abalone can be high, and indeed comparable with that of barnacles and other opportunistic colonizers of hard substrata. Estimates of more than 12 000 m^{-2} have been recorded (McShane & Smith, 1989b) and a mean recruitment of more than 1000 m^{-2} is not uncommon (Tanaka *et al.*, 1986; McShane *et al.*, 1988; McShane & Smith, 1991). Such high numbers are consistent with larval swarms and heavy settlement and can be explained by local dispersal of larvae. Larvae diffused and dispersed in open water would be unlikely to colonize in such high densities (see Strathmann, 1985). These recruitment data for *H. rubra* suggest that survival of recruits is favoured by retention of larvae near natal sites; there is little evidence of comparably high recruitment deriving from distant populations (McShane *et al.*, 1988; McShane & Smith, 1991). But recruitment in abalone is highly variable both spatially and temporally (McShane & Smith, 1991), as it is for marine invertebrates in general (Yoshioka, 1986).

The high variability in recruitment of abalone (McShane *et al.*, 1988; McShane & Smith, 1991) may reflect high variability in larval availability. High variability in larval availability with an associated direct influence on recruitment has been established for other species of marine invertebrates (Keough, 1983; Gaines *et al.*, 1985; Yoshioka, 1986; Gotelli, 1987). However, until the collection of larvae becomes reliable, this hypothesis cannot be tested for abalone. Nonetheless, annual variation in recruitment to several populations of *H. rubra* has been shown to be regional rather than local and to correlate with exogenous factors, for example temperature (McShane & Smith, 1991). An increase in larval survival due to a shorter larval life is one explanation for increased recruitment correlating with higher sea temperatures.

Localized dispersal of larvae because of attenuation of local water currents (McShane *et al.*, 1988) or presence of eddies (Tanaka *et al.*, 1986) can explain spatial variation in recruitment of abalone. Prince *et al.* (1988) found the highest numbers of recruits in sheltered habitat unexposed to violent wave action. Localized recruitment of *H. rubra* observed by Prince *et al.* (1987) is all the more likely because of the sheltered nature of the site; the presence of silt at their main study site is indicative of the calm conditions and the site's retentive capacity for neutrally buoyant particles such as abalone larvae. Similarly, McShane *et al.* (1988) found high recruitment only on reef surface protected from surface water currents by kelps and local reef topography. A further example is provided by the author's recent studies of recruitment of *H. rubra* on a coastal reef off a headland in southeast Australia. Recruitment on one side of the headland was lower than that on the other, where an easterly flowing current sets up an eddy (Fig. 11.2). Larval retention in the eddy can explain the observed patterns of recruitment and also the observed abundance of adult *H. rubra* (see also Tanaka *et al.*, 1986).

Further evidence of the governing influence of hydrodynamics is provided by another study of recruitment of *H. rubra* (McShane & Smith, 1991). Recruitment patterns in populations of *H. rubra* less than 1 km apart differed significantly. The fact that the relative size composition of recruits differed significantly indicated that recruitment in the two populations was asynchronous. A corollary is that dispersal of larvae was insufficient to mask differences in recruitment between these neighbouring populations. The populations of *H. rubra* studied by McShane & Smith (1991) were hydrodynamically isolated; one inhabited a sheltered reef in the lee of an exposed headland. Thus there is a correlation between hydrodynamic processes and recruitment of abalone that can account for the spatial variation in recruitment observed for abalone populations (Tomita *et al.*, 1977; Tanaka *et al.*, 1986; Prince *et al.*, 1987, 1988; McShane *et al.*, 1988; McShane & Smith, 1991).

A stock–recruitment relationship (cf Ricker, 1954; Beverton & Holt, 1957) is an attractive hypothesis to explain variation in recruitment in abalone populations. Indeed McShane *et al.* (1988) and Prince *et al.* (1988) did show that recruitment of *H. rubra* correlated with adult abundance, but McShane & Smith (1991) found that recruitment within a single population varied enormously over time despite no difference in adult abundance. On the other hand, Prince *et al.* (1988) proposed a stock–recruitment relationship for *H. rubra* on the basis of surveys of several populations; however, they surveyed recruits at least 2 months after settlement. McShane (1991) showed that post-settlement survival in populations of *H. rubra* is highly variable and that survival during the first two months can be less than 1%. Differential mortality could therefore account for the differences in recruitment in populations reported by Prince *et al.* and the observed low recruitment (mostly less than $2\,m^{-2}$). A more plausible explanation of the results from Prince's group is that regional hydrodynamics are the main determinant of recruitment.

Differences in post-settlement survival have been implicated as major determinants of population structure in benthic invertebrates (Keough & Downes, 1982). Because abalone settle only on crustose coralline algae, differential mortality related to survival on different substrata is unlikely to be important (cf Rowley,

1989, for sea urchins; Raimondi, 1988, for barnacles). Indeed, survival of abalone is probably enhanced by substratum specificity during settlement (see Young & Chia, 1981, 1982, for other organisms). The presence of concavities in the substratum and the abalone's acquisition of a shell colour similar to that of their surroundings means that predation by visual predators such as fish would be low at first (Garland *et al.*, 1985; Shepherd & Turner, 1985; Chapter 21 (this volume); but see also Osenberg & Mittelbach, 1989), yet post-settlement mortality in abalone can be very high (McShane, 1991). Grazing of the coralline surface by mobile grazers such as sea urchins and abalone has a 'bulldozing' effect on small sessile invertebrates (Dayton, 1971; Fletcher, 1987; Fletcher & Underwood, 1987; Miller & Carefoot, 1989) resulting in poor survival of recently-settled abalone recruits. By scraping the substratum with their radulae, marine snails such as adult abalone or limpets could remove tiny invertebrates such as abalone recruits (Dayton, 1971; Denley & Underwood, 1979; Miller & Carefoot, 1989). Similarly, the sea urchin *Centrostephanus rodgersii*, a voracious grazer abundant in temperate waters of the east coast of Australia (Shepherd, 1973b), could remove abalone recruits during its grazing activities (see Fletcher, 1987). Such mortality would be specific to juveniles and only applicable to those on exposed (i.e. non-cryptic) substrata. Older individuals would be immune to 'bulldozing', either because they are too large for grazers to displace or because the abalone had moved to cryptic habitat which provides refuge from mobile grazers (Shepherd & Turner, 1985). But the grazing activity that is important in maintaining coralline algae (Paine & Vadas, 1969; Ayling, 1981) is also important in preventing the smothering of recently-settled benthic invertebrates by epibenthic algae. Thus, paradoxically the grazing activity which provides the coralline surface suitable for settlement of abalone may well be a significant source of their mortality (Fletcher & Creese, 1985; Fletcher, 1987). Furthermore, predation of recruits by terebellid polychaetes and nematodes associated with coralline algae can be an important source of mortality (Morse *et al.*, 1979b). Shepherd & Breen (Chapter 21, this volume) suggest that the major predators of newly-settled abalone are nematodes, flatworms and certain species of polychaetes. Crabs predate on larger recruits (>4 mm shell length). Recruits can be removed by wave action (Cox, 1962; Denley & Underwood, 1979) or they may be smothered by deposition of sediment (Sainsbury, 1982). Sheltered habitat therefore favours retention of larvae near their natal site and protects recruits from wave and current shear by reef topography and seaweeds that attenuate water movement. Refuges have been shown to be important for new recruits of other sessile benthic organisms in avoiding predation (Connell, 1961; Paine, 1974; Creese, 1982; Lubchenco, 1983).

Space is an obvious limiting resource for benthic marine invertebrates (Roughgarden *et al.*, 1985), especially for abalone which colonize only coralline algae. But McShane *et al.* (1988) found that the density of recruits of *H. rubra* was not related to the area of coralline algae. The results of studies by Shepherd (1987) and McShane (1991) are consistent with those for other species of benthic marine invertebrate whose density-dependent mortality is influential only at high settlement densities (Connell, 1985; Davis, 1988; but see also McGuiness & Davis, 1989). Thus coralline 'space' might become limiting only at very high settlement densities. For example, the concavities in coralline algae offer protection from predators and the inadvertent 'bulldozing' of recruits by grazers. Evidence of this is provided by McShane (1991) who showed that early survival of recruits of *H. rubra* was related to density. Conversely, hypotheses of inverse density dependence for mortality of recruits of other benthic species are based on reduced susceptibility to predation at high settlement densities (e.g. for bryozoa, Buss, 1981; Keough, 1984), but this hypothesis has weak empirical support (Keough, 1986).

CONCLUSIONS

All the available evidence suggests that events in the early life history of abalone are as important in determining the population structure of abalone as they are for other species of benthic marine invertebrates. However, if post-settlement survival is density dependent, then variation in survival will influence observations of recruitment in abalone. Such variation is yet another factor that contributes to the chaotic stock–recruitment relationship for abalone.

The lecithotrophic larva of an abalone has only a relatively brief period during which it is competent to settle. Studies of regional hydrodynamics show that larvae would be transported away from habitat suitable for settlement unless water currents were sufficiently attenuated. Unlike some crustacean larvae, whose shape is conducive to orientation in moving water masses (Crisp & Barnes, 1954), the veliger larvae of abalone are weak swimmers which cannot 'select' habitat for settlement if water movement is violent. Water movement is frequently violent in the sublittoral coastal habitat favoured by abalone and there is little evidence to suggest that abalone larvae can behave other than as passively transported particles.

Larvae retained in discrete areas by current eddies or by the protection of local reef topography could complete their larval life cycle near their natal site (Fig. 11.1). Most of the larvae dispersed into the open sea would probably not encounter suitable habitat for settlement; yet the few that do, contribute to gene flow between populations. Evidence from studies of Australian populations of abalone (Chapter 3, this volume) indicates little inbreeding, and is consistent with some dispersal between populations (Grassle & Grassle, 1978; Behrens Yamada, 1989).

Localized dispersal of larvae implies that abalone stocks are composed of several discrete substocks (Sluczanowski, 1984). Removal of abalone by fishing will therefore affect substocks rather than the stock as a whole (Prince *et al.*, 1988; McShane & Smith, 1989b). Recruitment overfishing has been implicated in the decline of many abalone fisheries (Cicin-Sain *et al.*, 1977; Breen, 1980; Sluczanowski, 1984; Sloan & Breen, 1988). The broodstock for each discrete abalone substock could have a 'critical level' below which recruitment would consistently fail. Even so, the author's own studies have shown that there is considerable 'noise' in the form of temporal variation in recruitment; 'noise' that is characteristic of recruitment in marine invertebrates (Hancock, 1973; Keough, 1983; Yoshioka, 1986).

ACKNOWLEDGEMENTS

The author thanks Murray Smith for helpful discussions and technical assistance.

Dr Kerry Black enhanced my knowledge of coastal hydrodynamics and mechanisms of larval dispersal. Drs Darwin Evans and Scoresby Shepherd critically reviewed early drafts and helped improve the manuscript, as did comments from an anonymous reviewer.

REFERENCES

Akashige S., Seki T., Kan-no H. & Nomura T. (1982) Effects of gamma-aminobutyric acid and certain neurotransmitters on the settlement and the metamorphosis of the larvae of *Haliotis discus hannai* Ino. *Can. Transl. Fish Aquat. Sci.*, No 4882.

Andrew N.L. & Choat J.H. (1982) The influence of predation and conspecific adults on the abundance of juvenile *Evechinus chloroticus* (Echinoidea:Echinometridae). *Oecologia*, **54**, 80–7.

Ayling A.M. (1981) The role of biological disturbance in temperate subtidal encrusting communities. *Ecology*, **62**, 830–47.

Banse K. (1986) Vertical distribution and horizontal transport of planktonic larvae of echinoderms and benthic polychaetes in an open coastal sea. *Bull. Mar. Sci.*, **39**, 162–75.

Behrens Yamada S. (1989) Are direct developers more locally adapted than planktonic developers? *Mar. Biol.*, **104**, 403–11.

Bernstein B.B. & Jung N. (1979) Selective processes and co-evolution in a kelp canopy community in southern California. *Ecol. Monogr.*, **49**, 335–55.

Beverton R.J.H. & Holt S.J. (1957) On the dynamics of exploited fish populations. *Fish. Invest. (London)*, **19**, 1–533.

Bingham B.L. & Walters L.J. (1989) Solitary ascidians as predators of invertebrate larvae: evidence from gut analyses and plankton samples. *J. Exp. Mar. Biol. Ecol.*, **113**, 147–59.

Black K.P. (1989) The relationship of reef hydrodynamics to variations in numbers of planktonic larvae around coral reefs. *Proc. Int. Coral Reef Symp. Townsville 1988*, **2**, 125–30.

Black K.P. & Gay S.L. (1987) Eddy formation in unsteady flows. *J. Geophys. Res.*, **92**, 9514–22.

Breen P.A. (1980) Measuring fishing intensity and annual production in the abalone fishery of British Columbia. *Tech. Rep. Can. Fish. Res. Bd.*, No 947.

Breen P.A. & Adkins B.E. (1980) Spawning in a British Columbia population of northern abalone, *Haliotis kamtschatkana*. *Veliger*, **23**, 177–9.

Breitburg D.L. (1984) Residential effects of grazing: inhibition of competitor recruitment by encrusting coralline algae. *Ecology*, **65**, 1136–43.

Burke R.D. (1986) Pheromones and the gregarious settlement of marine invertebrate larvae. *Bull. Mar. Sci.*, **39**, 323–31.

Bushek D. (1988) Settlement as a major determinant of intertidal oyster and barnacle distributions along a horizontal gradient. *J. Exp. Mar. Biol. Ecol.*, **122**, 1–18.

Buss L.W. (1981) Group living, competition, and the evolution of cooperation in a sessile invertebrate, *Bugula turita*. *Science*, **213**, 1012–14.

Butman C.A. (1987) Larval settlement of soft-sediment invertebrates: the spatial scales of patterns explained by active habitat selection and the emerging role of hydrodynamic processes. *Oceanogr. Mar. Biol. Ann. Rev.*, **25**, 113–65.

Butman C.A. (1989) Sediment trap experiments on the importance of hydrodynamical processes in distributing settling invertebrate larvae in near-bottom waters. *J. Exp. Mar. Biol. Ecol.*, **134**, 37–88.

Caffey H.M. (1985) Spatial and temporal variation in settlement and recruitment of an intertidal barnacle. *Ecol. Monogr.*, **55**, 313–32.

Chia F.S., Buckland-Nicks J. & Young, C.M. (1984) Locomotion of marine invertebrate larvae: a review. *Can. J. Zool.*, **62**, 1205–22.

Cicin-Sain B., Moore J.E. & Wyner A.J. (1977) Management approaches for marine fisheries: the case of the Californian abalone. University of California Sea Grant College program. *Sea Grant Publication 54.*

Clarkson P.S. & Shepherd S.A. (1985) Distribution and feeding of archaeogastropods in a boulder habitat at West Island, South Australia. *J. Malac. Soc. Aust.*, **7**, 34–44.

Clavier J. & Richard O. (1986) Growth of juvenile *Haliotis tuberculata* (Mollusca:Gastropoda) in the natural environment. *J. Mar. Biol. Assoc. UK*, **66**, 497–503.

Connell J.H. (1961) The influence of interspecific competition and other factors on the distribution of the barnacle *Chthamalus stellatus*. *Ecology*, **42**, 710–23.
Connell J.H. (1972) Community interactions on marine rocky intertidal shores. *Ann. Rev. Ecol. Syst.*, **3**, 169–92.
Connell J.H. (1985) The consequences of variation in initial-settlement vs. post-settlement mortality in rocky intertidal communities. *J. Exp. Mar. Biol. Ecol.*, **93**, 11–45.
Cox K.W. (1962) California abalones, family Haliotidae. *Calif. Fish Game, Fish Bull.* **118**.
Creese R.G. (1982) Distribution and abundance of the acmaied limpet, *Patelloida latiotrigata*, and its interaction with barnacles. *Oecologia*, **52**, 85–96.
Crisp D.S. (1955) The behaviour of barnacle cyprids in relation to water movement over a surface. *J. Exp. Biol.*, **32**, 569–90.
Crisp D.S. (1974) Factors affecting settlement of marine invertebrate larvae. In *Chemoreception in Marine Organisms* (Ed. by P. Grant & A.M. Mackie), pp. 177–265. Academic Press, New York and London.
Crisp D.S. (1976) Settlement responses in marine organisms. In *Adaptations to Environment: Essays on the Physiology of Marine Animals* (Ed. by R.C. Newell), pp. 83–124. Butterworths, London.
Crisp D.J. & Barnes H. (1954) The orientation and distribution of barnacles at settlement with particular reference to surface contour. *J. Anim. Ecol.*, **23**, 142–62.
D'Asaro C.N. (1966) The egg capsules, embryogenesis, and early organogenesis of a common oyster predator, *Thais haemostoma floridana*. *Bull. Mar. Sci.*, **16**, 884–914.
Davis A.R. (1987) Variation in recruitment of the subtidal colonial ascidian *Podoclavella cylindrica* (Quoy & Gaimard): the role of substratum choice and early survival. *J. Exp. Mar. Biol. Ecol.*, **106**, 57–71.
Davis A.R. (1988) Effects of variation in initial settlement on distribution and abundance of *Podoclavella moluccensis* Sluiter. *J. Exp. Mar. Biol. Ecol.*, **117**, 157–67.
Dayton P.K. (1971) Competition, disturbance and community organization. The provision and subsequent utilization of space in a rocky intertidal community. *Ecol. Monogr.*, **41**, 351–89.
Dayton P.K. (1975) Experimental evaluation of ecological dominance in a rocky intertidal algal community. *Ecol. Monogr.*, **45**, 137–59.
Denley E.J. & Underwood A.J. (1979) Experiments on factors influencing settlement, survival and growth of two species of barnacles in New South Wales. *J. Exp. Mar. Biol. Ecol.*, **36**, 269–93.
Ebert E.E. & Houk J.L. (1984) Elements and innovations in the cultivation of red abalone, *Haliotis rufescens*. *Aquaculture*, **39**, 375–92.
Eckman J.E. (1983) Hydrodynamic processes affecting benthic recruitment. *Limnol. Oceanogr.*, **28**, 241–57.
Eckman J.E., Duggins D.O. & Sewell A.T. (1989) Ecology of understory kelp environments. I. Effects of kelps on flow and particle transport near the bottom. *J. Exp. Mar. Biol. Ecol.*, **129**, 173–87.
Fletcher W.J. (1987) Interactions among subtidal Australian sea urchins, gastropods, and algae: effects of experimental removals. *Ecol. Monogr.*, **57**, 89–109.
Fletcher W.J. & Creese R.G. (1985) Competitive interactions between co-occurring herbivorous gastropods. *Mar. Biol.*, **86**, 183–91.
Fletcher W.J. & Underwood A.J. (1987) Interspecific competition among subtidal limpets: effects of substratum heterogeneity. *Ecology*, **68**, 385–98.
Forster G.R., Potts G.W. & Swinfen R. (1982) Changes in the ormer populations of Guernsey and Jersey. *J. Mar. Biol. Assoc. UK*, **62**, 717–27.
Forward R.B. Jr., Wellins C.A. & Buswell C.U. (1989) Behavioural responses of larvae of the crab *Neopanope sayi* to hydrostatic pressure. *Mar. Ecol. Prog. Ser.*, **57**, 267–77.
Fretter V. & Manly R. (1977) Algal associations of *Tricolia pullus*, *Lacuna vincta* and *Cerithiopsis tubercularis* (Gastropoda) with special reference to the settlement of their larvae. *J. Mar. Biol. Assoc. UK*, **57**, 999–1017.
Gaines S.D., Brown S. & Roughgarden J. (1985) Spatial variation in larval concentration as a cause of spatial variation in settlement for the barnacle, *Balanus glandula*. *Oecologia*, **67**, 267–73.
Gaines S.D. & Roughgarden J. (1987) Fish in offshore kelp forests affect recruitment to intertidal barnacle populations. *Science*, **235**, 479–81.
Garland C.D., Cooke S.L., Grant J.F. & McMeekin T.A. (1985) Ingestion of the bacteria on and the cuticle of crustose (non-articulated) coralline algae by post-larval and juvenile abalone (*Haliotis rubra* Leach) from Tasmanian waters. *J. Exp. Mar. Biol. Ecol.*, **91**, 137–49.

Gerrodette T. (1981) Dispersal of the solitary coral *Balanophylia elegans* by demersal planular larvae. *Ecology*, **62**, 611–19.
Gotelli N.J. (1987) Spatial and temporal patterns of reproduction, larval settlement, and recruitment of the compound ascidean *Aplidium stellatum*. *Mar. Biol.*, **94**, 45–51.
Grassle J.F. & Grassle J.P. (1978) Life histories and genetic variation in marine invertebrates. In *Marine Organisms: Genetics, Ecology, and Evolution* (Ed. by B. Battaglia & J.A. Beardmore). *NATO Conf. Ser. (Ser. IV Mar. Sciences)*, **2**, 347–64.
Hadfield M.G. (1984) Settlement requirements of molluscan larvae: new data on chemical and genetic roles. *Aquaculture*, **39**, 283–98.
Hadfield M.G. & Scheur D. (1985) Evidence for a soluble metamorphic inducer in *Phestilla*: ecological, chemical and biological data. *Bull. Mar. Sci.*, **37**, 556–66.
Hahn K.O. (1989a) Gonad reproductive cycles. In *Handbook of Culture of Abalone and Other Marine Gastropods* (Ed. by K.O. Hahn), pp. 13–40. CRC Press, Florida.
Hahn K.O. (1989b) Induction of settlement in competent abalone larvae. In *Handbook of Culture of Abalone and Other Marine Gastropods* (Ed. by K.O. Hahn), pp. 101–12. CRC Press, Florida.
Hancock D.A. (1973) The relationship between stock and recruitment in exploited invertebrates. *Rapp. PV Reun. Comm. Int. Explor. Mer.*, **164**, 113–31.
Hannan C.A. (1984) Planktonic larvae may act like passive particles in turbulent near-bottom flows. *Limnol. Oceanogr.*, **29**, 1108–16.
Harrigan J.F. (1972) Behaviour of the planula larvae of the scleractinian coral *Pocillophora damicornis* L. *Am. Zool.*, **12**, 723.
Heck K.L. & McKoy E.D. (1978) Long distance dispersal and the reef-building corals of the eastern Pacific. *Mar. Biol.*, **48**, 349–65.
Hines A.H. (1979) The comparative reproduction ecology of three species of intertidal barnacles. In *Reproductive Ecology of Marine Invertebrates* (Ed. by S. Stancyk) pp. 213–34. University of South Carolina Press, Columbia, South Carolina.
Hines A.H. (1986) Larval problems and perspectives in life histories of marine invertebrates. *Bull. Mar. Sci.*, **39**, 506–25.
Hines A.H. & Pearce J.S. (1982) Abalone shells and sea otters: dynamics of prey populations in central California. *Ecology*, **63**, 1547–60.
Ino T. (1952) Biological studies of the propagation of the Japanese abalone (genus *Haliotis*). *Bull. Tokai Reg. Fish. Res. Lab.*, **5**, 29–102.
Jackson G.A. & Winant C.D. (1983) Effect of a 'kelp' forest on coastal currents. *Cont. Shelf Res.*, **2**, 75–80.
Jaeckle W.B. & Manahan D.T. (1989) Feeding by a 'nonfeeding' larva: uptake of dissolved amino acids from seawater by lecithrophic larvae of the gastropod *Haliotis rufescens*. *Mar. Biol.*, **103**, 87–94.
Johnson L.E. & Strathmann R.R. (1989) Settling barnacle larvae avoid substrata previously occupied by a mobile predator. *J. Exp. Mar. Biol. Ecol.*, **128**, 87–103.
Kennelly S.J. & Underwood A.J. (1984) Underwater microscopic sampling of a sublittoral kelp community. *J. Exp. Mar. Biol. Ecol.*, **76**, 67–78.
Keough M.J. (1983) Patterns of recruitment of sessile invertebrates in two subtidal habitats. *J. Exp. Mar. Biol. Ecol.*, **66**, 213–45.
Keough M.J. (1984) Kin-recognition and the spatial distribution of larvae of the bryozoan *Bugula neritina*. *Evolution*, **38**, 142–7.
Keough M.J. (1986) The distribution of the bryozoan *Bugula neritina* on seagrass blades: settlement growth and mortality. *Ecology*, **67**, 846–57.
Keough M.J. (1989) Benthic populations: is recruitment limiting or just fashionable? *Proc. 6th Int. Coral Reefs Symp.* (Ed. J.H. Choat *et al.*) Sixth International Coral Reef Symposium Executive Committee.
Keough M.J. & Downes B.J. (1982) Recruitment of marine invertebrates: the role of active larval choices and early mortality. *Oecologia (Berlin)*, **54**, 348–52.
Lee T.Y., Pyen C.K., Chin P. & Hong S.Y. (1978) Seed production and rearing of the abalone, *Haliotis discus hannai*. *Publ. Inst. Mar. Sci. Nat. Fish. Univ. Busan.*, **11**, 47–61.
Leighton D.L. (1972) Laboratory observations on the early growth of the abalone, *Haliotis sorenseni*, and the effect of temperature on larval development and settlement success. *Fish. Bull.*, **70**, 373–81.
Leighton D.L. (1974) The influence of temperature on larval and juvenile growth in three species of Southern California abalones. *Fish. Bull.*, **72**, 1137–45.

Lewin R. (1986) Supply-side ecology. *Science*, **234**, 25–7.

Lobel P.S. & Robinson A.R. (1986) Transport and entrapment of fish larvae by ocean mesoscale eddies and currents in Hawaiian waters. *Deep Sea Res.*, **33**, 483–500.

Lubchenco J. (1983) *Littorina* and *Fucus*: effects of herbivores, substratum heterogeneity, and plant escapes during succession. *Ecology*, **64**, 1116–23.

McGuinness K.A. & Davis A.R. (1989) Analysis and interpretation of the recruit–settler relationship. *J. Exp. Mar. Biol. Ecol.*, **134**, 197–202.

McShane P.E. (1991) Density dependent mortality of recruits of the abalone *Haliotis rubra* (Mollusca:Gastropoda). *Mar. Biol.* **110**, 385–9.

McShane P.E., Beinssen K.H.H., Smith M.G., O'Connor S. & Hickman N.J. (1986) Reproductive biology of blacklip abalone *Haliotis rubra* Leach from four Victorian populations. *Victorian Ministry for Conservation, Forests and Lands, Marine Science Laboratories, Tech. Rep. No. 55.*

McShane P.E., Black K.P. & Smith M.G. (1988) Recruitment processes in *Haliotis rubra* (Mollusca:Gastropoda) and regional hydrodynamics in southeastern Australia imply localized dispersal of larvae. *J. Exp. Mar. Biol. Ecol.*, **124**, 175–203.

McShane P.E. & Smith M.G. (1988) Measuring recruitment of abalone *Haliotis rubra* Leach (Gastropoda:Haliotidae) – comparison of a novel method with two other methods. *Aust. J. Mar. Freshwat. Res.*, **39**, 331–6.

McShane P.E. & Smith M.G. (1989a) Direct measurement of fishing mortality in abalone (*Haliotis rubra* Leach) off southeastern Australia. *Fish. Res.*, **8**, 93–102.

McShane P.E. & Smith M.G. (1989b) Victorian abalone fishery in good shape. *Aust. Fish.*, **48**, 20–5.

McShane P.E. and Smith M.G. (1991) Recruitment variation in sympatric populations of *Haliotis rubra* (Mollusca:Gastropoda). *Mar. Ecol. Prog. Ser.* (in press).

Meadows P.S. & Campbell J.I. (1972) Habitat selection by aquatic invertebrates. *Adv. Mar. Biol.*, **10**, 271–382.

Mileikovsky S.A. (1971) Types of larval development in marine bottom invertebrates, their distribution and ecological significance: a re-evaluation. *Mar. Biol.*, **10**, 193–213.

Mileikovsky S.A. (1973) Speed of active movement of pelagic larvae of marine bottom invertebrates and their ability to regulate their vertical position. *Mar. Biol.*, **23**, 11–17.

Miller K.M. & Carefoot T.H. (1989) The role of spatial and size refuges in the interaction between juvenile barnacles and grazing limpets. *J. Exp. Mar. Biol. Ecol.*, **134**, 157–74.

Morse A.N.C. & Morse D.E. (1984) Recruitment and metamorphosis of *Haliotis* larvae induced by molecules uniquely available at the surface of crustose red algae. *J. Exp. Mar. Biol. Ecol.*, **75**, 191–215.

Morse D.E. (1984) Biochemical and genetic engineering for improved production of abalones and other valuable molluscs. *Aquaculture*, **39**, 263–82.

Morse D.E. (1985) Neurotransmitter-mimetic inducers of larval settlement and metamorphosis. *Bull. Mar. Sci.*, **37**, 697–706.

Morse D.E., Hooker N., Duncan H. & Jensen L. (1979a) γ-amino butyric acid, a neurotransmitter, induces planktonic abalone larvae to settle and begin metamorphosis. *Science*, **204**, 407–10.

Morse D.E., Hooker N., Jensen L. & Duncan H. (1979b) Induction of larval abalone settling and metamorphosis by γ-amino butyric acid and its cogeners from crustose red algae. II: Application to cultivation, seed production and bioassays; principle causes of mortality and interference. *Proc. World Maricul. Soc.*, **10**, 81–91.

Morse D.E., Tegner M., Duncan H., Hooker N., Trevelyan G. & Cameron A. (1980) Induction of settling and metamorphosis of planktonic molluscan (*Haliotis*) larvae. III Signaling by metabolites of intact algae is dependent on contact. In *Chemical Signaling in Vertebrate and Aquatic Animals* (Ed. by D. Muller-Schwarze & R.M. Silverstein), pp. 67–86. Plenum Press, New York.

Mottet M.G. (1978) A review of the fishery biology of abalones. Washington Report. *Fish. Tech. Rep. No. 37.*

Newman G.G. (1967) Reproduction of the South African abalone, *Haliotis midae*. *Invest. Rep. Div. Sea Fish. S. Afr.*, **64**, 1–24.

Olson R.R. (1985) The consequences of short-distance larval dispersal in a sessile marine invertebrate. *Ecology*, **66**, 30–9.

Olson R.R. & Olson M.H. (1989) Food limitation of planktotrophic marine invertebrate larvae: does it control recruitment success? *Ann. Rev. Ecol. Syst.*, **20**, 225–47.

Osenberg C.W. & Mittelbach G.G. (1989) Effects of body size on the predator-prey interaction between pumpkinseed sunfish and gastropods. *Ecol. Monogr.*, **59**, 405–32.

Paine R.T. (1969) The *Pisaster–Tegula* interaction: prey patches, predator food preference, and intertidal community structure. *Ecology*, **50**, 950–61.
Paine R.T. (1974) Intertidal community structure: Experimental studies on the relationship between a dominant competitor and its principal predator. *Oecologia*, **15**, 93–120.
Paine R.T. (1980) Food webs: linkage, interaction strength and community infrastructure. *J. Anim. Ecol.*, **49**, 667–85.
Paine R.T. & Vadas R.L. (1969) The effects of grazing by sea urchins, *Stronglyocentrotus* spp., on benthic algal populations. *Limnol. Oceanogr.*, **14**, 710–19.
Pawlik J.R. (1986) Chemical induction of larval settlement and metamorphosis in the reef-building tube worm *Phragmatopoma californica* (Sabellariidae: Polychaeta). *Mar. Biol.*, **91**, 59–68.
Pennington J.T., Rumrill S.S. & Chia F.S. (1986) Stage-specific predation upon embryos and larvae of the Pacific sand dollar, *Dendraster excentricus*, by eleven species of common zooplankton predators. *Bull. Mar. Sci.*, **39**, 485–97.
Pond S. & Pickard G.I. (1983) *Introductory Dynamical Oceanography*. Pergamon, New York.
Poore G.C.B. (1972) Ecology of New Zealand abalones, *Haliotis* species (Mollusca: Gastropoda) 4. Reproduction. *NZ J. Mar. Freshwat. Res.*, **7**, 67–84.
Prince J.D. & Ford W.B. (1985) Use of anaesthetic to standardize efficiency in sampling abalone populations (genus: *Haliotis*: Mollusca:Gastropoda). *Aust. J. Mar. Freshwat. Res.*, **36**, 701–6.
Prince J.D., Sellers T.L., Ford W.B. & Talbot S.R. (1987) Experimental evidence for limited dispersal of haliotid larvae (genus: *Haliotis*: Mollusca: Gastropoda). *J. Exp. Mar. Biol. Ecol.*, **106**, 243–63.
Prince J.D., Sellers T.L., Ford W.B. & Talbot S.R. (1988) Confirmation of a relationship between the localized abundance of breeding stock and recruitment for *Haliotis rubra* Leach (Mollusca: Gastropoda). *J. Exp. Mar. Biol. Ecol.*, **122**, 91–104.
Quayle D.B. (1971) Growth, morphometry and breeding in the British Columbia abalone (*Haliotis kamtschatkana* Jones). *Tech. Rep. Can. Fish. Res. Bd, No 279*, 1–16.
Raimondi P.T. (1988) Rock type affects settlement, recruitment, and zonation of the barnacle *Chthamalus anisopoma* Pilsbury. *J. Exp. Mar. Biol. Ecol.*, **123**, 253–67.
Ricker W.E. (1954) Stock and recruitment. *J. Fish. Res. Bd Can.*, **11**, 559–623.
Roughgarden J., Iwasa Y. & Baxter C. (1985) Demographic theory for an open marine population with space-limited recruitment. *Ecology*, **66**, 54–67.
Rowley R.J. (1989) Settlement and recruitment of sea urchins (*Stronglyocentrotus* spp.) in a sea-urchin barren ground and a kelp bed: are populations regulated by settlement or post-settlement processes? *Mar. Biol.*, **100**, 485–94.
Sainsbury K.J. (1982) Population dynamics and fishery management of the paua, *Haliotis iris*. I. Population structure, growth, reproduction and mortality. *NZJ. Mar. Freshwat. Res.*, **16**, 147–61.
Saito K. (1981) The appearance and growth of 0-year old Ezo abalone. *Bull. Jap. Soc. Fish.*, **47**, 1393–401.
Scheltema R.S. (1974) Biological interactions determining larval settlement of marine invertebrates. *Thallasia Jugoslavica*, **10**, 263–96.
Scheltema R.S. (1986) On dispersal and planktonic larvae of benthic invertebrates: an eclectic overview and summary of problems. *Bull. Mar. Sci.*, **39**, 290–322.
Seki T. & Kan-no H. (1981a) Induced settlement of the Japanese abalone, *Haliotis discus hannai*, veliger by the mucous trails of the juvenile and adult abalones. *Bull. Tohoku Reg. Fish. Res. Lab.*, **43**, 29–36.
Seki T. & Kan-no H. (1981b) Observations on the settlement and metamorphosis of the veliger of the Japanese abalone *Haliotis discus hannai* Ino, Haliotidae, Gastropoda. *Bull. Tohoku Reg. Fish. Res. Lab.*, **42**, 31–9.
Shanks A.L. & Wright W.G. (1987) Internal-wave-mediated shoreward transport of cyprids, megalopae, and gammarids and correlated longshore differences on the settling rate of intertidal barnacles. *J. Exp. Mar. Biol. Ecol.*, **114**, 1–13.
Shepherd S.A. (1973a) Studies on southern Australian abalone (genus *Haliotis*) I. Ecology of five sympatric species. *Aust. J. Mar. Freshwat. Res.*, **24**, 217–57.
Shepherd S.A. (1973b) Competition between sea urchins and abalone. *Aust. Fish.*, **32**, 4–7.
Shepherd S.A. (1986) Studies on southern Australian abalone (genus *Haliotis*) VII. Aggregative behaviour of *H. laevigata* in relation to spawning. *Mar. Biol.*, **90**, 231–6.
Shepherd S.A. (1987) *Aspects of the biology of the abalone* Haliotis laevigata *and* Haliotis scalaris. PhD thesis, Deakin University.

Shepherd S.A. & Laws H.M. (1974) Studies on southern Australian abalone (genus *Haliotis*) II. Reproduction of five species. *Aust. J. Mar. Freshwat. Res.*, **25**, 49–62.

Shepherd S.A. & Turner J.A. (1985) Studies on southern Australian abalone (genus *Haliotis*). VI. Habitat preference and abundance and predators of juveniles. *J. Exp. Mar. Biol. Ecol.*, **93**, 285–98.

Slattery M. (1987) *Settlement and metamorphosis of red abalone* (Haliotis rufescens) *larvae: a critical examination of mucous, diatoms and γ-aminobutyric acid (GABA) as inductive substrates.* MA thesis, San Jose State University.

Sloan N.A. & Breen P.A. (1988) Northern abalone, *Haliotis kamtschatkana*, in British Columbia: fisheries and synopsis of life history information. *Can. Spec. Publ. Fish. Aquat. Sci.*, **103**.

Sluczanowski P.R. (1984) A management oriented model of an abalone fishery whose substocks are subject to pulse fishing. *Can. J. Fish. Aquat. Sci.*, **41**, 1008–14.

Spight T.M. (1975) Factors extending gastropod embryonic development and their selective cost. *Oecologia*, **21**, 1–16.

Steneck R.S. (1982) A limpet–coralline alga association: adaptations and defenses between a selective herbivore and its prey. *Ecology*, **63**, 507–22.

Steneck R.S. (1986) The ecology of coralline algal crusts: convergent patterns and adaptive strategies. *Ann. Rev. Ecol. Syst.*, **17**, 273–303.

Strathmann R.R. (1985) Feeding and nonfeeding larval development and life history evolution in marine invertebrates. *Ann. Rev. Ecol. Syst.*, **16**, 339–61.

Strathmann R.R., Branscomb E.S. & Vedderet K. (1981) Fatal errors in set as a cost of dispersal and the influence of intertidal flora on set of barnacles. *Oecologia*, **48**, 13–18.

Tanaka K., Tanaka T., Ishida O. & Oba T. (1986) On the distribution of swimming and deposited larvae of nursery ground of abalone at the southern coast of Chiba prefecture. *Bull. Jap. Soc. Sci. Fish.*, **52**, 1525–32.

Tanaka Y. (1978) Swimming behaviour of the abalone larvae, *Nordotis gigantea* (Gmelin), in an outdoor tank. *Bull. Tokai Reg. Res. Lab.*, **95**, 15–22.

Tegner M.J. & Butler R.A. (1985) Drift-tube study of the dispersal potential of green abalone (*Haliotis fulgens*) larvae in the southern California bight: implications for recovery of depleted populations. *Mar. Ecol. Prog. Ser.*, **26**, 73–84.

Thorson G. (1950) Reproduction and larval ecology of marine bottom invertebrates. *Biol. Rev.*, **25**, 1–45.

Thorson G. (1964) Light as an ecological factor in the dispersal and settlement of larvae of marine bottom invertebrates. *Ophelia*, **1**, 167–208.

Tomita K., Tajima K. & Kudo K. (1977) Morphological observations of the veliger and creeping larvae of an abalone *Haliotis discus hannai* from Rebun Island in the north-western Hokkaido. *Sci. Rep. Hokkaido Fish. Exp. Stn No 19*, 13–18.

Tutschulte T. & Connell J.H. (1981) Reproductive biology of three species of abalones (*Haliotis*) in southern California. *Veliger*, **23**, 195–206.

Underwood A.J. & Chapman M.G. (1989) Experimental analyses of the influences of topography of the substratum on movements and density of an intertidal snail, *Littorina unifasciata*. *J. Exp. Mar. Biol. Ecol.*, **134**, 175–96.

Underwood A.J. & Denley E.J. (1984) Paradigms, explanations and generalizations in models for the structure of intertidal communities on rocky shores. In *Ecological Communities: Conceptual Issues and Evidence* (Ed. by D.R. Strong, D. Simberloff, L. Abele & A.B. Thistle), pp. 151–80. Princeton University Press, Princeton.

Wethey D.S. (1984) Spatial pattern in barnacle settlement: day to day changes during the settlement season. *J. Mar. Biol. Assoc. UK*, **64**, 687–98.

Wethey D.S. (1986) Ranking of settlement cues by barnacle larvae. Influence of surface contour. *Bull. Mar. Sci.*, **39**, 393–400.

Wilson D.P. (1968) The settlement behaviour of the larvae of *Sabellaria alveolata* (L.). *J. Mar. Biol. Assoc. UK*, **48**, 387–435.

Yano I. & Ogawa Y. (1977) Effects of light, temperature and hydrostatic pressure on vertical distribution of abalone larvae, *Haliotis gigantea*. *Bull. Tokai Reg. Res. Lab.*, **91**, 19–27.

Yoshioka P.M. (1982) Role of planktonic and benthic factors in the population dynamics of the bryozoan *Membranipora membranacea*. *Ecology*, **63**, 457–68.

Yoshioka P.M. (1986) Chaos and recruitment in the bryozoan, *Membranipora membranacea*. *Bull. Mar. Sci.*, **39**, 408–17.

Young C.M. (1989) Larval depletion by ascideans has little effect on settlement of epifauna. *Mar.*

Biol., **102**, 481–9.
Young C.M. & Chia F.S. (1981) Laboratory evidence for delay of larval settlement in response to a dominant competitor. *Int. J. Invert. Reprod.*, **3**, 221–6.
Young C.M. & Chia F.S. (1982) Factors controlling spatial distribution of the sea cucumber *Psolus chitinoides*: settling and post-settling behaviour. *Mar. Biol.*, **69**, 195–205.

PART IV
FEEDING AND GROWTH

Chapter 12
The determinants and measurement of abalone growth

R.W. DAY and A.E. FLEMING *Zoology Department, University of Melbourne, Parkville, Victoria 3052, Australia*

ABSTRACT: Both mariculture and the management of wild fisheries require a knowledge of the growth rates of abalone (*Haliotis* species), partly because growth rates determine the time delay between settlement of larvae and recruitment to harvestable size classes. Sources of bias in the methods to measure growth in the field, which should receive more attention, are identified. Field and laboratory studies of many species of *Haliotis* have revealed extensive variation in growth rates. Both the quality and quantity of algal food appear to be important, and variations between sites, seasons and years may be due in part to the availability of algae. In our study of *Haliotis rubra*, much of the observed variation in growth does not arise from consistent differences between individuals and may be due to spatial patchiness in available food. Laboratory studies, mostly short-term studies on Japanese abalone, have focused largely on feeding trials using a single alga. Here, growth depends both on the nutritional value of the food and on the feeding rate. Abalone do not necessarily feed rapidly on the algae that they can convert efficiently into biomass. While such studies show that different algae support very different growth rates, growth may also be affected by other feeding parameters, such as the previous diet of the abalone. Furthermore, our longer-term studies show that single-species diets may sustain growth for short periods only. These results suggest that growth will be affected by long-term, rather than short-term differences in food supply between reefs, and the variety of algae available on a reef may be important.

RESUMEN: Tanto el cultivo marítimo como el manejo de las pesquerías requieren un conocimiento de las tasas de crecimiento del abulón (especies de *Haliotis*), en parte, porque las tasas de crecimiento determinan el tiempo de retraso entre el asentamiento de las larvas y el reclutamiento de las clases de tallas capturables. Las fuentes de los prejuicios en los métodos para medir crecimiento en el campo, son identificadas, las cuales deberían recibir más atención. Estudios de campo y de laboratorio de muchas especies de *Haliotis* han mostrado una gran variación en las tasas de crecimiento. Tanto la calidad como la cantidad de alimento algal parecen ser importantes, y las variaciones entre las estaciones, los lugares y los años puede ser debido, en parte, a la disponibilidad de algas. En nuestro estudio sobre *Haliotis rubra*, mucha de la variación observada en el crecimiento no surge de diferencias consistentes entre individuos, ésta puede deberse a las parcelaciones espaciadas de alimento disponible. Los estudios de laboratorio, principalmente de corta duración, sobre el abulón japonés se han enfocado bastante a ensayos alimenticios con una sola especie de alga. En estos estudios el crecimiento depende tanto del valor nutricional del alimento como de la tasa de alimentación. El abulón no necesariamente se alimenta rápido de algas que pueda convertir eficientemente en biomasa. Mientras estos estudios demuestran que diferentes algas soportan muchas diferentes tasas de crecimiento, éste puede ser tambien afectado por otros parámetros alimenticios, como la dieta previa del abulón. Además, nuestros ensayos de más larga duración, muestran que las dietas con una sola especie, pueden sustenar el crecimiento sólo por períodos cortos de tiempo. Estos resultados sugieren que el crecimiento será afectado por las diferencias a largo plazo más que a corto plazo en la provisión de alimento entre los arrecifes, pudiendo ser importante la variedad de algas disponibles en un arrecife.

INTRODUCTION

The rates of growth of abalone and the factors that affect growth are very important for mariculture and fishery management. Costs of mariculture are reduced by fast growth. In management, growth rates are important for yield-per-recruit and egg-per-recruit models, and affect the relation between future recruitment to the reproductive population and the size of the present breeding population, as they determine the time lag between spawning and reproductive recruitment. This time lag is itself important to the population's dynamics, and it also sets the period of pre-reproductive mortality and thus the proportion of juveniles surviving to reproductive recruitment. Furthermore, any density-dependent growth will fundamentally alter the stock–recruitment relation.

Variation in growth over time, between sites or between individuals at a site will also have important consequences for the stock-recruitment relation (Day & Leorke, 1986a). If variation between sites is significant, a management model based on data from some sites would be appropriate for these sites only. Similar concerns apply to variation over time as growth rates might be measured in an unusual year. Variation over time may affect some of the methods used to estimate growth rates, and individual variation in growth may also bias the estimates of growth parameters obtained (Sainsbury, 1980). These problems illustrate that not only average growth parameters but also the variation in these parameters must be known before reliable management models can be built.

In this chapter we review the problems associated with measuring growth rates of abalone and the factors that produce variation in growth rates. While this review is not exhaustive, we have attempted to document growth in the field of as many species as possible. We examine the methods used to measure growth, the form and the parameters of growth reported and especially the sources of variation in growth noted in the literature. The field studies point to a number of causes of variation in growth, and many laboratory studies have investigated the way in which environmental conditions affect growth (summarized by Hahn, 1989). Both lines of evidence suggest food is important. We have therefore critically assessed the methods used in the laboratory studies that investigated the effect of diet on growth in the light of our study of *H. rubra* in Victoria, Australia.

FIELD STUDIES

Table 12.1 summarizes published studies on growth rates in the field for 23 species of *Haliotis*. Growth of seeded juveniles is not included, and unfortunately we could not translate many Japanese papers. The parameter t_0 of von Bertalanffy (VB) growth has been omitted, as it is often not given or assumed to be zero, and is not very useful in comparing growth rates. Growth rates cannot be compared on the basis of K or L_∞ alone, as the estimates are correlated (Moreau, 1987). Ideally, confidence boundaries for K and L_∞ should be presented, on a graph of K versus L_∞ (Kimura, 1980). Kimura also provides a test for significant differences in VB growth. There are good reasons to use alternative parameters to describe VB growth (Schnute & Fournier, 1980; Ratkowsky, 1986; Francis, 1988) or the Gompertz model instead (see below). However, almost all studies fitted these

Table 12.1 Studies of abalone growth.

Species	Author	Sites	Method	Model	Parameters	Variation
H. australis	Poore, 1972b	1	GMI, T	von B	$K = 0.32$ $L_\infty = 87$	I
H. corrugata	Burge *et al.*, 1975	4	T	none?	$L_\infty = 155–169$	L
	Tutschulte & Connell, 1988	1	CS, T	linear	16 mm y^{-1}	S D
				von B (>70 mm)	$K = 0.09–0.11$ $L_\infty = 169–202$	not Y L G
	Guzmán del Próo (Chapter 24, this volume) (review table)	4	CS	von B	Male: $K = 0.15–0.36$ $L_\infty = 163–210$ Female: $K = 0.23–0.41$ $L_\infty = 159–186$	
		1	CS	von B	$K = 0.98$ $L_\infty = 119$	
H. cracherodii	Leighton & Bolootian, 1963	2	CF, T	none	26–30 mm y^{-1} for 1–2 y	L S
	Wright, 1975	1	T	none	$L_\infty \approx 125$	I
	Douros, 1985	2	T	none*	$K = 0.37$ $L_\infty = 117$ $K = 0.47$ $L_\infty = 139$	L
H. discus discus	Todoufuken . . . , 1972 (in Kojima *et al.*, 1977)	3	CF GMI GMI	von B	$K = 0.26$ $L_\infty = 177$ $K = 0.02$ $L_\infty = 744$ $K = 0.08$ $L_\infty = 319$	L
	Kojima *et al.*, 1977	2	GMI	von B	$K = 0.22$ $L_\infty = 188$ $K = 0.27$ $L_\infty = 167$	L I
H. discus hannai	Sakai, 1962a,b (in Kojima *et al.*, 1977)	4	GMI	von B	$K = 0.21$ $L_\infty = 139$ $K = 0.12$ $L_\infty = 237$ $K = 0.07$ $L_\infty = 369$ $K = 0.24$ $L_\infty = 134$	L Y
H. diversicolor aquatilis	Kim & Chung, 1985	1	GMI, CF, T	von B	$K = 0.21$ $L_\infty = 78$	S
H. d. diversicolor	Kim & Chung, 1985	1	GMI, CF, T	von B	$K = 0.18$ $L_\infty = 89$	S

Table 12.1 *Continued*

Species	Author	Sites	Method	Model	Parameters	Variation
H. d. supertexta	Oba *et al.*, 1968	1	CF	inc	22–28 mm at 1 y	S
	Peon, 1980	1	GMA	none	39 mm at 1 y	
	(in Chen, 1984)				54 mm at 2 y	
H. fulgens	Burge *et al.*, 1975	4	T	none?	$L_\infty = 161–169$	L
	Tutschulte, 1976;	1	CS, T	linear	15 mm y^{-1}	S D
	Tutschulte & Connell, 1988			von B	$K = 0.09–0.10$	
				(>90 mm)	$L_\infty = 179–205$	
	Guzmán del Próo (Chapter 24, this	1	T	von B	$K = 0.34$ $L_\infty = 189$	
	volume) (review table)	1	CF	von B	$K = 0.39$ $L_\infty = 126$	
		1	CF	von B	Male:	G
					$K = 0.38$ $L_\infty = 171$	
					Female:	
					$K = 0.37$ $L_\infty = 170$	
	Shepherd *et al.* 1991	1	T	von B	$K = 0.38$ $L_\infty = 183$	Not G
H. gigantea	Kurogane *et al.*, 1974	1	T	Gomp	$a = 0.64$ $L_\infty = 193$	
				von B	Length Inc:	
				(>70 mm)	$K = 0.41$ $L_\infty = 205$	
					Age length:	
					$K = 0.35$ $L_\infty = 221$	
H. iris	Poore, 1972b	2P	CF, T	inc	21 mm at 1 y	I L S Y
				(<50 mm)	51 mm at 2 y	
				von B	$K = 0.31$ $L_\infty = 146$	
	Sainsbury, 1982	1	CF, T	von B	$K = 0.17$ $L_\infty = 130$	I L S
				von B_I	$K = 0.16$ $L_\infty = 132$	
H. kamtschatkana	Quayle, 1971	3	T	von B	$K = 0.20$ $L_\infty = 137$	'Not L'
	(in Breen, 1986)				$K = 0.23$ $L_\infty = 130$	
					$K = 0.32$ $L_\infty = 124$	Y*
	Schnute & Fournier, 1980	13P	CS, T	von B	$K = 0.22$ $L_\infty = 133$	H
	Fournier & Breen, 1983	2	CS, T	von B	$K = 0.20$ $L_\infty = 92$	H
					$K = 0.25$ $L_\infty = 108$	

	Breen, 1986	6	T	von B	$K = 0.16$ $L_\infty = 123$ $K = 0.20$ $L_\infty = 95$ $K = 0.24$ $L_\infty = 114$ $K = 0.27$ $L_\infty = 124$ $K = 0.35$ $L_\infty = 114$ $K = 0.51$ $L_\infty = 101$	L Y*
H. lamellosa	Bolognari, 1953	1	CS, GMA	none	?14 mm at 1 y max L ≈ 70 mm	G
H. laevigata	Shepherd & Hearn, 1983	3	T, CF	von B_S	$K = 0.41$ $L_\infty = 131$ $K = 0.48$ $L_\infty = 138$ $K = 0.60$ $L_\infty = 148$	S G
	Shepherd, 1988	1	CF, T	linear (<60 mm)	20 mm y^{-1}	Not S
H. mariae	Sanders, 1982	18P	GMI, CS	von B	$K = 0.75$ $L_\infty = 119$	Not G
H. midae	Newman, 1968	1	T, CF	linear (<57 mm) von B	25 mm y^{-1} $K^* = 0.06$ $L_\infty = 224$	S
H. roei	Keesing & Wells, 1989	1	T	von B_S	Intertidal: $K = 0.76$ $L_\infty = 80$ Subtidal: $K = 0.64$ $L_\infty = 86$	H
H. rubra	Harrison & Grant, 1971	2	T	von B	Protected: $K = 0.22$ $L_\infty = 160$ Exposed: $K = 0.12$ $L_\infty = 175$	L
	Hamer, 1980	13	T	von B	$K = 0.24–0.35$ $L_\infty = 141–115$	L
	Shepherd & Hearn, 1983	3	T, CF	von B_S	$K = 0.32$ $L_\infty = 143$ $K = 0.34$ $L_\infty = 139$ $K = 0.41$ $L_\infty = 144$	L S
	McShane *et al.*, 1988b	3	T	von B	$K = 0.27$ $L_\infty = 133$ $K = 0.29$ $L_\infty = 117$ $K = 0.35$ $L_\infty = 121$	L

Table 12.1 *Continued*

Species	*Author*	*Sites*	*Method*	*Model*	*Parameters*	*Variation*
	Prince *et al.*, 1988a,b	1	CF, T	linear (>80 mm)	21 mm/y	
				von B	$K = 0.29$ $L_\infty = 140$	
	Prince, 1989	1	T	von B	$K = 0.47$ $L_\infty = 155$	
			A	Gomp	$a = 0.50$ $L_\infty = 155$	
	Nash *et al.* (in prep.)	5	A	von B	$K = 0.13–0.17$ $L_\infty = 127–146$	L
	Nash (Chapter 23, this volume)	6	A	Gomp	$a = 0.23–0.31$ $L_\infty = 124–173$	L
H. rufescens	Cox, 1962	3	T	none	raw data only	
	P Haaker, unpublished (in Tegner *et al.*, 1989)	1	CS (T)	von B	$K = 0.27$ $L_\infty = 201$	
	Tegner *et al.* (Chapter 26, this volume)	1	T	von B	$K = 0.27$ $L_\infty = 200$	I
	DeMartini, unpublished *(in Tegner et al.*, Chapter 26, this volume)	5P	T	von B	$K = 0.22$ $L_\infty = 192$	I
H. scalaris	Shepherd *et al.*, 1985, 1987	1	CF (T)	inc linear (<53 mm)	(during y 1) 14 mm y^{-1}	I S Y
H. sorenseni	Tutschulte, 1976; Tutschulte & Connell, 1988	1	CS	linear (<90 mm)	23 mm y^{-1}	
H. tuberculata	Forster, 1967	2	GMI, T	linear (<55 mm)	15–16 mm y^{-1}	H S Y
				von B	Subtidal:	
			T		$K = 0.39$ $L_\infty = 99$	
			GMI		$K = 0.27$ $L_\infty = 108$	
					Intertidal:	
			GMI		$K = 0.29$ $L_\infty = 119$	

	Hayashi, 1980	1	T (CF)	linear (<55 mm)	(Intertidal) >20 mm y^{-1}	H
				von B	$K = 0.45$ $L_\infty = 115$	
	Cochard, 1982	1	GMI	von B	$K = 0.25$ $L_\infty = 105$	H L S
	Clavier & Richard, 1985	2	CF GMA T	von B	$K = 0.37$ $L_\infty = 108$	L Y
				Gomp	$a = 0.53$ $L_\infty = 105$	
	Clavier & Richard, 1986	1	CF	inc	35 mm in 2 y	
H. virginea	Poore, 1972b	2	GMI	von B	$K = 0.32$ $L_\infty = 64$	L
					$K = 0.45$ $L_\infty = 62$	

Abbreviations: *Sites* P: data pooled. *Method* A: shells aged by spire layers. CF: Cohorts from length frequency data followed through time. CS: Growth of cohorts from one 'snapshot' set of data. GMA: length-at-age data from growth checks. GMI: increments in length from growth checks. T: tagging. *Model*. Gomp: Gompertz. Inc. growth rate increases with age. von B: von Bertalanffy. von B_I – allowing individual variation, von B_S – with seasonal variation, none*: von B parameters estimated from figure. *Parameters* a: Gompertz parameter as in Kaufmann (1981). K: Brody coefficient in y^{-1}. K*: based on shell breadth. L_∞ in mm. Reported variation in growth: D: depth. G: gender. H: habitat. I: individual variation. L: locality. S: seasons. Y: years. Y*: 2 estimates at the same site in different years.

parameters of VB growth. L_∞ values in the report by Kojima *et al.* (1977) for *H. discus discus* and *H. d. hannai* of 774 (corrected here to 744), 319 and 369 mm, associated with low values of K (Table 12.1), were queried by Kojima (pers. comm.). They were derived from previously published length-at-age data in which the average sizes of the oldest year classes were all less than 140 mm. Knight (1968) stresses that in such cases the K and L_∞ values are summaries of the data without biological meaning. The data may reflect false marks, omitted marks or no old animals in the samples (Kojima, pers. comm.).

Length-at-age data

Use of growth marks

In a few species of abalone, marks representing changes in calcification at the perimeter of the shell have been found. These allow the length at each mark to be measured and, if the period between marks is known, growth parameters can be estimated (see Table 12.1). This method has many advantages, as growth over a number of years can be measured without a long and costly field programme. However, Beamish and McFarlane (1983) emphasize that age marks should be validated for each population, and the validation must extend to the older age classes. While a number of studies have validated growth marks in some species, in other cases (e.g. Bolognari, 1953; Sanders, 1982) no verification was undertaken. Two important questions are (1) whether the marks occur consistently once every year and (2) whether the marks occur at the same time each year, so that the length increments between them consistently represent one year of growth.

Forster (1967) used three methods to address the first question in *H. tuberculata*. As a preliminary check, frequency histograms were used to show that the lengths at each growth mark formed separate cohorts. He checked for an extra mark in individuals recaptured a year after tagging, and compared Manzer and Taylor plots of length increments between growth marks with plots from tagging data. Growth marks for the year were absent in a few of the juveniles tagged, some of which showed no growth over the year. Many of the large abalone had not grown enough to identify a growth mark. Of those that had, most had produced one growth mark during the year, but 7% had no growth mark, while another 7% had produced two growth marks in a year. Forster also found that the disturbance of tagging could cause extra growth marks, and Cochard (1982) found these were frequently difficult to distinguish from annual marks. The lack of identifiable marks when no growth occurred would bias the estimated growth rate upward, especially for large animals, and Forster obtained a higher L_∞ and a lower K using growth marks than by using tagging data (Table 12.1).

Sakai (1960) reported that in shells of *H. discus hannai* the annual growth mark is characterized by a discontinuity in the nacre layer associated with external ridges, so that it can be distinguished from disturbance marks in sections through the shell (see photographs in Hayashi, 1955; Sakai, 1960). Using caged abalone, colour coded by the food supplied, he found an annual mark was formed in the same month in each abalone except one, in which it was formed in the following month. The timing of the marks corresponded to gonad maturation and spawning. In nature the most readily eaten brown algae wither just before the breeding period, resulting in a change in shell colour. Sakai recorded the colour around the second, third and fourth growth marks. Most occurred during or after the change in shell colour, but in a few the mark was in other colour bands, or its position could not be determined. Thus the timing of growth marks may vary occasionally, and some may not be visible.

Poore (1972b) sectioned shells of *H. australis* tagged one year previously. Growth marks in the nacre layer, like those in *H. discus hannai*, occurred in shells ≥40–50 mm. In most shells, marks appeared to have formed just before tagging and again a year later, but a few had not grown, and one showed an extra mark corresponding to a previous recapture. Poore could not compare increments between marks to tagging data as too few of the large animals tagged had grown by a measurable amount. Kojima (1975) collected *H. discus discus* at different times of year and measured the increment from the last growth mark (visible through the shell) to the edge of the shell. He found the increments increased during the year until the gonad maturation period in October–November. Individuals varied somewhat in the time when they formed new growth marks, in line with their stage of gonadal development.

These studies showed that growth marks in these species are approximately annual, but some missing or undetectable marks occur, especially when growth is slow, and will bias growth estimates. In particular, as older animals show little growth their age may be underestimated. Disturbance during tagging produced extra marks, both externally and in the nacre layer, so that extra growth marks might be produced in some years by natural disturbances such as storms. There was variation between animals in the timing of the growth marks and the timing may also vary between years. This would introduce extraneous variation in apparent growth if many marks on each shell are measured, and may bias results where one or two marks on each shell are measured.

While the growth mark in the nacre layer corresponded to gonad maturation and spawning in *H. discus hannai* (Sakai, 1960) and *H. d. discus* (Kojima, 1975), it corresponded to late autumn or early winter for *H. australis* (Poore, 1972b). A false mark occurs in *H. d. discus* where winter temperatures fall below 10°C (Kojima, pers. comm.). Kim & Chung (1985) found external marks in *H. diversicolor aquatilis* and *H. d. diversicolor* occurred in winter, not in the summer spawning period. Forster (1967) found the external marks in *H. tuberculata* corresponded to winter in juveniles, and to both winter and the spring spawning in adults.

A third important question is whether the previous sizes recorded using growth marks are a representative sample of the sizes of younger abalone at that time. As a result of size-selective sampling, or size-selective natural or fishing mortality, the mean sizes at previous ages, as estimated from growth marks on older animals, may be biased. This is 'Lee's phenomenon', well known for estimates based on rings on fish scales (Ricker, 1969; Gutreuter, 1987). Size-selective sampling or mortality may shift the estimated mean size of a year class without skewing the distribution (Ricker, 1969). The bias is severe when early growth marks on old animals are used, but may occur with any age-length data (Gutreuter, 1987).

The only study we have found addressing this question is by Kojima *et al.* (1977), who analysed commercial catches of *H. discus discus*. They found the mean length of 2-year-old animals was larger than the mean length at the second annual growth mark of older animals. Recruitment to the fishery starts at the 2+ year class, so that strong size-selective mortality would occur at this stage. Kojima *et al.* assumed all the size-selective mortality was due to fishing, and adjusted the mean lengths at each age using the age composition of the catch. Because fishing mortality will always be strongly size selective where size limits are used, samples for growth-mark data should ideally be taken from unfished areas, or from cohorts that have not begun to reach harvestable size. Natural mortality may be higher for smaller abalone (see Chapter 21) and if so, faster growing juveniles would be over-represented in older year classes. It would be instructive to estimate the strength of such size selection if several unfished cohorts of a species with annual growth marks could be sampled.

Growth marks can be used to determine the age of each shell, to provide length-at-age data, if the age when the first mark is formed is known. Forster's (1967) size-frequency data suggest that this is in the first year for *H. tuberculata*. Saito (1981) showed, by following cohorts, that *H. discus hannai* forms the first mark in December, at 13–14 months old, although after maturity in the second year (Sakai, 1962b), the marks occur in August to September (Sakai, 1960). Kojima (1975) showed, by rearing *H. d. discus*, that marks begin during the second year. Poore (1972b) was unable to age *H. australis*, as the age at which the first mark forms is not known.

Length-at-age data from growth marks may be used to calculate the mean lengths of year classes, or the individual lengths may be analysed. The year classes or individuals may be followed over the history of their growth (longitudinal data), or lengths at one time of a number of year classes or individuals may be obtained (cross-sectional data). Longitudinal data obtained from growth marks in shells from a single collection will be sensitive to bias, due to Lee's phenomenon (Ricker, 1969; Gutreuter, 1987), variation in the timing of marks and omitted or extra growth marks. Furthermore, use of repeated measurements from the same individuals leads to the problem of non-independence of errors in analyses (Day & Quinn, 1989). Cross-sectional length-at-age data would be less affected by the first set of problems, but Gulland (1964) points out that, in contrast to data for one cohort over time or the length increments of all animals over a fixed period, such data do not correspond with the growth of any animals at any time. Thus variation in growth between years or between year classes cannot be assessed. Ideally, length-at-age data should be obtained from shells collected over a number of years, to follow cohorts during their growth.

Use of shell layers

Muñoz Lopez (1976) estimated the age of abalone shells by grinding down the tip of the spire, through successive alternate layers of dark 'conchiolin' and white aragonite nacre. He found that the darker layer was deposited on the inner surface of shells of *H. corrugata*, *H. fulgens* and *H. rufescens* in winter, while white aragonite was formed in summer. He related the number of layers to shell length in *H. corrugata*, but no growth data from the area of collection were presented for comparison. Hayashi (1955, Plate II Fig. 5) illustrates similar layers in a section through the spire of *H. gigantea*. Prince *et al.* (1988b) related the spire layers to growth data for *H. rubra* at one site in southern Tasmania. After the first three 'minor' layers, the layers appeared to be annual and to correspond with the period of coldest water. As older animals do not grow appreciably, relating predicted sizes at each age to the number of layers does not verify the method for old animals. Francis (1988) also questions the use of tagging (length-increment) data to validate methods to determine ages (see below). Verification is also required for each area studied (Beamish & McFarlane, 1983). Prince *et al.* (1988b) noted that the method could not be applied to *H. rubra* from New South Wales, where water temperatures are higher.

Muñoz Lopez (1976), Prince *et al.* (1988b) and Nash *et al.* (in prep.) were unable to count nacre layers on shells with boring organisms in the spire. The layers are obliterated by the boreholes, and extra layers of conchiolin and nacre are deposited around the holes (Prince *et al.*, 1988b; R.W. Day, pers. obs.). Nash *et al.* (in prep.) found that between 53 and 67% of the shells could not be aged. More of the larger animals were severely bored, presumably because they were older. If so, more young than old animals may be sampled in the large size categories, which would lead to an overestimate of growth. Also, if boring animals infest slow growing abalone more often, or slow their growth, the growth parameters obtained will be seriously biased. Kojima & Imajima (1982) report that *H. diversicolor aquatilis* with many boring polychaetes had significantly lower body weights for their size, but Clavier (1992) found no difference for *H. tuberculata*. A higher proportion of bored shells was found in slower growing populations of *H. rubra* (Nash *et al.* in prep.) and *H. tuberculata* (Forster, 1967; Cochard, 1982).

Use of length-frequency data

The usual method used to obtain length-at-age data is the analysis of length-frequency data to distinguish cohorts. As the growth rates of juvenile abalone are very variable, even among siblings in the laboratory (Leighton, 1974; Momma, 1980), the cohorts that arise from bursts of settlement merge as they grow (see Day & Leorke, 1986a; Shepherd, 1988). Cohorts that arise close together during one spawning season will merge very rapidly and would probably be interpreted as one year class, although the distribution of lengths within this group might be far from normal. Such variation in settlement dates and growth rates of individuals will contribute to poor resolution of cohorts with methods based on normally distributed cohorts. Separation of cohorts could also be confounded by growth differences between the sexes (Hearn, 1986).

Many computer programs that obtain maximum likelihood estimates of the mean lengths of component cohorts have been described, most recently by MacDonald & Green (1985) and Fournier *et al.* (1990). These have replaced the graphical method of Cassie (1954). By optimizing the fit of distributions, the routines indicate when the data are inadequate by failing to converge, or by giving very large standard errors, whereas graphical methods do not provide these warnings (Grant *et al.*, 1987). The computer methods can be made more useful (produce less ambiguous results) by including biological assumptions, such as the form of the growth curve, as constraints to force a biologically reasonable fit (Schnute & Fournier, 1980; MacDonald, 1987). However, the result may depend on whether the biological assumptions are correct. As the growth of juvenile abalone has often been reported to be linear or increasing (Table 12.1), not declining as in VB growth, linear or Gompertz growth assumptions may be most appropriate.

Schnute & Fournier (1980) suggest that the relationship between size or age and the variation in size within cohorts is an important constraint in fitting cohorts. Their analysis of *H. kamtschatkana* data from Breen & Adkins (1979) suggested that the standard deviations in length of cohorts might be linearly related to the mean length of the cohort. In contrast, Fournier & Breen (1983) assumed that the variance would be related to the cohort age. In the strong isolated cohorts of *H. rubra*, followed in the field by Day & Leorke (1986a), variation within cohorts can be seen unobscured by overlapping cohorts. Regressions of the standard deviation of lengths against the mean length of cohorts (Fig. 12.1) provided the best fit to the data. When variances were regressed against age, the regressions explained less variation for both cohorts, and the deviations indicated non-linearity: most of the data fell below the line in the centre and above the line at the extremes. The relation between the standard deviation and age appeared linear for each cohort, but the slopes of the lines for the two cohorts differed. The spring cohort grew faster initially, and the standard deviation increased more rapidly than for the late summer cohort. In contrast, the slopes of standard deviation against mean length were similar for the two cohorts (Fig. 12.1). Although these data cover only the first two years of growth, they suggest that the coefficient of variation of length remains constant, as Schnute & Fournier (1980) suggested.

The above discussion of possible biases of cross-sectional data from growth marks also applies to data derived from shell layers or from analysis of cohorts. Two issues in particular are important. First, if size selective sampling or mortality has affected the cohorts for which age and mean size are determined, the means will not reflect true growth rates. This would occur when age and mean size are determined for older animals that have been subject to size-limited fishing. Second, a

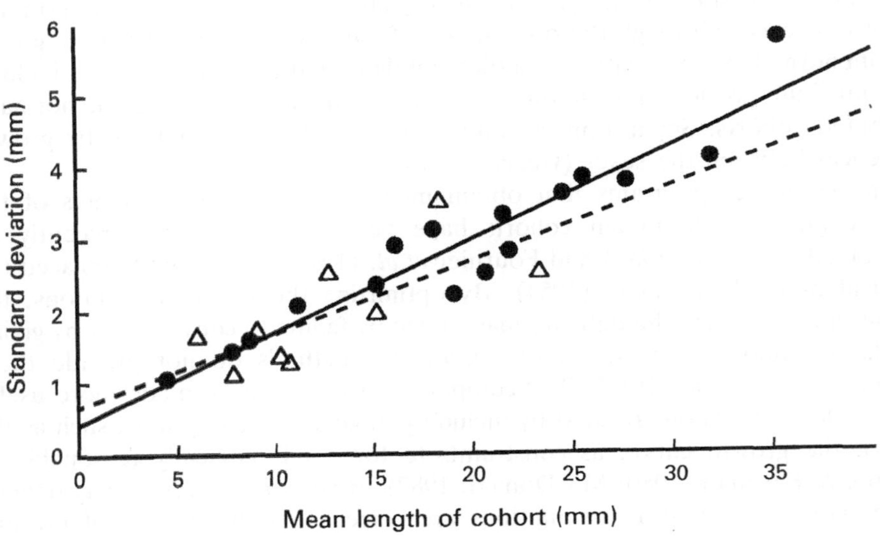

Fig. 12.1 The relationship between the standard deviation of the maximum length of juvenile abalone in two cohorts and the mean length of the cohorts. Circles and solid line: cohort of abalone born in early spring. Triangles and broken line: cohort of abalone born in late summer.

'snapshot' analysis of length frequencies as representing yearly cohorts assumes recruitment occurs every year at the same time and that the growth parameters do not vary between years. There are many reports of growth variation between years (Table 12.1) and variable recruitment between years (e.g. Poore, 1972b; Hayashi, 1980; Forster *et al.*, 1982; Sainsbury, 1982; Shepherd *et al.*, 1987; Tegner *et al.*, 1989), and strong recruitment may occur twice in one breeding season, producing two cohorts for one year (Day & Leorke, 1986a). Sexes may grow at different rates (Table 12.1), forming two overlapping cohorts. Where possible, the growth of cohorts should be followed over time or tagging should be used to verify the results (e.g. Shepherd & Hearn, 1983). A time sequence of length–frequency distributions will produce a more reliable separation of cohorts (MacDonald, 1987). Fournier *et al.* (1990) describe a programme to analyse such data, and claim that with long time series the presence of size selectivity in the recruitment of the first age class can be detected. However, the method still assumes that there is one cohort per year and that growth parameters are the same for all cohorts.

Length–increment data

Increments in length may be measured using growth marks, calculated for year classes by analysis of length–frequency distributions or measured in tagging studies. The increments in length between early growth marks on old shells may be biased due to Lee's phenomenon, as discussed for length-at-age data above. Increments between year classes should be based on longitudinal data, as for length-at-age data. Tagging, which is one of the most commonly used methods to estimate growth (Table 12.1), involves several assumptions that appear to have been seldom tested.

(1) The tagging process must not affect growth. Removal of abalone from the water, possibly even from the substratum, may affect growth. Sakai (1960), Forster (1967) and Poore (1972b) reported growth marks on the shell corresponding to previous handling. Shepherd & Hearn (1983) noted that abalone tagged in hot weather grew significantly less than others of the same size, tagged at other times. Long periods between marking and recapture will reduce this problem and tagging *in situ* is recommended (Sainsbury, 1982; McShane *et al.*, 1988b; Shepherd, 1988). Day & Leorke (1986b) left tagged abalone on caged rocks for 1 to 2 weeks and measured the initial size when the rocks were moved out of the cages onto the reef.

(2) The presence of the tag must not influence growth. McShane (1988) has reviewed the methods to tag abalone. Tags in pores (Cox, 1962; Poore, 1972b; Prince *et al.* 1988b) or notches in the shell edge (Leighton & Boolootian, 1963; Tutschulte & Connell, 1988) are likely to be more intrusive than labels glued to the outside of the shell. Wire tags threaded between two respiratory pores appear to reduce growth in *H. rubra* (Harrison & Grant, 1971; McShane *et al.*, 1988b). Tests are needed to determine whether the nylon rivets in pores described by Prince (1989, 1991) affect growth.

(3) The animals tagged and recaptured must be a random subset of the population. For example, if abalone that are cryptic or move further both grow more slowly and are less often recaptured, or if slow or fast growing animals suffer higher mortality during the tagging period, then the results will be biased. Hearn (1986) simulated the effects of size-limited fishing on tagging data and showed that growth appeared to change at the minimum harvestable size and that the effect is greatest when growth is very variable.

(4) As growth often varies seasonally (Table 12.1), the periods between tagging and recapture should be whole years if a standard growth equation is to be fitted. If the periods vary, the standard growth models should be modified as in Shepherd & Hearn (1983) and Appeldoorn (1987) to estimate seasonal parameters from tagging data. These modifications assume the seasonal variation in growth decreases linearly as the length approaches L_∞, so that the ratio of increments in any two seasons is the same for all lengths, but this assumption appears doubtful (Shepherd & Hearn, 1983; Hearn, 1986).

Consequences of individual variation

Many studies reported variation between individuals (Table 12.1). For example, Kojima *et al.* (1977) calculated K for individuals from growth marks in three areas. The ranges of K were 0.20–0.38, 0.25–0.49 and 0.30–0.66. This represents extremely variable growth. Francis (1988) pointed out that if endogenous individual variation is present, models fitted to length-at-age data and length-increment data are not the same. The VB length-at-age model predicts the mean length at a given age, whereas fitting the standard VB model to length-increment data predicts the mean increment at a given length. In the former model, L_∞ is the length the average individual would theoretically approach if it lived for a long time. Some might grow larger than L_∞ within their lifespan. In the latter model, L_∞ is in theory the length at which the mean increment is zero, or the maximum possible length of individuals. Francis (1988) notes that the fitted L_∞ is often below the maximum size observed and suggests this shows the model is inappropriate. The fitting procedure, which assumes a constant distribution around the mean increment at each length, is also likely to be inappropriate. When models that allow L_∞ to vary are used (Sainsbury, 1980; Kirkwood & Somers, 1984; Hearn, 1986), the mean L_∞ becomes the average size at which individuals are predicted to cease growing, that is the average maximum length individuals approach, which corresponds to the length-at-age model. The meaning of K is related to the meaning of L_∞ in each case. The fitted parameters summarize the data, and biological interpretations should not be extended beyond the range of the data (Knight, 1968).

As growth of abalone is usually measured in order to predict the length at each age for management models, length-at-age data should be fitted to provide age-based model predictions where possible. But there is an argument for the development of length-based management models if life-history parameters are more closely related to length than age, or known only in terms of length (Gutreuter, 1987). Differences between the sexes should also be considered. For example, if females grow larger, management based on average parameters may deplete female breeding stocks (Shepherd & Hearn, 1983; Hearn, 1986).

Tagging can usually only produce length-increment data for individuals, whereas length-frequency analysis only provides the mean increments in length for cohorts. Growth marks can be used to provide either. Both types of length-increment data are routinely used to estimate VB parameters, but Hancock (1965) stresses that they may produce different results, because smaller members of a cohort may grow faster (see also Ford, 1933). Hancock's Manzer and Taylor plots of cohorts of cockles showed that this 'compensatory growth' is less than the variation in growth between cohort means, so that smaller individuals in a cohort would remain smaller than their cohort fellows, and eventually reach different maximum sizes. The figures reported by Kojima *et al.* (1977) suggest that they found the same phenomenon for *H. discus discus*, but we have no translation of their paper. The 'compensatory growth' effect will bias the analysis of increments in length of individuals if the sampled individuals are largely from a few cohorts or from cohorts with closely spaced mean sizes and large standard deviations (e.g. older cohorts). The mean growth at each length would then largely reflect the differences in growth rate between large and small individuals within the same or similar cohorts. Many of the tagging studies in Table 12.1 were restricted to older abalone and may suffer from this problem. Analysis of the mean increment in length of cohorts (where possible) avoids the problems associated with individual variation.

Francis (1988) noted that results obtained using length-increment data (growth at each length) will be affected by mortality, as the individuals that have reached a given length will include young, fast growing animals and older animals that grew slowly. The proportion of older animals will depend on the rate of mortality. He demonstrated that even if the mean length at each age follows VB growth, the mean growth at each length will not fit a VB model. The mean age-based and length-based growth rates may be very different.

Sainsbury (1980) estimated the bias of VB parameters obtained by standard methods by fitting a model that included individual variation in the parameters, assuming that variation in K and L_∞ was independent. The bias is more serious for length-increment than for length-at-age data because the lengths at each age are constructed from the estimated parameters when length-increment data are analysed. Sainsbury showed that for length-at-age data the mean K will be underestimated using standard methods. Variation in the time of settlement of individuals in a year class or errors in aging individuals will increase this effect. Standard methods to fit the models to data also assume that the distribution of individual lengths has the same shape and spread at each age. As individuals vary in growth rate, the variance in length would increase with age, unless effects such as compensatory growth or size selective mortality are important, or if the coefficient of variation for K is large, in which case the variance may decrease slightly for the oldest abalone (Sainsbury, 1980). Therefore, the results will normally be sensitive to bias or chance variation in the data for older abalone, and relatively insensitive to such errors for the youngest animals.

Sainsbury (1980) and Hearn (1986) simulated tagging data with variation in K only, L_∞ only, independent variation in both and linked variation in both. Hearn found the estimates were always worse for constant tagging periods, especially over several years, than for variable periods. The value of K was always underestimated for tagging periods of 1 to 2 years, and L_∞ was overestimated, except for the case with linked variation in K and L_∞. Sainsbury (1980) found the predicted lengths for some ages may be much larger than they should be, because K is underestimated. *Haliotis iris* had a mean K of 0.16 and a coefficient of variation for K of 0.67, and the Fabens (1965) method underestimated the average length by 13% for some older year classes. For larger K (near 0.4), Sainsbury showed the largest bias will be for intermediate ages, whereas for smaller K (near 0.2) the bias increases with age. The size of the bias varies with the coefficient of variation of K.

Sainsbury (1980), Hearn (1986) and Francis (1988) considered inherent differences between individuals. However, Francis notes that if all the variation in growth is a result of environmental effects, so that past and future growth is not correlated, there will be no difference between the predicted growth based on age or length (assuming mortality does not depend on size or growth rate). It is clear from laboratory experiments that some variation in growth rates is endogenous. For example, Momma (1980) demonstrated that juveniles of different ages, but the same sizes, maintained the same relative growth rates. Morse (1984) suggested that variability in growth among siblings was a result of variation in growth hormone production. However, in the field, chance differences between individuals in food supply and other environmental factors are likely. It is important to determine how much variation in growth in the field is due to such chance differences and how much is inherent.

Day and Leorke (1986b) recorded the increments in length of *H. rubra* at approximately 3 month intervals over 18 months, using tags fixed to the shells. Abalone were measured *in situ* with minimum disturbance, and independent measurements on the same day established that measurement errors were small. If Ford-Walford plot regressions are used to predict the increment for each initial length in each season, the residuals from the regressions of the subset of abalone that were measured in several seasons may be used to measure correlations in growth between seasons (Table 12.2). All the correlations are very low, suggesting that endogenous variation plays a small part in the differences between individual growth rates over short periods. There was also little correlation between growth deviations in the autumns of two successive years. Growth was most variable in summer, when the mean growth was fastest. Similar observations on *H. iris* were interpreted by Sainsbury (1982) as indicating differences in the ability of individuals

Table 12.2 Correlations in growth between seasons: Pearson correlation coefficients for deviations from predicted growth of *H. rubra* measured repeatedly over 3-month periods at Point Cook, Victoria; numbers in brackets are sample sizes; variances of growth deviations are shown for the 19 abalone measured after all five periods.

	Variance	Autumn	Winter	Spring	Summer
Autumn 1983	2.47				
Winter 1983	0.99	0.17 (41)			
Spring 1983	1.56	0.12 (41)	0.15 (43)		
Summer 1983/4	4.36	0.30 (19)	0.23 (21)	−0.11 (22)	
Autumn 1984	3.33	0.18 (19)	0.16 (21)	0.01 (22)	0.20 (22)

to take advantage of good conditions for growth, but conditions may simply favour some animals by chance and not others. An example would be a patchy supply of nutritious drift algae in summer. Shepherd & Hearn (1983) calculated K values of 1.2 to -0.2 from *H. laevigata* recaptured repeatedly over years. As growth with zero or negative K values cannot persist, exogenous variation must be important unless these values are entirely due to measurement errors.

Other sources of variation

Seasonal variation in growth is very common (Table 12.1), but the underlying causes are usually confounded. For example, Keesing & Wells (1989) reported that the growth of *H. roei* is fastest in spring when water temperatures are rising, spawning is completed and food is abundant. Newman (1968) argued that because algal food did not vary seasonally, and growth was slow during the summer spawning season, seasonal growth in *H. midae* was influenced by spawning. But growth was slow in both summer and autumn, and Barkai & Griffiths (1986) found that the diet of *H. midae* at the same site included more preferred algal species in winter, so that food availability may be important here too. Laboratory evidence suggests all three factors are important. Low temperature slows growth in the laboratory (Chen, 1984). The growth of *H. discus hannai* is slowed in the laboratory during gonad maturation, at a time when temperatures are high (Sakai, 1960, 1962b). Ino (1952) found both low growth and reduced feeding during the gonad maturation period for adult *H. discus hannai*, but not for juveniles. Many laboratory studies have documented the importance of food to growth (Hahn, 1989).

The availability of food may override the effect of the reproductive cycle. Shepherd & Hearn (1983) found that the period of maximum growth of *H. laevigata* and *H. rubra* was one or two months after the period of maximum food supply, except at a site where food availability did not vary much. At this site, growth was highest in the non-reproductive period. Furthermore, the timing of both growth and reproduction may be an evolutionary response to the seasonal availability of food. For example, growth of *H. discus hannai* in the field varies with the abundance of the brown algae *Undaria* and *Laminaria* (Sakai, 1962a; Hirose, 1974). These preferred algae become scarce just before the spawning season when growth slows (Sakai, 1960).

Variation in growth between years is reported less often (Table 12.1), probably because many studies cover short periods. Shepherd & Hearn (1983) found no variation between years for *H. laevigata*. Sakai (1962a) concluded that annual variations in growth were related to changes in the supply of algae caused by the Oyashio current, rather than temperature. Decreased growth due to a lack of preferred algae in El Niño years has also been reported in California (Cox, 1962; Olsen, 1968). Tegner & Dayton (1987) and Dayton & Tegner (1989) found the 1982–4 El Niño suppressed supplies of *Macrocystis* and thus gonadal development in *H. fulgens*.

Table 12.1 shows that growth parameters are very variable between localities for many species. The most commonly suggested explanation for these differences is the supply of food (Leighton & Boolootian, 1963; Olsen, 1968; Hirose, 1974).

Shepherd & Hearn (1983) found that growth at various sites was correlated with the mean index of crop fullness at those sites. Often differences in growth between areas are correlated with the supply of drift algae. For example, Tutschulte & Connell (1988) found the growth of *H. corrugata* and *H. fulgens* was faster in shallow water where there was more drift. Poore (1972a,b) concluded that drift algae are the main food of *H. iris* and suggested that the composition of the drift, which varies between areas, affects the diet and growth. In general, preferred species give the fastest growth in laboratory feeding experiments (Ino, 1952; Sakai, 1962a; Kikuchi *et al.* 1967; Poore, 1972b) and Poore found the fastest growth in areas with more preferred algae.

Many authors speculate that the supply of drift, and thus growth, is higher in areas more exposed to wave action (Harrison & Grant, 1971; Cochard, 1982; Sainsbury, 1982; McShane *et al.*, 1988b). However, Shepherd (1973) suggests there is an optimum degree of water movement for feeding by *H. laevigata*. Breen (1980) suggested that growth of *H. kamtschatkana* in a wide range of habitats was related to the supply of food, but growth was lowest in the most exposed habitats. However, these habitats were dominated by either sea urchins or the unpalatable kelp *Pterygophora*, so that competition for food from urchins, or the species composition of the drift, may be important. In some areas there may be an absolute shortage of suitable algae, rather than a relative shortage of drift algae supplied to abalone by wave action.

Transplants between areas suggest that differences in growth are environmentally determined, but these studies all lack proper controls. Ino (1952) measured the growth in two areas ($n = 2$ and 4) of *H. discus hannai* transplanted as 'seed' from an area of slow growth in 1943, and Sakai (1962b) compared these with a control transplant conducted in 1957 ($n = 59$). The average growth of the transplants was greater than for the control, but no test is possible. Breen (1986) transplanted *H. kamtschatkana* from a moderately exposed area where the growth parameters were previously $K = 0.2$ and $L_\infty = 92$ mm to a sheltered area with *Macrocystis* beds. After one year, K was 0.27 and L_∞ was 124 mm. There were no controls. Henderson *et al.* (1988) transplanted *H. corrugata*, but most were apparently lost to poachers and those recovered had not grown, probably due to an El Niño event. Again, there were no controls. Emmett & Jamieson (1989) transplanted *H. kamtschatkana* collected from a series of headland and bay areas to two sheltered sites with *Macrocystis* 3 to 4 hours later. Only one headland pinnacle was used as a control and abalone were collected, tagged and replaced there within 2 to 3 hours. Over 9 months, growth was significantly faster at each transplant site than at the control site over the range of sizes transplanted. Better designed transplants are needed as well as experiments to show whether growth differences between sites are a result of food supply, as the circumstantial evidence suggests.

The effect of abalone density on growth in the field appears to have received little attention. High stocking densities have been found to reduce growth markedly in culture (Koike *et al.*, 1979; Chen, 1984). Clavier (1982) found maximum size was inversely related to density in *H. tuberculata* at several sites in France. Recruitment of *H. rubra* juveniles on corallines appears to be highest in sheltered habitats and reefs where larval retention may be high (McShane *et al.*, 1988a), and sheltered areas with high densities often have poor growth (W.J. Nash, pers.

comm.), but this may be due, for example, to a reduced food supply in sheltered areas. Breen (1980) found the largest abalone in areas of low density, and the highest densities of abalone below legal size in areas dominated by *Pterygophora*, where growth was slow. But high densities of abalone below harvestable size may simply be the result of slow growth. Douros (1985) reduced the density of *H. cracherodii* at one high-density site, and the removed abalone were transplanted to scraped rock surfaces at a low-density site to increase density. One high-density and one low-density site were used as controls, where abalone were removed and replaced to control for abalone transplantation. Growth was measured from April to December. Although abalone grew significantly faster at the low-density control site than the high-density one (see Table 12.1), no significant differences were detected between the reduced-density or the increased-density site and their respective control sites. However, most of the animals tagged were adults and the gonad volume in June was significantly larger at the reduced-density site than the control. Gonads were not examined at the low-control or increased-density sites. If this difference in gonad volume was due to the density at these sites, then density might be expected to affect the growth of juveniles, in which energy is not shunted to gonad development.

LABORATORY STUDIES

Various environmental parameters have been shown to affect growth. Low temperature reduces growth, at least in part by affecting feeding rates and feeding duration (Uki, 1981) and food absorption efficiency (Peck, 1989). Similarly, photoperiod affects feeding rates and conversion efficiencies (Ebert & Houk, 1984) and therefore growth. Reduced salinity and low oxygen concentration also reduce growth (Chen, 1984). Most laboratory studies have focused on the important effect of diet in determining growth rates (Hahn, 1989).

Since Ino (1952) first conducted his diet quality trials there have been many similar laboratory studies testing the dietary value of algae by offering them singly to abalone and measuring their growth. However, little consideration has been given to the suitability of this technique and its accuracy in testing diet quality. We now assess the methods used to measure the value of various diets for the growth of *Haliotis* in terms of growth indices used, duration of trials, number of animals per treatment and frequency of measurements. Laboratory trials on *H. rubra* are presented to demonstrate the importance of addressing these factors.

Ino (1952) fed *H. discus* single species of algae and measured their value to growth by the 'conversion coefficient' (total weight gain per total weight of food eaten) (Table 12.3). Ten years later, Sakai (1962a) fed groups of six abalone single species of algae for two months, measuring length and weight at the beginning and end of the experiment, and presented the dietary value of each algal species using Ino's index, calling it the 'efficiency of food conversion'. At about the same time, Leighton & Boolootian (1963) proposed the 'algal value quotient' (weight of alga consumed per weight gained by the abalone, i.e., the reciprocal of Ino's index) as a measure of food value. They fed a single group of abalone a series of test algae, changing the species of alga every one or two weeks and measuring the weight and length of the abalone at the time of diet change.

Table 12.3 Major publications on growth using various diets.

Authors	Growth value	Conditioned abalone	Duration of experiment	Measure of growth	Measure of variability	Sample size
Ino, 1952	conversion coefficient index	?	35–152 days	begin & end	no	5–15
Leighton & Boolootian, 1963	algal value quotient	no	1–2 weeks	begin & end	no	6
Kikuchi *et al.*, 1967	*Eisenia* equilavent	cultured	29 days	begin & end	no	?
Poore, 1972a	no index	no	6 months	?	no	?
Koike *et al.*, 1979	efficiency of food conversion	2-month conditioning period	112 days	2 weeks	no	150
Uki, 1981	*Eisenia* equilavent	cultured	30 days	begin & end	no	10
Ebert & Houk, 1984	food conversion efficiency	cultured	1 year	monthly	no	30
Uki *et al.*, 1986b	*Eisenia* equilavent	cultured – all fed *Eisenia*	30–35 days	begin & end	no	10–75
Leighton, 1990	no index	cultured	±140 days	40–50 days	differences between 2 groups	12

Both Sakai (1962a) and Leighton & Boolootian (1963) quoted the theoretical feeding rates required to promote growth for each alga of known dietary value. It is important to consider both feeding rate and dietary value, as both contribute to the growth value of the alga. If an alga of high dietary value is not eaten readily, it will not promote rapid growth rates. Kikuchi *et al.* (1967) introduced the '*Eisenia* equilavent' (growth in shell of *H. discus hannai* fed the test alga per growth in shell when fed *Eisenia*) as a measure that reflects the combined effects of food preference and dietary value. Thus *Eisenia*, which is readily eaten and promotes rapid growth, has a value of 1.00 and all test algae are evaluated in relation to it. This index also enables the standardization of growth values between experiments run at different times. Uki (1981) adopted the '*Eisenia* equilavent' to express the growth value (to both length and weight) of the alga tested, later naming it the 'relative growth index' (Uki *et al.*, 1986). Although this growth index is useful when conducting feeding experiments on *H. discus hannai*, it does not allow comparisons with other species of abalone. More recent investigations testing dietary quality on abalone other than *H. discus hannai* have used both the food conversion efficiency (FCE) and feeding rates to assess growth value (Koike *et al.*, 1979; Ebert & Houk, 1984).

Table 12.3 summarizes experimental designs in studies on the dietary value of algae to growth. The list is not comprehensive, but includes papers in which we could determine (or translate) the experimental procedure, to illustrate the lack of standardization and the shortcomings of many experiments. Hahn (1989) summarized results. The trials averaged approximately 30 days, ranging from one week (Leighton & Boolootian, 1963) to one year (Ebert & Houk, 1984). However, trials on *H. rubra* have shown that they should be long term to show true growth value, as growth may be influenced by factors other than food during short-term trials, and may not reflect the long-term effects of a single-species diet.

Both the previous diet of the test animals and the disturbance suffered by the animals when the trials are set up may influence growth in short-term trials. To investigate the influence of previous diet, groups of six abalone were fed one of three diets for 131 days and then all abalone were fed the red alga, *Jeannerettia* (Table 12.4). After 124 days on the same diet, the growth rates of groups with different dietary histories were significantly different. For those studies that used cultured animals, presumably all raised on the same diet, differences between treatments would not have been affected by the previous diet, but the initial growth in the trials, and thus the index of dietary value, may depend partly on the previous diet. Uki *et al.* (1986) state that the cultured abalone used in their trials were fed *Eisenia* prior to the trials. Other studies did not report the previous diet.

Table 12.4 Effect of previous diet on growth: *Haliotis rubra* were fed on one of the algae shown for 131 days and then fed the preferred alga *Jeannerettia lobata* for 124 days.

Previous diet	Mean growth (g)	Standard deviation	Sample size
Ecklonia radiata	5.75	3.60	6
Laurencia botryoides	1.40	2.00	6
Ulva sp.	7.50	1.50	6

In trials that use animals directly from the field the previous diet may vary between individuals and batches. For comparative results, all test animals should be fed a standard food before the trials begin, or only the growth in the latter part of the trials should be used.

Koike *et al.* (1979) noted reduced growth during the first 56 days of trials on cultured *H. tuberculata*. They concluded that the animals needed about two months to adapt to the changes of food and environment, and so used only the last phase of the trials to assess diet quality. Similarly, trials on *H. rubra* brought directly from the field, fed a variety of preferred algae and weighed regularly, have shown that growth may be reduced for at least 24 days after collection. For example, in Fig. 12.2(a) growth of the group 2 abalone over the first 24 days was 9 mg day^{-1}, but increased to 62 mg day^{-1} during the next 82 days. However, this effect did not always occur. The group 1 abalone in Fig. 12.2(a) were brought from the field one month before group 2 were collected and were fed the same diet. The mean growth of group 1 was 80 mg day^{-1} for the first 22 days and 65 mg day^{-1} for the next 27 days. These differences suggest that initial growth may be influenced by previous diet in the field rather than disturbance. Whatever the cause, short-term feeding trials may produce erroneous FCE values unless an initial period of adjustment to laboratory conditions and/or a standard diet is employed.

Another factor that requires long trial periods to test is the long-term effects on growth of a single-species diet. Trials on *H. rubra* using single species of dried algae have shown that abalone cease to grow after a period ranging from 50 to 200 days (Fig. 12.2(b), suggesting that these animals are not receiving all the nutrients necessary for growth. Duncan & Klekowski (1975) note that 'essential' substances may become limiting where animals are provided with only one type of food in long-term experiments.

In order to determine whether factors other than the test diet, such as previous diet or adjustment to laboratory conditions, affect growth during trials it is necessary to measure growth throughout the trial. As most of the feeding trials summarized in Table 12.3 measured weight and/or length only at the beginning and end of the trial period, no information on growth variation over time is available, and therefore the effect of these factors on the results is unknown. Only two papers recorded growth over time. Koike *et al.* (1979) identified the effect of disturbance on the first two months of growth as a result of recording growth every two weeks. As the trials were conducted for only 112 days, no information on the long-term effect of single-species diet on growth can be gained. Ebert & Houk (1984) recorded considerable variability over time of growth, feeding rates and conversion efficiencies, although they found no long term effect on growth during their year long trial using *Macrocystis*.

Assessment of the optimum duration of growth trials is important. The trial must be long enough to avoid the effects of previous diet and disturbance on the initial growth phase, and to demonstrate the effect of the test diet in the second phase. For *H. rubra*, trials longer than 100 days are required. The fact that some single algae did not support growth on a long-term basis raises doubts regarding the suitability of single-species trials for testing diet quality and the use of single species diets in mariculture. Although an alga may not support sustained growth when given singly, as part of a mixed diet it may be of great value, supplying

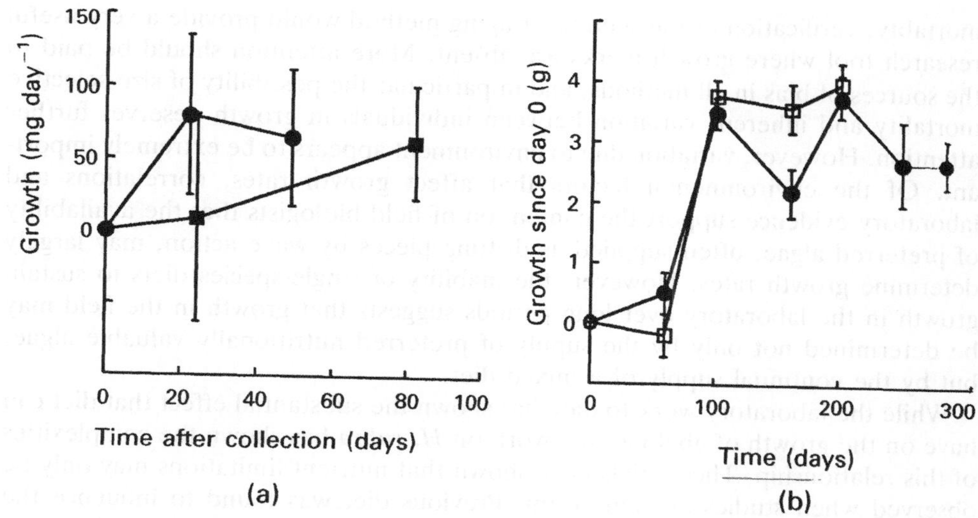

Fig. 12.2 (a) Growth increments in the laboratory of juveniles collected at two times. Means and standard deviations are shown. Squares: group 1 ($N = 85$), circles: group 2 ($N = 28$). (b) The long-term effect on growth rate of feeding abalone single species of dried algae. Means and standard errors are shown of the total growth from day 0. Circles: *Jeannerettia lobata* ($N = 5$), squares: *Macrocystis augustifolia* ($N = 4$).

essential nutrients to the diet. Koike *et al.* (1979) reported that offering mixed diets to *H. tuberculata* did not improve growth rates, but their value may be to sustain growth over time. Duncan & Klekowski (1975) state that it is rare for a species to feed on the same food for its whole life, which suggests that most species can only obtain the full range of required nutrients from a mixed diet. These results suggest that a good measure of the quality of diet in the field may be the availability of a regular supply of a mixture of algae.

The variability in growth and feeding rates between individuals was not reported in any of the papers (Table 12.3). Given the large variability of abalone growth, reported as early as 1963 by Leighton & Boolootian, it is difficult to interpret the significance of growth results without a knowledge of the size of the variation in growth between individuals. Monitoring the growth of 55 *H. rubra* fed a mixed diet of preferred algae for one year has shown variability in individual growth to be high. The coefficient of variation in growth for three month periods averaged 0.66. Over one year the mean growth was 61 mg day^{-1}, but some animals gained 117 mg day^{-1} while others grew at only 30 mg day^{-1}. This variability also indicates the need to use large sample sizes, as the effects of different diets on growth may be masked by the high variation.

CONCLUSIONS

A variety of methods have been used to measure growth rates of abalone in the field, but more studies are required to verify some of the methods used. As the determination of individual ages facilitates the measurement of both growth and mortality, verification of the shell layer aging method would provide a very useful research tool where growth marks are absent. More attention should be paid to the sources of bias in all methods, and in particular the possibility of size-selective mortality and inherent variation between individuals in growth deserves further attention. However, variation due to environment appears to be extremely important. Of the environmental factors that affect growth rates, correlations and laboratory evidence support the contention of field biologists that the availability of preferred algae, often supplied as drifting pieces by wave action, may largely determine growth rates. However, the inability of single-species diets to sustain growth in the laboratory over long periods suggests that growth in the field may be determined not only by the supply of preferred nutritionally valuable algae, but by the continual supply of a mixed diet.

While the laboratory work to date has shown the substantial effect that diet can have on the growth of abalone, the work on *H. rubra* has shown the complexities of this relationship. These trials have shown that nutrient limitations may only be observed when studies are long term. Previous diet was found to influence the growth of abalone when fed test diets, requiring a standardization of previous diet before testing begins. To identify the growth period that reflects the effect of the test diet, a number of measurements of growth during the test period are needed to determine the pattern of growth over time. Because individual growth rates are very variable, extensive replication within treatments is needed to ensure that this variation does not mask the effects of experimental diets. The variation in growth should be reported.

Correlative evidence suggests that food supply is important in determining growth rates in the field, but suitable replicated and controlled field experiments are needed. If food supply is important, then the role of competition for food from urchins and other abalone at high densities deserves further experimental investigation. Growth has been found to vary with density in culture, and may do so in the field. A knowledge of such density dependence is important for management, as it would change the results of yield per recruit and egg per recruit models, and the form of the stock–recruitment relation. Again, well designed field experiments are required.

ACKNOWLEDGEMENTS

This work was supported by grants to Rob Day from the Fishing Industry Research and Development Committee and the Australian Research Grants Scheme. The authors thank the Marine Science Laboratories of the Victorian Department of Conservation, Forests and Lands for the use of laboratory facilities. Travel was supported by the Department of Industry, Technology and Resources, and the Potter Foundation. The authors would like to thank the referees P.A. Breen, W.S. Hearn and Professor K. Sakata, and also H. Kojima, W.J. Nash and S.A. Shepherd for helpful criticism of the manuscript and discussions.

REFERENCES

Appeldoorn R.S. (1987) Modification of a seasonally oscillating growth function for use with mark-recapture data. *J. Cons. Int. Explor. Mer*, **43**, 194–8.

Barkai R. & Griffiths C.L. (1986) Diet of the South African abalone *Haliotis midae*. *S. Afr. J. Mar. Sci.*, **4**, 37–44.

Beamish R.J. & McFarlane G.A. (1983) The forgotten requirement of age validation in fisheries biology. *Trans. Am. Fish. Soc.*, **112**, 735–43.

Bolognari A. (1953) Richerche sulla sessualita di *Haliotis lamellosa* Lam. *Archo Zool. Ital. Torino*, **38**, 361–402 (in Italian).

Breen P.A. (1980) Measuring fishing intensity and annual production in the abalone fishery of British Columbia. *Can. Tech. Rep. Fish. Aquat. Sci.*, (947), 1–49.

Breen P.A. (1986) Management of the British Columbia fishery for northern abalone (*Haliotis kamtschatkana*). *Can. Spec. Pub. Fish. Aquat. Sci.*, **92**, 300–12.

Breen P.A. & Adkins B.E. (1979) A survey of abalone populations on the east coast of the Queen Charlotte Islands, August 1978. *Fish. Mar. Serv. MS Rep.* (1490).

Burge R., Schultz S. & Odemar M. (1975) *Draft Report on Recent Abalone Research in California with Recommendations for Management*. Resources Agency, Department of Fish and Game, California.

Cassie R.M. (1954) Some uses of probability paper in the analysis of size frequency distributions. *Aust. J. Mar. Freshwat. Res.*, **5**, 513–22.

Chen H.-C. (1984) Recent innovations in cultivation of edible molluscs in Taiwan, with special reference to the small abalone *Haliotis diversicolor* and the hard clam *Meretrix lusoria*. *Aquaculture*, **39**, 11–27.

Clavier J. (1982) (Initial data on natural ormer (*Haliotis tuberculata*) stocks in the St. Malo region.) Rapport d'étude, Association pour la Mise en Valeur du Littoral de la Côte d'Emeraude. (Translated into English in *Can. Transl. Fish. Aquat. Sci.*, (1983) (4928).

Clavier J. (1992) Infestation of *Haliotis tuberculata* shells by *Cliona celata* and *Polydora* sp. *Fish. Res. Pap. Dep. Fish. S. Aust.*, No. 24 (in press).

Clavier J. & Richard D. (1985) *Etudes sur les Ormeaux dans la Region de Saint-Malo*. Association pour la Mise en Valeur du Littoral de la Côte d'Emeraude (in French).

Clavier J. & Richard D. (1986) Growth of juvenile *Haliotis tuberculata* (Mollusca: Gastropoda) in their natural environment. *J. Mar. Biol. Assoc. UK*, **66**, 497–503.

Cochard J.-C. (1982) La croissance de l'ormeau *Haliotis tuberculata* L. en rade de Brest. *Haliotis*, **12**, 61–9 (in French).

Cox K.W. (1962) California abalones, family Haliotidae. *Fish. Bull. California*, **118**, 1–129.

Day R.W. & Leorke A. (1986a) Abalone catches – what factors affect them? *Aust. Fish.*, **45**, 32–6.

Day R.W. & Leorke A. (1986b) Growth rates of blacklip abalone (*Haliotis rubra*) in Port Phillip Bay. *Abstracts*, p. 127. Australian Marine Sciences Association conference, University of Tasmania, Hobart.

Day R.W. & Quinn G.P. (1989) Comparisons of treatments after an analysis of variance in ecology. *Ecol. Monogr.*, **59**, 433–63.

Dayton P.K. & Tegner M.J. (1989) Bottoms beneath troubled waters: benthic impacts of the 1982–1984 El Niño in the temperate zone. In *Global Ecological Consequences of the 1982–83 El Niño Southern Oscillation*. (Ed. by P.W. Glynn). Elsevier Press, Amsterdam.

Douros W.J. (1985) *Density, growth, reproduction and recruitment in an intertidal abalone: effects of intraspecific competition and prehistoric predation*. MSc thesis, University of California, Santa Barbara.

Duncan A. & Klekowski R.Z. (1975) Parameters of an energy budget. In *Methods for Ecological Bioenergetics*, IBP Handbook No 24 (Ed. by W. Grodzinski, R.Z. Klekowski & A. Duncan, pp. 97–148. Blackwell Scientific Publications, Oxford.

Ebert E.E. & Houk J.L. (1984) Elements and innovations in the cultivation of red abalone *Haliotis rufescens*. *Aquaculture*, **39**, 375–92.

Emmett B. & Jamieson G.S. (1989) An experimental transplant of northern abalone, *Haliotis kamtschatkana*, in Barkley Sound, British Columbia. *Fish. Bull. US*, **87**, 95–104.

Fabens A.J. (1965) Properties and fitting of the von Bertalanffy growth curve. *Growth*, **29**, 265–89.

Ford E. (1933) An account of the herring investigations conducted at Plymouth during the years from 1924 to 1933. *J. Mar. Biol. Assoc. UK*, **15**, 279–304.

Forster G.R. (1967) The growth of *Haliotis tuberculata*: results of tagging experiments in Guernsey 1963–65. *J. Mar. Biol. Assoc. UK*, **47**, 287–300.

Forster G.R., Potts G.W. & Swinfen R. (1982) Changes in the ormer populations of Guernsey and Jersey. *J. Mar. Biol. Assoc. UK*, **62**, 717–27.

Fournier D.A. & Breen P.A. (1983) Estimation of abalone mortality rates with growth analysis. *Trans. Am. Fish. Soc.*, **112**, 403–11.

Fournier D.A., Sibert J.R., Majkowski J. & Hampton J. (1990) MULTIFAN a likelihood-based method for estimating growth parameters and age composition from multiple length frequency data sets illustrated using data for southern bluefin tuna (*Thunnus maccoyii*). *Can. J. Fish. Aquat. Sci.*, **47**, 301–17.

Francis R.I.C.C. (1988) Are growth parameters estimated from tagging and age–length data comparable? *Can. J. Fish. Aquat. Sci.*, **45**, 936–42.

Grant A., Morgan P.J. & Olive P.J.W. (1987) Use made in marine ecology of methods for estimating demographic parameters from size/frequency data. *Mar. Biol.*, **95**, 201–8.

Gulland J.A. (1964) Manual of methods of fish population analysis. *FAO Fish. Biol. Tech. Pap.*, (40).

Gutreuter S. (1987) Considerations for estimation and interpretation of annual growth rates. In *The Age and Growth of Fish* (Ed. by R.C. Summerfelt & G.E. Hall), pp. 115–26. Iowa State University Press, Ames, Iowa.

Hahn K.O. (1989) Nutrition and growth of abalone. In *Handbook of Culture of Abalone and Other Marine Gastropods*, (Ed. by K.O. Hahn), pp. 135–56. CRC Press, Boca Raton.

Hamer G.D. (1980) Growth of *Haliotis rubra* in N.S.W. *J. Malac. Soc. Aust.*, **4**, 256–7.

Hancock D.A. (1965) Graphical estimation of growth parameters. *J. Cons. Perm. Int. Explor. Mer.*, **29**, 340–51.

Harrison A.J. & Grant J.F. (1971) Progress in abalone research. *Tasmanian Fish. Res.*, **5**, 1–10.

Hayashi I. (1980) Structure and growth of a shore population of the ormer, *Haliotis tuberculata*. *J. Mar. Biol. Assoc. UK*, **60**, 431–7.

Hayashi T. (1955) Studies on the age determining characters of shellfish. I. Observations by sectioning of the shell (1). *Bull. Hokkaido Reg. Fish. Res. Lab.*, (12), 43–8 (in Japanese).

Hearn W.S. (1986) *Mathematical models for evaluating marine fisheries*. PhD thesis, University of New South Wales.

Henderson K.C., Parker D.O. & Haaker, P.L. (1988) The survival and growth of transplanted adult pink abalone, *Haliotis corrugata* at Santa Catalina Island. *Calif. Fish Game*, **74**, 82–6.

Hirose T. (1974) On the growth of the abalone, *Haliotis discus hannai* Ino, especially concerned with the food condition in nature. *Bull. Tohoku Reg. Fish. Res. Lab.*, (33) 87–94 (in Japanese).

Ino T. (1952) Biological studies on the propagation of the Japanese abalone (genus *Haliotis*). *Bull. Tokai Reg. Fish. Res. Lab.*, (5), 1–102 (in Japanese).

Kaufmann K.W. (1981) Fitting and using growth curves. *Oecologia (Berl.)*, **49**, 293–9.

Keesing J.K. & Wells F.E. (1989) Growth of the abalone *Haliotis roei* Gray. *Aust. J. Mar. Freshwat. Res.*, **40**, 199–204.

Kikuchi S., Sakurai Y., Sasaki M. & Ito T. (1967) Food values of certain marine algae for the growth of the young abalone, *Haliotis discus hannai*. *Bull. Tohoku Reg. Fish. Res. Lab.*, (27), 93–100 (in Japanese).

Kim J.-W. & Chung S.-C. (1985) On the growth of abalones, *Sulculus diversicolor diversicolor* (Reeve) and *S. diversicolor aquatilis* (Reeve), in Cheju Island. *Bull. Mar. Res. Res. Inst. Cheju Nat. Univ.*, (9) 39–50 (in Chinese).

Kimura D.K. (1980) Likelihood methods for the von Bertalanffy growth curve. *Fish. Bull. US*, **77** 765–76.

Kirkwood G.P. & Somers I.F. (1984) Growth of two species of tiger prawn, *Penaeus esculentus* and *P. semisulcatus*, in the western Gulf of Carpentaria. *Aust. J. Mar. Freshwat. Res.*, **35**, 703–12.

Knight W. (1968) Asymptotic growth: an example of nonsense disguised as mathematics. *J. Fish. Res. Brd. Can.*, **25**, 1303–7.

Koike Y., Flassch J. & Mazurier J. (1979) Biological and ecological studies on the propogation of the ormer, *Haliotis tuberculata* (L.). II Influence of food and density on the growth of juveniles. *La Mer*, **17**, 43–52.

Kojima H. (1975) A study on the growth of the shell of *Haliotis discus discus* (Gastropoda: Haliotidae) – I, Age and growth. *Aquaculture*, **23**, 61–6 (in Japanese).

Kojima H. & Imajima M. (1982) Burrowing polychaetes in the shells of the abalone *Haliotis diversicolor aquatilis* chiefly on the species of *Polydora*. *Bull. Jap. Soc. Sci. Fish.*, **48**, 31–5 (in Japanese).

Kojima H., Nakahisa Y., Tanimoto H. & Isibasi K. (1977) A study on the stock of Japanese black abalone, *Haliotis discus discus* in Tokushima Prefecture I. Growth of shells. *Bull. Tokai Reg. Fish. Res. Lab.*, (90), 25–37 (in Japanese).

Kurogane K., Sagara J., Hamada S. & Tanaka Y. (1974) On the growth of tagged abalone, *Haliotis gigantea* Gmelin, in the Manazuru fishing port. *Bull. Tokai Reg. Fish. Res. Lab.*, (77), 7–19 (in Japanese).

Leighton D.L. (1974) The influence of temperature on larval and juvenile growth in three species of southern California abalones. *Fish. Bull. US*, **72**, 1137–45.

Leighton D.L. (1990) Abalone (Genus *Haliotis*) mariculture on the North American Pacific Coast. *Fish. Bull. US*, **87**, 689–702.

Leighton D.L. & Boolootian R.A. (1963) Diet and growth in the black abalone, *Haliotis cracherodii*. *Ecology*, **44**, 227–38.

MacDonald P.D.M. (1987) Analysis of length–frequency distributions. In *The Age and Growth of Fish*, (Ed. by R.C. Summerfelt & G.E. Hall), pp. 371–84. Iowa State University Press, Ames, Iowa.

MacDonald P.D.M. & Green P.E.J. (1985) *User's guide to program MIX: an interactive program for fitting mixtures of distributions*. Ichthus Data Systems, Hamilton, Ontario.

McShane P.E. (1988) Tagging abalone. In *Australian Society for Fish Biology Tagging Workshop* (Ed. by D.A. Hancock), pp. 65–7. Australian Government Publishing Service, Canberra.

McShane P.E., Black K.P. & Smith M.G. (1988a) Recruitment processes in *Haliotis rubra* (Mollusca: Gastropoda) and regional hydrodynamics in southeastern Australia imply localized dispersal of larvae. *J. Exp. Mar. Biol. Ecol.*, **124**, 175–203.

McShane P.E., Smith M.G. & Beinssen K.H.H. (1988b) Growth and morphometry in abalone (*Haliotis rubra* Leach) from Victoria. *Aust. J. Mar. Freshwat. Res.*, **39**, 161–6.

Momma H. (1980) Studies on the variation of the abalone I. On the growth of the different aged young abalone. *Aquaculture*, **28**, 142–55 (in Japanese).

Moreau J. (1987) Mathematical and biological expression of growth in fishes: recent trends and further developments. In *The Age and Growth of Fish* (Ed. by R.C. Summerfelt & G.E. Hall), pp. 81–113. Iowa State University Press, Ames, Iowa.

Morse D.E. (1984) Biochemical and genetic engineering for improved production of abalones and other valuable molluscs. *Aquaculture*, **39**, 263–82.

Muñoz Lopez T. (1976) Resultados preliminares de un método para determinar edad en abulones (*Haliotis* spp.) de Baja California. *Memorias del Simposío sobre Recursos Pesqueros Masivos de México Ensenada B.C. del 28 al 30 de Septiembre, 1976*, pp. 281–301 (in Spanish).

Nash W., Sellers T., Talbot S., Cawthorn A. & Ford W. (in prep) The population biology of abalone (*Haliotis* species) in Tasmania. I. Blacklip abalone (*H. rubra*) from the north coast and the Furneaux group of islands. *Dept. Sea Fish. Tasmania Tech. Rept.*

Newman G.G., (1968) Growth of the South African abalone *Haliotis midae*. *S. Afr. Sea Fish. Branch Invest. Rept*, **67**, 1–24.

Oba T., Sato H., Tanaka K. & Toyama T. (1968) Studies on the propagation of an abalone, *Haliotis diversicolor supertexta* – III. On the size of the one year old specimen. *Bull. Jap. Soc. Sci. Fish.*, **34**, 457–61.

Olsen D.A. (1968). Banding patterns in *Haliotis* – II. Some behavioral considerations and the effect of diet on shell coloration for *Haliotis rufescens*, *Haliotis corrugata*, *Haliotis sorenseni*, and *Haliotis assimilis*. *Veliger*, **11**, 135–9.

Peck L.S. (1989) Feeding, growth and temperature in the ormer, *Haliotis tuberculata* (L.). *Prog. Underwat. Sci.*, **14**, 95–107.

Poore G.C.B. (1972a) Ecology of New Zealand abalones, *Haliotis* species (Mollusca: Gastropoda). 1. Feeding. *NZ J. Mar. Freshwat. Res.*, **6**, 11–22.

Poore G.C.B. (1972b) Ecology of New Zealand abalones, *Haliotis* species (Mollusca; Gastropoda): 3. Growth. *NZ J. Mar. Freshwat. Res.*, **6**, 534–59.

Prince J.D. (1989) *The fisheries biology of the Tasmanian stocks of* Haliotis rubra. PhD thesis, University of Tasmania.

Prince J.D. (1991) A new technique for tagging abalone. *Aust. J. Mar. Freshwat. Res.*, **42**, 101–6.

Prince J.D., Sellers T.L., Ford W.B. & Talbot S.R. (1988a) Recruitment, growth, mortality and population structure in a southern Australian population of *Haliotis rubra* (Mollusca: Gastropoda). *Mar. Biol.*, **100**, 75–82.

Prince J.D., Sellers T.L., Ford W.B. & Talbot S.R. (1988b) A method for ageing the abalone *Haliotis rubra* (Mollusca: Gastropoda). *Aust. J. Mar. Freshwat. Res.*, **39**, 167–75.

Quayle D.B. (1971) Growth, morphometry and breeding in the British Columbia abalone (*Haliotis kamtschatkana* Jonas). *Tech. Rep. Can. Fish. Res. Bd*, **279**, 1–19.

Ratkowsky D.A. (1986) Statistical properties of alternative parameterizations of the von Bertalanffy growth curve. *Can. J. Fish. Aquat. Sci.*, **43**, 742–7.

Ricker W.E. (1969) Effects of size-selective mortality and sampling bias on estimates of growth, mortality, production, and yield. *J. Fish. Res. Brd. Can.*, **26**, 479–540.

Sainsbury K.J. (1980) Effect of individual variability on the von Bertalanffy growth equation. *Can. J. Fish. Aquat. Sci.*, **37**, 241–7.

Sainsbury K.J. (1982) Population dynamics and fishery management of the paua, *Haliotis iris* I. Population structure, growth, reproduction and mortality. *NZ J. Mar. Freshwat. Res.*, **16**, 147–61.

Saito K. (1981) The appearance and growth of 0-year-old Ezo abalone. *Bull. Jap. Soc. Sci. Fish.*, **47**, 1393–400.

Sakai S. (1960) On the formation of the annual ring on the shell of the abalone, *Haliotis discus* var. *hannai* Ino. *Tohoku J. Agric. Res.*, **11**, 239–44.

Sakai S. (1962a) Ecological studies on the abalone, *Haliotis discus hannai* Ino. I. Experimental studies on the food habit. *Bull. Jap. Soc. Sci. Fish.*, **28**, 766–79 (in Japanese).

Sakai S. (1962b) Ecological studies on the abalone, *Haliotis discus hannai* Ino. III. Study on the mechanism of production of the abalone in the region of Onagawa Bay. *Bull. Jap. Soc. Sci. Fish.*, **28**, 891–8 (in Japanese).

Sanders M.J. (1982) Preliminary stock assessment for the abalone taken off the south east coast of Oman. Report No. FI:DP/RAB/80/015/3, Food and Agriculture Organization of the United Nations, Rome.

Schnute J. & Fournier D. (1980) A new approach to length–frequency analysis: growth structure. *Can. J. Fish. Aquat. Sci.*, **37**, 1337–51.

Shepherd S.A. (1973) Studies on southern Australian abalone (genus *Haliotis*) I. Ecology of five sympatric species. *Aust. J. Mar. Freshwat. Res.*, **24**, 217–57.

Shepherd S.A. (1988) Studies on southern Australian Abalone (genus *Haliotis*) VIII. Growth of juvenile *H. laevigata*. *Aust. J. Mar. Freshwat. Res.*, **39**, 177–83.

Shepherd S.A., Clarkson P.S. & Turner J.A. (1985) Studies on southern Australian abalone (genus *Haliotis*) V. Spawning, settlement and early growth of *H. scalaris*. *Trans. R. Soc. S. Aust.*, **109**, 61–2.

Shepherd S.A., Guzmán del Próo S.A., Turrubiates J. Belmar J., Baker J.L. & Sluczanowski P.R. (1991) Growth, size at sexual maturity and egg-per-recruit analysis of the abalone *Haliotis fulgens* in Baja California. *Veliger*, **34**, 324–30.

Shepherd S.A. & Hearn W.S. (1983) Studies on southern Australian abalone (genus *Haliotis*) IV. Growth of *H. laevigata* and *H. ruber*. *Aust. J. Mar. Freshwat. Res.*, **34**, 461–75.

Shepherd S.A., Mower A.G.J. & Hill K. (1987) Studies on southern Australian abalone (genus *Haliotis*) IX. Growth of *H. scalaris*. *Trans. R. Soc. S. Aust.*, **112**, 93–5.

Tegner M.J., Breen P.A. & Lennert C.E. (1989) Population biology of red abalones, *Haliotis rufescens*, in southern California and management of the red and pink, *H. corrugata*, abalone fisheries. *Fish. Bull. US*, **87**, 313–39.

Tegner M.J. & Dayton P.K. (1987) El Niño effects on Southern California kelp forest communities. *Adv. Ecol. Res.*, **17**, 243–79.

Todoufuken suishi isoneshigen kenkyu gurupu (1972) Isone shigen to sono zosyoku I-awabi. *Nippon suisan shigen hogo kyokai, Suisan Zoyosyoku Sosyo*, (24) (in Japanese).

Tutschulte T.C. (1976) The comparative ecology of three sympatric abalones. PhD dissertation, University of California, San Diego.

Tutschulte T.C. & Connell J.H. (1988) Growth of three species of abalones (*Haliotis*) in southern California. *Veliger*, **31**, 204–13.

Uki N. (1981) Feeding behaviour of experimental populations of the abalone, *Haliotis discus hannai*. *Bull. Tohoku Reg. Fish. Res. Lab.*, (43), 53–8 (in Japanese).

Uki N., Sugiura M. & Watanabe T. (1986) Dietary value of seaweeds occurring on the Pacific coast of Tohoku for growth of the abalone *Haliotis discus hannai*. *Bull. Jap. Soc. Sci. Fish.*, **52**, 257–66. (in Japanese).

Wright M.B. (1975) Growth in the black abalone, *Haliotis cracherodii*. *Veliger*, **18**, 194–9.

Chapter 13
Food preferences of three Australian abalone species with a review of the algal food of abalone

S.A. SHEPHERD *Department of Fisheries, 135 Pirie Street, Adelaide, South Australia 5000, Australia*

P.D. STEINBERG *School of Biological Sciences, University of Sydney, New South Wales 2006, Australia*

ABSTRACT: This study examined, by correlation analysis, algal characteristics which affect consumption rates or feeding preferences of abalone. In Australia, three species of abalone consistently ate two red algae (*Pterocladia capillacea* and *Asparagapsis armata*) and one brown alga (*Lobospira bicuspidata*) at higher rates than fucoid or laminarian brown algae (generally uneaten). Consumption rates of these algae by the three Australian abalone species were negatively correlated with both algal toughness and phenolic levels, but we could not distinguish the relative importance of each factor.

Although food preferences or diets of abalone elsewhere in the world vary from place to place, some patterns are evident. Leathery, brown macrophytes (fucoids, kelps) are less palatable to Australian abalone than to species elsewhere. This could be due to higher phenolic levels in the Australian species, but we hypothesize that there may be little selection pressure for Australian abalone to evolve preferences for such algae, because red algae are abundant in their habitat. World-wide, the feeding preferences of abalone are not easily explained by simple models based on deterrence (by chemical defences) or morphology. We suggest that differences in digestibility of red and brown algae may also be important.

RESUMEN: Este estudio examinó por analisis de correlación, las características de las algas que afectan la tasa de consumo o las preferencias alimenticias del abulón. En Australia tres especies de abulón consistentemente comieron dos algas rojas, (*Pterocladia capillacea* y *Asparagopsis armata*) y una alga café (*Lobospira bicuspidata*) en tasas mas altas que las fucoides o laminarias algas cafés (generalmente incomibles). Las tasas de consumo de estas algas de las tres especies Australianas de abulón fueron correlacionadas negativamente tanto con la dureza algal como con los niveles fenólicos pero no pudimos distinguir la importancia relativa de cada factor.

Aunque las preferencias de alimento o dietas del abulón en cualquier parte del mundo varía de un lugar a otro, algunos patrones son evidentes.

Laminarias y fucoides cafés son menos apetecibles para el abulón Australiano que para especies de otros lugares. Esto podría deberse a niveles fenólicos más altos en las especies Australianas, pero sostenemos la hipótesis de que puede haber poca presión de selección para el abulón Australiano para desarrollar preferencias por tales algas, porque las algas rojas son abundantes en su habitat.

En todo el mundo las preferencias alimenticias del abulón no son fácilmente explicadas por simples modelos basados en disuación (por defensas químicas) o morfología. Sugerimos que las diferencias en digestibilidad de algas rojas y cafés pueden también ser importantes.

INTRODUCTION

Although the diet of most exploited species of abalone in the world has been described, little is known of the factors which affect their choice of algae in nature. Some factors have been identified including (a) the presence in algae of chemical metabolites, (b) algal morphology (toughness) and (c) nutritive value.

Chemical metabolites include stimulants (reviewed by Sakata & Ina, Chapter 14, this volume) and polyphenolics, which are important deterrents to herbivory (Geiselman & McConnell, 1981; Steinberg, 1984, 1985, 1988; Hay & Fenical, 1988). A better understanding of these factors is especially important for the development of optimal diets for abalone culture, which is now expanding rapidly in many parts of the world.

In this chapter we measure the consumption rates of three common Australian species of abalone in relation to two factors, polyphenolic levels and algal toughness. We then review the literature on diet and food preferences of abalone around the world and consider whether their diet is affected by the same factors in algae. Lastly, we speculate on the adaptive response of abalone to evolved polyphenolic levels in algal floras in different parts of the world.

MATERIALS AND METHODS

The consumption rates of algae by three Australian abalone species, *Haliotis roei* Gray, *Haliotis laevigata* Donovan and *Haliotis rubra* Leach, were studied at West Island, South Australia (35°37′S, 138°35′E). Animals of uniform size were collected from depths of 1–5 m, and feeding experiments were carried out in cages of about 0.25 m^3 capacity fitted with 5 mm mesh in the summer of 1986 (sea temperatures 20°C) at a depth of 4 m. *Haliotis laevigata* and *H. rubra* were of shell length (SL) 120–130 mm and *H. roei* were of 50–70 mm SL (in one experiment 80–100 mm). They were then placed in the cages after collection, without being removed from the water, and were starved for 24 hours before use. New sets of abalone were used for each experiment to avoid conditioning them to particular algae.

The algae selected for the feeding trials are all common, either epilithic or as drift in the habitats of these abalone at West Island (Shepherd & Womersley, 1970). The species selected were brown algae: *Ecklonia radiata* (C.Ag.) J.Ag., *Cystophora moniliformis* (Esper) Womersley and Nizamuddin, *Seirococcus axillaris* (R.Br.) Grev., *Sargassum fallax* Sonder, *Lobospira bicuspidata* Aresch., and red algae: *Pterocladia capillacea* (Gmel.) Born. & Thur. and *Asparagopsis armata* Harvey. These algae vary considerably in morphology and secondary chemistry (Table 13.1).

Herbivores may respond differently to algae according to whether or not they are offered a choice (Vadas, 1977; Steinberg, 1988). Even when they have a choice it is not clear what array of algal food is perceived by them in the field. So we performed two kinds of experiments: no choice (only one species of alga per cage) and choice (more than one species per cage). In the former experiments algae were blotted with a paper towel to remove excess water and about 100 g of the alga, tied in several bundles, were weighed out into each cage. There were three abalone per cage and three replicates for each alga. Abalone were allowed to feed for 48 hours and then the remaining algae were removed, blotted and re-weighed. Three control cages for each experiment gave estimates of weight loss due to factors other than grazing.

In the choice experiments, about 25 g of each of five species of algae were weighed out and placed into each of four replicate cages with abalone and into

Table 13.1 Algal secondary chemistry and morphology.

Species of algae	Polyphenolic levels[1]	Other secondary metabolites	Functional form group[2]
Ecklonia radiata	7.46 ± 0.86	none known	leathery macrophyte
Sargassum fallax	11.46 ± 0.99	terpenoids[3]	leathery macrophyte
Seirococcus axillaris	4.64 ± 1.43	none known	leathery macrophyte
Cystophora moniliformis	12.16 ± 1.01	terpenes[4,5]	leathery macrophyte
Lobospira bicuspidata	1.13 ± 0.11	none known	corticated macrophyte
Phyllospora comosa	2.67 ± 0.48	none known	leathery macrophyte
Macrocystis pyrifera	0.62 ± 0.08	none known	leathery macrophyte
Pterocladia capillacea	none	none known[6]	corticated macrophyte
Asparagopsis armata	none	unknown but halogenated ketones in *A. taxiformis*[6]	corticated macrophyte

[1] Steinberg, 1989.
[2] Steneck & Watling, 1982.
[3] Van Altena & Steinberg, unpublished.
[4] Van Altena, 1987.
[5] Kazlauskas *et al.*, 1978.
[6] Fenical, 1975.

control cages without abalone. Four *H. laevigata*, four *H. rubra* and eight *H. roei* were used in two sets of experiments. We used five species of brown algae in one set and two species of red algae together with *Ecklonia*, *Cystophora* and *Lobospira* in the other set.

Results of the no-choice experiments were tested by (a) t-tests comparing the amount of algae eaten in the experimental cages with the amount lost in the controls (Schiel, 1982) and (b) one-way analysis of variance for all species of algae together with comparison of the means with the Student-Newman-Keuls test after correction for mean losses in controls. ANOVA is not appropriate for the multiple-choice experiments, because the amounts of algae consumed are not independent of all other species. So here we simply present the results to demonstrate consistency with the other experiments.

Phenolic levels for most algae are derived from Steinberg (1989); for *Lobospira* we used the method of Ragan & Jensen (1977) and Steinberg (1985, 1989). Data for other secondary metabolites (Table 13.1) are taken from the literature.

RESULTS

No-choice experiments

The quantity of each algal species eaten by *H. laevigata*, *H. rubra* and *H. roei* in the no-choice experiments is ranked from lowest to highest in Fig. 13.1. Consumption of different algae is highly significantly different for all three species of abalone, as determined by single-factor ANOVA ($P < 0.001$ for all three experiments).

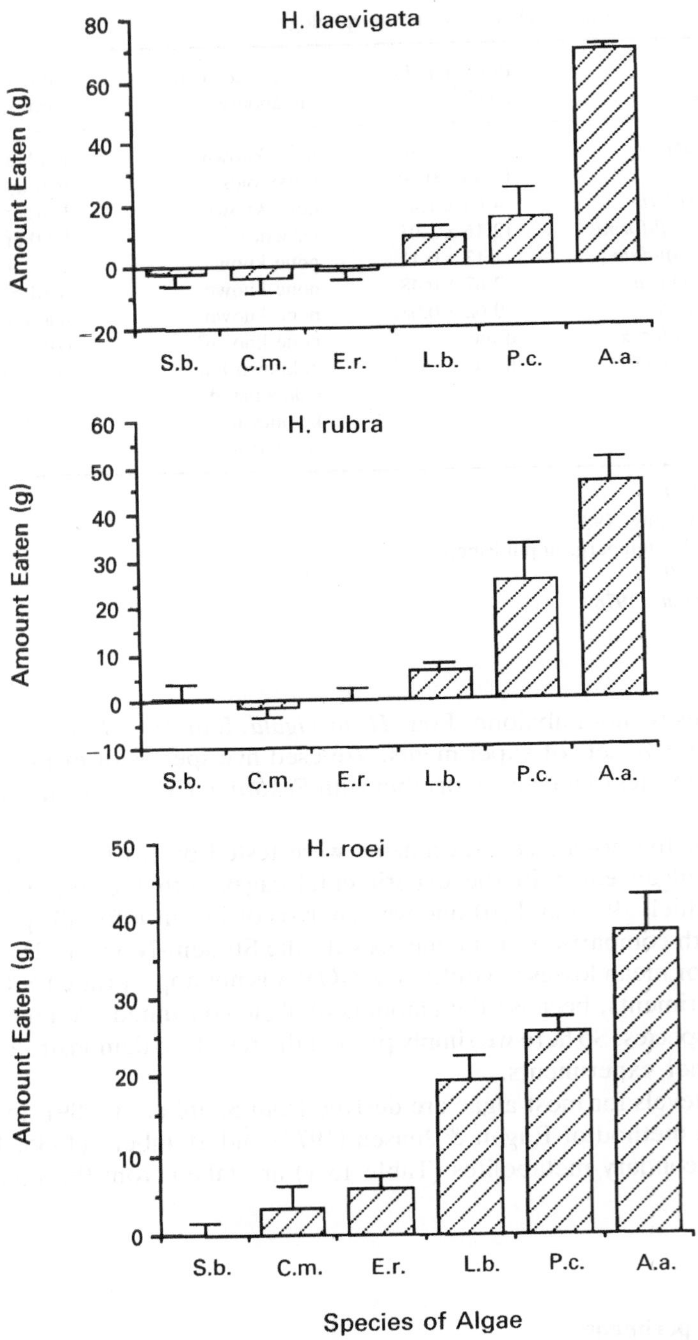

Fig. 13.1 Amounts of algae (g wet weight) consumed in no-choice (one alga per cage) experiments ($N = 3$) on three Australian species of abalone. S.b. = *Sargassum fallax*; C.m. = *Cystophora moniliformis*; E.r. = *Ecklonia radiata*; L.b. = *Lobospira bicuspidata*; P.c. = *Pterocladia capillacea*; A.a. = *Asparagopsis armata*. Error bars are standard errors.

Except for *H. roei*, which consumed small but detectable amounts of *Ecklonia radiata*, none of the large leathery brown algae (*E. radiata*, *Sargassum fallax*, *Cystophora moniliformis*) were consumed by the abalone (loss in weight of these algae from experimental cages was not significantly different from the loss of algae in control cages, as determined by t-tests). The other three species of algae were all eaten by all three species of abalone, except for *Pterocladia capillacea* by *H. laevigata*, where large losses from the control cages made it difficult to distinguish loss due to consumption and loss from other factors.

The relative consumption of algae was very similar for all three species of abalone (Fig. 13.1 and SNK tests: $P<0.05$). *H. laevigata* consumed *Asparagopsis* at a greater rate than all other species (which, except for *Lobospira*, were not consumed). The pattern of algal consumption for *H. rubra* was:

Asparagopsis $>$ *Pterocladia* $>$ all other species, and for *H. roei* was:

Asparagopsis, *Pterocladia*, *Lobospira* $>$ all other species

where species sharing an underline are not significantly different.

Multiple-choice experiments

In the experiments in which five species of brown algae were offered, the only alga eaten by each of the three abalone species was *Lobospira* ($P<0.05$) (Fig. 13.2). The loss of other species was not significantly different from the controls.

In the experiments in which two species of red algae and three species of brown algae were offered, *Cystophora* and *Ecklonia* were again not eaten, but *Lobospira*, *Pterocladia* and *Asparagopsis* were each consumed in proportions which are qualitatively in agreement with the earlier results (Fig. 13.3). This suggests that the feeding rates of these abalone on different algae are consistent across different arrays of algal species.

Algal characteristics

The phenolic levels, presence of other secondary metabolites and the functional form (Steneck & Watling, 1982) of the algae used in the feeding experiments are given in Table 13.1. The four brown algae, *Ecklonia*, *Cystophora*, *Sargassum* and *Seirococcus* are tough, leathery macrophytes with high (4.5–12% dry weight) phenolic content (Steinberg, 1989), whereas *Lobospira* has a low phenolic content and is a delicate species of low toughness. The red algae lack polyphenolics, are corticated and low in toughness. However, *Asparagopsis armata* may contain halogenated ketones characteristic of other members of the genus (Fenical, 1975; Ragan & Glombitza, 1986).

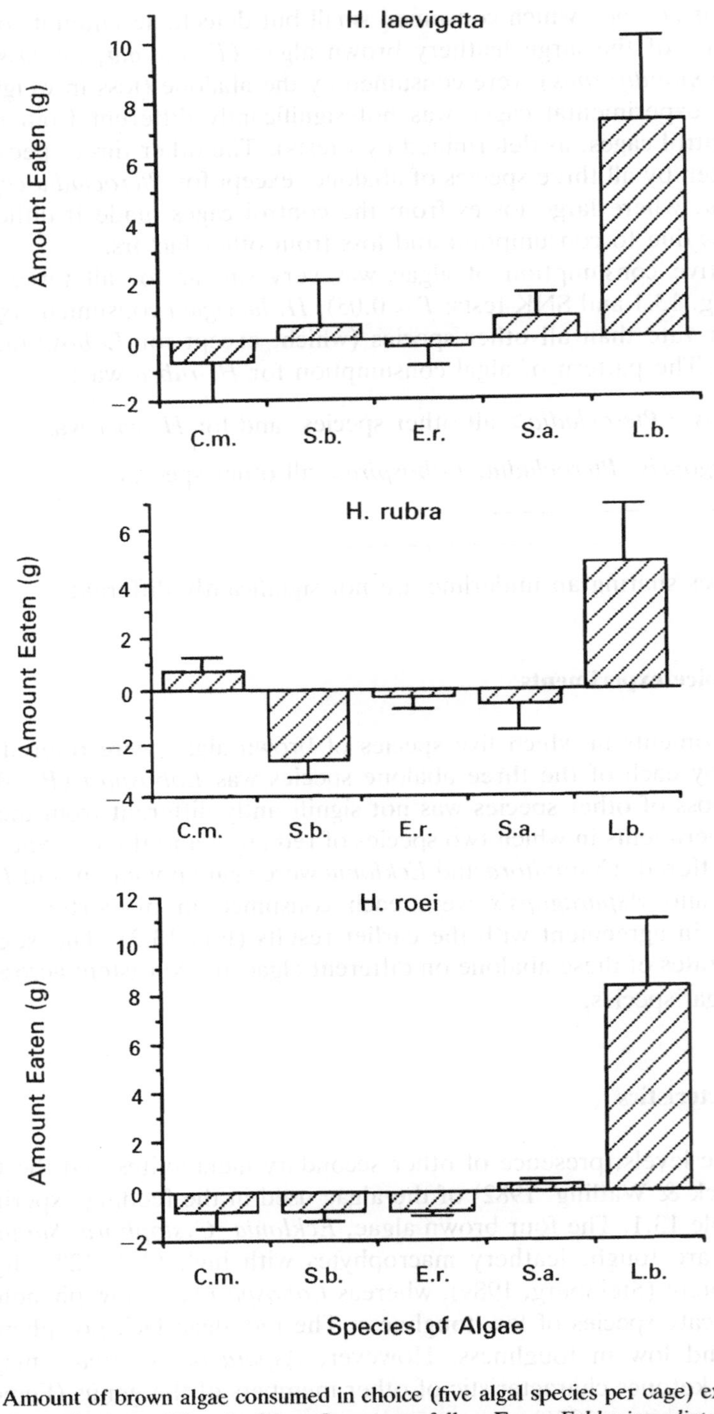

Fig. 13.2 Amount of brown algae consumed in choice (five algal species per cage) experiments (N = 4). C.m. = *Cystophora moniliformis*; S.b. = *Sargassum fallax*; E.r. = *Ecklonia radiata*; *S.a.* = *Seirococcus axillaris*; L.b. = *Lobospira bicuspidata*. Error bars are standard errors.

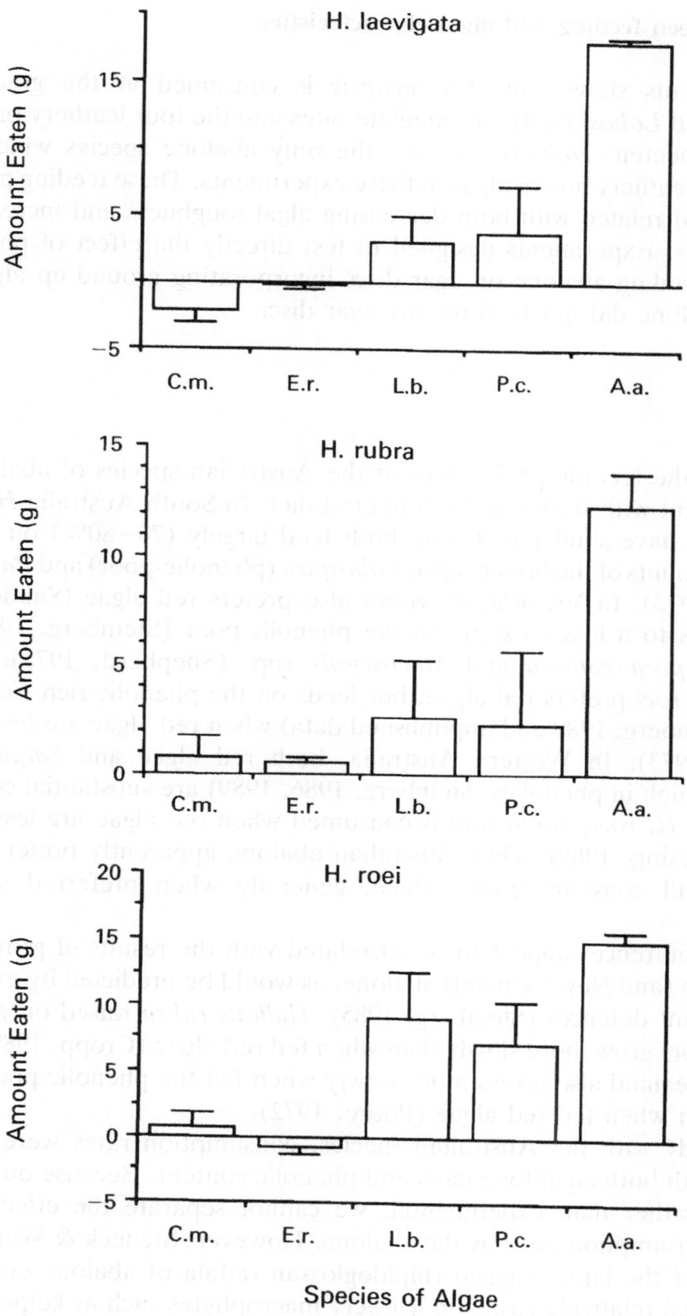

Fig. 13.3 Amounts of algae consumed in choice (five algal species per cage) experiments ($N = 4$). C.m. = *Cystophora moniliformis*; E.r. = *Ecklonia radiata*; L.b. = *Lobospira bicuspidata*; P.c. = *Pterocladia capillacea*; A.a. = *Asparagopsis armata*. Error bars are standard errors.

Relation between feeding and algal characteristics

Our experiments show that *Asparagopsis* is consumed at the greatest rate, *Pterocladia* and *Lobospira* at intermediate rates and the four leathery brown algae are largely uneaten. *Haliotis roei* was the only abalone species which ate any phenolic-rich leathery brown algae in these experiments. These feeding preferences are strongly correlated with both decreasing algal toughness and increasing algal phenolic levels. Experiments designed to test directly the effect of phenolics on feeding (by feeding abalone on agar discs incorporating ground-up algae) failed when the abalone did not feed on any agar disc.

DISCUSSION

Our data on the feeding preferences of the Australian species of abalone are in good agreement with studies of their natural diet. In South Australia *H. laevigata* and *H. rubra* have similar diets and both feed largely (70–80%) on red algae, with small amounts of the brown algae *Lobospira* (phenolic-poor) and *Pachydictyon* (Shepherd, 1973). In Victoria, *H. rubra* also prefers red algae (Sanders, 1981), but also feeds to a lesser extent on the phenolic-poor (Steinberg, 1989) brown algae *Phyllospora comosa* and *Macrocystis* spp. (Shepherd, 1975). In South Australia, *H. roei* prefers red algae, but feeds on the phenolic rich *Ecklonia* and *Zonaria* (Steinberg, 1989 and unpublished data) when red algae are less abundant (Shepherd, 1973). In Western Australia, both red algae and *Sargassum* spp. (presumably high in phenolics, Steinberg, 1986, 1989) are substantial components of the diet of *H. roei*; *Sargassum* is consumed when red algae are less abundant (Wells & Keesing, 1989). Thus Australian abalone apparently prefer to eat red algae, but will consume brown algae, generally when preferred species are uncommon.

Feeding preferences appear to be correlated with the results of growth studies for Australian (and New Zealand) abalone, as would be predicted by evolutionary models of plant defences (Steinberg, 1985). *Haliotis rubra* raised on *Macrocystis* (phenolic-poor) grow more slowly than when fed red algae (Cropp, 1989). *Haliotis iris* in New Zealand also grows more slowly when fed the phenolic-poor *Lessonia variegata* than when fed red algae (Poore, 1972).

In our study with the Australian species, consumption rates were negatively correlated with both algal toughness and phenolic content. Because our study was correlative, rather than experimental, we cannot separate the effects of these factors on consumption rates by the abalone. However, Steneck & Watling (1982) suggested that the large, rugged rhipidoglossan radula of abalone enables these animals to feed relatively easily on leathery macrophytes such as kelps or fucoids. Moreover, Winter (1989) and Winter & Estes (in prep.) concluded that most of the variation in feeding preferences for the North American *H. rufescens* was not due to variation in toughness among the algae studied (although differences in consumption of preferred species could be ascribed to variation in toughness among these algae). These results suggest that variation in phenolic content among algae may play a role in determining food choice or consumption rates of the Australian species.

This conclusion, however, does not seem to apply generally to abalone elsewhere in the world (see Table 13.2 for a summary of information on other abalone species). Although red algae are often high in terms of preference, brown algae are more commonly highly preferred by abalone elsewhere in the world than in Australasia (Table 13.2). In the northeast Pacific, a number of phenolic-poor kelp species are highly preferred by several *Haliotis* spp. The kelp *Undaria* is highly preferred by the Japanese abalone *H. discus hannai*, the phenolic-rich (Anderson & Velimirov, 1982; but see Tugwell & Branch, 1989) kelp *Ecklonia maxima* is preferred by the South African species *H. midae* and phenolic-rich *Ecklonia radiata* is eaten by the Omani species *H. mariae* (Barratt *et al.*, 1986). These studies suggest that leathery (Steneck & Watling, 1982) macrophytic brown algae such as kelps are often highly palatable to abalone, and cast further doubt on the

Table 13.2 Algal food of abalone species from different parts of the world; an asterisk indicates a preferred species as determined by feeding trials or inferred from field data.

Species and references to feeding studies	Habitat and algal dominants	Phenolic-poor brown algae	Algal food Phenolic-rich brown algae	Other algae
Northeast Pacific				
Haliotis kamtschatkana Paul *et al.*, 1977; Ueda & Okada, 1939	Shallow subtidal (to 5 m) *Laminaria* spp. *Alaria* *Agarum*	*Laminaria* spp.* *Alaria**		Diatoms* Red algae* *Ulva** *Enteromorpha**
Haliotis rufescens Cox, 1962 Leighton, 1966 Olsen, 1968	Depth range 0–40 m *Macrocystis* *Nereocystis* *Laminaria* *Agarum* *Pelagophycus*	*Macrocystis** *Egregia** *Pelagophycus* *Laminaria* spp.* *Nereocystis*		Red algae
Haliotis corrugata, *Haliotis fulgens* Leighton, 1966 Tutschulte & Connell, 1988	Depth range 5–40 m *Macrocystis* *Laminaria* spp. *Agarum*	*Macrocystis** *Eisenia** *Egregia*		Red algae*
Haliotis cracherodii Leighton & Boolootian, 1963 Stein, 1984	Intertidal and shallow subtidal *Egregia*, *Eisenia* *Macrocystis* *Pelvetia*	*Egregia* *Eisenia** *Macrocystis**	*Pelvetia fastigiata*	sulphur bacteria
Northeast Atlantic				
Haliotis tuberculata Crofts, 1929 Culley & Peck, 1981	Depth range 0–10 m Commonly intertidal *Laminaria* spp.			*Ulva** *Palmaria** *Enteromorpha* Red algae*

Table 13.2 *Continued*

Species and references to feeding studies	Habitat and algal dominants	Algal food Phenolic-poor brown algae	Phenolic-rich brown algae	Other algae
Northwest Pacific				
Haliotis discus hannai Sakai, 1962 Uki *et al.*, 1986 Nie & Yan, 1985 Kikuchi *et al.*, 1967 Tomita, 1972	Depth range 0–10 m	*Undaria** *Laminaria* spp.		*Ulva** *Enteromorpha*
Southwest Pacific				
Haliotis iris Sinclair, 1963 Tunbridge, 1967 Poore, 1972 Dutton & Tong, 1981	Depth range 0–10 m *Ecklonia* *Carpophyllum* *Macrocystis*	*Macrocystis* *Lessonia*	*Ecklonia* *Carpophyllum* *Blossevillea*	Red algae*
Northwest Indian				
Haliotis mariae Barratt *et al.*, 1986	Depth range 0–10 m *Ecklonia radiata* *Sargassopsis* *Sargassum*		*Ecklonia** *Sargassopsis*	Red algae Green algae
Southern Australia				
Haliotis roei Shepherd, 1973 Wells & Keesing, 1989 (this chapter)	Intertidal, shallow subtidal *Sargassum* *Ecklonia* *Cystophora* spp.		*Ecklonia* *Sargassum* spp.	Red algae*
Haliotis laevigata Shepherd, 1973 (this chapter)	Depth range 5–40 m *Ecklonia* *Cystophora* spp. *Sargassum* spp.	*Lobospira* *Pachydictyon*		Red algae*
Haliotis rubra Shepherd, 1973, 1975, (this chapter) Prince, 1989	Depth range 1–30 m *Ecklonia* *Macrocystis* *Phyllospora*	*Macrocystis* *Phyllospora*		Red algae*
South Africa				
Haliotis midae Newman, 1969 Barkai & Griffiths, 1986	Depth range 0–10 m *Ecklonia maxima* *Laminaria pallida*	*Laminaria*	*Ecklonia**	Red algae*

adequacy of toughness as an explanation for feeding preferences of abalone. The high phenolic content of these species also suggests that polyphenolics are not consistently deterrent to abalone.

Why do brown algae appear to be more palatable to abalone in parts of the world other than Australasia? Estes & Steinberg (1988) presented an evolutionary/ecological model of plant–herbivore interactions. They argued that in the north Pacific, predation (notably by sea otters) historically kept the intensity of herbivory at a low level. This led to weakly evolved plant–herbivore interactions so that there was little selective pressure for algae to evolve chemical defences against herbivores. Conversely, in the temperate southern hemisphere, the intensity of herbivory was greater in the absence of sea otters and more tightly coupled plant–herbivore interactions would have evolved. They predicted more substantial chemical defences to have evolved in algae and the herbivores to have co-evolved a greater tolerance of plant defences.

Although the prediction (Estes & Steinberg, 1988) of greater levels of chemical defences in Australasian waters is borne out (Steinberg, 1989), Australasian abalone are not more tolerant of brown algae, phenolic-rich or otherwise, than are abalone elsewhere in the world. This may be due to differences in the relative abundances of alternative algae. Red algae are abundant as drift or living plants in nearly all abalone habitats in Australia (Shepherd & Womersley, 1970, 1971, 1976, 1981). If abalone have a general tendency to feed on red algae (Table 13.2), then there may simply be no selection pressure on Australian abalone to evolve preferences for brown algae. Conversely, in North America, where abalone feed more readily on brown algae, particularly as drift (Tutschulte & Connell, 1988), the dominant component of this drift is typically brown algae such as *Macrocystis pyrifera* (Gerard, 1976). Thus feeding preferences of these different species may simply reflect differences in the relative abundances of algae in their habitats.

World-wide, the feeding biology of abalone species is not easily explained by simple models based on either secondary metabolites (e.g. Estes & Steinberg, 1988) or morphology (e.g. Steneck & Watling, 1982). The relative abundance of different algae in different regions is one possible explanation, and differences due to taxonomy of the algae may also be important. In particular, red and brown algae differ in many respects. Their storage polysaccharides are quite different (Bold & Wynne, 1981), and Montgomery & Gerking (1980) suggest that difficulties in digesting brown algal storage polysaccharides may explain why many fish avoid brown algae. Abalone may have similar digestive difficulties. Thus much of the variation in feeding by abalone throughout the world, particularly, in their fondness for red algae, may reflect a variation in the ability to digest certain kinds of algal products.

ACKNOWLEDGEMENTS

The authors are grateful to A. Mower for field assistance and thank J.A. Estes, F. Winter, R.W. Day and A.E. Fleming for their criticism of the manuscript.

REFERENCES

Anderson R.J. & Velimirov B. (1982) An experimental investigation of the palatability of kelp bed algae to the sea urchin *Parechinus angulosus*. *Mar. Ecol.* (PSZNI) **3**, 357–73.

Barkai R. & Griffiths C.L. (1986) Diet of the South African abalone *Haliotis midae*. *S. Afr. J. Mar. Sci.*, **4**, 37–44.

Barratt L., Ormond R.F.G. & Wrathall T.J. (1986) *Ecological studies of southern Oman kelp communities Part I.* Tropical Marine Research Unit, University of York, York.

Bold H.C. & Wynne M.J. (1981) *Introduction to the Algae*, 2nd edn. Prentice-Hall, Englewood Cliffs, NJ.

Cox K.W. (1962) California abalones, family Haliotidae. *Fish. Bull., No. 118*, Dept Fish & Game (Resource Agency of California).

Crofts D.R. (1929) *Haliotis. Liverpool Marine Biology Committee Memoirs*, **29**, 1–74.

Cropp R.A. (1989) Abalone culture in Tasmania. *Dept Sea Fisheries, Tasmania Technical Report*, **37**.

Culley M.B. & Peck L.S. (1981) The feeding preferences of the ormer *Haliotis tuberculata* (L). *Kieler Meeresforsch. Sonderh.*, **5**, 570–2.

Dutton S. & Tong L.J. (1981) Food preferences of paua. *Catch 81*, 15 March.

Estes J.A. & Steinberg P.D. (1988) Predation, herbivory and kelp evolution. *Paleobiology*, **14**, 19–36.

Fenical W. (1975) Halogenation in the Rhodophyta. *J. Phycol.*, **11**, 245–59.

Geiselman J.A. & McConnell O.J. (1981) Polyphenols in the brown algae *Fucus vesiculosus* and *Ascophyllum nodosum*: chemical defenses against the herbivorous snail *Littorina littorea*. *J. Chem. Ecol.*, **7**, 1115–33.

Gerard V.A. (1976) *Some aspects of material dynamics and energy flow in a kelp forest in Monterey Bay, California.* PhD thesis, University of California, Santa Cruz.

Hay M.E. & Fenical W. (1988) Marine plant–herbivores interactions, the ecology of chemical defense. *Ann. Rev. Ecol. Syst.*, **19**, 111–45.

Kazlauskas R., Murphy P.T. & Wells R.J. (1978) Two derivatives of farnesylacetone from the brown alga *Cystophora moniliformis*. *Experientia*, **34**, 156–7.

Kikuchi S., Sakurai Y., Sasaki M. & Ino T. (1967) Food values of certain marine algae for the growth of the young abalone *Haliotis discus hannai*. *Bull. Tohoku Reg. Fish. Res. Lab.*, No. 27, 93–100.

Leighton D.L. (1966) Studies on food preference in algivorous invertebrates of southern California kelp beds. *Pac. Sci.*, **20**, 104–13.

Leighton D.L. & Boolootian R.A. (1963) Diet and growth in the black abalone *Haliotis cracherodii*. *Ecology*, **44**, 227–38.

Montgomery W.L. & Gerking S.D. (1980) Marine macroalgae as foods for fishes: an evaluation of potential food quality. *Envir. Biol. Fish.*, **5**, 143–53.

Newman G.G. (1969) Distribution of the abalone *Haliotis midae* and the effect of sea temperature on productivity. *Invest. Rep. Div. Sea Fish. S. Afr.*, **74**.

Nie Z. & Yan J. (1985) Studies on food habit of adult abalone *Haliotis discus hannae* Ino. *J. Fish. China*, **9**, 19–27.

Olsen D.A. (1968) Banding patterns in *Haliotis*–II. Some behavioural considerations and the effect of diet on shell colouration for *Haliotis rufescens*, *Haliotis corrugata*, *Haliotis sorenseni* and *Haliotis assimilis*. *Veliger*, **11**, 135–9.

Paul A.J., Paul J.M., Hood D.W. & Neve R.A. (1977) Observations on food preferences, daily ration requirements and growth of *Haliotis kamtschatkana* Jonas in captivity. *Veliger*, **19**, 303–9.

Poore G.C.B. (1972) Ecology of New Zealand abalones, *Haliotis* species (Mollusca, Gastropoda) 1. Feeding. *NZ J. Mar. Freshwat. Res.*, **6**, 11–22.

Prince J.D. (1989) *The fisheries biology of* Haliotis rubra *in Tasmanian waters.* PhD thesis, University of Tasmania.

Ragan, M.A. & Glombitza K.W. (1986) Phlorotannins, brown algal polyphenols. *Progr. Phycol. Res.*, **4**, 129–241.

Ragan M.A. & Jensen A. (1977) Quantitative studies on brown algal polyphenols. I. Estimation of absolute polyphenol content of *Ascophyllum nodosum* (L) and *Fucus vesiculosus* (L). *J. Exp. Mar. Biol. Ecol.*, **34**, 245–58.

Sakai S. (1962) Ecological Studies on the abalone, *Haliotis discus hannai* Ino–I. Experimental studies on the food habit. *Bull. Jap. Soc. Sci. Fish.*, **28**, 766.

Sanders S. (1981) *Feeding ecology of blacklip abalone* Haliotis rubra *Leach, 1814 at two locations in Port Phillip Bay, Victoria*. BSc (Hons) thesis, University of Melbourne.

Schiel D.R. (1982) Selective feeding by the echinoid, *Evechinus chloroticus*, and the removal of plants from subtidal algal stands in northern New Zealand. *Oecologia (Berl.)*, **54**, 379–88.

Shepherd S.A. (1973) Studies on southern Australian abalone (genus *Haliotis*). 1. The ecology of five sympatric species. *Aust. J. Mar. Freshwat. Res.*, **24**, 217–57.

Shepherd S.A. (1975) Distribution, habitat and feeding habits of abalone. *Aust. Fish.*, **34**(1), 1–4.

Shepherd S.A. & Womersley H.B.S. (1970) The sublittoral ecology of West Island, South Australia. I. Environmental features and the algal ecology. *Trans R. Soc. S. Aust.*, **94**, 105–38.

Shepherd S.A. & Womersley H.B.S. (1971) Pearson Island Expedition 1969–7. The subtidal ecology of benthic algae. *Trans R. Soc. S. Aust.*, **95**, 155–67.

Shepherd S.A. & Womersley H.B.S. (1976) The subtidal algal and seagrass ecology of St Francis Island, South Australia. *Trans R. Soc. S. Aust.*, **100**, 177–91.

Shepherd S.A. & Womersley H.B.S. (1981) The algal and seagrass ecology of Waterloo Bay, South Australia. *Aquat. Bot.*, **11**, 305–71.

Sinclair M. (1963) Studies on the paua, *Haliotis iris* Martyn in the Wellington District 1945–46. *Zool. Publ. Victoria University of Wellington*, No. 35.

Stein J.L. (1984) Subtidal gastropods consume sulphur-oxidizing bacteria: evidence from coastal hydrothermal vents. *Science*, **223**, 696–8.

Steinberg P.D. (1984) Algal chemical defense against herbivores: allocation of phenolic compounds in the kelp *Alaria marginata*. *Science*, **223**, 405–7.

Steinberg P.D. (1985) Feeding preferences of *Tegula funebralis* and chemical defenses in marine brown algae. *Ecol. Monogr.*, **55**, 333–49.

Steinberg P.D. (1986) Chemical defenses and the susceptibility of tropical marine algae to herbivores. *Oecologia*, **69**, 628–30.

Steinberg P.D. (1988) The effects of quantitative and qualitative variation in phenolic compounds on feeding in three species of marine invertebrate herbivores. *J. Exp. Mar. Biol. Ecol.*, **120**, 221–37.

Steinberg P.D. (1989) Biogeographical variation in brown algal polyphenolics and other secondary metabolites, comparison between temperate Australasia and North America. *Oecologia*, **78**, 373–82.

Steneck R.S. & Watling L. (1982) Feeding capabilities and limitations of herbivorous molluscs: a functional group approach. *Mar. Biol. (Berl.)*, **68**, 299–319.

Tomita K. (1972) Experiments on the food selectivity and digestion of abalone seed *Haliotis discus hannai*. *J. Hokkaido Fish. Scient. Inst.*, **29**(4), 17–23.

Tugwell S. & Branch G.M. (1989) Differential polyphenolic distribution among tissues in the kelps *Ecklonia maxima*, *Laminaria pallida* and *Macrocystis angustifolia* in relation to plant-defence theory. *J. Exp. Mar. Biol. Ecol.*, **129**, 219–30.

Tunbridge B.R. (1967) Feeding habits of paua (*Haliotis iris* Martyn). *Fish. Technical Report*, No. 20, NZ Marine Dept.

Tutschulte T.C. & Connell J.H. (1988) Feeding behaviour and algal food of three species of abalones (*Haliotis*) in southern California. *Mar. Ecol. Prog. Ser.*, **49**, 57–64.

Ueda S. & Okada Y. (1939) Studies on the food of useful gastropoda in Japan. 1. Two species of Japanese Abalone, *Haliotis gigantea* Gmelin and *H. kamtschatkana* Jonas. *Bull. Jap. Soc. Sci. Fish.*, **8**, 51.

Uki N., Sugiura M. & Watanabe T. (1986) Dietary value of seaweeds occurring on the Pacific coast of Tohuku for growth of the abalone *Haliotis discus hannai*. *Bull. Jap. Soc. Sci. Fish.*, **52**, 257–66.

Vadas R.L. (1977) Preferential feeding: an optimization strategy in sea urchins. *Ecol. Monogr.*, **47**, 337–71.

Van Altena I.A. (1987) Terpenoids from the brown alga *Cystophora moniliformis*. *Aust. J. Chem.*, **41**, 49–56.

Wells F.E. & Keesing J.K. (1989) Reproduction and feeding in the abalone *Haliotis roei* Gray. *Aust. J. Mar. Freshwat. Res.*, **40**, 187–97.

Winter F.C. (1989) Brown algal polyphenolics deter grazing by the California red abalone. 1st International Symposium on Abalone Biology, Fisheries and Culture, La Paz, Mexico (abstr.).

Chapter 14
Algal feeding stimulants for abalone

KANZO SAKATA and KAZUO INA *Research Laboratory of Marine Biological Science, Faculty of Agriculture, Shizuoka University, Mochimune, Shizuoka 421–01, Japan*

ABSTRACT: This chapter describes a simple and reliable bioassay procedure (Avicel plate method) for chemical studies of feeding behaviour of marine herbivorous gastropods which was used to screen algal feeding stimulants for young *Haliotis discus*. Young abalone showed feeding preferences for different algal methanol extracts. From an extract of the brown alga *Undaria pinnatifida*, one of the most effective ones, digalactosyldiacylglycerol (DGDG) and phosphatidylcholine (PC) were isolated and identified as phagostimulants for the abalone.

The Avicel plate method has been shown to be applicable to other herbivorous gastropods, such as the turban shell *Turbo cornutus* and the topshell *Omphalius pfeifferi*. The authors have also established a simple phagostimulant assay, using filter paper, for another herbivorous gastropod, the sea hare *Aplysia juliana*. Isolation and identification of gastropod feeding stimulants from the green alga *Ulva pertusa* are also described briefly. Several active principles have been isolated and identified. DGDG and a novel glycerolipid, 1,2-diacylglyceryl-4′-*O*-(*N*,*N*,*N*-trimethyl)-homoserine(DGTH or DGTS, but the authors prefer the former), and its lyso-form, 1-monoacylglyceryl-4′-*O*-(*N*,*N*,*N*-trimethyl)-homoserine (MGTH), were found to be effective for all the gastropods tested.

RESUMEN: Estamos interesados en los estudios químicos de comportamiento alimentario, de gasterópodos herbívoros marinos y tuvimos éxito en establecer un método de bioensayo simple y confiable; metodo de placa de Avicel; que nos permite probar estimulantes alimenticios de origen algal con juveniles de *Haliotis discus*. Los abulones jóvenes mostraron diferentes preferencias alimenticias por los extractos metanólicos de algas. Del extracto del alga parda *Undaria pinnatifida* se obtuvo del degalactosildiacilglicerol (DGDG) y la fosfatidilcolina (PC) aislados por primera vez como fagoestimulantes para el abulón.

Recientemente el método Avicel probó ser aplicable a otros gasterópodos herbívoros como *Turbo cornutus* y *Omphalius pfeifferi*. Establecimos también una prueba simple para fagoestimulantes usando papel filtro para otros gasterópodos herbívoros, como *Aplysia juliana*. El extracto metanólico del alga verde *Ulva pertusa*, uno de los más efectivos para todos los gasterópodos, fue fraccionado y sujeto a los métodos de bioensayo arriba mencionados. Se aislaron e identificaron varios principios activos. Es muy interesante que el DGDG y un nuevo glicerolípido el 1,2-diacylgliceril-4′-*O*-(*N*,*N*,*N*-trimetil)-homoserina (DGTH) y su lisoforma, 1-monoacilgliceril-4′-*O*-(*N*,*N*,*N*-trimetil)-homoserina (MGTH) hayan resultado ser comúnmente efectivos para todos los gasterópodos probados.

INTRODUCTION

Chemoreception has been proposed to play an important role in the feeding behaviour of the abalone *Haliotis discus hannai*, and feeding preferences for some kinds of brown and red algae have been observed (Kikuchi *et al.*, 1967; Uki, 1981). Uki *et al.* (1986) showed that brown algae, including *Undaria pinnatifida* and *Eisenia bicyclis*, are the most suitable food for abalone growth. In contrast, Australian abalone feed chiefly on red algae, and a particular species, *H. laevigata*, fe seagrass (Shepherd, 1975). As feeding is essential for animals, selection and ingestion of appropriate foods can be expected to have an adaptive or evolutionary basis. Table 14.1 lists chemical substances concerned with chemoreception in various kinds of feeding behaviour, based on definitions by Mackie (1982) and Lindstedt (1971). In all cases, the animal responds to a stimulus involving a chemical gradient.

Recognition of active substances is actually quite difficult. Feeding arrestants can be involved in the response to attractants in some cases. Feeding incitants and stimulants are also difficult to distinguish unless appropriate bioassay procedures are used.

Harada and colleagues studied the attractant effect of seaweeds (Harada & Kawasaki, 1982) and the algal constituents (Harada *et al.*, 1984), but the chemical basis of the results was not clear. The anticipated shortage of fishery resources in the future will increase the need to establish efficient culture methods for commercially valuable gastropods, including abalone. For this reason, feeding stimulants are likely to be more important than feeding attractants for seed production and culture. These considerations prompted us to study feeding stimulants for the abalone *H. discus*, as well as other useful gastropods.

MATERIALS AND METHODS

Test animals

Young abalone (*H. discus*, 15–25 mm, 1–2 years old), kindly supplied by the Shizuoka Prefecture Fish Farming Center, were kept in an experimental aquarium (Fig. 14.1). The animals were fed the brown alga *E. bicyclis* and were starved for at least one day before the bioassay.

Feeding-stimulant assay

Figure 14.1 shows a diagrammatic view of an experimental aquarium (a commercially available dark blue plastic planter) with a shelter made of black polyvinyl

Table 14.1 Chemical substances involved in chemoreception in different feeding behaviours.

Substance	Effect
Arrestant	orientation toward the apparent source possibly over long distances
Attractant	cessation of locomotion close to the apparent source
Repellent	orientation away from the apparent source
Incitant	evocation of the initiation of feeding (tasting)
Suppressant	inhibition of the initiation of feeding
Stimulant	promotion of ingestion and continuation of feeding
Deterrent	prevention of continued feeding or accelerated termination of feeding

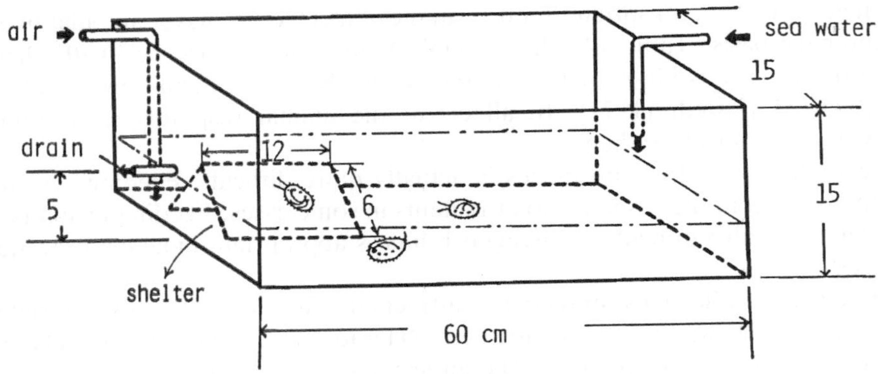

Fig. 14.1 Diagram of an experimental aquarium for the feeding-stimulant assay for young *Haliotis discus*.

chloride plate for the nocturnal abalone to rest under during the day. Between 10 and 20 young abalone were used for the assay.

Algal methanol extracts or sample solutions (40–100 μl) were applied with a microsyringe on to the sample zone (about 25–30 mm in diameter, drawn with a pair of compasses) on an Avicel plate, which was made as follows. Avicel SF (crystalline cellulose, Asaki Kasei Co) was coated to a thickness of 0.25 mm on glass plates (5 × 20 or 10 × 20 cm) in the conventional way for thin-layer chromatography (TLC) and dried at 120°C for 1 hour. The prepared Avicel plate was set on the bottom of the test aquarium after sunset (Fig. 14.2).

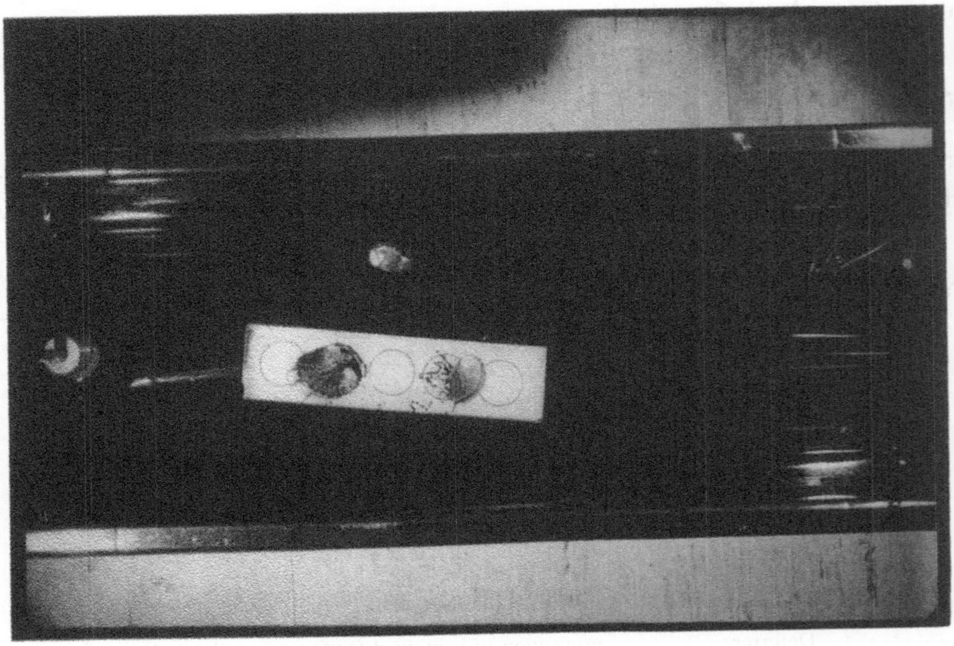

Fig. 14.2 An overview of the experimental aquarium of *Haliotis discus* during a bioassay. This photograph was taken at 23.00 hours. Two sample zones have just been bitten off by three abalone.

The seawater supply was stopped during the assay. By comparing the number of characteristic feeding traces left on the plate the next morning (Fig. 14.3), the feeding-stimulant activity of each sample was judged as:

++ biting traces are observed over almost all the sample zone and the activity can be judged clearly; for example, 50- and 25-mg equivalent extracts of the air-dried alga in Fig. 14.3;

\+ many more biting traces are found in the sample zone than outside in a unit area; 13-mg equivalent sample zone in Fig. 14.3;

± a few more biting traces are observed in the sample zone than outside, and it is desirable to repeat the assay to judge the activity;

− no or nearly the same number of the biting traces are inside the sample zone as are outside: for example, reference zones in Fig. 14.3.

Isolation of the feeding stimulants

The air-dried alga *U. pinnatifida* Harvey, (94 g) harvested in May, was extracted continuously with ether by using a Soxhlet extractor (Fig. 14.4). The methanol extracts were obtained after the residue had been soaked in methanol (500 ml) for several days. The extraction process was repeated four times. Clear evidence of feeding-stimulant activity was not found in the ether extracts, even at a dose equivalent to 0.1 g of the dried material, but was very apparent in the methanol extracts at a dose equivalent to 0.016 g. The methanol of the extracts was evaporated. The residue was combined with water (100 ml), extracted with ethyl acetate (60 ml, four times) and then with butanol (60 ml, four times). The ethyl acetate extracts ($\frac{1}{10}$ of the extracts) were concentrated and chromatographed on a Sephadex LH-20 column (internal diameter 2×40 cm) with methanol as an elution solvent (each fraction, 10 g = 12.8 ml). The butanol extracts ($\frac{1}{10}$ of the extracts) were fractionated in the same way as the ethyl acetate extracts. In both

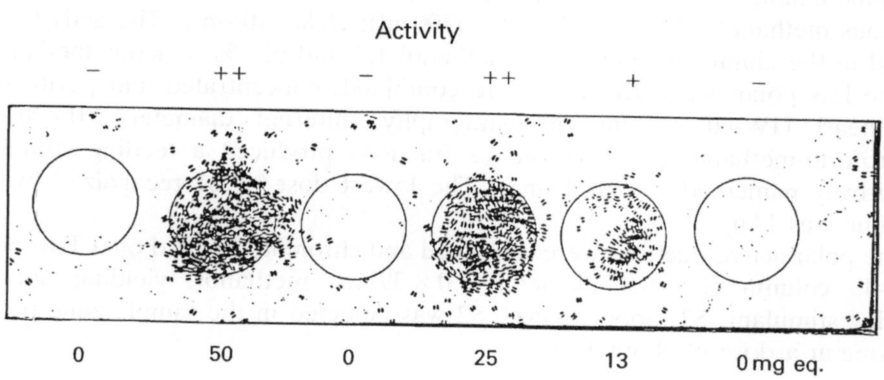

Fig. 14.3 Typical assay results obtained by using young *Haliotis discus*. Methanol extracts equivalent to the indicated amount of air-dried *Undaria pinnatifida* were absorbed in each sample zone on the Avicel plate. See text for details of experiment.

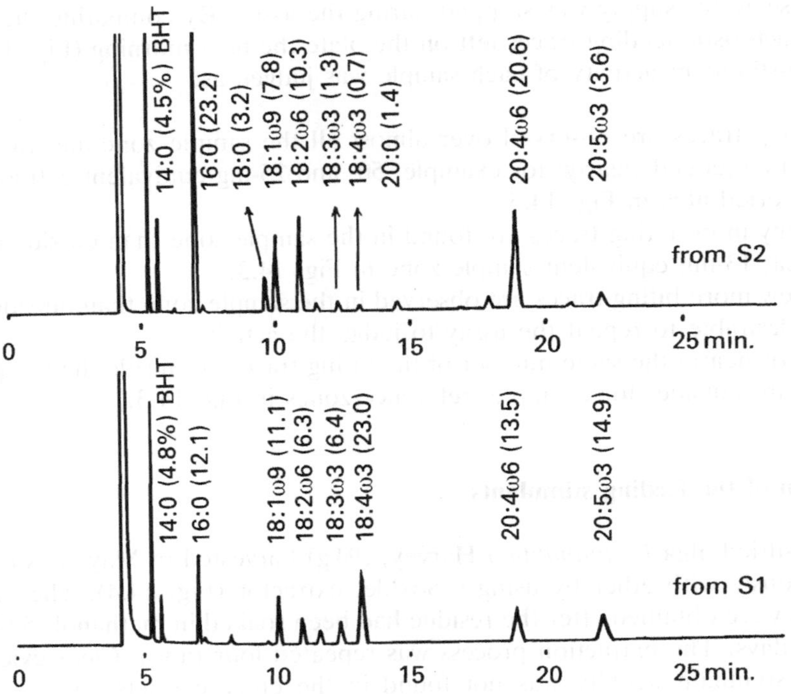

Fig. 14.4 Isolation procedure for S1 (DGDG) and S2 (PC) from the methanol extracts of the brown alga *Undaria pinnatifida*.

cases, fractions 5 to 7 stimulated feeding. The remaining extracts were fractionated on a larger column to yield the active fractions.

The combined active fractions were concentrated and chromatographed on a silica gel column (Wako-gel C-300, internal diameter 2.5 × 14 cm, packed with a mixture of chloroform-methanol (20:1)). The elution was performed with chloroform-methanol: 20:1 (60 ml), 5:1 (120 ml), 1:2 (120 ml); methanol (160 ml); aqueous methanol: 5% (100 ml), 10% (100 ml); 20% (100 ml). The activity was found in the eluates of chloroform-methanol 1:2 and of 5% aqueous methanol.

The less polar active fractions were combined, concentrated and purified by Toyopearl HW-40 column chromatography (internal diameter 2.0 × 46 cm, chloroform-methanol 1:2). The active fractions produced a feeding stimulant tentatively named S1 (about 10 mg). The lowest dose of S1 recognized by the abalone was 14 μg.

The polar active fractions were combined and chromatographed on a Toyopearl HW-40 column (internal diameter 2.0 × 39 cm, methanol, yielding another feeding stimulant, S2 (about 58 mg). S2 was detected in the sample zone by the abalone at a dose of about 10 μg.

Thin-layer chromatography

Thin-layer chromatography of S1 and S2 was performed on plates (5 × 20 or 10 × 20 cm) coated with silica gel GF_{254} (Merck), 0.25 mm thick, by using the solvent systems shown in Table 14.2. S1 was detected as a single spot by I_2 vapour, 0.02% $KMnO_4$ and Anthrone reagents. S2 was detected by I_2 vapour, 0.02% $KMnO_4$ and Dittmer-Lester and Dragendorff reagents.

Identification of the feeding stimulants S1 and S2

S1

A mixture of S1 (0.5 mg), methanol (0.5 ml) and methanolic *N* KOH (three drops) was left at room temperature for 1 hour and passed through a small column of Dowex 50W (H^+). The effluent was evaporated to dryness. Part of the residue was dried and trimethylsilylated by using a mixture (30 µl) of bis-trimethylsilyl acetamide/trimethylchlorosilane/pyridine (50/10/100, v/v) at 80°C for 5 minutes for gas-liquid chromatography (GLC) and gas chromatography-mass spectroscopy (GC-MS). A known sample of DGDG was treated in the same way as S1. Analytical data were identical, confirming that S1 is DGDG.

The hexane extract was analyzed by GLC to clarify the fatty acid composition (Fig. 14.5).

^{1}H and ^{13}C nuclear magnetic resonance (NMR) and fluid desorption-mass spectroscopy (FD-MS) spectra measurements of S1 also confirmed the structure (Sakata & Ina, 1985).

S2

S2 was treated in the same way as S1, except for reaction conditions (40°C, 30 minutes). After neutralization with dilute HCl followed by addition of butylated hydroxy toluene (BHT) (about 3 µg), the mixture was evaporated to dryness. The residue was triturated with hexane, yielding the extract for fatty acid analysis by GLC (Fig. 14.5).

^{13}C-NMR data of S2 (22.5 MHz, DMSO-d_6, δ: 173.0, 71.3 (d), 66.3 (t), 63.2 (t), 59.0 (t), 53.9 (q) and FD-MS spectra of perhydrogenated S2 confirmed that S2 is PC (Sakata & Ina, 1985).

RESULTS AND DISCUSSIONS

We maintained young *H. discus* (shell lengths about 1.5–3.00 cm) in an aquarium (Fig. 14.1) and observed their feeding behaviour. They begin to search for food

Table 14.2 Thin-layer chromatography of S1 and S2; chromatography plate: silica gel GF_{254} (Merck), 0.25 mm.

Solvent system	Rf value		
	S1	DGDG*	S2
Diisobutyl ketone/acetic acid/water (40/25/4, v/v)	0.29	0.28	0.25
chloroform/methano/ammonium hydroxide/water (48/34/4/2, v/v)	0.47	0.46	0.58
butanol/acetic acid/water (4/1/2, v/v)	0.60		0.46
propanol/water (35/15, v/v)	0.79	0.79	0.34

* Known sample from spinach.

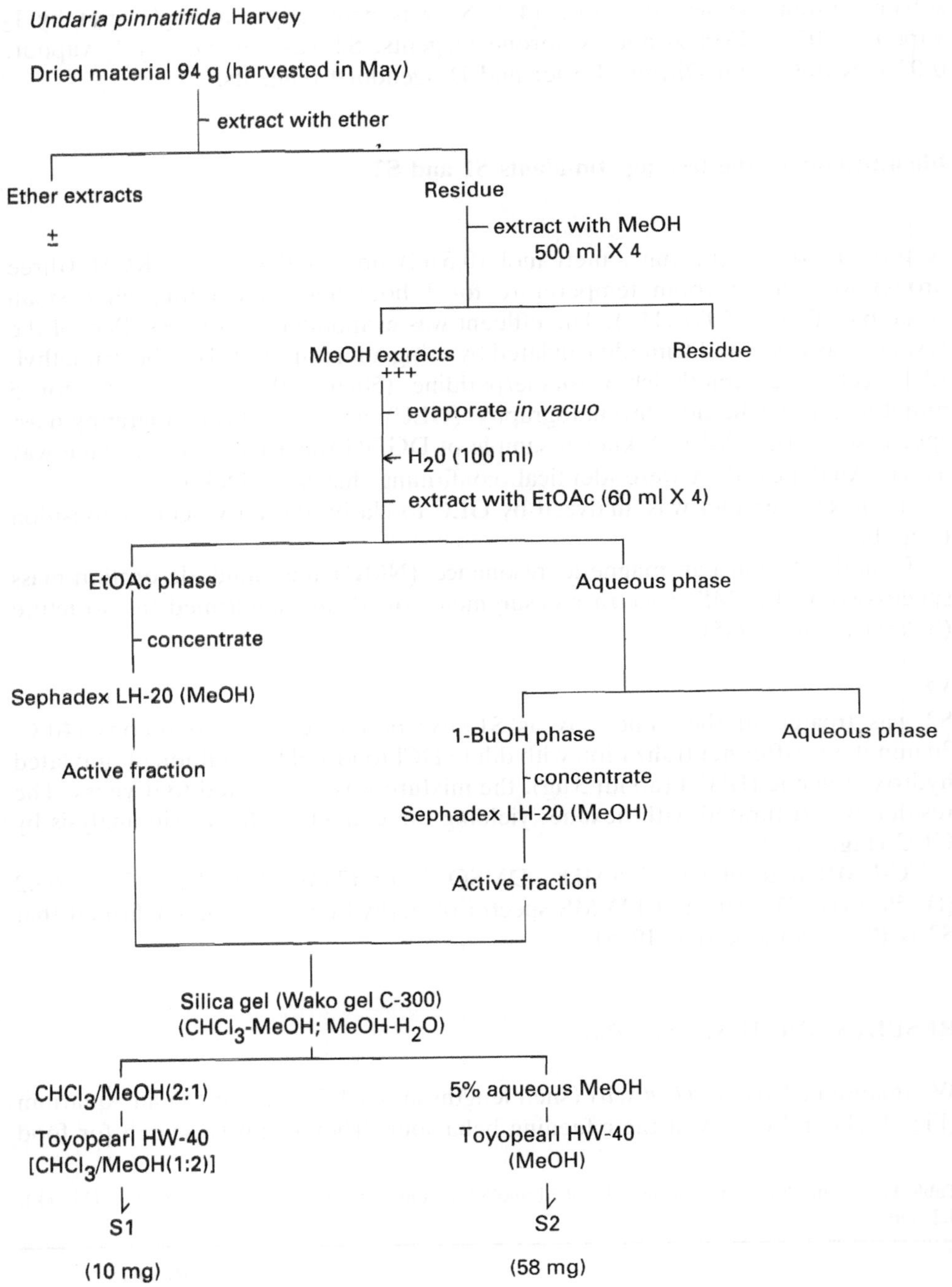

Fig. 14.5 Gas-liquid chromatograms of fatty acid methyl esters from S1 and S2. Instrument, Hitachi gas chromatograph Model 163 (FID); column, PEG-HT G-SCOTT fused silica capillary column, internal diameter 0.25 mm × 50 m; oven temperature, 210°C; carrier gas, N_2, 1.5 ml/min; injection and detection temperature 240°C. Values in parentheses are area percentages of each peak.

after sunset. Careful observation of their feeding behaviour led to establishment of a simple and reliable bioassay procedure (Avicel plate method) for chemical study of feeding preferences (Sakata *et al.*, 1984).

Figure 14.3 shows the results of a typical overnight Avicel plate bioassay. Assessment of the feeding activity was easily achieved. The abalone seem to creep on to the test plates in search of food, leaving a pair of their characteristic feeding traces here and there. They bite off the Avicel when they come across a sample zone where feeding stimulants have been absorbed. This assay method is applicable to the identification of feeding stimulants rather than attractants.

The extract of *U. pinnatifida*, one of the most phagostimulatory algae, was fractionated as shown in Fig. 14.4. Application of this assay procedure to each fraction allowed us to isolate two active principles, tentatively named S1 and S2. Analysis by TLC, including colouring reactions, suggested that S1 and S2 are complex glycerolipids.

Spectroscopic and chemical studies unambiguously revealed that S1 and S2 are DGDG and TC, respectively, having saturated/unsaturated fatty acid moieties composed of C_{14}–C_{20} chains (Fig. 14.5; Sakata & Ina, 1983, 1985). Both S1 and S2 were detected by the young abalone at doses of less than 20 μg (Table 14.3).

This is the first report to have clarified the feeding-stimulant activity of the glycerolipids, DGDG and PC, which are rather lipophilic compounds. Feeding attractants and stimulants of marine organisms elucidated previously were all water-soluble compounds such as amino acids and nucleosides (Kohn, 1961; Lehnhoff & Lindstedt, 1974; Leighton, 1966; Mackie, 1982; Sakata, 1988, 1989).

CONCLUDING REMARKS

Simple feeding-stimulant assay procedures have been established for other marine herbivorous gastropods and algal feeding stimulants for these animals have been isolated and identified. Comparison of the phagostimulants for *H. discus* with those of other herbivorous gastropods is quite interesting. Thus we briefly describe the bioassay procedures and the algal feeding stimulants for them.

Young sea hares, *Aplysia juliana*, were kept in a test aquarium (32 × 61 × 18 cm) with a drain to maintain a water depth of 5 cm. Filter papers with

Table 14.3 Feeding-stimulant activities of the glycerolipids isolated from the green alga *Ulva pertusa*; results are given as the minimal effective dose (MED) in micrograms (MED = minimal sample amount/sample zone necessary to show positive activity); these values are preliminary.

Isolate	*Haliotis discus*	*Turbo cornutus*	*Omphalius pfeifferi*	*Aplysia juliana*
DGDG	14–28	15–25	18–23	800*
DGTH	10–20	24	<10	100
MGTH	23–50	25–50	<23	
SQDG	>300	20–40	<20	>1000
PC†	<10	15	<15	<50

* in the case of *Aplysia*, minimal sample amount/filter paper (Toyo filter paper No 2, 5.5 cm in diameter).

† isolated from the brown alga *Undaria pinnatifida*

absorbed sample solutions were weighed and placed at the bottom of the test aquarium at the beginning of the assay. The sea hares moved around; some of them showed their typical feeding stimulatory responses, and some of them reached the filter paper and started eating it. One fraction was obviously quite stimulatory (Fig. 14.6). The number of sea hares at each filter paper was counted after 10, 20 and 30 minutes. Assessment of activity was based on the total number of sea hares aggregated on each filter paper (Sakata *et al.*, 1986).

The turban shell ('sazae' in Japanese) *Turbo cornutus* and the top shell ('bateira' in Japanese) *Omphalius pfeifferi* are both marine herbivorous gastropods of commercial value in Japan whose delicate taste is highly esteemed. The Avicel plate method was recently found to be applicable to chemical studies of feeding stimulants for these gastropods (Sakata *et al.*, 1988a). We have now established feeding-stimulant assay procedures for four different species of marine herbivorous gastropods.

Among several methanolic algal extracts, the *Ulva pertusa* extract was the most stimulatory for all gastropods tested. Chromatographic purification of the extracts yielded several glycerolipids as phagostimulants: DGDG, DGTH, MGTH and 6-sulphoquinovosildiacylglycerol (SQDG) (Fig. 14.7) (Sakata *et al.*, 1985, 1988b). Although SQDG is present in the brown algae *E. bicyclis* and *U. pinnatifida*, it was not isolated as a feeding stimulant for *H. discus* in our previous study (Sakata & Ina, 1985). This time, we found that SQDG was a feeding stimulant for the

Fig. 14.6 An overview of a test aquarium for *Aplysia juliana* during the bioassay. At the beginning of the assay, 10 test animals were distributed along the edge of the aquarium bottom. Filter papers (5.5 cm in diameter) with absorbed aliquots of sample solutions of the green alga *Ulva pertusa* were set with sinkers on the bottom of the aquarium. This photograph was taken 20 minutes after the beginning of the assay.

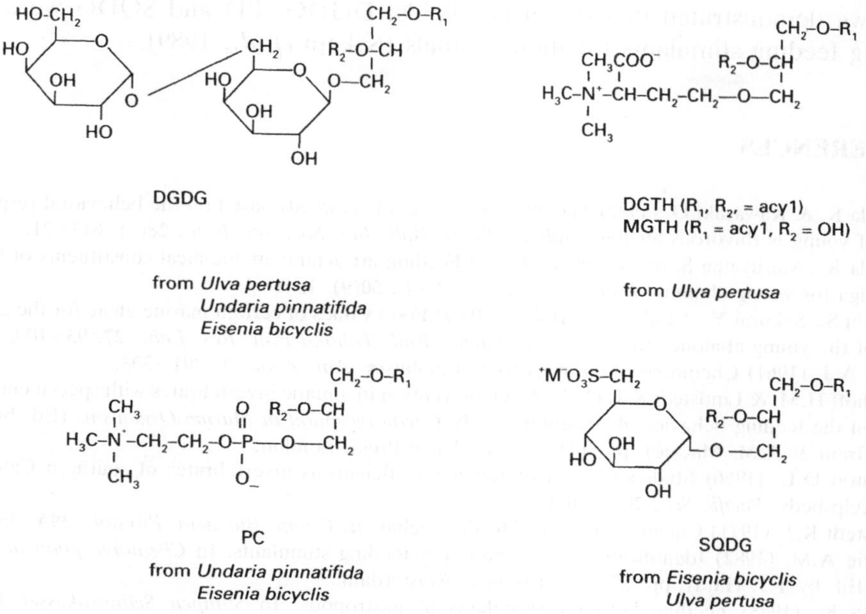

Fig. 14.7 Algal feeding stimulants isolated and identified. R_1 and R_2 are saturated/unsaturated fatty acids composed of C_{14} to C_{22}.

turban shell *T. cornutus* and the top shell *O. pfeifferi*. This suggests that the food preference of these two herbivorous gastropods is similar, but slightly different from that of *H. discus*. The phagostimulatory activity of the organic extracts of this green alga can, therefore, be attributed to the glycerolipids DGDG, DGTH, MGTH and SQDG (Sakata *et al.*, 1988b). DGDG, DGTH, MGTH and PC were effective for all the gastropods tested. SQDG was not as active as the others with *H. discus*, although the young abalone were quite responsive to amounts of PC, DGDG and DGTH as small as 20–30 µg.

We have tried to isolate feeding stimulants from attached diatoms, which are widely used as natural diets for postlarval abalone, by using smaller *H. discus* (8–15 mm shell length). One of the active principles we isolated was, to our surprise, SQDG (Sakata *et al.*, pers. obs.), because SQDG showed much lower feeding stimulatory activity in somewhat larger *H. discus* in our preliminary experiment (Table 14.3). We are re-examining the phagostimulatory activity of SQDG.

The wide occurrence of DGDG and PC in the plant kingdom strongly suggests that marine herbivorous gastropods could be stimulated to feed not only on any algae but also on terrestrial plants. The abalone *H. discus hannai*, however, reportedly shows quite different feeding preferences (Kikuchi *et al.*, 1967). Some algal species are not eaten by them, even though the algae are not particularly tough. Suppressants (feeding deterrents) may be present in these particular species of algae. The Avicel plate method is now being used to study this interesting phenomenon. We have found that the Avicel plate method may be successfully applied to feeding-stimulant studies of the sea urchin *Strongylocentrotus intermedius* and we demonstrated that the glycerolipids, DGDG, PD and SQDG were also strong feeding stimulants for these animals (Sakata *et al.*, 1989).

REFERENCES

Harada K. & Kawasaki O. (1982) The attractive effect of seaweeds based on the behavioral responses of young herbivorous abalone *Haliotis discus*. *Bull. Jap. Soc. Sci. Fish.*, **28**(5), 617–21.

Harada K., Maruyama S. & Nakano K. (1984) Feeding attractants in chemical constituents of brown alga for young abalone. *Bull. Jap. Soc. Sci. Fish.*, **50**(9), 1541–4.

Kikuchi S., Sakurai Y., Sasaki M. & Itoh T. (1967) Food values of certain marine algae for the growth of the young abalone, *Haliotis discus hannai*. *Bull. Tohoku Fish. Res. Lab.*, **27**, 93–100.

Kohn A.J. (1961) Chemoreception in gastropod molluscs. *Am. Zool.*, **1**, 291–308.

Lehnhoff H.M. & Lindstedt K.J. (1974) Chemoreception in aquatic invertebrates with special emphasis on the feeding behavior of coelenterates. In *Chemoreception in Marine Organisms* (Ed. by P.T. Grant & A.M. Mackie), pp. 143–75. Academic Press, London.

Leighton D.L. (1966) Studies of food preference in algivorous invertebrates of southern California kelp beds. *Pacific Sci.*, **20**(1), 104.

Lindstedt K.J. (1971) Chemical control of feeding behavior. *Comp. Biochem. Physiol.*, **39A**, 553–81.

Mackie A.M. (1982) Identification of the gustatory feeding stimulants. In *Chemoreception in Fishes* (Ed. by T.J. Hara), pp. 275–91. Elsevier, Amsterdam.

Sakata K. (1988) Feeding behavior stimulants of gastropods. In *Shinpen Seibutsukassei Tennen Busshitsu* (Ed. by S. Shibata), pp. 138–44. Ishiyaku Shuppan, Tokyo.

Sakata K. (1989) Feeding attractants and stimulants for marine herbivorous gastropods. In *Marine Bioorganic Chemistry*, Vol. 3 (Ed. by P. Scheuer), pp. 115–29. Springer-Verlag, Heidelberg.

Sakata K. & Ina K. (1983) Digalactosyldiacylglycerols isolated from a brown alga as effective phagostimulants for a young abalone. *Agric. Biol. Chem.*, **47**(12), 2957–60.

Sakata K. & Ina K. (1985) Digalactosyldiacylglycerols and phosphatidylcholines isolated from a brown alga as effective phagostimulants for a young abalone. *Bull. Jap. Soc. Sci. Fish.*, **51**(4), 659–65.

Sakata K., Itoh T. & Ina K. (1984) A new bioassay method for a young abalone, *Haliotis discus* Reeve. *Agric. Biol. Chem.*, **48**(2), 425–9.

Sakata K., Kato K., Ina K. & Machiguchi Y. (1989) Glycerolipids as potent feeding stimulants for the sea urchin, *Strongylocentrotus intermedius*. *Agric. Biol. Chem.*, **53**(5), 1457–9.

Sakata K., Sakura T. & Ina J. (1988a) Algal phagostimulants for marine herbivorous gastropods. *J. Chem. Ecol.*, **14**(5), 1405–16.

Sakata K., Sakura T., Kamiya Y. & Ina K. (1988b) A simple feeding stimulant assay for marine herbivorous gastropods, the turban shell *Turbo cornutus* and the top shell *Omphalius pfeifferi*. *Nippon Suisan Gakkaishi*, **54**(10), 1715–18.

Sakata K., Tsuge M. & Ina K. (1986) A simple bioassay for feeding-stimulants for the young sea hare *Aplysia juliana*. *Mar. Biol.*, **91** 509–11.

Sakata K., Tsuge M., Kamiya Y. & Ina K. (1985) Isolation of a glycerolipid (DGTH) as a phagostimulant for a sea hare, *Aplysia juliana*, from a green alga, *Ulva pertusa*. *Agric. Biol. Chem.*, **49**(6), 1905–7.

Shepherd S.A. (1975) Distribution, habitat and feeding habits of abalone. *Aust. Fish.*, **34**(1), 12–15.

Uki N. (1981) Food value of marine algae of order Laminariales for growth of the abalone, *Haliotis discus hannai*. *Bull. Tohoku Fish. Res. Lab.*, **42**, 19–29.

Uki N., Sugiura M. & Watanabe T. (1986) Dietary value of seaweeds occurring on the Pacific coast of Tohoku for growth of the abalone *Haliotis discus hannai*. *Bull. Jap. Soc. Sci. Fish.*, **52**(2), 257–66.

Chapter 15
Feeding attraction activity of fragrant and pungent spice extracts in black abalone, *Haliotis discus* (Studies on the Feeding Attractants for Fishes and Shellfishes XVII)

KATSUHIKO HARADA *Laboratory of Biochemistry, Department of Food Science and Technology, Shimonoseki University of Fisheries, Yamaguchi 759–65, Japan*

ABSTRACT: The feeding-attraction activity of fragrant and pungent spices in young black abalone, *Haliotis discus*, was estimated statistically on the basis of exploratory behaviour of the gastropod. Of 18 specimens tested among the odorous spices, anise, *Pimpinella anisum*, basil, *Ocimum basilicum*, caraway, *Carum carvi*, marjoram, *Majorana hortensis*, and peppermint, *Mentha piperita*, were found to be highly attractive to abalone. Among four specimens of the acrid spices, strong attractants were black pepper, *Piper nigrum*, and mustard, *Synapis alba*. Of eight specimens of odour-corrective or odour-masking spices, onion, *Allium cepa*, and oregano, *Origanum vulgare*, strongly stimulated the exploratory behaviour. Marjoram was highest in attraction activity. The relationship between the attractive effect and various spice characteristics is also discussed.

RESUMEN: Con el objeto de determinar la atracción alimentaria que ejercen materiales fragantes y picantes, de especias que no tienen oportunidad alguna de ser proporcionadas como pienso a los animales acuáticos, se estimó estadísticamente, sobre la base del comportamiento alimentario y exploratorio, la atracción que ejercen algunas especias sobre juveniles de abulón negro (*Haliotis discus*) colocados en un tanque experimental. La muestra experimental fue el extracto obtenido después de agitar la especia por 16 horas en agua a temperatura ambiente. Entre las 30 especias se incluyeron 18: olorosas, cuatro picantes y ocho inodoras. Un total de 18 compuestos; nueve olorosos, tres picantes y seis inodoros, atrajeron en cierta medida al abulón negro. Las nueve especias con fuerte actividad atrayente fueron: cinco olorosas (anis, basilio, alcaravea, mejorana y menta), dos picantes (pimienta y mostaza) y dos inodoras, (cebolla y orégano). En una prueba comparativa de cada combinación seleccionada arbitrariamente, tres especies entre estas nueve especias (basilio, pimienta y mejorana) mostraron evidentemente gran actividad atrayente. Más aún, en la combinación de las tres especias anteriores, la mejorana mostró ser la mas atrayente. La actividad atrayente de la mejorana dependió claramente de la concentración usada. La relación entre el efecto atrayente y la característica de la especias también se discute. Se señala especialmente la posibilidad de que las especias picantes en el caso de utilizarlas como aditivos alimenticios en el pienso, desempeñen un triple, es decir, como estimulantes y antisepticos, además de atrayentes.

INTRODUCTION

Many studies on feeding attractants for aquatic animals have dealt with the attractive effect of constituents and compounds in the natural food of fishes and shellfishes (Harada, 1987). Among the various substances, amino acids and related compounds have received much attention.

In contrast, only a few studies deal with the attractive effect of constituents and compounds in food substances foreign to aquatic animal environments (Michelson, 1960; Takei, 1967; Goldschmeding & Jager, 1973; Harada & Kawasaki, 1982). A study of favourite food attractants for finfishes (Takei, 1967) reports that anise oil (obtained from certain spices) is favoured by rainbow trout, *Salmo gairdneri*, but not carp, *Cyprinus carpio*, or yellowtail, *Seriola quinqueradiata*. In commercial and sport fishing in Japan, spices such as garlic and onion are added to some baits. On the basis of these findings, it is presumed that spices may also be effective as feeding attractants for other aquatic animals. Furthermore, spices may be optimal additives of artificial diets from the standpoint of their antimicrobial and antiseptic actions (Kanda *et al.*, 1971). Spices may have an appetizing effect distinct from their water extracts obtained from each of 30 specimens of 28 species of spices; the herbivorous young black abalone, *Haliotis discus*, was used as the test animal.

MATERIALS AND METHODS

Test animals

Young black abalone (average shell length 2.7 cm) were kindly supplied by Yamaguche Prefecture Gaikai Fisheries Experimental Station. The maintenance and the numbers of individuals used in the experiment were the same as described in a previous report (Harada, 1987). Briefly, the abalone were fed overnight once every 4 days with green algae, *Ulva pertusa*, their most common food.

Estimate of the attraction index

The estimate of attraction activity in terms of the attraction index α (AIα) was described by Harada (1987). To summarize, in this series of experiments a test sample (7.5 ml) as described below was applied to a crumpled gauze sheet (25 × 25 cm) and placed at the S position in the compartment of the test tank, as shown in Fig. 15.1. Gauze without a test sample was used as a control. The experiments were carried out with a group of 50 individuals twice a day on the third and fourth days after starvation. The experiments were conducted at 10:00 and 15:00 hours, following a design similar to the Latin square method. Three test samples with a control were placed at each S position, and the septum or perforated screen was immediately lifted. Each batch of abalone that entered the compartments (A, B, C and D) from the residence, left the compartments, or remained in the compartments was counted for 20 minutes at 2-minute intervals. Four series of numbers counted every 2 minutes in each test sample were summed. The numbers summed were integrated successively with time. The numbers subtracted during the first 2 minutes from the integrated ones in each successive time were used as the time-course data to estimate AIα. The attraction index can be presented as the coefficient α of a logistic curve $y = g/(1 + e^{-r(x-a)})$. Where the logistic curve is $Y_x = \Sigma\ y_x$ the calculation method is as follows. The y_x shows the number of abalone which remained in the test compartment $x - 1$ to x (two minute interval) after the start of the experiment. r was estimated from the coefficient of the regression of reciprocal Y_x on reciprocal of Y_{x-1}, while a and g were estimated

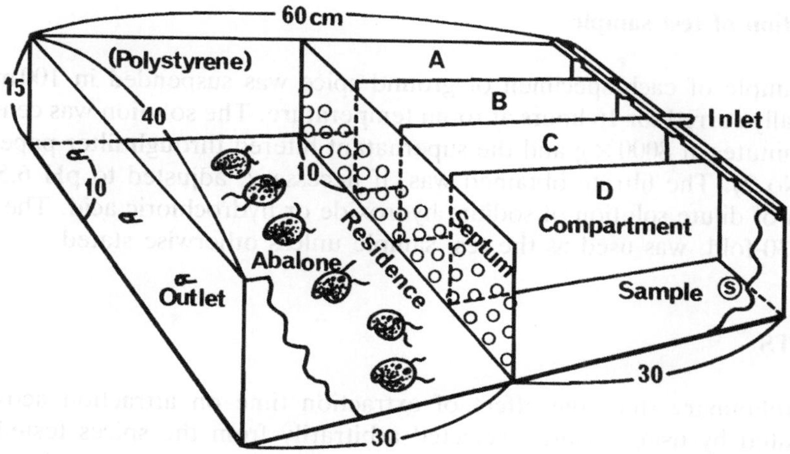

Fig. 15.1 A test tank used in the estimate of the attraction index.

from the coefficient and constant of the regression line of reciprocal of Y_x on exp $[-rx]$ $(x = 1 \text{ to } 10)$ (weighted by x). The degree of significance was established with 0.100 by the chi-squared test. The coefficient g, r and a indicate the maximal remaining number in unlimited time, inclination and average remaining time, respectively.

Fragrant and pungent material

Thirty specimens of spices were kindly supplied by Shiono Koryo Kaisha:

Odorous spices (18): allspice (*Pimenta officinalis*, family Myrtacea), anise (*Pimpinella anisum*, Umbelliferae), basil (*Ocimum basilicum*, Labiatae), caraway (*Carum carvi*, Umbelliferae), cardamon (*Elettaria cardamomum*, Zingiberaceae), celery seed (*Apium graveolens*, Umbelliferae), cinnamon (*Cinnamomum zeylanicum*, Lauraceae), clove (*Syzygium aromaticum*, Myrtaceae), coriander (*Coriandrum sativum*, Umbelliferae), cumin (*Cuminum cyminum*, Umbelliferae), dill seed (*Anethum graveolens*, Umbelliferae), fennel (*Foeniculum vulgare*, Umbelliferae), mace (*Myristica fragrans*, Myristicaceae), marjoram (*Majorana hortensis*, Labiatae), nutmeg (*Myristica fragrans*, Myristicaceae), peppermint (*Mentha piperita*, Labiatae), star anise (*Illicium verum*, Magnoliacea), and tarragon (*Artemisia dracunculus*, Compositae).

Acrid spices (4): Black pepper (*Piper nigrum*, Piperaceae), ginger (*Zingiber officinale*, Zingiberaceae), mustard (*Synapis alba*, Cruciferae) and white pepper (*Piper nigrum*, Piperaceae).

Odour-corrective or odour-masking spices (8): garlic (*Allium sativum*, Liliaceae), bay leaves (*Laurus nobilis*, Lauraceae), onion (*Allium cepa*, Liliaceae), oregano (*Origanum vulgare*, Labiatae), rosemary (*Rosmarinus officinalis*, Labiatae), sage (*Salvia officinalis*, Labiatae), savory (*Satureia hortensis*, Labiatae) and thyme (*Thymus vulgaris*, Labiatae).

Preparation of test sample

A 3 g sample of each specimen of ground spice was suspended in 100 ml water and usually stirred for 16 hours at room temperature. The solution was centrifuged for 10 minutes at 8000 × *g* and the supernatant filtered through filter paper (Toyo Roshi No 2). The filtrate obtained was, if necessary, adjusted to pH 6.5 by the addition of dilute solution of sodium hydroxide or hydrochloric acid. The filtrate, diluted 30-fold, was used as the test sample unless otherwise stated.

RESULTS

As a preliminary step, the effect of extraction time on attraction activity was investigated by using allspice, selected arbitrarily from the spices tested (Table 15.1). The attraction activity increased with increased extraction time. Accordingly, the extraction time was fixed at 16 hours in the preparation of test samples, because adequate attraction activity was found in 16-hour extracts.

Table 15.2 shows the attraction activity of each of the spices, including odorous, acrid and odour-corrective ones. Among 18 odorous specimens (experiments 2–7), nine specimens were measurably effective in attraction. Anise (experiment 2) basil (experiment 2), caraway (experiment 3), marjoram (experiment 7) and peppermint (experiment 7) were the most potent. Among four acrid specimens (black pepper in experiment 8 and white pepper in experiment 9), all specimens except ginger (experiment 8) were attractive. Black pepper (experiment 8) and mustard (experiment 8) were especially high in attraction activity. Finally, among eight odour-corrective specimens (garlic in experiments 9–11), five specimens were attractive. Onion (experiment 9) and oregano (experiment 10) appeared the most potent.

To determine the strongest attractants among the attractive specimens indicated, we selected nine specimens with high attraction activity and compared their attraction activity (Table 15.3). First, three series (experiments 12–14) of arbitrarily chosen combinations of three of the nine specimens were investigated. Basil (experiment 12), marjoram (experiment 13), and black pepper (experiment 14) were most effective. In the next experiment (experiment 15), the attraction activity of these three spices was compared. The attraction activity of marjoram was highest.

Table 15.1 Attraction activity of allspice extract over different extraction times.

Experiment number	Extraction time (hours)	Attraction activity (AIα)*
	0	–
	8	2.3
1	16	4.6
	24	9.6

* The α of attraction index (AIα) is the coefficient of a logistic curve $y = g/(1 + e^{-r(x-a)})$ in the remaining timecourse for 20 minutes.

Table 15.2 Attraction activity of the spices under test.

Experiment number	Spice	Attraction activity (AIα)*	Experiment number	Spice	Attraction activity (AIα)*
2	Control	2.6	3	Control	4.7
	Allspice	4.7		Caraway	8.4
	Anise	10.1		Cardamon	2.8
	Basil	12.2		Celery seed	3.9
4	Control	5.8	5	Control	4.5
	Cinnamon	8.5		Dill seed	1.5
	Clove	5.2		Fennel	5.8
	Coriander	5.2		Nutmeg	5.4
6	Control	5.4	7	Control	0.5
	Mace	5.0		Cumin	3.7
	Star anise	2.5		Marjoram	7.9
	Tarragon	4.9		Peppermint	6.3
8	Control	6.0	9	Control	6.6
	Black pepper	11.6		White pepper	9.5
	Ginger	6.9		Garlic	5.7
	Mustard	10.7		Onion	9.8
10	Control	4.0	11	Control	5.6
	Oregano	8.7		Bay leaves	5.5
	Rosemary	6.3		Savory	7.3
	Sage	6.1		Thyme	6.2

* Definition as in Table 15.1.

Table 15.3 A comparison of the activity indices of nine spices selected for their high attraction activity.

Experiment number	Spice	Attraction activity (AIα)*	Experiment number	Spice	Attraction activity (AIα)*
12	Control	4.0	13	Control	5.1
	Anise	7.4		Marjoram	9.3
	Basil	9.6		Mustard	6.3
	Caraway	7.2		Peppermint	6.6
14	Control	3.2	15	Control	4.6
	Black pepper	10.8		Basil	5.5
	Onion	6.6		Black pepper	5.4
	Oregano	7.4		Marjoram	12.0

* Definition as in Table 15.1.

The relationship between attraction activity and concentration is shown in Fig. 15.2 for potent marjoram. The attraction activity was examined for three concentrations (0.05, 0.1 and 1%) diluted from the original filtrate (3%). The attraction activity increased appreciably as the concentration increased.

DISCUSSION

The tested spices are from terrestrial plants, which do not occur in the sea. Aquatic animals have no opportunity to feed on them. However, a terrestrial plant, lettuce, *Lactuca sativa* var. *longifolia*, obviously attracts a snail, *Australorbis glabratus* (Michelson, 1960), and clearly stimulates a pond snail, *Lymnaea stagnalis* (Goldschmeding & Jager, 1973). Furthermore, a similar finding has also been reported for the black abalone. The abalone are strongly attracted by some algae, such as the brown alga *Ishige okamurai* ('Ishige' in Japanese) and the red alga *Grateloupia filicina* ('Tsurutsuru' in Japanese), which are not distributed, as a rule, in the natural enviroment of abalone (Harada & Kawasaki, 1982). From these studies, it appeared possible that the organisms might accept non-marine plant products.

An attractant elicits a response such as orientation towards or increase in receptivity to the apparent source; a stimulus promotes ingestion and continuation of feeding (Lindsted, 1971). In some fishes and shellfishes, an attractant is also a stimulant (Harada, 1987). In this context, the attraction activity of spices was examined in black abalone as part of a study on the feeding attractants of fishes and shellfishes.

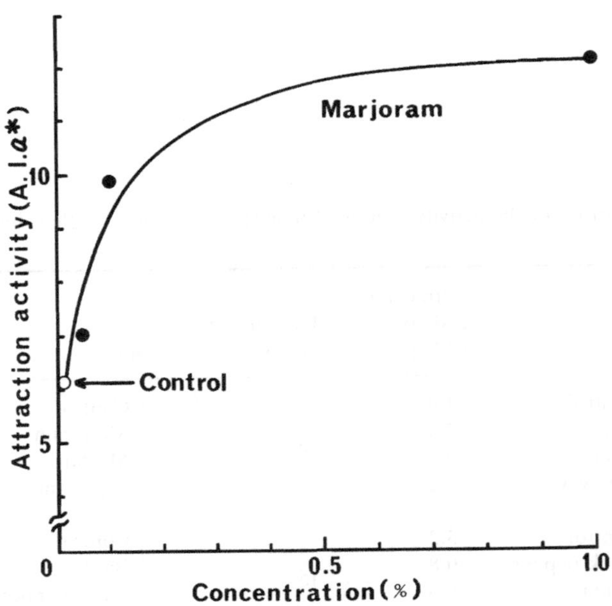

Fig. 15.2 Effect of concentration of the test spice marjoram on attraction activity, measured in terms of the attraction index AIα, as in Table 15.1.

Table 15.4 A summary of the attraction effects and characteristics (from Ohra, 1976) of the spices tested.

Spice	Attraction effect	Characteristics						
		Odorous	Acrid	Odour	Bitter	Sweet	Antiseptic	Appetizing
Odorous								
Allspice	**	**	*	**	**			
Anise	***	***						
Basil	***	***			**			
Caraway	***	***		**	**			
Cinnamon	*	***	**			**		
Cumin	**	**				**		
Fennel	*	***		**		**		
Marjoram	***	***			**			
Peppermint	***	***		*			*	
Acrid								
Black pepper	***		***				**	**
Mustard	***		***				**	**
White pepper	**		***				**	**
Odour-corrective or Odour-masking								
Onion	***	**	**			**	**	**
Oregano	***	**		***	**			
Rosemary	*	**		***				
Sage	*			***	**			
Savory	*	**						
Thyme	*	**		***	**			**

*, **, ***: weak, moderate and strong, respectively.

Table 15.4 indicates the extent of attraction activity obtained from our experiments, and the general characteristics of spices (Ohta, 1976). The magnitude of the attraction is based on the comparison between the attraction indices of spice and control; i.e., 'strong' is much higher and 'weak' is only slightly higher than the value for the control. Among the characteristics, the antiseptic and the appetizing actions are defined as the inhibition of the growth of microorganisms in foods and the stimulation of appetite for foods, respectively. Among 30 specimens of spices, 18 had some attraction activity: nine odorous spices, three acrid spices and six odour-corrective spices. The most active spices were nine species including anise and four others among the odorous, black pepper and mustard in the acrid and onion and oregano in the odour-corrective. These nine spices also attracted the abalone in the time-course observations.

When the relationship between attraction and the characteristics of the spices is considered, about half the attractive spices belong to the family Labiatae. Furthermore, many attractive spices are odorous and odour-corrective. The attractive acrid spices show most promise for the production of artificial diets. Among the four acrid spices, three were strong or moderate in attraction activity. In addition, they have appetizing and antiseptic properties (other spices lack these qualities in general). Thus, the attractive acrid spice, if used as a food additive, may play three roles: attractant, stimulant and antiseptic.

REFERENCES

Goldschmeding J.T. & Jager J.C. (1973) Feeding responses to sucrose with pond snail *Lymnaea stagnalis* after nerve section and tentacle amputation. *Neth. J. Zool.*, **23**, 118–24.

Harada K. (1987) Favorite and disliked constituents for aquatic animals. *Kansui*, **276**, 25–33.

Harada K. & Kawasaki K. (1982) The attractive effect of seaweeds based on the behavioral response of young herbivorous abalone *Haliotis discus*. *Nippon Suisan Gakkaishi*, **48**, 617–21.

Hasegawa T. (1979) Chemistry of spices. *Food Nutr.*, **32**, 267–75.

Kanda T., Yamamoto T. & Saito H. (1971) Antimicrobial and antiseptic properties of spice. *Food Indust.*, **14**, 73–6.

Lindsted K.J. (1971) Chemical control of feeding behavior. *Comp. Biochem. Physiol.*, **39A**, 553–81.

Michelson E.H. (1960) Chemoreception with the snail *Australorbis glabratus*. *Am. J. Trop. Med. Hyg.*, **9**, 480–7.

Ohta S. (1976) Spices. In *Food Seasoning* (Ed. by S. Ohta), pp. 202–21. Yukishobou, Tokyo.

Takei M. (1967) Studies on fish favorite foods. I. *Bull. Tokai Reg. Fish. Res. Lab.*, **49**, 119–29.

PART V
DISEASES

Chapter 16
Investigations into a mortality among populations of the California black abalone, *Haliotis cracherodii*, on the central coast of California, USA

J.R. STEINBECK *Tenera Environmental Services, 586 Higuera St. Suite 200, San Luis Obispo, California 93401, USA*
J.M. GROFF, C.S. FRIEDMAN, T. McDOWELL and R.P. HEDRICK
Department of Medicine, School of Veterinary Medicine, University of California, Davis, California 95616, USA

ABSTRACT: In June 1988 several moribund black abalone, *Haliotis cracherodii*, were found in Diablo Cove, a semi-protected cove located in San Luis Obispo County in central California. Diablo Cove is the site of the cooling water discharge for Pacific Gas and Electric Company's Diablo Canyon Power Plant. Moribund animals found in the cove were characterized by a body mass smaller relative to shell size than normal. Studies which were initiated to investigate the geographical extent of the condition in the Diablo Canyon area showed that it was restricted to areas closest to and most influenced by the power plant thermal discharge. Laboratory experiments also showed a correlation between higher seawater temperatures and increased mortality. Animals collected during the studies were sent to the Fish Disease Laboratory at the University of California, Davis, where histological examination revealed massive infections of the kidney by a coccidian-like protozoan. A correlation between infections with the coccidian and mortality of abalone in Diablo Cove, however, has yet to be established. The potential role of the coccidian and environmental factors in the observed mortality are under investigation.

RESUMEN: En junio de 1988, varios abulones negros (*Haliotis cracherodii*) se encontraron moribundos en la Ensenada del Diablo, una ensenada semi-protegida localizada en el condado de San Luis Obispo en el centro de California. La Ensenada del Diablo es el sitio de descargo de aguas de enfriamiento de la planta de energía eléctrica del Cañon del Diablo y de la Compañía de Gas del Pacífico. Los animales moribundos encontrados en la ensenada se caracteirizaban por que la relación de la masa del cuerpo respecto del tamaño de la concha era mas pequeña que lo normal. Los estudios que se iniciaron para investigar la extensión geográfica de la condicíon en el área del Cañón del Diablo, demonstraron que ese fenómeno se restringió a las áreas mas cercanas y mas influenciadas por la descarga termal de la planta eléctrica. Experimentos de Laboratorio demostraron también una correlación entre las temperaturas más altas del agua de mar y el incremento de la mortalidad. Los animales recolectados durante nuestros estudios fueron enviados al Laboratorio de Enfermedades de Peces de la Universidad de California, Davis, en donde un examen histológico reveló infecciones masivas del riñón ocasionadas por un protozoario del tipo de los coccidios. Sin embargo, todavía no se ha establecido una correlación entre las infecciones por el coccidio y la mortalidad del abulón en la Ensenada del Diablo. El papel potencial del coccidio y los factores del medio ambiente en la mortalidad observada, están bajo investigación.

INTRODUCTION

The black abalone, *Haliotis cracherodii*, is a conspicuous member of the intertidal community in areas of central California where sport harvest is limited by restricted public access. Diablo Cove is a semi-protected cove located in San Luis Obispo County in central California and is the site of the cooling water discharge for Pacific Gas and Electric Company's Diablo Canyon Power Plant. Restricted public access and the resulting high abundance have allowed abalone to be the subject of a number of studies at Diablo Canyon (Chapter 18, this volume). In June 1988 several moribund animals were found in the cove. These mortalities were characterized by a smaller than normal body mass relative to shell size, a syndrome which has been referred to as atrophied foot condition or withering syndrome (WS) (Chapter 17, this volume). Gross examination of the animals did not reveal any likely cause of the condition. All live afflicted animals from Diablo Cove kept in aquaria in the laboratory at the plant site eventually died. Recently high mortality rates and WS have been documented as affecting populations of black abalone in southern California (Chapter 19, this volume). The condition was first observed in the spring of 1987, and has been especially evident on Santa Cruz Island where monthly mortality rates have averaged 6% through 1988, and population densities at some sites have been reduced by greater than 90% (Tissot, 1988; Chapter 17, this volume).

In contrast to the mass mortalities observed on the Channel Islands, only isolated occurrences of WS have been reported from mainland California. Diablo Cove has been the only reported mainland site where WS and mass mortality have been reported to occur on a scale similar to the Channel Islands. This chapter presents results from field studies, which were part of the power plant's marine environmental monitoring programme, that document the decline of the abalone population in Diablo Cove. Results from field studies which were initiated to investigate the geographical extent of the condition in the Diablo Canyon area are also presented.

Additional studies were undertaken to investigate the aetiology of the condition and the role the raised seawater temperatures in Diablo Cove may have had in the decline. The most rapid and dramatic mortality rates occurred in areas closest and most influenced by the power plant thermal discharge. Results from a laboratory experiment designed to investigate the role of temperature in the decline are presented.

Moribund and grossly healthy abalone recovered from the field during our studies were sent to the Fish Disease Laboratory at the University of California, Davis (UCD), where histological examination revealed massive infection of the kidneys by a coccidian-like protozoan. The coccidians are protozoans belonging to the phylum Apicomplexa, class Sporozoa, subclass Coccidia (Levine *et al.*, 1980). All of the members of the phylum are parasitic. Coccidians have been known to infect the kidneys of several species of marine molluscs including the whelk *Buccinum undulatum* and the limpet *Patella vulgata* (Lauckner, 1980). Morado *et al.* (1984) recently described a coccidian from the kidney of the littleneck clam, *Protothaca staminea*, that has a number of stages very similar in appearance to those found infecting abalone. A preliminary description of this infection will be presented here.

MATERIALS AND METHODS

Following the discovery of WS in Diablo Cove, qualitative surveys were conducted to investigate the extent of the condition in surrounding coastal areas. This method was designed to assess the condition of a large number of animals over a wide area and involved locating individual animals and attempting to detach them manually from the substrate. Animals in varying stages of WS can be easily detached from the substrate. This type of survey was conducted three times over the period from June 1988 to July 1989 and assessed the condition of abalone over 15 km of coastline. The abalone population inside Diablo Cove has been censused since 1981 using a set of 10 fixed transects (NDC1–5 and SDC1–5) which encircle the cove at approximately the +0.3 m mean lower low water (MLLW) tidal level (Fig. 16.1). Up to 100 randomly selected 1 m^2 quadrats were sampled per survey.

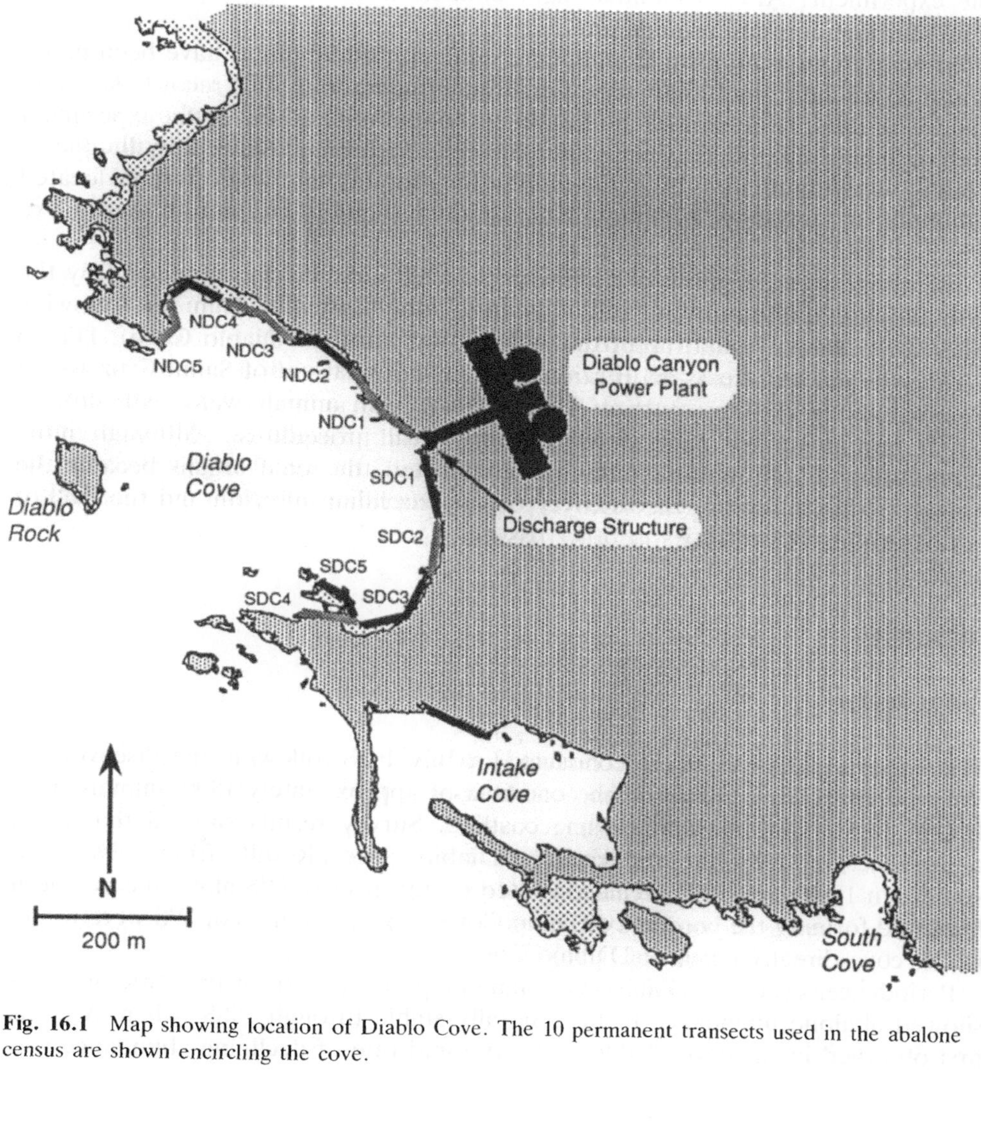

Fig. 16.1 Map showing location of Diablo Cove. The 10 permanent transects used in the abalone census are shown encircling the cove.

Two randomly generated coordinates were used in selecting each quadrat, with one coordinate designating the horizontal position along the transect and the second designating the vertical orientation of the quadrat relative to the transect. The objective of this procedure was to sample approximately 10% of the total area in Diablo Cove between the −0.3 m and +1.5 m MLLW tide level.

A laboratory study was conducted to assess the possible effects of elevated seawater temperatures on the susceptibility of abalone to WS. Grossly healthy abalone were collected from seven areas outside Diablo Cove. The collection areas ranged from directly outside Diablo Cove to approximately 3.5 km distant. Two groups of 60 animals each were acclimated for two weeks to two temperature treatments, 16 and 18°C. A third group was held at ambient seawater temperature (10–14°C) throughout the experiment. The laboratory has a flow-through seawater system with the intake located outside Diablo Cove. Animals were held in separate tanks and fed in excess *Macrocystis pyrifera*, the giant kelp, throughout the experiment. After 6 months, eight animals from each group were sent to UCD for histological examination.

Seawater temperatures in Diablo Cove and surrounding areas have been monitored continually since 1980 using Sea Data Corporation. TR2 recorders. These data were used to determine the treatment temperatures used in the experiment and in documenting the changes from ambient conditions produced by the plant's cooling water discharge. Data reported in this chapter were from stations located subtidally at a depth of 3.3 m in north and south Diablo Cove and in South Cove (Fig. 16.1).

Histological investigation of approximately 100 animals was conducted by the Fish Disease Laboratory at UCD. Animals were collected from the following sites: Año Nuevo Island (approximately 250 km north of Diablo Cove), Diablo Cove and adjacent areas and Frazier Point at the west end of Santa Cruz Island (approximately 130 km south of Diablo Cove). All animals were sent alive to UCD and processed using standard histological procedures. Although initial examination of all organ systems was conducted, the renal organs became the primary focus following the discovery of the coccidian infection and the lack of any apparent abnormalities in other tissues.

RESULTS

Field studies

The initial qualitative survey conducted in July 1988 following the discovery of WS in Diablo Cove assessed the condition of approximately 1300 animals from Diablo Cove and the surrounding coastline. Survey results showed that gross occurrences of WS were restricted to Diablo Cove. Results from subsequent surveys in 1988 and 1989 revealed isolated occurrences of WS along the northern headland forming the borders of Diablo Cove. No occurrences of WS were found in the cove directly north of Diablo Cove.

Periodic census of the Diablo Cove abalone population from 1982 through 1989 showed abalone abundance to be generally stable through 1988, when WS was first observed in the cove (Table 16.1). A rapid rate of declining abundance was

Table 16.1 Numbers of black abalone counted along fixed transects located inside Diablo Cove.

	Survey (month/year)						
Transect	08/82	08/83	07/85	08/88	01/89	07/89	% decline
North 1	80	ns	26	21	0	0	100
North 2	221	50	177	48	28	22	90
North 3	194	194	192	186	89	10	95
North 4	110	46	90	23	15	1	99
North 5	197	79	56	113	4	16	92
South 1	15	2	1	0	1	0	100
South 2	29	28	27	31	17	6	79
South 3	38	16	10	16	5	1	97
South 4	188	75	151	181	114	58	69
South 5	52	ns	73	237	149	129	**

ns – not sampled.
** – 248% increase.

recorded at most of the cove transects from 1988 through 1989 (Table 16.1). Although total percentage declines between 1982 and 1989 were similar among most of the transects, the declines at transects NDC2 through NDC4 were most dramatic because of the historically higher densities at those transects. These transects contain some of the better abalone habitat in the cove, consisting of large immovable boulders overlaying bedrock with large cracks. Transects in south Diablo Cove showed more variable rates of decline. Transect SDC4 showed the lowest overall decline of any of the transects and SDC5 showed an increase in abundance (Table 16.1).

Subtidal temperature monitoring has shown that since the power plant started operation in 1985, seawater temperatures in the north portion of the cove can average up to 2 to 4°C warmer than the south cove and up to 11°C warmer than surrounding ambient waters (Fig. 16.2). The more variable rates of decline in south Diablo Cove can be attributed to the cooler temperatures there.

Temperature experiment

The percentage survivorship in the three temperature groups of abalone after 6 months is shown in Table 16.2. The higher percentage mortality in the 18°C treatment is consistent with the result that mortality rates are higher in the warmer regions of Diablo Cove. Many of the abalone which died in the 16 and 18°C treatments showed shrinkage of body mass similar to moribund animals found in Diablo Cove and on Santa Cruz Island. Animals used in the experiment were held at the test temperatures past the 6-month duration of the study. Mortality continued to occur in the 18°C treatment and the mortality rate increased in the 16°C treatment.

Histological examination of eight randomly selected live animals from each of the three temperature groups conducted following completion of the experiment showed a negative correlation between temperature-related mortality and coccidian

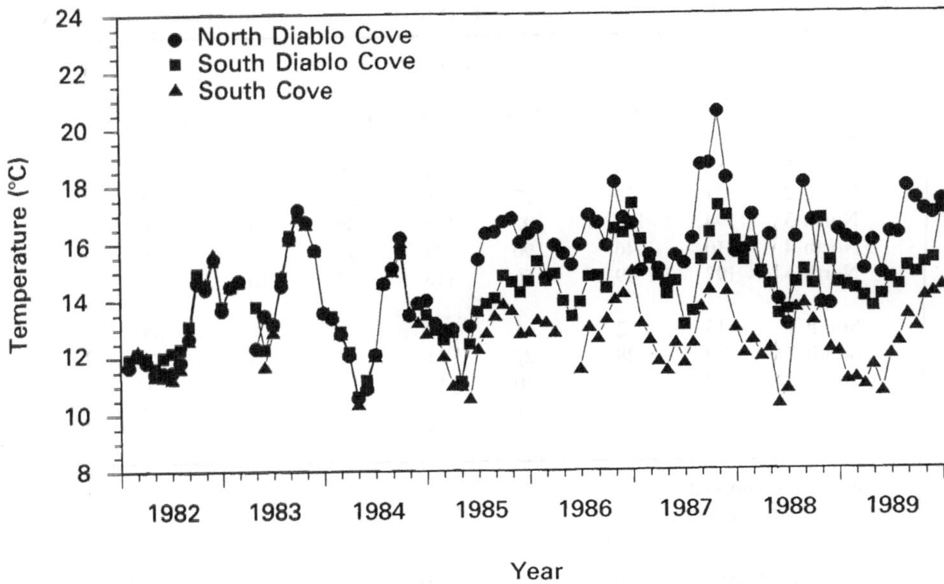

Fig. 16.2 Mean monthly seawater temperatures from 1982 through 1989 measured at depths of −3.3 m mean lower low water at two stations in Diablo Cove and at South Cove, a location approximately 1 km south of Diablo Cove. The power plant began commercial operation in April 1985.

Table 16.2 Survival of black abalone at three seawater temperatures at the Pacific Gas & Electric Company laboratory at Diablo Canyon from December 1988 through June 1989; the control group was held at ambient seawater temperature (10–15°C).

	Treatment		
	Control	16°C	18°C
No. of animals/treatment	42	60	60
No. of mortalities	0	4	23
% survivorship	100	93	62

infection intensity (Table 16.3). None of these animals exhibited any symptoms of WS. Moderate to very heavy infections were found in 100% of the abalone from the ambient seawater control group. Sixty-three per cent of the abalone in the 16°C treatment contained slight to moderate infections and 25% of the abalone held at 18°C showed only slight infections.

Coccidian infection

Abalone have paired kidney organs which are dissimilar in structure. Infection by the coccidian occurred in either kidney or in both simultaneously. Light infections did not change the normal histological appearance of the kidneys, whereas severe

Table 16.3 Presence of renal coccidiosis in black abalone held in three temperature treatments for approximately 6 months; ambient seawater temperature generally varied from 10 to 15°C.

	Treatment		
	Control	16°C	18°C
No. of specimens examined	8	8	8
No. infected	8	5	2
% infected	100	63	25

infections rendered epithelial and stromal cells nearly unrecognizable (Figs 16.3, 16.4). In a few abalone the left kidney had undergone such extensive destruction that no apparently functional kidney tissue could be found (Fig. 16.4). Overall, infection intensity in the right kidney appeared to be lower than that observed in the left kidney.

Infected kidney epithelial cells underwent hypertrophy during development of the parasite. Generally, histological examination of infected kidney did not reveal any host response to the parasite. However, an inflammatory response consisting of an infiltration of host haemocytes was observed in the left kidney of a few abalone. Some infected cells apparently ruptured, leaving only the basal portion of the cell attached to the basal membrane. Infected tissue contained numerous developmental stages of the coccidian. Many of the stages were similar in form to those described infecting the littleneck clam by Morado *et al.* (1984).

Black abalone infected with the parasite were found from all collection sites. No quantitative measure of infection intensity was used but, generally, infection levels were lowest in collections from Frazier Pt on Santa Cruz Island and from inside Diablo Cove. Animals from these areas would range from being apparently free of the parasite to mildly infected. Some of the highest infection intensities were found in individual abalone from areas with grossly healthy populations, that is, Año Nuevo Island. These areas also appeared to have more consistent levels of infection among collected individuals.

DISCUSSION

The results in this chapter document a mass mortality of black abalone in Diablo Cove, San Luis Obispo County, California, that is similar to occurrences described from the southern California Channel Islands (Tissot, 1988; Chapter 19, this volume). A majority of the mortalities observed in the field and laboratory showed significant and rapid reduction of body mass described by Haaker *et al.* (Chapter 17, this volume) as withering syndrome. This condition accounted for population reductions in north Diablo Cove of greater than 90% between 1982 and 1989 (Table 16.1). These values are similar in magnitude to declines reported from Santa Cruz Island (Tissot, 1988; Chapter 17, this volume).

In south Diablo Cove the effects of WS were more variable, being lower in areas least affected by the power plant's discharge plume. The predominant direction of the discharge plume trajectory is towards the north channel of the

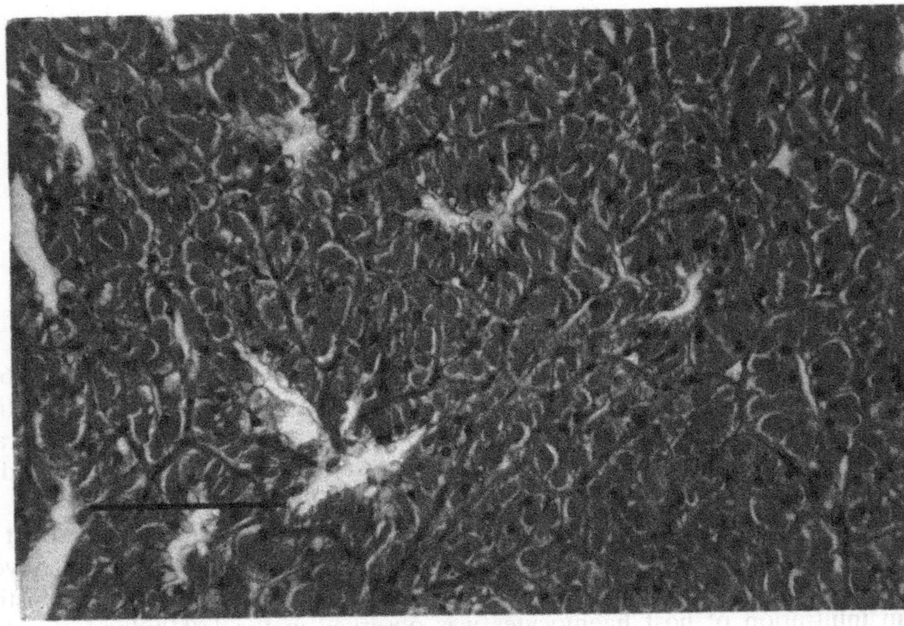

(a)

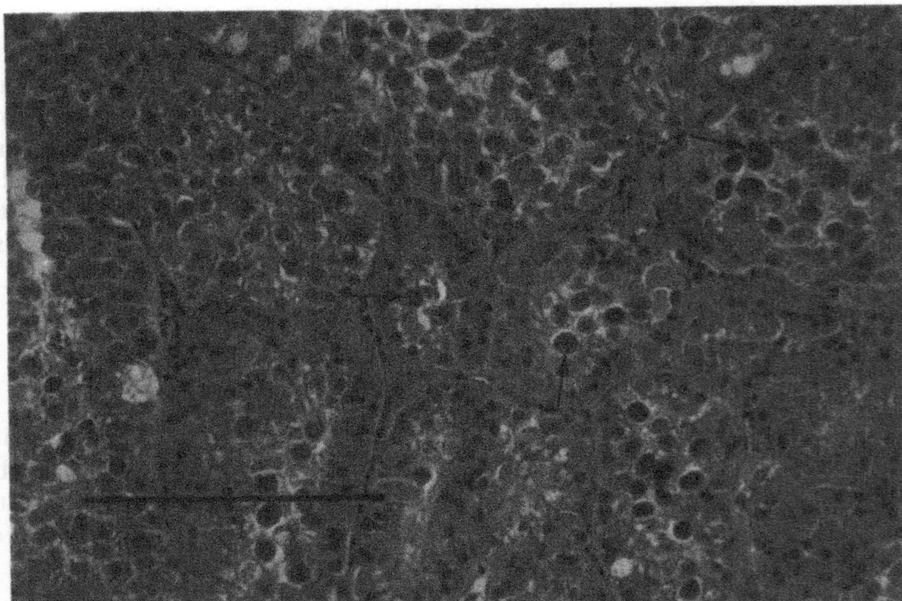

(b)

Fig. 16.3 Normal (a) and infected (b) tissue from right kidney. Coccidian noted by arrows on bottom photograph (scale bar represents 100 μm).

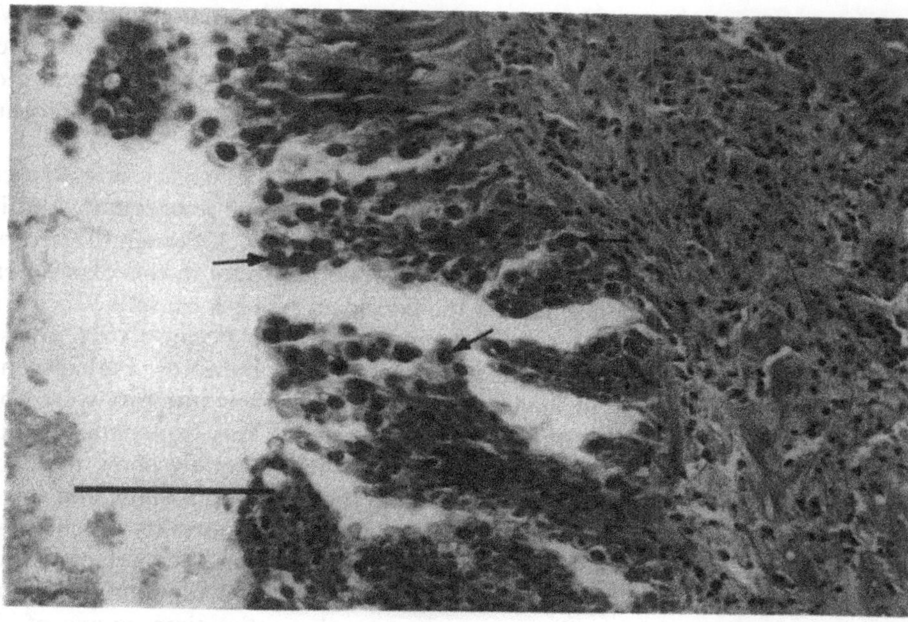

Fig. 16.4 Infected tissue from left kidney. Coccidian noted by arrows (scale bar represents 100 μm).

cove, therefore that area of the cove is warmer than the southern portion. This would appear to be the reason that total population declines are more consistent between transects in north Diablo Cove (Table 16.1). Transects SDC4 and SDC5 at the south point of the cove are areas least affected by the discharge. The trajectory of the plume out of the north channel can actually induce a countercurrent of cooler oceanic water which flows into the cove along the south point near transects SDC4 and SDC5. These factors may account for the increase in abalone recorded at SDC5 over the duration of the study (Table 16.1).

Studies at Diablo Canyon through 1982 showed abalone to have an inherent thermal tolerance level exceeding the temperature ranges recorded in Diablo Cove during plant operation. The 96-hour ET50 (median effective temperature) value for black abalone acclimated to 11.5°C was 26.1°C and the value for animals acclimated to 16°C was 27.4°C (Hines *et al.*, 1980). Growth studies conducted for 90 days showed that optimal growth in black abalone occurred at approximately 18°C (Pacific Gas & Electric Company, 1982). No mortality was recorded from a treatment held at approximately 24°C during the same experiment. Studies on other species of abalone from southern California show that even non-littoral species should survive the thermal regime experienced in Diablo Cove (Leighton, 1974).

The high mortality in the 18°C treatment from the experiment described in this chapter suggests that a change in the thermal tolerance of the Diablo Canyon population of black abalone has occurred between 1982 and 1988. The available evidence of decreased population densities in Diablo Cove, the high incidence of WS and the positive relationship between temperature and mortality indicate that observed effects were related to the elevated seawater temperatures in Diablo

Cove resulting from power plant operation. The factor (or factors) causing WS and the mass mortality was presumably involved in the decreased thermal tolerance because no changes in the population were observed during the three years of power plant operation between spring 1985, when power plant operation began, and summer 1988, when WS was first observed.

The relationship between the coccidian and the mass mortality is equivocal given the negative correlation between infection intensity and temperature and its natural occurrence in areas with grossly healthy populations. Although the inverse relationship between infection intensity and abalone mortality in this experiment could be interpreted as evidence against the coccidian being a causative factor in the mass mortality, other interpretations remain possible because only grossly healthy survivors from the experiment were sampled for parasites. Perhaps only animals without the infection or having the infection at a low intensity were able to survive the stress induced at the higher temperature. This explanation would also be consistent with current field collections which generally show infection intensities to be highest in areas with outwardly healthy populations and lower in animals from Diablo Cove and Santa Cruz Island where mortality was highest. To gain a better understanding of the correlation between the coccidian- and temperature-related mortality described here, an experiment randomly sampling individuals for parasites from all treatments over the entire time course of the experiment would need to be conducted, with sampling of all mortalities. Unfortunately, the absence of any *in situ* test for the presence of the parasite would necessitate the sacrifice of a large number of animals from an already severely depressed population.

Infection by the coccidian was found among black abalone from all collections with some of the most intense infections occurring in areas with grossly healthy populations (i.e., Año Nuevo Island). This, and the results of the experiment described here, would indicate that the presence of even severe infection alone does not appear to cause WS or mortality. The absence of infected animals or reduced infection intensities in Diablo Cove may be due to the direct effect of temperature on the parasite. Temperature effects on parasites and host defence have been shown for a number of vertebrate and invertebrate host–parasite relationships (Kinne, 1980). It is more likely that the isolated areas of high mortality are disturbed habitats due either to elevated temperature, as is the case in Diablo Cove, or to another factor such as changes in food availability, as has been suggested for the Channel Islands (Tissot, 1988). Under normal conditions, a host–parasite relationship may exist that is not fatal to the abalone. In contrast, this relationship may be upset in disturbed habitats that place additional stress on the diseased animals, thus causing mortality of both agent and host.

The parasite stages seen in the black abalone kidney in our study closely resemble those identified in littleneck clams. In the clams, all stages of the coccidian are observed and this may indicate that the entire life cycle of the parasite occurs in this host (Morado *et al.*, 1984). The parasite in clams was believed to be the cause of mortality among populations in Sequim Bay, Washington (Morado *et al.*, 1984). Most coccidians, however, have an alternative host at some stage in the life cycle. A comprehensive study of the coccidian in black abalone, including its distribution in different abalone species and potential effects on the host, is the subject of current research at both UCD and the California Department of Fish and Game.

ACKNOWLEDGEMENTS

The authors thank the personnel at the Diablo Canyon Biology Laboratory for assistance in specimen collection and Brian Tissot for his assistance in the collections from Santa Cruz and Año Nuevo Islands. The authors also thank the personnel at the Fish Disease Laboratory, University of California, Davis, for assistance in specimen preparation, Jay Carroll and Allan Fukuyama for their comments on drafts of this chapter and the Pacific Gas & Electric Company for financial support and use of their facilities.

REFERENCES

Hines A., Anderson S. & Brisbin M. (1980) Heat tolerance in the black abalone, *Haliotis cracherodii* Leach, 1814: Effects of temperature fluctuation and acclimation. *Veliger*, **23**, 113–18.

Kinne O. (1980) General aspects. In *Diseases of Marine Animals, Vol I: General Aspects, Protozoa to Gastropoda*, (Ed. by O. Kinne), pp. 13–73. Wiley, New York.

Lauckner G. (1980) Diseases of Mollusca: Gastropoda. In *Diseases of Marine Animals, Vol I: General Aspects, Protozoa to Gastropoda* (Ed. by O. Kinne), pp. 311–424. Wiley, New York.

Leighton D.L. (1974). The influence of temperature on larval and juvenile growth in three species of southern California abalones. *Fish. Bull.*, **72**, 1137–45.

Levine N.D., Corliss J.O., Cox F.E.G., Deroux G., Grain J., Honigberg B.M, Leedale G.F., Loeblich III A.R., Lom J., Lynn D., Merinfeld E.G., Page F.C., Poljansky G., Sprague V., Vavra J. & Wallace F.G. (1980) A newly revised classification of the Protozoa. *J. Protozool.*, **27**, 37–58.

Morado J.F., Sparks A.K. & Reed S.K. (1984) A coccidian infection of the kidney of the native littleneck clam, *Protothaca staminea. J. Invert. Pathol.*, **43**, 207–17.

Pacific Gas & Electric Company (1982) *Compendium of Thermal Effects Laboratory Studies* Vol. 2, pp. 63–80. TERA Corporation, Berkeley.

Tissot B.N. (1988) Mass mortality of black abalone in southern California. *Am. Zool.*, **28**, 69A (abstr.).

Chapter 17
Mass mortality and withering syndrome in black abalone, *Haliotis cracherodii*, in California

PETER L. HAAKER, DAVID O. PARKER and HEIDI TOGSTAD
California Department of Fish and Game, Marine Resources Division, Long Beach, California 90802, USA
DANIEL V. RICHARDS and GARY E. DAVIS *US National Park Service, Channel Islands National Park, Ventura, California 93001, USA*
CAROLYN S. FRIEDMAN *California Department of Fish and Game, Fish Disease Laboratory, Rancho Cordova, California 95670, USA*

ABSTRACT: Accelerated mortality of black abalone, *Haliotis cracherodii*, was observed at several intertidal sites at the California Channel Islands. Since 1986, densities on Anacapa, Santa Cruz, Santa Barbara and Santa Rosa Islands have been the most affected, with up to 99% of the abalone lost at several sites. All sizes of abalone were affected. Black abalone at these sites were discoloured and weakened, and the soft tissues were atrophied and non-responsive to stimuli. We call these symptoms withering syndrome (WS). Pesticides, heavy metals pathology, food and harvest were considered as possible causes of WS. A coccidian-like protozoan was found in nephridia of both moribund and apparently healthy abalone, but its relation to the accelerated mortality is uncertain. No single factor has been identified as the causative agent.

RESUMEN: Se observó una mortalidad acelerada del abulón negro, *Haliotis cracherodii*, en diferentes sitios intermareales en las Islas Channel en California. Desde 1986, las densidades en Anacapa, Santa Cruz, Santa Bárbara e Islas Santa Rosa, han sido las más afectadas con una pérdida del abulón hasta del 99% en varios sitios. Todos los tamaños fueron afectados. El abulón negro en estos sitios estaba descolorido y debilitado y los tejidos suaves estaban atrofiados y no respondían al estímulo. A estos síntomas los denominamos Síndrome de Deterioro (SD). Los pesticidas, la patología provocada por los metales pesados, el alimento y la captura se consideraron como posibles causas del Síndrome de Deterioro. Protozoarios del tipo de los coccidios fueron encontrados en los nefridios, tanto en abulones moribundos como en los aparentemente sanos, pero su relación con la mortalidad acelerada es incierta. Ningún factor específico ha sido identificado como agente causante de este Síndrome.

INTRODUCTION

In 1986, we found evidence of unusual mortality of black abalone, *Haliotis cracherodii*, in permanent intertidal transect sites at the Channel Islands National Park off the coast of southern California. At low tide, exposed black abalone attach tightly to the substrate and are difficult to remove without implements, but at several sites we found individuals that were loosely attached and hanging from the rocks, partly decomposed abalone lying in tidepools and unusually high numbers of fresh shells of all sizes.

Examination of these abnormal abalone revealed discoloration of the epipodium and mantle: the colour was blue-grey, grey or purple, rather than the normal black colour, the soft tissue mass was flaccid and atrophied, so that the animal did not fill its shell and there was little or no response to palpation. We call the presence of these symptoms withering syndrome (WS). Inspection of Channel Islands study sites revealed the presence of WS in black abalone at some, but not all, locations. At sites where WS occurred, there were indications of black abalone population decline. Replicate density surveys in permanently marked plots at the sites confirmed a decline in numbers of abalone.

Mass mortalities of abalone in natural populations are known and have been attributed to starvation and disease (Lester & Davis, 1981; Tanaka *et al.*, 1986). Parasites and pathogenic organisms were implicated in abalone mortality in hatcheries (Ebert & Houk, 1989; Hahn, 1989a,b). In California, occurrences of WS-like symptoms in the late 1950s were reported for pink abalone, *H. corrugata* (MacGinitie & MacGinitie, 1966) and black abalone (Cox, 1962; Young, 1964) and were related to starvation.

The present black abalone mortality problem disrupted a valuable commercial fishery. These mass mortalities attracted the interest of government, the scientific community and the public who are concerned about the fishery, public health considerations and the future of the black abalone resource, as well as effects on other abalone. This chapter describes unusual declines and the presence of WS in black abalone populations in California. It documents the efforts to discover the cause of the declines, to delineate its extent and to find a relationship between the declines and WS.

METHODS

Information about declines in black abalone populations was collected at 11 intertidal study sites within the Channel Islands National Park (Fig. 17.1). These sites are part of the natural resources monitoring programme which was established in 1985 by the National Park Service (NPS) at Channel Islands National Park (Richards, 1988; Richards & Davis, 1988). They are located at West (or Cat Rock) and Middle Anacapa Island, Johnsons Lee, Ford Point, Fossil Reef, East Point and Northwest (NW) Talcott on Santa Rosa Island, Crook Point, Harris Point and Otter Harbor on San Miguel Island and at the sea lion rookery on Santa Barbara Island (Fig. 17.1). The sites are visited in the spring (March–April) and the fall (October–December).

At each site, except East Point which only has one, five permanently marked, measured areas of representative black abalone habitat have been established. The total area of permanently marked transects at each site is: Anacapa Island, West, 11 m^2 and Middle, 7.5 m^2; Santa Rosa Island, Johnsons Lee, 8.8 m^2, Ford Point, 14.4 m^2, NW Talcott, 38.9 m^2, East Point, 54 m^2, and Fossil Reef, 26.9 m^2; San Miguel Island, Harris Point, 26.5 m^2, Otter Harbor, 20.6 m^2, and Crook Point, 11.1 m^2; and Santa Barbara Island, 31.3 m^2. Abalone in the plots are counted and measured, without removal, during each visit. Sizes are divided into three groups: juveniles, <44 mm, adult, 44–126 mm and (sport) legal, >126 mm (Richards & Davis, 1988; Davis, 1989). Size distribution based on these groups is expressed as a percentage of the total number sampled at each site (Table 17.1). The mean density for the five transects, ± the standard error, was determined for each visit to a site (Fig. 17.2).

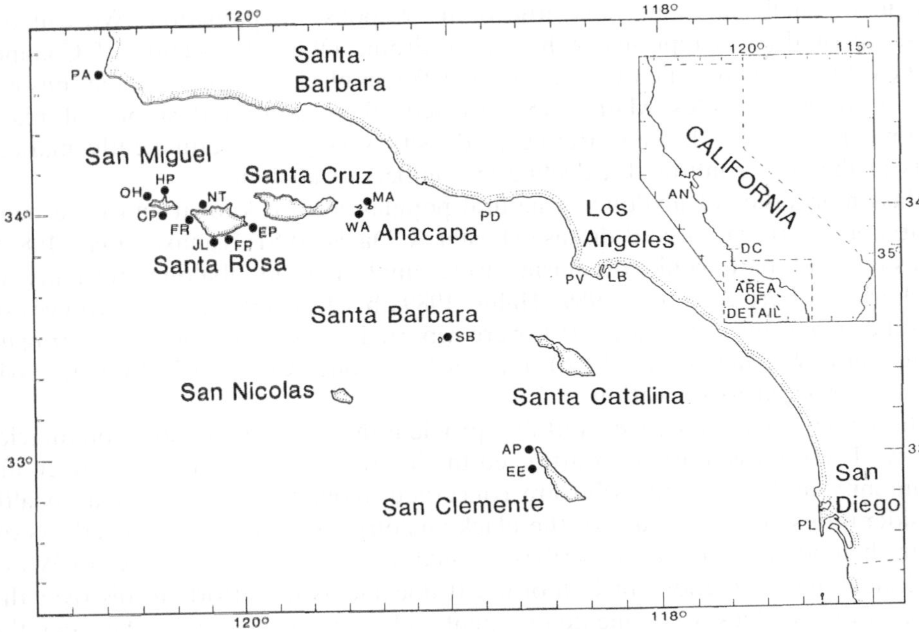

Fig. 17.1 The California Channel Islands and the mainland coast of southern California. The inset map shows sites in central California.

Additional black abalone tagging sites were established in 1988 at Eel Point and near the airport on San Clemente Island and at a mainland site at Point Arguello (Fig. 17.1). At these locations there are no permanently marked transects.

In 1988, freshly dead abalone shells which had accumulated within more or less discrete locations were collected, measured and destroyed during repetitive visits, to find the rate of shell accumulation. Fresh shells are those that have a clean, shiny, nacreous interior without any marine encrustation. These surveys were conducted at West Anacapa Island, Fossil Reef and Harris Point.

Initial efforts to identify the cause of WS included submitting samples of WS-affected abalone to the CDFG Fish Disease Laboratory and the CDFG Pesticide Laboratories at Rancho Cordoba, the University of California School of Veterinary Medicine, Davis, and to R.J.G. Lester of the University of Queensland Department of Parasitology, Queensland, Australia.

To determine whether food availability could be implicated in the occurrence of WS, 52 individuals, 35 of which appeared robust and healthy and 17 of which had symptoms of WS, were collected at Johnsons Lee and transported live to Cabrillo Marine Museum, San Pedro, California. The animals were individually tagged, placed together in a holding tank and fed kelps, *Macrocystis pyrifera* and *Egregia laevigata*, *ad libitum*. The abalone were monitored at approximately 3-day intervals and weighed weekly. Dead animals were removed upon discovery.

Other investigation focused upon the delineation of WS in California abalone populations and identification of its cause. Abalone collected from coastal and island locations throughout California, in addition to the Channel Island sites, were sent to the CDFG Disease Laboratory. Pericardial, kidney and other tissues

Table 17.1 Size distribution (percentage) of juvenile (J), adult (A), and legal (L) black abalone at the Channel Islands.

	West, AI				Middle, AI				Johnsons Lee, SRI			
	n	J(%)	A(%)	L(%)	n	J(%)	A(%)	L(%)	n	J(%)	A(%)	L(%)
1985	544	3	76	22	1078	7	81	12	370	2	72	26
1986	377	3	70	27	895	12	76	13	793	3	67	30
1987	256	1	76	23	406	7	80	14	515	2	69	29
1988	97	0	79	21	74	1	89	9	243	2	72	26
1989	32	0	66	34	8	0	100	0	122	2	70	28
	Ford Point, SRI				N.W. Talcott, SRI				East Point, SRI			
	n	J(%)	A(%)	L(%)	n	J(%)	A(%)	L(%)	n	J(%)	A(%)	L(%)
1985	342	0	82	18	—	—	—	—	—	—	—	—
1986	525	4	81	15	483	2	67	30	58	3	91	5
1987	162	1	76	23	915	2	67	32	114	0	92	8
1988	27	0	67	33	885	1	67	32	86	0	87	13
1989	10	0	70	30	698	0	68	31	18	6	89	6
	Fossil Reef, SRI				Harris Point, SMI				Otter Harbor, SMI			
	n	J(%)	A(%)	L(%)	n	J(%)	A(%)	L(%)	n	J(%)	A(%)	L(%)
1985	—	—	—	—	818	12	75	13	1073	6	60	34
1986	—	—	—	—	943	10	79	12	952	2	59	39
1987	—	—	—	—	742	9	78	13	952	1	64	34
1988	660	4	64	32	830	7	80	13	885	3	64	34
1989	1029	7	65	28	792	7	86	8	842	3	62	35
	Crook Point, SMI				Santa Barbara Is							
	n	J(%)	A(%)	L(%)	n	J(%)	A(%)	L(%)				
1985	796	2	77	23	311	5	82	13				
1986	647	1	74	25	332	2	89	10				
1987	263	2	76	22	332	3	83	14				
1988	340	4	77	19	204	2	88	9				

AI – Anacopa Island.
SRI – Santa Rosa Island.
SMI – San Miguel Island.

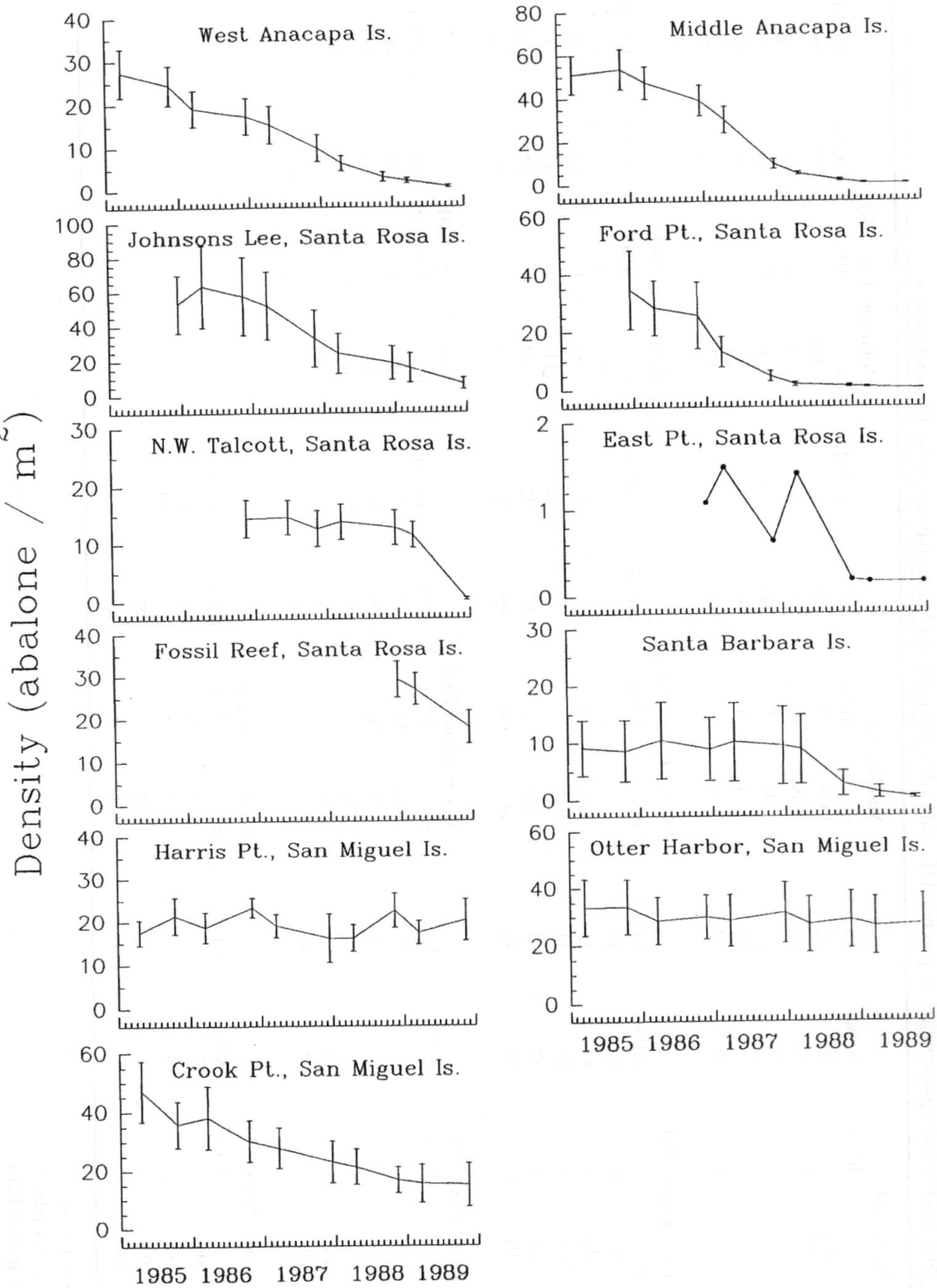

Fig. 17.2 The mean density (±se) of black abalone in permanent transects at the California Channel Islands from 1985–9. At East Point there is only one transect, thus no standard error can be determined.

were sampled for histopathological screening from black abalone. Left and right kidney tissue was excised from abalone, fixed in Davidson's solution (Shaw & Battle, 1957) and processed for routine paraffin histology. De-paraffinized, 5–7 μm sections were stained with Harris' Haematoxylin and Eosin (Luna, 1968) and were examined under light microscopy.

RESULTS

Declines in the black abalone density were first identified in 1986 at Anacapa Island (Fig. 17.2). The declines seriously affected the more abundant adult group (44–126 mm), but the cryptic juvenile (<44 mm) and the legal (>126 mm) groups similarly declined (Table 17.1).

At Anacapa and Santa Rosa Islands, black abalone were symptomatic for WS. Individuals were seen foot-side up in tidepools in various states of decomposition. Abalone, normally firmly attached, could be knocked from the rocks while they were being measured. Unusually high numbers of fresh shells were seen throughout the sites.

The decline in numbers and the symptoms of WS observed in 1986 were not seen at every location. The density at sites on Santa Barbara and San Miguel Islands was stable during this period (Fig. 17.2).

Through 1987, populations at NW Talcott and Fossil Reef appeared healthy. During subsequent visits to these sites, it became apparent that abalone declined there as well (Fig. 17.2).

In 1988, all sites except Harris Point, Otter Harbor and NW Talcott were undergoing population decline (Fig. 17.2). Visits in the following year revealed dramatically reduced numbers of black abalone at Anacapa, the south sites on Santa Rosa and Santa Barbara Islands. At most other sites abalone declines continued.

The number of shells collected repeatedly at three sites (Fig. 17.3) revealed trends which closely duplicated events at the respective sites. Where abalone stocks declined to low levels, as on Anacapa Island, the number of fresh shells also declined. Where populations were stable, as at Harris Point, the number of fresh shells was stable as well. Where rates of declines were accelerating, as at Fossil Reef, the number of shells on the beach increased (Figs 17.2, 17.3).

Pesticide analysis did not detect chlorinated hydrocarbons (<5 ppb), but heavy metals were found. Mean dry weight concentrations of lead, chromium, copper, mercury and zinc were 0.81, 3.25, 50.59, 0.15 and 95.27 ppm, respectively, but no significance was attributed to their presence (Pesticide Laboratory Report L-89–88, J. Harrington, pers. comm.).

Analysis of black abalone by R.J.G. Lester (pers. comm.) indicated that there was no infestation by the parasite *Perkinsus* sp., which is a problem in abalone populations in South Australia (Lester & Davis, 1981).

Initial samples of black abalone sent to the Fish Disease Laboratory revealed no unusual histopathology (Report dated 21 January 1981, Don Manzer, CDFG Senior Fish Pathologist, pers. comm.). Subsequent analysis of black abalone from Santa Rosa and Santa Cruz Islands conducted by R.P. Hedrick at the University of California, Davis, School of Veterinary Medicine, revealed a coccidian-like

Table 17.2 Locations from which black abalone have been tested for the presence of a coccidian parasite and the occurrence of WS, + indicates presence, – indicates absence, n/d is not determined and i indicates an area with insignificant numbers of abalone to detect WS.

	Collection date	Parasite	WS
Island			
Anacapa			
West	Nov 88	+	+
Middle	Nov 88	+	+
Santa Rosa			
Johnsons Lee	Dec 88	+	+
	Jan 89	+	+
Ford Pt	Dec 88	+	+
NW Talcott	Dec 88	+	+
East Pt	Dec 88	+	+
Fossil Reef	Dec 88	+	+
	Jan 89	+	+
San Miguel			
Harris Pt	Mar 89	+	+
Otter Harbor	Mar 89	+	–
Crook Pt	Mar 89	+	–
Santa Barbara	n/d	n/d	–
Santa Catalina			
Brick Pt	Feb 89	+	–
San Clemente			
Eel Pt	Oct 88	+	–
Airport	Oct 88	+	–
Coast			
Ana Nuevo	Oct 88	+	–
Diablo Canyon	Jul 88	+	+
Pt Conception			
Pt Arguello	Nov 88	+	–
Government Pt	Nov 88	+	–
Palos Verdes	n/d	n/d	i
Long Beach Harbor	Sep 88	–	i
San Diego	May 89	n/d	i

protozoan in both the left and right nephridia. Additional tissue samples taken from black abalone at various locations in California revealed the presence of the coccidian-like parasite in the kidneys of black abalone at most locations (Table 17.2).

Although the kidneys contained large numbers of parasites, most animals lacked an inflammatory response. Histological examination of infected kidney tissue revealed massive destruction of host tissue and replacement with the coccidian-like parasites. Black abalone which appeared healthy harboured large numbers of this parasite, whereas some abalone with WS exhibited only moderate parasite infestation.

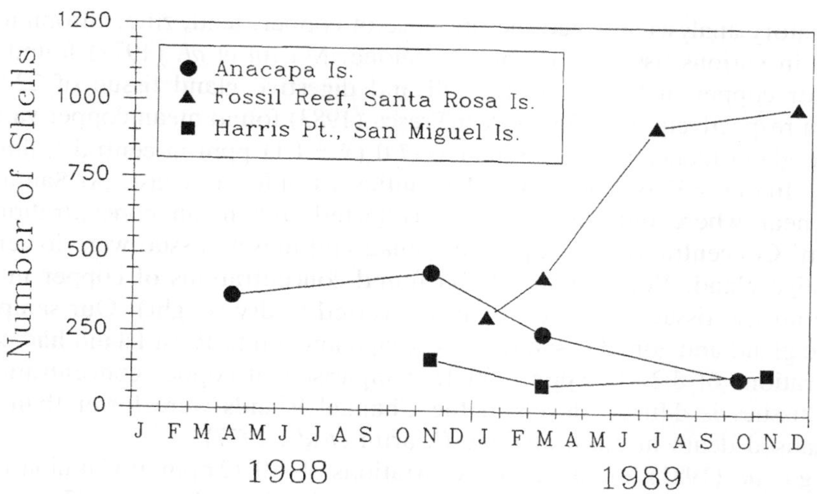

Fig. 17.3 Counts of black abalone shells collected at three sites at the Channel Islands. Shells were collected from areas of natural accumulation near sample locations.

Of the 52 abalone taken to Cabrillo Marine Museum, two individuals, one WS-affected and one apparently healthy, died in transit. Within 30 days all 16 WS-affected and 17 (50%) of the apparently healthy abalone died. Seven more abalone died by the end of the experiment about four months later.

DISCUSSION

The density plots (Fig. 17.2) provided convincing evidence that a progressive mortality of black abalone occurred at the Channel Islands, including Santa Cruz Island (B. Tissot, pers. comm.). However, not every Channel Island location experienced mortality. Harris Point and Otter Harbor on San Miguel Island and two sites at San Clemente Island (Fig. 17.1) did not decline, and as of December 1989, we have received no reports of the problem at San Nicholas Island.

On the south central coast, at Government Point, Point Arguello, and Diablo Canyon (Fig. 17.1 inset), we found moderate numbers of black abalone. Accelerated mortality was observed only at Diablo Cove, the site of a power plant cooling water discharge (Anonymous, 1989). Populations of abalone outside the Diablo Cove area are to date unaffected.

The few abalone analyzed for pesticide pollution yielded nothing significant, thus further costly tests and analyses were not indicated.

Heavy metals pollution, especially copper which is known to affect molluscs, was considered as a cause of mortalities. At Diablo Canyon Power Generating Station an accidental release of copper into nearshore waters killed large numbers of abalone (Martin *et al.*, 1977). This event and the sinking, in 1988, of the *PAC Baroness*, a loaded copper ore ship, near Point Conception led to concern that copper leaching from the wreck might be responsible for observed mortalities at the Channel Islands. However, because the mortalities are far removed from the wreck site, occurred over an extensive area and predated the wreck, we believe it is unlikely that copper from the wreck caused the observed mortalities.

Laboratory analysis detected the presence of copper, lead, zinc, chromium and mercury in various tissues from black abalone. Martin *et al*. (1977) found LD_{50} values for copper in black abalone gill and digestive gland tissue of 22.9 and 95.5 ppm respectively. In 1973–4, VenTresca (1983) found mean copper values in digestive gland tissues ranging from 1.1–7.0 ($\bar{x} = 4.1$) ppm in central California, and 6.6–104 ($\bar{x} = 32.6$) ppm along the southern California coast. At Santa Cruz Island, near where our samples were collected, the mean concentration was 10.9 ppm. Concentrations of copper in gonad and muscle tissue were lower than in digestive gland. Young *et al*. (1981) found concentrations of copper in black abalone muscle tissue of 1.7–6.7 ppm (converted to dry weight). Our samples of digestive gland and gonad tissue from Anacapa and Santa Rosa Island had copper concentrations of 8.2–12.5 ppm. While comparison of copper concentrations of various tissues is difficult, levels at the Channel Islands were lower than levels which caused death in black abalone (Martin *et al*., 1977).

Young *et al*. (1981) found lead concentrations of <0.12 ppm at Catalina Island, which compare closely with our findings. Stewart & Schultz-Baldes (1976) reported lead concentrations in red abalone from San Diego that were an order of magnitude higher than the levels we found in black abalone at Anacapa and Santa Rosa Islands. They also found that concentrations of lead in red abalone were non-lethal at levels two orders of magnitude higher than we observed in wild populations of black abalone.

Young *et al*. (1981) reported ranges (ppm wet weight) for zinc (4.3–11), chromium (0.04–0.10) and mercury (<0.01–0.06) in black abalone from Santa Catalina Island. Our data found ranges of 11.6–14.2, 0.34–0.49 and 0.03 ppm, for black abalone from Anacapa and Santa Rosa Islands, respectively. Few data are available on background levels or lethal concentrations of these metals for black abalone. Laboratory analyses by others failed to implicate pesticides, heavy metals, inorganic compounds or radioisotopes in abalone mortalities (Diablo Canyon Power Plant Marine Environmental Monitoring program, Annual Report, April 1988). From the available information we conclude that heavy metals and pesticides are not a direct causal factor.

Sport and commercial harvest was not directly implicated in the declines. Few abalone larger than 126 mm (sport legal size) occur at any site, which can be attributed to harvest (Table 17.1), and relative percentages remained constant at most sites. The continued appearance of fresh shells at the sites is also evidence against harvest, as abalone are legally required to be landed alive.

Occurrences of WS-like symptoms have been reported from southern California. In 1956, black abalone at Palos Verdes 'were shrunken' (Cox, 1962). An exchange of abalone between Palos Verdes and a healthy stock at Santa Catalina Island resulted in the recovery of the former and the starvation of the latter (Cox, 1962; Young, 1964). At the same time, pink abalone, *H. corrugata*, near Laguna Beach were 'shrunken within their shells and the flesh was not firm'. The condition was attributed to starvation resulting from the local seaweeds being killed by ocean discharges (MacGinitie & MacGinitie, 1966).

Quantity and quality of food have been implicated in abalone declines in California (Cox, 1962; Young, 1964; MacGinitie & MacGinitie, 1966; Tanaka *et al*., 1986). Our experimental results and other studies (Diablo Canyon Power Plant Marine Environmental Monitoring program, Annual Report, April 1988) suggest that the observed mortality could not be reversed by simply providing food, but food together with other factors may be important (Chapter 19, this volume).

Parasites are known to affect abalone elsewhere. In South Australia the red lip abalone, *H. rubra*, may become infected with high levels of a protozoan parasite, *Perkinsus olseni*, which infects muscle and haemolymph, causing discoloured pustules on the foot (Lester & Davis, 1981). This genus is also known to infect oysters on the east coast of North America. Extensive outbreaks of this parasite seriously impact commercial abalone beds in Australia. The symptoms, however, are not similar to those of WS, and this organism was not found in samples of black abalone.

The discovery of a coccidian parasite and its ability to damage kidney tissue focused investigation of the cause of WS. We found the parasite throughout the range of the black abalone (Table 17.2), but it was not virulent everywhere it occurred. Preliminary findings indicated that there was a very low correlation of the presence and the numbers of the parasite in the kidneys with the appearance of WS. These findings cast doubt that the parasite is a direct single cause of WS.

The search for the cause of accelerated mortality has thus far been fruitless. We have eliminated single factors, i.e. fishery harvest, parasites, pollution and lack of food, which have caused problems in abalone and closely related forms. There are other, probably many, factors we have not addressed. For example, the Diablo Canyon observations seem to implicate temperature. Having examined a number of the most likely single factors, we conclude that a combination of environmental and biological factors is involved.

ACKNOWLEDGEMENTS

The authors thank Ron Hedrick, Robert J.G. Lester and Don Manzer for their assistance and examination of black abalone. The authors are also grateful to Brian Tissot and an anonymous reviewer for their helpful comments and suggestions.

REFERENCES

Anonymous (1989) Diablo Canyon Power Plant Marine Environmental Monitoring program, Annual Report, April 1988. Pacific Gas and Electric Company, Avila Beach.

Cox K.W. (1962) California abalones, family Haliotidae. *Calif. Fish Game, Fish. Bull.*, **118**.

Davis G.E. (1989) Design of a long-term ecological monitoring program for Channel Islands National Park, California. *Nat. Areas J.*, **9**, 80–9.

Ebert E.E. & Houk J.L. (1989) Abalone cultivation methods used at the California Department of Fish and Game's Marine Resources Laboratory. In *Handbook of Culture of Abalone and other Marine Gastropods* (Ed. by K.O. Hahn), pp. 239–54. CRC Press, Boca Raton.

Hahn K.O. (1989a) Biotic and abiotic factors affecting the culture of abalone. In *Handbook of Culture of Abalone and other Marine Gastropods* (Ed. by K.O. Hahn), pp. 113–34. CRC Press, Boca Raton.

Hahn K.O. (1989b) Nutrition and growth of abalone. In *Handbook of Culture of Abalone and other Marine Gastropods* (Ed. by K.O. Hahn), pp. 135–56. CRC Press, Boca Raton.

Lester R.J.G. & Davis G.H.G. (1981) A new *Perkinsus* species (Apicomplexa, Perkinsea) from

abalone *Haliotis ruber*. *J. Invert. Pathol.*, **37**, 181–7.
Luna L.G. (1968) *Manual of Histologic Staining Methods of the Armed Forces Institute of Pathology*. 3rd edn. McGraw-Hill, New York.
MacGinitie N. & MacGinitie G.E. (1966) Starved abalones. *Veliger*, **8**, 313.
Martin M., Stephensen M.D. & Martin J.H. (1977) Copper toxicity experiments in relation to abalone deaths observed in a power plant's cooling waters. *Calif. Fish Game*, **63**, 95–100.
Martin M., Stephensen M.D., Smith D. & Goetzl J. (1986) Pilot chemical surveillance program at Channel Islands National Park and National Marine Sanctuary. *Calif. Dept. Fish Game, Mar. Poll. Studies Lab. Mussel Watch Survey Oct. 85*.
Richards D.V. (1988) Rocky intertidal ecological monitoring in Channel Islands National Park, California 1986–1987. *Chan. Is. Nat. Park Nat. Sci. Rpts.*, CHIS-88-001.
Richards D.V. & Davis G.E. (1988) *Rocky Intertidal Communities Monitoring Handbook*. Channel Islands National Park, Ventura.
Shaw B.L. & Battle H.I. (1957) Growth and microscopic anatomy of the digestive tract of the oyster, *Crassostrea virginica* (Gmelin). *Can. J. Zool.*, **35**, 325–47.
Stewart J. & Schultz-Baldes M. (1976) Long-term accumulation in abalone (*Haliotis* spp.) fed on lead-treated brown algae (*Egregia laevigata*). *Mar. Biol.*, **36**, 19–24.
Tanaka K., Ishida O. & Tanaka T. (1986) On the lost weight of abalone in the area around Mera Shoal; in the southern coast of the Bósó Peninsula, especially concerning the habitat conditions. *Bull. Jap. Sea Reg. Fish. Res. Lab.*, **36**, 49–57.
VenTresca D.A. (1983) *Concentration of selected trace elements in the black abalone*, Haliotis cracherodii *Leach, 1814, collected between Pillar Point, California and Punta Banda, Mexico*. MA thesis, California State University, San Francisco.
Young D.R., Moore M.D., Jan T.K. & Eganhouse R.P. (1981) Metals in seafood organisms near a large California municipal outfall. *Mar. Poll. Bull.*, **12**, 134–8.
Young P.H. (1964) Some effects of sewer effluent on marine life. *Calif. Fish Game*, **50**, 33–41.

Chapter 18
Aspects of the biology of the black abalone (*Haliotis cracherodii*) near Diablo Canyon, central California

JAMES B. BLECHA and JOHN R. STEINBECK *Tenera Environmental, 586 Higuera St., Suite 200, San Luis Obispo, California 93401, USA*

DAVID C. SOMMERVILLE *Pacific Gas & Electric Company, Biology Laboratory, PO Box 117, Avila Beach, California 93424, USA*

ABSTRACT: Black abalone (*Haliotis cracherodii*) have been monitored from 1976 to the present as part of the marine environmental studies at Pacific Gas & Electric Company's Diablo Canyon Power Plant. Four permanent stations were established at which individually tagged abalone are periodically relocated and measured. The resulting data were analyzed to estimate growth rates and measure movement from permanent bench marks. In addition to studies at permanent stations, periodic area-wide population censuses have been conducted along the intertidal zone of Diablo Cove, the receiving water body for the plant's thermal discharge. Growth rates of tagged individuals were computed using a modified von Bertalanffy algorithm. Growth varied among stations. In the study area, black abalone reached a shell length of 40 mm during their third year. Sizes greater than 100 mm required in excess of 20 years of growth. Movement of tagged individuals was quite variable. In general, large (>100 mm shell length) individuals tended to be sedentary, rarely moving from the site where they were originally tagged. Animals in an intermediate size range (40–80 mm) were most active, often moving several metres in a single tidal cycle. Abundance estimates inside Diablo Cove remained relatively constant over the first three years of power plant operation. A recruitment event occurred in 1985–6. At this time the power plant was in full commercial operation, discharging water approximately 10°C above ambient temperature. Since June 1988, there has been a marked decrease in abundance of black abalone inside Diablo Cove due to mortality associated with an atrophied foot condition first observed in the Channel Islands off southern California. Observations of disease-associated effects on the distribution of black abalone are presented.

RESUMEN: El abulón negro (*Haliotis cracherodii*) ha sido monitoreado desde 1976 hasta la fecha como parte del estudio del medio ambiente marino en las Compañías de Gas del Pacífico y la Planta de energía eléctrica del Cañón del Diablo. Las tasas de crecimiento de los individuos marcados en las 4 estaciones fueron computadas usando un algoritmo modificado de von Bertalanffy. El crecimiento varió entre las estaciones. El abulón negro alcanzó una concha de 40 mm de longitud durante su tercer año. Las tallas más grandes de 100 mm requirieron mas de 20 años de crecimiento. El movimiento de los individuos marcados fue bastante variable. En general, los individuos grandes (>100 mm de longitud de la concha) eran sedentaros, raramente se movieron del lugar donde fueron marcados. Los animales en un rango de tamaño intermedio (40 a 80 mm) eran más activos, a menudo moviéndose varios metros en un solo ciclo mareal. Una abundancia de estimaciones dentro de la Ensenada del Diablo permanecieron relativamente constantes durante el funcionamiento de la planta eléctrica. Un evento de reclutamiento ocurrió en 1985–6. En este período la planta de energía eléctrica estaba en su plena operación comercial, descargando agua aproximadamente 10°C sobre la temperatura ambiente. Desde junio de 1988, ha habido un marcado decrecimiento en la abundancia del abulón negro dentro de la Ensenada del Diablo debido a la mortalidad asociada con una condición de atrofia del pie, primeramente observada en las Islas Channel. Observaciones de los efectos de la enfermedad en la distribución del abulón negro son presentadas.

INTRODUCTION

The black abalone, *Haliotis cracherodii* Leach, is one of seven species of abalone common to California and the only species which lives primarily in the intertidal zone. Its optimal habitat is the rocky intertidal zone in areas with protective crevices and moderate to heavy surf action. In such habitats, densities may range from 25–40 per m^2 on the central California mainland (pers. obs.), to over 120 per m^2 in the Channel Islands (Douros, 1987).

While not as extensively studied as the other species of California abalone, black abalone have recently been the subject of studies to determine the cause of the mass mortalities associated with an atrophied foot condition, first observed on the Channel Islands offshore southern California in 1987 (Tissot, 1988a; Chapters 16, 17 and 19, this volume).

Black abalone have been the subject of special studies as part of an extensive programme begun in 1976 to monitor the effects of the Diablo Canyon Power Plant on the nearshore marine environment. Field studies monitored growth, movement, distribution, abundance and recruitment. These studies were designed to develop a baseline which would be used to evaluate the potential effects of an increase in seawater temperature resulting from the discharge of the power plant. This chapter presents a preliminary examination of growth and movement studies carried out between 1979 and 1984, and distribution and abundance studies conducted from 1981 to the present. Following one year of start-up testing, Diablo Canyon Power Plant began commercial operation in May 1985.

LOCATION

Diablo Cove (35°11.3′N, 120°48.8′W) is located on a rocky headland midway between the cities of Morro Bay and Avila Beach, in central California (Fig. 18.1). This area is open coast, subject to heavy surf action. Winter storms create waves which average 6 to 9 m in height (trough to crest), with occasional waves in excess of 9 m. The annual mean seawater temperature is 12°C, with a range from 9°C (April through May) to 16°C (October through November). The nearshore environment is influenced by the southerly flow of the colder California Current from late winter through late summer, and the northerly flow of the warmer Davidson Current during the autumn and winter months. Strong upwelling is common from February to July.

The area around Diablo Cove is prime habitat for *Haliotis* species. It lies within the historical area for the commercial fishery for red abalone, *H. rufescens*, which averaged over 1800 tons per year between 1965 and 1974 (Morris *et al.*, 1980). The area is well suited for field studies because abalone are abundant, study sites are inaccessible to the public and the normal heavy wave action prevents extensive predation by the sea otter (*Enhydra lutris*).

METHODS

Four areas, approximately 40 to 50 m^2 each, were selected for growth and movement studies (Fig. 18.1). Two areas, south Diablo Cove (SDC) and north

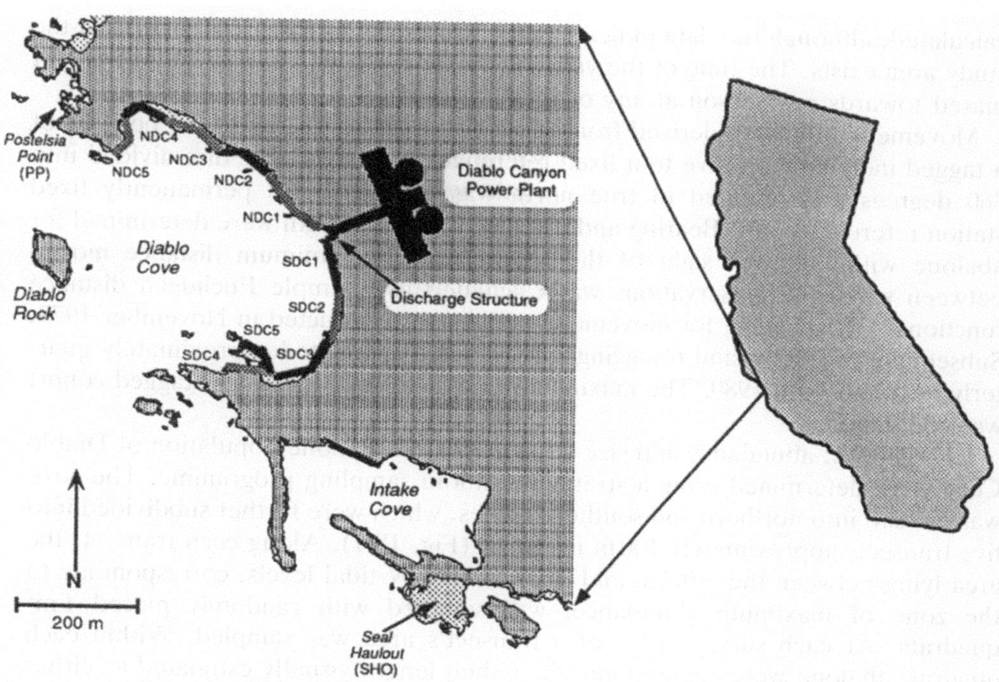

Fig. 18.1 Map of study area, showing growth and movement study areas, and fixed transect locations.

Diablo Cove (NDC) were relatively protected from wave action. Postelsia Point (PP) and Seal Haulout (SHO) were located on the headlands to the north and south, respectively, of Diablo Cove in areas normally subject to heavy wave action.

Growth and movement in the field were determined from successive measurements of tagged animals. All abalone were tagged without removal from the substratum. A numbered 5 × 15 mm tag was secured to a length of stainless steel wire passed through adjacent respiratory pores. Growth was defined as the change in shell length between successive recaptures of tagged individuals. Shell length was measured *in situ* using calipers. A minimum time interval of two months between shell length measurements was used to calculate growth curves.

Growth was assumed to fit the Brody-Bertalanffy model (Ricker, 1975) which produces a growth curve described by the equation:

$$\Delta L = (1 - e^{-K\Delta t})\ (L_{\infty} - L_1)$$

where K is the growth constant, L_1 is the initial shell length, L_{∞} is the mean asymptotic shell length, Δt is the time interval (days) from capture to recapture and ΔL is the growth over this interval. This function has been used to describe the growth of other *Haliotis* species (Forster, 1967; Shepherd & Hearn, 1983; Tutschulte & Connell, 1988). The non-linear least squares method of Fabens (1965), as modified by T.A. Ebert (pers. comm.), was used to estimate the parameters of the Brody-Bertalanffy growth curve. This method is designed specifically for incremental growth measurements from mark-recapture studies, in which absolute age at any time is not known. Seasonal growth was not calculated, although the data indicate that a seasonal component of growth in the study area exists. The time of the year over which growth was calculated was not biased towards any season at any of the stations.

Movement data were derived from successive determinations of the position of a tagged individual relative to a fixed reference point. A plastic disc divided into 360 degrees and oriented to true north was attached to a permanently fixed station reference point. Bearing and distance from the origin were determined for abalone within line of sight of the apparatus. The minimum distance moved between successive observations was calculated as a simple Euclidean distance function. Initial tagging for movement studies was conducted in November 1978. Subsequent recapture and retagging surveys were conducted approximately quarterly, through June 1980. The maximum time at liberty of the first tagged cohort was 606 days.

Distribution, abundance and size structure of the abalone population of Diablo Cove were determined using a stratified random sampling programme. The cove was divided into northern and southern halves, which were further subdivided into five transects approximately 100 m in length (Fig. 18.1). Along each transect, the area lying between the −0.3 m and +1.3 m MLLW tidal levels, corresponding to the zone of maximum abundance, was sampled with randomly placed 1 m^2 quadrats. At each survey, 10% of a transect's area was sampled. Within each quadrat, abalone were counted and their shell length visually estimated as either small (<50 mm), medium (51–100 mm) or large (>100 mm).

RESULTS

Over 1100 abalone were tagged between 1979 and 1984 at the four stations. Growth data were derived from 173 recaptures at SDC, 190 recaptures at NDC, 228 recaptures at SHO and 169 recaptures at PP. Recapture data contained negative ΔL values, an indication of, among other factors, the variability of *in situ* measurements. Other than obvious mensural errors, all ΔL values, positive and negative, were used to estimate the growth parameters and generate growth curves.

Mean shell lengths and standard deviations of the initial sizes (size at tagging) of animals recaptured at the four areas are presented in Table 18.1. With the exception of NDC, the observed difference in mean shell lengths among recaptures at the study sites is relatively small and not considered to be a significant variable affecting the estimates of the growth parameters. Habitat differences among stations may have accounted for some of the size differences seen. For example, at NDC fewer small animals were measured due to their inaccessibility.

The time increment between recaptures varied between 60 and 90 days for sizes below 50 mm, and between 150 and 200 days for sizes greater than 60 mm. These results reflect the differential success of recapture between the two size classes. Some of the possible reasons for this difference are discussed in the section on movement.

Estimated K values, asymptotic lengths and estimated age at the asymptotic length minus 1 mm are presented for the four study sites in Table 18.1. Growth curves generated by the method of Fabens (1965) are shown in Fig. 18.2. Size at a

Table 18.1 Tagging data and von Bertalanffy growth parameter estimates, by station.

Station	Mean initial size (mm)	sd	K	sd	L_{∞}	sd	Age (y) at (L_{∞} − 1.0 mm)	L_{max}
South Diablo Cove	64.7	28.9	0.196	0.018	122.7	3.9	25	145
North Diablo Cove	81.4	23.5	0.118	0.018	112.0	5.2	36	136
Seal Haulout	67.9	20.5	0.200	0.016	100.3	2.2	24	130
Postelsia Point	73.2	24.3	0.231	0.043	97.5	5.4	20	128

known age is required for the model and since these data are lacking, a size of 0.0 mm at time $T = 0$ was assumed.

All movement data presented here are from a population of 416 tagged animals at the SHO station. Movement was divided into long term and short term. Long-term movement was defined as a change in location of an abalone between two, not necessarily successive surveys. If an abalone was in one position during one survey, and another position during another survey 154 days later, it 'moved' in the 154-day period. Of course, it could have moved many times during that time period, and at any time. Short-term movement was that which occurred over a period of successive survey days; positions of tagged abalone were recorded on day one and rechecked on the next two days.

Long- and short-term movement data are summarized in Table 18.2. The percentage of individuals which moved over the long term was greatest in the size range of 55–65 mm shell length. The percentage of recaptures was greatest among the larger, less active size classes. For both types of movement, the most

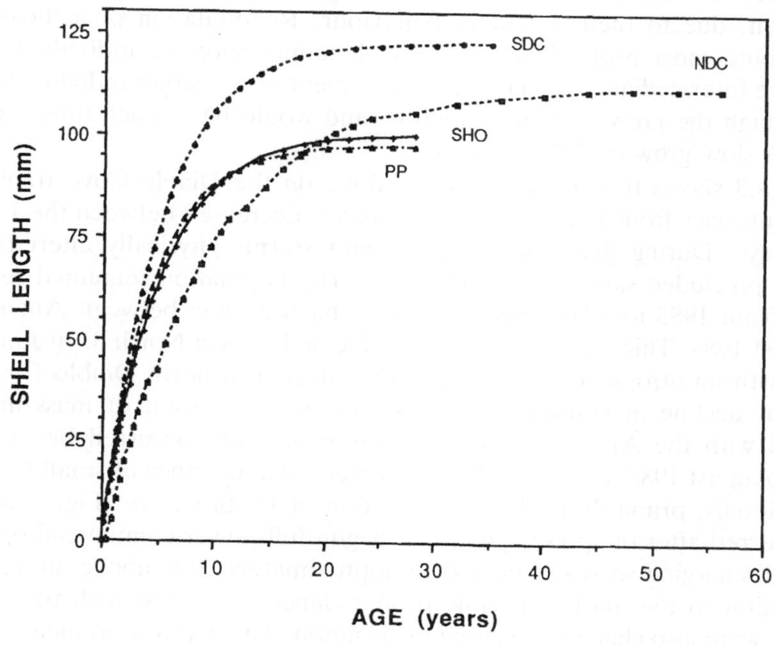

Fig. 18.2 von Bertalanffy growth curves, by station.

Table 18.2 Long-term and short-term movement data.

	Size class midpoint (mm)							
	≤35	45	55	65	75	85	95	≥105
Number tagged	23	51	43	38	36	42	23	28
Number recaptured	12	24	23	21	29	32	16	23
Long term								
Number moved	4	11	17	15	13	13	3	2
Percentage moved	33	46	74	71	45	41	19	9
Short term								
Percentage moved	–	18	25	23	33	7	9	–
Mean distance moved per day (cm)	–	83	189	387	278	54	119	–

active individuals were those between 45 and 75 mm shell length. The average distance moved per day was also greatest in the intermediate (55–75 mm) size classes. In fact, some individuals in the size range for 60–69 mm were observed to move over 7 m in a 24-hour period. Long-term movement data suggest that, given enough time, any size abalone may move.

The effect of movement behaviour on repopulation of habitats is illustrated in Fig. 18.3. These data are the result of a field manipulation begun in September 1980 and monitored for 5 years. In this experiment, all abalone were removed from an isolated crevice which was subsequently monitored for recolonization. The replacement cohort ranged in size from 45 to 95 mm, representative of the most active size classes. Replacement of larger size classes did not occur by immigration, due to their sedentary behaviour. Repopulation of a 'favourable' habitat would most probably occur through immigration by individuals in the more active intermediate size classes. Replacement of the larger individuals would occur through the growth of these recruits and would be a much longer process due to the slow growth of the species.

Table 18.3 shows the abundance of abalone on the Diablo Cove transects at irregular intervals from 1982 to 1989. Abundance decreased between the 1982 and 1983 surveys. During this time, severe winter storms physically altered station relief and precluded sampling two transects. The population remained relatively constant from 1983 to 1988, then showed a sharp decline between August 1988 and August 1989. This trend is most noticeable at Transect North 3. In June 1988, abalone with an atrophied foot (AF) were collected in north Diablo Cove. The subsequent decline in abundance in this area was the result of mass mortality associated with the AF condition. A recruitment event occurred between July 1985 and August 1988, as reflected in the increased abundance of small (<50 mm) size individuals, primarily in the south portion of Diablo Cove (Fig. 18.4). This event occurred after the power plant had begun full-power commercial operation and was discharging seawater heated to approximately 10°C above ambient.

In addition to the marked decline in abundance associated with the onset of AF, there were also changes in spatial distribution. The variance to mean ratio for the quadrat data pooled over the north Diablo Cove transects (Table 18.4) shows

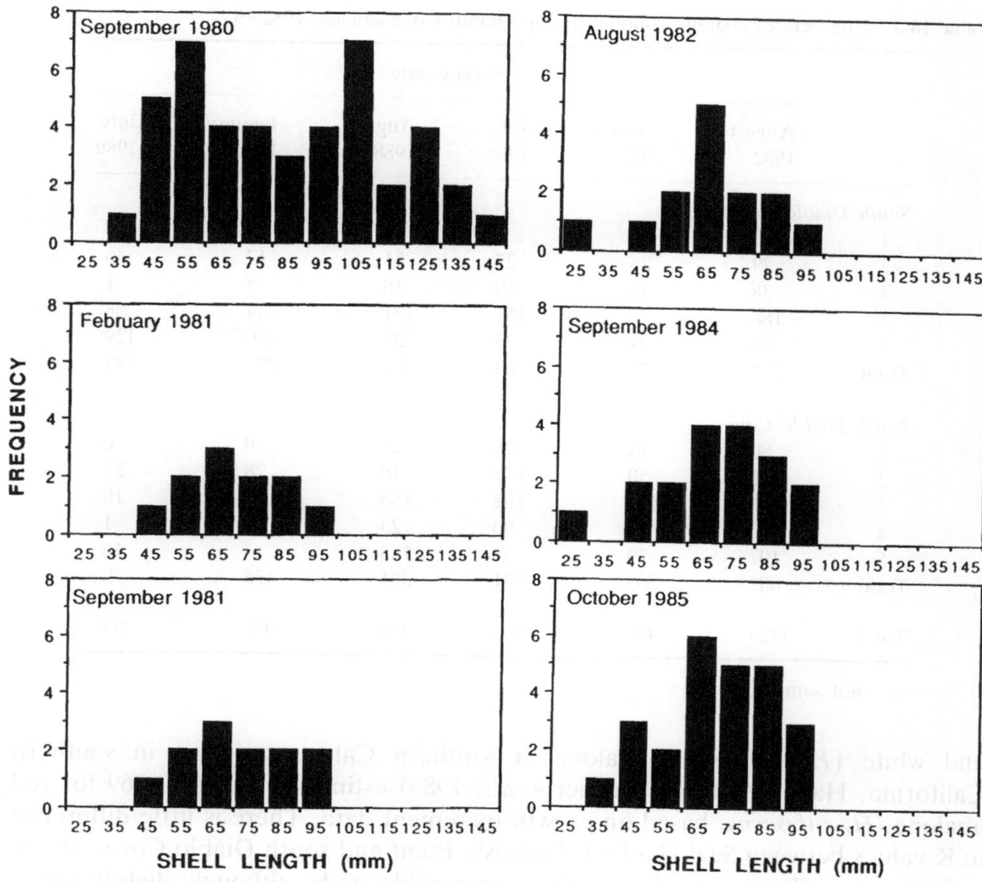

Fig. 18.3 Size frequency of abalone repopulating a crevice cleared in 1980 and monitored through 1985.

a decreasing trend. Kolmogorov-Smirnov tests, used to determine whether the distribution of individuals corresponds to a Poisson (random) distribution, are significant for the two surveys in 1989, indicating a change from a spatially clumped to a random distribution.

DISCUSSION

The objective of these studies was to assess the potential effects of the operation of the power plant on various aspects of the biology of the black abalone. Studies of growth, movement and abundance during plant operation are ongoing. This discussion is, therefore, intended to summarize the results of growth, movement and abundance studies conducted to date.

Estimates for the growth parameter K determined in this study (Table 18.1) are similar to those for other species of abalone from California. Tutschulte & Connell (1988) found K values of 0.08 to 0.15, depending on experimental treatment and method of calculation, for green (*H. fulgens*), pink (*H. corrugata*)

Table 18.3 Numbers of abalone counted during Diablo Cove census, 1982–9.

	Survey date					
	August 1982	August 1983	July 1985	August 1988	January 1989	July 1989
South Diablo Cove						
1	15	2	2	0	1	0
2	29	28	27	31	17	6
3	38	16	10	16	5	1
4	188	75	151	181	114	58
5	52	ns	73	237	150	129
Total	322	121	263	465	287	194
North Diablo Cove						
1	80	ns	26	21	0	0
2	221	50	177	48	28	22
3	194	194	192	186	89	10
4	110	46	90	23	15	1
5	197	79	56	113	4	16
Total	802	369	541	391	136	49
Total	1124	490	804	856	423	243

ns = not sampled.

and white (*H. sorenseni*) abalone in southern California. Also in southern California, Haaker (cited in Tegner *et al.*, 1989) estimated a K of 0.269 for red abalone, *H. rufescens*, based on growth increment data. There is little difference in K values between Seal Haulout, Postelsia Point and south Diablo Cove. The K estimates for *H. cracherodii* are also comparable with, although slightly lower

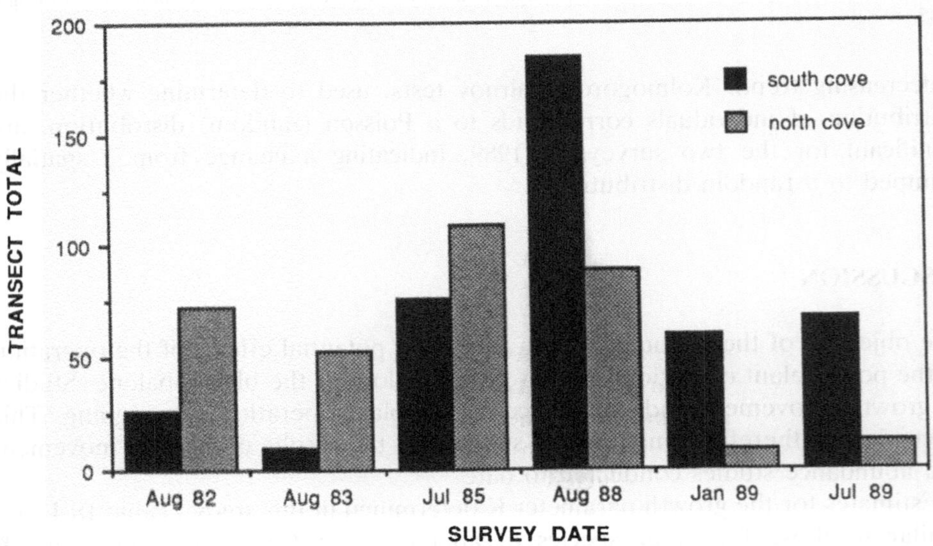

Fig. 18.4 Abundance of small (<50 mm) individuals on the north and south Diablo Cove transects between 1982 and 1989.

Table 18.4 Summary statistics for pooled data from north Diablo Cove transects 2–4.

	Survey date					
	August 1981	August 1983	July 1985	August 1988	January 1989	July 1989
Number of quadrats surveyed (N)	572	609	664	650	673	640
Number of quadrats with abalone	138	80	140	123	67	29
Total number of abalone observations in N	722	366	516	385	119	49
Mean number of abalone per quadrat	1.26	0.60	0.78	0.59	0.18	0.08
Standard deviation	4.19	2.39	2.56	1.95	0.72	0.39
Variance/mean	13.9	9.6	8.4	6.5	2.8	1.9
K–S *d*-value	0.48	0.32	0.33	0.26	0.06*	0.03*

* Denotes significant Kolomogorov-Smirnov *d*-value at 0.05 level of significance.

than, those for other *Haliotis* species presented in Shepherd & Hearn (1983) and Sloan & Breen (1988).

A comparison of the asymptotic length and the maximum measured size at each station shows the measured maximum length to exceed the asymptotic length by over four standard deviations in all cases. If the estimated value for L_∞ (and K) represents a distributional mean for individual abalone (Sainsbury, 1980), then the degree to which measured L_{max} differs from the asymptotic length would be an indication of the variability in growth rate between individual black abalone. In areas or populations where there is little variability in growth rates between individuals, the asymptotic length and measured L_{max} may be nearly equal. Therefore a high degree of variability in growth among individual black abalone is indicated and appears to be most pronounced at wave-exposed sites and less at protected sites.

Shell length alone cannot be considered to be a good indication of the age of black abalone. They appear to be long-lived, generally in excess of 30 years. For each study site, 20 or more years were required to reach the asymptotic length. One individual, 100 mm in shell length, at SHO showed essentially no change in length from 1976 to 1988. Wright (1975) calculated a time of 51 years for black abalone from Santa Cruz Island, southern California, to grow from 90 to 130 mm shell length. Similarly, Tutschulte & Connell (1988) reported times ranging from 33 to 57 years for pink, white and green abalone in southern California to reach a size of L_∞, where asymptotic lengths ranged from 169 to 221 mm.

The following growth model for the study area is proposed. Growth is generally regular during the first two years, regardless of location. During the third year, the abalone reach a size of 40 mm. Thereafter, growth becomes more irregular and episodic, influenced in part by locality. For abalone in areas with strong wave action, individual growth is variable and they generally will not grow as large as abalone in areas with less surf action. It appears that those black abalone which grow to the sport legal fishing size of 127 mm are the exception, rather than the rule, in this area.

The abalone in this study exhibited distinct, size-related activity patterns. Small individuals (<40 mm) are cryptic by nature, moving about actively within the protective confines of crevices. This behaviour pattern provides a defence against predation. Various predators, including *Pycnopodia helianthoides*, *Pachygrapsus crassipes*, *Hemigrapsus* spp. and *Cancer* spp., as well as gulls and oyster catchers, have been observed to capture small abalone in the study area. Intermediate sizes (55–75 mm) are the most active, both as a group and individually. This same pattern was noted by Tissot (1988b), as it affected immigration and emigration. Larger size classes are mostly sedentary. Individuals above 105 mm were observed to be predictably stationary, and were commonly recaptured due to their conspicuous nature. This size class was most frequently found in very open, exposed locations, while intermediate and small sizes preferred the protection of cracks and crevices. The change in movement behaviour with growth suggests that a size is reached which provides refuge from many predators, permitting free movement into and ultimately colonization of open, less protected locations.

Movement is generally thought to be related to feeding. Shepherd (1986) found that movement was related to crevice availability and quality. He suggested that crevice-seeking behaviour is largely affected by predation pressure and factors associated with feeding. Bergen (1971) reported a relationship between size and movement in *Haliotis cracherodii*: large individuals remained in place for long periods of time, while smaller animals appeared to leave their scars (depressions in the rock) to graze at night, returning to the same spot. Our studies of movement show neither evidence of diel movement nor homing. Repeated sampling trips to the same areas and familiarity with the study area have, however, suggested that there are 'preferred' stopping-over sites. There are, within the sampling station areas, relocatable locations in which an abalone will invariably be found, usually a member of the more active size classes (55–75 mm). These sites are most often small ledges which, although not well protected, seem to represent a preferred stopping site as the tide ebbs. These sites almost always contain abalone, although the same individual is rarely present on successive days. Most published accounts of movement in abalone have reported little or no movement. Cox (1962) noted that an adult black abalone was observed to remain in the same spot for two years. We have observed abalone in the same location for over 12 years. These were large (>100 mm) individuals in open, exposed locations. This seems to confirm the notion of attainment of a safe size of refuge at some stage of growth.

Based on extrapolation, the abalone population within Diablo Cove ranged between 7000 and 11 000 from 1982 to 1988. The persistence of their abundance and the observed recruitment event in the southern portion of the cove suggest that a healthy, viable population existed in the cove during the first three years of power plant operation. Between 1988 and 1989 the population decreased to less than 2300 individuals as a result of mortality associated with the AF condition. Although the exact cause of this condition is unknown at present (Chapter 16, this volume), the similarity of the symptoms and timing of its occurrence suggest that it is the same disease which affected the black abalone population on the Channel Islands (Chapters 17 and 19, this volume).

Since the onset of the AF condition, the spatial distribution of abalone in Diablo Cove has changed from a clumped to a random distribution. If this condition is contagious, then transmission by proximity and subsequent mortality could result in a random distribution which is a relict of the previously clumped distribution. The change in distribution may also be related to a change in the behaviour of afflicted individuals. In the field, AF-afflicted individuals have frequently been observed out in the open, on top of boulders, in cobble beds and in other atypical habitats. The cause may be an active mechanism such as a change in behaviour, a passive mechanism such as movement by wave action due to a diminished ability to attach to the substratum or a combination of both. In any case, the result would be a redistribution of individuals which are normally clumped around preferred microhabitat features.

ACKNOWLEDGEMENTS

This project was funded by Pacific Gas & Electric Company. J. Carroll, C. Ehrler, A. Fukuyama, J. Kelly, R. Kimura, S. Krenn and R. Massengill participated in data collection. T.A. Ebert provided the computer program used to fit the growth curves and provided valuable discussion on the subject of growth. F. Steinert and D. Innis assisted with graphics production. The authors wish to thank M.J. Tegner, B.N. Tissot and an anonymous reviewer for their editorial guidance.

REFERENCES

Bergen M. (1971) *Growth, feeding and movement in the black abalone*, Haliotis cracherodii *Leach, 1814*. MA thesis, University of California, Santa Barbara.

Cox K.W. (1962) California abalones, family Haliotidae. *Calif. Dept Fish Game. Fish Bull.*, **188**.

Douros W.J. (1987) Stacking behavior of an intertidal abalone: an adaptive response or a consequence of space limitation? *J. Exp. Mar. Biol. Ecol.*, **108**, 1–14.

Fabens A.J. (1965) Properties and fitting of the von Bertalanffy growth curve. *Growth*, **29**, 265–89.

Forster G.R. (1967) The growth of *Haliotis turberculata*: results of tagging experiments in Guernsey 1963–5. *J. Mar. Biol. Assoc. UK*, **47**, 287–300.

Morris R.H., Abbott D.P. & Haderlie E.C. (1980) *Intertidal Invertebrates of California*. Stanford University Press, Stanford.

Ricker W.E. (1975) Computation and interpretation of biological statistics of fish populations. *Bull. Fish. Res. Bd. Can.*, **191**, 1–382.

Sainsbury K.J. (1980) Effect of individual variability on the von Bertalanffy growth function. *Can. J. Fish. Aquat. Sci.*, **37**, 241–7.

Shepherd S.A. (1986) Movement of the Southern Australian abalone *Haliotis laevigata* in relation to crevice abundance. *Aust. J. Ecol.*, **11**, 295–302.

Shepherd S.A. & Hearn W.S. (1983) Studies on southern Australian abalone (genus *Haliotis*). IV. Growth of *H. laevigata* and *H. ruber*. *Aust. J. Mar. Freshwat. Res.*, **34**, 461–75.

Sloan N.A. & Breen P.A. (1988) Northern abalone, *Haliotis kamtschatkana* in British Columbia: fisheries and synopsis of life history information. *Can. Spec. Publ. Fish. Aquat. Sci.*, **103**, 1–46.

Tegner M.J., Breen P.A. & Lennert C.E. (1989) Population biology of red abalones, *Haliotis rufescens*, in southern California and management of the red and pink, *H. corrugata*, abalone fisheries. *Fish. Bull. US*, **87**, 313–39.

Tissot B.N. (1988a) Mass mortality of black abalone in southern California. *Am. Zool.*, **28**, 69A (abstr.).

Tissot B.N. (1988b) Morphological variation along intertidal gradients in a population of black

abalone *Haliotis cracherodii* Leach 1814. *J. Exp. Mar. Biol. Ecol.*, **117**, 71–90.
Tutschulte T.C. & Connell J.H. (1988) Growth of three species of abalones (*Haliotis*) in southern California. *Veliger*, **31**, 204–13.
Wright M. (1975) Growth in the black abalone, *Haliotis cracherodii*. *Veliger*, **18**, 194–9.

Chapter 19
Abalone population declines and fishery management in southern California

GARY E. DAVIS and DANIEL V. RICHARDS *US National Park Service, Channel Islands National Park, 1901 Spinnaker Dr, Ventura, California 93001, USA*
PETER L. HAAKER and DAVID O. PARKER *California Department of Fish and Game, 330 Golden Shore, Suite 50, Long Beach, California 90802, USA*

ABSTRACT: From 1985 to 1989, black abalone, *Haliotis cracherodii*, populations declined by up to 96% in abundance on four California Channel Islands. Similarly, island populations of red abalone, *H. rufescens*, declined by 97% to 98% and pink abalone, *H. corrugata*, declined by 51% to 94% between 1983 and 1989. The cause(s) of the declines are not clear, but we suggest that fishery management practices have altered the structure and function of coastal ecosystems and that the observed abalone population crashes were precipitated in a destabilized system by the extreme 1982–4 El Niño event. Modern civilization has dramatically reduced natural predator populations in coastal ecosystems and selectively removed a few well marketed species. These actions apparently facilitated accrual of abnormally high densities of unharvested prey species and shifted natural controls on herbivore populations from predation to competition, and perhaps disease. The 1982–4 El Niño event abruptly altered kelp abundance and intensified competition for food between abalones and other drift kelp consumers. A dramatic crash in abalone populations followed. This sudden change from the apparent status quo disrupted a multi-million dollar abalone fishery and initiated discussions of ecologically based approaches to fishery management.

RESUMEN: De 1985 a 1989, abulón negro, *Haliotis cracherodii*, poblaciones declinaron de 82% a 96% en su abundancia en cuatro de las Islas Channel de California. Poblaciones del abulón rojo, *H. rufescens*, declinaron de 97% a 98%, y de abulón amarillo, *H. corrugata*, declinaron de 51% a 94%. Las causas de la declinación no son claras todavía, pero sugerimos un posible guión con evidencia disponible: que los cambios sociales y las prácticas modernas en la administración del recurso alteraron la estructura y función del ecosistema costera y que el colapso observado en la población de abulón en este sistema desestabilizado se precipitó por el evento de 'El Niño' de 1982–84. La civilización moderna redujo drásticamente la depredación natural sobre abulones. Estas acciones en conjunto favorecieron densidades excepcionalmente altas de abulón ($125\,m^{-2}$), incrementando y desviando los controles naturales sobre las poblaciones herbívoras, de la depredación a las enfermedades y competencia, fundamentalmente por alimento y espacio. Un evento extremo como 'El Niño' de 1982–84 deterioró la base alimentaria del abulón. Estos eventos intensificaron la competencia por alimento entre abulones y otros consumidores de arribazones de algas. El dramático colapso de la población del abulón continuó. Este cambio súbito del aparente statu quo impulsó el interés por salud pública, en una pesqueria multimillionaria de abulón, económicamente desorganizada e inició discusiones acerca de una nueva administración de las pesquerías, basada en la ecología.

INTRODUCTION

Dramatic fluctuations in marine invertebrate populations are not uncommon, and it appears that they are being observed and reported with increasing frequency world-wide. Some are clearly associated with extreme natural events, like the hermatypic coral mortality resulting from hurricanes (Stoddart, 1963; Connell, 1978) and the rapid passage of atmospheric cold fronts (Davis, 1982), but causes of many of these population fluxes are unknown. Outbreaks of the sea star *Acanthaster planci* in the Pacific in the 1960s and 1970s were initially attributed to human activities (Chesher, 1969; Endean, 1977), although some indirect evidence for natural causes was advanced early in those investigations (Frankel, 1977). Nearly 15 years were required to develop a convincing argument based on correlation and natural history that these outbreaks could be caused by natural events (Birkeland, 1982; Walbran *et al.*, 1989). Both the effects and causes of mysterious coral 'bleaching' and sea urchin 'massacres' in the Western Atlantic and Caribbean, and sea star depletion in the eastern Pacific are currently under investigation (Dungan *et al.*, 1982; Roberts, 1987; Lewin, 1988). These kinds of large scale community changes probably result from a combination of endogenous and exogenous factors (Pickett & White, 1985), and may well be symptoms of anthropogenic impacts on global environmental processes.

In this chapter, we describe drastic declines of populations of black, *Haliotis cracherodii*, red, *H. rufescens*, and pink, *H. corrugata*, abalone that once supported a multi-million dollar fishery in southern California. We suggest that the population declines were prompted by an extreme natural event, the 1982–4 El Niño, in an ecosystem destabilized by chronic human activity.

History of the California abalone fishery

These abalone inhabit rocky intertidal and shallow subtidal communities from southern Oregon to Baja California, Mexico (Frey, 1971; Ault, 1985). As adults, they subsist largely on drifting fronds of giant kelp, *Macrocystis pyrifera*, and other algae (Cox, 1962; Leighton & Boolootian, 1963; Douros, 1987).

Abalone were harvested by aboriginal peoples in California for thousands of years (Cox, 1962; Muratto, 1984). Spanish missionaries removed the Chumash people from the Channel Islands in the early 19th century (Johnson, 1982), leaving island abalone populations largely unharvested for some 25 years. The modern fishery was developed in the 1850s and grew to an annual harvest of nearly 2000 metric tons (mt) by 1879 (Cox, 1962; Cicin-Sain *et al.*, 1977). In 1900, sale of abalone collected in the littoral zone was prohibited. Annual commercial harvests of all species grew to about 1600 mt by 1935, were briefly suspended during World War II and then quickly soared to and remained at about 2000 mt during 1950–70 (Fig. 19.1). Recreational harvests also grew dramatically during the 1950s and 1960s with the advent of neoprene wet suits and SCUBA. By 1970, the recreational harvest was estimated to be 1360–1820 mt, equalling or exceeding the commercial harvest (Frey, 1971).

Fishery management was based on closed seasons, minimum sizes, bag limits, and gear restrictions designed primarily to protect reproductive stocks for 6–8

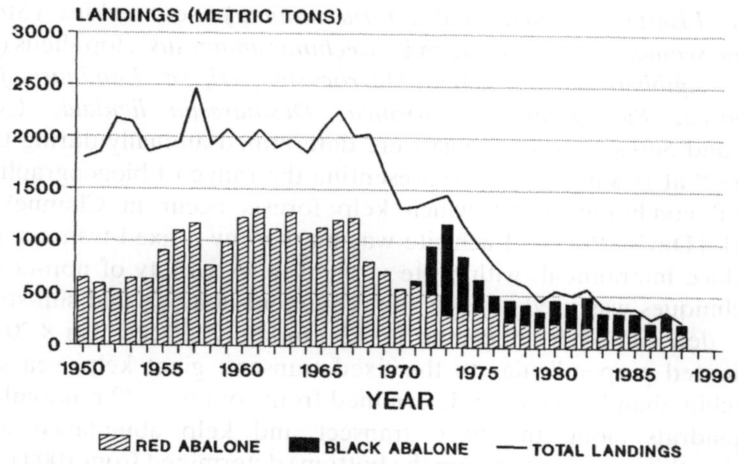

Fig. 19.1 California commercial abalone landings 1950–88 in shell.

years prior to exploitation (Chapter 27, this volume). However, these safeguards apparently failed and subtidal stocks were depleted in the late 1960s, in part reflecting sea otter recovery (VanBlaricom & Estes, 1988). Black abalone became a major component of the commercial harvest, comprising 32 to 60% of the total landings from 1972 to 1988. Coastal development, pollution and harvest had depleted abalone stocks along the mainland coast of southern California by the mid-1970s (Haaker *et al.*, 1986, Tegner, 1989), but even with commercial harvest, populations around the offshore islands remained exceptionally dense and appeared generally healthy until recently. New areas were legally opened for harvest in 1984, but this did not increase or stabilize commercial landings, which continued to fall, reaching a 73-year low of 253 mt in 1988 (Fig. 19.1).

METHODS

The California Channel Islands lie 20–100 km off the coast in two groups of four, stretching from Point Conception to San Diego. The four northern islands sit astride the boundary of two major biogeographical provinces (Seapy & Littler, 1980; Murray *et al.*, 1980). The Oregonian province to the north and west is a cold-water, boreal system, whereas the Californian province to the south and east is a warmer, temperate system. The region around the northern islands is characterized by both high biological diversity, a result of contributions from two provinces, and high biological productivity, the result of persistent upwelling off the mainland coast at Point Conception, local upwellings from deep basins south of the islands and eddies in the California Current system (Owen, 1980).

The number and sizes of black abalone were determined each spring and autumn during 1985–9 in 40 fixed plots, each approximately 3 m^2, in area established in a stratified random fashion at 10 intertidal sites on San Miguel, Santa Rosa, Anacapa and Santa Barbara Islands. Abundance, distribution and size frequencies of abalone (*H. rufescens* and *H. corrugata*), sea stars (*Pycnopodia helianthoides*, *Pisaster giganteus* and *Patiria miniata*), sea urchins (*Strongylocentrotus franciscanus*, *S. purpuratus*, and *Lytechinus anamesus*), top shells (*Astraea undosa* and *A. gibberosa*) and kelps (*Macrocystis pyrifera*, *Laminaria farlowii*, *Eisenia arborea*, *Pterygophora californica*, *Desmarestia ligulata*, *Cystoseira osmundacea* and *Sargassum muticum*) were determined annually during the summers of 1982–9 at 16 subtidal sites representing the range of biogeographical and environmental conditions under which kelp forests occur in Channel Islands National Park (Davis, 1989a). Each site was marked by a fixed transect 100 m in length to reduce interannual, within-site variability. A variety of non-destructive sampling techniques was used at each site: emergent abalone and sun-star abundances were determined from counts in 12 randomly placed 3 m × 20 m band transects oriented perpendicular to the fixed transect; giant kelp, sea star, top shell, and urchin abundances were determined from counts in 20 randomly placed 1 m × 2 m quadrats along the fixed transect and kelp abundance was also measured as benthic cover (<1 m from the bottom) determined from 1000 randomly placed points within 1.5 m of the fixed transect.

RESULTS

The abundance of black abalone of all sizes in the fixed plots declined by 96% at Anacapa Island, 90% at Santa Barbara Island, 83% at Santa Rosa Island and 39% at San Miguel Island from 1985 to 1989 (Table 19.1). Initial densities in these plots were exceptionally high and represented the bulk of the populations at these sites, not mean densities of the rocky intertidal zone, since areas without abalone were excluded by the initial stratification. Intertidally, unusually large numbers of black abalone shells accumulated rapidly in surge channels and on cobble beaches (Chapter 17, this volume). Red abalone abundance declined by 98% at the three Santa Rosa Island sites, and fell from 450 ha^{-1} to less than 14 ha^{-1} (the lower limit of detectability) at the four Santa Cruz Island sites between 1983 and 1988 (Table 19.2). Pink abalone populations at the four Santa Cruz Island sites virtually disappeared, dropping from a mean density of 250 ha^{-1} to less than 14 ha^{-1}. At Anacapa Island, where two of the three sites were protected from harvest, pink

Table 19.1 Abundance of black abalone, *Haliotis cracherodii*, in fixed plots, California Channel Islands (mean number of abalone $m^{-2} \pm 1$ se).

Year/season	Santa Barbara	Anacapa	Santa Rosa	San Miguel
	N = 5	*N* = 10	*N* = 15	*N* = 20
1985 Spring	9 ± 5	51 ± 10	no data	33 ± 6
1985 Autumn	9 ± 5	51 ± 11	44 ± 11	30 ± 4
1986 Spring	10 ± 17	44 ± 10	46 ± 13	28 ± 5
1986 Autumn	9 ± 5	37 ± 8	41 ± 13	28 ± 3
1987 Spring	10 ± 7	29 ± 6	32 ± 12	25 ± 4
1987 Autumn	9 ± 7	12 ± 2	18 ± 9	16 ± 5
1988 Spring	9 ± 6	7 ± 1	13 ± 7	21 ± 4
1988 Autumn	3 ± 2	3 ± 1	10 ± 5	23 ± 4
1989 Spring	1 ± 1	2 ± 0.4	8 ± 5	20 ± 4

Table 19.2 Abundance of red and pink abalone, *Haliotis rufescens* and *H. corrugata*, at Santa Rosa, Santa Cruz and Anacapa Islands (number of emergent abalone per hectare ± 1 se).

	Red abalone		Pink abalone	
Year	Santa Rosa	Santa Cruz	Santa Cruz	Anacapa
1983	1042 ± 408	450 ± 181	250 ± 29	358 ± 106
1984	633 ± 217	117 ± 50	108 ± 22	494 ± 64
1985	534 ± 210	167 ± 68	208 ± 21	560 ± 56
1986	278 ± 84	153 ± 75	188 ± 20	389 ± 41
1987	70 ± 38	0	31 ± 11	430 ± 39
1988	21 ± 16	0	0	272 ± 32
1989	9 ± 4	0	0	357 ± 75

abalone abundance increased between 1983 and 1985 to a high of 560 ha^{-1}, but then declined by 52% in the next three years. The size structure of red and pink abalone populations remained relatively stable during the declines. Juveniles, sublegal adults and legally harvestable adults all declined equally within the size range encountered (14–208 mm). The size structure of the black abalone populations also remained virtually unchanged during the population declines (Chapter 17; this volume).

The remaining black abalone were in poor condition, especially at Anacapa and Santa Rosa Islands. Compared to apparently normal individuals at San Miguel Island, median sized abalone (85 mm) were 9% and 42% lighter at southeastern Santa Rosa and Anacapa Islands, respectively. Compared to a mainland population from Point Dume measured in 1958 (Leighton & Boolootian, 1963), black abalone on the northern Channel Islands in 1988 were 36% lighter, but the use of this population as a baseline for the islands is confounded by geographic variation in size and shape (B. Tissot, pers. comm.) and the probable effects of the 1957–9 El Niño (Cox, 1962). In 1987, 10–15% of the intertidal black abalone at Anacapa Island and the southeastern Santa Rosa sites were extremely weak, unable to adhere firmly to the rock and easily removed by hand. No evidence of pathogens or contamination by pesticides or heavy metals was found in black abalone at the islands (Chapter 17, this volume). No information on the condition of surviving red and pink abalones is available.

The abundance of abalone food (primarily kelps), competitors (unharvested sea urchins and *Astrea*) and predators (sea stars) also varied considerably among islands and annually from 1982 through 1989 (Table 19.3). In 1981 and 1982, all 16 subtidal sites exhibited virtually complete surface canopies of *Macrocyctis pyrifera* (95–100% cover). Major winter storms in early 1983 removed 85–90% of this canopy at the study sites, exposing the sea floor to light. Kelp recruitment the following spring (1983) was extremely heavy. Mean benthic phaeophyte cover increased from 14% in 1982 to 55% in 1983. However, those recruits failed to survive beyond their first year, and cover dropped back to 15% for the next two years. Abundance of canopy and detritus-producing adult *M. pyrifera* declined at all locations after 1982. By 1986, adult *M. pyrifera* were virtually absent at Santa Rosa, Santa Cruz and Santa Barbara Islands and greatly reduced at Anacapa and San Miguel Islands. Pulses of kelp recruitment were evident in 1986 (San Miguel

Table 19.3 Abundance of selected kelp forest organisms in Channel Islands National Park, 1982–7, expressed as density or percentage cover ± 1 se.

Species/island	1982	1983	1984	1985	1986	1987	1988	1989
Giant kelp (number $10\,m^{-2}$)								
Macrocystis pyrifera								
San Miguel Island	10 ± 2	8 ± 2	6 ± 1	2 ± <1	22 ± 3	2 ± <1	6 ± 1	6 ± 2
Santa Rosa Island	7 ± 2	37 ± 8	10 ± 2	2 ± 1	1 ± <1	<1 ± <1	52 ± 4	48 ± 4
Santa Cruz Island	8 ± 1	32 ± 2	4 ± 1	17 ± 2	0	0	3 ± <1	1 ± <1
Anacapa Island	23 ± 2	110 ± 9	27 ± 4	15 ± 3	56 ± 7	12 ± 3	6 ± 1	20 ± 4
Santa Barbara Island	6 ± 1	2 ± 1	<1 ± <1	2 ± 1	<1 ± <1	0	0	1 ± <1
Brown algae (% benthic cover)								
Phaeophyta								
San Miguel Island	5 ± 1	59 ± 5	8 ± 3	5 ± 2	50 ± 3	8 ± 2	48 ± 6	13 ± 2
Santa Rosa Island	8 ± 3	68 ± 13	16 ± 3	9 ± 3	3 ± 2	2 ± 2	54 ± 10	53 ± 4
Santa Cruz Island	23 ± 3	43 ± 7	21 ± 3	4 ± 2	<1 ± <1	<1 ± <1	<1 ± <1	2 ± <1
Anacapa Island	18 ± 5	60 ± 7	16 ± 4	42 ± 8	45 ± 8	38 ± 8	15 ± 6	38 ± 7
Santa Barbara Island	16 ± 4	45 ± 9	14 ± 4	16 ± 6	5 ± 1	1 ± 1	1 ± <1	3 ± 1
Top snails (number $10\,m^{-2}$)								
Astrea undosa (*A. gibberosa* at San Miguel Island)								
San Miguel Island	0	1 ± <1	<1 ± <1	2 ± <1	2 ± <1	0	<1 ± <1	0
Santa Rosa Island	0	<1 ± <1	<1 ± <1	2 ± <1	2 ± <1	<1 ± <1	0	0
Santa Cruz Island	2 ± <1	5 ± 1	13 ± 1	14 ± 1	27 ± 2	26 ± 2	22 ± 2	16 ± 1
Anacapa Island	2 ± <1	2 ± <1	4 ± 1	7 ± 2	19 ± 2	24 ± 3	19 ± 2	18 ± 2
Santa Barbara Island	9 ± 1	2 ± 1	21 ± 2	48 ± 5	36 ± 4	31 ± 17	7 ± 1	8 ± 1
Unharvested sea urchins (number $10\,m^{-2}$)								
Strongylocentrotus purpuratus and *Lytechinus anamesus*								
San Miguel Island	55 ± 8	28 ± 7	46 ± 18	70 ± 18	229 ± 48	79 ± 24	42 ± 13	26 ± 7
Santa Rosa Island	121 ± 1	128 ± 28	168 ± 36	294 ± 34	337 ± 29	212 ± 31	11 ± 6	54 ± 9
Santa Cruz Island	44 ± 12	59 ± 14	48 ± 19	124 ± 26	408 ± 60	372 ± 40	296 ± 42	373 ± 19
Anacapa Island	20 ± 5	11 ± 3	10 ± 3	18 ± 7	156 ± 59	138 ± 41	159 ± 54	93 ± 17
Santa Barbara Island	314 ± 52	256 ± 35	76 ± 16	304 ± 43	952 ± 86	788 ± 87	809 ± 77	838 ± 64

Sea stars (number 10 m^{-2})								
Patiria miniata and *Pisaster giganteus*								
San Miguel Island	52 ± 8	40 ± 5	8 ± 2	7 ± 2	16 ± 4	15 ± 4	17 ± 4	23 ± 3
Santa Rosa Island	44 ± 5	18 ± 4	4 ± 1	5 ± 1	12 ± 2	17 ± 4	12 ± 3	22 ± 2
Santa Cruz Island	4 ± 1	1 ± 1	1 ± <1	2 ± <1	2 ± 1	5 ± 1	4 ± 1	8 ± 1
Anacapa Island	<1 ± <1	<1 ± <1	<1 ± <1	0	<1 ± <1	1 ± <1	3 ± 1	2 ± <1
Santa Barbara Island	1 ± 1	1 ± <1	1 ± <1	<1 ± <1	1 ± 1	1 ± <1	3 ± 2	1 ± <1
Pycnopodia helianthoides (number ha^{-1})								
San Miguel Island	no data	166 ± 29	25 ± 9	56 ± 20	167 ± 45	305 ± 47	125 ± 40	438 ± 56
Santa Rosa Island	no data	0	6 ± 3	23 ± 8	32 ± 12	981 ± 119	1273 ± 102	1625 ± 131
Santa Cruz Island	no data	4 ± 2	0	0	0	0	0	0
Anacapa Island	no data	0	0	0	0	0	0	0
Santa Barbara Island	no data	0	0	0	0	0	0	0

and Anacapa), 1988 (Santa Rosa) and 1989 (Santa Rosa and Anacapa). Extensive surface canopy was once again present at Santa Rosa and Anacapa Islands in 1989, but very patchy or absent at the other islands.

While abalone populations fell at all study sites during 1982–9, other kelp consumer populations increased remarkably. Sea urchin abundance increased at all locations after 1984, generally peaking in 1986, but the patterns were quite different along the biogeographical gradient. At San Miguel and Santa Rosa Islands, the elevated densities (23–34 m^{-2}) survived only three years and dropped back to 1–5 m^{-2} by 1988. The increases in urchin densities at Santa Cruz Island were similar to San Miguel and Santa Rosa Islands in magnitude, up from 4–6 m^{-2} in 1982–4 to 41 m^{-2} in 1986, but the elevated densities persisted through 1989. At Anacapa Island, the increase was more gradual, the peak density lower (16 m^{-2}), and abundance declined in 1989. Santa Barbara Island revealed yet another pattern with a decline preceding the general 1984–6 increase, followed by a sharp increase from 7 m^{-2} to 95 m^{-2}, and sustained high densities through 1989.

Populations of the unharvested herbivorous snail, *Astrea undosa*, increased by an order of magnitude, from 0.2 m^{-2} to 3–5 m^{-2}, between 1982–3 and 1985–6 at the three islands where it occurred (Table 19.3). At Santa Cruz and Anacapa Islands, densities began to decline slightly in 1988, but at Santa Barbara Island the population density dropped back to less than 1 m^{-2}.

Juvenile abalone and urchin predators include the sunflower star, *Pycnopodia helianthoides*, bat star, *Patiria miniata*, and giant spined sea star *Pisaster giganteus* (B. Tissot & G. Davis, pers. obs). Bat star and giant spined sea star populations declined sharply in 1982 and 1983, remained at low levels for the next two to three years and began gradually increasing in 1986 and 1987. Of particular note was the dramatic and steady increase in sunflower star populations at San Miguel and Santa Rosa Islands from 6–25 ha^{-1} in 1984 to 438–1625 ha^{-1} in 1989.

DISCUSSION

Several factors, both natural and anthropogenic, appear to have combined in complex ways to weaken abalone, accelerate their mortality and cause subsequent declines. Fishery harvests may have contributed to the population declines, but the persistence of stable proportions of legally harvestable abalone in populations undergoing such drastic drops indicates that harvest was not the primary factor. Descriptions of abalone dynamics alone will not adequately explain recent abalone declines. Nor will observations of abalone alone define the ecological relationships among producers, consumers and environmental factors that drive the dynamics of kelp forests and adjacent rocky intertidal communities of which abalone are a part. We hypothesize that a series of ecological interactions, mediated by anthropogenic actions, are responsible for the abalone declines observed in the Channel Islands in the 1980s. Consequently, we believe that restoration of abalone stocks and a productive fishery will require ecological management, not just additional single-species restrictions on seasons and harvest (Tegner *et al.* 1989).

The following discussion of factors probably responsible for the recent declines is summarized in Fig. 19.2. The process which led to recent abalone population declines may have begun in the early 19th century when Spanish missionaries

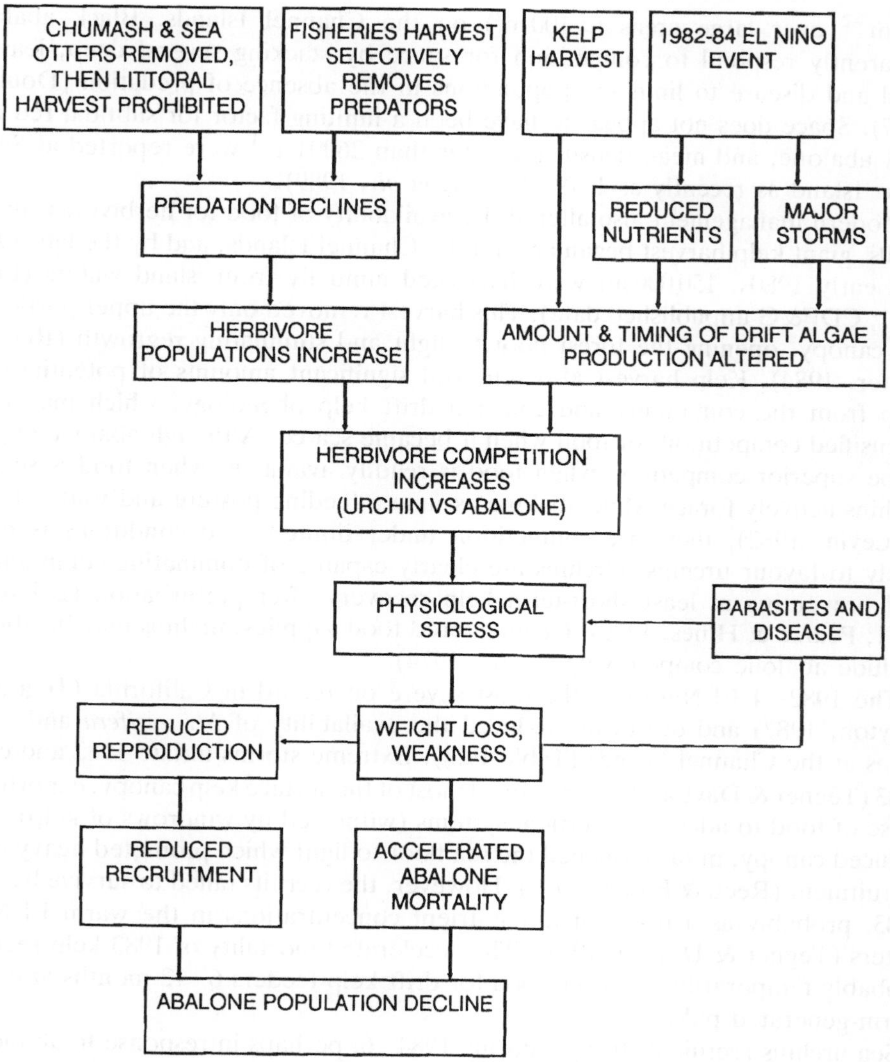

Fig. 19.2 Schematic summary of hypothetical process leading to mass mortality and population decline of black abalone, *Haliotis cracherodii*, at the California Channel Islands during 1983–9.

removed the Chumash people from the Channel Islands, thus ending over 7000 years of aboriginal abalone and sea urchin harvest (Johnson, 1982). Russian and American fur hunters removed sea otters, *Enhydria lutris*, another major urchin and abalone predator, from southern California by the end of the 19th century (VanBlaricom & Estes, 1988).

Fisheries in southern California continue to remove substantial numbers of abalone and urchin predators, such as spiny lobster, *Panulirus interruptus*, and California sheephead, *Semicossyphus pulcher* (Tegner, 1980; Tegner & Levin, 1983). This reduced predation could allow herbivores, including urchins and abalone, to increase to the point that disease and competition would limit their populations. Black abalone reached remarkable population densities of more than

$125\,m^{-2}$ over large areas ($>100\,m^2$) on the Channel Islands. Black abalone apparently respond to competition for space by stacking several deep, leaving food and disease to limit the populations in the absence of predation (Douros, 1987). Space does not appear to have been a limiting factor for subtidal red and pink abalone, and mean densities greater than $2600\,ha^{-1}$ were reported at Santa Rosa Island as recently as 1979 (Tegner, *et al.*, 1989).

Modern management also altered the availability of food for herbivores. In the 1950s, giant kelp harvest began around the Channel Islands, and by the late 1970s and early 1980s, 150 000 mt were harvested annually from island waters (Frey, 1971; CDF&G unpublished data). This harvest removed only the upper portion of the canopy, opening the forest floor to light and stimulating regrowth (Reed & Foster, 1984). Kelp harvest also removed significant amounts of potential drift kelp from the community and changed drift kelp phenology, which may have intensified competition for food when it became scarce. Although abalone appear to be superior competitors when food is readily available, when food is scarce, urchins actively forage while abalone assume a feeding posture and wait (Tegner & Levin, 1982), therefore competition under limited food conditions is more likely to favour urchins. Urchins are clearly capable of dominating communities and preventing at least short-term kelp recovery after perturbation (Leighton, 1971; Pearse & Hines, 1979). Given altered food supplies, urchins may be able to exclude abalone competitively (Paine, 1974).

The 1982–4 El Niño was the most severe on record in California (Tegner & Dayton, 1987) and drastically reduced the availability of *M. pyrifera* and other kelps at the Channel Islands (Table 19.3). Extreme storms in late 1982 and early 1983 (Tegner & Dayton, 1987) removed most of the surface kelp canopy, exporting a pulse of food to adjacent intertidal systems (witnessed by windrows of kelp). The reduced canopy, in turn, opened the sea floor to light which promoted heavy algal recruitment (Reed & Foster, 1984). However, the recruits failed to survive beyond 1983, probably as a result of low nutrient concentrations in the warm El Niño waters (Tegner & Dayton, 1987). The accelerated mortality of 1983 kelp recruits probably temporarily enhanced food for drift kelp feeders 6–12 months after the storm-generated pulse.

Sea urchins recruited strongly during 1984–6, perhaps in response to abundant food and low predation. Red urchin harvests also reduced direct competition with purple and white urchins (Tegner & Dayton, 1981). Grazing by the greatly increased white and purple urchin populations reduced survival of subsequent algal recruits, and slowed or eliminated development of canopy and subsequent drift kelp production (Lawrence, 1975; Pearse & Hines, 1979). Subtidal urchin consumption and low production of drift kelp may also have further reduced kelp availability to intertidal black abalone, as well as subtidal abalone.

Abalone declines were most rapid and extreme at Anacapa, Santa Cruz and Santa Rosa Islands, where both kelp availability was severely disrupted and abalone were initially abundant. These factors increased intraspecific competition and could have facilitated disease transmission. At San Miguel Island (and San Nicholas Island) in the boreal zone, abalone were abundant, but kelp recovered relatively quickly from the winter storms of 1982–3 and subsequent El Niño conditions, and abalone populations did not collapse. Initially, abalone were not as severely affected at Santa Barbara Island in the temperate Californian zone, where kelp was greatly reduced by the 1982–3 events, but initial abalone population densities were much lower, consequently competition for food may not have been as acute and opportunities for disease transmission were greatly reduced. Southern California mainland populations in the Californian zone, already thinned by harvest and pollution, apparently were less dramatically affected by this event, and mainland populations in the Oregonian zone at Point Arguello were not affected (Chapter 17, this volume).

The declines reported here apparently resulted from dense abalone populations that were stressed by a natural disruption of food. Similar abalone weight loss and mortality were attributed to starvation in Japan (Tanaka *et al.*, 1985), and food-induced stress seems to be a reasonable possibility at the Channel Islands. Population control appears to have shifted from predation to competition for food, and perhaps to disease, apparently as a result of human activities which altered ecosystem structure by reducing predator populations. Furthermore, intense and extensive sea urchin grazing has accentuated the effects of natural perturbations to the system by extending the recovery time. Southern California kelp canopies were relatively stable for the first 60 years of the 20th century, recovering quickly following earlier El Niño events (North, 1971). The extended recovery (at least 7 years) in the temperate and transition zones along the Channel Islands following the 1982–4 event (Table 19.3) may be the response of a destabilized system in which benthic areas alternate between sea urchin-dominated barrens and dense, kelp-dominated forests, rather than open kelp forest mosaics.

Models have been proposed to explain urchin barrens and how such an alternating pattern might occur (Ebeling *et al.*, 1985; Harrold & Reed, 1985). However the details are resolved, there are strong indications that stochastic physical environmental factors, such as storms, now appear to be the principal agents of change in this system, rather than the biological components which previously buffered the system against extreme, long-lasting fluctuations. The reduction of this buffering capacity and subsequent shift of control from probabilistic biological factors to unmanageable deterministic physical factors, paradoxically, will make the system more predictable but less manageable. The effects of uncontrollable storms will be easier to predict and the outcomes of competitive interactions more certain to foretell, but control through traditional biological measures, such as harvest and size limits, will become increasingly ineffective.

Under these circumstances, fishery managers need to explore ecosystem-based management approaches for the future (Davis, 1989b) and seek better understanding of the ecological interactions driving the systems. It appears that management actions regarding abalone harvest and predators were taken without adequate knowledge of the linkages among system components, and therefore without knowledge of their consequences. For example, without knowing that food could become a major limiting factor for abalone, the connection between kelp harvest and subsequent abalone stock levels could not be considered when setting kelp harvest constraints. Without a better understanding of the relationships among kelps, urchins, abalone and their respective predators, abalone management based solely on abalone biology is doomed to be a series of expensive and highly uncertain experiments repeated often as long as a few abalone survive.

REFERENCES

Ault J.S. (1985) Species profiles: life histories and environmental requirements of coastal fishes and invertebrates (Pacific Southwest) – black, green, and red abalones. *US Fish Wildl. Serv. Biol. Rep. 82(11.32)*, US Army Corps of Engineers, TR EL-82–4.

Birkeland C. (1982) Terrestrial runoff as a cause of outbreaks of *Acanthaster planci* (Echinodermata: Asteroidea). *Mar. Biol.*, **69**, 175–85.

Chesher R.H. (1969) Destruction of Pacific corals by the sea star *Acanthaster planci*. *Science*, **165**, 280–3.

Cicin-Sain B., Moore J.E. & Wyner A.J. (1977) Management approaches for marine fisheries: the case of the California abalone. *Univ. Calif. Sea Grant Pub.*, 54, IMR 77–101.

Connell J.H. (1978) Diversity in tropical rain forests and coral reefs. *Science*, **199**, 1302–10.

Cox K.W. (1962) California abalones, family Haliotidae. *Calif. Dept Fish Game Fish. Bull.*, **118**, 1–133.

Davis G.E. (1982) A century of natural change in coral distribution at the Dry Tortugas: a comparison of reef maps from 1881 and 1976. *Bull. Mar. Sci.*, **32**(2), 608–23.

Davis G.E. (1989a) Design of a long-term ecological monitoring program for Channel Islands National Park, California. *Natural Areas J.*, **9**, 80–9.

Davis G.E. (1989b) Harvest refugia: the next stage in California fishery management. *Calif. Coop. Fish. Invest. Repts.*, **30**, 53–8.

Dayton P.K. & Tegner M.J. (1984) The importance of scale in community ecology: a kelp forest example with terrestrial analogs. In *Novel Approaches to Interactive Systems* (Ed. by P.W. Price, C.N. Slobodchikoff & W.S. Gaud), pp. 457–81. Wiley, New York.

Douros W.J. (1987) Stacking behavior of an intertidal abalone: an adaptive response or a consequence of space limitation? *J. Exp. Mar. Biol. Ecol.*, **108**, 1–14.

Dungan M.L., Miller T.E. & Thompson D.A. (1982) Catastrophic decline of a top carnivore in the Gulf of California rocky intertidal zone. *Science*, **216**, 989–91.

Ebeling A.W., Laur D.R. & Rowley R.J. (1985) Severe storm disturbances and reversal of community structure in a southern California kelp forest. *Mar. Biol.*, **84**, 287–94.

Endean R. (1977) *Acanthaster planci* infestations of reefs of the Great Barrier Reef. In *Proceedings of the Third International Coral Reef Symposium* Ed. by D.L. Taylor, pp. 185–92. Rosenstiel School of Marine and Atmospheric Science, University of Miami.

Frankel E. (1977) Previous *Acanthaster* aggregations in the Great Barrier Reef. In *Proceedings of the Third International Coral Reef Symposium* Ed. by D.L. Taylor, pp. 201–8. Rosenstiel School of Marine and Atmospheric Science, University of Miami.

Frey H.W. (Ed.) (1971) *California's Living Marine Resources and their Utilization*. The Resources Agency, Department of Fish and Game, State of California.

Haaker P.L., Henderson K.C. & Parker D.O. (1986) California abalones. *California Marine Resources Leaflet*, **11**.

Harrold C. & Reed D.C. (1985) Food availability, sea urchin grazing, and kelp forest community structure. *Ecology*, **66**, 1160–9.

Johnson J.R. (1982) *An ethnohistoric study of the island Chumash*. MA thesis, University of California, Santa Barbara.

Lawrence J. (1975) On the relationships between marine plants and sea urchins. *Oceanogr. Mar. Biol. Ann. Rev.*, **13**, 213–86.

Leighton D.L. (1971) Grazing activities of benthic invertebrates in kelp forests. *Nova Hedwigia*, **32**, 421–53.

Leighton D. & Boolootian R.A. (1963) Diet and growth in black abalone, *Haliotis cracherodii*. *Ecology*, **44**, 227–38.

Lewin R. (1988) Sea urchin massacre is a natural experiment. *Science*, **239**, 867.

Muratto M.J. (1984) *California Archeology*. Academic Press, New York.

Murray S.N., Littler M.M. & Abbott I.A. (1980) Biogeography of the California marine algae with emphasis on the southern California Islands. In *The California Channel Islands: Proceedings of a Multidisciplinary Symposium* (Ed. by D.M. Power), pp. 325–39. Santa Barbara Museum of Natural History, Santa Barbara.

North W.J. (1971) The biology of giant kelp beds (*Macrocystis*) in California. *Beihefte Zur Nova Hedwigia Heft*, 32.

Owen R.W. (1980) Eddies of the California Current system: physical and ecological characteristics. In *The California Channel Islands: Proceedings of a Multidisciplinary Symposium*. (Ed. by D.M. Power), pp. 237–63. Santa Barbara Museum of Natural History, Santa Barbara.

Paine R.T. (1974) Ecological determinism in the competition for space. *Ecology* **65**(5), 1339–48.

Pearse J.S. & Hines A.H. (1979) Expansion of a Central California kelp forest following the mass mortality of sea urchins. *Mar. Biol.*, **51**, 83–91.

Pickett S.T.A. & White P.S. (1985) *The Ecology of Natural Disturbance and Patch Dynamics*. Academic Press, San Diego.

Reed D.C. & Foster M.S. (1984) The effects of canopy shading on algal recruitment and growth in a giant kelp forest. *Ecology*, **65**(3), 937–48.

Roberts L. (1987) Coral bleaching threatens Atlantic reefs. *Science*, **238**, 1228–9.

Seapy R.R. & Littler M.M. (1980) Biogeography of rocky intertidal macroinvertebrates of the southern California Islands. In *The California Channel Islands: Proceedings of a Multidisciplinary Symposium*. (Ed. by D.M. Power), pp. 307–24. Santa Barbara Museum of Natural History, Santa Barbara.

Stoddart D.R. (1963) Effects of hurricane Hattie on the British Honduras reefs and cays, October 30–31, 1961. *Atoll Res. Bull.*, **95**, 1–142.

Tanaka K., Ishida O. & Tanaka T. (1985) On the lost weight of abalone in the area around Mera Shoal on the southern coast of the Boso Peninsula, especially concerning the habitat conditions. *Japan Sea Regional Fisheries Research Laboratory Achievement A*, No 430.

Tegner M.J. (1980) Multispecies considerations of resource management in southern California kelp beds. *Can. Tech. Rep. Fish. Aquat. Sci.*, **954**, 125–43.

Tegner M.J. (1989) The California abalone fishery: production, ecological interactions, and prospects for the future. In *Scientific Approaches to Management of Invertebrate Stocks* (Ed. by J.F. Caddy), pp. 401–20. Wiley, New York.

Tegner M.J., Breen P.A. & Lennert C.E. (1989) Population biology of red abalones, *Haliotis rufescens*, in southern California and management of red and pink, *H. corrugata*, abalone fisheries. *Fishery Bull. US*, **87**, 313–39.

Tegner M.J. & Dayton P.K. (1981) Population structure, recruitment and mortality of two sea urchins (*Strongylocentrotus franciscanus* and *S. purpuratus*) in a kelp forest. *Mar. Ecol. Prog. Ser.*, **5**, 255–68.

Tegner M.J. & Dayton P.K. (1987) El Niño effects on southern California kelp forest communities. *Adv. Ecol. Res.*, **17**, 243–79.

Tegner M.J. & Levin L.A. (1982) Do sea urchins and abalones compete in California kelp forest communities? In *International Echinoderms Conference, Tampa Bay* (Ed. by J.M. Lawrence) pp. 265–71. A.A. Balkema, Rotterdam.

Tegner M.J. & Levin L.A. (1983) Spiny lobsters and sea urchins: analysis of a predator–prey interaction. *J. Exp. Mar. Biol. Ecol.*, **73**, 125–50.

VanBlaricom G.R. & Estes J.A. [Eds] (1988) *The Community Ecology of Sea Otters. Ecology Series 65*. Springer-Verlag, New York.

Walbran P.D., Henderson R.A., Jull A.J.T. & Head M.J. (1989) Evidence from sediments of long-term *Acanthaster planci* predation on corals of the Great Barrier Reef. *Science*, **245**, 847–50.

PART VI
FISHERY BIOLOGY

Chapter 20
A review of models used for stock assessment in abalone fisheries

PAUL A. BREEN *Ministry of Agriculture & Fisheries, Fisheries Research Centre, Box 297, Wellington, New Zealand*

ABSTRACT: The application of stock assessment models to abalone fisheries is briefly reviewed and aspects of the assumptions for each method are discussed. Because many abalone fishery managers rely heavily on a minimum legal size to maintain yield efficiency and egg production, the most common assessments are based on yield- and egg-per-recruit models. Several authors have used Leslie-DeLury techniques to estimate stock size, but these methods and surplus–production analysis depend heavily on the relation between catch per unit effort (CPUE) and stock size. For abalone, CPUE is not a good index of abundance and should be treated with extreme caution in assessments. Stock–recruit relations are of over-riding importance for abalone fishery management, but abalone stock definition is difficult and these relations are not understood. Stochastic simulation modelling and multi-species assessments are productive directions for future work. Cohort analysis, stock reduction analysis and Deriso's method have been used sparingly for abalone; their use could be expanded.

RESUMEN: La aplicación de modelos de evaluación de los stocks a las pesquerías del abulón es revisada someramente y se discuten los supuestos para cada método. Debido a que muchos empresarios de pesquerías de abulón confían mucho en un tamaño legal mínimo para mantener un rendimiento y producción de huevos eficientes, las evaluaciones más comunes están basadas en modelos de rendimiento y huevos por recluta. Varios autores han usado las técnicas de Leslie-DeLury para estimar el tamaño de los stocks, pero este método y el análisis de producción excedente dependen considerablemente de la relación entre la captura por unidad de esfuerzo (CPUE) y el tamaño del stock. Para el abulón, la CPUE no es un buen índice de abundancia y debe ser tratado con extrema precaución en las evaluaciones. Las relaciones entre reclutamiento y el stock son de suma importancia para el manejo de las pesquerías del abulón pero la definición de stock en abulón es difícil y estas relaciones no están bien comprendidas. Los modelos de simulación estocástica y las evaluaciones multiespecíficas son direcciones productivas para un trabajo futuro. El análisis cohorte el análisis de reducción poblacional y el método de Deriso han sido usados escasamente para el abulón; su uso podría ser expandido.

INTRODUCTION

Most abalone fisheries are 'managed' in some way by a regulatory agency. Restrictions can be simple, for instance, minimum legal size (MLS) limits are very common, or they can be very complicated, as seen in systems of limited entry, royalties and individual quotas. For a manager there exists some desirable output from the fishery. This is only rarely specified as explicitly and carefully as it should be. Sustainable yield, sustainable economic yield, recreational diver satisfaction, the number of viable commercial fishing units and aboriginal access to historic fishing are all possible (and conflicting) outputs that might be important to a manager. The underlying premise of management is that the output of interest can be increased or even maximized by restricting the fishery in some way.

The purpose of stock assessment is to predict the relation between management action and observed result, at least under a set of simplifying assumptions. For this purpose it is desirable (but by no means essential) to have estimates of stock size. This chapter reviews both methods of estimating stock size and models of population dynamics used to make predictions.

In the absence of a formal stock-assessment approach, abalone fishery management once proceeded in one of two ways. First, where possible it drew on a long period of essentially experimental management. Ino (1966) describes the effect of a variety of restrictions imposed in Ikura, Fukui Prefecture, Japan. There were no restrictions from 1909–18, a total closure in 1918–19, then various seasonal and area closures that apparently tripled the output of the fishery by 1933. A similar analysis for a different area is provided by Nonaka & Fushimi (1972).

Second, managers apparently made restrictions in many abalone fisheries for 'intuitive' reasons. MLS restrictions are a good example: they have been a feature of most abalone fisheries for many decades, but the basis for their imposition is undocumented or prompted by informal models, as will be shown below.

Abalone biologists were relatively slow to apply modern stock assessment techniques. Mottet's (1978) review of 'fishery management' is essentially a list of regulatory tools; formal assessments were simply unavailable. Application of the Leslie-DeLury approach to estimate stock size first attracted attention in fisheries in 1947 (Ricker, 1975) but was not applied to abalone until about 1965 (Ino, 1966). The yield-per-recruit approach, pioneered by Baranov in 1918 (Ricker, 1975), Thompson & Bell (1934), Ricker (1945), and Beverton & Holt around 1953 (Ricker, 1975), was first applied to abalone around 1977 by Doi *et al.* (1977) and Sainsbury (1977). Surplus production analysis, developed by Schaefer (1954), was first attempted for abalone by Harrison (1983).

Why were biologists slow to apply such techniques for abalone? There are five possible contributing reasons. First, invertebrate fisheries during the period 1950–75, when stock-assessment procedures became widespread for finfish species, tended to be less important and less valuable than finfish. Except in Japan and California, abalone fisheries were localized and relatively unimportant until the development of strong export markets in the mid-1970s.

Second, abalone differ from other exploited marine populations in important ways, and these differences may have prevented direct application of standard stock-assessment techniques developed for finfish. This is especially true of techniques dependent on catch per unit effort (CPUE), a topic discussed at length below.

Other differences are numerous. Abalone are essentially sessile; this and their tendency to aggregate makes them more vulnerable to exploitation than other types of exploited invertebrates. For instance, Beinssen (1979) estimated from diver experiments with simulated abalone that a diver harvesting Australian blacklip abalone, *Haliotis rubra*, could exert an instantaneous rate of fishing mortality $F = 0.80 - 1.05$. In many abalone species growth and mortality rates in the exploited portion of the population are low (Harrison, 1986). The combination of locally high fishing intensity with low growth and natural mortality rates causes local populations to respond rapidly to fishing, in turn creating assessment problems where biologists arrive some time after fishing has begun. Only a few species of abalone can be aged, so growth and mortality estimates can often be obtained only from tagging or from polymodal length frequency analysis. This may have delayed the application of age-structured models to abalone fisheries.

Third, recruitment or settlement of abalone is often erratic (see review in Breen, 1986 and Tegner *et al.* 1989). (In this chapter, 'recruitment' is used in the sense of Ricker, 1975, as recruitment to the vulnerable stock, and the term 'settlement' is used in the sense of 'recruitment to the population'. The latter is, of course, a product of both larval settlement from the plankton and subsequent survival. In other contexts the term 'settlement' may not be appropriate (e.g. Rowley, 1989) to use in this sense.) Shepherd (1990) found that settlement of Australian greenlip abalone, *H. laevigata*, was log-normally distributed: over 20 years a few years of good settlement interspersed with many poor years. Early stock-assessment procedures were based on the assumption of equilibrium conditions.

Fourth, stock-recruitment functions for abalone populations are unknown; this has precluded stock-recruit analyses. Few populations have been harvested for long enough and in such a way that this function can be examined. Some authors suggest that abalone exist in very small, localized stocks (e.g. Prince *et al.*, 1987). If so, then management approaches and stock assessments must address exploitation on a very fine scale.

Fifth, abalone fisheries seem especially prone to unpredictable collapse. Examples include the California fishery for several species (Burge *et al.*, 1975; Cicin-Sain *et al.*, 1977; see Tegner, 1989), the British Columbia fishery for northern abalone, *H. kamtschatkana* (Breen, 1980, 1986; Sloan & Breen, 1988) and the Channel Islands fishery for ormers, *H. tuberculata* (see review in Breen, 1986). Whereas many finfish species have been assessed for many years, abalone fisheries have rarely persisted long enough to generate reasonable time series of data.

The variety of techniques that have now been applied to abalone world-wide is great. This review lists applications of standard and non-standard approaches and discusses their strengths and limitations based on the assumptions involved. In some cases, only the English abstracts of papers in Japanese were available. Although it is difficult to glean a complete understanding from these, examples are included to encourage translation.

MODELS USED IN ABALONE FISHERY ASSESSMENTS

Critical size and yield-per-recruit models

The weight of a 'cohort' of abalone (those spawned in the same period) will be influenced simultaneously by growth and mortality throughout its lifespan. Initial cohort weight is very small (the weight of a large number of fertilized eggs); eventually the cohort weight is zero when all members have died. At some stage in between, the cohort weight reaches a maximum. That maximum weight could be taken at that point by a fishery exercising infinitely high fishing mortality.

The age of a cohort at its maximum biomass, denoted 'critical age' or T_{mb}, is related to the instantaneous rate of natural mortality M and the growth parameters (see Alverson & Carney, 1975). The simplest yield-per-recruit (YPR) analysis involves the calculation of T_{mb}. The size at T_{mb} ('critical size') would be the MLS that maximized YPR at an infinitely high fishing mortality rate. This kind of analysis was reported by Breen (1980), Hamer (1983) and S.A. Schultz (unpublished data). Doi *et al.* (1977) calculated critical age and size for two populations of Mexican *H. corrugata* by actually calculating the estimated biomass of each cohort and reading T_{mb} from the resulting curve.

It is impractical for fisheries to remove an entire cohort just when it reaches critical size. A common simplifying assumption is that recruitment is constant from year to year. The yield taken from a single cohort over its lifespan is then the same as the yield from all cohorts in a single year under equilibrium conditions. Under this assumption, YPR can be examined at various combinations of F (instantaneous fishing mortality rate) and either the length (l_c) or age (t_c), at which animals become vulnerable to the fishery. A review of traditional methods is provided by Mohn (1986).

Isibasi & Kojima (1979) used the Beverton-Holt (1957; Beverton, 1954) method to analyse YPR for Japanese black abalone, *H. discus discus*. They describe a modification of the method to model seasonal fishing. The authors estimated independently that fishing mortality was very high ($F = 2.0$) in this fishery. In their figures, the authors show the impact of varying estimated M, F and t_c. The YPR isopleths were relatively flat for $F > 1.0$, so little increased yield would result from reducing fishing mortality. However, estimated YPR would increase substantially if t_c were increased from 2.5 years to 4 years, assuming $M = 0.24$. Accordingly, an increased MLS was recommended for this species.

Harrison (1983) used the Beverton-Holt (1966) YPR tables, using an M/K ratio of 1.0, to evaluate YPR in *H. rubra* in Tasmania. He held l_c constant at 125 mm and examined the effect of varying F. Yields increased little with increasing F for $F > 0.50$.

Sluczanowski (1984a, 1986) developed a YPR model along the same general lines as Beverton & Holt (1957) and examined YPR for South Australian *H. laevigata*. Whereas YPR models usually vary natural mortality rate M, t_c or l_c, and F, Sluczanowski assumed that individual reefs could be fished at high rates over short periods and then left alone for some period P. He varied P instead of F; thus the model essentially examines 'pulse fishing', a management strategy suggested for California species by Wyner *et al.* (1977). When l_c was varied, YPR varied with a maximum near the present MLS. Variation of P had little effect on YPR.

Clavier & Richard (1985) used the Beverton-Holt (1957) model for estimating YPR in *H. tuberculata* near St. Malo, France. They estimated recruitment in each of a number of areas by examining length frequencies from stratified surveys (Clavier & Richard, 1986a) and using growth estimates (Clavier & Richard, 1986b) to identify individuals that had recruited within the past year. This procedure was also used by Breen (1980, 1986) to estimate recruitment strength in British Columbia *H. kamtschatkana*. Clavier & Richard (1985) then multiplied YPR estimates and recruitment estimates to obtain absolute yield estimates for each area. They used a variety of values for the economic threshold of abalone density to determine which areas contained commercially viable populations. The explicit assumption used for this procedure is that recruitment would remain at levels observed during population surveys of the pre-exploited stock. This procedure is potentially valuable: it gives the manager an estimate of the annual sustainable yield at equilibrium under a variety of values for F and MLS.

Breen (1986) used the method of Beverton & Holt (1957) to analyse YPR for *H. kamtschatkana* in British Columbia. He found current MLS to be slightly larger than optimum at the estimated current level of *F*. The YPR isopleths suggested that, for an MLS of 110 mm or greater in this species, YPR was independent of *F*. Based on this pattern and ignoring concerns other than YPR, a manager with limited resources could set MLS and ignore *F*. Conversely, at $F < 0.20$ the YPR changed little as MLS varied.

Sainsbury (1977, 1982b) analysed YPR for New Zealand black-footed paua, *H. iris*. His model was based on a projection matrix model of the population. Growth was described with a mean growth increment for animals within a given size class, but also by a variance associated with each mean so individual growth rates varied. Natural mortality could be made size specific, making a more realistic model. YPR was calculated from a yield matrix assuming a fixed recruitment pattern. Sainsbury concluded that the existing MLS was too large and that a higher YPR would result from a smaller MLS in the modelled population. Sainsbury (1982b) also used his model to examine population structure under various conditions. He concluded that the structure observed in the field could not have been produced under equilibrium conditions and thus that recent settlement had varied.

Doi *et al.* (1977) also used their own age-structured model to calculate YPR for Mexican *H. corrugata*. They estimated the instantaneous rate of total mortality (*Z*) for each of several years from length frequencies, using the method of Doi (1975). (Marín (1981) also used this method to estimate survival and mortality in *H. corrugata* in Bahia Tortugas, Mexico.) With an estimate of natural mortality rate they calculated fishing mortality rate and exploitation rate; with estimates of fishery yields they could estimate stock size. Their paper discusses the problem of estimating the availability to the fishery of partially recruited age classes. The authors developed YPR isopleths by varying MLS and *F*.

A Leslie matrix model (Leslie, 1945; Usher, 1972) was used to examine yields from a Japanese abalone species by Watanabe *et al.* (1982). Without a translation of their paper, I am unable to describe their results further.

Tegner *et al.* (1989) used a Ricker (1945, 1975) model to analyse YPR in California red abalone, *H. rufescens*, and pink abalone, *H. corrugata*. Schiel & Breen (1991) also used this model, which is easy to construct on a spreadsheet, for *H. iris*.

Hamer (1983) described the use of a YPR approach in the New South Wales fishery for *H. rubra*, but gave no details as to how YPR estimates were used to obtain estimates of yield from the fishery.

Sanders (1982) conducted YPR analysis for the abalone *H. mariae* on the southeast coast of Oman. He concluded that yields would increase if either *F* were increased or MLS were decreased.

The requirements for YPR modelling are straightforward: an estimate of natural mortality rate *M*, estimates of growth parameters and the relation between length and weight.

It is essential to consider seriously the assumptions made in YPR analysis and to evaluate how estimates might be different if assumptions are violated. Hancock (1979) presents a lucid discussion of this issue for invertebrate fisheries, especially with respect to Beverton & Holt's (1957) model.

The Beverton-Holt model assumes that M is age independent, at least for the portion of the population recruited to the fishery. Not much is known about how M varies with age in abalone. Prince *et al.* (1988a) suggest that M decreases with age in Tasmanian *H. rubra*, based on the polymodal length frequency analysis MIX (MacDonald & Pitcher, 1979; MacDonald & Green, 1988). The curve of the logarithm of abundance versus age was concave upwards rather than straight, suggesting that M decreases with age. Hines & Pearse (1982) suggested that *H. rufescens* and flat abalone, *H. walallensis*, in California have type III survivorship curves (Deevey, 1947), where mortality decreases with age. Appeldoorn (1988) found decreasing M with increasing age in conch, *Strombus gigas*, reviews evidence for the phenomenon in other marine invertebrates, and discusses the effect on YPR analysis.

Conversely, Sainsbury (1982a) concluded that M was independent of age for *H. iris* larger than 30 mm; Shepherd *et al.* (1982) made the same conclusion for *H. laevigata* and *H. rubra* in South Australia. Vetter (1988) presents a general discussion of non-constant M and reviews the sensitivity of YPR and other models to estimated M.

Most models assume that F is age independent. This could be a problem where recruitment occurs before maturity, as abalone reproductive behaviour makes them more vulnerable to fishing (Shepherd, 1986). Usually it is assumed that the fishery causes no mortality to sublegal individuals, but the mortality caused when abalone are removed from the bottom for measuring and replaced can be considerable. The problem is well documented only for *H. rufescens* (Burge *et al.*, 1975; Cicin-Sain *et al.*, 1977; Tegner *et al.*, 1989). The mortality caused by cutting the foot of removed animals and poor handling can be assumed to vary directly with F and incorporated in YPR analyses.

Third, the Beverton-Holt (1957) model requires the von Bertalanffy (1938) growth model parameters. See Roff (1980) for criticisms of this model, Sainsbury (1980) for a discussion of how individual variability in growth leads to problems in accurate parameter estimation and Francis (1988) for a discussion of the difference between these growth parameters estimated from ageing and tagging. The Ricker (1945, 1975) and other models are more flexible.

Hancock (1979) considered that YPR models assume a constant recruitment, or recruitment that varies independently of stock size. This is true only if one tries to relate total yield directly to YPR. If the stock–recruitment relation were known, then it could be combined with YPR modelling to evaluate overall yield (e.g. Beverton & Holt, 1957; Shepherd, 1982; Mohn, 1986). Bannister & Addison (1986) carried out simulations along these lines for European lobsters, *Homarus gammarus*. Their results underscored the need to base management decisions on more than simple YPR analyses.

Implicitly YPR models assume that growth rates are independent of stock size. If growth rates are instead density dependent, YPR would be greater at lower stock sizes (higher F values) than simple models predict. Clavier (1982) showed an inverse relation between maximum size and density among *H. tuberculata* at several sites in France, suggesting density dependence in growth. This idea needs to be confirmed experimentally, with abalone growth measured at the same site at different densities. For Dungeness crabs, *Cancer magister*, Botsford (1981) explored the effect of increased growth rate when population size was decreased by fishing and found a significant effect in model populations. Shepherd & Hearn (1983) demonstrated sex-dependent growth rates in South Australian *H. rubra* and *H. laevigata*; such differences must be handled correctly in YPR models.

Similarly, natural mortality rates might vary with density. Not much is known about this for adult abalone (Chapter 21, this volume). The phenomenonally high densities of natural abalone populations (e.g. Douros, 1987) might suggest this effect is normally weak.

Egg-per-recruit models

YPR analysis gives estimates of maximum yield when recruitment is assumed to be constant, varies independently of stock size or can be measured directly. The method is, however, incapable of addressing any effect that fishing may have on subsequent recruitment. Sainsbury (1982b) states that YPR

> '... provides no indications as to how sustainable a given level of exploitation might be; i.e. it considers the ultimate yield from a given number of recruits but not the ultimate number of progeny produced by them.'

Maintenance of reproductive capacity has long been recognized as essential to sustain abalone fisheries. This may have been the single most important consideration in establishing MLS regulations. For Japanese abalone species Ino (1966) states:

> 'The purpose of [a minimum legal size] is to allow each abalone to engage in at least one reproduction, so as to ensure that a sufficient number of new abalones are produced. The determination of minimum shell length depends on the size at which an average abalone's reproductive organs grow to maturity.'

Bissell & Hubbard (1968) argue for *H. rufescens*:

> 'since abalones of four to five inches [100–150 mm] are known to spawn there is no apparent shortage of eggs and larvae in their environment, and there seems little danger of this mollusc becoming extinct because of lack of spawning activity.'

This sort of argument is widespread: Mace (1984) applies it to Australian abalone. Similarly, for *H. rufescens* Tegner (1989) describes the California management philosophy:

> 'The [California Dept. of Fish & Game] assumed that an appropriate size limit, large enough to allow several spawnings of these highly fecund animals but small enough that the size would be achieved within a reasonable number of years following recruitment [to the population], could maintain the stocks.'

There are problems with using overly simple and informal models to determine whether the MLS will maintain sufficient reproductive capacity. Fecundity is usually related to body weight, and far fewer eggs are produced by a newly-matured individual than by an older female. If natural mortality rates are low, much of a population's egg production will arise from large, old females; conversely, the production of a fished population depends on reproduction by comparatively young animals. The extent to which egg production is reduced by fishing must be determined analytically: *ad hoc* models such as those above are inadequate to support their claims.

One approach is to examine the egg production associated with various combinations of MLS and *F*. The relation between egg production and subsequent recruitment or settlement is of course unknown, but the scale of reduction in egg production caused by fishing can be useful in determining a management strategy. This method has been used for homarid (Campbell, 1986) and palinurid (Annala & Breen, 1989) lobsters and for fishes (Prager *et al.*, 1987). By analogy with YPR analysis, this technique is called egg-per-recruit (EPR) analysis, although the absolute estimate of eggs-per-recruit is not usually examined.

Doi *et al.* (1977) examined EPR for Mexican *H. corrugata*, varying *F* at a constant MLS. They considered 'optimum' an EPR value of 50% the value in the virgin population, but gave no supporting information. There is no way to choose the appropriate value of EPR until the egg production/recruitment function is understood and stochastic modelling has explored the effects of unpredictable recruitment variability.

Sluczanowski's (1984a, 1986) model is described above. This model also estimated equilibrium biomass and EPR at various values of MLS and the interval *P* between fishing South Australian *H. laevigata* on a reef. Although the current MLS would yield a YPR value near the maximum, it yielded a relatively low EPR:

> 'The present legal minimum length of 130 mm achieves a high yield efficiency for all periods, but leads to an egg production efficiency between 16 and 36% on the most optimistic assumptions ....'

Sluczanowski estimated that, under the worst assumptions, EPR could be as low as 1% of virgin levels. When *P* (the period between fishing on a reef) was increased, EPR increased greatly without much sacrifice in YPR. This approach suggests that, if MLS is used as the major management tool (as it is in most abalone fisheries), both YPR and EPR must be examined.

For *H. kamtschatkana* in British Columbia, Breen (1986) used a simple model to analyse EPR (an error affects an equation and the appearance of two figures, but does not change the conclusions: details are available from the author.) Like YPR, EPR is insensitive to changes in *F* when the MLS is greater than 110 mm. Nash (Chapter 23, this volume) used a similar model to examine EPR in *H. rubra* in Tasmania.

Breen (1986) suggested combining the results of YPR and EPR quantitatively. He transformed absolute YPR into relative YPR, i.e. YPR/YPR_{max} where YPR_{max} is the YPR at the optimum combination of t_c and *F*. EPR, when expressed as a percentage of EPR in the virgin population, could be transformed into a number between 0 and 1. He called this egg production value (EPV). This could be done by assuming that an EPR less than some percentage such as 20% is unacceptable (i.e. EPV = 0) and that above some percentage such as 50% EPR would be sufficient to sustain the population (i.e. EPV = 1). EPV can be calculated for intermediate EPR values by some simple arbitrary function. Relative YPR and EPV can then be combined by multiplication to give a relative strategy value for each combination of t_c and *F*.

Sluczanowski's approach was also adopted by Tegner *et al.* (1989) for their analysis of California *H. rufescens* and *H. corrugata*. This model was modified to include some variability in length about the mean length-at-age. As in most YPR and EPR models, this model assumes that mean length-at-age is described by a von Bertalanffy growth curve. However, Tegner *et al.*'s (1989) model allows lengths to be normally distributed around their mean. This incorporation of variability removes the unrealistic discontinuities (such as those in Breen, 1986) from the EPR isopleths. The authors conclude that the declines observed in the California fishery cannot be explained on the basis of simple recruitment overfishing, because egg production at equilibrium should have remained at a high percentage of virgin levels.

This conclusion was qualified by several considerations. First, the analysis is sensitive to changes in the von Bertalanffy growth parameters, so the results may not be applicable to abalone in different locations growing at different rates. Second, fishing mortality on sublegal abalones (Cicin-Sain *et al.*, 1977) has the effect of reducing the effective MLS. This effect can easily be incorporated into YPR models. Third, the analysis ignores spatial complexity: abalone might have very small unit stocks, as discussed below, and serial depletion could occur even with an acceptable overall value of *F*. Aggregation for spawning might be an important ecological adaptation (Shepherd, 1986), and divers in turn concentrate on aggregations. This spatial complexity has two consequences: effective *F* on a small spatial scale may be higher than the effort level would suggest, and fertilization efficiency may be greatly reduced at low abalone densities.

Finally, Tegner *et al.* (1989) note that models such as those described above are all based on deterministic equilibria, whereas real populations (especially those of abalone) have highly variable settlement and recruitment patterns. Such variation can significantly reduce the response of exploited populations below expectations (see, for example, Archibald *et al.*, 1983, describing rebuilding probabilities for depleted stocks). Where abalone populations are driven by occasional strong year classes, then recruitment overfishing may be more likely than simple models suggest.

The data required for EPR analysis are simple: an estimate of *M*, growth parameters, the size at 50% maturity and the relation between fecundity and length or weight. The assumptions and associated problems that apply to YPR models also apply to these models. In addition, EPR analysis assumes that the fecundity–length relation and the onset of maturity are independent of stock size. Variation in these estimates has apparently not been examined in abalone, but density-dependent variation in growth rate suggests that density-dependent variation in fecundity may well occur. In lobsters, reduced density may decrease the size at onset of maturity, for example in the Hawaiian spiny lobster *Palinurus marginatus* (Polovina, 1989).

An implicit assumption is that the reproductive potential of an egg is independent of spawning female size. For American lobsters, *Homarus americanus*, Attard & Hudon (1987) suggest that eggs from larger females have a greater potential for survival. This effect requires modifications to EPR analysis.

Leslie and DeLury models

These models estimate the original biomass B_0 of a fished population at some point (e.g. in the virgin state or at the beginning of a season). They use the catch over some period beginning at that point and the decline in CPUE over the same period. CPUE is assumed proportional to stock size. The catchability, q, is also estimated. The Leslie (Leslie & Davis, 1939) and DeLury (1947) methods are reviewed by Ricker (1975).

If CPUE is indeed an index of stock size (see below), then declining CPUE indicates a declining stock. Whether it also indicates a stock declining to lower than desirable levels cannot be determined without much more analysis. Newman (1967), for instance, used declining CPUE in the South African fishery for perlemoen, *H. midae*, to be an indication of overfishing. This conclusion is unwarranted without more evidence.

Ino (1966) describes use of the DeLury model to estimate the pre-fishery abalone biomass and catchability for one area in Japan (Shimoda Cho, Taushi) by the Shizuoka Prefectural Marine Products Research Station. This is apparently the first documented stock assessment for any abalone species.

For *H. discus discus* in Tokushima Prefecture, Kojima *et al.* (1978) used daily catch records and the DeLury method to estimate the pre-season biomass and q for each year from 1962 to 1966. Results suggested that finite annual exploitation rates (A of Ricker, 1975) were 0.85–0.90.

Kojima *et al.* (1978) and Kojima & Isibasi (1985) modified the Leslie and DeLury models to incorporate recruitment to the fishery between years and during the fishing season in order to improve the estimates of catchability from several years' catch and effort data. Translations of these papers would be very useful.

Geaghan & Castilla (1986) used Leslie and DeLury models to estimate population size and catchability for loco, *Concholepas concholepas*, a commercially harvested Chilean muricid gastropod. They also modified the basic model, which assumes that natural mortality and recruitment are negligible over the period of analysis, to include these two parameters. The modified model can be used where more than one season's data are available, but prior estimates of some parameters are required. These authors found that recruitment was a major component of the annual catch, so that the fishery was highly vulnerable to recruitment failure.

Hirayama *et al.* (1989) used a modified DeLury model to estimate B_0 and q for *H. discus discus* in a small fishery at Kisakata, Japan. They had daily records of catch for two seasons. They modified the model to take into account the effect on CPUE of catching other species and used multiple linear regression to evaluate the model.

Tanaka (1988) used a Leslie-DeLury method to estimate the stock sizes of *H. discus discus* in several regions within Chiba Prefecture. Data required for this technique are time series of catch and effort data. The basic model assumes that natural mortality, movements and recruitment are negligible; this is violated if the period over which the estimate is made is long.

Failure of the assumption that CPUE is proportional to stock size is more serious. Geaghan & Castilla examined this problem for Chilean loco. They found that divers return to previously fished beds in a very short time (2–4 weeks), and this combined with a high rate of exploitation makes CPUE an apparently good reflector of stock size. For most abalone species, CPUE is probably not a good index. This will be discussed below in the context of surplus-production analysis.

Stock reduction models

If the amount by which a stock has decreased in a given time is known, the landings are known and some assumptions can be made about natural mortality and recruitment, then it should be possible to calculate the original stock size and reconstruct its exploitation history. Kimura & Tagart (1982) described a rigorous model, refined by Kimura *et al.*, (1984) and Kimura (1985, 1988). The simple model simultaneously estimates stock biomass B_i and fishing mortality rate F_i for each year i of the period being considered, and a constant recruitment biomass R. The more elaborate versions incorporate a stock-recruitment function.

Breen (1986) used the Kimura & Tagart (1982) model to assess *H. kamtschatkana* stocks over the period 1975–83. Kimura & Tagart suggest looking for solutions of the equation system where annual recruitment is at the virgin population replacement level. Using realistic estimates of stock reduction, and M estimated from the length frequency method of Fournier & Breen (1983), credible estimates of both F_i and R could not be obtained. This and other evidence suggested that recruitment was lower than replacement levels during 1975–83 and that the stock would have declined by about a third during this time even without fishing.

Stock-reduction analysis is a valuable tool for abalone fisheries, where biologists must often make an assessment for a new fishery with a short time series of fishery data and little biological information. Stock biomass is difficult to estimate directly (for example with biomass surveys), but relative abundance estimates are much easier to obtain. Catches are generally known from government statistics. With an estimate of natural mortality, the decline in relative abundance from two surveys and the catch data, the biologist can estimate absolute original biomass, fishing mortality and recruitment with this technique.

The model assumes that recruitment is constant over the period for which the estimate is made. Failure of this assumption affects the estimate of biomass directly. If the estimate of stock reduction is biased, this would also affect the biomass estimate directly.

Surplus-production models

These models estimate the equilibrium-state relation between biological production and fishing intensity, where 'production' is the difference between increases in recruited biomass caused by recruitment and growth and decreases caused by fishing and natural mortality. This type of model was originally developed by Schaefer (1954); for reviews see Schnute (1977, 1985) and Fogarty & Murawski (1986).

Harrison (1983) used this method for *H. rubra* in Tasmania. First, however, Harrison considered several factors that might affect nominal effort as a unit of real effort. He conducted a questionnaire survey of divers to estimate changes in diver fishing power over time. From this information he developed a time series of estimated diver power for use in standardizing the time series of effort. Harrison also attempted to standardize effort using weather information, but he found no correlation between CPUE and 'good' weather indices.

From the catch and standardized effort data, Harrison used the Fox (1970) model to estimate surplus production for the whole of Tasmania and for each of several areas within the state. A problem with the analysis appears to be a lack of 'contrast' in the data. Catch at the highest level of effort in Harrison's data was also the maximum catch in all but one area, so the 'dome' of the dome-shaped production function was not well defined. This is an inherent problem with surplus-production analysis – in order to obtain good estimates of the relevant population parameters the observer must have data from both lightly and heavily fished situations.

Hamer (1983) briefly describes use of the Schnute (1977) model for analysis of *H. rubra* in New South Wales. Tanaka (1988) fitted a surplus production curve for a population of Japanese *H. discus discus*, but no details are available in English.

Sluczanowski (1984b) plotted catch against effort for South Australian *H. rubra* and *H. laevigata* and used the method of Schnute (1977). The curve shows a simple linear increase in catch with increasing effort from 1967–84; Sluczanowski considered that surplus-production analysis was of little use in this case and perhaps for abalone generally. His reasons were as follows:

(1) The time series of data is often too short for assessment of abalone fisheries. The lag between spawning and recruitment is 4–5 years in South Australian species (much longer in many others), so long series of data are required.
(2) Diver power changes over time, so time series of effort must be standardized properly to be meaningful.
(3) Handling time is a significant fraction of diver time, and so must be included in the effort estimates.
(4) To obtain a valid production function, one must use data from several levels of sustained effort, both in the declining and recovery phases of the stock. It is difficult to obtain such data from abalone fisheries, where collapses are more common than rebuilding.
(5) Abalone have a 'clumped' distribution. Because divers search for 'clumps' and the search time is not proportional to numbers of individuals, catch rates will not decline at the same rate as the reduction in abundance of clumps.

Sluczanowski concluded that 'models which deal with catch and effort data offer almost no guidance for managers of the [South Australian] abalone fishery until more data are available'. This work, together with the studies by Harrison (1983) and Hamer (1983) are also discussed by Mace (1984).

Rocha & Arreguin-Sanchez (1987) used a surplus-production model to analyse fisheries for Mexican *H. cracherodii* and *H. fulgens* on the Baja Peninsula. (The author has seen only the English abstract from this paper.)

The data required for surplus-production analysis are time series of catches and effort. Early versions of the method contained assumptions about the shape of the production function and assumed that observed data were from equilibrium conditions, but later models relax these constraints. The method assumes a unit stock, but in abalone fisheries the definition of a unit stock is not clear. This problem is discussed further below, in the context of stock-recruitment models.

Real fishing power per unit of nominal effort may change over time, as discussed above. For instance, diver effort per day may increase because of competition between divers or in response to management actions (Harrison, 1983; Breen, 1986). When this happens, or when learning increases diver efficiency, effort standardization (e.g. Harrison, 1983) is necessary.

This is just one problem with the important assumption that CPUE is a valid index of stock size (i.e. that catchability is constant). The assumption implies that divers operate in a random selection of locations containing fish – a clear oversimplification of reality. Hilborn & Walters (1987) explored the consequences of divers choosing to fish in areas with higher catch rates, modelling the South Australian abalone fishery. In their simulation, CPUE decreased at a much slower rate than abundance in a developing fishery. This paper emphasizes the need to consider diver behaviour explicitly as part of the fishery system, and discusses several possible diver strategies. Prince (1990) explores the effect of price on diver behaviour in the Tasmanian fishery for *H. rubra*.

If divers can apportion their effort among species based on relative abundance, CPUE for the species of interest can be affected. Kojima *et al.* (1978) describe the situation in Tokushima Prefecture, where four species of abalone are fished. When abalone are scarce, divers take the gastropod *Turbo cornutus*. The authors demonstrate a mathematical relation between CPUEs for the abalone and the gastropod.

The assumption of constant catchability also requires that fish re-disperse after fishing. This assumption fails in several ways for abalone. When an abalone stock is being 'fished down' in a new fishery, abalone divers can progressively discover and exploit unharvested beds; their CPUE remains high as the stock declines (Breen, 1980; Harrison, 1983). That effect occurs whatever the local dispersion pattern may be. Aggregative behaviour further contributes to the problem (see above; Sluczanowski, 1984b; Chapter 22, this volume). Finally, abalone may actually re-aggregate after fishing rather than redisperse (Chapter 22, this volume), exacerbating the problem. Breen (1986) estimated that the northern British Columbia stock of *H. kamtschatkana* declined by 50% or more during 1975–83; a period in which CPUE (catch per diver day) declined by only 7%.

The weather can obviously influence CPUE. Harrison (1983) attempted unsuccessfully to standardize CPUE for Tasmanian *H. rubra* from weather data. He suggested that the available meteorological data might not be appropriate reflectors of the effect of weather on abalone fisheries. Kojima & Isibasi (1985) found a good relation between barometric pressure and the diving hours per day of three divers operating in a small area in Japan.

Problems with CPUE in abalone fisheries are serious, and stock assessments based on CPUE must be used extremely cautiously.

Cohort analysis

Age-based cohort analysis uses a time series of catch data and age structures to estimate the time series of fishing mortality rates and population sizes (see Pope, 1972). Age structures require a verified aging procedure. Rings in *H. tuberculata* have been verified by growth studies (Forster, 1967; Hayashi, 1980). Verification is also reported in Japanese *H. discus discus* (Kojima *et al.*, 1977) and the Korean *H. diversicolor diversicolor* and *H. d. aquatilis* (Kim & Cheung, 1985). Inoue & Oba (1980) suggest that shell circuli are annual for Japanese *H. gigantea*. Prince *et al.* (1988b) describe a method for aging *H. rubra*, based on the method of Muñoz Lopez (1976).

For many abalone species aging is not possible. Jones (1974, 1981, 1984) describes a length-based cohort analysis, where individuals are assigned an age based on the growth curve. Castilla & Jerez (1986) applied this method to *Concholepas concholepas*. Sainsbury (1982b), however, suggests that this method 'is unsuited to species such as *H. iris* which exhibit a high degree of individual variability in growth'.

Cohort analysis requires annual catch data and also annual estimates of the proportion of the catch in each age or length class. The analysis is sensitive to the assumed value of *M*.

Stock-recruit models

There are no abalone species for which we know the relation between stock and recruitment or settlement. The longest time series of appropriate data is that of Shepherd (1990), who found no relation between stock size and subsequent 2-year-old abundance of *H. laevigata* in an unfished population in South Australia.

It is often suggested that the frequently-observed collapse of abalone fisheries (see, for example, Chapter 24, this volume) is caused by recruitment failure, and recruitment or settlement failure is well documented. Examples of such failure include California *H. rufescens* (Cicin-Sain *et al.*, 1977), New Zealand *H. iris* (Sainsbury, 1977, 1982a), Guernsey *H. tuberculata* (Hayashi, 1980) and Canadian *H. kamtschatkana* (Breen, 1980, 1986; Boutillier *et al.* 1985; Sloan & Breen, 1988). Recruitment failures can have several causes, probably most of which are direct or indirect environmental effects (see Tegner *et al.*, 1989, for discussion). These effects are not generally well understood, and the same environmental events may have different effects on different abalone species (Dayton & Tegner, 1990).

However, it is not clear whether recruitment failure in abalone has ever resulted from recruitment overfishing. Harrison (1986) contends that recruitment overfishing has never been convincingly demonstrated in an abalone fishery, and I concur (but see Chapter 24, this volume). The four species cited above experienced decreased recruitment or settlement independently of the fisheries. Tegner *et al.* (1989) described the classic model of recruitment overfishing in abalone fisheries: abalone populations often contain accumulations of large, old individuals in whom much of the population reproductive potential is vested. Fishing can quickly remove these accumulations and drastically reduce the population's reproductive capacity. However, these authors analysed EPR for California *H. rufescens* and *H. corrugata* and concluded that the model cannot explain what happened in the California fisheries.

In addition to the large environmental effects on settlement and recruitment, a major uncertainty in the elucidation of stock–recruit patterns in abalone is the problem of defining a unit stock. This depends on fully understanding dispersal (McShane *et al.* 1988). Abalone have swimming larvae with stage lengths on the order of 1–2 weeks. If the larvae are truly planktonic then larvae spawned at one place could settle significant distances away; thus a functional 'stock' could comprise the abalone living along tens or hundreds of kilometres of coast.

Prince *et al.* (1987) suggest that veliger larvae of Tasmanian *H. rubra* may not be planktonic, but instead remain close to the bottom. Prince *et al.* (1988c), and McShane *et al.* (1988) in southeastern Australia tested this idea by removing adults from experimental plots and measuring the abundance of post-larval settlement with the techniques of Prince & Ford (1985) and McShane & Smith (1988) respectively. Abundance of newly-settled abalones was positively correlated with adult density, as predicted by the hypothesis under test. This result is not a 'confirmation' of the hypothesis, as other explanations have not been disproven. For instance, larvae from distant spawnings may settle preferentially where adult density is higher, as in experiments conducted by Tutschulte (1976) on *H. corrugata* and green abalone, *H. fulgens*. But these studies are valid tests of the hypothesis. If the Prince *et al.* (1987) hypothesis is correct, stocks may be smaller than originally supposed. Tegner (Chapter 34, this volume) suggests a scale of hundreds of metres for *H. fulgens*. Additional support for the idea is presented by McShane *et al.* (1988), who conducted plankton surveys to search for larvae and conducted current measurements and modelling. Even if larvae disperse widely, but settling is gregarious, there are important management implications (Prince *et al.* 1988c).

In contrast, Day & Leorke (1986) reported high variability in the abundance of several-weeks-old *H. rubra* that was unrelated to the breeding abalone density on small reefs in Victoria.

If stocks are highly localized, then local recruitment overfishing and sequential recruitment collapse on small reefs is a serious management problem (Sluczanowski, 1984a, 1986; Harrison, 1986). Lack of understanding of stock–recruit patterns is the single biggest gap in our understanding of abalone fishery biology. Shepherd (1984) suggested:

> 'With such limitations to knowledge the only safe way to manage the fishery is 'on the worst case' i.e. to assume that a sensitive stock/recruitment relation exists. It follows that a principal management goal must be to maintain optimal egg production and a principal research goal to determine what that is. While these aspects remain unknown, the fishery should be managed in a way which past experience has shown to be 'safe' in respect of egg production.'

Economic analyses

Fishery management is often more concerned with economic returns than with yield from the fishery. Waugh (1980, 1984) examined the economics of the fishery for *H. rubra* in New South Wales. He used a Beverton-Holt yield equation as the basis for the biological model and made assumptions about the pattern of recruitment. His approach then follows that of Gordon (1953, 1954). He concluded that *F* should be decreased or MLS increased in order to increase economic yield from the fishery.

Harrison (1983) extended his surplus–production analysis to examine revenue as a simple function of catch and total costs as a simple function of fishing effort, as in the Gordon model. He estimated that maximum average economic yield would be derived from an effort level lower than that corresponding with maximum average yield. Harrison discussed the economic position of the fishery with reference to this simple model and to trends in market price.

Hamer (1983) extended his YPR analysis to an economic analysis. He first converted the YPR to fishery yield using assumptions not stated, then converted fishery yield to revenue in dollars, assuming a constant unit price. Cost of fishing as a function of effort was estimated from real data. Surplus can be defined as the difference between revenue and costs. Hamer estimated the 'breakeven' point or no-surplus equilibrium, where revenue equals costs, and the point of greatest surplus.

Mace (1984) presented a Gordon analysis of the abalone fisheries of New South Wales, South Australia, Tasmania and Victoria. He compared equilibrium revenue, estimated from a surplus–production model, with a constant cost per unit effort. Mace does not make clear what data he used. His analysis suggested that New South Wales was at the open access equilibrium, where revenue equals costs, and that Victoria was nearby. Tasmania and South Australia, with limited entry, appeared to be operating near maximum economic yield.

Rocha & Arreguin-Sanchez (1987) also used a Gordon analysis for the Baja California abalone fishery. They concluded that the fishery was providing some net economic return in two places, but elsewhere the fishery was close to the no-surplus equilibrium or beyond.

A more sophisticated approach was used by Deacon (1989) to examine the effects of policy decisions on the fisheries for *H. rufescens* and *H. corrugata* in California. His biological model is age structured. Parameters were fitted to real catch and effort data from the fishery with a non-linear estimation procedure. For *H. corrugata*, Deacon examined the economic rent from the fishery under varying MLS and effort conditions. He confirmed the suggestion by Burge *et al.* (1975) that increasing the MLS in 1970 decreased fishery yield; he supported the limited entry recommended by those authors. Deacon discussed the 'optimal-path problem' of finding a set of fishing efforts that provides the best net present benefit from the fishery as the stock responds to management, but provided no results from the abalone fishery.

Limited entry does not limit effort in abalone fisheries unless further restrictions are imposed (Harrison, 1986). Breen (1986) describes spectacular effort increases in the British Columbia fishery for *H. kamtschatkana* after entry limitation. One would expect effort to increase in limited-entry fisheries until the no-surplus equilibrium was reached.

Miscellaneous models

Energy budgets are useful for examining the role of abalone in their ecological community and estimating their production potential. Barkai & Griffiths (1988) constructed individual and population energy budgets for South African *H. midae* based on estimates of feeding and growth rates, fecundity, absorption efficiency, respiration and excretion rates and population surveys. Abalone populations at two sites consumed a large proportion of the total primary production of kelp. The estimated annual ratio of somatic production to abalone biomass was 13% and 15% at the two sites. Models such as this can give useful indications of the scale of annual production relative to abalone biomass in new fisheries for which little other information is available.

CONCLUSIONS

Although abalone biologists were slow to adopt modern quantitative stock-assessment tools, a wide variety of approaches have now been applied to abalone populations.

Yield- and egg-per-recruit techniques are among the easiest to apply, require only easily obtained data and are of immediate application where managers rely heavily on a minimum legal size limit to maintain yield efficiency and egg production.

Methods involving CPUE, such as surplus-production analysis for estimating sustainable yields and Leslie-DeLury methods for estimating biomass, should be used extremely cautiously for abalone because the underlying assumptions are very likely to be violated. Largely overlooked techniques widely used in finfish assessments include length-based cohort analysis (Jones, 1974, 1981, 1984) and the Deriso (1980) method (mentioned in passing by Hamer, 1983).

Experimental harvesting as a direct approach to estimating population responses has also been overlooked. (Clavier, 1982) suggested for *H. tuberculata* (in translation):

> 'Such an approach offers the possibility of comparing, *a priori*, the data obtained through theoretical estimation of stock potential with those obtained from actual harvesting. It may also provide data on the rate of recolonisation after harvesting, and on the feasibility of repopulation by spreading seed-ormers in the natural environment ... Only an evaluation of this type will enable us to elucidate the economic interest of an ormer fishery and to decide on the advisability of ormer harvesting over a large area.'

The key problem in abalone population dynamics lies in understanding settlement and recruitment. The problem of stock definition is tangled with this problem. Abalone populations often show highly variable, unpredictable recruitment. This variation affects the conclusions made from most stock assessments. If enough data are available so that the pattern of recruitment variability can be described, then assessments should explore the results of stochastic simulations rather than relying on deterministic models and equilibrium conditions. See Walters (1986) for a discussion dealing with both unpredictability and measurement uncertainty.

Most important for stock assessment to be useful in developing effective fishery management is an elucidation of the relation between present stock and future recruitment. Modelling such as described by Bannister & Addison (1986) can explore the implications of various stock–recruit relations. Such modelling is essential to the development of management strategy when the stock–recruit relation and the role of extraneous variability are unknown.

Also central to effective management is the question of stock structure. Whether abalone populations disperse widely or instead are highly localized (e.g. Prince *et al.*, 1987) has profound implications for management. If unit stocks for abalone are very small, then appropriate regulation of effort and fishing mortality is probably impossible on an individual stock basis. In that situation, fishery data and modelling results could be highly misleading if used in the context of 'too large' an area. Such small unit stocks would need to be managed without effective local control of fishing effort or fishing mortality: the importance of having an appropriate MLS would be greatly magnified.

Finally, nearly all abalone assessments have been based on single-species approaches. Yet abalone populations are dependent on, and vary with, other elements in their ecological community (see, for instance, Tanaka *et al.*, 1986; Tegner, 1980, 1989). Ino (1966) describes the decline of abalone as a result of kelp harvesting in Japan, and suggests a model for determining how much algal volume should be retained to support abalone. Conversely, sea urchin fisheries may increase the standing stock of algae (for example in the California fishery for red sea urchins, *Strongylocentrotus franciscanus*, M.J. Tegner, pers. comm.). It is also possible that sea urchin removal may result in altered algal species composition (Vadas, 1968), with consequences in abalone growth. Multi-species approaches to abalone stock assessment will become increasingly more important with increasing interest in the harvest of marine algae, sea urchins and other grazers and predators from coastal production systems.

ACKNOWLEDGEMENTS

This chapter is dedicated to the author's former teacher, Norm Wilimovsky, on his retirement. The author is grateful for constructive criticism to Rob Day, Ray Hilborn, Jeremy Prince, Scoresby Shepherd, Phillip Sluczanowski and Mia Tegner.

REFERENCES

Alverson D.L. & Carney M.J. (1975) A graphic review of the growth and decay of population cohorts. *J. Cons. Int. Explor. Mer*, **36**(2), 133–43.

Annala J.H. & Breen P.A. (1989) Yield- and egg-per-recruit analyses for the New Zealand rock lobster, *Jasus edwardsii*. *NZ J. Mar. Freshwat. Res.*, **23**(1), 93–105.

Appeldoorn R.S. (1988) Ontogenetic changes in natural mortality rate of queen conch, *Strombus gigas* (Mollusca: Mesogastropoda). *Bull. Mar. Sci.*, **42**(2), 159–65.

Archibald C.P., Fournier D.A. & Leaman B.M. (1983) Reconstruction of stock history and development of rehabilitation strategies for Pacific Ocean perch in Queen Charlotte Sound, Canada. *N. Am. J. Fish. Manag.*, **3**, 283–94.

Attard J. & Hudon C. (1987) Embryonic development and energetic investment in egg production in relation to size of female lobster (*Homarus americanus*). *Can. J. Fish. Aquat. Sci.*, **44**, 1157–64.

Bannister R.C.A. & Addison J.T. (1986) Effect of assumptions about the stock–recruitment relationship on a lobster (*Homarus gammarus*) stock assessment. *Can. J. Fish. Aquat. Sci.*, **43**(11), 2353–9.

Barkai R. & Griffiths C.L. (1988) An energy budget for the South African abalone *Haliotis midae* Linnaeus. *J. Molluscan Stud.*, **54**(2), 43–51.

Beinssen K. (1979) The estimation of fishing mortality of abalone based on experiments to measure the rate of reef coverage by the fishing fleet. *Rapp. PV Réun. Cons. Int. Explor. Mer*, **175**, 27–9.

Beverton R.J.H. (1954) Notes on the use of theoretical models in the study of the dynamics of exploited fish populations. *US Fish. Lab., Beaufort, NC, Misc. Contrib.*, **2**.

Beverton R.J.H. & Holt S.J. (1957) On the dynamics of exploited fish populations. *UK Min. Agric. Fish. Food Fish. Invest. (Ser. 2)*, **19**, 1–533.

Beverton R.J.H. & Holt S.J. (1966) Tables of yield functions for fisheries assessment. *FAO Fish. Tech. Pap.*, **38**.

Bissell H. & Hubbard F. (1968) *Report on the sea otter, abalone and kelp resources in San Luis Obispo and Monterey Counties and proposals for reducing the conflict between the commercial abalone industry and the sea otter*. Mimeo report to the California Senate, California Department of Fish and Game, Sacramento, California.

Botsford L.W. (1981) The effects of increased individual growth rates on depressed population size. *Am. Nat.*, **117**(1), 38–63.

Boutillier J.A., Carolsfeld W., Breen P.A., Farlinger S. & Bates K. (1985) Abalone resurvey in the southeast Queen Charlotte Islands, July 1984. *Can. MS Rep. Fish. Aquat. Sci.*, **1818**.

Breen P.A. (1980) Measuring fishing intensity and annual production in the abalone fishery of British Columbia. *Can. Tech. Rept Fish. Aquat. Sci.*, **947**.

Breen P.A. (1986) Management of the British Columbia fishery for northern abalone (*Haliotis kamtschatkana*). In *North Pacific Workshop on Stock Assessment and Management of Invertebrates* (Ed. by G.S. Jamieson & N.F. Bourne). *Can. Spec. Publ. Fish. Aquat. Sci.*, **92**, 300–12.

Burge R., Schultz S. & Odemar M. (1975) *Draft report on recent abalone research in California with recommendations for management*. Mimeo report, California Department of Fish and Game.

Campbell A. (1986) Implications of size and sex regulations for the lobster fishery of the Bay of Fundy and southwestern Nova Scotia. In *North Pacific Workshop on Stock Assessment and Management of Invertebrates* (Ed. by G.S. Jamieson & N.F. Bourne). *Can. Spec. Pub. Fish. Aquat. Sci.*, **92**, 126–32.

Castilla J.C. & Jerez G. (1986) Artisanal fishery and development of a data base for managing the loco, *Concholepas concholepas*, resource in Chile. In *North Pacific Workshop on Stock Assessment and Management of Invertebrates* (Ed. by G.S. Jamieson & N. Bourne). *Can. Spec. Publ. Fish. Aquat. Sci.*, **92**, 133–9.

Cicin-Sain B., Moore J.E. & Wyner A.J. (1977) Management approaches for marine fisheries: The case of the California abalone. *University of California Sea Grant Program, Sea Grant Publ.*, **54**.

Clavier J. (1982) (Initial data on natural ormer (*Haliotis tuberculata*) stocks in the St Malo region.) Rapport d'étude, Association pour la Mise en Valeur du Littoral de la Côte l'Emeraude. (Translated from French in *Can. Transl. Fish. Aquat. Sci.*, **4928**, (1983))

Clavier J. & Richard O. (1985) *Etudes sur les Ormeaux dans la Région de Saint-Malo*. Association pour la Mise en Valeur du Littoral de la Côte d'Emeraude, Dinard, Bretagne.

Clavier J. & Richard O. (1986a) Estimation du stock naturel d'ormeaux (*Haliotis tuberculata*) dans la région de Saint-Malo. *Rev. Trav. Inst. Pêches Marit.*, **48**(3–4), 95–105.

Clavier J. & Richard O. (1986b). Growth of juvenile *Haliotis tuberculata* (Mollusca: Gastropoda) in their natural environment. *J. Mar. Biol. Assoc. UK*, **66**, 497–503.

Day R.W. & Leorke A. (1986) Abalone catches – what factors affect them? *Aust. Fish.*, **45**(10), 32–6.

Dayton P.K. & Tegner M.J. (1990) Bottoms below troubled waters: Benthic impacts of the 1982–1984 El Niño in the temperate zone. In *Ecological Consequences of the 1982–83 El Niño to Marine Life* (Ed. by P.W. Glynn). Elsevier, Amsterdam.

Deacon R.T. (1989) An empirical model of fishery dynamics. *J. Env. Econ. Manag.*, **16**, 167–83.

Deevey E.S. (1947) Life tables for natural populations of animals. *Q. Rev. Biol.*, **22**, 283–314.

DeLury D.B. (1947) On the estimation of biological populations. *Biometrics*, **3**, 145–67.

Deriso R.B. (1980) Harvesting strategies and parameter estimation for an age-structured model. *Can.*

J. Fish. Aquat. Sci., **37**, 268–83.

Doi T. (1975) Análisis matemático de poblaciones pesqueras. Compendio para uso práctico. *Instituto Nacional de Pesca, México, Serie Información*, INP/SI:m12.

Doi T., Guzmán del Próo S.A., Marín Aceves V., Ortiz Quintillana M., Camacho Acosta J. & Muñoz Lopez T. (1977) Análisis de la poblacion y diagnostico de la pesqueria de abulón amarillo (*Haliotis corrugata*) en el area de Punta Abreojos e Islas Cedros, B.C. Direccion. *General del Instituto Nacional de Pesca, Serie Cientifica*, **18**.

Douros W.J. (1987) Stacking behavior of an intertidal abalone: An adaptive response or a consequence of space limitation? *J. Exp. Mar. Biol. Ecol.*, **108**(1), 1–14.

Fogarty M.J. & Murawski S.A. (1986) Population dynamics and assessment of exploited invertebrate stocks. In *North Pacific Workshop on Stock Assessment and Management of Invertebrates* (Ed. by G.S. Jamieson & N.F. Bourne). *Can. Spec. Publ. Fish. Aquat. Sci.*, **92**, 228–44.

Forster G.M. (1967) The growth of *Haliotis tuberculata*: Results of tagging experiments in Guernsey. *J. Mar. Biol. Assoc. UK*, **47**, 287–300.

Fournier D.A. & Breen P.A. (1983) Estimation of abalone mortality rates with growth analysis. *Trans. Am. Fish. Soc.*, **112**, 403–11.

Fox W.W. (1970) An exponential surplus-yield model for optimizing exploited fish populations. *Trans. Am. Fish. Soc.*, **99**, 80–8.

Francis R.I.C.C. (1988) Are growth parameters estimated from tagging and age-length data comparable? *Can. J. Fish. Aquat. Sci.*, **45**(6), 936–42.

Geaghan J.P. & Castilla J.C. (1986) Use of catch and effort data for parameter estimates for the loco (*Concholepas concholepas*) fishery of central Chile. In *North Pacific Workshop on Stock Assessment and Management of Invertebrates* (Ed. by G.S. Jamieson & N. Bourne). *Can. Spec. Publ. Fish. Aquat. Sci.*, **92**, 168–74.

Gordon, H.S. (1953) An economic approach to the optimum utilization of fishery resources. *J. Fish. Res. Bd Canada*, **10**, 442–57.

Gordon H.S. (1954) The economic theory of a common property resource. *J. Pol. Econ.*, **62**, 124–42.

Hamer G.T. (1983) NSW abalone stock assessment shows effort should be reduced. *Aust. Fish.*, **42**(8), 7–11.

Hancock D.A. (1979) Population dynamics and management of shellfish stocks. *Rapp. PV. Réun. Cons. Int. Explor. Mer*, **164**, 113–31.

Harrison A.J. (1983) The Tasmanian abalone fishery. *Tasmanian Fish. Res.*, **1**.

Harrison A.J. (1986) Gastropod fisheries of the Pacific with particular reference to Australian abalone. In *North Pacific Workshop on Stock Assessment and Management of Invertebrates* (Ed. by G.S. Jamieson & N.F. Bourne). *Can. Spec. Publ. Fish. Aquat. Sci.*, **92**, 14–22.

Hayashi I. (1980) Structure and growth of a shore population of the ormer, *Haliotis tuberculata*. *J. Mar. Biol. Assoc. UK*, **60**, 431–7.

Hilborn R. & Walters C.J. (1987) A general model for simulation of stock and fleet dynamics in spatially heterogeneous fisheries. *Can. J. Fish. Aquat. Sci.*, **44**(7), 1366–9.

Hines A.H. & Pearse J.S. (1982) Abalones, shells, and sea otters: Dynamics of prey populations in central California. *Ecology*, **63**, 1547–60.

Hirayama N., Yamada S., Kikuchi H. & Yamada J. (1989) A modification of DeLury method to abalone fishery. *Nippon Suisan Gakkaishi*, **55**(3), 409–16.

Ino T. (1966) [The abalone science and its propagation in Japan.] *Nippon Suisan Shigen Hogo Kyokai, Suisan Zoyoshoku Sosho*, **11**. [Translated from Japanese in *Fish. Res. Bd Can. Trans. Ser.*, 1078, (1968).]

Inoue M. & Oba T. (1980) Circuli on the shell as an age determination and growth of the Japanese abalone, *Haliotis gigantea*. *Bull. Kanagawa Prefecture Fish. Exp. Stn*, **1**, 107–13.

Isibasi K. & Kojima H. (1979) A study on the stock of Japanese black abalone, *Haliotis discus discus* in Tokushima Prefecture – III. Analysis on abalone fisheries management. *Bull. Tokai Reg. Fish. Res. Lab.*, **98**, 65–75. [Partial translation from Japanese in *Can. Trans. Fish. Aquat. Sci.*, 4866, (1982).]

Jones R. (1974) Assessing the long term effort and mesh size from length composition data. *Cons. Int. Explor. Mer*, CM1974/F:33.

Jones R. (1981) The use of length composition data in fish stock assessment (with notes on VPA and cohort analysis). *FAO Fish. Circ.*, **734**.

Jones R. (1984) Assessing the effects of changes in exploitation pattern using length composition data. *FAO Fish. Tech. Pap.*, **256**, 1–118.

Kim J.-W. & Cheung S.-C. (1985) On the growth of abalones, *Sulculus diversicolor diversicolor* (Reeve) and *S. diversicolor aquatilis* (Reeve) in Cheju Island. *Bull. Mar. Res. Res. Inst., Jeju Nat. Univ.*, **9**, 39–50.

Kimura D.K. (1985) Changes to stock reduction analysis indicated by Schnute's general theory. *Can. J. Fish. Aquat. Sci.*, **42**, 2059–60.

Kimura D.K. (1988) Stock-recruit curves as used in the stock-reduction analysis model. *J. Cons. Perm. Int. Explor. Mer*, **44**, 253–8.

Kimura D.K., Balsiger J.W. & Ito D.H. (1984) Generalized stock reduction analysis. *Can. J. Fish. Aquat. Sci.*, **41**, 1325–33.

Kimura D.K. & Tagart J.V. (1982) Stock reduction analysis, another solution to the catch equations. *Can. J. Fish. Aquat. Sci.*, **39**, 1467–72.

Kojima H. & Isibasi K. (1985) The estimation of population parameters on the stock of Japanese abalone, *Haliotis discus discus* in Tokushima Prefecture. *Bull. Tokai Reg. Fish. Res. Lab.*, **118**, 11–20.

Kojima H., Nakahisa Y. & Tanimoto H. (1977) A study on the stock of Japanese black abalone, *Haliotis discus discus* in Tokushima Prefecture – I. Growth of shells. *Bull. Tokai Reg. Fish. Res. Lab.*, **90**, 25–37.

Kojima H., Nakahisa Y., Tanimoto H. & Isibasi K. (1978) A study on the stock of Japanese black abalone, *Haliotis discus discus* in Tokushima Prefecture – II. Abalone fishery and its population. *Bull. Tokai Reg. Fish. Res. Lab.*, **93**, 45–62.

Leslie P.H. (1945) On the use of matrices in certain population mathematics. *Biometrika*, **33**(3), 183–212.

Leslie P.H. & Davis D.H.S. (1939) An attempt to determine the absolute number of rats on a given area. *J. Anim. Ecol.*, **8**, 94–113.

MacDonald P.D.M. & Green P.E.J. (1988) *User's Guide to Program Mix 2.3: An Interactive Program for Fitting Mixtures of Distributions.* Icthus Data Systems, Hamilton, Ontario.

MacDonald P.D.M. & Pitcher T.J. (1979) Age-groups from size-frequency data: A versatile and efficient method of analysing distribution mixtures. *J. Fish. Res. Bd Can.*, **36**, 987–1001.

McShane P.E., Black K.P. & Smith M.G. (1988) Recruitment processes in *Haliotis rubra* (Mollusca: Gastropoda) and regional hydrodynamics in southeastern Australia imply localized dispersal of larvae. *J. Exp. Mar. Biol. Ecol.*, **124**, 175–203.

McShane P.E. & Smith M.G. (1988) Measuring recruitment of abalone *Haliotis rubra* Leach (Gastropoda: Haliotidae) – comparison of a novel method with two other methods. *Aust. J. Mar. Freshwat. Res.*, **39**, 331–6.

Mace J.T. (1984) *Management of the abalone fisheries of south-eastern Australia.* New Zealand Ministry of Agriculture and Fisheries, Nelson, NZ, unpublished report.

Marín A.V. (1981) Parametros poblacionales y diagnostico de la pesqueria de abulón amarillo (*Haliotis corrugata*) en Bahia de Tortugas, Baja California Sur, Mexico. *Cienc. Pesq.*, **1**(2), 68–79.

Mohn R.K. (1986) Generalizations and recent usages of yield per recruit analysis. In *North Pacific Workshop on Stock Assessment and Management of Invertebrates* (Ed. by G.S. Jamieson & N.F. Bourne). *Can. Spec. Publ. Fish. Aquat. Sci.*, **92**, 318–25.

Muñoz Lopez T. (1976) Resultados preliminares de un metodo para determinar edad en abulónes (*Haliotis* spp.) de Baja California. *Memorias del Symposio sobre Recursos Pesqueros Masivos de Mexico Ensenada BC del 28 al 30 de Septiembre, 1976*, 281–301.

Mottet M.G. (1978) A review of the fishery biology of abalones. *Wash. State Dept Fish. Tech. Rep.*, **37**.

Newman G.C. (1967) Abalone research in South Africa. *Fish. Bull. S. Afr.*, **4**, 28–34.

Nonaka M. & Fushimi H. (1972) Ecological studies on the rocky shore resources off Shizuoka Prefecture – X. On the effects of the regulation of catch amount of abalone off Toji, southern Izu Peninsula. *Bull. Shizuoka Prefecture Fish. Exp. Stn*, **5**, 51–6. [Translated from Japanese in *Can. Trans. Fish. Aquat. Sci.*, 4933, (1983).]

Polovina J.J. (1989) Density dependence in spiny lobster, *Panulirus marginatus*, in the northwestern Hawaiian Islands. *Can. J. Fish. Aquat. Sci.*, **46**(4), 660–5.

Pope J.G. (1972) An investigation of the accuracy of virtual population analysis using cohort analysis. *Int. Comm. Northwest Atl. Fish. Res. Bull.*, **9**, 65–74.

Prager M.H., O'Brien J.F. & Saila S.B. (1987) Using lifetime fecundity to compare management strategies: A case history for striped bass. *N. Am. J. Fish. Manag.*, **7**, 403–9.

Prince J.D. (1988a) Recruitment, growth, mortality, and population structure in a Southern Australian population of *Haliotis rubra* (Mollusca: Gastropoda). *Mar. Biol.*, **100**(1), 75–82.

Prince J.D. (1988b) A method for ageing the abalone *Haliotis rubra* (Mollusca: Gastropoda). *Aust. J. Mar. Freshwat. Res.*, **39**(2), 167–76.

Prince J.D. (1988c) Confirmation of a relationship between the localized abundance of breeding stock and recruitment for *Haliotis rubra* Leach (Mollusca: Gastropoda). *J. Exp. Mar. Biol. Ecol.*, **122**, 91–104.

Prince J.D. (1990) *The fisheries biology of the Tasmanian stocks of* Haliotis rubra. PhD thesis, University of Tasmania, Hobart.

Prince J.D. & Ford W.B. (1985) Use of anaesthetic to standardize efficiency in sampling abalone populations (genus *Haliotis*; Mollusca: Gastropoda). *Aust. J. Mar. Freshwat. Res.*, **36**, 701–6.

Prince J.D., Sellers T.L., Ford W.B. & Talbot S.R. (1987) Experimental evidence for limited dispersal of haliotid larvae (genus *Haliotis*; Mollusca: Gastropoda). *J. Exp. Mar. Biol. Ecol.*, **106**, 243–63.

Ricker W.E. (1945) A method of estimating minimum size limits for obtaining maximum yield. *Copeia*, **1945**(2), 84–94.

Ricker W.E. (1975) Computation and interpretation of biological statistics of fish populations. *Bull. Fish. Bd Can.*, **191**.

Rocha E. & Arreguin-Sanchez F. (1987) Diagnostico de la pesqueria del abulón (*Haliotis* spp.) de la Peninsula de Baja California, México. *Investigaciones Marinas Cicimar*, **3**(2), 65–77.

Roff D.A. (1980) A motion for the retirement of the Von Bertalanffy function. *J. Fish. Res. Bd Can.*, **37**, 127–9.

Rowley R.J. (1989) Settlement and recruitment of sea urchins (*Strongylocentrotus* spp.) in a sea-urchin barren ground and a kelp bed: are populations regulated by settlement or post-settlement processes? *Mar. Biol.*, **100**, 485–94.

Sainsbury K.J. (1977) *Population dynamics and fishery management of the abalone*, Haliotis iris. PhD thesis, Canterbury University, Christchurch, NZ.

Sainsbury K.J. (1980) The effects of individual variation on the von Bertalanffy growth equation. *Can. J. Fish. Aquat. Sci.*, **37**, 241–7.

Sainsbury K.J. (1982a) Population dynamics and fishery management of the paua, *Haliotis iris*. I. Population structure, growth, reproduction, and mortality. *NZ J. Mar. Freshwat. Res.*, **16**, 147–61.

Sainsbury K.J. (1982b) Population dynamics and fishery management of the paua, *Haliotis iris*. II. Dynamics and management as examined using a size class population model. *NZ J. Mar. Freshwat. Res.*, **16**, 163–73.

Sanders M.J. (1982) *Preliminary stock assessment for the abalone taken off the south east coast of Oman, March 24 to April 2.* FAO FI-DP/RAB/80/015/3, Rome.

Schaefer M.B. (1954) Some aspects of the dynamics of commercial marine fisheries. *Bull. Inter-Am. Trop. Tuna Comm.*, **1**, 25–56.

Schiel D.R. & Breen P.A. (1991) Population structure, ageing and fishing mortality of the New Zealand abalone *Haliotis iris*. *Fishery Bull.*, **89**, 681–91.

Schnute J. (1977) Improved estimates from the Schaefer production model: Theoretical considerations. *J. Fish. Res. Bd Can.*, **34**, 583–603.

Schnute J. (1985) A general theory for analysis of catch and effort data. *Can. J. Fish. Aquat. Sci.*, **42**, 414–29.

Shepherd J.G. (1982) A versatile new stock-recruitment relationship for fisheries, and the construction of sustainable yield curves. *J. Cons. Int. Explor. Mer*, **40**(1), 67–75.

Shepherd S.A. (1984) Biology. In *An assessment of the South Australian abalone resource*, pp. 2–13. S. Aust. Dept. Fish., unpublished report.

Shepherd S.A. (1986) Studies on Southern Australian abalone (genus *Haliotis*). VII. Aggregative behaviour of *H. laevigata* in relation to spawning. *Mar. Biol.*, **90**, 231–6.

Shepherd S.A. (1990) Studies on Southern Australian abalone (genus *Haliotis*). XII. Long term recruitment and mortality dynamics of an unfished population. *Aust. J. Mar. Freshwat. Res.*, **41**, 475–92.

Shepherd S.A. & Hearn W.S. (1983) Studies on Southern Australian abalone (genus *Haliotis*). IV. Growth of *H. laevigata* and *H. ruber*. *Aust. J. Mar. Freshwat. Res.*, **34**, 461–75.

Shepherd S.A., Kirkwood G.P. & Sandland R.L. (1982) Studies on Southern Australian abalone (genus *Haliotis*). III. Mortality of two exploited species. *Aust. J. Mar. Freshwat. Res.*, **33**, 265–72.

Sloan N.A. & Breen P.A. (1988) Northern abalone, *Haliotis kamtschatkana*, in British Columbia: Fisheries and synopsis of life history information. *Can. Spec. Publ. Fish. Aquat. Sci.*, **103**.

Sluczanowski P.R. (1984a) A management-orientated model of an abalone fishery whose substocks are subject to pulse fishing. *Can. J. Fish. Aquat. Sci.*, **41**, 1008–14.

Sluczanowski P.R. (1984b) The fishery. In *An assessment of the South Australian abalone resource*, pp. 14–49. S. Aust. Dept Fish., unpublished report.

Sluczanowski P.R. (1986) A disaggregate model for sedentary stocks: The case of South Australian abalone. In *North Pacific Workshop on Stock Assessment and Management of Invertebrates* (Ed. by G.S. Jamieson & N.F. Bourne). *Can. Spec. Publ. Fish. Aquat. Sci.*, **92**, 393–401.

Tanaka K. (1988) Study on the abalone culture in the coast of Awa Region, Chiba Prefecture. *Bull. Jap. Sea Reg. Fish. Res. Lab.*, **38**, 21–132.

Tanaka K., Ishida O. & Tanaka T. (1986) On the lost weight of abalone around Mera Shoal in the southern coast of the Boso Peninsula, Japan. I. On the distribution of abalone. *Bull. J. Reg. Fish. Res. Lab.*, **36**, 49–58.

Tegner M.J. (1980) Multispecies considerations of resource management in southern California kelp beds. In *Proceedings of the Workshop on the Relationship between Sea Urchin Grazing and Commercial Plant/Animal Harvesting*, (Ed. by J.D. Pringle, G.J. Sharp & J.F. Caddy). *Can. Tech. Rep. Fish. Aquat. Sci.*, **954**, 125–43.

Tegner M.J. (1989) The California abalone fishery: Production, ecological interactions, and prospects for the future. In *Marine Invertebrate Fisheries: Their Assessment and Management* (Ed. by J.F. Caddy), pp. 401–20. Wiley, New York.

Tegner M.J., Breen P.A. & Lennert C.E. (1989) Population biology of red abalones (*Haliotis rufescens*) in southern California and management of the red and pink (*H. corrugata*) abalone fisheries. *Fish. Bull.*, **87**(2), 313–39.

Thompson W.F. & Bell F.H. (1934) Biological statistics of the halibut fishery. 2. Effect of changes in intensity upon total yield and yield per unit of gear. *Rep. Int. Fish. Comm.*, **6**.

Tutschulte T.C. (1976) *The comparative ecology of three sympatric abalones*. PhD thesis, University of California, San Diego.

Usher M.B. (1972) Developments in the Leslie matrix model. In *Mathematical Models in Ecology*, (Ed. by J.N.R. Jeffer), pp. 29–60. Blackwell, Oxford.

Vadas R.L. (1968) *The ecology of* Agarum *and the kelp bed community*. PhD thesis, University of Washington, Seattle.

Vetter E.F. (1988) Estimation of natural mortality in fish stocks: A review. *Fish. Bull.*, **86**(1), 25–43.

von Bertalanffy L. (1938) A quantitative theory of organic growth. *Hum. Biol.*, **10**, 181–213.

Walters C.J. (1986) *Adaptive Management of Renewable Resources*. MacMillan, New York.

Watanabe S., Matsunaga R. & Fushimi H. (1982) Age-structured matrix model including catch and restocking. *J. Tokyo Univ. Fish.*, **68**(1–2), 15–23.

Waugh G. (1980) *Management of fisheries: theoretical analysis and applications to prawn and abalone fisheries in Australia*. PhD thesis, University of New South Wales.

Waugh G. (1984) *Fisheries Management: Theoretical Development and Contemporary Applications*. Westview, Boulder, Colorado.

Wyner A.J., Moore J.E. & Cicin-Sain B. (1977) Politics and management of the California abalone fishery. *Mar. Pol.*, **1**, 326–39.

Chapter 21
Mortality in abalone: its estimation, variability and causes

S.A. SHEPHERD *Department of Fisheries, 135 Pirie Street, Adelaide, South Australia*

P.A. BREEN *Ministry of Agriculture and Fisheries, Fisheries Research Centre, Box 297, Wellington, New Zealand*

ABSTRACT: The principal methods of measuring natural mortality (*M*) are: analysis of age- and size-frequency distributions, mark-recapture experiments, examination of death assemblage, population models and the use of correlations between *M* and other life parameters. The methods are described and compared. Estimates of *M* for abalone range from about 0.05 to more than 1.00. There is a weak trend for *M* to increase with decreasing latitude (i.e. from cold to warm temperature waters). Seasonal and inter-annual variations in *M* have rarely been examined. One study has shown wide variations of both kinds, suggesting that *M* may be much more variable than has been assumed. The predators, parasites and environmental hazards of abalone are also reviewed.

RESUMEN: Los principales métodos para estimar la mortalidad natural (*M*) son: análisis de la distribución de frecuencias; experimentos de marca-recaptura; análisis de las conchas, modelos poblacionales junto con la correlación de *M* y otros parámetros del ciclo de vida. Se describen y comparan estos métodos. Las estimaciones de *M* para el abulón oscilan entre 0.05 y ligeramente por arriba de 1. Existe una tendencia débil de *M* a aumentar conforme decrece la latitud (es decir, desde aguas frías hacia aguas templadas). Existen pocas observaciones sobre variaciones estacionales e inter-anuales de *M*. Un estudio ha mostrado amplias variaciones de los dos tipos, sugiriendo que *M* puede presentar una mayor variabilidad, de la que se había supuesto. También se comenta sobre los depredadores del abulón.

INTRODUCTION

Modern, non-equilibrium theories of populations have greatly extended, if not replaced, the classical view that populations are stable. The physical and biological influences on populations are not constant from year to year or from place to place, so there is substantial temporal and spatial variation in the parameters that describe the dynamics of these populations. The natural mortality rate, *M*, is among the most important of such parameters but remains poorly understood.

Ideally, management of abalone population requires knowledge of *M* for all phases of the life history: larval, juvenile and adult. The relation between a stock and recruits depends on knowing *M* during the larval and juvenile phases, and nearly all models used to describe the dynamics of abalone stocks require estimates of *M* in the adult phase.

Given that *M* is likely to vary in time and space, exact knowledge is an unattainable goal. Instead, we need to understand the scale and extent of its variability and the conditions or factors that cause variation. Furthermore, we need to appreciate what effects deviations from the *M* estimates used in models are likely to have on the results.

Because M is one of the most difficult of population parameters to measure, this review considers first the methodology. Next we review published studies of mortality in abalone and examine various hypotheses that may account for variability in M. Finally we review the classes of predators reported for abalone from settlement to adulthood, the harmful commensals and parasites of abalone and environmental hazards. We use the standard terminology of Ricker (1975); the reader is referred to that work and Vetter (1988) for a broader discussion of mortality in fish stocks.

METHODS USED TO DERIVE *M*

M can be estimated in five ways:

(1) by analysing single, or sequences of, age- or size-frequency distributions,
(2) by performing mark/recapture experiments,
(3) by examining the death assemblage,
(4) by fitting population models, that usually include catch and effort data, and
(5) by using correlations between M and other life-history parameters.

Where there is fishing mortality of the ages, sizes or marked individuals used, then methods (1) and (2) will measure the total mortality rate, Z.

Age/size-frequency distributions

Abalone shells of some species can be aged by interpreting growth checks or layers of nacre in the shell (Chapter 12, this volume). Where the method is valid, length-frequency distributions can be converted to age-frequencies (for example Chapter 23, this volume). A number of methods allow estimation of M from the age-frequency distribution or 'catch curve' (see Ricker, 1975, and Gulland, 1983, for comprehensive discussions). A common method is that of Chapman & Robson (1960). This method assumes a stable age structure (i.e. constant recruitment and mortality rates) and a stable population size. Stable age structures are rare, but if variations in cohort strength are random and not excessive, M can still be estimated. Ricker (1975) suggests combining catch curves from successive years.

Different cohorts are not always equally available to sampling, and usually smaller abalone are less available than larger ones (Shepherd, 1990). For this reason it is important to include only that part of the age-frequency distribution whose components are equally available. At the other end of the distribution, growth checks may become increasingly obscured, eroded or obliterated by boring organisms (Chapter 12, this volume).

Age determination has an element of imprecision: different workers may estimate age differently from the same shell material, and repeated determinations by the same reader may also differ (Beamish & Fournier, 1981). If aging errors are unbiased, then the method of Chapman & Robson (1960) is robust to aging errors. However, Barlow (1984) showed that improper exclusion of age classes from the analysis leads to biased estimates of mortality. For several reasons, therefore, the age classes included in a catch-curve analysis must be chosen very carefully.

Where abalone cannot be aged, numerous methods to estimate M have been developed by using size-frequency distributions and a knowledge of the growth rate. Because the size distribution of animals taken or measured is obviously related to growth and mortality rates, M can be estimated from size parameters of the catch, if growth rates are known. Most of these methods require that growth be adequately described by the Brody-Bertalanffy growth equation and that M be constant over the size range being examined. The assumption that the population has a stable age structure, referred to above, also applies. The simplest of these methods are listed in Table 21.1. Smith (1972) described another method and estimated mortality of *Haliotis rufescens* to be $M = 0.05$, and for *H. cracherodii* $Z = 1.0$.

Fournier & Breen (1983), following Schnute & Fournier (1980), developed a non-linear procedure to estimate simultaneously the parameters of the growth equation and M or Z from a size–frequency distribution. Because the number of solutions for a given size–frequency distribution is infinite, Fournier & Breen (1983) imposed constraints on parameter values in the light of prior biological knowledge; they placed bounds on the Brody coefficient, K, and upper and lower limits on the standard deviations of lengths of individuals in the first and last age classes. The growth parameters, K and L_∞, can be estimated from field experiments,

Table 21.1 Formulae based on size-frequency distributions and the Brody-Bertalanffy growth parameters for estimating M.

Equation	Parameters	Comment
(1) $Z = K(L_\infty - l)/(l - l_c)$ (Beverton & Holt, 1957; and see Gulland, 1983)	Z: total mortality; K, L_∞: von Bertalanffy growth parameters; l: mean length of animals in sample; l_c: length at first capture	Used by Sainsbury (1982) for *H. iris* and found satisfactory. Formulation is seldom used; but Le Guen (1971) shows it compares well with other *L*/*F* distribution-based methods
(2) $Z = \frac{n}{n+1} K/\ln(l - l_c/L_\infty) - \ln(l - l/L_\infty)$ (Ssentongo & Larkin, 1973)	As above	Used by Sainsbury (1982), but tended to give high values. But see Hoenig *et al.* (1983)
(3) $Z = -gs + K$ (Ricker, 1975; Van Sickle, 1977)	g: mean annual increment over range of sizes examined; s: slope of regression of natural logarithm of length frequency on length	Used by Breen (1980) with variable results

and used to force estimates of *M*. Breen & Fournier (1984) provide guidelines on use of the method. They point out that the model only applies to unfished populations, populations fished for a long time at a constant rate or virgin populations fished for a very short time. Fournier & Breen (1983) performed analyses on simulated data and found their method reasonably robust to recruitment variability and random individual variation in growth rate.

Instead of analysing a single size–frequency distribution it may be possible to examine the same cohort, or group of cohorts, over time. The methods of Ebert (1986) and Damm (1986) examine changes in mean length of the catch over time to estimate growth and mortality. Other methods use more of the length–frequency data. If the cohorts are distinct the growth rate can be obtained by plotting the modal progressions over time. If the growth rate is known, a sequence of polymodal size–frequency distributions can be decomposed into year classes and the results treated as equivalent to an age–frequency distribution. (The Fournier & Breen method combines the two steps in a single analysis.) A popular polymodal analysis is the MIX program (MacDonald & Green, 1988) and practical advice on its use is given by MacDonald (1987) and Grant *et al.* (1987) (see Chapter 12, this volume). Sainsbury's (1982) model is a more automated version of the above procedures, useful when the modes of size–frequency distributions are indistinct; it assumes neither a stable age nor size structure (i.e. a constant recruitment).

It is rarely possible to distinguish older cohorts in size–frequency distributions (e.g. Prince *et al.*, 1988, and Fig. 21.2), but they can be combined and the decline in total numbers measured over time. Shepherd (1990) used this technique, after separating fully- from partially-recruited age classes by knife-edge selection, and found that the resulting estimates of *M* agreed well with those obtained by mark/recapture experiments. Tutschulte (1976) used both a life-table method (after conversion from length to age based on polymodal analysis) and a comparison of estimated age structures in the same population in two successive years for *H. fulgens* and *H. corrugata* to calculate *M*.

Most of the methods discussed above assume a stable age structure and a constant *M* for all ages. If this is not the case, then mortality must be estimated by following the abundance of cohorts through time. The length-based method MULTIFAN (Fournier *et al.*, 1990) allows simultaneous analysis of size–frequency distributions observed at different times.

Mark/recapture experiments

In this section we discuss variations of mark/recapture techniques that have been found useful in abalone studies. For a more general discussion of mark/recapture the reader is referred to the reviews of Cormack (1968) and Seber (1973, 1986).

Cormack (1964) described a method for estimating *M* in which animals are marked and individually identified and located on subsequent visits without recapture. Assumptions of the method are that *M* is constant and that the probability of sighting an individual is independent among visits. The method was used by Shepherd *et al.* (1982) to estimate *M* for two abalone species at two sites over several years. Because there was evidence that the second assumption might be violated, a model of Sandland & Kirkwood (1981) was used to test for significant dependence between sightings. A problem of the technique is that if the abalone is a cryptic species, sighting probabilities may be variable, and the standard errors of the estimates consequently large. Low or variable sighting probabilities will always cause problems by increasing 'noise'; for example, Shepherd *et al.* (1991) found low sighting probabilities of *H. fulgens* of about 10% and could obtain no significant estimate of M.

Interpretation of the results of transplant, outplant and mark/recapture studies is difficult when, as is usual, recovery of both live animals and empty shells is incomplete. Thus the rates of mortality and emigration cannot be separated. Typically, some transplanted individuals emigrate from the study area or become so cryptic that they are not recovered. If they subsequently die, their shells may not be recovered either. Shells of animals who die within the study area may be crushed or carried off by predators, or may simply disappear. This problem has been handled in various ways by different authors. For instance, Henderson *et al.* (1988) recovered alive only 5% of their original transplant of *H. corrugata* in California and they recovered another 18% as dead shells. They concluded that the remainder were probably taken illegally by recreational divers. Ebert & Ebert (1988), in discussing the large percentage of missing *H. rufescens* in their study, suggested using the dead-shell recovery as an estimate of mortality. However, they listed several reasons why dead-shell recovery must also be incomplete, and they concluded that estimated survival (calculated as 1.0 − the mortality estimate) 'seems inordinately high'.

In evaluating a transplant of *H. kamtschatkana* in British Columbia, Emmett & Jamieson (1989) considered that dead shells could be used to estimate the lowest possible mortality, under the assumption that all dead shells had been recovered; and conversely that the greatest possible mortality could be calculated from the number of living abalone recovered, under the assumption that all living animals had been recovered. In this way they 'bracketed' the real mortality rate. In this example the minimum and maximum estimates were widely different (0.27 and 1.42 respectively at one site). In situations where a smaller proportion of abalone remains 'unaccounted for' the two estimates would be closer together and more useful. This approach has the advantage that both live and dead shells are taken into consideration, and thus is more objective than studies where either live or dead recoveries are chosen as the basis for the mortality estimate.

If possible, the mortality rate at the beginning of a transplant should be estimated separately from the subsequent mortality, as the former is likely to be substantially higher because of stress, injury and predator interest. Inoue (1976) suggests that numbers of surviving transplanted *H. sieboldii* and *H. gigantea* decreased for about 30 days, then stabilized. Tegner & Butler (1989) list a number of opportunist predators and scavengers attracted to abalone transplants, but which are not ecologically important predators of abalone otherwise. Momma (1972) reported that a high proportion of abalone injured during transplant are subsequently eaten. Tegner & Butler (1989) recommend monitoring transplants for dead shells in the early stages as an effective assessment procedure, although they list several mechanisms preventing the complete recovery of dead shells. They suggest that shells demonstrating no growth can be considered transplant mortalities.

An ingenious method which overcomes many of the above problems is that of

Beinssen & Powell (1979). Individuals are marked and placed in the cells of a grid at successive time intervals. At the conclusion of the experiment, the whole grid is exhaustively searched and the cell of recovery of marked abalone is noted. Alternatively, marked animals can be placed in the grid at the beginning of the experiment and the grid searched at successive times without removal of marked animals. The former procedure is better because it disturbs the habitat less and does not increase density as much. The data enable an estimate to be obtained of the probability of a marked abalone remaining within the grid. This information is then used to weight the ratio of number of recaptures to number marked, from which M is derived by regression analysis. Care must be taken in setting up the grid so that the diagonal of the grid is larger than the maximum expected movement of abalone. If the grid is too small, the probability of animals staying in the grid will be overestimated and consequently M will be overestimated. Shepherd & Godoy (1989) conducted searches outside the grid to reduce the bias from underestimating the extent of movement beyond the grid. Beinssen & Powell (1979) also assumed random movements within the grid, but failure to meet this assumption can be allowed for in the experimental design and analysis (Shepherd & Godoy, 1989).

Both methods allow simultaneous measurement of M, growth and movement – an important consideration in light of the need to maximize gains for a given cost. However, Beinssen & Powell's technique appears to have advantages in versatility and cost efficiency over other mark/recapture methods. Problems likely to be encountered are migration of marked animals beyond the maximum estimated and low or variable sighting probabilities. If there is uncertainty on either matter, a pilot experiment is a prerequisite and well worth the effort.

Tags, tag-loss and methods of marking

Reviews of tagging abalone are given by Mottet (1978) and McShane (1989). The most satisfactory modern methods of marking are with numbered plastic or metal tags either (a) glued to the shell with cyanoacrylate (superglue) or epoxy, or (b) fixed to a pore hole with a nylon rivet (Prince, 1991) or stainless-steel wire. Method (a) requires marking to be done out of water, so may incur marking mortality due to exposure or unsatisfactory replacement on the bottom (see Shepherd & Godoy, 1989). Method (b) can be used underwater in order to minimize marking mortality, but is not suitable for animals less than about 50 mm long; the use of wire may also interrupt growth (Chapter 12, this volume) but otherwise may be quite satisfactory (Emmett & Jamieson, 1989). For hatchery-reared juveniles, tagging with diet manipulation (Olsen, 1968) is simple and effective; alternatively, colour-coded spots of epoxy can be used, providing easier recognition underwater (R. Day, pers. comm.). Tarr (1986) overcame the problem of overgrowth of the tag in *H. midae* by using copper discs (set in epoxy) which inhibit settlement of epibiota.

A tag-less method is simply to file a notch in the growing edge of the shell (Leighton & Boolootian, 1963; Tegner & Butler, 1989); a variant procedure is to inscribe a number with a fine drill into the shell. However, this type of mark is more likely to become overgrown or obscured than more prominent tags.

Tag loss will almost always occur, and Wetherall (1982) reviews methods of estimating the loss rate from double-marking experiments. Shepherd & Godoy

(1989) found that only about 1% of tags glued with superglue were lost; however Prince (1991) reported a 10–14% loss of riveted tags. Apparently the elevated pores of *H. rubra* can be broken by the animal rubbing its shell against rock, so dislodging the tag; this was not a problem with *H. laevigata*, whose pore-holes are not elevated.

Estimate *M* from the death assemblage

Conceptually this method (see review by Powell *et al.*, 1989) is elegant because predators might be expected to leave the shell intact, or eject it whole or fractured. Systematic collection of dead shells, if aged or subjected to length–frequency analysis, should then give an estimate of *M*. Problems that always need to be addressed are emigration of the living abalone, removal of dead shells by predators (Tegner *et al.*, 1989; Shepherd, 1990) and disappearance of shells by dissolution, breakage or burial.

The method appears to have been successfully applied in four studies. Tegner & Butler (1985) found that dead shells of 40–80 mm shell length (SL) gave an estimate of *M* for that size range of abalone. They cited earlier work by Schmitt and colleagues showing that marked dead shells disappeared at the rate of 29% in 4 months, and considered transport of shells from the site by predators unimportant. Tutschulte (1976) used data on the growth rate of serpulids on the inside of dead shells to infer date of death; he calculated *M* for three successive 4-month periods for *H. fulgens* and *H. corrugata* for ages 1–7 years. Hines & Pearse (1982) marked dead shells to establish that shells were not transported by storms out of the study site and concluded that the production rate of fresh empty shells gives the best estimate of *M*. However, Shepherd (1990) found a high disappearance rate of shells during predation; small ones were consumed by wrasses and large ones by stingrays. He found that dead-shell collections were more useful for estimating the proportion of abalone taken by different predators, rather than for measuring *M* itself.

Population models

It has long been known that catch rates are poor measures of abundance of abalone (reviewed by Sloan & Breen, 1988, and in Chapter 20, this volume). For this reason, classical fisheries models that depend on the assumption that catch rates reflect stock abundance must be used with great caution for abalone. There is a need for more realistic models that take into account the aggregative behaviour of abalone and diver behaviour (see review in Chapter 20, this volume). Several models have been applied to abalone and have yielded estimates of *M*. Doi *et al.* (1977) and Marín (1981) plotted yield against biomass for the virgin stock and, assuming a maximum longevity of 20 years, chose the most likely of several curves calculated with different values of *M*. The method is largely intuitive and its assumptions have not been validated.

Geaghan & Castilla (1986) used a modified DeLury model to estimate simultaneously the catchability, natural mortality and recruitment of the muricid gastropod *Concholepas concholepas*. This model was fitted to time series of two or more years of catch and effort data from each of a number of areas. The authors concluded that the method appeared to give reasonable estimates of some parameters, but cautioned that the three parameter estimates are not independent. Use of this method to determine M might require a prior estimate of at least one of the other two parameters. Deacon (1989) applied an age-structured model of the general class developed by Deriso (1980) and Schnute (1987) to catch and effort data for *H. corrugata* and *H. rufescens* for the period 1951–84. He obtained estimates of $M = 0.35$ for *H. corrugata* and $M = 1.239$ for *H. rufescens*. Deacon attributed the higher mortality rate for *H. rufescens* to the spread of sea otters into the red abalone range after 1968.

Some authors have used models to estimate the total mortality rate Z. Kojima *et al.* (1978) and Kojima & Isibasi (1985) estimated Z for *H. discus discus* with a modified Leslie-DeLury method, using catch and effort time series; their estimates ranged from 1.9 to 2.2, reflecting a very intense fishery. Hamer (1983) estimated Z from growth and size–frequency distributions for *H. rubra* in New South Wales, using the formulae in Table 21.1 (G. Hamer, pers. comm.). Tauti (1981, 1984) and Tauti *et al.* (1984) used a polymodal separation method described by Tanaka in 1956 (reference unavailable) to calculate the survival rate of *H. discus* spp. The ratio of 5-year old (recruits) to 4-year old abalone (pre-recruits) gave an estimated $Z = 1.79$.

Correlation of M with life-history parameters

A number of authors have noticed that M correlates with life-history parameters for many fish stocks (reviewed by Vetter, 1988). These correlations may suggest trends and hypotheses for testing. They may also be useful if a crude estimate of M is needed for a preliminary stock assessment, where better data are absent. However, it is dubious to estimate M from trends seen in other taxonomic groups.

There are few accounts of such correlations for molluscs and none for abalone. Hoenig (1983) examined 28 molluscan stocks and found that M was inversely correlated with longevity. This seems obvious because a population with a low mortality rate should contain more older individuals than would a population with a high mortality rate. Some authors have estimated M crudely from the estimated longevity of the species (before fishing or when fishing is light) using the simpler version of Hoenig's (1983) argument. With an estimate of maximum age one can assume that some small percentage of an original cohort will reach that age, then the equation

$$N_t = N_0 \exp(-Mt)$$

can be rearranged to solve for M:

$$M = -(\ln(N_t/N_0)/t)$$

where t is taken to be the estimated longevity and N_t/N_0 is the proportion assumed to reach the maximum age, usually 0.01 but sometimes 0.05 (e.g. Anonymous, 1990).

Berthou *et al.* (1985) estimated the longevity of *H. tuberculata* in Brittany to be

15 years. Based on 5% of a cohort reaching this age (and with other evidence), they suggested M is of the order of 0.20. Murray & Akroyd (1984) estimated that the longevity of *H. iris* is 60 years; this would imply $M = 0.08$. Sloan & Breen (1988) estimated the longevity of *H. kamtschatkana* to be 50 years, based on observations of sponge infestation of shells. This gives $M = 0.09$.

Alverson & Carney (1975) present a method for estimating M from the Brody growth coefficient K and critical age (the age at which a cohort reaches its maximum biomass). Schultz (unpublished data) estimated critical age as 25% of maximum age, for which he had an estimate for *H. rufescens*, and obtained $M = 0.01$ to 0.05.

Others (e.g. Roff, 1986) have suggested that M is correlated with the Brody growth coefficient, K. Data for *H. rubra* and *H. laevigata* (Fig. 21.1) show a trend of increasing M with increasing K. However, for *H. rubra* the relationship is too diffuse, and for *H. laevigata* there are too few data points to be completely convincing.

Conclusions

There are both easy and sophisticated methods, based on analysis of length–frequency distributions, which give estimates of M. They are based on models subject to numerous restrictive assumptions and their value depends on whether

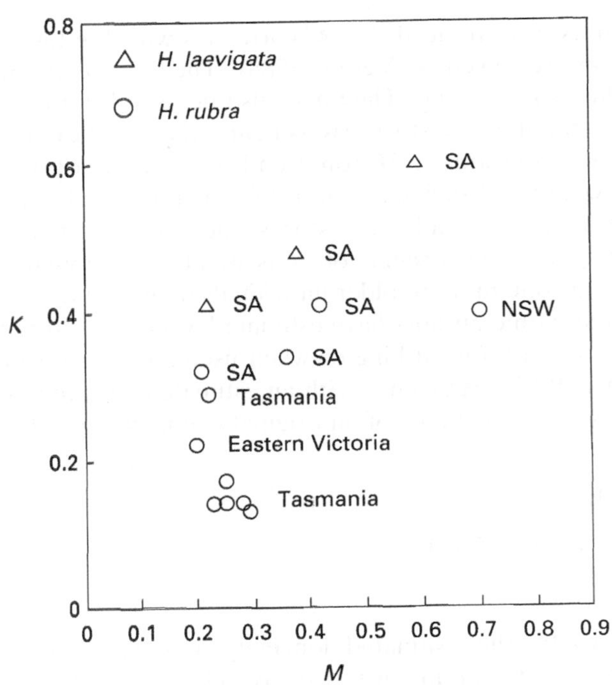

Fig. 21.1 Plot of Brody growth coefficient, K, against M for populations of *H. laevigata* and *H. rubra* in southeastern Australia.

these are met. Aging techniques need critical examination. They require validation (Beamish & MacFarlane, 1983), which is not easy, and may contain bias due to errors or unreadable shells and restrictive assumptions may still apply. Subject to these caveats, they show promise for some species.

Mark/recapture experiments using the Beinssen-Powell design are likely to produce the most accurate estimates of *M* for a given amount of effort. Concurrent dead shell collections can be obtained with little additional effort and will enhance an understanding of the causes of *M*. In such experiments the optimal size of the plot and grid depends *inter alia* on the extent of abalone movement, and the accuracy of estimates of *M* on sighting probabilities, movement and numbers tagged. Little guidance can be found in the literature on any of these matters, so pilot experiments seem essential. There is also much scope for simulation studies to improve the experimental design.

VARIABILITY OF *M*

In this section we tabulate the studies in which *M* has been derived for abalone, first for seed transplants (Table 21.2) and then for wild stocks (Table 21.3). The estimates for seed transplants are generally crude and are based on the recognition of the distinctive shell colour of hatchery-raised seed compared with that of wild seed. Values for *M* of 0.7–1.5 for seed 6–12 months old are common.

Table 21.2 Estimated instantaneous rates of natural mortality (*M*) of abalone seed transplants.

Species with reference	Length range (mm)	M (y^{-1})	Comment
Haliotis discus hannai Saito (1979)	10–20	0.6–0.7	Placed on natural and artificial reefs
Momma *et al.* (1980)	15 21	1.8 1.1	Placed in concrete crib filled with boulders. (S.A. Shepherd, pers. obs.)
Miyamoto *et al.* (1982)	<10 >22	1.7 0.9–1.1	After 491 days in natural habitat
Saito (1984)	Not known	0.7–1.0	On natural reefs
Haliotis discus discus Sakemoto *et al.* (1982) (cited by Tegner & Butler, 1985)	15–50	*ca* 1.0	On concrete artificial reefs
Kojima (1981)	8–18 8–18	0.67 1.3–1.5	Confined in netting cage in channel
H. rubra (R. Whyte & R.W. Whyte, pers. comm.)	9–22	0.5 $month^{-1}$	Released into boulder habitat; handling mortality was 20%
H. rufescens Tegner & Butler (1985)	45–71	0.4–0.5	Estimated from dead shell collection; released on natural reefs in *Macrocystis* forest

Table 21.3 Estimated instantaneous rates of natural mortality (*M*) for different species of abalone.

Species	Author	M (y^{-1})	Comment
H. corrugata	Tutschulte (1976)	0.20	>7 years old
		0.70	Aged 1–7 years
	Doi *et al.* (1977)	0.35–0.43	Adults
	Marín (1981)	0.36	Adults
	Deacon (1989)	0.35	Adults
H. fulgens	Tutschulte (1976)	0.53	Aged 2–7 years
	Olsen (1971) (cited in Sainsbury, 1982)	0.07	Adults
H. iris	Sainsbury (1982)	0.10	Adults
H. kamtschatkana	Sloan & Breen (1988) Breen (1980, 1986)	0.15–0.20	Adults
	Olsen (1984) (cited in Sloan and Breen, 1988)	0.36	Aged 1–3 years
	Emmett & Jamieson (1989)	0.15–0.27	Adults, from dead shell collection analysis
		0.48–2.51	Adults, from tag recaptures
H. laevigata	Shepherd (1987)	3.5–4.6	Aged 0–6 months; West Island
		0.2–1.1 (mean 0.64)	Aged 6 months–2$\frac{1}{2}$ years; West Island
	Shepherd *et al.* (1982)	0.22	>3 years old; Tiparra Reef
	Shepherd *et al* (1982)	0.22–0.38	>3 years old; West Island
	Shepherd (1990a)	0.38	>3 years old; mean of 17 years data at West Island
	Shepherd (1986)	0.59*	Aged 2–5 years; Waterloo Bay (3 years of data)
	Shepherd (unpublished data)	0.27	Aged 2–5 years; Waterloo Bay; 87 tagged animals released for 1 year
H. mariae	Sanders (1982)	0.6	Adults; fishery on Omani coast
H. rubra	Beinssen & Powell (1979)	0.20	Adults; a stunted stock in eastern Victoria
	Hamer (unpublished data)	0.7	Adults; several localities in NSW
	Nash (Chapter 23, this volume)	0.19–0.22	Two sites in Bass Strait
	Prince *et al.* (1988)	0.70	Aged 1 year (see Fig 21.2)
	Day & Leorke (1986)[†]	0.91	Aged 0.5–1$\frac{1}{2}$ years
		0.81	Aged 2 years
	McShane (1990)	6.6–10.2	Settlement to 5 months (3 sites)
	Shepherd (unpublished data)	0.42	Analysis of subset of data from Shepherd & Hearn (1983) aged 2–4 years; Waterloo Bay
	Shepherd *et al.* (1982)	0.21	Tiparra Reef; aged 2–6 years
		0.36	West Island; aged 2–5 years

H. scalaris	Shepherd & Godoy (1989)	1.1	Aged 1–3 years
H. rufescens	Tegner *et al.* (1989)	0.15	Emergent adults
	Hines and Pearse (1982)	0.3–1.0	>2 years old in presence of sea otters (lower estimate from data in mapped plots; higher estimate from dead shell collections)
	Deacon (1989)	1.3	Adults (in presence of sea otters)
H. tuberculata	Clavier & Richard 1985	1.7	0+ age class
		0.26	Aged 1–2 years
		0.14	Aged 4 years
		0.26	Aged 10–12 years

* Later analyses suggested a tag loss rate of 0.04, giving a value for *M* of 0.55.
† Recalculated from survival values by R.W. Day (pers. comm.).

Mortality rates estimated from seed transplants should not be extrapolated to wild populations. One reason for this caution is the high initial mortality previously described. Another is that hatchery seed may behave differently from wild abalone and may be more susceptible to predation (Breitburg, 1984 (in Tegner & Butler, 1989); Saito, 1984; Schiel & Welden, 1987; Tegner & Butler, 1989).

Estimates of *M* for wild stocks are given in Table 21.3. Great variability occurs related to age, habitat, density and other environmental differences. Here we present evidence of correlations between *M* and age, density and latitude.

Age

It is usually assumed in fishery models that *M* is constant in the emergent, and later exploited, phase of the abalone's life history. Failure of this assumption affects the results of yield-per-recruit and other models (Chapter 20, this volume). Some authors (Tutschulte, 1976; Hines & Pearse, 1982; Prince *et al.*, 1988b) have suggested that *M* decreases in older age classes, but without strong evidence. Shepherd *et al.* (1982) found no significant difference in *M* for *H. laevigata* over a size range of 60–150 mm, and Sainsbury (1982) concluded that *M* was independent of age for *H. iris* larger than 30 mm. Similarly, data by Nash (Chapter 23, this volume) for Babel I populations disclosed no such trend. On the other hand, Clavier & Richard (1985) provided evidence that *M* increases among the oldest (10–12 year) age classes (Table 21.3). At present, the assumption of constant *M* during the adult phase of the life history appears to be a reasonable one.

The change in *M* through all age classes has been seldom examined. The best data available are for *H. tuberculata*, *H. rubra* and *H. laevigata* (Fig. 21.2). Measurement of *M* from, or soon after, the time of settlement is the most difficult to obtain and, so far, has been estimated only by McShane (1990), and McShane *et al.* (1990). However, Shepherd (1987) and Prince (1989) both obtained data from an early post-settlement stage when abalone were about 1 mm long and 2–4 weeks old.

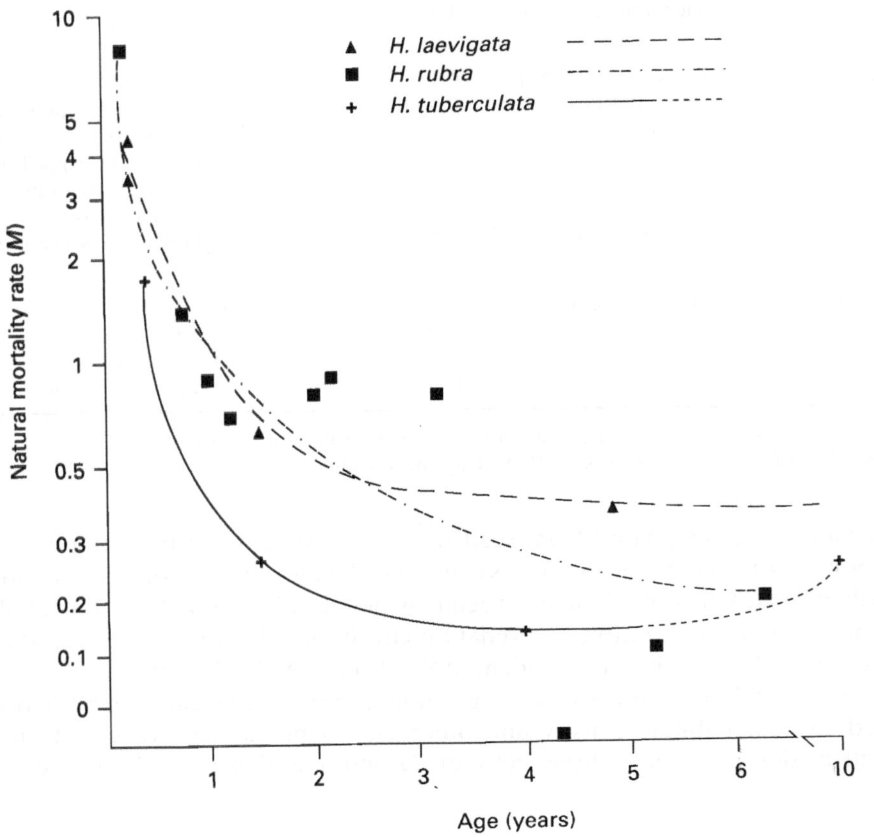

Fig. 21.2 Plots of M against age for *H. laevigata* in South Australia, *H. rubra* in Victoria and Tasmania and *H. tuberculata* in France. Data are given in Table 21.3.

Density

Density-dependent mortality, although well documented for fish stocks, is not well known among abalone. Shepherd (1990) found that M for *H. laevigata* varied from year to year in accordance with changing adult density during a 18-year study, although the addition of data for 3 more years (Fig. 21.3) caused the regression to become non-significant ($0.05 > P < 0.1$); he attributed the variable M to the behaviour of stingrays, the predominant predator of adults.

Peck & Culley (1990) suggested the density-dependent control of recruitment by adult *H. tuberculata*, whose browsing activity, at high densities, smothered recently-settled juveniles. Their data are limited and open to other interpretations, but the question deserves more attention.

There is some evidence for density-dependent M among juvenile abalone. Shepherd (1987 and unpublished data) followed the survival of newly settled *H. laevigata* and *H. scalaris* for 7 years and found high M (3.5–4.5) in years when settlement was heavy. However in other years M was lower, but independent of density. He concluded that these species of abalone conformed to the model

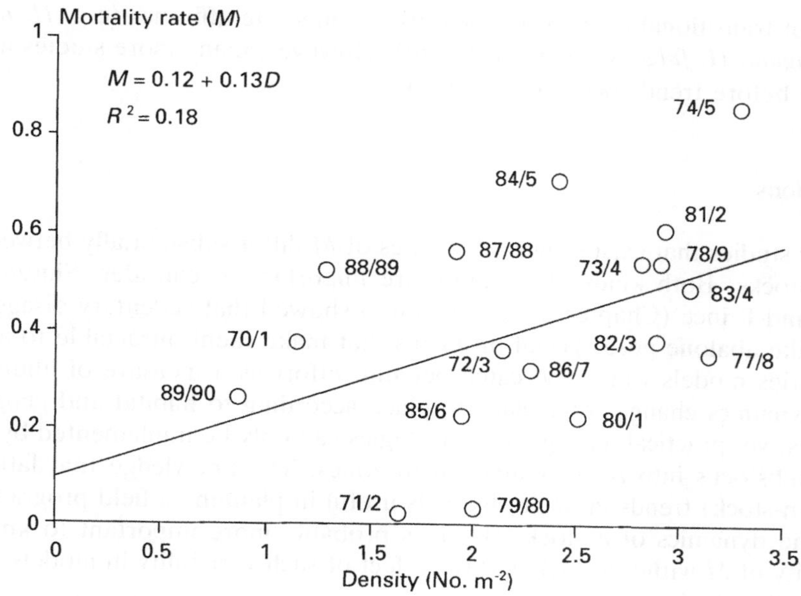

Fig. 21.3 Plot of M against density for *H. laevigata* at West I., South Australia (updated from Shepherd, 1990).

proposed by Connell (1985); when settlement is light, M for juveniles is independent of density, but when settlement is heavy, M is density-dependent. The question has practical implications of some importance, because if stock–recruitment relations are to be elucidated it is necessary to know how much recruitment variability is due to pre-settlement events and how much to post-settlement mortality. The limited evidence available suggests that strong settlements are reflected later in above-average abundances.

Latitude

Abalone occur over a range of latitudes, especially in the northeastern and southwestern Pacific. They have a range of habitats and are subject to predation by different suites of predators. It might be expected that predation should vary along a temperature gradient. Vermeij (1978 and references therein) considered that predation is more effective and more prevalent in tropic than in temperate seas. This appears to be true at least for some molluscs (Palmer, 1979). However, Choat (1982) warns that any conclusions about latitudinal trends must take geographical faunal differences into account. For example, myliobatid rays, which are most abundant in temperate waters, could account proportionally for just as much mortality in abalone as other predators in warmer waters. The best data for abalone comes from the numerous studies on emergent *H. rubra* between 35° and 42°S, where M declines from about 0.7 to 0.2 concomitantly with the growth parameter K (Fig. 21.1). Other evidence can be found in Table 21.3. Low M values of 0.1–0.2 seem typical for species such as *H. kamtschatkana* and *H. iris* in cool temperate waters, whereas higher values of 0.3–0.6 are common among species of transitional cool–warm or warm–temperate affinities (e.g. *H. mariae*, *H. corrugata*, *H. fulgens* and *H. laevigata*). However, many more studies need to be done before trends become convincing.

Conclusions

The few studies that exist suggest that rates of *M* differ substantially between and within stocks. Both kinds of variation are important to consider. Sluczanowski (1986) and Prince (Chapter 22, this volume) showed that sedentary disaggregate stocks like abalone pose special problems that make them intractable to analysis by fisheries models which use catch per unit effort as a measure of abundance. Their dynamics change from place to place according to habitat and geographic variables, yet practical management strategies can only be implemented by aggregating substocks into larger management zones. The knowledge that latitudinal (between-stock) trends in *M* are likely is useful in planning a field programme to study the dynamics of a stock. But it is probably more important to know the variability of *M* within a stock and the effect of such variability in models used to describe the stock.

Sluczanowski (1984) found that higher estimates of *M* will lead to lower estimates of the optimal age at first capture, which in turn is critical in order to maintain an appropriate egg production. He emphasized the need for accurate estimates of *M*. However, differences in *M* made little difference to egg-per-recruit analyses. Other studies of fish stocks (reviewed by Vetter, 1988) show that errors in estimates of *M* propagate into roughly equal errors in estimates of yield-per-recruit, but with the sign reversed. Because there are likely to be density-dependent effects and between-habitat variation, it seems important for analysis of a given stock to do experiments that not only give accurate and representative estimates of *M*, but explore other likely causes of variation. Preferably two independent methods should be used at a number of locations, each replicated and over several years.

CAUSES OF MORTALITY

Isolating the effect of different classes of predators from other causes of mortality is not easy. Studies of the death assemblage (see above) provide the best evidence of predation and its causes for abalone over about 50 mm long, when the shell is generally strong enough to resist breakdown and dissolution, at least for a time. With decreasing size of shell the value of the method decreases as shells are increasingly fragile. In this section we summarize information on causes of mortality from (a) catastrophic natural events, disease, handling and pollution and (b) predators of abalone from the newly settled phase to adult size.

Catastrophe, disease, handling and pollution

Episodic high or low temperatures, low salinities, storm, disease and pollution may each cause increased mortality in abalone.

The survival of abalone larvae may vary with water temperature (Chapter 11, this volume). Hayashi (1980) and Forster *et al.* (1982) suggested that larval *H. tuberculata* had a decreased survival in below-average temperatures, resulting in poor ormer settlement in exceptionally cold years. Beaudry (1983) examined larval survival in *H. kamtschatkana* and found an inverse relation between temperature and survival over the range 14–21°C. Elevated temperatures can be a stress that promotes bacterial infection in intensive abalone culture (Elston & Lockwood, 1983).

The warm East Australia current occasionally penetrates to the west of Tasmania, elevating sea temperatures to over 18°C and causing high mortality to *H. rubra*, possibly from depletion of oxygen in surface waters (Winstanley, 1972). A physiological problem known as 'blue spot', caused by abnormally high water temperatures in the wild or by poor holding conditions after capture, also occurs (Thrower, 1977). In March 1989 large numbers of perlemoen (*H. midae*) were washed ashore at Stony Point, South Africa, when dinoflagellate blooms of *Gymnodinium nagasakiense* were sweeping the coast (R. Tarr, pers. comm.) Apparently the dinoflagellate releases a muscle relaxant causing molluscs, dependent on muscular attachment to the rock, to release their hold. A combination of extreme low tides and warm offshore winds occasionally causes high mortality of intertidal *H. roei* in Western Australia (Hodgkin, 1959; F.E. Wells, pers. comm.).

Heavy storms can also cause mortality. Sainsbury (1982) and Tegner & Butler (1989) observed that sudden changes in sediment level during storms can bury large numbers of abalone. Cox (1962) reported that storms in northern California can cause high mortality to abalone through crushing by dislodged boulders, freshwater runoff, silt and debris. Similarly, Bonnot (1930) reported mortality to *H. corrugata* and *H. rufescens* from spring freshets. Thompson (1914) reported mortality of *H. kamtschatkana* caused by winter storms in British Columbia. Such storms are often accompanied by heavy rainfall, which may reduce surface salinities and cause shallow-living abalone to be dislodged. Unusually severe storms on the New South Wales coast in 1974 rolled boulders and dislodged abalone, causing high mortalities of juvenile and adult *H. rubra* (G. Hamer, pers. comm.).

Larvae of *H. discus hannai* demonstrated tolerance to only a narrow range of salinities, from 31.8–33.4‰ (Nie & Chen, 1984). The range of salinities tolerated appeared to increase as the larvae developed.

Starvation may occur. Ino (1966) suggested that heavy harvesting of brown algae during World War I caused a reduction in abalone populations in Ibaraki Prefecture, Japan. However, Sakai (1962) found a high correlation between Japanese abalone landings and harvest of *Undaria pinnatifida*; he suggested that algal growth varies with the strength of the Kurile current and that abalone abundance varies with it. Tanaka *et al.* (1986) suggested that high mortalities of abalone seen in part of Japan in 1975–80 may have been caused by 'withering' of the brown algal communities in that area and their replacement by red algae. The weight of abalone in the affected area was 20% less than normal, suggesting a shortage of food. In contrast, Tegner (1989) considers that harvesting of *Macrocystis pyrifera* in California can have little effect on abalone. Schmitt & Connell (1984) refer mortality in a transplant of *H. rufescens* to starvation caused by storm damage to kelp availability (Tegner & Butler, 1989).

The dynamics of large-scale environmental events are becoming better understood, and could have major effects on abalone. The El Niño of 1957–9 caused

Macrocystis forest recession in California, and abalone growth was reported to stop (Cox, 1962). The El Niño of 1982–4 was accompanied by storms that destroyed standing *Macrocystis* beds (Dayton & Tegner, 1984; Tegner & Dayton, 1987) and led to the starvation of *H. fulgens*, reduced gonad growth and reduced recruitment; the recruitment of *H. rufescens* was also suppressed (Dayton & Tegner 1989).

There are two reports of mass mortality from disease. A coccidian-like parasite affects the black abalone *H. cracherodii* in southern California (Chapter 16, this volume), and the protozoan parasite *Perkinsus olseni* affects the South Australian species *H. rubra* and *H. laevigata* (Lester *et al.*, 1988; O'Donoghue *et al.*, 1991). Other diseases have been described in culture situations. Elston & Lockwood (1982, 1983) describe bacterial infections causing morbidity or moribundity in intensive cultures of *H. rufescens*. Hatai (1982) describes a fungus causing lesions in a live-storage facility for *H. sieboldii* in Japan. Bower (1987a,b) describes a protist pathogen causing mass mortalities to young individuals in intensive cultures of *H. kamtschatkana* and *H. rufescens*.

The effect of handling on abalone survival is important in transplant and commercial contexts. Ino (1966) presents a table showing that average survival of a Japanese abalone in air is inversely related to air temperature. At −0.8°C the average survival was 5 days, but Ino cautions that this will probably vary with species, humidity and other conditions, and the physiological condition of the abalones used. James & Olley (1970, 1974) state that the optimum temperature for holding *H. rubra* in air is 6°C. At this temperature the survival rate after 40 hours was 82%. These authors present evidence suggesting that smaller abalone are better able to survive in air than larger ones. Whang & Chung (1977) found experimentally that seedlings of *H. discus* spp. survived better in the shade than in the sun, and best when covered with moist gauze. The thermal tolerance of *H. cracherodii* was examined by Hines *et al.* (1980). The response to heat stress was abrupt; there was only about 1°C difference between no mortality and total mortality.

Bar cutting, the injury caused to sublegal abalone during removal and replacement, can also cause serious mortality. Burge *et al.* (1975) considered this a serious problem for *H. rufescens*, and Tegner *et al.* (1989) concluded that the pattern seen in their size–frequency distributions of *H. rufescens* reflected this source of mortality. Even where the foot is not cut, the practice of divers taking abalone to the surface for measurement, and then not replacing them quickly on an appropriate substratum, can cause sublegal mortality, as observed for *H. kamtschatkana* (P.A. Breen, unpublished data). Regulations in some countries require the abalone to be measured and replaced underwater, but they are difficult to enforce.

The only known report of mass mortality from pollution is that of Martin *et al.* (1977), who identified copper in the cooling water system of a nuclear power plant in California as the responsible agent for abalone deaths in the vicinity of the discharge. Hunt & Anderson (1989) developed sensitive toxicity tests to determine the toxicity of sewage effluents and a reference toxicant, zinc, on larvae of *H. rufescens*.

Predators, harmful commensals and parasites

Predators

Studies of the predators of abalone have been carried out in several parts of the world. Clavier & Richard (1985) list the predators of *H. tuberculata* on the French Brittany coast; earlier notes were given by Crofts (1929) and Stephenson (1924). The predators of Californian species are known from the classic work of Cox (1962) and the later review of Tegner & Butler (1989). Sloan & Breen (1988) review the predators of *H. kamtschatkana* in Canada. In Japan lists of predators are given by Ino (1966), Inoue (1976), and Hayashi (1988). In South Australia, Shepherd (1973) lists the known predators; Mower & Shepherd (1988) and Shepherd (1990) provide further data.

Rather than reproduce these lists, we synthesize information on the predators which have been found to be important in the three types of habitat occupied by abalone. These are:

(a) the crustose coralline habitat of newly-settled abalone,
(b) the crevice or under-boulder habitat of cryptic juveniles or small abalone and
(c) the cave or open habitat of large or adult abalone.

The definitions of these habitats are not precise, but are adequate for us to show broad phyletic patterns of predators in relation to habitat type.

Crustose coralline habitat

Very little is known about the predators of very small abalone. In principle, any predator that can capture plankton to about 0.5 mm size is a potential predator of settling abalone larvae. This would include terebellid annelids, anemones and stony corals. Other hunting predators which might take small abalone include polyclad flat worms, some nematodes and eunicid, glycerid and polynoid annelids. Of the above groups, Morse *et al.* (1979) found that terebellids were very active predators of settling larvae, but they also implicated other annelids and small (<10 mm) nematodes. Shepherd & Gray (1986) found eggs and gastropod larvae in the gut of a common anemone *Anthothoe albocincta* and considered it a likely predator of abalone larvae. J. Turrubiates (pers. comm.) noted that an unidentified polyclad flatworm ate many small abalone in a hatchery. The effect of such predators on settlement strength of abalone is unknown. Shepherd & Turner (1985) could find no detectable difference between the numbers settling on substrata from which predators had been removed and the numbers on substrata where the density of predatory anemones was about 12 m^{-2}. On the other hand, Morse *et al.* (1979) found that predators could consume the majority of settling larvae.

Cryptic habitat

The importance of shelter for juveniles is shown by studies of the relation between substratum type and survival of transplanted seed. Tanaka (1988) suggested that the relation between shell height of transplants and crevice space is critical for survival. Similarly, Shepherd (1973) found that *H. roei* lived only in crevices too small for its crab predators to enter. Ebert & Ebert (1988) used a protective 'seeding module' which is transplanted with acclimated seed, and found that the procedure reduced injury to the seed, the interest of opportunistic predators and subsequent predation. In laboratory experiments with predatory fishes and lobsters, Murray & Akroyd (1984) found that predation on *H. iris* decreased when rocks were provided for shelter.

The spine canopy of sea urchins may also provide a refuge for small abalone. Juveniles (3–10 mm) of the South African species *H. midae* commonly shelter under the sea urchin *Parechinus angulosus* (Tarr, 1989), Californian species under the sea urchin *Strongylocentrotus franciscanus* (Tegner & Dayton, 1977), and transplanted Japanese *H. discus* under the sea urchin *Anthocidaris crassispina* (Kojima, 1981). In the last example, seed survival was higher in the presence of sea urchins, although that was not the only difference between sites.

The recorded predators of cryptic abalone can be divided into four groups: (a) gastropods and octopuses, (b) crabs and lobsters, (c) sea-stars and (d) fishes.

(a) Numerous neogastropods have been reported as predators, but almost invariably the authors note that they attack only aged, weakened or stressed abalone, by inserting the proboscis into the soft tissue. The genera involved are *Nassarius* and *Buccinum* (Clavier, 1983; Clavier & Richard, 1985), *Kelletia* (Tegner & Butler, 1989) and *Pleuroploca* (Shepherd, 1973). However, some drilling species also attack healthy abalone in order to feed on them, without necessarily causing death (Table 21.4). The escape response of some abalone to such predators may reduce the incidence of attack (Parsons & Macmillan, 1979).

Octopuses generally are not important predators of cryptic abalone. An exception is California, where Tutschulte (1976) and Tegner & Butler (1985, 1989) considered them major predators of small and middle-sized abalone. Small, cryptic octopuses may also take small abalone (Shepherd, 1973) but predation is probably minor.

(b) The importance of crabs as abalone predators is well established. Although there is an upper limit to the size of prey they can consume, crabs may select a smaller size that can be more easily handled (reviewed by Vermeij, 1978; Lau 1987). Most abalone taken by crabs are small (<50 mm SL) but large crabs can take abalone up to 200 mm SL (Tegner *et al.*, 1989). Crab predation is often recognizable because the abalone shells are chipped at the margin (Clavier & Richard, 1985; Hayashi, 1988; Mower & Shepherd, 1988), but some crabs consume the shell as well (Tegner & Butler, 1989; Schiel & Welden, 1987). Lobsters can also be important predators of abalone, but the only experimental study is that of Clavier & Richard (1985). Because large fish prey on both abalone and predatory crabs they may compel the two to share a common refuge. Shepherd (1973) describes how the period of feeding activity of fish, crabs and abalone is staggered to minimize interactions. Momma (1980) found in laboratory experiments that feeding behaviour increased the exposure of abalone to predation. From field observations he suggested that abalone react in two different ways to decreased food availability. Some forage for food and become more vulnerable to predators; others remain cryptic and lose weight.

The major genera of decapod crustacea implicated as predators by observation or experiment are: *Macropipus*, *Carcinus*, *Homarus* and *Palaemon* (Clavier & Richards, 1985), *Loxorhynchus*, *Taliepus*, *Cancer* and *Panulirus* (Tegner & Butler, 1989; Schiel & Welden, 1987) *Plagusia*, *Megametope*, *Paguristes*, *Lomis*, *Necto-*

Table 21.4 Animals that bore into abalone shells or drill through the shell to feed, and other parasites of abalone.

Species	Reference	Comment
Sponges		
Cliona lobata	Forster (1967)	70–90% of infested shells
Cliona vastifica *Cliona euryphylla*	Shepherd (1973)	Infestation length- and habitat-dependent.
Cliona celata californiana	MacGinitie & MacGinitie (1949), Hansen (1970) J. DeMartini (pers. comm.)	Infestation common; at one site infestation of *H. rufescens* from 5% at 120 mm SL to >50% at 180 mm
Cliona celata	Clavier & Richard (1985)	60–80% of shells 80–100 mm SL infested
Annelids		
Polydora monilaris *Polycirrus* sp. *Nereis kerguelensis*	Sinclair (1963)	*H. iris* often heavily infested
Polydora spp. Syllidae (5 species) Eunicidae (1 species)	Kojima & Imajima (1982)	Frequency of occurrence from 2–36% in shells
Polydora spp. *Ceratonereis* sp. *Dodecaceria* sp. *Lysidice natalense* *Typosillis prolifera*	Shepherd (1973)	*Polydora* spp. are the primary, and other species secondary, borers
Polydora spp.	Clavier & Richard (1985)	Frequency of occurrence 77% in shells >80 mm SL
Bivalve molluscs		
Lithodomus sp.	Crofts (1929)	Infestation common
Pholadidex parva *Penitella conradi* *Lithophaga subula*	Cox (1962)	These species bore into shells; incidence unknown
Ectoparasitic gastropods		
Odostomia evalea phanea	Cox (1962)	Apparently sucks blood from the region of the mantle
Ceratostoma nuttalli	Tegner & Butler (1989)	Occasional predator of juveniles
Thais baileyana *Thais orbita* *Thais clavigera*	Shepherd (1973) F.E. Wells (pers. comm.) Inoue (1976)	Thaids drill through shell to feed on the foot
Internal parasites		
Cercaria cf. brachyura	Crofts (1929)	A trematode that infests viscera and ctenidia
Echinocephalus pseudouncinatus	Cox (1962)	Often encysts foot and gonad of shallow water species
Digenian trematode (fam. Opecoilidae?)	Keesing (1984)	1–4% infestation of gonads of *H. roei* at 2 sites
Digenian trematode (fam. Allocreadiidae)	Harrison & Grant (1971)	3% infestation of gonads of *H. rubra* at one site

carcinus, *Ozius* and *Jasus* (Shepherd, 1973; Mower & Shepherd, 1988; Riley, 1988), *Pachygrapsus* (J. Turrubiates, pers. comm.) and *Gaetice*, *Thalamitra*, *Etisus*, *Cancer*, *Pugettia* and *Charybdis* (Shibui, 1971; Inoue, 1976; Kojima, 1981; Hayashi, 1988).

(c) Sea-stars are seldom important predators of small abalone. This is probably because abalone have a well-developed escape response to them (Montgomery, 1967; Parsons & Macmillan, 1979). There are two exceptions. Tegner *et al.* (1989) gave evidence of significant predation by sea-stars (genera *Pycnopodia* and *Astrometis*) that share the crevice habitat of *H. rufescens*, and McShane & Smith (1986) recorded high juvenile abalone mortality where the sea-star *Coscinasterias* was abundant.

(d) The fish predators of small abalone are those that fossick or search visually for prey in cryptic habitat (Choat, 1982). Like crabs, they may select the size of prey according to their ability to remove them (Shepherd & Turner, 1985). The species recorded are mostly labrids and in the genera *Labrus*, *Symphodus*, *Blennius* (Clavier & Richard, 1985), *Pseudolabrus*, *Pictilabrus*, (Shepherd, 1973), *Sparus*, *Gymnothorax*, *Semicossyphus*, *Lateolabrax*, *Goniistius*, *Halichoeres*, *Ditrema*, *Limanda*, *Ceratocottus*, *Microstomus*, *Pseudolabrus*, *Astroconger*, *Anguilla*, *Stephanolepis* and *Platichthys* (Ino, 1966; Inoue, 1976).

Non-cryptic habitat

The predators of emergent abalone are octopuses, sea-stars, large fish including eels, rays and sharks, mammals such as sea otters and seals, and shore-feeding birds (see Mottet, 1978).

Octopuses are a major predator of abalone in South Africa (R. Tarr, pers. comm.) and Japan (Kojima, 1988). Ino (1966) reported an inverse relation between abalone catches and octopus abundance, implying a strong predator–prey link. The mode of attack by octopuses apparently differs between species and individuals. Some species drill a characteristic hole (Pilson & Taylor, 1961; Ambrose & Nelson, 1983) in order to inject a toxin, or may physically pull the abalone off the substratum. Kojima (1988) found experimentally that 14% of shells were drilled and suggested that drilling occurred only if removal by force failed. Similarly, Tegner & Butler (1985) found that about 33% of shells were drilled and suggested that only small abalone can be removed by force; those longer than 125 mm SL are not taken at all. *Octopus flindersi* does not drill but smothers the abalone until the foot muscle relaxes: possibly a toxin is involved (Shepherd, 1990). Ino (1966) suggested that the octopus used its beak to break the shell before injecting a salivary toxin. Other papers reporting octopuses (genus *Octopus*) as predators of abalone are by Stephenson (1924), Shepherd (1973), Hartwick *et al.* (1981), Ambrose (1982) and Tegner & Butler (1989).

Sea-stars are generalist predators that capture emergent abalone or scavenge stressed or moribund individuals. Authors do not always distinguish the two and some of the records given below may refer to scavenging. Except in the instances cited previously, sea-stars do not seem to be serious predators. The following predatory genera have been implicated for abalone: *Asterias* and *Orthasterias* (Crofts, 1929; Forster, 1962), *Marthasterias* (Tarr, 1986), *Coscinasterias* (Shepherd,

1973; Hayashi, 1988), *Pycnopodia*, *Astrometis* and *Pisaster* (Cox, 1962; Hines & Pearse, 1982; Tegner & Butler, 1989). The last named authors state that *Pisaster* is a scavenger only.

Large fish, rays and small sharks capture abalone by variously ramming and fracturing the shell, crushing the shell with the teeth or prising the animal from the substratum (Shepherd, 1973). Most authors cite myliobatid rays as serious predators, and Shepherd (1990) estimated that they were responsible for up to 94% of the mortality of one abalone species at one site. The main predatory genera are *Scorpaenichthys* (O'Connell, 1953; Cox, 1962; Tegner & Butler, 1989), *Anarrhichthys* (Emmett & Jamieson, 1989; Sloan & Breen, 1988), *Raja* (Crofts, 1929), *Conger* and *Scylliorhinus* (Clavier & Richard, 1985), *Achoerodus* (Shepherd, 1973), *Heterodontus* (McLaughlin & O'Gower, 1971), *Myliobatis* (Tegner & Butler, 1989; Tarr, 1986; Shepherd, 1990; Inoue, 1976) and probably large species of the fish genera cited as predators of cryptic abalone.

Of the mammals, sea otters are a significant predator whose importance is still a matter of controversy (Cox, 1962, Hines & Pearse, 1982; review by Tegner, 1989). Chumash Indians were also major predators of *H. cracherodii*, but their decline and the depletion of sea-otter populations allowed the development of very high intertidal densities of *H. cracherodii* (Douros, 1987). Seals (e.g. *Neophoca*) also take abalone (S.A. Shepherd, unpublished data) but their importance as predators is not known. Lastly, shore-birds, notably the oyster-catcher (*Haematopus* spp.) and gulls (*Larus*), take abalone from the intertidal zone (Crofts, 1929; S.A. Shepherd, unpublished data) and may be important in some places.

Harmful commensals and parasites

The most common animals inhabiting abalone shells are species of the boring sponge genus *Cliona* (Table 21.4). These species bore into and weaken the shell, so that the abalone more easily falls prey to other predators (Forster, 1967) or breakage by rolling rocks. Reported infestation rates are generally length dependent, becoming high on adults, but can vary markedly between species, habitat and place. Boring annelids of the family Spionidae (reviewed by Blake & Evans, 1973) and boring molluscs of the family Pholadidae often bore into the shell of some abalone species (Table 21.4). Other species of annelid also inhabit and enlarge the galleries made by polydorids. In addition, ectoparasitic gastropods drill through the shell to feed on the abalone foot. All of these animals, to a greater or lesser extent, weaken the shell and render the abalone more susceptible to predation. If the shell is penetrated, the abalone produces nacre on the inside of the shell, so diverting energy from growth and reproduction. Kojima & Imajima (1982) found that the body weight of an abalone whose shell was entered by 10 or more *Polydora* was significantly lower than that of an abalone with fewer or no *Polydora* in the shell. On the other hand, Clavier & Richard (1985) found no significant difference between bored shells and those without borers. However, the latter authors noted that their samples did not include examples of extreme infestation which were likely to suffer physiological stress.

Abalone are occasionally hosts of trematodes that complete their life cycle in a vertebrate host (Fretter & Graham, 1962; Hyman, 1951). The few infestations reported (Table 21.4) occur in the muscular foot or gonad, which in serious cases can be completely destroyed. However, the incidence of infestation appears to be low, and in any event does not appear to affect the health of the animal.

ACKNOWLEDGEMENTS

The authors are especially grateful for the criticism of Drs N.L. Andrew, R.W. Day and M.J. Tegner, as well as discussions with many colleagues.

REFERENCES

Alverson D.L. & Carney M.J. (1975) A graphic review of the growth and decay of population cohorts. *J. Cons. Int. Explor. Mer*, **36**, 133–43.

Ambrose R.F. (1982) *Octopus predation and community structure of subtidal rocky reefs at Santa Catalina Island, California*. PhD dissertation, University of California, Los Angeles.

Ambrose R.F. & Nelson B.V. (1983) Predation by *Octopus vulgaris* in the Mediterranean. P.S.Z.N.I. *Mar. Ecol.*, **4**, 251–61.

Barlow J. (1984) Mortality estimation: biased results from unbiased ages. *Can. J. Fish. Aquat. Sci.*, **41**, 1843–7.

Beamish R.J. & Fournier D.A. (1981) A method for comparing the precision of a set of age determinations. *Can. J. Fish. Aquat. Sci.*, **38**, 982–3.

Beamish R.J. & MacFarlane G.A. (1983) The forgotten requirement for age validation in fisheries biology. *Trans. Am. Fish. Soc.*, **112**(6), 735–43.

Beaudry C.G. (1983) Survival and growth of the larvae of *Haliotis kamtschatkana* Jonas at different temperatures. *J. Shellfish Res.*, **3**, 109–10.

Beinssen K. & Powell D. (1979) Measurement of natural mortality in a population of blacklip abalone *Notohaliotis ruber*. *Rapp. PV Reun. Cons. Int. Explor. Mer*, **175**, 23–6.

Berthou P., Beurier J.P., Buestel D. & Clavier J. (1985) Etat actuel des connaissances sur les stocks naturels d'ormeau: Eléments halicutiques et juridiques. *La Pêche Maritime*, **1286**, 288–94.

Beverton R.J.H. & Holt S.J. (1957) On the dynamics of exploited fish populations. *UK Min. Ag. Fish. Food Fish. Invest. (Ser. 2)*, **19**, 1–533.

Blake J.A. & Evans J.W. (1973) *Polydora* and other genera as borers in mollusc shells and other calcareous substrates. *Veliger*, **15**, 235–49.

Bonnot P. (1930) Abalones in California. *Calif. Fish Game*, **16**, 15–23.

Bower S.M. (1987a) *Labyrinthuloides haliotidis* (new species: Protozoa: Labyrinthomorpha), a pathogenic parasite of small juvenile abalone in a British Columbia, Canada mariculture facility. *Can. J. Zool.*, **65**, 1996–2007.

Bower S.M. (1987b) Pathogenicity and host specificity of *Labyrinthuloides haliotidis* (Protozoa: Labyrinthomorpha), a parasite of juvenile abalone. *Can. J. Zool.*, **65**, 2008–12.

Breen P.A. (1980) Measuring fishing intensity and annual production in the abalone fishery of British Columbia. *Can. Tech. Rep. Fish Aquat. Sci.*, No. 947.

Breen P.A. & Fournier D.A. (1984) A user's guide to estimating total mortality rates from length frequency data with the method of Fournier and Breen. *Can. Tech. Rep. Fish. Aquat. Sci.*, No. 1239.

Burge R., Schultz S. & Odemar M. (1975) *Draft report on recent abalone research in California with recommendations for management*. Mimeo, California Department of Fish and Game.

Chapman D.G. & Robson D.S. (1960) The analysis of a catch curve. *Biometrics*, **16**, 354–68.

Choat J.H. (1982) Fish feeding and the structure of benthic communities in temperate waters. *Ann. Rev. Ecol. Syst.*, **13**, 423–49.

Clavier J. (1983) (Initial data on natural ormer (*Haliotis tuberculata*) stocks in the St Malo region.) Association pour la Mise en Valeur du Littoral de la Côte l'Emeraude. Study report. (Translated from French in *Can. Transl. Fish. Aquat. Sci.*, **4928**, (1983).)

Clavier J. & Richard O. (1985) *Etudes sur les Ormeaux dans la Région de Saint-Malo*. Association

pour la Mise en Valeur du Littoral de la Cote D'Emeraude, Dinard, France.
Connell J.H. (1985) The consequences of variation in initial settlement vs. post-settlement mortality in rocky intertidal communities. *J. Exp. Mar. Biol. Ecol.*, **93**, 11–45.
Cormack R.M. (1964) Estimates of survival from the sighting of marked animals. *Biometrika*, **51**, 429–38.
Cormack R.M. (1968) The statistics of capture-recapture methods. *Oceanogr. Mar. Biol. Ann. Rev.*, **6**, 455–506.
Cox K.W. (1962) California abalones, Family Haliotidae. *Calif. Dept Fish Game, Fish. Bull.*, No. 118.
Crofts D.R. (1929) *Haliotis. LMBC Memoirs on Typical British Plants and Animals*, **29**.
Damm U. (1986) Some modifications of Ebert's method to calculate growth and mortality parameters from average lengths in a population. In *Length-based Methods in Fisheries Research. Vol. I. ICLARM Conference Proceedings 13* (Ed. by D. Pauly & G. Morgan). ICLARM, Manila, Philippines and Kuwait Institute for Scientific Research, Safat, Kuwait.
Day R.W. & Leorke A.E. (1986) Abalone catches – what factors affect them. *Aust. Fish.*, **45**(10), 32–6.
Dayton P.K. & Tegner M.J. (1984) Catastrophic storms, El Niño, and patch stability in a southern California kelp community. *Science*, **224**, 283–5.
Dayton P.K. & Tegner M.J. (1989) Bottoms beneath troubled waters: benthic impacts of the 1982–1984 El Niño in the temperate zone. In *Global Ecological Consequences of the 1982–83 El Niño-Southern Oscillation* (Ed. by P.W. Glynn). Elsevier, Amsterdam.
Deacon R.T. (1989) An empirical model of fishery dynamics. *J. Env. Econ. Manag.*, **16**, 167–83.
Deriso R.B. (1980) Harvesting strategies and parameter estimation for an age-structured model. *Can. J. Fish. Aquat. Sci.*, **37**, 268–82.
Doi T., Guzmán del Próo S.A., Marín V., Ortiz M., Camacho J. & Muñoz T. (1977) Análisis de la población y diagnostico de la pesquería de abalon amarillo (*Haliotis corrugata*) en el area de Punto Abreojos e Isla Cedros, B.C. *Inst. Nacional de Pesca. Ser. Cientifica*, No. 18.
Douros W.J. (1987) Stacking behavior of an intertidal abalone: An adaptive response or a consequence of space limitation? *J. Exp. Mar. Biol. Ecol.*, **108**, 1–14.
Ebert T.A. (1986) Estimating growth and survival parameters by nonlinear regression using average size in catches. In *Length-based Methods in Fisheries Research. Vol. I. ICLARM Conference Proceedings 13* (Ed. by D. Pauly & G. Morgan). ICLARM, Manila, Philippines and Kuwait Institute for Scientific Research, Safat, Kuwait.
Ebert T.B. & Ebert E.E. (1988) An innovative technique for seeding abalone and preliminary results of laboratory and field trials. *Calif. Fish Game*, **74**, 68–81.
Elston R. & Lockwood G.S. (1982) Pathogenesis and microbiological characterization of a bacterial infection of cultured juveniles of the red abalone *Haliotis rufescens* Swainson. *J. Shellfish Res.*, **2**, 94–5.
Elston R. & Lockwood G.S. (1983) Pathogenesis of vibriosis in cultured juvenile red abalone, *Haliotis rufescens* Swainson. *J. Fish Dis.*, **6**(2), 111–28.
Emmett B. & Jamieson G.S. (1989) An experimental transplant of prerecruit abalone in Barkley Sound, British Columbia. *Fish. Bull.*, **87**, 95–104.
Forster G.R. (1962) Observations on the ormer population of Guernsey. *J. Mar. Biol. Assoc. UK*, **42**, 493–8.
Forster G.R. (1967) The growth of *Haliotis tuberculata*: results of tagging experiments in Guernsey 1963–65. *J. Mar. Biol. Assoc. UK*, **47**, 287–300.
Forster G.R., Potts G.W. & Swinfen R. (1982) Changes in the ormer populations of Guernsey and Jersey. *J. Mar. Biol. Assoc. UK*, **62**, 717–27.
Fournier D.A. & Breen P.A. (1983) Estimation of abalone mortality rates with growth analysis. *Trans. Am. Fish. Soc.*, **112**, 403–11.
Fournier D.A., Sibert J.R., Majkowski J. & Hampton J. (1990) MULTIFAN a likelihood-based method for estimating growth parameters and age composition from multiple length frequency data sets illustrated using data for southern bluefin tuna (*Thunnus maccoyii*). *Can. J. Fish. Aquat. Sci.*, **47**, 301–17.
Fretter V. & Graham A. (1962) *British Prosobranch Molluscs*. Ray Society, London.
Geaghan J.P. & Castilla J.C. (1986) Use of catch and effort data for parameter estimates for the loco (*Concholepas concholepas*) fishery of central Chile. In *North Pacific Workshop on Stock Assessment and Management of Invertebrates* (Ed. by G.S. Jamieson & N. Bourne). *Can. Spec. Publ. Fish. Aquat. Sci.*, **92**, 168–74.

Grant A., Morgan P.J. & Olive P.J.W. (1987) Use made in marine ecology of methods for estimating demographic parameters from size-frequency data. *Mar. Biol.*, **95**, 201–8.

Gulland J.A. (1983) *Fish Stock Assessment: A Manual of Basic Methods. FAO/Wiley Series on Food and Agriculture Vol. 1.* Wiley, Chichester.

Hamer G.T. (1983) NSW abalone stock assessment shows effort should be reduced. *Aust. Fish.*, **42**(8), 7–11.

Hansen J.C. (1970) Commensal activity as a function of age in two species of California abalones. *Veliger*, **13**, 90–4.

Harrison A.J. & Grant J.F. (1971) Progress in abalone research. *Tasmanian Fish. Res.*, **5**, 1–10.

Hartwick B., Tullock L. & MacDonald S. (1981) Feeding and growth of octopus doflcini (Wulker). *Veliger*, **24**, 129–38.

Hatai K. (1982) On the fungus *Haliphthoros milfordensis* isolated from temporarily held abalone (*Haliotis sieboldii*). *Fish Pathol. (Tokyo)*, **17**(3), 199–204.

Hayashi I. (1980) Structure and growth of a shore population of the ormer, *Haliotis tuberculata. J. Mar. Biol. Assoc. UK*, **60**, 431–7.

Hayashi I. (1988) Habitat preference, diurnal activity pattern and predators of the artificially bred juvenile abalone *Haliotis discus discus. Venus*, **47**, 104–20.

Henderson K.C., Parker D.O. & Haaker P.L. (1988) The survival and growth of transplanted adult pink abalone, *Haliotis corrugata* at Santa Catalina Island. *Calif. Fish Game*, **74**(2), 82–6.

Hines A., Anderson S. & Brisbin M. (1980) Heat tolerance in the black abalone, *Haliotis cracherodii* Leach, 1814: effects of temperature fluctuation and acclimation. *Veliger*, **23**, 113–18.

Hines A.H. & Pearse J.S. (1982) Abalones, shells and sea otters: dynamics of prey populations in central California. *Ecology*, **63**, 1547–60.

Hodgkin E.P. (1959) Catastrophic destruction of the littoral fauna and flora near Fremantle January 1959. *West. Aust. Nat.*, **7**, 6–11.

Hoenig J.M. (1983) Empirical use of longevity data to estimate mortality rates. *Fish. Bull.*, **82**, 898–903.

Hoenig J.M., Lawing W.D. & Hoenig N.A. (1983) On estimating the total mortality rate from mean age, mean length or median length. *Proc. Int. Council for Exploration of the Sea*. Copenhagen, Denmark.

Hunt J.W. & Anderson B.S. (1989) Sublethal effects of zinc and municipal effluents on larvae of the red abalone *Haliotis rufescens. Mar. Biol.*, **101**, 545–52.

Hyman L.H. (1951) *The Invertebrates – Platyhelminthes and Rhynchocoela*, Vol. 2. McGraw-Hill, New York.

Ino T. (1966) (The abalone science and its propagation in Japan.) *Nippon Suisan Shigen Hogo Kyokai, Suisan Zoyoshoku Sosho*, **11**. 104pp. (Translated from Japanese in *Fish. Res. Bd Can. Trans Ser.*, **1078**, (1968).)

Inoue M. (1976) Awabi (Abalone) Suisan Zoyoshoku Deeta Bukku (Marine Aquaculture Data Book) Suisan Shuppan. (Translated from Japanese by M.G. Mottet, 1978, in *Wash. Dept Fish. Tech. Rep.*, **37**.)

James D.G. & Olley J. (1970) Moisture and pH changes as criteria of freshness in abalone and their relationship to texture of the canned product. *Food Technol. Aust.*, **22**(7), 350–7.

James D.G. & Olley J. (1974) The abalone industry in Australia. In *Fishery Products* (Ed. by R. Kreuzer), pp. 238–42. Fishing News International, Surrey.

Keesing J. (1984) *Reproductive biology of the abalone* Haliotis roei *Gray 1827 in south-western Australia.* BSc (Hons) thesis, Murdock University.

Kojima H. (1981) Mortality of young Japanese black abalone *Haliotis discus discus* after transplantation. *Bull. Jap. Soc. Sci. Fish*, **47**, 151–9.

Kojima H. (1988) Hole-drilling predation by *Octopus vulgaris* on abalone. *La Mer*, **26**, 115–19.

Kojima H. & Imajima M. (1982) Burrowing polychaetes in the shells of the abalone *Haliotis diversicolor aquatilis* chiefly of species of *Polydora. Bull. Japan. Soc. Sci. Fish.*, **48**, 31–5.

Kojima H. & Isibasi K. (1985) The estimation of population parameters on the stock of Japanese abalone, *Haliotis discus discus* in Tokushima Prefecture. *Bull. Tokai Reg. Fish. Res. Lab.*, **118**, 11–20.

Kojima H., Nakahisa Y., Tanimoto H. & Isibasi K. (1978) A study on the stock of Japanese black abalone, *Haliotis discus discus* in Tokushima Prefecture – II. Abalone fishery and its population. *Bull. Tokai Reg. Fish. Res. Lab.*, **93**, 45–62.

Lau C.J. (1987) Feeding behaviour of the Hawaiian slipper lobster, *Scyllarides squamosus*, with a review of decapod crustacean feeding tactics on molluscan prey. *Bull. Mar. Sci.*, **41**, 378–91.
Le Guen J.C. (1971) Dynamiques des populations de *Pseudotolithus (Fonticulus) elongatus* (Bowd, 1825). Poissons-Sciaenidae. Cah. ORSTOM. Ser. *Oceanography*, **9**, 3–84.
Leighton D.L. & Boolootian R.A. (1963) Diet and growth in the black abalone, *Haliotis cracherodii*. *Ecology*, **44**, 227–38.
Lester R.J.G., Goggin C.L. & Sewell K.B. (1988) *Perkinsus olseni* and other *Perkinsus* infections from Australian molluscs. *3rd Internat. Colloq. Pathol. Marine Aquacul.*, Gloucester Pt, VA.
MacDonald P.D.M. (1987) Analysis of length–frequency distributions. In *Age and Growth of Fish* (Ed. by R.C. Summerfelt & G.E. Hall), pp. 371–84. Iowa State University Press, Ames.
MacDonald P.D.M. & Green P.E.J. (1988) *Program Mix 2.3: An interactive program for fitting mixtures of distributions*. Icthus Data Systems, Hamilton, Ontario.
MacGinitie G.E. & MacGinitie N. (1949) *Natural History of Marine Animals*. McGraw-Hill, New York.
McLaughlin R.H. & O'Gower A.K. (1971) Life history and underwater studies of a heterodont shark. *Ecol. Monogr.*, **41**, 271–89.
McShane P.E. (1989) Tagging abalone. In *Aust. Soc. Fish Biology Tagging Workshop* (Ed. by D.A. Hancock), pp. 65–7. Australian Government Publishing Service, Canberra.
McShane P.E. (1990) *The fisheries ecology of Victorian abalone*. PhD thesis, Latrobe University.
McShane P.E., Black K.P. & Smith M.G. (1990) Recruitment processes in *Haliotis rubra* (Mollusca: Gastropoda) and regional hydrodynamics in south-eastern Australia imply localised dispersal of larvae. *J. Exp. Mar. Biol. Ecol.*, **124**, 175–203.
McShane P. & Smith M. (1986) Starfish vs abalone in Port Philip Bay. *Aust. Fish.*, **45**(4), 16–18.
Marín V. (1981) Parametros poblacionales y diagnostico de la pesquería de abulón amarillo (*Haliotis corrugata*) en Bahia Tortugas, B.C.S. *Ciencia Pesquería. Inst. Nal. Pesca Depto Pesca Mexico*, **1**, 67–79.
Martin M., Stephenson M.D. & Martin J.H. (1977) Copper toxicity experiments in relation to abalone deaths observed in a power plant's cooling waters. *Calif. Fish Game*, **63**(2), 95–100.
Miyamoto T., Saito K., Motoya S. & Kawamura K. (1982) Experimental studies on the release of the cultured seeds of abalone, *Haliotis discus hannai* Ino in Oshoro Bay, Hokkaido. *Sci. Repts Hokkaido Fisheries Experimental Station*, No. 24, 59–89. (English abstract, figures and tables.)
Momma H. (1972) Studies on the release of the seed disk abalone – I. Behaviour immediately after release. *Bull. Jap. Soc. Sci. Fish.*, **38**(7), 671–6.
Momma H. (1980) On the abalone behaviour under the influence of predator. *Seibutsugaka Kenkyu*, **16**(1–2), 60–9.
Momma H., Kobayashi K. & Kato T. (1980) On the artificial propagation method of abalone and its effects on rocky shore. I. Remaining ratio of the artificial seed abalone (*Haliotis discus hannai* Ino) on latticed artificial reefs. *Suisan Zoshoku* (The Aquaculture), **28**, 59–65. (Translation by M. Mottet, State of Washington, Department of Fisheries.)
Montgomery D.H. (1967) Responses of two haliotid gastropods (Mollusca), *Haliotis assimilis* and *H. rufescens* to the forcipulate asteroids (Echinodermata) *Pycnopodia helianthoides* and *Pisaster ochraceus*. *Veliger*, **9**, 359–68.
Morse D.E., Hooker N., Jensen L. & Duncan H. (1979) Induction of larval abalone settling and metamorphosis by gamma-aminobutyric acid and its congeners from crustose red algae: II: Applications to cultivation, seed-production and bioassays; principal causes of mortality and interference. *Proc. World Maricul. Soc.*, **10**, 81–91.
Mottet M.G. (1978) A review of the fishery biology of abalones. *Wash. State Dept Fish. Tech. Rep.*, No. 37.
Mower A.G.J. & Shepherd S.A. (1988) The crab fauna of West Island: their abundance, diet and role as predators of abalone. *Trans R. Soc. S. Aust.*, **112**, 83–6.
Murray T. & Akroyd J.-A. (1984) The New Zealand paua fishery. *NZ Fish. Int. Rep.*, No. 5.
Nie Z.-Q. & Chen W.-H. (1984) Studies on rearing conditions of abalone, *Haliotis discus hannai*, Ino. 2. The effects of salinity and inorganic nutrients on the development of fertilized eggs and living of larvae. *Mar. Fish. Res.*, **6**, 41–8.
O'Connell C.P. (1953) The life history of the cabezon. *Calif. Fish Game Bull.*, **93**, 1–76.
O'Donoghue P.J., Phillips P.H. & Shepherd S.A. (1991) Infections by *Perkinsus* (Protozoa: Apicomplexa) in abalone from South Australian waters. *Trans R. Soc. S. Aust.*, **115**, 77–82.

Olsen D. (1968) Banding patterns of *Haliotis rufescens* as indicators of botanical and animal succession. *Biol. Bull.*, **134**, 139–47.

Palmer A.R. (1979) Fish predation and the evolution of gastropod shell sculpture: experimental and geographic evidence. *Evolution* **33**, 697–713.

Parsons D.W. & Macmillan D.L. (1979) The escape response of abalone (Mollusca, Prosobranchia, Haliotidae) to predatory gastropods. *Mar. Behav. Physiol.*, **6**, 65–82.

Peck L.S. & Culley M.B. (1990) Structure and density of *Haliotis tuberculata* populations around the coasts of Jersey, Channel Isles. *J. Mar. Biol. Assoc. UK*, **70**, 67–75.

Pilson M.E.Q. & Taylor P.B. (1961) Hole drilling by octopus. *Science*, **134**, 1366–8.

Powell E.N., Staff G.M., Davies D.J. & Callender W.R. (1989) Macrobenthic death assemblages in modern marine environments: formation, interpretation and application. *Rev. Aquat. Sci.*, **1**, 555–90.

Prince J.D. (1989) *The fisheries biology of the Tasmanian stocks of* Haliotis rubra. PhD thesis, University of Tasmania.

Prince J.D. (1991) A new technique for tagging abalone. *Aust. J. Mar. Freshwat. Res.*, **42**, 101–106.

Prince J.D., Sellers T.L., Ford W.B. & Talbot S.R. (1988) Recruitment, growth, mortality, and population structure in a southern Australian population of *Haliotis rubra* (Mollusca: Gastropoda). *Mar. Biol.*, **100**(1), 75–82.

Ricker W.E. (1975) Computation and interpretation of biological statistics of fish populations. *Bull. Fish. Res. Bd Can.*, **191**.

Riley S.P. (1988) *Predation of juvenile abalone*, Haliotis rubra *by the crab* Plagusia chabrus: *an application of the theory of optimal foraging*. BSc (Hons) thesis, University of Tasmania.

Roff D.A. (1986) Predicting body size with life history model. *Bioscience*, **6**, 316–23.

Sainsbury K.J. (1982) Population dynamics and fishery management of the paua, *Haliotis iris* I. Population structure, growth, reproduction, and mortality. *NZ J. Mar. Freshwat. Res.*, **16**, 147–61.

Saito K. (1979) Studies on propagation of Ezo abalone, *Haliotis discus hannai* Ino – I. Analysis of the relationship between transplantation and catch. *Bull. Jap. Soc. Sci. Fish.*, **45**, 695–704.

Saito K. (1984) Ocean ranching of abalones and scallops in northern Japan. *Aquaculture*, **39**, 361–73.

Sakai S. (1962) Ecological studies on the abalone, *Haliotis discus hannai* Ino. IV. Studies on the growth. *Bull. Jap. Soc. Sci. Fish.*, **28**(9), 899–904.

Sakemoto J., Tanaka K. & Kobayashi K. (1982) (On the utilization of the marine environment for the intermediate culture of laboratory-reared abalone seed.) *Chiba Ken Suisan Shikenjo Kenkyu Hokoku* (Chiba Pref. Fish. Res. Lab. Reports), **40**, 123–30. (Translation by M. Mottet, State of Washington, Department of Fisheries.)

Sanders M.J. (1982) *Preliminary Stock Assessment for the Abalone taken off the South East Coast of Oman*. FAO, Rome.

Sandland R.L. & Kirkwood G.P. (1981) Estimation of survival in marked populations with possible dependent sighting probabilities. *Biometrika*, **68**, 531–41.

Schiel D.R. & Welden B.C. (1987) Responses to predators of cultured and wild red abalone, *Haliotis rufescens*, in laboratory experiments. *Aquaculture*, **60**, 173–88.

Schmitt R.J. & Connell J.H. (1984) Field evaluation of an abalone enhancement program. In *California Sea Grant College Program 1980–82 Biennial Report*. Institute of Marine Resources, University of California, La Jolla.

Schnute J. (1987) A general fishery model for a size-structured fish population. *Can. J. Fish. Aquat. Sci.*, **44**, 924–40.

Schnute J. & Fournier D. (1980) A new approach to length–frequency analysis: growth structure. *Can. J. Fish. Aquat. Sci.*, **37**, 1337–51.

Seber G.A.F. (1973) *The Estimation of Animal Abundance*. Griffin, London.

Seber G.A.F. (1986) A review of estimating animal abundance. *Biometrics*, **42**, 267–92.

Shepherd S.A. (1973) Studies on southern Australian abalone (genus *Haliotis*) I. Ecology of five sympatric species. *Aust. J. Mar. Freshwat. Res.*, **24**, 217–57.

Shepherd S.A. (1986) Movement of the southern Australian abalone *Haliotis laevigata* in relation to crevice abundance. *Aust. J. Ecol.*, **11**, 295–302.

Shepherd S.A. (1987) *Aspects of the biology of the abalone* Haliotis laevigata *and* Haliotis scalaris. PhD thesis, Deakin University.

Shepherd S.A. (1990) Studies on southern Australia abalone (genus *Haliotis*) XII. Long term recruitment and mortality dynamics. *Aust. J. Mar. Freshwat. Res.*, **41**, 475–92.

Shepherd S.A. & Godoy C. (1989) Studies on southern Australian abalone (genus *Haliotis*) XI. Movement and natural mortality of juveniles. *J. Malac. Soc. Aust.*, **10**, 87–95.
Shepherd S.A. & Gray J.D. (1986) Food of the anemone *Anthothoe albocincta* at West Island, South Australia. *Trans. R. Soc. S. Aust.*, **109**, 191–2.
Shepherd S.A., Guzmán del Próo S.A., Turrubiates J., Belmar J., Baker J.L. & Sluczanowski P.R. (1991) Growth, size at sexual maturity and egg-per-recruit analysis of the abalone *Haliotis fulgens* in Baja California. *Veliger*, **34**, 324–30.
Shepherd S.A., Kirkwood G.P. & Sandland R.L. (1982) Studies on southern Australian abalone (genus *Haliotis*). III Mortality of two exploited species. *Aust. J. Mar. Freshwat. Res.*, **33**, 265–72.
Shepherd S.A. & Turner J.A. (1985) Studies on southern Australian abalone (genus *Haliotis*) VI. Habitat preference, abundance and predators of juveniles. *J. Exp. Mar. Biol. Ecol.*, **93**, 285–98.
Shibui T. (1971) Experimental studies on the predatory animals of young abalones. *Bull. Jap. Soc. Sci. Fish.*, **37**(12), 1173–6.
Sinclair M. (1963) Studies on the paua *Haliotis iris* Martyn in the Wellington District 1945–46. *Zool. Publ. Victoria Univ. Wellington*, No. 35.
Sloan N.A. & Breen P.A. (1988) Northern abalone, *Haliotis kamtschatkana*, in British Columbia: fisheries and synopsis of life history information. *Can. Spec. Publ. Fish. Aquat. Sci.*, **103**.
Sluczanowski P.R. (1984) A management orientated model of an abalone fishery whose substocks are subject to pulse fishing. *Can. J. Fish. Aquat. Sci.*, **41**, 1008–14.
Sluczanowski P.R. (1986) A disaggregate model for sedentary stocks: the case of the South Australian abalone. In *North Pacific Workshop on Stock Assessment and Management of Invertebrates* (Ed. by G.S. Jamieson & N. Bourne). *Can. Spec. Publ. Fish. Aquat. Sci.*, **92**, 393–401.
Smith S.V. (1972) Production of calcium carbonate on the mainland shelf of southern California. *Limnol. Oceanogr.*, **17**(1), 28–41.
Ssentongo G.W. & Larkin P.A. (1973) Some simple methods of estimating mortality rates of exploited fish populations. *J. Fish. Res. Board Can.*, **30**, 695–8.
Stephenson T.A. (1924) Notes on *Haliotis tuberculata*. *Mar. Biol. Assoc. UK*, **13**, 480–95.
Tanaka K. (1988) Study on the abalone culture in the coast of Awa Region, Chiba Prefecture. *Bull. Jap. Sea Reg. Fish. Res. Lab.*, **38**, 21–132.
Tanaka K., Ishida O. & Tanaka T. (1986) On the lost weight of abalòne around Mera Shoal in the southern coast of the Boso Peninsula, Japan. I. On the distribution of abalone. *Bull. Jap. Sea Reg. Fish. Res. Lab.*, **36**, 49–58.
Tarr R.J.Q. (1986) *Bird Island Abalone Report*. Sea Fisheries Research Institute, South Africa.
Tarr R.J.Q. (1989) Abalone. In *Oceans of Life off Southern Africa* (Ed. by A.I.I. Payne & R.J.M. Crawford), pp. 62–9. Vlaeberg, Cape Town.
Tauti M. (1981) Growth rate, age composition, apparent survival rate and fishing exploitation rate of Japanese black abalone *Haliotis discus discus* in the artificial abalone reefs at Akabane, Jogashima. *Bull. Kanagawa Pref. Fish. Exp. Stn*, **3**, 51–7.
Tauti M. (1984) Growth, age composition, survival rate and recapture rate of Japanese black abalone, *Haliotis discus* Reeve, at Nagai. *Bull. Kanegawa Pref. Fish. Exp. Stn*, **6**, 17–21.
Tauti M., Kanasugi S. & Chikayama M. (1984) Growth, age composition, survival rate and recapture rate of Japanese black abalone, *Haliotis discus* Reeve, at Kamoi, Yokosuka City, Kanagawa Prefecture. *Bull. Kanagawa Pref. Fish. Exp. Stn*, **6**, 51–5.
Tegner M.J. (1989) The California abalone fishery: Production, ecological interactions, and prospects for the future. In *Marine Invertebrate Fisheries: Their Assessment and Management* (Ed. by J.F. Caddy), pp. 401–20. Wiley, New York.
Tegner M.J., Breen P.A. & Lennert C.E. (1989) Population biology of red abalones, *Haliotis rufescens*, in southern California and management of the red and pink, *H. corrugata* abalone fisheries. *Fish. Bull.*, **87**, 313–39.
Tegner M.J. & Butler R.A. (1985) The survival and mortality of seeded and native red abalones, *Haliotis rufescens*, on the Palos Verdes Peninsula. *Calif. Fish Game*, **71**, 150–63.
Tegner M.J. & Butler R.A. (1989) Abalone seeding. In *Handbook of Culture of Abalone and Other Marine Gastropods* (Ed. by K.O. Hahn), pp. 157–82. CRC Press, Boca Raton.
Tegner M.J. & Dayton P.K. (1977) Sea urchin recruitment patterns and implications of commercial fishing. *Science*, **196**, 324.
Tegner M.J. & Dayton P.K. (1987) El Niño effects on southern California kelp forest communities. *Adv. Ecol. Res.*, **17**, 243–79.
Thompson W.F. (1914) Report on the abalone of British Columbia (*Haliotis giganteus* Chemnitz). In

Report of the British Columbia Commissioner for Fisheries for 1913, pp. 126–30.
Thrower S.J. (1977) Blue spot in abalone – possible causes and solutions. *Aust. Fish.*, **36**(10), 6–7, 13, 20.
Tutschulte T.C. (1976) *The comparative ecology of three sympatric abalones*. PhD thesis, University of California, San Diego.
Van Sickle J. (1977) Mortality rates from size distributions. *Oecologia*, **27**, 311–18.
Vermeij G.J. (1978) *Biogeography and Adaptation: Patterns of Marine Life*. Harvard University Press, Cambridge.
Vetter E.F. (1988) Estimation of natural mortality in fish stocks: a review. *Fish. Bull.*, **86**, 25–43.
Wetherall J.A. (1982) Analysis of double-tagging experiments. *Fish. Bull.*, **80**, 687–701.
Whang H.C. & Chung K.O. (1977) Study on relationship between exposure time and mortality of young abalone, *Haliotis discus* (Reeve). *Bull. Fish. Res. Dev. Agency, Busan*, **18**, 123–9.
Winstanley R.H. (1972) Abalone and rock lobster mortality and abnormal water temperatures. *Tasmanian Fish. Res.*, **6**, 21.

Chapter 22
Using a spatial model to explore the dynamics of an exploited stock of the abalone *Haliotis rubra*

JEREMY D. PRINCE *PO Box 209, Leederville, Western Australia 6007*

ABSTRACT: A dynamic spatial model of a stock of *Haliotis rubra* is described in this chapter. The biological parameters for this model, including growth, movement, mortality and recruitment and fishery parameters, are drawn from the results of recent research in Tasmania, Australia. The results and the inferences drawn from them are discussed briefly. The model's simulations of the long-term dynamics of an exploited abalone population suggest that, except under extremely high levels of exploitation, catch per unit effort (CPUE) is likely to be an inaccurate indicator of stock abundance. This is partly due to the handling time required by divers when catching abalone, but principally because of the movement patterns of abalone and the visual nature of the fishery. In essence, abalone form and reform sedentary 'schools' which are targeted by divers and this causes the catchability of an abalone stock to increase with declining stock density.

RESUMEN: Este artículo describe un modelo dinámico de la distribución espacial de un stock de *Haliotis rubra*. La información biológica para este modelo, incluyendo crecimiento, mortalidad y reclutamiento, y la información de la pesquería son los resultados de un estudio recientemente acabado en Tasmania, Australia. Estos resultados y sus inferencias son discutidos de una manera general. Cuando el modelo simula el dinámico de larga duración de población de abulón explotada los resultados indican que, con excepción de niveles extremadamente altos de explotación, la captura por unidad de esfuerzo (CPUE) es un indicador inexacto de stock de abundancia. Esto es en parte debido el tiempo que los buzos necesitan para recoger abulón. Pero principalmente debido a los movimientos del abulón y al carácter visual de la pesquería. Esencialmente, el abulón forma bancos sendatarios y los buzos se concentran en ellos. Este hecho produce que la capturabilidad del abulón aumente cuando el stock de abundancia disminuye, lo que a su vez produce que la CPUE se mantenga estable.

INTRODUCTION

It is rarely possible to measure the actual abundance of an exploited stock directly and, as a consequence, it is more usual for an index of abundance to be used. Many methods for estimating the potential yield of a stock use catch per unit effort (CPUE) as an index of stock abundance (Gulland, 1977), and the assumption that CPUE reflects stock abundance is probably the most common assumption made in the field of fisheries biology. It is widely recognized that this assumption is not always valid; Russell (1931) reported that the allocation of effort by fishermen can affect this relationship and Paloheimo & Dickie (1964) documented the difficulty of using this assumption with schooling fish. Hancock (1979) cautioned that the catchability of many invertebrates can be extremely variable, which can make CPUE an unreliable index of invertebrate stock abundance. Hilborn & Walters (1987) presented evidence suggesting that this assumption may also be tenuous in spatially heterogeneous fisheries.

The relationship between CPUE and stock abundance in abalone has been discussed in the literature. It has been suggested that several factors could affect this relationship:

- Changes in diver efficiency, due to learning, improvement of equipment and technique and competition amongst divers (Harrison, 1983; Breen, 1986);
- the progressive discovery and exploitation of sedentary beds of abalone (Harrison, 1983);
- variation in catchability associated with different bottom types (Harrison, 1986).

However, little serious study has been conducted on this aspect of the fishery dynamics of abalone stocks. The work of Beinssen (1979), who quantified the time taken to handle abalone as they were collected, stands alone. Beinssen found that within the Victorian (Australia) abalone fishery, there could have been a three-fold increase in the efficiency of divers because the proportion of time spent handling abalone had decreased with stock density.

Despite the questions raised about the relationship between CPUE and stock abundance, the assumption that CPUE can be used as an index of stock abundance continues to be used, either quantitatively or qualitatively, in Australia (Ward, 1986).

In 1983 the author initiated a study of the fishery biology of *Haliotis rubra* (Leach) in Tasmanian waters (southeastern Australia). At that time the commercial abalone divers of Tasmania stated that they did not believe that the stable catch rates observed in the fishery indicated the actual state of the stock. They believed that the stock was declining rapidly. Consequently, a primary aim of this study was to examine how the fishery dynamics of this species could affect the relationship between CPUE and stock abundance within the fishery. To this end, a study was undertaken which examined both the exploited abalone and the exploiting diver. Various aspects of abalone biology were documented, including growth, mortality, fecundity, larval and post-larval movement and feeding. The fishing patterns, behaviour and fishing power of the divers were also studied. The results were used to model the dynamics of an exploited population of *H. rubra*.

The purpose of this model and the aim of this chapter is to examine the potential for interaction between the biology of abalone and the fishing patterns of abalone divers. The principal purpose of this chapter is to examine the implications that these potential interactions have for our understanding of the relationship between CPUE and stock abundance within a single stock of abalone.

The results on which the model is based are published elsewhere (Prince *et al.* 1987, 1988a,b; Prince, 1989) and will only be summarized here, readers are referred to these sources for more detailed presentation and discussion of this work.

DESCRIPTION OF THE MODEL

Physical structure of the modelled reef

The model describes the dynamics of an abalone population similar to the population studied by the author (Prince, 1989) at a site in southern Tasmania, George III

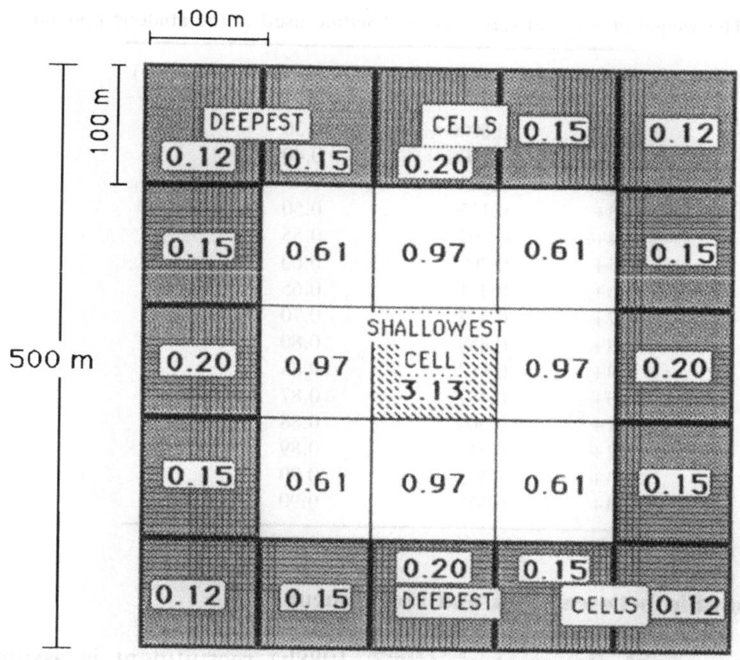

Fig. 22.1 The hypothetical reef structure for the abalone population model. Figures in bold type indicate typical breeding stock densities (abalone m^{-2}) produced by the movement parameters of the model when no fishing pressure is applied.

Rock (43°31′S, 146°58′E). The reef has been modelled as a square made up of 25 cells, each 100 × 100 m. The central cell is the shallowest area of the reef and a depth gradient exists between the central and outermost cells (Fig. 22.1).

General population structure

The model follows 15 year classes, with the oldest year class being an accumulation of older year classes which are assumed to have the same body weight and the same mortality rates. The growth parameters used within the model are those observed at George III Rock and the model uses an average live weight for each year class (Table 22.1). The survivorship table (Table 22.1) used by the model is based on mortality data gathered at this and another site (Prince *et al.*, 1988a). Mortality is assumed to decline from approximately 70% per annum in the youngest age class to 10% in the oldest age classes (Table 22.1).

Maturation of *H. rubra* in Tasmania occurs at between 5 and 10 years of age with approximately 50% of the population being mature at 7+. Upon maturity, the abalone move out of cryptic habitats and become vulnerable to fishing (Prince *et al.* 1988a; Prince, 1989). For simplicity, maturity and recruitment to the fished population is assumed to be knife-edged, with all abalone maturing and entering the fishable population during their eighth year of life (7+).

Table 22.1 The weight at age and survivorship schedule used in the abalone population model.

Year	Weight (kg)	Survival ($annum^{-1}$)
0+	0.0006	0.30
1+	0.0008	0.40
2+	0.043	0.45
3+	0.115	0.50
4+	0.207	0.55
5+	0.298	0.60
6+	0.371	0.65
7+	0.423	0.70
8+	0.459	0.80
9+	0.482	0.85
10+	0.497	0.87
11+	0.506	0.88
12+	0.512	0.89
13+	0.515	0.90
14+	0.517	0.90

The relationship between stock and recruitment

In accordance with Prince *et al.* (1987, 1988b) recruitment is assumed to be localized, i.e. the level of recruitment within each 100 × 100 m cell is assumed to be proportional to the level of spawning stock within that cell and independent of the recruitment or spawning stock in surrounding cells. Fecundity in *H. rubra* is approximately proportional to the body weight of mature individuals (Nash *et al.*, in prep.) so spawning stock abundance was estimated in terms of biomass of mature animals.

In the absence of any published data to indicate the form of the relationship between stock and recruitment for abalone, a curve was created for use with the model. The curve incorporates an inflexion at low densities and an absolute decline in recruitment abundance at higher densities, implying that population fecundity will be favoured by aggregation at low densities and by dispersion at high densities. Figure 22.2 shows this stock–recruitment curve (curve A) compared with the equilibrium replacement line (curve B). The equilibrium replacement line indicates the relationship between any constant level of recruitment and the abundance of breeding stock in an unfished population that it would produce. Where recruitment falls below the equilibrium replacement line the stock will tend to decline towards the point where the two curves intersect. When recruitment is above this line, stock abundance will tend to increase towards the next intersection of the curves.

Movement

Two types of movement were observed and measured within the population of *H. rubra* at George III Rock in southern Tasmania (Prince, 1989). The first is a relatively random foraging behaviour, the magnitude of which is inversely related to food availability. This type of movement might have been expected to lead to

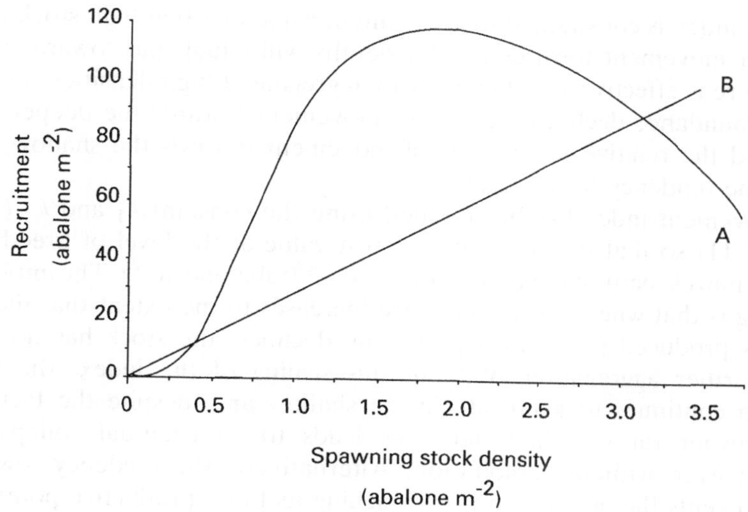

Fig. 22.2 The relationship between breeding stock (abalone m^{-2}) and recruitment density (abalone m^{-2}) used for the abalone population model (curve A) and the equilibrium replacement line (curve B).

the dispersion of abalone away from areas of low food density and to their aggregation in areas of high food density. In contrast to this expectation, within the studied area over 50% of the population was found to be aggregated in the shallowest region which had the poorest supply of food and which comprised only about 14% of the total area. This aggregation was apparently explained by the second type of movement which was a general tendency within the abalone population to move towards shallower depths. Movement was also observed to be size related within that population.

For simplicity, within this model, it has been assumed that movement only occurs amongst the mature population, and that it is independent of age or size amongst these animals. Movement has been modelled as simple directional diffusion. The maximum and annual rate of diffusion in any direction has been assumed to be 0.24, implying that during any year 24% of the abalone in any 100×100 m cel will move out of the cell in one of the four cardinal directions. This maximum rate approximates the highest rate of movement observed in the abalone population at George III Rock, which was movement towards shallower water. The food-related rate of movement and the tendency to aggregate within the shallowest area have been described by multiplying movements which are neutral or negative with regard to depth, by a movement index that is $\leqslant 1$:

$$\text{number moving} = \text{number in cell} \times \text{movement index} \times 0.24 \qquad (22.1)$$

Within the model it is assumed that food availability provides a mechanism by which the rate of dispersion is related to population density. This is described by making the movement index proportional to the density of breeding stock:

$$\text{movement index} = k_1 + (k_2 \times \text{breeding stock abundance}) \qquad (22.2)$$

where k_1 and k_2 are arbitrary constants used to scale the equation. Since the movement index is constrained to a maximum value of 1.0 at high stock densities, the rate of movement towards greater depths will equal that towards shallower depths, there is effectively no tendency to aggregate at high densities. In contrast, as stock abundance declines the rate of movement towards the deeper cells will decline and the relative importance of movement towards the shallow cells will increase the tendency to aggregate.

The movement index has been scaled using the constant k_1 and k_2 ($k_1 = 0.24$ and $k_2 = 0.11$) so that it reaches its greatest value at the level of breeding stock which maximizes recruitment to the stock (1.0–2.0 abalone m^{-2}). The implication of this scaling is that when stock abundance increases to the extent that the number of recruits produced per spawning abalone declines, the stock has no tendency towards further aggregation. Without this scaling of the index, the modelled population continues to aggregate in the shallow area despite the fact that the recruit/spawner ratio declines and this leads to an eventual collapse of the population even without exploitation. Alternatively, the tendency towards dispersion prevents the population from reaching its full reproductive potential until a uniform density has been established in every cell.

These movement parameters cause the model to establish a density gradient over the cells at all but the highest densities, with the central cell having the highest density (Fig. 22.1). Within the model there is no provision for diagonal movement between cells, which results in the apparently anomalous concentration of abalone in corner cells. The degree of aggregation and patchiness created by this movement model, with approximately 50% of the stock being found in 16% of the area, is similar to the patchiness observed in the population at George III Rock, Tasmania, where approximately 50% of the stock was found in 14% of the reef area (Prince, 1989).

The model has a minimum spatial resolution of 100 m × 100 m and this prevents the description of aggregations below this scale. In the situation where the abundance of spawning stock in the central cell falls below the lower intersection with the cohort decay line (Fig. 22.2) this would prevent any recovery of the stock, since the stock continues to decline towards the cohort decay curve. To circumvent this situation, and to give the population the ability to increase from any level of abundance, the model was modified with respect to the stock-recruitment relationship of the central cell. At low stock abundance the stock in the central cell is assumed to collapse into an area smaller than 100 m × 100 m so as to achieve a density that will maximize the recruit/spawner ratio. In this situation the abundance of recruitment to the central cell is estimated on the basis of the area the remaining abalone will occupy if they optimize their density.

Fishery dynamics

The relationship between catch and effort has been described by the equation:

$$\text{CPUE} = d\ P/(1 + h\ d\ P) \tag{22.3}$$

where CPUE is the catch rate of abalone in numbers per diving hour, d is the density of abalone in numbers per square metre, P is the fishing power of the divers in square metres effectively swept per hour and h is the handling time required to capture each abalone. The fishing power of divers was measured and found to be 1128 m^2 h^{-1} (Prince, 1989). The handling time of 5.1 s $abalone^{-1}$ measured by Beinssen (1979) in the Victorian fishery has been used in this model.

The effect of handling time is to flatten the functional response of catch to effort expended (Fig. 22.3) at densities > 1.5 abalone m^{-2}, or equivalent to catch rates of approximately 250 kg h^{-1}. However, within each 100 m × 100 m cell at stock densities < 1.5 abalone m^{-2} a relatively direct relationship should exist between catch rate and stock abundance.

Abalone divers do not allocate their effort randomly. Using their general experience they can predict with considerable accuracy the locations of abalone aggregations even within an area in which they have no prior knowledge (Prince, 1989). Within this model it is assumed that divers direct each unit of effort to the most preferable cell. This is achieved by using an index of preference which can be based on any number of factors to which a numerical weighting can be assigned. Two factors have been used in this model, the primary factor is the abundance of abalone (D), however abundance has been multiplied by a secondary value which is used to represent the attractiveness of shallower areas compared to deeper areas (F):

$$\text{Preference index} = D \times F \tag{22.4}$$

During the experimental programme, commercial divers were observed to create and maintain a catch rate differential of 15–30 kg h^{-1} between depths of 3–5 m and 12–18 m, reflecting their preference for fishing in shallower depths. The weighting factor (F) used in this model has been scaled so that fishing will create and maintain a difference of approximately this magnitude between the central and peripheral cells.

The structure of the model is such that the model assesses the relative attractiveness of each of the 25 cells before allocating a block of effort (5 hours). After allocating the effort to the most attractive cell, the model estimates the effect of

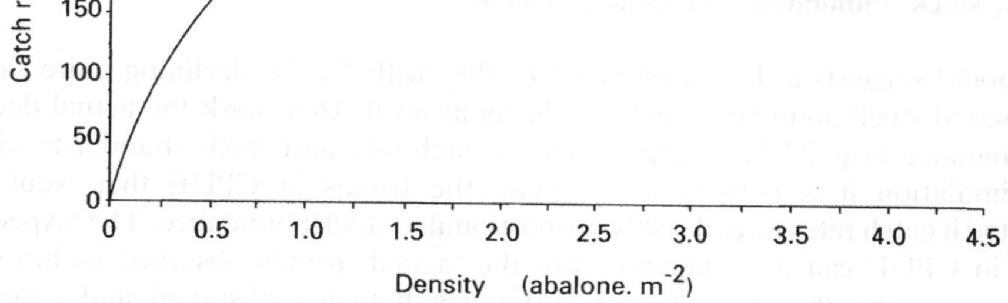

Fig. 22.3 The functional response of catch rate (kg h^{-1}) to stock density (abalone m^{-2}) used in the abalone population model.

that effort and re-assesses the relative merits of each cell before allocating the next unit of effort.

A listing of the model in Microsoft Basic can be obtained from the author.

SIMULATING THE DYNAMICS OF AN EXPLOITED STOCK

The model was used to simulate fishing the theoretical stock with various levels of constant effort over the time span of many years. With the parameters built into the model the stock will sustain a constant 185 h annum^{-1}. Figure 22.4A–C shows some examples of the dynamics exhibited by the model under a range of exploitation rates which eventually lead to collapse. In this figure the plotted values for recruitment denote the relative strength of the youngest year class. Stock abundance is the level of stock resulting from recruitment to the fished population, natural mortality and fishing mortality. The plotted value for fishing mortality is the finite rate of fishing mortality or the proportion of the available stock removed by fishing in each year. With the exception of fishing mortality, all values have been scaled in order to facilitate graphing.

While the level of effort applied clearly determines the speed with which the stock declines, a pattern is evident through the different simulations. During the first phase of these declines the accumulated virgin stock is fished down. During this period stock abundance and CPUE decline rapidly, although the decline in CPUE is much less than the decline in abundance.

The dome shaped stock–recruitment curve produces significantly improved recruitment as a consequence of fishing down the virgin biomass. This improvement in recruitment stabilizes stock size to a certain degree, which in turn stabilizes CPUE. Fishing mortality also stabilizes during this second phase. The duration of this phase, and its absolute level of stability, are determined by the level of effort applied; however the effect of declining recruitment, (recruitment occurs in the eighth year) eventually leads to a decline in stock abundance, which in turn affects recruitment. Recruitment failure accelerates the decline in stock abundance and eventually CPUE begins to decline; however this 'index of stock abundance' responds some time after the 'actual stock' has begun responding to the excessive level of effort. Fishing mortality escalates rapidly during the final phase of the fishery.

CPUE, stock abundance and fishing mortality

The model suggests a degree of hyperstability with CPUE declining more slowly than actual stock abundance and stabilizing at levels that mask the actual decline in abundance (Fig. 22.4). Using the initial catch rate and stock abundance within any simulation it is possible to estimate the trends in CPUE that would be expected if catch rates were directly proportional to stock abundance. The 'expected' trend in CPUE can be compared with the 'actual' trends observed within each simulation using the mean square difference between observed and expected values. Figure 22.5 shows the mean difference between expected and observed CPUE over a range of fixed levels of effort which resulted in the eventual collapse

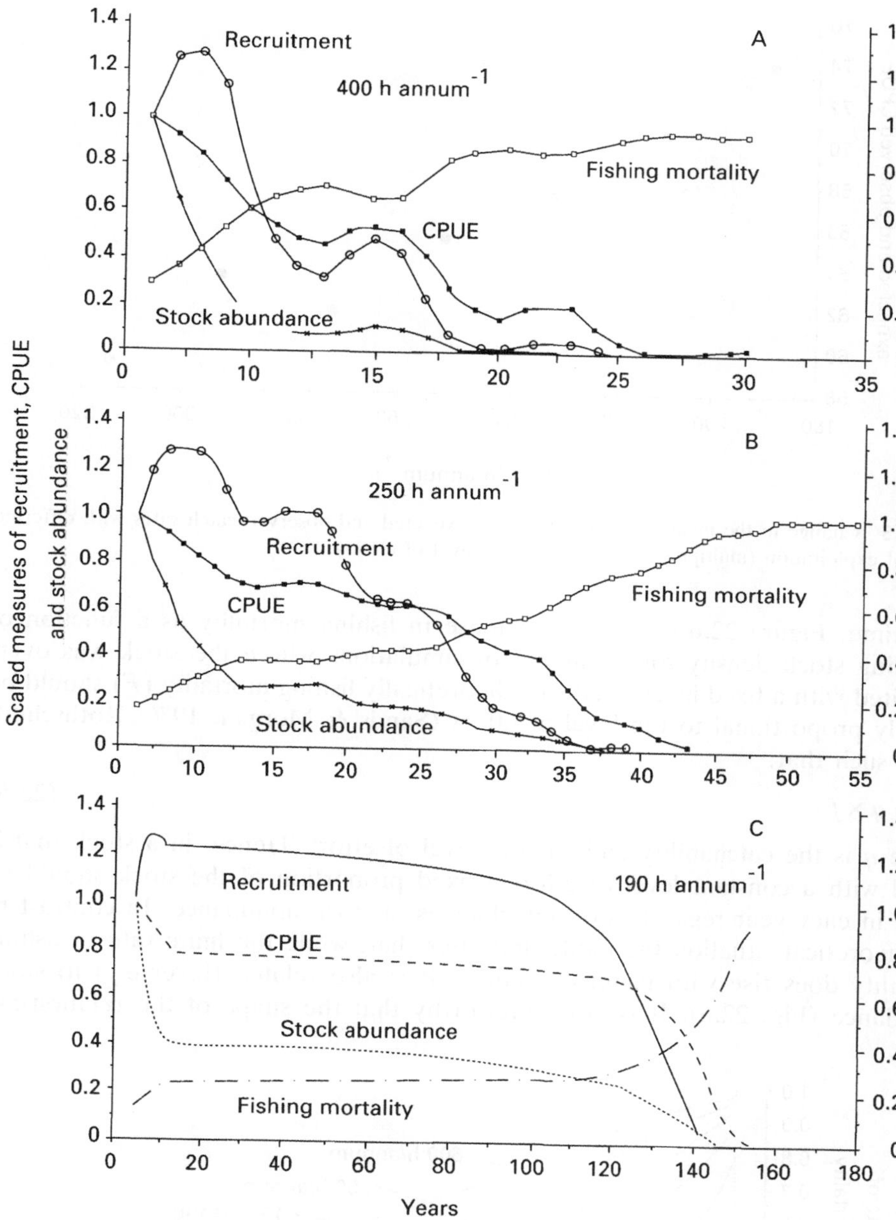

Fig. 22.4 Simulated exploitation of an abalone stock with different constant levels of effort 400 (A), 250 (B), 190 h annum^{-1} (C), showing trends in recruitment, CPUE, stock abundance (all in scaled units) and fishing mortality (proportion of available biomass caught each year).

of the stock. From this figure it can be seen that the model predicts that 'actual' CPUE will become less useful as an index of abundance as effort approaches the level that can be sustained by the stock.

Another notable feature of Fig. 22.4 is that the proportion of the stock being removed by the fixed level of effort (the finite level of fishing mortality) increases

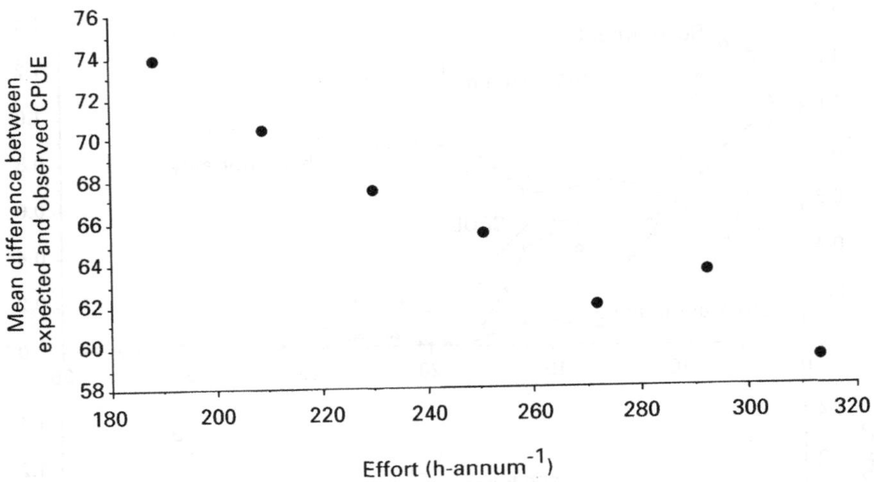

Fig. 22.5 Change in the mean difference between expected and observed catch rates with differing levels of exploitation (multiples of the sustainable level of effort).

with time. Figure 22.6 shows this increase in fishing mortality as a function of declining stock density for a number of simulations where the stock was over-exploited with a fixed level of effort. Theoretically fishing mortality (F) should be directly proportional to the level of effort (Sands & Morgan, 1976; Rothschild, 1977) such that:

$$F = q \times f \tag{22.5}$$

where q is the catchability and f is the level of effort. Hence, in a stock that is fished with a constant level of effort a fixed proportion of the stock should be taken in eacy year regardless of any changes in stock abundance. In contrast to the theoretical situation the model indicates that, while the finite rate of fishing mortality does rise with increasing effort, it is also related (inversely) to stock abundance (Fig. 22.6). It is also noteworthy that the shape of the relationship

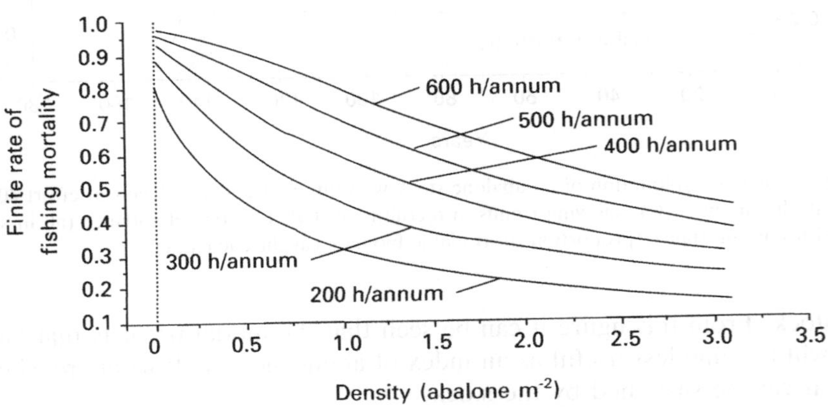

Fig. 22.6 The relationship between stock abundance (abalone m^{-2}) and the finite rate of fishing mortality within simulations using fixed effort levels of 200, 300, 400, 500 and 600 h annum^{-1}.

between CPUE and stock abundance becomes less linear as effort decreases, explaining the observation that CPUE becomes less useful as an index of stock abundance as effort declines.

As the level of effort is held constant during these simulations it is obvious from equation 22.5 that the catchability of the abalone within the modelled population must vary with effort and density. Using equation 22.5 an estimate of the catchability coefficient (q.est) can be crudely estimated using effort (*f*) and the finite rate of fishing mortality (*fF*):

$$q.est = fF/f \qquad (22.6)$$

When this equation is used to convert the data in Fig. 22.6 so that catchability, rather than the finite rate of fishing mortality, is shown as a function of density for each simulation, the variability of catchability with changing levels of both effort and density becomes more evident (Fig. 22.7).

DISCUSSION

At high stock densities (>1.5 abalone m^{-2}) the variability of catchability could be attributed to the effect of handling time, with the proportion of effort being expended on searching rather than handling, increasing as density declines. However, within the model catchability exhibits most variability at densities of <1.0 abalone m^{-2} (Fig. 22.7) and handling time has little impact on the linearity of the relationship between stock abundance and CPUE at these densities (Fig. 22.3). The variability in catchability observed within this model is primarily due to the aggregative behaviour of abalone and the ability of the divers to target these aggregations. As stock abundance declines the modelled population displays an increasing tendency to aggregate and the distribution of abalone over the reef becomes increasingly heterogeneous. Because divers target these aggregations the

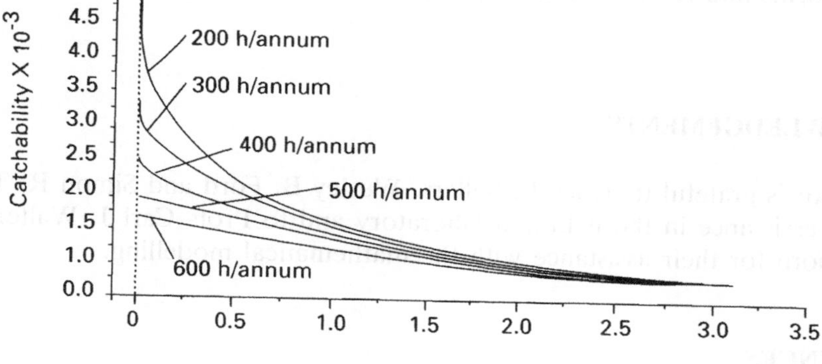

Fig. 22.7 The relationship between stock density (abalone m^{-2}) and the estimated coefficient of catchability (finite rate of fishing mortality/effort) in simulations using fixed effort levels of 200, 300, 400, 500 and 600 h $annum^{-1}$.

catchability of the abalone or the relative efficiency of the divers is increased.

At higher levels of effort divers will tend to catch the abalone before they can re-aggregate and the effect of aggregation will be less pronounced. This can be seen in Figs 22.6 and 22.7, where the relationship between the finite rate of fishing mortality or catchability and stock abundance is more linear and flatter with higher levels of effort. Hence, within this model it is the aggregative nature of abalone, together with the ability of divers to target aggregations, which causes CPUE to become less reliable as an index of stock abundance at low exploitation rates. The performance of this model suggests that the nature of the relationship between CPUE and stock abundance will be determined by the history of exploitation within the fishery. In this context it should be noted that the simulations performed with the model and displayed here, with fixed levels of effort being applied to the stock, represent the simplest possible patterns of exploitation. When simulations are performed with this model in which the level of effort is allowed to vary, the variability exhibited by the estimated coefficient catchability can become extreme, and the relationship between CPUE and stock abundance more complex and tenuous.

Sluczanowski(1984) suggested that the clumped distribution of abalone could affect the relationship between CPUE and stock abundance. However, he apparently saw the distribution of abalone primarily affecting the relationship through its interaction with the searching time of divers, because he stated '. . . catch rates will not sufficiently reflect a reduction in the number of such "clumps" because the search time required to check the existence of a "clump" is not in proportion to the number of individuals which it once contained'. This comment is typical of the published discussion about the relationship between CPUE and stock abundance which has concentrated on the activity of the divers and and treated the distribution of the abalone as a passive feature of the stock. These simulations suggest that it is the activity of the abalone, i.e. the movement patterns observed within the abalone population at George III Rock, which has the major influence on the relationship between CPUE and stock abundance. Hence the view that abalone are sedentary organisms which are simply picked up by divers is too simplistic. It may be more useful to conceptualize them as a super-organism which actively forms and reforms sedentary schools which are targeted by the abalone divers.

ACKNOWLEDGEMENTS

The author is grateful to Tracy L. Sellers, Wesley B. Ford and Simon R. Talbot for their assistance in the field and laboratory and to Profs Carl J. Walters and Ray Hilborn for their assistance with the mathematical modelling.

REFERENCES

Beinssen K. (1979) Fishing power of divers in the abalone fishery of Victoria, Australia. *Rapp. P.-v. Reun. Cons. Int. Explor. Mer*, **175**, 20–2.

Beverton R.J.H. & Holt S.J. (1957) On the dynamics of exploited fish populations. *UK Min. Ag. Fish., Fish. Invest. (Ser. 2)*, **19**, 1–533.

Breen P.A. (1986) Management of the British Columbia fishery for northern abalone (*Haliotis kamschatkana*). *Can. Spec. Publ. Fish. Aquat. Sci.*, **92**, 300–12.

DeLury D.B. (1947) On the estimation of biological populations. *Biometrics*, **3**, 145–67.

Gulland J.A. (1977) The analysis of data and development of models. In *Fish Population Dynamics* (Ed. by J.A. Gulland), pp. 67–95. Wiley, New York.

Hancock D.A. (1979) Population dynamics and management of shellfish stocks. *Rapp. P.-v. Reun. Cons. Int. Explor. Mer*, **175**, 8–19.

Harrison A.J. (1983) The Tasmanian abalone fishery. *Tas. Fish. Res.*, No. 26.

Harrison A.J. (1986) Gastropod fisheries of the Pacific with particular reference to Australian abalone. *Can. Spec. Publ. Fish. Aquat. Sci.*, **92**, 14–22.

Hilborn R. & Walters C.J. (1987) A general model for simulation of stock and fleet dynamics in spatially heterogeneous fisheries. *Can. J. Fish. Aquat. Sci.*, **44**, 1366–9.

Palohcimo J.E. & Dickie L.M. (1964) Abundance and fishery success. *Rapp P.-v. Reun. Cons. Int. Explor. Mer*, **155**, 152–63.

Prince J.D. (1989) *The fisheries biology of the Tasmanian stocks of* Haliotis rubra. PhD thesis, University of Tasmania.

Prince J.D., Sellers T.L., Ford W.B. & Talbot S.R. (1987) Experimental evidence for limited dispersal of haliotid larvae (*Haliotis*; Mollusca: Gastropoda). *J. Exp. Mar. Biol. Ecol.*, **106**, 243–63.

Prince J.D., Sellers T.L., Ford W.B. & Talbot S.R. (1988a) Recruitment, growth, mortality and population structure in a southern Australian population of *Haliotis rubra* (Mollusca: Gastropoda). *Mar. Biol.*, **100**, 75–82.

Prince J.D., Sellers T.L., Ford W.B. & Talbot S.R. (1988b) Confirmation of a relationship between the localized abundance of breeding stock and recruitment for *Haliotis rubra* Leach (Mollusca: Gastropoda). *J. Exp. Mar. Biol. Ecol.*, **122**, 91–104.

Ricker W.E. (1954) Stock and recruitment. *J. Fish. Res. Bd Can.*, **11**, 559–623.

Rothschild B.J. (1977) Fishing effort. In *Fish Population Dynamics* (Ed. by J.A. Gulland), 96–115, Wiley, New York.

Russell E.S. (1931) Some theoretical considerations on the 'overfishing' problem. *J. Cons. Int. Explor. Mer*, **6**, 3–20.

Sanders M.J. & Morgan A.J. (1976) Fishing power, fishing effort, density, fishing intensity and fishing mortality. *J. Cons. Int. Explor. Mer*, **37**, 36–40.

Shepherd S.A. (1986) Studies on southern Australian abalone (genus *Haliotis*) VII. Aggregative behaviour of *H. laevigata* in relation to spawning. *Mar. Biol.*, **90**, 231–6.

Sluczanowski P.R. (1984) The fishery. In *An assessment of the South Australian abalone resource*, pp. 14–49. S. Aust. Dept Fish., unpublished report.

Ward T.J. (1986) Abalone biology and fisheries in Australia: a review of research. *Aust. Fish. Serv., Canberra, Aust. Fish. Pap.*, No. 86–12.

Chapter 23

An evaluation of egg-per-recruit analysis as a means of assessing size limits for blacklip abalone (*Haliotis rubra*) in Tasmania

WARWICK J. NASH *Division of Sea Fisheries, Department of Primary Industry, Crayfish Point, Taroona, Tasmania 7053, Australia*

ABSTRACT: Several characteristics of blacklip abalone (*Haliotis rubra*) stocks in Tasmania suggest a means by which size limits appropriate for sustainable fishing may be determined.

(1) There is a latitudinal trend in growth rate: abalone in the south grow faster, and to a larger size, than those in the north.

(2) Sexual maturity is attained at a given age, not a given size. The slower-growing northern stocks therefore reach sexual maturity at a smaller size than their southern counterparts.

(3) There is currently a single Tasmania-wide legal minimum size limit for abalone. Hence, the slow-growing northern stocks are able to breed for a greater number of years before they reach catchable size than the southern stocks. There is therefore greatest protection of the egg production in the northern stocks, and least protection in the southern stocks.

(4) Since, as argued here, the northern stocks are underexploited at this size limit and the southern stocks are vulnerable to overfishing, it follows that some geographically intermediate stocks, with intermediate growth rates, are receiving an optimal level of protection by the current size limit, assuming relatively even distribution of fishing mortality over the area in question. If these 'ideally-fished' stocks can be identified, the 'ideal' level of protection of the egg production can be determined by egg-per-recruit analysis.

(5) Egg-per-recruit analysis can then be used to determine, for other areas, the size limit which provides the same level of protection of egg production as in the 'ideally-fished' area. The validity of this approach relies on (i) being able to identify 'ideally-fished' stocks, and (ii) being able to estimate M for fished stocks. Neither of these requirements could be met: no satisfactory method for identifying 'ideally-fished' stocks was found, and the intended method of determining M for fished stocks was shown to be invalid. Other limitations of egg-per-recruit analysis in assessing size limits are also discussed.

RESUMEN: Varias características en las poblaciones del abulón de holán negro (*Haliotis rubra*) en Tasmania indican factores por los cuales puede ser determinada una talla mínima apropieda para una pesca sostanida:

(1) El abulón de holán negro de Tasmania presenta un crecimiento variable, con una tendencia latitudinal en la tasa de crecimiento: los abulones en el sur crecen más rápido y alcanzan mayor talla que los abulones en el norte.

(2) La madurez sexual es alcanzada a determinada edad, no a una talla determinada. Las poblaciones del norte, de crecimiento más lento, alcanzan la madurez sexual a talles menores que su contraparte del sur.

(3) Existe en la actualidad en toda Tasmania, una sola talle minima legal de captura para el abulón. Por lo tanto, las poblaciones del norte, de crecimiento más lento, pueden reproducirse por un mayor número de años antes que alcancen la talla de captura, en comparación a las poblaciones del sur. Por esto mismo, la producción de huevos de las poblaciones norteñas se encuentra mejor protegida que la del sur.

(4) Como se ha argumentado aquí, la población del norte está sub-explotada en cuanto a límite de talla y la población del sur es vulnerable a una sobre-pesca, suce de que algunas poblaciones que se encuentran intermedias geográficamente, con tasas de crecimiento intermedias, están recibiendo un nivel óptimo de protección mediante el límite de talla establecido, asumiendo una distribución relativamente igual de mortalidad por pesca en el área en cuestión. Si se pudiesen identificar estas poblaciones 'Idealmente pescables', el nivel 'ideal' de protección de la producción de huevos puede terminarse por medio del análisis de reclutamiento de huevos.

(5) Este tipo de análisis podría usarse entonces en la determinación, para otras áreas, de la talla-límite que asegure el mismo grado de protección a la producción de huevos, como en la área 'ideal' de pesca. La validez de este estudio depende de: (i) poder identificar la población de la 'pesca ideal', y (ii) poder estimar *M* para la población pescada. Ninguno de estos requerimientos pudo ser satisfecho: ningún método satisfactorio para identificar la población 'idealmente pescable' fue encontrado y los métodos intentados para determinar *M* para la población pescada demostraron ser nulos. Otras limitaciones del análisis de reclutamiento de huevos en la evaluación de los límites de talla son también discutidos.

INTRODUCTION

The larval stage of abalone species is relatively short (approximately 4 to 10 days: Hooker & Morse, 1985). The dispersal capacity of the larvae from the point of spawning is therefore limited. Prince *et al.* (1987, 1988b) provided experimental evidence that dispersal of blacklip abalone (*Haliotis rubra*) larvae in Tasmania may be measured on a scale of tens or hundreds of metres.

If these findings are representative throughout this species' range, then the management implications are profound. They suggest that there are many (perhaps hundreds of) essentially unit stocks around Tasmania, between which exchange rates of larvae are ecologically insignificant. Hence, replenishment of stocks subject to localized intensive fishing will occur only slowly, since recruitment of larvae from outside the area will be minor.

There are two management options available to protect multiple small unit stocks which are subject to variable levels of fishing pressure brought about by accessibility, proximity to population centres and boat ramps and prevailing weather conditions. The first is to manage each small unit stock separately. This is not a feasible option because information on catch size, fishing mortality and changes in stock abundance would be necessary for each of the multitude of stocks around Tasmania. This would be impossible to attain.

The second option is to provide blanket protection for all (or at least many) of the unit stocks simultaneously by protecting a suitable proportion of the egg production from being fished. This can be done by the imposition of appropriate minimum size limits. The means of determining appropriate size limits for *H. rubra* stocks in Tasmania is the main subject of this chapter, the objective being to determine what proportion of the egg production must be protected to conserve the stock.

In an evaluation of the possible causes of the sharp and continuing decline of abalone stocks in California, Tegner *et al.* (1989) used a similar approach to determine the level of egg production conserved by the size limits there. They concluded that these levels appeared adequate, and suggested that the decline in stocks was caused by factors other than overfishing (such as predation pressure and environmental factors). The adequacy of size limits alone to manage the Californian abalone fisheries was therefore questioned. Since minimum size limits are regarded as an important component of the management strategy of the abalone fishery in Tasmania (Harrison, 1986), an evaluation of the levels of egg production afforded by the size limit regulation in Tasmania was considered important. The initial results of this work are presented here.

Haliotis rubra stocks in Tasmania appear well suited to the determination of optimal size limits, because of three factors:

(1) There is a fairly even latitudinal trend in growth of *H. rubra* in Tasmania, with abalone in the south of the state growing faster, and to a larger size, than those in the north;
(2) there is a single, uniform size limit of 132 mm maximum shell diameter in force throughout the state;
(3) onset of sexual maturity is age-related, not size-related (Nash *et al.*, in prep.).

Fast-growing abalone in southern waters attain sexual maturity after they have reached legal size. Since *H. rubra* emerges from the sub-boulder (or otherwise cryptic) habitat at the onset of sexual maturity (Nash *et al.*, in prep.), this means they are vulnerable to the fishery as soon as they mature. These stocks therefore receive little protection of the egg production at all. At the other extreme, *H. rubra* stocks from along the northern coast of Tasmania grow so slowly and reach such a small maximum size that they reach sexual maturity several years before they reach the legal size limit. Some of the northern stocks are so stunted that they do not even reach the size limit. These stocks, therefore, are well protected by the current size limit.

Since there is a continuum of growth rates latitudinally, with slow-growing northern stocks being extremely protected (overprotected) and fast-growing southern stocks being only lightly protected, it follows that, somewhere between these two extremes, the stocks have just the right degree of protection of egg production. The task is to identify these optimally-fished stocks. This was approached by discussions with divers and cursory surveys of the abalone stock state-wide.

Not unexpectedly, what appear to be optimally-fished stocks were generally located in the northeast and northwest of the state, geographically intermediate between the slow-growing northern and fast-growing southern stocks. The levels of egg production conserved in these areas were determined by egg-per-recruit (EPR) analysis.

In this chapter, the validity of this approach to determining appropriate size limits is assessed and, more generally, the validity of the assumptions underlying EPR analysis is evaluated.

METHODS

Sampling of abalone populations

Abalone stocks have so far been sampled from the islands in Bass Strait north of Tasmania, and northeast, west and southwest coastal waters. The four sites described in this chapter (Fig. 23.1) are representative of the regions from which they were taken. Full descriptions of the population profiles taken around Tasmania will appear in the *Technical Report Series* of the Tasmanian Division of Sea Fisheries.

At least 200 abalone were collected at each site. From experimental truncation of larger data sets, a sample size of 200 is considered to be the minimum number for accurate determination of the parameters of growth, maturation and mortality. Searching was initially for the emergent abalone, which constitute the fraction of the population normally taken by the commercial abalone diver. Care was taken to ensure that all emergent abalone seen were collected, to avoid size- (or age-)

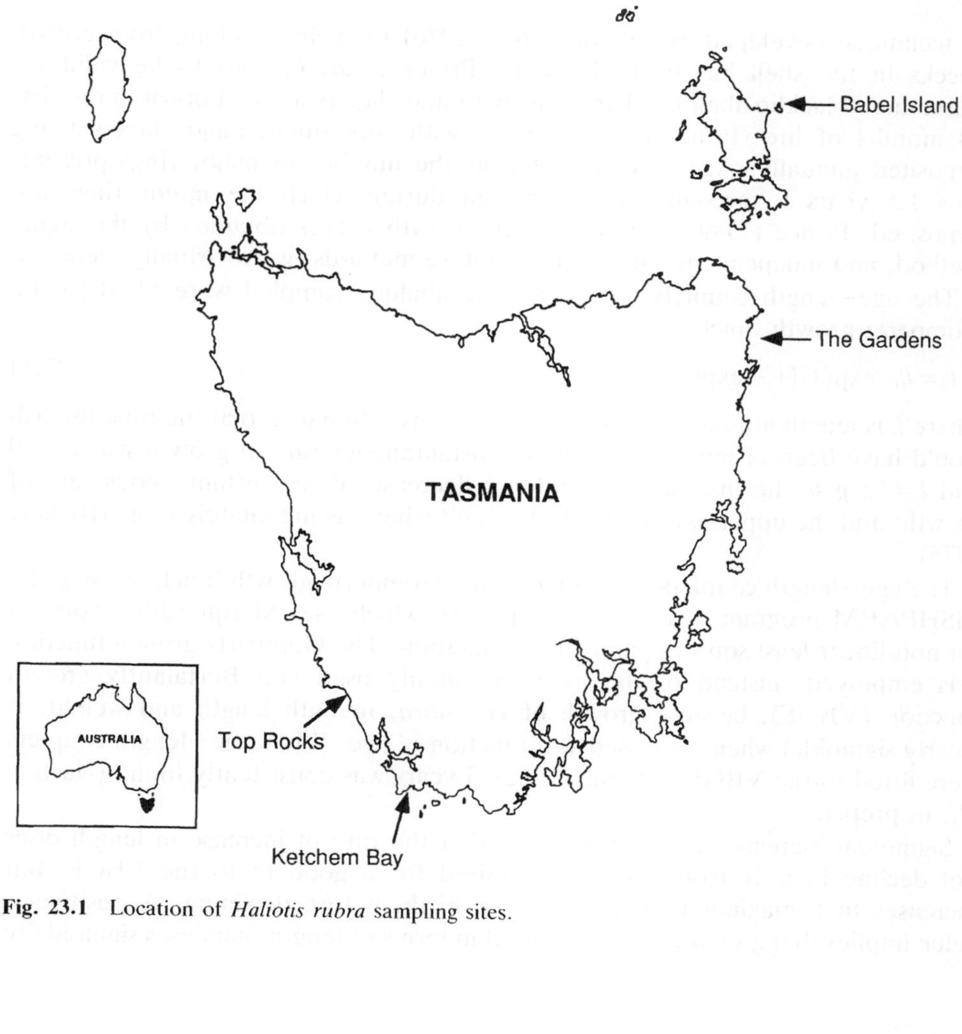

Fig. 23.1 Location of *Haliotis rubra* sampling sites.

related bias. When the desired sample size of emergent abalone was attained, divers searched under boulders and in crevices to obtain sufficient juveniles for accurate determination of the age–size relationship. Abalone were frozen as soon as possible after sampling, and maintained at −30°C until they were processed.

Measurements taken

The following measurements were made from each abalone: sex, whole weight, meat weight, visceral weight, maximum diameter, minimum diameter, vertical shell height, shell weight (after cleaning and drying to constant weight in a 60°C oven) and count of the number of growth checks (rings) in the shell. The cross-sectional area of the conical appendage of the visceral mass, containing both gonad and digestive gland tissue, was traced for estimation of fecundity (see below).

Growth rates and aging

A technique developed by Muñoz-Lopez (1976) for aging abalone from growth checks in the shell has been shown by Prince *et al.* (1988a) to be valid for Tasmanian blacklip abalone. Three distinct minor layers are laid down in the first 18 months of life (Prince *et al.*, 1988a), with subsequent major layers being deposited annually. Age was calculated as the number of major rings present, plus 1.5 years to account for the period during which the minor rings are deposited. Prince (1989) showed that the growth curves obtained by this aging method, and independently by mark/recapture methods, were virtually identical.

The age–length couplets for all ageable abalone sampled were fitted to the Gompertz growth function:

$$l_t = l_0 \cdot \exp(G(1 - \exp(-gt))) \tag{23.1}$$

where l_t is length at time t, measured from a conventional zero at the time the fish would have been of length l_0; Gg is the instantaneous rate of growth when $t = 0$ and $l = l_0$; g is the instantaneous rate of decrease of the instantaneous rate of growth; and the upper asymptote $(L_\infty) = l_0 \cdot e^G$ when t is indefinitely large (Ricker, 1975).

The age–length couplets were fitted to the Gompertz growth function using the FISHPARM program of Prager *et al.* (1987), which uses Marquardt's algorithm for non-linear least squares parameter estimation. The Gompertz growth function was employed, instead of the more commonly used von Bertalanffy growth function (VBGF), because growth of *H. rubra*, in both length and weight, is clearly sigmoidal when expressed as a function of age. When age–length couplets were fitted to the VBGF, a t_0 value of 2–3 years was consistently found (Nash *et al.*, in prep.).

Sigmoidal increase in length indicates that the rate of increase in length does not decline linearly from birth, as required for a good fit to the VBGF, but increases to a maximum some time after birth, before declining. A positive t_0 value implies that growth, when measured in terms of length, increases sigmoidally with age. The Gompertz growth function is suitable for fitting growth to a sigmoid curve. Descriptions of the Gompertz growth function and its properties may be found in Yamaguchi (1975, 1977), Haukioja & Hakala (1979) and Kaufmann (1981).

Onset of sexual maturation

Abalone were considered sexually mature if gonad tissue could be detected; presence of gonad was equated with sexual maturity. It is unlikely that spent (fully spawned) gonad was misidentified as immature, since *H. rubra* are in a ripe or semi-ripe condition throughout the year (Harrison & Grant 1971; Division of Sea Fisheries, unpublished data). Furthermore, the gonad of *H. rubra* remains clearly identifiable as ovary or testis following artificially-induced spawning (pers. obs.).

Fecundity determination

Gonad volume is a good measure of fecundity. Newman (1967) and Tutschulte & Connell (1981) have shown that gonad volume is closely related to fecundity, measured as number of ripe eggs in the gonad. Although this method does not measure absolute fecundity (number of eggs), this is not necessary for the EPR analysis, since only the relationship between fecundity and age (or fecundity and length) is required.

Gonad volume was determined by measuring the cross-sectional area of the gonad. This measure, taken at the base of the excised conical appendage, is closely and linearly related to gonad volume (Nash *et al.*, in prep.), and is easily obtained.

Cross-sections of the excised conical appendages were traced onto transparent acetate (plastic) sheets. These were then transferred to an Apple Macintosh computer via an optical scanner, and the area of both gonad and digestive gland measured by a computer program, written in BASIC, which counts the number of pixels in each area. The number of pixels was then converted to mm^2.

It is assumed that fecundity as measured by gonad cross-sectional area is proportional to the number of eggs released, over all individuals and sites.

Mortality

Instantaneous total mortality (Z) was calculated by catch-curve analysis: the frequency of each age class (logarithmically transformed) was plotted against age, and Z estimated as the negative slope of the regression line, fitted to all fully-recruited age classes. The first fully-recruited age class was defined as the one immediately after the modal age. This does not mean that, after age at full recruitment, all abalone were catchable; it means that they were then of equal catchability. The rate of decline in frequency then is a measure of Z.

Strictly speaking, this definition of full recruitment applies only to unfished populations; the first age at full recruitment to the fishery should be the first age for which the members of that age class will not grow larger than the legal size limit. It was found, however, that variation in individual length-at-age was so great that very few (if any) age classes fit this criterion. Since there was no noticeable change in the rate of decline of abundance of the age classes after the modal age (no increase in rate of decline as legal size was reached), all age classes following the modal age were used to calculate Z.

Since the stunted abalone stocks in northern coastal waters are subject to very light (or zero) fishing pressure, the total mortality here was an estimate of instantaneous natural mortality (M). In the absence of independently-derived estimates of M, the value of Z obtained for these stunted stocks was used as an estimate of M elsewhere around Tasmania. Values of Z obtained for these stunted stocks ranged from 0.164 to 0.278 (mean value 0.216) (Nash *et al.*, in prep.). In the interpretation of the EPR analyses below, a value of M of 0.2 was used, except where this was obviously invalid, as outlined below.

For the purpose of this chapter, natural mortality was treated as constant over the range of age classes examined in this model. Prince *et al.* (1988c) found that M declined with age until about 3 years old, after which it was approximately constant up to age 7 years, the maximum age for which abundance and mortality were measured. The effect of non-constant M on EPR results and their interpretation is discussed below.

Egg-per-recruit analysis

Data required for the EPR analyses were the Gompertz growth parameters (l_0, G and g), maximum age, instantaneous natural mortality coefficient M, the length–weight relationship

$$\text{weight} = a\cdot(\text{length})^b \tag{23.2}$$

and the length–fecundity relationship

$$E_t = a\cdot(\text{length})^b \tag{23.3}$$

where fecundity (E_t) was expressed as the gonad cross-sectional area. It is assumed that spawning occurs once per year and that all eggs are released.

In addition, an age-related selection coefficient was used to adjust F over the range of ages at which maturation occurred. This was done because, as is demonstrated below, abalone become vulnerable to capture at the onset of sexual maturity. The proportion of each age class that is vulnerable to fishing therefore increases from 0 in immature age classes to 1 in those age classes which contain only sexually mature individuals. The selection coefficient increases monotonically from 0 to 1 in the partially mature age classes.

To construct EPR isopleth diagrams, F was varied systematically from 0.0 to 2.0 in increments of 0.25. The minimum size at first capture was varied between the areas studied, since the size range suitable for, for example, the stunted population was inappropriate for the population with rapid growth.

EPR was calculated as

$$\sum_{t=t_R}^{t\lambda} E_t \cdot \exp(-Z_t) \cdot N_{t-1} \tag{23.4}$$

and expressed as a percentage of the egg production in the absence of fishing (i.e. $Z = M$):

$$\mathrm{EPR} = 100 \cdot \left\{ \sum_{t=t_R}^{t\lambda} E_t \cdot \exp(-Z_t) \cdot N_{t-1} \right\} \Big/ \left\{ \sum_{t=t_R}^{t\lambda} E_t \cdot \exp(-M) \cdot N_{t-1} \right\} \tag{23.5}$$

Z_t was calculated as

$$Z_t = M + F \cdot q_t \tag{23.6}$$

where q_t is the age-specific selection coefficient, calculated as the proportion of each age class that was sexually mature; q_t ranged from 0 for age classes that were wholly sexually immature to 1 for those age classes containing only sexually mature animals; t_R is age at recruitment to the fishery and $t\lambda$ is longevity (in years), measured as the maximum age of abalone in a sample.

RESULTS

The age–length plots demonstrate the increase in maximum size from north to south (Fig. 23.2 A,D,G). The L_∞ values calculated from the Gompertz growth parameters range from 124 mm maximum shell length for the northern Babel Island (site 1) population to 172.6 mm for the southern Ketchem Bay sample (Table 23.1). Growth rate shows a similar trend, although not so pronounced. Growth in all populations sampled to date exhibits a sigmoid relationship when expressed as length versus age.

In contrast to this variation in growth around the state, the onset of maturation follows a similar pattern at all sites. Maturation commences at about 6–7 years of age and increases steadily until, by age 12–13 years, 100% of the age class is sexually mature (Fig. 23.2 B,E,H,K). Age at 50% maturity ranges from 7.5 years for the Babel Island sites to 10.2 years for the Gardens site 10 population

Table 23.1 Parameter values used in the egg-per-recruit analyses for *Haliotis rubra*. The maximum asymptotic length (L_∞) is also shown.

Site	l_0	G	g	a	b	a'	b'	L_∞
Babel Island site 1	0.164	6.628	0.314	1.450×10^{-4}	2.977	5.392×10^{-5}	3.000	124.0
Babel Island site 2	1.461	4.564	0.234	7.286×10^{-5}	3.125	4.464×10^{-5}	3.035	140.2
Gardens site 7	1.210	4.852	0.273	1.176×10^{-4}	3.008	6.274×10^{-11}	5.641	154.9
Gardens site 10	0.449	5.783	0.292	5.806×10^{-5}	3.147	3.285×10^{-4}	2.574	145.8
Top Rocks	0.574	5.608	0.288	5.170×10^{-5}	3.203	7×10^{-9}	4.485	156.5
Ketchem Bay	5.218	3.499	0.231	2.522×10^{-5}	3.341	1×10^{-9}	5.198	172.6

l_0, G, g: parameters of the Gompertz growth function (equation (23.1)).
a, b: constants in the length–weight relationship (equation (23.2)).
a', b': constants in the length–fecundity relationship (equation (23.3)).

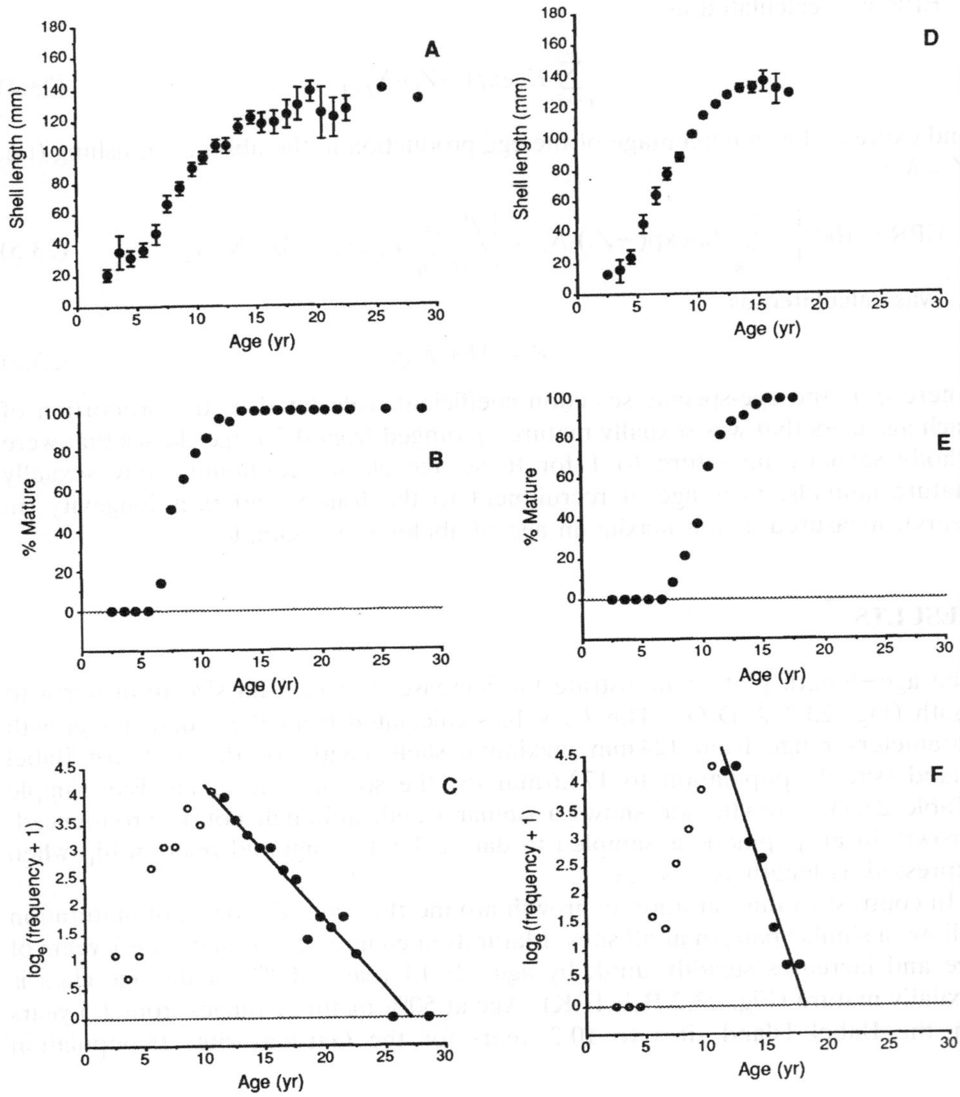

Fig. 23.2 Growth, maturation and mortality of *Haliotis rubra* from Babel Island site 1 (A–C), The Gardens site 10 (D–F), Top Rocks (G–I) and Ketchem Bay (J–L). (A,D,G,J) Age–length relationship; (B,E,H,K) age at maturity; (C,F,I,L) age–frequency plots, from which Z was calculated as –(slope of the regression line fitted to the solid data points (●)). Error bars in (A), (D), (I) and (J) are the 95% confidence limits about the mean.

(Table 23.2); the corresponding lengths are 64 and 110 mm respectively. Not all the difference in length maturity between these two sites can be explained by differences in age: the mean increase in length between ages 7.5 and 10.2 years for the Gardens site 10 is 32 mm (calculated with the Gompertz growth parameter values in Table 23.1), whereas the difference in length at maturity between these two sites is 46 mm (Table 23.2). The greater variation in length at maturity than age at maturity is also shown by the coefficients of variation (defined as 100 × standard deviation/$\bar{x}$ (Table 23.2). Onset of maturation is therefore primarily age

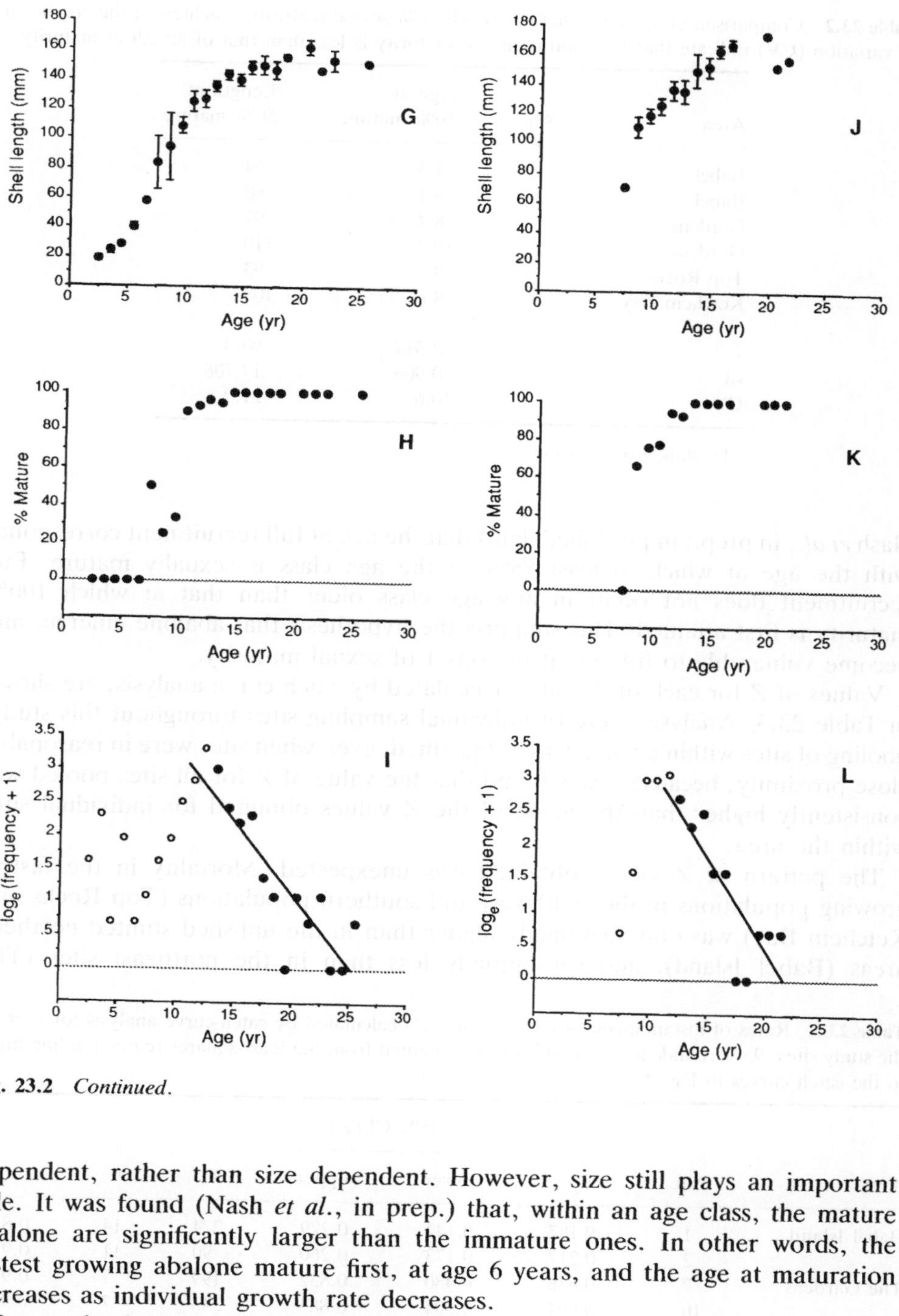

Fig. 23.2 *Continued.*

dependent, rather than size dependent. However, size still plays an important role. It was found (Nash *et al.*, in prep.) that, within an age class, the mature abalone are significantly larger than the immature ones. In other words, the fastest growing abalone mature first, at age 6 years, and the age at maturation increases as individual growth rate decreases.

It can also be seen from Fig. 23.2 that an age class becomes fully recruited to the fishery when it becomes fully sexually mature: using the results for Babel Island site 1 as an example, the 11-year class is the first fully-recruited age class (Fig. 23.2C), and the proportion of abalone within this age class that has reached sexual maturity is very nearly 100%. It has been found consistently (Fig. 23.2;

Table 23.2 Comparison of age and length at which 50% sexual maturity is achieved; the coefficients of variation (CV) indicate that variation in age at maturity is less than that of length at maturity.

Area	Site	Age at 50% mature	Length at 50% mature
Babel	1	7.5	64
Babel	2	8.1	68
Gardens	7	8.4	97
Gardens	10	10.2	110
Top Rocks		8.3	93
Ketchem Bay		8.6	108
$\bar{x}$		8.517	90.0
sd		0.906	19.708
CV		10.6	21.9

sd = standard deviation

Nash *et al.*, in prep.; unpublished data) that the age at full recruitment corresponds with the age at which at least 85% of the age class is sexually mature. Full recruitment does not occur in any age class older than that at which 100% maturity is first attained. This supports the hypothesis that abalone emerge, and become vulnerable to fishing, at the onset of sexual maturity.

Values of Z for each of the sites, calculated by catch curve analysis, are shown in Table 23.3. Analyses were of individual sampling sites throughout this study; pooling of sites within an area was not justified, even when sites were in reasonably close proximity, because it was found that the value of Z for all sites pooled was consistently higher than the mean of the Z values obtained for individual sites within the area.

The pattern of Z values obtained was unexpected. Mortality in the faster-growing populations in the southwest and southern populations (Top Rocks and Ketchem Bay) was only marginally higher than in the unfished stunted northern areas (Babel Island), and substantially less than in the northeast sites (The

Table 23.3 Rates of instantaneous total mortality (Z) calculated by catch-curve analysis for each of the study sites. 95% confidence limits (CL) were obtained from the least-squares regression line fitted to the catch curves in Fig. 23.2.

			95% CL(Z)				
Area	Site	Z	lower	upper	N	n	r^2
Babel Island	1	0.187	0.144	0.229	224	14	0.834
	2	0.217	0.173	0.260	90	11	0.925
The Gardens	7	0.656	0.480	0.832	199	7	0.948
	10	0.694	0.504	0.884	343	7	0.946
Top Rocks	—	0.218	0.130	0.305	143	13	0.732
Ketchem Bay	—	0.271	0.116	0.427	109	10	0.671

N = number of agable abalone sampled at each site.
n = number of data points used to calculate Z in Fig. 23.3.
r^2 = coefficient of determination of the regression line.

Gardens) (Table 23.3). The Z value of 0.218 obtained for the fished Top Rocks site is virtually the same as that for the unfished Bass Strait sites, where $\bar{Z} = 0.216$. Hence, the state-wide application of a single M value of 0.2, obtained from the stunted northern stocks, is not valid. It is certain that F is substantially higher than 0.018 for the Top Rocks sample, because catch and effort statistics show that about 25–35% of the total fishing effort is of, and a slightly larger proportion of the total Tasmanian catch is taken from, the west and southwest coastal regions of Tasmania annually (Sea Fisheries Division, unpublished data).

Parameter values for the length–weight and length–fecundity relationships (equations (23.2) and (23.3)), used in the EPR analysis, are shown in Table 23.1. The EPR isopleths obtained are illustrated for the Babel Island site 1 and The Gardens site 7 populations in Fig. 23.3, using $M = 0.2$. For the purpose of comparison of EPR between sites, a range of M values was used for both the Top Rocks and Ketchem Bay sites, and 0.2 for all other sites. The M values used for the Top Rocks and Ketchem Bay sites almost certainly encompass the true value of M. For the four fished sites (The Gardens sites 7 and 10, Top Rocks and Ketchem Bay), the current size limit, in conjunction with the F value determined as $Z - M$, gives the levels of egg production conserved shown in Table 23.4. Egg per recruit was moderate to high (40–47%) at all sites (Table 23.4).

The sensitivity of EPR analysis to variation in M can be seen from Table 23.4 for both the Top Rocks and Ketchem Bay populations. M values of 0.05, 0.10 and 0.15 were used for Top Rocks, and 0.10, 0.15 and 0.20 for Ketchem Bay. For the Top Rocks site, EPR values of 40, 56 and 74% were obtained for $M = 0.05$, 0.10 and 0.15 respectively; for the Ketchem Bay site, EPR values of 44, 58 and 74% were obtained when $M = 0.10$, 0.15 and 0.20 respectively.

Between-site differences in EPR were then compared in two other ways: first, by determining the proportion of egg production at the present minimum legal size that would be conserved at moderate to high levels of F ($F = 1$ to 2); and

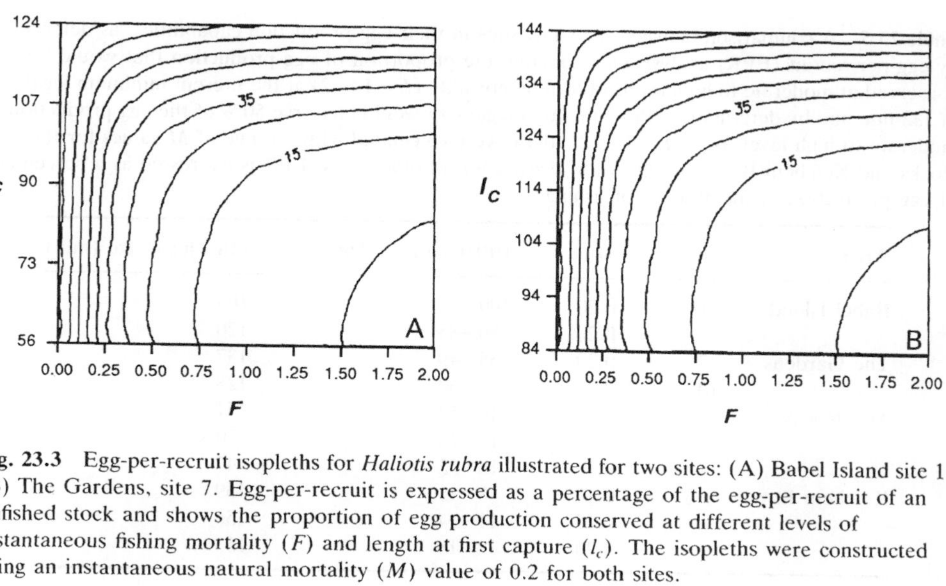

Fig. 23.3 Egg-per-recruit isopleths for *Haliotis rubra* illustrated for two sites: (A) Babel Island site 1; (B) The Gardens, site 7. Egg-per-recruit is expressed as a percentage of the egg-per-recruit of an unfished stock and shows the proportion of egg production conserved at different levels of instantaneous fishing mortality (F) and length at first capture (l_c). The isopleths were constructed using an instantaneous natural mortality (M) value of 0.2 for both sites.

Table 23.4 Egg-per-recruit (EPR) conserved at the four fished sample sites at the present combination of size limit (132 mm) and fishing mortality (F); $M=0.2$ for the two Gardens sites, whereas a range of M values, likely to encompass the true M, is given for the Top Rocks and Ketchem Bay sites, where M is almost certainly less than 0.2.

Site	Z	M	F	EPR(%)
Gardens 7	0.656	0.2	0.456	58
Gardens 10	0.694	0.2	0.494	70
Top Rocks	0.218	0.05	0.168	40
Top Rocks	0.218	0.10	0.118	56
Top Rocks	0.218	0.15	0.068	74
Ketchem Bay	0.271	0.10	0.171	44
Ketchem Bay	0.271	0.15	0.121	58
Ketchem Bay	0.271	0.20	0.071	74

EPR is expressed as a percentage of egg production in the absence of fishing.
$F=Z-M$.
Z values were obtained from Table 23.3.

second, by determining what minimum size limit would conserve 50% of the egg production at similarly high levels of F. The latter comparison was made because a conservative approach to management of abalone fisheries might be to maintain egg production at 50% of the egg production of the unfished population or greater (Breen, 1986). These comparisons are shown in Table 23.5.

By the first comparison, the proportion of the egg production conserved at the present size limit of 132 mm ranges from 80 to 100% at the Babel Island sites in the north to 8–18% (depending on the value of M) at the Ketchem Bay site in the south (Table 23.5). By the second comparison, the minimum size limit at which 50% of the egg production is conserved ranges from 103 mm at Babel Island site 1 to 153–159 mm at Ketchem Bay (depending on the value of M). By

Table 23.5 A comparison between the study sites in their egg production capabilities, as determined by egg-per-recruit (EPR) analysis by estimating the proportion of egg production that would be conserved at moderate to high levels of fishing mortality ($F=1$ to 2) at the present minimum legal size of 132 mm and by determining what minimum legal size would conserve 50% of the egg production at moderate to high levels of F ($F=1$ to 2); these were calculated with a range of M values for the Top Rocks and Ketchem Bay sites, and with $M=0.2$ for all other sites; EPR is expressed as a percentage of egg production in the absence of fishing.

Area	Site	M	EPR(%) at 132 mm	Min. legal size (mm)
Babel Island	1	0.20	100	103
	2	0.20	80–85	120
The Gardens	7	0.20	35–40	137
	10	0.20	55–60	128
Top Rocks		0.05	10–12	151
		0.10	13–16	150
		0.15	18–22	147
Ketchem Bay		0.10	8–11	159
		0.15	11–14	156
		0.20	14–18	153

both comparisons, there is a major difference between populations in levels of protection of egg production at high levels of fishing mortality, with the reduction at the southwest sites (Top Rocks and Ketchem Bay) being extreme.

DISCUSSION

Three features of the *H. rubra* stocks in Tasmania suggested a means by which optimal levels of egg production might be determined. These are (i) a latitudinal cline in growth rate and maximum size; (ii) maturation primarily by age rather than size; and (iii) a single legal minimum size limit throughout Tasmania. It seemed probable that some geographically intermediate stock received an optimal level of protection of egg production.

The successful application of this approach relies on (i) the accurate estimation of M for the fished populations subjected to EPR analysis and (ii) an ability to identify 'ideally-fished' populations. Since the total mortality (Z) of the southwestern and southern populations that were sampled is only slightly higher than the estimated natural mortality rate of the unfished stunted northern Babel Island populations (Table 23.3), it is clear that M varies considerably between populations; the application of a single M value derived from stunted populations to other, fished populations is therefore not justified.

This is particularly important because EPR analyses are sensitive to variation in M, as shown in Table 23.4, and reported by other workers (Tegner *et al.*, 1989; Annala & Breen, 1989). Nevertheless, by carrying out EPR analyses with a range of M values appropriate for each sampling site, it is possible to determine the likely confidence limits on the levels of egg production protection provided by the current size limit. The considerable variation in EPR that is caused by varying M, however, demonstrates that the estimate of EPR may be seriously inaccurate unless M can be determined with reasonable accuracy. The need for reliable site-specific estimates of M is clear.

There is no satisfactory set of criteria by which an ideally-fished population may be identified. For practical purposes, the advice of experienced abalone divers must be relied upon to identify populations which maintain high densities of pre-recruits despite reasonably intense fishing. Although The Gardens sites, geographically intermediate between the northern stunted stocks and the southern fast-growing stocks, were identified by abalone divers as productive stocks subjected to high fishing pressure, there were several sites elsewhere around Tasmania that were also regarded as highly productive. This suggests that factors other than levels of egg production influence the productivity of the stock. One such factor may be the rate of recruitment (that is, survival rate between the egg stage and recruitment to the fishery). If there is variation in larval survival rates between localities, then the proportion of egg production that must be conserved to maintain commercial densities of abalone will also vary: areas with high recruitment rates will sustain commercial densities at a lower level of egg production than areas with low recruitment rates. Hence, conserving 50% of the egg production may be adequate for a stock with high recruitment rates, whereas a much higher level of protection may be needed for stocks with low recruitment rates.

Evidence for variable recruitment rates between *H. rubra* populations in

Tasmania does exist: near The Gardens sample site in the northeast of Tasmania (Fig. 23.1), major differences in abundance were observed between an exposed site (where abundance was relatively low) and a more sheltered embayment (where abundance was much higher), despite close similarity of age–length and age–fecundity relationships and mortality rates (unpublished data). These observations strongly support the notion that the recruitment rate to the more sheltered site was higher.

At current levels of fishing mortality, these analyses suggest that the proportions of egg production conserved are reasonably high at all sites (Table 23.4). If *F* were to increase substantially, however, major reductions in egg production in the southwestern sites (Top Rocks and Ketchem Bay) would occur (Table 23.5). These populations are particularly sensitive to increases in *F* because of their fast growth, large maximum size and low *M*.

As an index of fecundity, gonad volume represents the potential fecundity only; unless all eggs in the gonad are released at spawning, the realized fecundity will be less. This is likely to be so (*e.g.* Chapter 8, this volume). It is not clear that the use of potential fecundity will yield the same age–relative fecundity relationship as the use of the realized fecundity, because the experience in abalone hatcheries suggests that younger, smaller abalone are more vigorous spawners, and spawn more frequently, than larger, older abalone (T. Lewis, K. Hahn, pers. comms). If this is true for wild abalone as well, the relationship between age and fecundity would be 'flatter' (less steep) than the volumentric index of fecundity, used here, would suggest. The effects of this flattening on the EPR analyses would be to reduce the sensitivity of the relationship between minimum legal size and egg per recruit; in other words, a higher size limit would not afford the increase in protection of egg production that the isopleths in Fig. 23.3 and EPR values in Table 23.4 indicate. The effect of this flattening of the age–fecundity relationship EPR has yet to be explored quantitatively; it may help explain the decline of abalone stocks in California that has occurred despite the apparent high degree of protection of the egg production by the size limits there reported by Tegner *et al.* (1989).

A further potential source of error in the estimate of fecundity may arise if the relationship between gonad volume and number of eggs spawned is not linear. There is no direct evidence that this is the case, although Han *et al.* (1989) demonstrated a reasonably close linear relationship between abalone body weight and number of eggs spawned following induction of spawning in the laboratory. The induction method used was exposure of the abalone to air for one hour followed by immersion in UV-irradiated seawater. On the other hand, Ault (1985) found that the relationship between total weight and number of eggs released for laboratory-conditioned and field-conditioned *H. rufescens* had an exponent of 1.88 and 1.49 respectively, when spawning was induced using the hydrogen peroxide method of Morse (1977). The effect of an exponential relationship of this sort on the form of the EPR curve would be the reverse of that caused by greater spawning vigour of young abalone, described in the previous paragraph; that is, protection of egg production would be particularly sensitive to changes in the minimum legal size limit. Linearity of the gonad volume–fecundity relationship, used in this study, may still exist if the total weight–fecundity relationship is exponential, if gonad volume increases exponentially with total weight. Clearly, further work is needed to clarify the length–fecundity and/or the weight–fecundity relationships of *H. rubra*.

The analyses carried out here assume that the relationship between number of growth layers in the shell and age, established by Prince *et al.* (1988a) at one site in southeast Tasmania, is applicable throughout Tasmanian waters. Various analyses (unpublished data) have been conducted, however, to test hypotheses that more than one shell layer is deposited annually, or that shell layer deposition is irregular; so far the results are consistent with the findings of Prince *et al.* (1988a).

Ability to determine the age of an abalone from growth rings in the shell is affected by the degree to which the spire of the shell is damaged by boring organisms (particularly sponges). The proportion of a sample that was agable varied greatly around Tasmania, and was least in the northern samples (Fig. 23.4). Since shell damage is likely to increase with time, it follows that the proportion of abalone that is impossible to age is likely to increase with age. If this were true, as suggested by Fig. 23.4, it would introduce an age-related bias into the estimation of the age–length relationship and, more importantly, into the determination of Z by catch-curve analysis. Z would tend to be overestimated because of the under-representation of older animals in the agable sample. Despite this, however, the highest Z values were found for The Gardens populations, which had the lowest proportion of non-agable shells (Table 23.3, Fig. 23.4). Overestimation of Z is most likely to have been at the Babel Island and Top Rocks sites, where the proportion of large abalone that was not agable was particularly high (Fig. 23.4).

Serious declines of some of the world's abalone stocks have occurred. In California, stocks have declined seriously (Burge *et al.*, 1975; Tegner *et al.*, 1989) despite the belief that appropriate size limits alone could maintain stocks (Heath, 1925; Cox 1962). Similarly, abalone stocks have declined in both Japan and the European Channel Islands, despite the fact that size limits have been an important component of the management strategies there (Bossy & Culley, 1976; Mottet, 1978). In some of these cases, the decline may have been caused by the inappropriate selection of size limits. It can be seen from Fig. 23.3 that there is a large area on each graph where high fishing pressure, combined with low size limits, leads to a low egg production. If minimum size limits fall in this region, then stock declines may occur because of recruitment failure (Breen, 1986). Since the size at which yield-per-recruit is maximized is often less than that at which a substantial proportion of the egg production is conserved (Breen, 1986), abalone fisheries that have been managed from yield-per-recruit considerations may be in danger of collapse through recruitment failure. Sluczanowski (1984, 1986) showed that a slight increase in size limit may substantially increase the conservation of egg production with only a slight loss in yield.

In an analysis of Californian abalone stocks, Tegner *et al.* (1989) showed that current legal minimum size limits conserve a substantial proportion of the egg production, and concluded that the decline in stocks there may have been for reasons other than overfishing. Similarly, Breen (1986) argued that the decline of the British Columbia fishery for *H. kamtschatkana* was not primarily caused by overfishing, since the time from larval settlement to recruitment to the fishery for this species is at least 6 years, and the fishery had been operating for only 8 years.

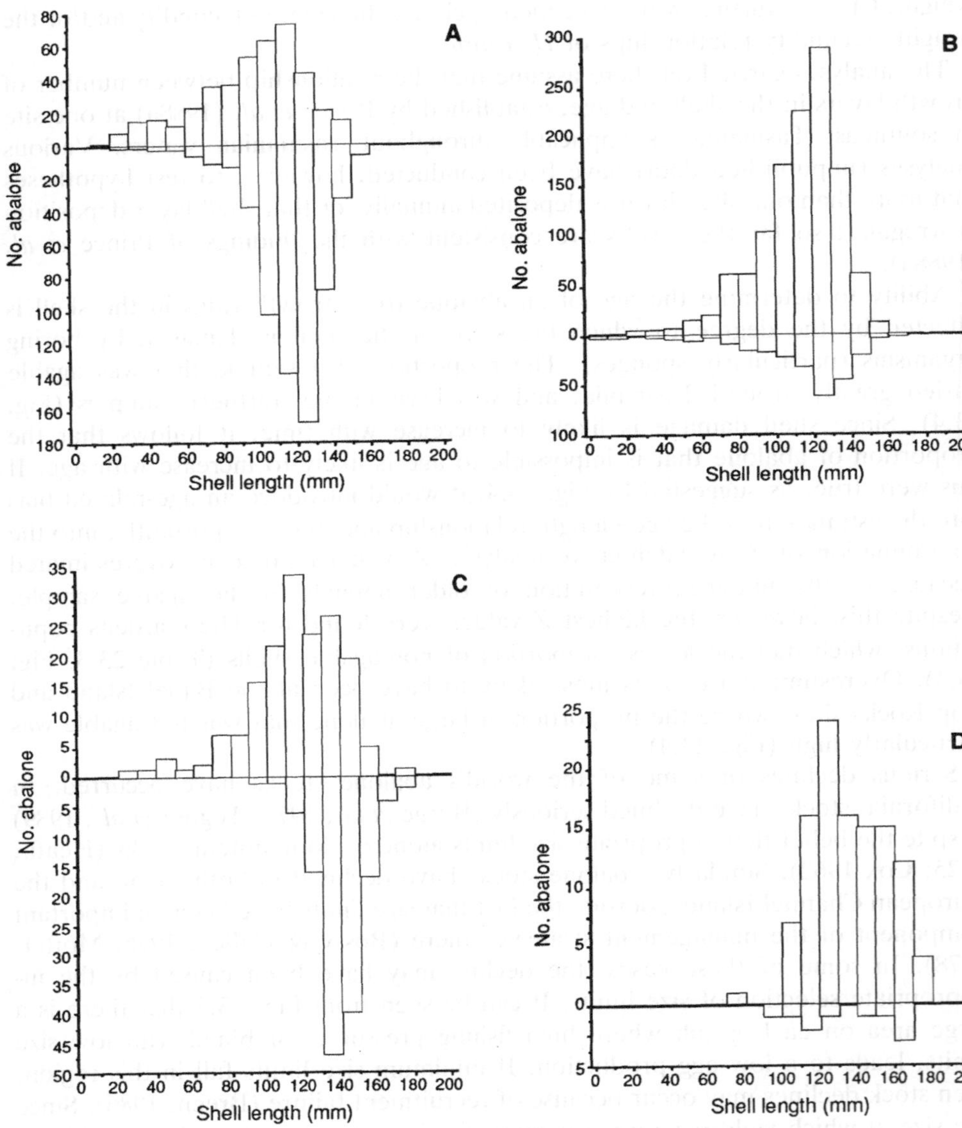

Fig. 23.4 Length composition of agable versus unagable *Haliotis rubra* from (A) Babel Island (all sites combined); (B) The Gardens (all sites combined); (C) Top Rocks; (D) Ketchem Bay. The agable fraction is above the X-axis.

This suggests that recruitment success was sporadic, and that occasional heavy recruitment in some years was insufficient to sustain the fishery.

If abalone recruitment rates fluctuate substantially due to natural causes alone, as Hayashi (1980), Forster *et al.* (1982) and Sainsbury (1982), for example, have concluded, then the stocks will be particularly in danger of overfishing and recruitment failure if fishing pressure is sustained during the periods of natural depletion. Management of these species to conserve egg production will not prevent these stocks from declining, and may not reduce the risk of recruitment failure, since the EPR approach to fishery management ignores spatial and ecological factors, such as the reduced fertilization rate that may occur at low abalone densities (Tegner *et al.*, 1989), although the apparent aggregation of abalone to spawn (Breen & Adkins, 1980; Shepherd, 1986) may help to maintain fertilization rates when abundances are low.

Despite the uncertainty of *M*, the results presented here suggest that, as for the abalone fisheries in California (Tegner *et al.*, 1989), conservation of egg production in *H. rubra* in Tasmania is reasonably high. If commercial abalone divers' accounts of recent stock declines in Tasmania are correct, then it would appear that either (i) those declining stocks had lower recruitment rates, and hence less resilience to fishing; (ii) the declines were for reasons other than overfishing; or (iii) there is a direct relationship between stock size and recruitment to the fishery even at quite high levels of abundance. These explanations are not mutually exclusive. If abalone stocks have site-specific recruitment rates (that is, survival rates between the egg stage and recruitment to the fishery), as some of our results suggest (unpublished data), then those with the lowest recruitment rates will be most vulnerable to overfishing at any given legal minimum size limit. The declines in abundance observed by divers in some areas may not reflect a more widespread state of overfishing, but rather represent an inevitable consequence of the development of a fishery, where the least productive stocks cannot be sustained at virtually any level of fishing.

Tegner *et al.* (1989) suggested that the decline of *H. rufescens* in California may have been exacerbated by the nutrient-depleting effects of El Niño events. Similar climatic effects on fisheries have been reported in southeastern Australia by Harris *et al.* (1988), who demonstrated a relationship between mean annual wind strength, water temperature and the size of the annual catch of the rock lobster *Jasus novaehollandiae* in Tasmania. If the low nutrient content of marine algae during periods of warm water influx found in California (Tegner & Dayton, 1987) exists also in Tasmania during warm periods, then low survival rates of *H. rubra* may also occur in Tasmania when seawater temperatures become elevated.

Unfortunately, there is no information available on the stock–recruitment relationship for *H. rubra* in Tasmania. Bannister & Addison (1986) have argued that a direct relationship at high levels of stock abundance would be unstable, since a population would tend to collapse as fishing effort increased. However, the fact that abalone fisheries around the world have collapsed indicates that this possibility cannot be ruled out.

If abalone recruitment rates to the fishery do fluctuate greatly because of natural (non-fishery-related) factors, how might fisheries of these species be safely managed? One approach might be to monitor recruitment rates of (say) one-year old abalone. If observed abundances of these young age classes closely index subsequently recruitment to the fishery, then, theoretically, annual catch could be adjusted accordingly. In practice, however, this approach would not work, because the individual variation in both growth rate (Shepherd & Hearn, 1983; Nash *et al.* in prep.) and age at onset of sexual maturation (Fig. 23.2) indicates that the members of a cohort will enter the fishery over a number of years. The problem is compounded if the numerous unit stocks around the state are subject to a variety of environmental conditions at any one time; this may be the case, given the variety of habitats and hydrological conditions around Tasmania. Hence, some areas may be experiencing poor conditions for recruitment while others are unaffected.

The best approach to management therefore appears to be to conserve egg production, even at the cost of some potential yield (Sluczanowski, 1984, 1986; Tegner *et al.*, 1989), to implement long-term monitoring of abalone abundance at representative sites around the state and to develop guidelines for responding to any observed long-term changes in abundance that might occur.

ACKNOWLEDGEMENTS

The author thanks Tony Wurf, Robbie Lowden and Jack Bock, who offered their vessels, crew and time to survey the more remote coastal waters of Tasmania. Tracy Brothers, Simon Talbot, Wes Ford, James Bridley, Andrew Cawthorn, Martine Nash-Lamaury, Melissa Millington and Natalie Wickham provided invaluable assistance in the field and laboratory. Iain Larner wrote the BASIC computer program to calculate gonad area from the cross-sectional tracings. This work was supported by the Division of Sea Fisheries Abalone Research Fund, and by the sale of the abalone meats obtained from the survey samples. The author is grateful for helpful discussions with Jeremy Prince, Keith Sainsbury, Tony Harrison and Paul McShane, and to Paul Breen, Rob Day, Scoresby Shepherd and Philip Sluczanowski for reviewing this manuscript.

REFERENCES

Annala J.H. & Breen P.A. (1989) Yield- and egg-per-recruit analyses for the New Zealand rock lobster, *Jasus edwardsii*. *NZ J. Mar. Freshwat. Res.*, **23**, 93–105.

Ault J.S. (1985) Some quantitative aspects of reproduction and growth of the red abalone, *Haliotis rufescens* Swainson. *J. World Maricult. Soc.*, **16**, 398–425.

Bannister R.C.A. & Addison J.T. (1986) Effect of assumptions about the stock–recruitment relationship on a lobster (*Homarus gammarus*) stock assessment. *Can. J. Fish. Aquat. Sci.*, **43**, 2353–9.

Bossy S. & Culley M. (1976) Channel Islands 'ormer' fishery faces uncertain future. *Fishing News Int.*, Sept. 1976, 55–60.

Breen P.A. (1986) Management of the British Columbia fishery for northern abalone (*Haliotis kamtschatkana*). *Can. Spec. Publ. Fish. Aquat. Sci.*, **92**, 300–12.

Breen P.A. & Adkins B.E. (1980) Spawning in a British Columbia population of northern abalone, *Haliotis kamtschatkana*. *Veliger*, **23**, 177–9.

Burge R., Schultz S. & Odemar M. (1975) Draft report on recent abalone research in California, with recommendations for management. *Calif. Fish Game Comm. Calif. Dept Fish Game*, unpublished report.

Cox K.W. (1962) California abalones, family Haliotidae. *Calif. Dept Fish Game Fish. Bull.*, **118**.

Forster G.R., Potts G.W. & Swinden R. (1982) Changes in the ormer populations of Guernsey and Jersey. *J. Mar. Biol. Assoc. UK*, **62**, 717–27.

Han S.J., Lee J.U., Kim B.G. & Baek J.H. (1989) A study on the spawning experiment of abalone, *Haliotis discus* Reeve, reared in a temperature-controlled indoor tank. *Bull. Nat. Fish. Dev. Agency*, **43**, 173–80.

Harris G.P., Davies P., Nunez M. & Meyers G. (1988) Interannual variability in climate and fisheries in Tasmania. *Nature*, **333**, 754–7.

Harrison A.J. (1986) Gastropod fisheries of the Pacific with particular reference to Australian abalone. *Can. Spec. Publ. Fish. Aquat. Sci.*, **92**, 14–22.

Harrison A.J. & Grant J.F. (1971) Progress in abalone research. *Tasmanian Fish. Res.*, **5**, 1–10.

Haukioja E. & Hakala T. (1979) Asymptotic equations in growth studies – an analysis based on *Anodonta piscinalis* (Mollusca, Unionidae). *Ann. Zool. Fenn.*, **16**, 115–22.

Hayashi I. (1980) Structure and growth of a shore population of the ormer, *Haliotis tuberculata*. *J. Mar. Biol. Assoc. UK*, **60**, 431–7.

Heath H. (1925) Abalones have sufficient protection. *Calif. Fish Game*, **11**, 138–9.

Hooker N. & Morse D.E. (1985) Abalone: The emerging development of commercial cultivation in the United States. In *Crustacean and Mollusk Aquaculture in the United States* (Ed. by J.V. Huner & E.E. Brown), pp. 365–413. AVI, Westport.

Kaufmann K.W. (1981) Fitting and using growth curves. *Oecologia (Berl.)*, **49**, 293–9.

Morse D.E. (1977) Hydrogen peroxide induces spawning in molluscs with activation of prostaglandin endoperoxide synthetase. *Science*, **196**, 298–300.

Mottet M.G. (1978) A review of the fishery biology of abalones. *Washington State Dept. Fish. Tech. Rep.*, **37**.

Muñoz-Lopez T. (1976) Resultados preliminares de un método para determinar edad en abulones (*Haliotis* spp.) de Baja California. In *Memorias del Simposio sobre Recursos Pesqueros Masivos de México Ensenada. B.C. del 28 al 30 de Septiembre, 1976*, pp. 281–301.

Nash W.J., Sellers T., Talbot S. in prep. The population biology of abalone (*Haliotis* species) in Tasmania. I. Blacklip abalone (*H. rubra*) from the north coast and the Furneaux group of islands. *Tasmanian Div. Sea Fish., Tech. Rep. Series*.

Newman G.G. (1967) Reproduction of the South African abalone *Haliotis midae*. *Rep. S. Afr. Div. Sea Fish., Invest. Rep.*, **64**.

Prager M.H., Saila S.B. & Recksiek C.W. (1987) FISHPARM: A microcomputer program for parameter estimation of nonlinear models in fishery science. *Old Dominion University Research Foundation Technical Report*, 87–10.

Prince J.D. (1989) *The fisheries biology of the Tasmanian stocks of* Haliotis rubra. PhD dissertation, University of Tasmania.

Prince J.D., Sellers T.L., Ford W.B. & Talbot S.R. (1987) Experimental evidence for limited dispersal of haliotid larvae (genus *Haliotis*; Mollusca: Gastropoda). *J. Exp. Mar. Biol. Ecol.*, **106**, 243–63.

Prince J.D., Sellers T.L., Ford W.B. & Talbot S.R. (1988a) A method for ageing the abalone *Haliotis rubra* (Mollusca: Gastropoda). *Aust. J. Mar. Freshwat. Res.*, **39**, 167–75.

Prince J.D., Sellers T.L., Ford W.B. & Talbot S.R. (1988b) Confirmation of a relationship between the localized abundance of breeding stock and recruitment for *Haliotis rubra* Leach (Mollusca: Gastropoda). *J. Exp. Mar. Biol. Ecol.*, **122**, 91–104.

Prince J.D., Sellers T.L., Ford W.B. & Talbot S.R. (1988c) Recruitment, growth, mortality and population structure in a southern Australian population of *Haliotis rubra* (Mollusca: Gastropoda). *Mar. Biol.*, **100**, 75–82.

Ricker W.E. (1975) Computation and interpretation of biological statistics of fish populations. *Bull. Fish. Res. Bd Can.*, **191**.

Sainsbury K.J. (1982) Population dynamics and fishery management of the paua, *Haliotis iris*. II. Dynamics and management as examined using a size class population model. *NZ J. Mar. Freshwat. Res.*, **16**, 163–73.

Shepherd S.A. (1986) Studies on southern Australian abalone (genus *Haliotis*). VII. Aggregative behaviour of *H. laevigata* in relation to spawning. *Mar. Biol.*, **90**, 231–6.

Shepherd S.A. & Hearn W.S. (1983) Studies on southern Australian abalone (genus *Haliotis*). IV. Growth of *H. laevigata* and *H. rubra*. *Aust. J. Mar. Freshwat. Res.*, **34**, 461–75.

Sluczanowski P.R. (1984) A management-oriented model of an abalone fishery whose substocks are subject to pulse fishing. *Can. J. Fish. Aquat. Sci.*, **41**, 1008–14.

Sluczanowski P.R. (1986) A disaggregate model for sedentary stocks: the case for South Australian abalone. *Can. Spec. Publ. Fish. Aquat. Sci.*, **92**, 393–401.

Tegner M.J., Breen P.A. & Lennert C.E. (1989) Population biology of red abalones, *Haliotis rufescens*, in southern California and management of the red and pink, *H. corrugata*, abalone fisheries. *Fish. Bull.*, **87**, 313–39.

Tegner M.J. & Dayton P.K. (1987) El Niño effects on southern California kelp forest communities. *Adv. Ecol. Res.*, **17**, 243–79.

Tutschulte T.C. & Connell J.H. (1981) Reproductive biology of three species of abalones, *Haliotis*, in southern California. *Veliger*, **23**, 195–206.

Yamaguchi M. (1975) Estimating growth parameters from growth rate data: problems with marine sedentary invertebrates. *Oecologia (Berl.)*, **20**, 321–32.

Yamaguchi M. (1977) Shell growth and mortality rates in the coral reef gastropod *Cerithium nodulosum* in Pago Bay, Guam, Mariana Islands. *Mar. Biol.*, **44**, 249–63.

PART VII
CASE STUDIES OF ABALONE FISHERIES

Chapter 24
A review of the biology of abalone and its fishery in Mexico

SERGIO A. GUZMÁN DEL PRÓO *Instituto Politécnico Nacional, México City, México*
(English translation by S.A. Shepherd)

ABSTRACT: The earliest statistical records for the Mexican abalone fishery in Baja California date from 1923, when the catch was 1721 tonnes. It rose to 6000 t in 1950, but the fishery has since collapsed. The current catch is about 600 t per year taken by 19 fishing cooperatives.

Research has been conducted since 1966, mainly on the life history, abundance and population dynamics of *Haliotis fulgens* and *H. corrugata*. There are good data on catch and fishing effort, geographic and vertical distribution, growth and mortality rates and length frequency, species and sex composition from landings mainly for central Baja California. Studies of fecundity, size at maturity and juvenile and adult ecology have commenced only in the last five years. Little progress has been made on the artificial culture and reseeding of juveniles or barrel culture.

RESUMEN: El registro de captura más antiguo de la pesquería mexicana de abulón (*Haliotis* spp.) en Baja California se remonta al año de 1929. En ese año se capturaron 1721 t, alcanzándose una producción record de 6000 t en 1950, sin embargo, actualmente es una pesquería colapsada con 500 a 600 t/año; con capturas realizadas por 19 cooperativas pesqueras. Las investigaciones sobre estas especies se inician en 1966. Los primeros trabajos fueron descriptivos de la pesquería y sobre su biología. En los últimos 15 años la investigación se ha concentrado sobre la biología de *H. corrugata* y *H. fulgens* y su dinámica poblacional, sin embargo el conocimiento logrado aún es parcial y limitado a algunas áreas de la península. Existe una información relativamente amplia sobre captura y esfuerzo de las principales zonas de captura. Aspectos tales como su distribucion geográfica y vertical, crecimiento, mortalidad y composición de la captura por especies tallas y sexos para la zona central de Baja California, parecen estar ya bien cubiertos. Otros, como fecundidad y madurez gonádica han sido abordados con relativa intensidad en los últimos años, sin embargo, la falta de estudios histológicos sistemáticos impiden concluir con certeza sobre el número y épocas precisas de desove de cada especie. Temas como la ecología de adultos, y juveniles, cultivos experimentales, dispersión larval, etc., son líneas de investigación que apenas se inician, mientras tanto el cultivo piloto de estas especies y su resiembra o engorda de juveniles en canastas y barriles concentran actualmente el interés general de los laboratorios de cultivo del gobierno y de las Cooperativas. Se concluye que es todavía necesario realizar investigaciones sobre temas selectos de la ecología y la biología de éstas especies, pero concentrando la atención sobre aquellos factores bióticos y abióticos que determinan el éxito del reclutamiento de juveniles y la abundancia final de reclutas al stock adulto.

INTRODUCTION

The papers of de Buen (1960), Chapa (1963) and Ortiz (1966) provide the earliest knowledge of the fishery and biology of abalone in Baja California. Subsequent studies have mainly addressed the population dynamics of the various species, with the goal of arresting the decline and facilitating the recovery of abalone

Table 24.1 The scientific and Mexican and Californian common names for the Mexican species of abalone.

Species	Common name	
	Mexican	Californian
Haliotis corrugata Gray 1828	yellow abalone	pink
H. fulgens Philippi 1854	blue abalone	green
H. cracherodii Leach 1817	black abalone	black
H. rufescens Swainson 1822	red abalone	red
H. sorenseni Bartsch 1940	chinese abalone	white
H. assimilis Dall 1878	striped abalone	threaded

stocks which from the early 1970s began to decline sharply. Unfortunately this decline continued until 1985, when production fell to 400 t a year.

In view of the great value of abalone to the local economy of Baja California, and the intense interest shown by fishermen's cooperatives to establish mariculture facilities, a review of current knowledge and status of the biology and ecology of *Haliotis* in Baja California is overdue. This chapter reviews only studies carried out in Mexico.

The taxonomy of the Mexican species of abalone was reviewed by Cox (1962). The common (Mexican and Californian) and scientific names of the species are listed in Table 24.1. Additional subspecies are *H. fulgens turveri*, described for specimens from Magdalena Bay, and *H. fulgens guadalupensis* and *H. cracherodii californiensis* for material from Guadalupe Island, (see Cox, 1962; Talmadge, 1964).

DISTRIBUTION

The distribution of the main species, *H. fulgens*, *H. corrugata* and *H. cracherodii*, extends throughout the greater part of Baja California from the Islas Coronado in the north to Isla Margarita in the south (Robles, 1970; Muñoz, 1976a). They are most abundant in the central coast of the peninsula from Cedros to Punta Asunción, although before the 1970s *H. corrugata* was abundant as far south as Punta Abreojos.

H. rufescens occurs between Islas Coronado and Punta Baja in Bahía del Rosario. Cox (1962) states that it occurs to Bahía Tortugas, but here it is quite rare. *Haliotis sorenseni* occurs occasionally in Cedros, Punta Eugenia and Bahia Tortugas (Fig. 24.1).

In general, these species share the same habitat, especially *H. fulgens* with *H. corrugata* and *H. rufescens* with *H. sorenseni*. The species have different but overlapping vertical distribution patterns, (Fig. 24.2) and surveys (Guzmán del Próo *et al.*, 1976) show that maximum densities usually occur at slightly different depths, implying the evolution of different niches (Leighton, 1974).

In summary, the depth distributions of the species are:

H. corrugata – (3–) 10–20 (–30) m;

H. cracherodii – intertidal (–3) m;

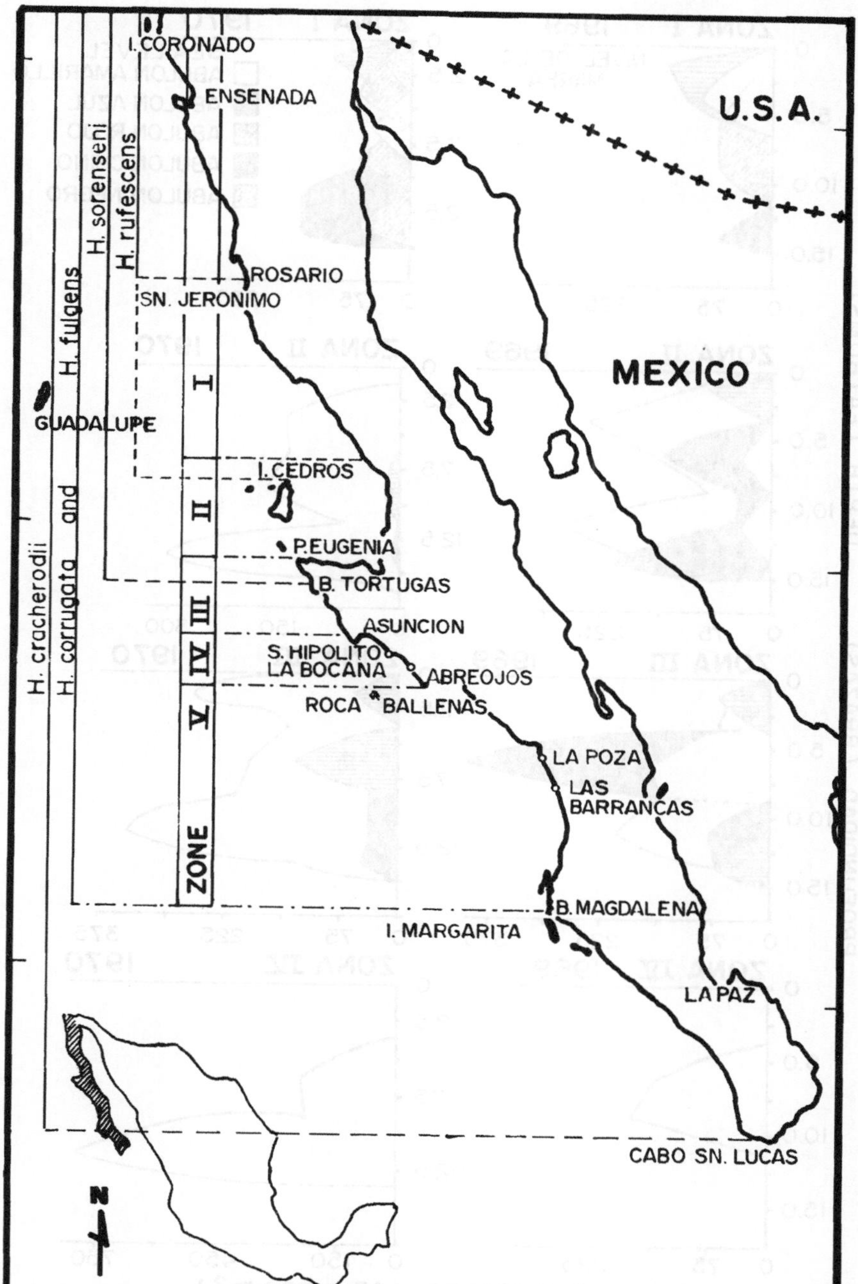

Fig. 24.1 Distribution of five abalone species in Baja California and the fishery management zones (from Robles, 1970, Muñoz, 1976a, and Guzmán del Próo *et al.*, 1976).

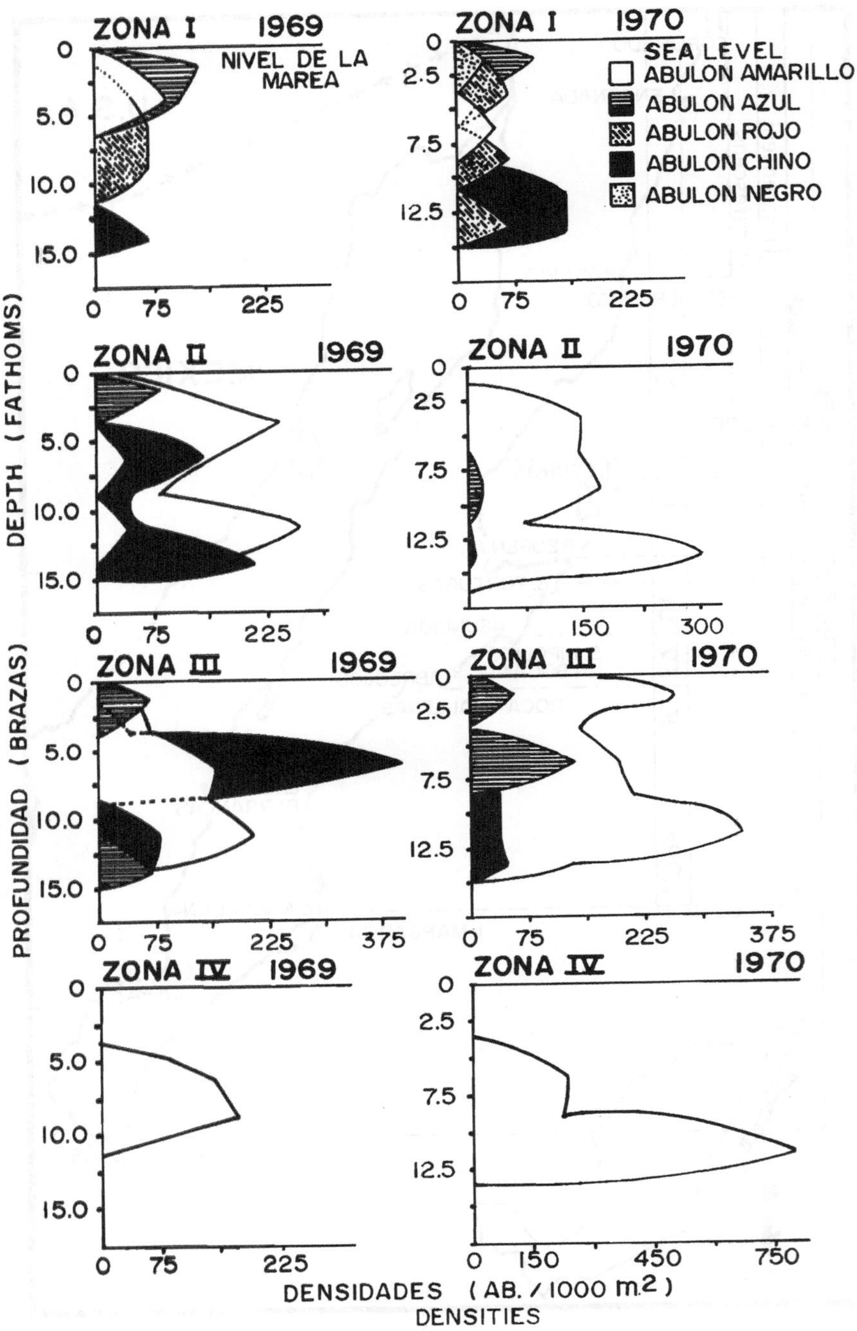

Fig. 24.2 Densities and depth distribution of abalone species in Baja California (after Guzmán del Próo *et al.*, 1976). Abulón amarillo = *H. corrugata*; abulón azul = *H. fulgens*; abulón rojo = *H. rufescens*; abulón chino = *H. sorenseni*; abulón negro = *H. cracherodii*.

H. fulgens – (3–) 8–15 (–20) m;
H. rufescens – (4–) 8–18 (–25) m; and
H. sorenseni – (12–) 15–25 (–30) m.

BIONOMY AND LIFE CYCLE

Haliotis species are dioecious. There is only one account of hermaphroditism (M. Ortíz, pers. comm.). In almost all cases the sexes are easily distinguishable as an individual approaches sexual maturity; ovaries are grey-green and testes cream or beige to rose coloured.

The sex ratio, observed in samples from commercial catches (Guzmán del Próo & Castro, 1976; Andrade, 1980), varies between zones and years. In some cases the ratio deviates from the normal 1:1 (male:female) ratio and may be as much as 1:1.8 for *H. corrugata* and 1:2.8 for *H. fulgens*. Andrade (1980) found in Zones III and IV (Fig. 24.2) a ratio of 1:1 for *H. corrugata* and *H. fulgens*, in the shell length (SL) range 100 to 180 mm. In Natividad it was 1:2 and in Bahía Asunción 1:1.8 for *H. fulgens*.

The observations on maturity and spawning of the two most important species, *H. corrugata* and *H. fulgens*, are summarized in Figure 24.3 and described below.

Haliotis corrugata

According to Cardenas *et al.* (1971) and Andrade (1980) there is no sexual activity from February until June (except for small peaks of maturity in April seen in Natividad, La Bocana and Abreojos). From July until December the gonads are generally full and ripe, with peaks in October and November. Between October and January a significant fraction of gonads is spent, thus indicating the main period of spawning.

Haliotis fulgens

The pattern described for *H. corrugata* applies also to *H. fulgens*. From June to September the gonads are mature. However, in Zone III the gonads are mature in October and November and in Zones IV and V south to La Poza, gonads are mature until December (León & Winkler, 1980). Thus, spawning occurs from September to December in these zones. In the southern part of this region the period of maturity and spawning appears to expand, suggesting multiple spawnings. On the basis of the occurrence of spent gonads, November, December, February and May appear to be the months of spawning.

Ortiz (1966) and Sevilla (1971) note that the reproductive season in Cedros is between June and September and, according to Sevilla, with minor spawning in October. Sevilla hypothesized that more than one episode of spawning could occur during the year.

All the above studies, except that of Sevilla (1971), measured changes in the gonad index, but they lack histological data to confirm the visual observations of gonad maturity. Further histological studies are needed since *Haliotis* spp. in the central and southern part of the peninsula may spawn several times a year. In that situation the gonad index has limitations because it cannot distinguish between

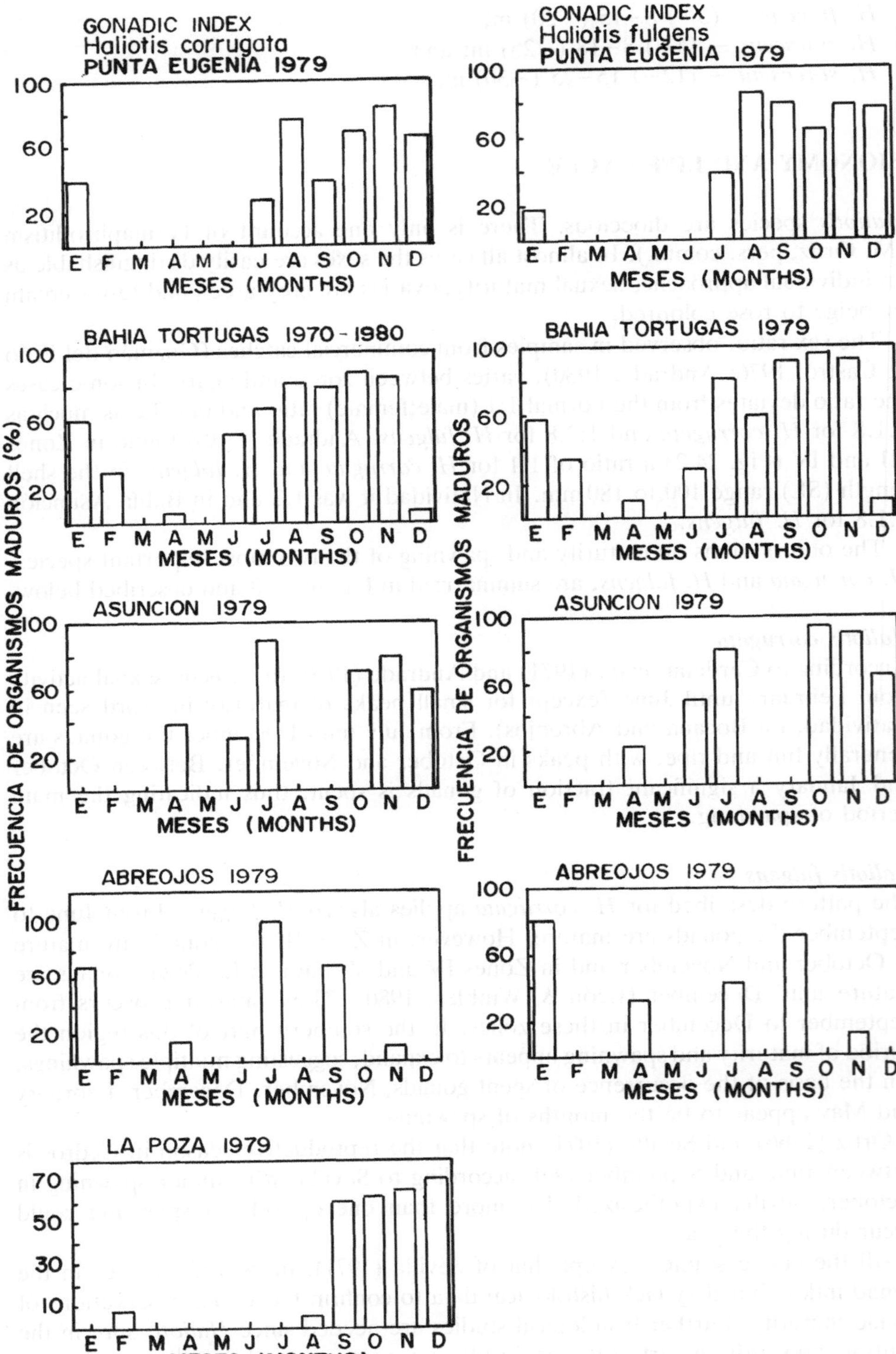

Fig. 24.3 Percentage frequency of sexually mature gonads of *H. corrugata* and *H. fulgens* throughout the year at different sites in Baja California (after Andrade, 1980, and León & Winkler, 1981).

the resting and immature stages of the gonad, nor between the mature and spawning stages.

For *H. rufescens*, there are partial observations from indices of maturity for Bahía del Rosario. According to Molina (1983), maturity of the gonads is attained in August, with a maximum between September and November; spawning is more or less continuous from late September to mid-November. For *H. cracherodii*, González & Ortíz (1986), citing Ortíz Q., M. (unpublished data), state that gonads are mature from May to July, with spawning from July to September at I. Cedros.

Size of sexual maturity

Few studies mention the age or size of sexual maturity, although this is of critical importance for management. Ortíz & González (1986) found that 100% of the population of *H. rufescens* was mature at 150 mm shell length (SL) and of *H. cracherodii* at 140 mm SL. However Molina (1983) gave 170 mm SL for *H. rufescens* in Bahía Rosario, but compare with the study of Giorgi & DeMartini (1977) who gave 125 mm SL for the same species in Mendocino, California. Guzmán del Próo *et al.* (1980) summarized previous work (e.g. Muñoz & Camacho, 1976) and showed that the size of sexual maturity, defined as the size at which 50% of the population show mature gonads, decreases from north to south, suggesting sexual precocity in the southern populations. For *H. corrugata* the average size is 129 mm, 138 mm, 132 mm, 120 mm and 102 mm at Rosalita, Cedros, Tortugas, Abreojos and La Poza respectively. For *H. fulgens* the figures are 145 mm, 152 mm, 141 mm, 125 mm and 117 mm for the same sites. For *H. rufescens* the figures are 169 mm for Bahía del Rosario, and for *H. cracherodii* it is 120 mm at the same place.

Fecundity

Information on fecundity is available for *H. rufescens*, *H. cracherodii* and *H. corrugata*. For *H. rufescens*, values of 6.4 to 30 million eggs at 165 mm SL are given. From that size fecundity decreases and becomes more variable (Molina, 1983). More recently, Ortíz & González (1986) reported values of 0.56 million at 136 mm SL and a maximum of 28.7 million at 212 mm SL for *H. rufescens*.

For *H. cracherodii*, González & Ortíz (1986) reported values of 0.8 million for individuals of 106 mm SL to 12.1 million for those of 140 mm SL. With increasing length fecundity declines. Lastly, for a small sample of *H. corrugata* at I. Benito they gave values of 3.8 to 9.5 million eggs for individuals 123 and 149 mm SL. Table 24.2 presents equations derived by these authors.

Larval phase

There is little published information on larval stages of the Mexican species of abalone. The first experiments to rear larvae of *H. rufescens* at Sauzal BC were by

Table 24.2 Equations relating fecundity (F) in numbers of eggs and shell length (L) in mm for two species of abalone in Baja California; equations are of the form $F = a\ L^b$.

Species	a	b	Source
H. rufescens	1.68×10^{-14}	9.3	Molina (1983)
	7.06×10^{-7}*	5.79*	Ortiz & González (1986)
H. cracherodii	4.20×10^{-14}	6.72	González & Ortíz (1986)

* Values have been recalculated from these authors' data.

Cota (1971). Later Kikuchi & Uki (1976) set out the main problems in rearing larvae at Sauzal. The problems were mostly logistic and operational, since both authors believed that the natural conditions for cultivation were more favourable in Baja California than in Japan. Of later studies, only the work of Searcy & Salas (1990) has been published. This gives information about the larval development of *H. rufescens* and *H. fulgens*, with emphasis on the ecology and rearing of juveniles for reseeding reefs in northern Baja California.

Adult phase

Longevity

No data on longevity are available. Some estimates have been made by fitting growth curves to data on growth from which an age corresponding to L_∞ can be obtained. These estimates are based on the separation of modes of length–frequency distributions following the methods of Cassie, Battacharia and others. By this method, Guzmán del Próo & Marín (1976) and Doi *et al.* (1977) calculated a maximum age of 20 years for *H. fulgens* and *H. corrugata*.

Muñoz (1976b) was the first to describe a method of determining the age of the abalone shell. By rasping the spire of the shell, layers of nacre are exposed and, if validated, can give an estimate of age (Chapter 12, this volume). Preliminary studies with this method for *H. corrugata* gave a maximum age of 21 years at 179.5 mm SL (Muñoz, 1976b).

Competitors

There are no studies in Baja California about the competitors of abalone. The sea urchins *Strongylocentrotus purpuratus* and *S. franciscanus* occur throughout Baja California in the same habitat as abalone and also feed on drift algae; however there is no evidence of competition. If that were to exist it would show up on the northwest coast of Baja California between Xatay and El Rosario Bay where *S. franciscanus* is so abundant that it has supported an important fishery since 1972. In the central part of the peninsula the abundance of this sea urchin decreases markedly, while herbivorous gastropods such as *Astrea undosa*, which co-exist on the same reefs as abalone and are the subject of another fishery, increase in abundance. *Astrea* is a potential competitor of *H. fulgens* and *H. corrugata* but, as yet, there is no experimental evidence either for or against this hypothesis.

Maximum size

Given the present conditions of over-exploitation of almost all the abalone reefs of Baja California, the size of capture today is much lower than the maximum size in virgin populations. However, collections of abalone divers of Baja California include very large shells of *H. fulgens* exceeding 230 mm SL (I. Olachea, pers. comm.).

Growth

The growth rate of *H. corrugata* and *H. fulgens* has been estimated at Cedros, Tortugas, Abreojos and La Poza (Table 24.3). Most of these studies are based on analysis of length–frequency distributions with adjustment of identified age groups to conform with the von Bertalanffy growth model (Guzmán del Próo & Marín, 1976; Doi *et al.*, 1977; Marín, 1981). Other estimates have been obtained by mark/recapture experiments (Guzmán del Próo & Salas, unpublished data; Shepherd *et al.*, 1991). Results obtained by different methods generally agree and suggest that the growth rate of these species increases from north to south in the peninsula (Guzmán del Próo, 1989; Shepherd *et al.*, 1991). The values of *K* range from around 0.2 for *H. corrugata* at Cedros to 0.98 at La Poza, with slight differences between the sexes. The latter estimate is high and needs to be verified by mark/recapture experiments. For *H. fulgens*, *K* ranges from 0.34 at Asuncion (0.35 at Bahía Tortugas) to 1.39 at La Poza.

Table 24.3 Parameters of the Brody-von Bertalanffy growth curve for *H. fulgens* and *H. corrugata* at various sites in Baja California; the estimates for the first site listed (by Guzmán del Próo & Salas) were obtained by mark/recapture experiments and fitted by the Gulland and Holt procedure; estimates for all other sites were by analysis of length–frequency distributions and use of Ford-Walford plots.

K	Sex	t_0 (mm)	L_∞ (mm)	Locality	Source
H. fulgens					
0.342	mf	–	189	Asunción	Guzmán del Próo & Salas (unpublished data)
0.368	f	0.05	170	Abreojos	Guzmán del Próo & Marín (1976)
0.384	m	0.10	171	Abreojos	Guzmán del Próo & Marín (1976)
1.386	mf	0.70	126	La Poza	Guzmán del Próo *et al.* (1980)
H. corrugata					
0.150	m	0.16	210	Cedros	Doi *et al.* (1977)
0.233	f	0.10	186	Cedros	Doi *et al.* (1977)
0.406	f	0.17	159	B. Tortugas	Marín (1981)
0.337	m	0.13	170	B. Tortugas	Marín (1981)
0.363	m	0.21	173	Asunción	Guzmán del Próo *et al.* (1980)
0.341	f	0.02	174	Asunción	Guzmán del Próo *et al.* (1980)
0.336	m	0.46	163	Abreojos	Doi *et al.* (1977)
0.372	f	0.60	170	Abreojos	Doi *et al.* (1977)
0.981	fm	1.56	119	La Poza	Guzmán del Próo *et al.* (1980)

Table 24.4 Length–weight relations for *H. fulgens*, *H. corrugata* and *H. rufescens* from different parts of Baja California; equations are of the form $W = a\ L^b$, where total weight, W, is expressed in g and shell length, L, in mm.

	H. fulgens		*H. corrugata*		*H. rufescens*	
Locality	*a*	*b*	*a*	*b*	*a*	*b*
Bahía Rosario					1.66×10^{-4}	3.02
Cedros I.	2.46×10^{-6}	3.84	1.07×10^{-5}	3.57		
Bahía Tortugas	2.72×10^{-5}	3.43	4.62×10^{-5}	3.25		
Asunción			1.20×10^{-4}	3.07		
Abreojos			3.19×10^{-5}	3.34		
La Poza	5.76×10^{-6}	3.67	1.29×10^{-4}	3.03		

Length–weight relations

Biometric relations have been obtained from commercial landings at various places (Table 24.4). These and other relations with meat weight are available and are useful for converting catch landings to live weight or checking legal minimum sizes of shelled abalone.

POPULATION STRUCTURE

Commercial landings from the main fishing grounds have been sampled more or less regularly since 1973. Available data are length–frequency distributions of species by sex (Guzmán del Próo *et al.*, 1973; Ortíz & Marín, 1976), but most remain unpublished.

Species composition of catches

Until 1976 abalone catches in the central region (Zones II–IV) were dominated by *H. corrugata* (62–98%). According to De Buen (1960), this dominance existed at least since 1957 because ratios of *H. corrugata*/*H. fulgens* were 70:30 in Natividad and 98:2 in Puerto Alcatraz near Bahía Magdalena.

In the northern region (Zone I), *H. fulgens* and *H. rufescens* have been the dominant species taken (27–60%), although on occasions *H. corrugata*, *H. sorenseni* and *H. cracherodii* are also taken (Guzmán del Próo & Castro, 1976). In the southern region (Zone V), *H. corrugata* remains the dominant species. León & Winkler (1981) report an overall ratio of 62:38 for *H. corrugata*/*H. fulgens*. The catch composition has changed notably in the central region in recent years. Since 1984, *H. fulgens* has dominated the catch from Cedros to Abreojos, comprising 48–63%, while *H. corrugata* has become of secondary importance (SEPESCA, 1987) (Table 24.5). This change seems to be linked with the serious decline in the populations of *H. corrugata* in the last 15 years.

Table 24.5 Percentage composition of abalone species in the commercial catch in Baja California in different years.

Zone	Year	*H. corrugata*	*H. fulgens*	*H. rufescens*	*H. sorenseni*	*H. cracherodii*	Source
I	1965	38	15	23			Ortíz (1966)
(Ensenada to Sta. Rosaliita)	69	24	60	16			Guzmán del Próo (1976)
	70	13	27	27	27	7	Guzmán del Próo (1976)
	84					16	SEPESCA (1987)
II	1965	75	25				Ortíz (1966)
(I. Cedros to Natividad)	68	83	14		3		Guzmán del Próo (1976)
	69	79	7		13		Guzmán del Próo (1976)
	70	80			20		Guzmán del Próo (1976)
	73	46	48		2	4	Guzmán del Próo & Castro (1976)
	74	49	49		2		Guzmán del Próo & Castro (1976)
	75	67	30		3		Guzmán del Próo & Castro (1976)
	76	62	38				Guzmán del Próo & Castro (1976)
	84	36	48		0.5	8	SEPESCA (1987)
III							
(Pt Eugenia to Pt Prieta)	69	85	3	12			Guzmán del Próo (1976)
	70	89	8		3		Guzmán del Próo (1976)
	73	52	46				Guzmán del Próo & Castro (1976)
	74	67	30				Guzmán del Próo & Castro (1976)
	75	46	53				Guzmán del Próo & Castro (1976)
	76	66	34				Guzmán del Próo & Castro (1976)
	84	36	55				SEPESCA (1987)
IV							
(Bocana to Abreojos)	68	67	33				Guzmán del Próo (1976)
	69	100	0				Guzmán del Próo (1976)
	70	100	0				Guzmán del Próo (1976)
	73	93	6				Guzmán del Próo & Castro (1976)
	74	91	8				Guzmán del Próo & Castro (1976)
	75	98	2				Guzmán del Próo & Castro (1976)
	76	92	8				Guzmán del Próo & Castro (1976)
	84	36	63				SEPESCA (1987)
V							
(S. Ignacio to I. Margarita)	68	58	42				Guzmán del Próo (1976)
	69	35	65				Guzmán del Próo (1976)
	70	40	60				Guzmán del Próo (1976)
	80	61	38				León and Winkler (1981)

Size and age composition

Data on samples from commercial catches have been published for each zone by Ortíz & Marín (1976), but more recent information has not been published.

Abundance (density)

The first attempt to estimate abundance was by surveys of the whole west coast of Baja California between 1968 and 1970. Random transect lines were put down and five replicate quadrats (14.4 m^2 each) were sampled per transect. Only emergent abalone larger than 10 cm were recorded and no attempt was made to record juvenile abundance. There were 306 transects along the length of the coast (Guzmán del Proó *et al.*, 1976). Estimates of density and biomass from this type of survey (Table 24.6) have high variance in most cases. The variability is related to habitat differences and the different zones in which the surveys were done. Semi-exposed rocky bottom, with dense beds of *Macrocystis pyrifera*, had the highest abalone densities. The highest densities were between Cedros and San Hipólito. *Haliotis corrugata* was numerically dominant on the whole coast. North of Cedros and south of San Hipólito, abalone beds were more dispersed and densities were lower.

Similar surveys of the coast were carried out between 1977 and 1988 with the purpose of allotting catch quotas to each co-operative in the region (INP, unpublished data). In general, the survey data (Table 24.6) are not very consistent. In some cases the changes in density appear dramatic, even from one year to the next, as occurred in Zone III (Table 24.6). The density of *H. corrugata* doubled in Zone III and almost tripled in Zone IV from 1969 to 1970. However, those values are not reflected in the catch data for the same years. Otherwise, from 1970 to

Table 24.6 Mean densities (numbers m^{-2}) of abalone in different zones in Baja California from surveys in different years; values given are based on data of (1) Guzmán del Próo *et al.* (1976); (2) Pineda & Muñoz (1977); (3) Pineda *et al.* (1978); (4) Ortíz *et al.* (1985).

Zone	Species	1968 (1)	1969 (1)	1970 (1)	1977 (2)	1978 (3)	1985 (4)
I	*H. corrugata*		0.062	0.046			
	H. fulgens		0.124	0.069			
	H. sorenseni		0.069	0.138			
	H. rufescens		0.062	0.069			
II	*H. corrugata*		0.168	0.162			
	H. fulgens		0.055	0.023			
	H. sorenseni		0.109	0.104			
III	*H. corrugata*		0.123	0.206	0.054	0.056	
	H. fulgens		0.069	0.095	0.029	0.027	
	H. sorenseni		0.149	0.038	—	—	
IV	*H. corrugata*	0.181	0.155	0.408		0.124	
	H. fulgens	0.106	—	—	—	0.021	
V	*H. corrugata*	0.208	0.138	0.069			0.048
	H. fulgens	0.208	0.138	0.208	—	—	0.310

1978 and 1985, the two main species declined steeply in density, coinciding with the severe decline in overall catch recorded for those years when the fishery practically collapsed. It seems that only in the long term is there any relation between the changes in density and the catch. In any event, given the patchy distribution of abalone and the high variance obtained in the censuses, the density data appear to be a poor measure of true abundance.

Recruitment

There are no data on recruitment, because no surveys have searched for abalone less than 10 cm SL. Specific long-term programmes, which set out to quantify periodically the juvenile fraction on abalone reefs, are critical for an understanding of the population dynamics of abalone in Baja California.

Mortality

The instantaneous total mortality rate Z has been estimated from monthly length-frequency data obtained at several places on the peninsula. Values obtained for *H. corrugata* are: 1.04 at Cedros, 0.61 at Abreojos and 0.48 at Tortugas. Natural mortality was estimated from curves of the theoretical biomass of abalone in a virgin stock and its relation with different levels of survival and longevity (Doi *et al.*, 1977; Marín, 1981). The estimates by this method (e.g. $M = 0.43$ in Abreojos and 0.35 in Cedros) are consistent with other studies (see review in Chapter 21, this volume).

EXPLOITATION AND MANAGEMENT

History of the fishery

Three distinct periods in the fishery can be recognized: 1860–1900, 1900–39 and 1939 to the present. In the earliest period, migrant Chinese established themselves in Baja California around 1860. Abalone were taken in the intertidal zone predominantly from Is Coronado south to Ensenada. The efficiency of the Chinese fishermen was a matter of concern to the Mexican government which imposed a licence and a fee of US$60 per boat, and set up a consulate in San Diego to maintain a register of the catch (Cox, 1962).

In 1900, California, USA, prohibited the capture of abalone down to low tide mark. That measure eliminated the Chinese fishermen. Their place was taken by Japanese who were the first to dive with a system that earned them the name 'sake barrel divers'. It is likely that the same system was used in Baja California. From this time until World War II the commercial fishery was dominated by Japanese. From about 1929 diving suits came to be used by the Japanese. By then the fishery in Baja California was substantial, exceeding 3350 t (Fig. 24.4).

At the beginning of World War II the Japanese were displaced from the abalone industry and at the end of the war the fishery was ceded exclusively to

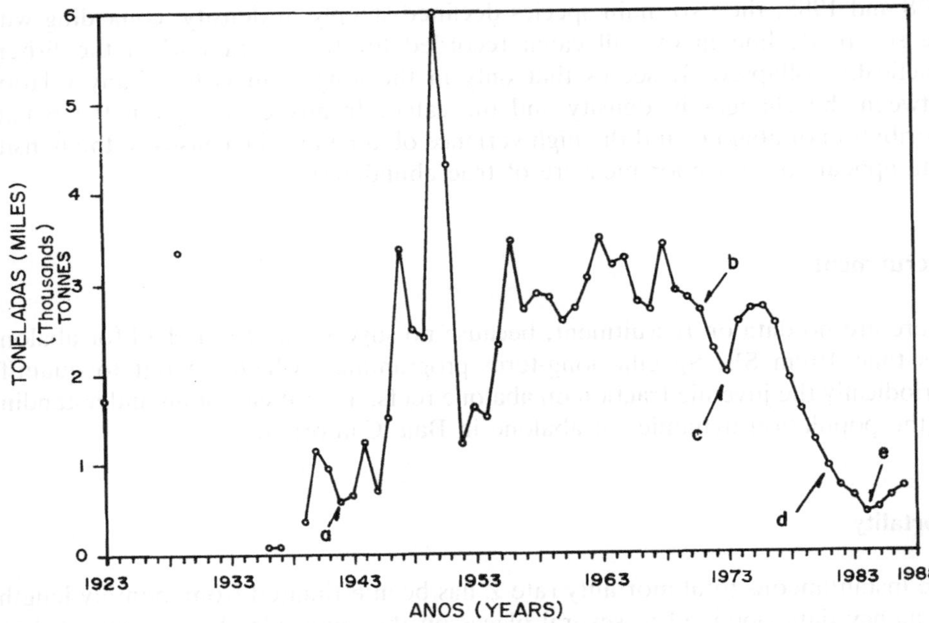

Fig. 24.4 Total production of abalone (live weight) in Baja California from 1923 to 1988. Data for 1923 and 1929 from Crocker (1931). Arrows indicate the time of critical management changes: a, exploitation by Mexican fishermen's cooperatives replaces exploitation by Chinese and Japanese; b, closed season changed from winter to summer; c, quotas are introduced; d, closed seasons increased and new size limits imposed; e, size limits enforced by landing abalone in the shell.

fishermen's co-operatives by the Mexican government. This situation exists to the present day. There is no sport fishery for any abalone species. From 1929 until the present, the abalone fishery has concentrated in Zones II to IV of Baja California, which have been historically the most productive part of the peninsula. In Zone I, the Is Coronados, S. Jeronimo and I. Guadalupe have been the most productive. In the south the regions of La Poza, Las Barrancas and I. Margarita are the most productive. However, the total production of the south and north combined has never amounted to more than 15–20% of the total production of Baja California. Thus, the central region (Zones II–IV) has always been the mainstay of the fishery. From 1970 the fishery began to decline, and fell to a low of 400 t in 1984. Since then, there has been a slight increase (Fig. 24.4).

Areas of abalone reefs

Between Is Coronados and I. Margarita the coastline is divided into 19 sectors corresponding to those ceded to the 19 fishermen's co-operatives. These sectors in turn are aggregated into the five zones delimited by the Secretaria de Pesca for the purpose of control and management (Fig. 24.1). Muñoz (1976a) shows the location of 244 fishing grounds in their zones for the central coasts of Baja California (from Cedros to P. Abreojos).

Present fishery

There are about 500 fishermen in the abalone fishery and 180 of them are divers (SEPESCA, 1987). Until 1987, 162 boats were used. Each boat normally carries a crew of three (diver, rower and deckhand). The boats are made of wood or fibreglass, are about 6–7 m long and powered by a 40–50 hp motor. Divers use hookah gear and neoprene wetsuits. Dry suits, still used by some old divers, had fallen into disuse by 1970.

Divers work 3–5 hours a day, including travelling time between the port and the reefs. Until 1974 no information was obtained on diver effort, but from that date the number of days fished (termed 'mareas') in the season was recorded. From 1984 the Secretaria de Pesca has tried to introduce diver return forms to provide more information on fishing effort. Information now sought is the number of divers and diving hours for each boat. Apparently this programme has not produced better data because divers have seldom conscientiously filled out the returns. Data on catch per unit effort available from Zones II and III (Fig. 24.5) indicate a continuous decline from 1958 to 1984, from 205 to 18 kg/boat per day (SEPESCA, 1987).

Management

Three distinct periods of management of the Mexican abalone fishery can be identified: from 1926–72, 1973–80 and 1981 to the present. In the first period size limits (SL) were largely based on those in force in California, USA, i.e.

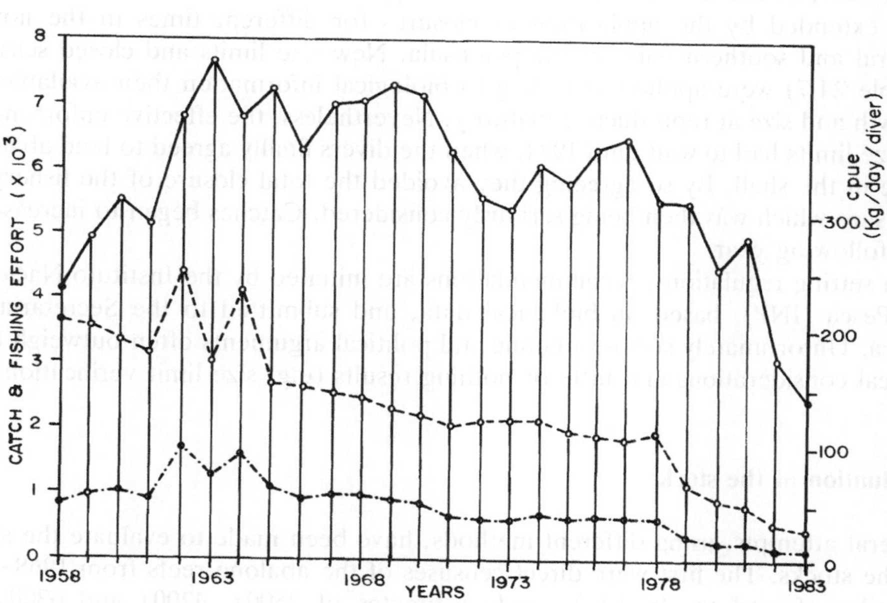

Fig. 24.5 Total abalone catch (meat weight) -·-·-, total effort (fishing days) ——— and catch per unit effort (kg/diver/day) - - - for Zones II, III and IV.

Table 24.7 Minimum legal sizes (SL in mm) and closed season (x) for abalone species in Baja California.

Species	Zones I, II	Zone III	Zone IV	Zone V
H. corrugata	140	135	130	110
H. cracherodii	120	120	120	120
H. fulgens	150	145	140	120
H. rufescens	165	165	165	165
Closed season				
1 July–30 Nov	x			
1 Aug–31 Dec		x	x	
1 Sept–31 Jan				x

H. corrugata – 150 mm, *H. fulgens* – 160 mm, *H. rufescens* – 175 mm and *H. cracherodii* – 120 mm. A closed season from 15 January to 15 March was imposed. Even then, a regulation existed requiring abalone to be brought ashore in the shell to enable checking of the size limit, but this regulation was not enforced. In 1973 the closed season was changed from winter to summer (1 July to 31 August) to improve protection of the spawning population. At the same time a system of catch quotas for each co-operative was introduced and a second attempt was made to require landing of abalone in the shell. Fishermen argued that the difficulties of transporting live animals in the boat, shrinkage of the meat and reduction in quality of the product were serious obstacles. Their arguments once more prevailed and abalone continued to be shelled at sea.

In 1981, as the near collapse of the fishery became evident, the closed season was extended by the application of closures for different times in the north, central and southern parts of the peninsula. New size limits and closed seasons (Table 24.7) were applied according to biological information then available on growth and size at reproductive maturity. Nevertheless, the effective enforcement of size limits had to wait until 1984, when the divers finally agreed to land abalone alive in the shell. By so agreeing they avoided the total closure of the fishery, a measure which was then being seriously considered. Catches began to increase in the following year.

In setting regulations, recommendations are initiated by the Instituto Nacional de Pesca (INP), based on biological data, and submitted to the Secretaria de Pesca. Unfortunately socio-economic and political arguments often outweigh biological considerations and little or nothing results (e.g. size limit verification).

Evaluation of the stock

Several attempts, using different methods, have been made to evaluate the state of the stocks. The first were direct censuses of the abalone reefs from 1968–70, already referred to. In 1969, crude estimates of 2500 t, 4200 t and 6300 t of abalone meat (of all sizes), including all species, were made for Zones I, II and III respectively (see Guzmán del Próo *et al.*, 1976). These figures, although they reflect or coincide with the levels of abundance of each of the zones, appear to have been over-estimates in as much as they were calculated for the entire rocky bottom of each zone, instead of being limited to the area of the known abalone-bearing reefs, which have a considerably smaller area.

Recent estimates based on surveys conducted specifically on abalone reefs have produced more realistic values (Pineda *et al.*, 1978; Pineda & López, 1978; Ortíz *et al.*, 1985), but all are estimates at one point in time and for small areas only. Other methods have also been used. Doi *et al.* (1977) and Marín (1981) used analytical models and Rocha & Arreguín-Sanchez (1987) used a surplus-production model. The former group of models, that were based on what Doi called a life cycle model, indicated that the *H. corrugata* fishery in Abreojos was showing signs of overfishing as early as 1975. However, the model failed to recognize a similar situation at Cedros and Tortugas.

Socioeconomic aspects

Many small towns of the central Baja California coast are economically dependent on the abalone and spiny lobster fisheries. To a lesser extent, the more recent algal, sea-urchin and conch fisheries provide an economic base for human coastal populations from El Rosario to Bahia Magdalena. The landed value of all these benthic fisheries was 33 billion Mexican pesos (US$11.5 million) in 1987 (SEPESCA, 1988), representing about 40% of the value of all fisheries on the Baja California Pacific coast.

Poaching is a problem mainly in Zones I and V. The intermediate zones have more isolated and more permanent settlements, with more difficult access to the grounds due to the fewer and rougher roads, and, therefore, can be more effectively controlled and protected by the fishermen's co-operatives.

DISCUSSION

Studies on abalone in México have been in progress for nearly two decades. The general focus of these studies has been related to the fisheries, especially monitoring of the catch (measuring catch composition and fishing effort) and surveys to estimate abundance. Complementary studies have considered growth, reproduction and fecundity. However, some fundamental aspects of the biology of wild populations have not been addressed. There have been few experimental, biological or ecological studies in the field. Practically all studies have been done by biologists of the INP and these have invariably addressed ad hoc management questions (such as quotas, closures and other controls). For this reason there has been a lack of continuity of research, information obtained has been fragmentary and many questions about basic biology have been poorly answered, if at all.

Regarding aquaculture, there have been more blunders than advances since the cultivation programme began in 1970. The most critical problem, apparently still unresolved, is to achieve high survival of larvae to the point of settlement and then to the juvenile stage. The very low production of juveniles for reseeding is proof of this.

It is thus necessary to re-appraise all abalone research in México to determine what needs to be pursued and where the priorities lie. The wide distribution of the fishery and the difficulty of access to many fished areas create logistic problems and increase costs. All of these will have to be considered in order to make the best use of human and financial resources. This can be achieved by concentrating on the critical information needs for particular species and by choosing representative locations for study. Within such a framework aspects which need to be addressed are:

(a) the reproductive cycles of *H. fulgens*, *H. corrugata* and *H. cracherodii* and, in particular, the timing, frequency and intensity of spawning and the factors that affect it;
(b) the identification and classification of the Mexican stocks in terms of growth rates, and an understanding of their population dynamics;
(c) the long-term recruitment of juveniles and pre-adults to fishable stocks and the identification of environmental and habitat factors which determine cohort strength;
(d) the food and feeding of juvenile and adult abalone and, in particular, the effect of food availability and quality on growth rate and reproductive success;
(e) the predators and parasites of abalone, especially if programmes to restock abalone reefs are to be undertaken.

A better understanding of critical population parameters (e.g. recruitment, growth and mortality) of stocks and of their variability within and between zones will lead to a more realistic application of existing models, such as the egg-per-recruit model, and hence better management. Ultimately, better models (see Chapter 20, this volume) incorporating a stock–recruitment relationship need to be developed.

ACKNOWLEDGEMENTS

The author is grateful to Drs M.J. Tegner, S.A. Shepherd and K. Karpov for valuable criticism and suggestions for improving the manuscript.

REFERENCES

Andrade P.M. (1971) *Algunos aspectos sobre madurez gonadal en* Haliotis fulgens *y* H. corrugata. Tesis licenciatura. UABC Ensenada, México.

Andrade P.M. (1980) Estudio biológico y prospección de los bancos de las Cooperativas de la zona central de Baja California. Informe Técnico mecanografiado. *Inst. Nal de Pesca*, 1–78.

Cárdenas B.R., Andrade P.M. & Villaseñor A. (1971) Estudio gonadal del abulón comercial en el litoral ocidental de la Península de Baja California. Fed. Reg. de Soc. Coop. de la Ind. Pesqu. de Baja California. *Informe Mecanografiado*, 1–24.

Chapa S.H. (1963) Los abulones. Importante recurso pesquero de Mexico. *Secretaría de Ind. y Com. Dir. Gral. de Pesca México Trab. Div.*, **IV**(34), 1–17.

Cota F.I. (1971) *Fecundación artificial y desarrollo embrionario de* Haliotis fulgens *Phillip: 1845 y* Haliotis rufescens *Swainson 1822 en condiciones de acuario*. Tesis profesional, Univ. Aut. Baja California, Mexico.

Cox K.W. (1962) California abalones, family Haliotidae. *Calif. Fish Game, Fish. Bull.*, **118**, 1–133.

Croker R.S. (1931) Abalones. *Calif. Fish Game, Fish. Bull.*, **30**, 58–72.

De Buen F. (1960) Abulones de Baja California. Moluscos del género *Haliotis*. *Univ. Chile, Rev. Biol. Mar.*, **10**(1–3), 201–7.

Doi T., Guzmán del Próo S.A., Marín V., Ortiz M., Camacho J. & Muñoz T. (1977) Análisis de la población y diagnóstico de la pesquería de abulón amarillo *Haliotis corrugata* en el área de Abreojos e Isla de Cedros B. C. *Inst. Nal de Pesca Ser. Cient.*, **18**, 1–17.

Giorgi A.E. & DeMartini J.D. (1977) A study of the reproductive biology of the red abalone, *Haliotis rufescens* Swainson, near Mendocino, California. *Calif. Fish Game*, **63**, 80–94.

González J.,G. & Ortíz Q.,M. (1986) Estudios sobre madurez, índice gonadal y fecundidad en abulón negro *Haliotis cracherodii* (Mollusca: Gasteropoda) a partir de muestras colectadas en la Isla de Cedros, Baja California, en Junio 1982. *Inst. Nal de Pesca, CRIP Ensenada Doc. Tec. Inf.*, **1**, 54–72.

Guzmán del Próo S.A. (1989) Problemas y perspectivas de las investigaciones pesqueras sobre los abulones de México. *Memorias del Seminario México–Australia sobre Ciencias Marinas*, 297–305. CINVESTAV Mérida, México.

Guzmán del Próo S.A. & Castro A.C. (1976) Un sistema de registro de la composición de la captura comercial de abulón (*Haliotis* spp.). Resultados de cuatro temporadas (1972–1976) en Baja California. *Memorias del Primer Simposium sobre Recursos Pesqueros Masivos de México. Vol. Esp. sobre Abulón/Langosta*, **2**, 1–36.

Guzmán del Próo S.A. & López S.F. (1992) Crecimiento de *Haliotis fulgens* (Mollusca: Gastropoda) en Bahía Asunción. BCS, México. *Inv. Mar. CICIMAR* (in press).

Guzmán del Próo S.A. & Marín A.V. (1976) Resultados preliminares sobre el crecimiento de abulón amarillo y azul (*Haliotis corrugata* y *H. fulgens*) en Punta Abreojos, B.C. *Inst. Nal de Pesca Ser. Cient.*, **17**, 1–11.

Guzmán del Próo S.A., Marín A.V. & Castro C. (1976) Estructura y abundancia de la población de abulón (*Haliotis* spp.) de Baja California, en 1968–1970. *Memorias del Primer Simposium Nacional de Recursos Pesqueros Masivos de México. Vol. Esp. Abulón/Langosta*, 219–78.

Guzmán del Próo S.A., Ortíz M. & Marín A.V. (1973) La explotación de abulón en la temporada 1972–1973. Primeros análisis sobre composición de la captura. *Inst. Nal de Pesca, SI*, **7** 1–52.

Guzmán del Próo S.A., Pineda B.J., Molina J., Uribe F., López F., Aguilar R., Andrade M., León G. Marín V. & Castro C. (1980) Análisis de la pesquería del abulón de Baja California. Fundamentos biológicos para un nuevo régimen de explotación del recurso. *Depto. Pesca. Inst. Nal de Pesca Doc. Tec. Int.*, 1–294.

Kikuchi S. & Uki N. (1976) Informe técnico sobre el cultivo y pesquería del abulón en México. *Inst. Nal de Pesca, Ser. Inf.*, 96.

Leighton D.L. (1974) The influence of water temperature on larval and juvenile growth in three species of southern California abalones. *Fish. Bull.*, **72**, 1137–45.

León C.G. & Winkler L.N. (1981) *Informe final de actividades realizadas en el programa Abulón-Langosta durante 1980*. Inst. Nal de Pesca, CRIP La Paz BCS Mecanografiado.

Marín V. (1981) Parámetros poblacionales y diagnóstico de la pesquería del abulón amarillo (*Haliotis corrugata*) en Bahía Tortugas B.C.S. *Ciencia Pesquera. Inst. Nal de Pesca México*, **1**(2), 67–79.

Molina J. (1983) Estimaciones de fecundidad en *Haliotis rufescens* del Bajo, Bahía Rosario, B.C. *Ciencia Pesquera. Inst. Nal de Pesca México*, **4**, 77–95.

Muñoz L.,T. (1976a) Clave artificial, diagnosis de las especies de híbridos de *Haliotis* spp. (Mollusca: Gastropoda) en Baja California, México. *Memorias del Primer Simposio Nacional de Recursos Pesqueros Masivos de México, Vol. Esp. Abulón/Langosta*, **2**, 81–130.

Muñoz L.,T. (1976b) Resultados preliminares de un método par determinar edad en abulones (*Haliotis* spp.) de Baja California. *Memorias del Primer Simposio Nacional de Recursos Pesqueros Masivos de México, Vol. Esp. Abulón/Langosta*, **1**, 281–301.

Muñoz L.,T. (1976c) Los bancos abuloneros de la parte central de Baja California. *Memorias del Primer Simposium Nacional de Recursos Pesqueros Masivos de México, Vol. Esp. Abulón/Langosta*, **1**, 161–84.

Muñoz L.,T. & Camacho A.J. (1976) Ensayo para la determinación de la primera edad de madurez de acuerdo a muestreos biométricos y cortes de concha de *H. corrugata* en Punta Abreojos, B.C. Sur. *Informe de Desarrollo. Mecanografiado. Inst. Nal de Pesca*, EIPc–209, 30 Abril 1976.

Ortíz Q.,M. (1966) Informe preliminar de las investigaciones sobre la biología y pesca del abulón comercial de las Islas de Cedros, Benitos y Guadalupe. Baja California. *Dir. Gral. Pesca, INIBP, Trab. Div.*, **XI**(109), 1–42.

Ortíz Q.,M. & González A.J. (1986) Notas sobre madurez, índice gonadal y fecundidad de tres especies de abulón *Haliotis* (Mollusca:Gasteropoda), a partir de colectas en el Rosario, Islotes Benitos y Bahía Asunción B. C. *Inst. Nal Pesca CRIP Ensenada Doc. Tec. Inf.*, **1**, 17–41.

Ortíz Q.,M. & Marín A.V. (1976) Composición de la captura comercial de abulón en la temporada 1972–1973. *Memorias del Primer Simposium Nacional de Recursos Pesqueros Masivos de México, Vol. Esp. Abulón/Langosta*, **1**, 185–218.

Ortíz Q.,M. Turrubiates J., Léon C.G. & Gonzalez J.G. (1985) Prospección y evaluación cuantitativa de los bancos abuloneros de la costa occidental de Baja California. *Secretaría de Pesca, Inst. Nal de Pesca, Informe Técnico. Mecanografiado nov. 1985.*

Pineda B.J. & López S.F. (1978) Densidad y abundancia de abulón (*Haliotis* spp.) en las áreas de las cooperativas Progreso y Punta Abreojos. *Depto de Pesca, Inst. Nal de Pesca. Informe de campo. Mecanografiado.*

Pineda B.J., López S.,F. & Guzmán del Próo S.A. (1978) Densidad y abundancia de los bancos abuloneros de las Cooperativas California San Ignacio y Leyes de Reforma durante Abril de 1978. *Departamento de Pesca, Inst. Nal de Pesca. Informe Técnico de Campo. Mecanografiado.*

Pineda B.J. & Muñoz L.,T. (1977) Densidad y abundancia de abulón en el área de la Cooperativa Leyes de Reforma. *Departamento de Pesca, Inst. Nal de Pesca, Informe Técnico de Campo. Mecanografiado.*

Robles Pacheco J.M. (1970) *Contribución al conocimiento de la biología pesquera del abulón en Baja California.* Tesis profesional, Univ. Aut. Baja California. Mecanografiado.

Rocha E. & Arreguín-Sanchez F. (1987) Diagnóstico de la pesquería de abulón (*Haliotis* spp.) de la península de Baja California, México. *Inv. Mar. CICIMAR*, **3**(2), 65–77.

Searcy B.R. & Salas Garza A.E. (1990) Investigaciones sobre el cultivo de abulón en la Universidad Autonoma de Baja California. México. *Serie Cientifica (UABCS, México)* **1**, 44–50.

SEPESCA (Secretaria de Pesca) (1987) *Pesquerías Mexicanas: Estrategias para su Administración.* Secretaria de Pesca, México.

SEPESCA (1988) *Anuario Estadístico de Pesca 1987.* Secretaría de Pesca, México.

Sevilla M.L. (1971) Desarrollo gonádico del abulón azul *Haliotis fulgens* Philippi. *Rev. Soc. Mex. Hist. Nat.*, **32**, 129–39.

Shepherd S.A., Guzmán del Próo S.A., Turrubiates J., Belmar J., Bahar J. & Sluczanowski P.R. (1991) Growth, size at sexual maturity and egg-per-recruit analysis of the abalone *Haliotis fulgens* in Baja California. *Veliger* **34**, 324–30.

Talmadge R.R. (1964) The races of *Haliotis fulgens* Philippi, Mollusca, Gastropoda. *Trans. San Diego Soc. Nat. Hist.*, **13**, 371–6.

Chapter 25

Alternativas cientificas para la recuperación de los recursos abuloneros de la Baja California, México: administración, biológico-pesquera y acuacultura

(Scientific alternatives for the recovery of abalone resources of Baja California, Mexico)

RODOLFO RAMÍREZ-GRANADOS *Laboratorio de Ecología, Escuela Nacional de Ciencias Biológicas, IPN Apartado Postal 26375, México DF*

ABSTRACT: Present information shows the poor results obtained through the administrative regulations based on fishery biology studies, as well as the minimum amount of larvae produced by culture in the last decade. Failure to apply effectively the recommendations derived from ecological research, has caused depletion of stocks of the more important species, which are now at a very low level. The author supports the opinion that rehabilitation by stock management is extremely doubtful. However, it is necessary to maintain at least the present abundance of the stocks as the reproductive stocks for aquaculture, since this technique offers the most promising method for restoring the abalone stocks. This alternative will be effective in the short term only if aquaculture is practised under a continuous and well designed experimental and ecological research programme involving the firm collaboration of the sectors which have been involved in the exploitation, marketing, study and administration of the abalone resources. A plan is proposed to undertake this task.

RESUMEN: Las investigaciones biológico-pesqueras que se analizan en este trabajo muestran los escasos o nulos resultados obtenidos, tanto en lo referente a las reglamentaciones administrativas basadas en estudios biológico-pesqueros, como en la mínima producción de larvas y juveniles lograda mediante el cultivo, por lo menos en la última década. Al no ser acatadas en forma efectiva, las plausibles recomendaciones derivadas de la investigación biológico-pesquera, las poblaciones de las especies principales han decaído a niveles de abundancia mínimos. Su restablecimiento por ese camino es en extremo dudoso. Si acaso, sólo puede esperarse que no se lleve a cabo su exterminio total. De esa manera, a partir de las existencias sobrevivientes, puede insistirse en la alternativa que en este trabajo se considera mas viable: el cultivo, la cual será efectiva a plazo más corto si se realiza en forma continua, basada en diseños experimentales adecuados, tanto en laboratorio como en condiciones naturales, asi como con la colaboración estrecha de los sectores implicados en la producción, comercialización, investigación y administración de los recursos abuloneros aun disponibles.

INTRODUCCION

En trabajos presentados en este mismo Simposio y otros que aquí se citan, se hace historia de la pesquería de abulón en Baja California y de las circunstancias que han influído en su ascenso y decadencia. Paralelamente se hace mención de los estudios realizados a lo largo de dos líneas de investigación principales: el estudio de las pesquerías y el de los aspectos bioecológicos de los abulones, con vistas a lograr, respectivamente, la estabilización de los recursos que originaron y sostuvieron importante pesquería, y la creación de una biotecnología de cultivo, adecuada a las condiciones naturales de los bancos de la costa occidental bajacaliforniana.

La revisión de las aportaciones hechas por los estudios biológico-pesqueros a la administración de los recursos abuloneros, conduce a una conclusión poco satisfactoria (Chapa, 1973; Guzmán del Próo 1974; Doi *et al.*, 1977; Aldana-Aranda, 1979; INP, 1980; López Salas y Ortiz Quintanilla, 1986; Ortíz Quintanilla y González Avilez, 1986; Rocha y Arreguín, 1987), lo que plantea la necesidad de proponer alternativas que conduzcan a resultados positivos.

Entre las diversas alternativas que pueden ser sugeridas, la que parece más factible y eficaz es la de emprender el cultivo de las especies de abulón más agotadas en momentos en que aún es factible su recuperación. La condición indispensable es lograr que, a través de las medidas administrativas que se apliquen, se conserven las existencias a partir de las cuales se emprenderán nuevamente los cultivos en forma más intensiva, contínua y eficiente. El logro de este propósito haría realidad las metas que ahora son imposibles: (a) Promover la instalación de granjas para el engorde de juveniles producidos en México, (b) Llevar a cabo la repoblación de bancos sobre-explotados, liberando en condiciones adecuadas las formas juveniles excedentes de los cultivos, cuya porción mayor se dedicaría al engorde.

Tal esquema, relativamente simple, debe ser el punto de partida no sólo de la rehabilitación de los bancos sino también de la creación de una línea de comercialización de perspectivas insospechadas; pero es indispensable, como antecedente, llevar a cabo en forma creativa, sistemática, contínua, la investigación científica de los aspectos más esenciales de la bioecología de los abulones para crear las biotecnologías de cultivo más eficientes.

ANTECEDENTES

Estudios biológico-pesqueros

Los estudios biológico-pesqueros realizados a lo largo del litoral occidental bajacaliforniano, complementados con la información obtenida de las cooperativas regionales, permitieron formular los enfoques teóricos de Doi *et al.* (1977) y el de Rocha y Arreguín, 10 años más tarde. El primero, con indudable carácter preventivo, se elaboró en una etapa en la que aún podía actuarse con mayor posibilidad de éxito. El segundo, bajo condiciones deplorables del recurso, propone medidas correctivas de aplicación urgente. En esas condiciones la investigación biológico-pesquera sólo puede contribuír a evitar el agotamiento total de las existencias si se logra el acatamiento estricto de las últimas reglamentaciones. Este objetivo es absolutamente indispensable para que la alternativa del cultivo pueda ser aplicada.

Estudios bioecológicos para desarrollar biotecnologías de cultivo

La acuacultura, debido a sus grandes logros a escala mundial y a sus avances científicos a través de la investigación, ha surgido en años recientes como una biotecnología de importancia indudable, tanto para atender a los problemas alimentarios del hombre como para proteger, rehabilitar e incrementar la productividad de los organismos sujetos a cultivo. Por esas razones se considera que es a través de la acuacultura como podrá lograrse en un plazo menor y en forma más segura la recuperación de las poblaciones y especies de abulón sometidas a un régimen de explotación insensato, que las ha puesto al borde del agotamiento total. Las investigaciones que se requieren para crear las tecnologías de cultivo son de carácter bioecológico, lo que significa que deben estudiarse los procesos biológicos de los organismos a cultivar, frente a las condiciones ambientales en que proliferan. La fundamentación ecológica para desarrollar los métodos de cultivo es absolutamente necesaria, igual para lograr un monocultivo que un policultivo mediante la aplicación de métodos de campo y de laboratorio, tanto de observación como experimentales y analíticos.

Cultivo de abulones en México

En el caso de México, los intentos de cultivo de los abulones en Baja California han seguido un curso un tanto azaroso, debido a circunstancias ajenas a la investigación científica propiamente dicha. Las primeras iniciativas surgen cuando se realizan los estudios del desarrollo gonádico de *H. fulgens* y *H. corrugata*, no sólo para determinar una época de veda más congruente con su proceso reproductivo, sino también para conocer la época más propicia para la obtención de gametos, con vistas a la producción masiva de 'semilla'. Después de los primeros intentos, que condujeron incluso a la construcción del primer criadero en El Sauzal, tuvo lugar una época de escasa actividad experimental en la que se dió prioridad a la investigación biológico-pesquera. Es hasta fines de la década de los 70 y principio de la presente cuando, de acuerdo con Salas Garza y Searcy Bernal (Capitulo 41, en estas memorias), 'se inician' los trabajos de cultivo, lo cual se hace siguiendo parcialmente las directrices de los investigadores de otros países, sobre todo de los japoneses, que para entonces ya disponían de métodos sistematizados de cultivo, después de alrededor de 35 años de investigación de los aspectos esenciales para fundamentar esa metodología, a partir de Murayama (1935) citado en Shibui (1972) quien desarrolló el método de fertilización artificial.

Aspectos relevantes del cultivo en Japón

En el Japón, el curso de los trabajos experimentales se apegó a la información indicativa resultante de la investigación ecológica y biológico-pesquera que condujo a la selección de especies a cultivar, sobre la base de criterios y características bioecológicas y comerciales de la especie elegida. Una vez hecha la selección, la atención se enfocó hacia los aspectos considerados más importantes del proceso reproductivo con el objetivo preciso de lograr la producción masiva, primero de las larvas y después, de las siguientes etapas ontogénicas. La operación incluyó la selección y mantenimiento de los reproductores previo conocimiento de la temporada de reproducción y de los factores que influyen en ella o la determinan (IG,

TEA, etc.). Los avances mayores de esta etapa se refieren al logro de los métodos de inducción, que ahora son parte esencial de la producción masiva de larvas. La inducción a la liberación de gametos se realiza actualmente en forma sistemática y eficiente, ya que se dispone de varios procedimientos a través de los cuales se producen, sin problema, cantidades considerables de larvas planctónicas (Imai, 1967; Shibui, 1972; Uki, 1989). Estas, sin embargo, se enfrentan al problema de sobrevivencia, tanto en el curso de su etapa planctónica como, sobre todo, al momento de asentarse y convertirse en larva bentónica. Aunque los involucrados en el cultivo de abulones conocen este crucial problema, pocos son quienes han logrado superarlo.

Problemas de la investigación bioecológica de los abulones en México

Atendiendo a lo observado en el caso de otras especies, se considera que la superación de este problema en México es el punto clave, en primer lugar para la producción masiva de larvas bentónicas y, a continuación, de formas juveniles tempranas en condiciones de ser conducidas con relativa facilidad al proceso de engorde, única realización que justificaría y apoyaría el paso a la producción comercial de ejemplares para repoblación o de 'medallones', ejemplares juveniles para el comercio y consumo directos. Los obstáculos científicos y técnicos, en la etapa a la que se hace referencia, al menos de acuerdo a lo observado en otros casos de acuacultura, son de los siguientes tipos:

Desconocimiento de organismos nocivos

No se ha investigado lo relativo a la presencia y abundancia de microorganismos, algunos de los cuales pueden ser patógenos, parásitos o depredadores de las larvas. La técnica japonesa para inducir la reproducción mediante el uso de agua irradiada con UV, influye seguramente en la eliminación de ese problema y a ello debe su éxito, al menos en parte. La desinfección del agua en esta etapa es de particular importancia y puede lograrse también mediante el uso de antibióticos, aunque ignoramos si en el caso de los abulones, hayan sido utilizados alguna vez.

La alimentación

Es otro aspecto crucial no sólo en esta etapa, sino a lo largo del desarrollo ontogénico de los abulones y, en general, de todas las especies bajo cultivo. El proceso alimentario actúa como factor limitante en el desarrollo de la acuacultura, bien sea por razones de calidad, cantidad, disponibilidad y eficiencia ecológica en el aprovechamiento, como por razones económicas de costo de producción de las dietas balanceadas; si bien, se puede recurrir a la producción de alimento vivo, también mediante cultivo, para atender a la necesidad de algas microscópicas, diatomeas y macroalgas, sobre todo de las formas larvarias. Este tema, al que le han concedido particular interés los investigadores japoneses, no ha sido aquilatado debidamente en nuestro país, excepto en algunos casos en que se han hecho observaciones de campo al respecto, pero ninguna de carácter experimental, mediante el establecimiento de ceparios y la realización de pruebas de selectividad en el laboratorio.

En el Japón se han realizado, incluso, trabajos de 'reforestación' de macroalgas, para proveer a las necesidades de juveniles repoblados. Ortíz-Quintanilla (1966), quien se refiere a hábitos alimentarios de 3 especies y Guzmán del Próo *et al.* (1972), informan acerca de las macroalgas que abundan en torno a los bancos de abulón y que, supuestamente, pueden servirles de alimento; sin embargo, queda de manifiesto que este tema amerita investigaciones a fondo que conducirán al abatimiento de la mortalidad de larvas y juveniles y confirmarán la viabilidad biológica y económica de los proyectos comerciales de engorde.

La buena alimentación juega también papel primordial en el caso de los reproductores, ya que implica altos costos de mantenimiento. En los intentos de cultivo realizados en México, este paso se suprime y se recurre a reproductores silvestres (Mazón Suástegui, *et al.*, en este Symposio.), lo cual puede anular las ventajas del cultivo semicerrado, ya que da entrada a factores abióticos y bióticos que es conveniente controlar.

Se discute aún lo referente a la alimentación durante la etapa larvaria planctónica, pues hay estudios y opiniones que niegan que la larva se alimente en esa etapa; sin embargo, Shibui (1972), demostró que en *H. discus hannai*, durante la etapa planctónica, las larvas no se alimentan y que la presencia de diatomeas y flagelados las condicionan para su comportamiento alimentario en la etapa siguiente de su desarrollo como larvas bentónicas, después de asentarse en el fondo, momento importante en el que se alimentan activamente, lo cual influye en la sobrevivencia. Es necesario investigar si lo dicho es válido para todas las especies.

Enfermedades, parásitos y depredadores

En la costa bajacaliforniana, no así en la californiana, hay un vacío de conocimiento en lo que se refiere a enfermedades, parásitos, depredadores y competidores, debido a que la acción de estos factores tiende generalmente a subestimarse, aunque hay evidencias de que puede ser muy importante: cuando el agua en los estanques de cría no se cambia o desinfecta proliferan protozoarios y nemátodos depredadores no identificados (Obs. pers.). Asimismo, en los bancos del norte de la península, los erizos *Strongylocentrotus* sp. forman densas poblaciones en areas antes ocupadas, supuestamente, por abulones, según información proporcionada por viejos buzos de escafandra. Este hecho no se ha comprobado científicamente, aunque debe tomarse en cuenta cuando se determinen las interrelaciones de las poblaciones de abulones con las poblaciones que las circundan. De esta manera se explicaría con mejor fundamento por qué es tan dificil la rehabilitación de bancos sobreexplotados, ya que además de la distancia cada vez mayor entre reproductores, que dificulta el encuentro de los gametos, si hay producción suficiente de larvas para llegar al asentamiento, las que lo logren encontrarán la competencia por espacio y alimento de otros organismos. Aún cuando no fuera así, es necesario investigar cuál es la verdadera situación al respecto.

Cambios ecológicos diferentes

Se desconoce también la influencia que tienen sobre las poblaciones de abulón y sus respuestas metabólicas, los cambios ecológicos originados con cierta periodicidad por el fenómeno conocido como El Niño, pues si bien se sabe de sus efectos sobre las poblaciones de peces, las comunidades planctónicas, los sargazales y otras unidades ambientales, no se ha intentado averiguar los que pudiera tener sobre comunidades sedentarias como las que forman los abulones.

DISCUSION

La alternativa de cultivo, revisada someramente, para ser útil bajo nuevas condiciones de trabajo, debe complementarse con la rica información procedente de otros países (Japón, EUA, Francia, Australia), pero la transferencia tecnológica deberá consistir en un proceso de adecuación a las condiciones y necesidades nacionales, no a la aplicación automática de recetas que hacen persistir al empirismo y el subdesarrollo científico. La información científica de otros países será complementaria de la nacional y servirá para reorientarla o reforzarla, sobre todo en situaciones tan críticas como la que ahora enfrentan los sectores regionales en relación con los recursos abuloneros.

Ante la falta o insuficiencia de los resultados, es necesaria la restructuración de la investigación bajo lineamientos más estrictos y condiciones organizativas que aseguren su continuidad y garanticen la consecusión parcial o total de los objetivos más esenciales. La breve revisión aquí presentada, permite al menos confirmar que es necesaria la realización de un Plan en situación de emergencia, a través del cual y con la participación de todos los sectores, se complementen y refuercen las acciones positivas y se supriman o corrijan las negativas, para lograr la recuperación de los recursos abuloneros. Las expectativas de que las existencias abuloneras de la región se repongan a partir de medidas administrativas, biológico-pesqueras, depende de diversas circunstancias, no todas confiables o predecibles con algún grado de seguridad. Hablando objetivamente, dependen de que los bancos se mantengan a una densidad que posibilite su recuperación, aún considerando las deficientes condiciones actuales. En la mayor parte de los casos, se trata de un verdadero trabajo de rescate; recuérdese que la implementación de medidas proteccionistas no se ha hecho efectiva ni siquiera en la época de mayor extracción. Lo anterior no significa que deban abandonarse esas medidas, sino sostenerse para asegurar las existencias a partir de las cuales se realizarán los cultivos.

Al analizar los problemas del cultivo de los abulones en México, Salas Garza y Searcy (1990), reconocen que radican en 'la falta de una biotecnia eficiente para la producción de "semilla", lo que impide satisfacer o atender a la necesidad de repoblar los bancos naturales, así como la de emprender el engorde comercial en cautiverio'. Ambos hechos indican que debe ponerse particular énfasis en la investigación a través de la cual se resuelvan los problemas científicos que han impedido hasta ahora, la producción masiva de crías para el logro de los objetivos fundamentales de la investigación, iniciada hace 25 años. Aunque es indudable que ha habido esfuerzos muy meritorios, es obvio que los resultados están lejos de satisfacer las necesidades. Los propios investigadores reconocen que la mortalidad en la etapa de asentamiento supera el 95%, por falta de un inductor adecuado e infestación de bacterias y protozoarios, lo cual revela un diseño experimental inadecuado, ya que los mismos obstáculos han sido superados en otros países. Por otra parte, las escasas repoblaciones de los bancos naturales tampoco han rendido resultados positivos, por lo que la recuperación no podría lograrse utilizando densidades de siembra poco significativas, sujetas a la dispersión y al desconocimiento de la tasa de depredación existente en los bancos. En tales condiciones, las alternativas planteadas por otros autores se contradicen con los resultados negativos de las investigaciones: no hay producción suficiente de juveniles para emprender repoblaciones, ni proyectos comerciales de engorde, a menos que se extraiga la producción natural de reclutas, seguramente escasa, lo que agravaría la situación de las existencias; o bien, como ya alguien ha sugerido, recurrir a la compra de juveniles, en las instituciones extranjeras que ya saben producirlos, lo que daría lugar a una nueva dependencia.

Si la investigación propia, debidamente dirigida y apoyada para darle seguridad y continuidad, resuelve este problema central de la situación abulonera, sistematizando la producción masiva de juveniles, cesarán las especulaciones carentes de fundamento científico y por tanto, de posibilidades de éxito. Sin decir que no haya otros problemas científicos de importancia, puede afirmarse que en el aspecto particular mencionado, radica el éxito de la tarea de recuperación más trascendente entre todas las que se presentan a escala regional. Este tema debe situarse con carácter absolutamente prioritario en los nuevos esquemas de investigación abulonera. Se considera que el mayor esfuerzo deberá dirigirse hacia los trabajos de cultivo, aún cuando hasta la fecha no hayan producido resultados que incidan sobre las condiciones generales de los bancos ni sobre las actividades económicas que continuarán siendo potenciales, en tanto no se logre la producción masiva de juveniles.

CONCLUSIONES Y RECOMENDACIONES

El análisis crítico pero constructivo, de lo hecho hasta ahora, debe servir de pauta a lo que deberá hacerse en el futuro inmediato, incorporando o adaptando como se ha dicho, a la experiencia nacional positiva, los resultados igualmente positivos, logrados en otros países y de los cuales carecemos.

Plan de trabajo y sectores implicados

La grave situación actual de los recursos abuloneros, amerita atención especial a través de la realización de un Plan de Trabajo y la creación de un instrumento de Acción para ponerlo en práctica, en los que se conjuguen los esfuerzos reorganizados de todos aquellos que han participado en la intención por darles una administración científica que asegure su aprovechamiento permanente. El Plan de Trabajo debe incluír las acciones que se consideren vitales o indispensables, a la luz de la situación diagnosticada, debidamente evaluadas para asignarles la prelación que ameriten. La ejecución deberá estar a cargo de los sectores nacionales que han sido:

Usufructuarios: Sector Social, Cooperativas.
Beneficiarios: Sector Privado, Empresas enlatadoras y fileteadoras.
Investigadores: Sector Académico.
Administradores: Sector Público.

Todos estos Sectores organizados en un Consejo Técnico o Comisión, con facultades ejecutivas y con medios suficientes para poner en marcha un Programa Regional, básicamente de Investigaciones Bioecológicas, en cuyo contenido se plantee con claridad y energía no sólo la necesidad sino la obligatoriedad de salvar los recursos abuloneros, por lo que significan como fuente de trabajo y alimentación y también como patrimonio inalienable de los mexicanos.

Estructura Institucional Regional para la elaboración y ejecución del Plan de Trabajo, que conduzca a la rehabilitación a mediano plazo de los recursos abuloneros de la Baja California

Los sectores involucrados deberán crear un organismo regional que se aboque a la tarea principal de rehabilitar los deteriorados recursos abuloneros en que abundaba la costa occidental de la Baja California. El organismo deberá asumir la estructura que los sectores consideren conveniente, en tanto que el funcionamiento estará regido por un consejo interno (integrado por los representantes sectoriales), el cual decidirá democráticamente los objetivos y los métodos del programa de rehabilitación, así como gestionará los recursos y demás apoyos logísticos que se requerirán para su realización. Deberán crearse grupos de trabajo científico en lugares estratégicos, para la realización de las investigaciones fundamentales y de las tareas necesarias que mantengan la continuidad y aseguren, en lo posible, el cumplimiento de los objetivos acordados. La restauración de los recursos abuloneros tendrá que ser una labor de responsabilidad colectiva, no sujeta al libre albedrío de los sectores sino la resultante de los acuerdos a que lleguen, atendiendo prioritariamente al interés nacional. La participación de los pescadores cooperativistas es quizás la más comprometida, ya que son ellos, por ley, los usufructuarios de lo que era un vasto recurso, reducido ahora a mínima expresión por una explotación excesiva y desordenada. El sector académico está obligado a dar máxima atención a los problemas científicos y también máxima orientación al sector social, para lograr su cooperación decidida y el acatamiento de las disposiciones que emanen de la investigación científica. En el caso probable de que el recurso se rehabilite totalmente, no debe volver a permitirse la explotación irracional que ha puesto a punto de extinción un recurso regional invaluable.

BIBLIOGRAFIA

Aldana-Aranda D. (1979) Aspectos biológicos y ecológicos a considerar para lograr una administración racional del recurso abulonero en México. *Informe para obtener el título de Biólogo*. ENCB, IPN.

Chapa S.H. (1973) Datos para la integración de un programa de desarrollo pesquero para el Noroeste de México II. *Gaceta Cooperativa* No. 38, 22–6.

Doi T., Guzmán del Próo S.A., Marín V.A., Ortíz Q.M., Camacho A.J. y Muñoz L.T. (1977) Análisis de la población y diagnóstico de la pesquería de abulón amarillo (*Haliotis corrugata*) en el área de Punta Abreojos e Isla de Cedros, B.C. *Dpto. Pesca: Dir. Gral. del INP. Ser. Cien.*, **18**.

González Avilés J.G., Ortíz Quintanilla M. (1986) Estudios sobre madurez índice gonadal y fecundidad en abulón negro, *Haliotis cracherodii* (Mollusca: Gastropoda), a partir de muestras colectadas en la Isla de Cedros, Baja California, México, en junio de 1982. Contr. Biol. y Tecnol-Pesq. CRIP Ensenada, INP, *SEPES, Doc. Tecn. Inf.*, **1**, 54–72.

Guzmán del Próo S.A. (1974) La pesquería de abulón. *Gaceta Cooperativa*, **52**, 14–17.

Guzmán del Próo S.A., de la Campa J.S., Pineda B.J. (1972) Flora macroscópica asociada a los bancos de abulón (*Haliotis* sp.) en algunas áreas de la costa occidental de Baja California, México. *Mem. IV Cong. Nac. Ocean. (México)*, **257**, 263.

Imai T. (1967) Mass production of molluscs by means of rearing the larvae in tanks. *Venus, Jpn. J. Malacol.*, **25**, 3 y 4, 159–67.

Inst. Nac. de la Pesca (INP) (1980) Análisis de la Pesquería de Abulón en Baja California y Fundamentos para un nuevo régimen de explotación del recurso. *Instituto Nacional de Pesca. Doc. Tec. Int.*

López Salas F., Ortíz Quintanilla M. (1986) Informe de la temporada de pesca 1984–1985 de abulón (*Haliotis* sp.) en la costa noroccidental de la península de Baja California, México. *Contrib. Biol. y Tecnol. Pesq. CRIP Ensenada, INP, SEPES, Doc. Técn. Int.*, **3**, 80–93.

Ortíz Quintanilla M. (1966) Informe preliminar de las investigaciones sobre la biológía y pesca del abulón comercial de las islas de Cedros, Benitos y Guadalupe, Baja California. *Dir. Gral. Pesca., Trab. Div.*, **XI**(109), 1–42.

Ortíz Quintanilla M., González Avilés J.G. (1986) Notas sobre madurez, índice gonadal y fecundidad en tres especies de abulón, *Haliotis* spp. (Mollusca: Gastropoda), a partir de colectas en El Rosario, Islotes Benitos y B. Asunción, Baja California, México. *Contrib. Biol. y Tecnol. Pesq. CRIP Ensenada, INP, SEPES, Doc. Tecn. Inf.* **1**, 17–41.

Rocha E., Arreguín-Sánchez F. (1987) Diagnóstico de la pesquería de abulón (Haliotis sp.) de la península de Baja California, México. *Inv. Mar. CICIMAR*, **3**, 2, 65–77.

Salas Garza A., Searcy Bernal R. (1990) Problemas y alternativas del cultivo de abulón en México. *Serie Cientifica* (UABCS, México) **1**, 51–8.

Shibui T. (1972) On the normal development of the eggs of Japanese abalone, *Haliotis discus hannai* Ino, and ecological and physiological studies of its larvae and young, *Bull. Iwate Pref. Fish. Exp. Stat.*, **2**, 1–69.

Uki N. (1989) Abalone seedling production and its theory. *Int. J. Aq. Fish. Technol.*, **1**, 3–15.

Chapter 26
The California red abalone fishery: a case study in complexity

MIA J. TEGNER *Scripps Institution of Oceanography, La Jolla, California 92093–0201, USA*
JOHN D. DeMARTINI *Humboldt State University Marine Laboratory, Trinidad, California 95521, USA*
KONSTANTIN A. KARPOV *California Department of Fish and Game, 19160 S. Harbor Dr., Fort Bragg, California 95437, USA*

ABSTRACT: The red abalone, *Haliotis rufescens* (Swainson), is historically the most important species in the California abalone fishery by virtue of its large size and distribution along the entire coast. Red abalone range through three regions varying considerably in food availability, predators and predation pressure and the physical characteristics of their habitat; these factors affect population structure, growth, reproduction and behaviour. The commercial fishery is largely restricted to southern California, sea otter (*Enhydra lutris*) predation has precluded most fishing in central California, and the north coast has been reserved for recreational fishermen since 1945. Thus both biogeography and politics complicate management of this haliotid. Management of commercial and recreational fisheries has been based primarily on size limits plus a daily bag limit for sport fishermen. In spite of continued market demand, commercial landings declined precipitously in the early 1970s after many years of relative stability. Limited entry was instituted in 1977, but the downward trend has continued unabated, and now recreational take on the north coast is apparently larger than the commercial harvest. These factors plus rapidly rising prices for abalone have put additional pressure on stocks of the north coast, and the advent of a sea-urchin fishery in this region has apparently led to a major poaching problem. While culture is technically feasible, stock enhancement measures have yet to be proven effective. Thus for the near term, the future of the red abalone fishery will depend upon better management of existing stocks.

RESUMEN: El abulón rojo, *Haliotis rufescens* (Swainson), es históricamente la especie más importante en la pesquería comercial, debido a su gran tamaño y a que se distribuye a todo lo largo de la costa del Estado de California. Los abulones rojos se distribuyen en tres regiones, las cuales varián considerablemente en disponibilidad de alimento, depredadores, presíon de depredación y en las caracteristicas físicas de su habitat. Estos factores afectan la estructura de la población, el crecimiento, la reproducción y su conducta. La pesquería comercial está en gran medida restringida al Sur de California; la depredación ejercida por la nutria marina (*Enhyra lutris*) ha evitado que se pesque más en el Centro de California y la Costa Norte ha sido reservada a la pesca deportiva desde 1945; así, la biogeografiá y la política complican la administración de este haliotido. La administración de la pesquería comercial y deportiva se ha basado en tamaño mínimo y además, para los pescadores deportivos, en un número limitado. A pesar de una contínua demanda en el mercado, las descargas comerciales disminuyeron precipitadamente al inicio de la decada de los 70s, después de muchos años de relativa estabilidad. En 1977 se restringió la incorporación de nuevos pescadores, pero la tendencia a la baja se ha mantenido, y en la actualidad, la captura deportiva de la Costa Norte aparentemente es mayor que la captura comercial. Estos factores, además de una elevación rápida de los precios del abulón, han sumado presiones adicionales a los stocks de la Costa Norte; asímismo, el desarrollo de la pesquería de erizos en esta zona, aparentemente ha contribudo a un agravamiento del problema de pesca furtiva. En tanto que el cultivo es técnicamente factible, las medidas de mejoramiento de poblaciones aún tienen que probar su efectividad; por lo que a corto plazo, el futuro de la pesquería del abulón rojo dependerá de una mejor administración de las poblaciones existentes.

INTRODUCTION

The red abalone, *Haliotis rufescens* (Swainson), is historically the most important species in the California abalone fishery by virtue of its size and distribution throughout the state. Reported to 298.6 mm (R. Warner, pers. comm.), it is the largest member of the genus (Cox, 1962). With a geographical range extending over 16° of latitude, from Coos Bay, Oregon, to Bahia Tortugas, Baja California (Cox, 1962), this species inhabits a range of ecosystems varying widely in physical and biological makeup. These factors affect population dynamics and fishery management. An evaluation of sport and commercial fisheries for red abalone in light of recent research is long overdue; the last major treatment was published by Cox in 1962. Since that time a variety of factors affecting the fishery have changed, including the expansion of predator populations, increased fishing power and growth of sport utilization. These factors, coupled with continued commercial demand, apparently led to the major decline in commercial landings in recent years. The diminishing supply and continued pressure on the resource have also been blamed for an increase in illegal commercial harvest in areas of Northern California reserved for the recreational fishery since 1945.

Here we review the biology of red abalone, how it changes regionally within California and its effects on fishery management. We focus on northern California, where most animals are now taken, and the object of recent research. An historical review is provided of commercial landings with an emphasis on changes since the Cox (1962) report. We contrast commercial fishing trends in south-central California to the San Francisco area, evaluating recent changes in landings and market value. The effort and landing estimates from the 1985–8 sport fishing assessment in northern California are presented to allow comparison with levels of commercial harvest in the rest of the state. We conclude with a discussion of the continuing challenges facing managers of the red abalone fishery.

RED ABALONE BY REGION

Biogeographic differences and present management regulations effectively divide the state into three regions (Fig. 26.1). A major predator, the sea otter (*Enhydra lutris*), precludes abalone fishing within its current established range in Central California (US Fish and Wildlife Service, 1987). The commercial fishery is restricted to southern California, including the Channel Islands, and in the portion of northern California from San Mateo County to Point Lobos in San Francisco County and the Farallon Islands, immediately north of the present sea otter range. The north coast, from Marin County to the Oregon border, is reserved for sport fishing. Latitudinal gradients in temperature, light and oceanographic conditions further affect the distribution of red abalone and their habitat in patterns which roughly follow this same division into three regions.

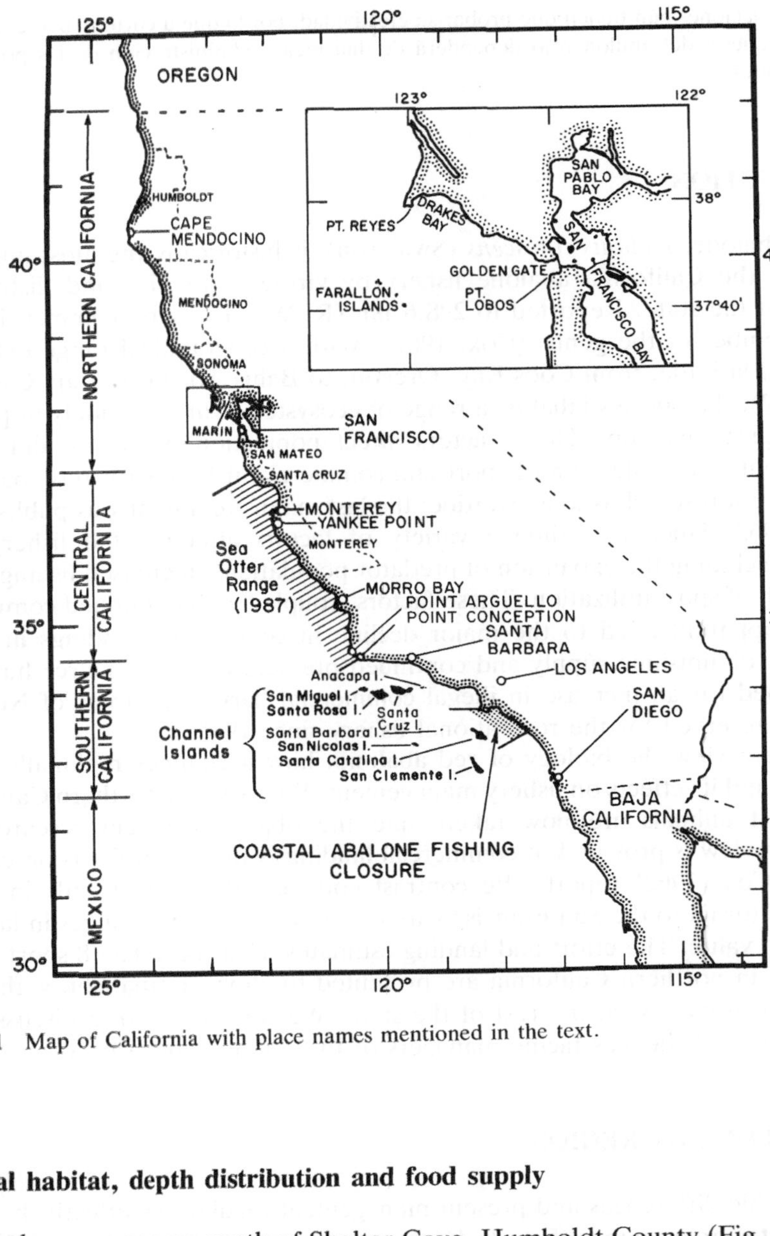

Fig. 26.1 Map of California with place names mentioned in the text.

Physical habitat, depth distribution and food supply

Red abalone are sparse north of Shelter Cove, Humboldt County (Fig. 26.1). The coastline in this region is subject to very severe ocean conditions, major rivers contribute much freshwater and sediment, and the temperature range is well below the optimum for this species (Leighton, 1974). Red abalone occur from the lower intertidal to no deeper than 6–7 m, the lower depths of their food supply in this area. The abundance of food species is highly seasonal, and standing crops are low relative to sites further south. While reduced salinities or sand abrasion may be harmful to abalone, the low abundances of other grazers suggest limitation by food supply.

The major sport fishery of northern California occurs in Mendocino and Sonoma

Counties (Fig. 26.1). This coastline is also highly exposed, but the waters are less turbid and the photic zone is deeper. Red abalone are most abundant from the lower intertidal to about 7–8 m, but are found in depths to 25 m where appropriate drift algae are supplied by surge channels (Deacon, 1977; Parker *et al.*, 1988; J.D. DeMartini pers. obs.). These observations and the migration patterns between deep and shallow water in response to highly seasonal and depth-limited abundance of algae reported by Ault and DeMartini (1987) indicate that the availability of algal drift sets the lower depth limit.

One of the authors (J.D. DeMartini, unpublished data) has studied feeding behaviour on the north coast, focusing on a site in Mendocino County. Red abalone food consisted primarily of annual brown algae (*Alaria marginata* and *Costaria costata* accounted for 58% of the feeding observations, N = 552) and virtually all observations involved drift. Food species generally started appearing by mid-spring, although the intensity of initial growth was extremely variable. These algal species started dying back by late summer and by mid-autumn virtually no food was available, a condition which persisted to the following spring. During the season of algal availability, the presence of drift and feeding behaviour both correlated with rough seas. The seasonality of food production and its temporal and spatial variability within this season favour opportunistic feeding, and abalone were observed consuming 18 species of macroalgae.

Marin through Santa Cruz Counties (Fig. 26.1) are all characterized by turbid waters; Marin and San Francisco Counties are strongly impacted by the outflow from San Francisco Bay. Major ecological changes begin in Santa Cruz County. A perennial and copious drift producer, *Macrocystis pyrifera*, becomes the dominant kelp; while there is an annual cycle of abundance, drift kelp is available virtually all year (Gerard, 1976). From Monterey County south, the photic zone deepens and algal productivity is very high. Red abalone are distributed from the lower intertidal to the outer edge of kelp forests.

Ambient temperatures strongly affect abalone distributions in southern California, where changes in current patterns create warmer environments. Red abalone are most abundant along the northwestern Channel Islands and along mainland sites where upwelling produces cooler temperatures, notably the areas north and west of Santa Barbara, the Palos Verdes Peninsula and San Diego. The distribution of red abalone becomes deeper with increasing ambient temperatures and deeper kelp forests to the south; off San Diego their minimum depth is about 12 m and their range extends to the outer edge of the kelp forest, about 25 m. *Macrocystis* is both the major source of drift in Southern California (Tegner & Dayton, 1991) and the preferred food of this haliotid (Leighton, 1966).

Competitors, predators and mortality rates

Two echinoids, the red (*Strongylocentrotus franciscanus*) and the purple (*S. purpuratus*) sea urchins, are other abundant, conspicuous invertebrates specializing on drift, and these species also range throughout the state. The apparent sea-urchin population explosion in southern California in the 1960s may have partially resulted from reduced competition with abalone (North & Pearse, 1970). While the negative impact of red abalone on the growth rate of red urchins is apparently not strong (Tegner & Levin, 1982), sea urchins clearly have a detrimental effect on other herbivores when food is limiting. These echinoids attack attached plants when drift becomes scarce; *en masse* they are capable of destroying entire stands of kelp, eliminating drift production and preventing regeneration of the vegetation. A red-urchin fishery, which began in the early 1970s and has now spread throughout California, has reduced the incidence of destructive grazing episodes by this species, but purple uchins remain a problem in some areas (reviewed by Tegner, 1989). Prior to the 1980s, red urchins dominated rocky bottoms below 6–7 m in Mendocino and Sonoma Counties apparently limiting algae to shallower depths. With the advent of the urchin fishery, there has been substantial algal recruitment to depths of about 20 m, expanding abalone habitat and food supply (J.D. DeMartini & K.A. Karpov, pers. obs.).

Sea otters dramatically affect the distribution, abundance and size structure of abalone populations (Hines & Pearse, 1982). Hunted almost to extinction in the eighteenth and nineteenth centuries, otters have slowly expanded their range since they were protected in 1911. Within the established otter range (Fig. 26.1), abalone are restricted to crevices and thus are highly dependent on the physical structure of the habitat; abalone are virtually absent from sedimentary rock substrates which do not provide crevices. Hines & Pearse (1982) estimate a very high turnover rate (1–3 years) in a kelp forest near Monterey; this estimate depends also on high rates of recruitment and growth. Sea otters have been transplanted to San Nicolas Island in southern California in an effort to secure the status of this mammal potentially threatened by oil spills in central California. However, most of the transplanted otters have left the island, and the future of this effort is not clear (E. Ebert, pers. comm.).

Outside the sea-otter range, predation is less important. Hundreds of hours of observations on the north coast have yielded few examples of predation; most of these involved a variety of asteroids feeding on large abalone (J.D. DeMartini, pers. obs.). Accordingly, Smith (1972) estimated an annual turnover rate (finite mortality rate) of 0.05 based on a length–frequency distribution from Mendocino County. Southern California supports a variety of abalone predators which collectively prey upon a wide size range of animals, but none are as dominant as the otter. These include asteroids, octopuses, crustaceans and fishes (Cox, 1962; Hines & Pearse, 1982; Tegner & Butler, 1985, 1989; Tegner *et al.*, 1989). Several of these predators, e.g. the spiny lobsters (*Panulirus interruptus*), are generally not found north of Point Conception, a major biogeographic boundary. Tegner *et al.*, (1989) estimated a natural mortality of 0.15 from five years of length–frequency distributions on Santa Rosa Island.

Growth rates, reproduction, recruitment and population structure

Long-term tagging programmes have been carried out in two areas, on Santa Rosa Island in southern California (Haaker *et al.*, 1986, unpublished data; Tegner *et al.*, 1989) and at several locations within Mendocino and Sonoma Counties in northern California (J.D. DeMartini, unpublished data). A total of 809 animals tagged at Santa Rosa Island between 1978 and 1981 yielded von Bertalanffy growth parameters of $K = 0.270$ and $L_\infty = 200.1$ mm. At this growth rate, 8.0

years are required to reach a sport minimum size of 178 mm and 15.0 years for a commercial minimum legal size of 197 mm. In the north, 1797 animals were tagged and recovered for 16 one-year intervals between 1971 and 1977 at five sites. The modal values were $K = 0.217$, $L_\infty = 192.4$, and time to sport minimum legal size was 11.8 years. Both set of estimates are based on considerable variability; the range for time to sport minimum legal size was 5.4–10.8 and 7.9–25 years for Santa Rosa Island and the north coast, respectively. Nevertheless, the data suggest that there are important differences in growth rates between southern and northern California.

Reproduction on the north coast is tightly linked to the onset of algal production (Giorgi & DeMartini, 1977; Ault, 1985). Ault's study clearly demonstrated the importance of the food regime in achieving sexual maturity, in apparently hastening gamete development and in increasing fecundity. Ault estimated that 120 days are required for the completion of the reproductive cycle in the field. This plus the seasonality of food production suggests that spawning in northern populations is temporally limited. Leighton (1974) found that red abalone collected from Estero Bay in Central California could be induced to spawn in the laboratory every month of the year. Similarly Price (1974), who studied animals collected near San Diego in Southern California where food was never limiting, detected spawning during at least three periods of the year. The size of the gonads in Price's study remained high throughout the year, suggesting that spawning may occur continually at a low level in this region.

Reproductive patterns are likely to affect recruitment rates and population structure. Recruitment rates were measured on Santa Rosa Island (Tegner *et al.*, 1989) and inferred to be high in central California to account for the high productivity observed (Hines & Pearse, 1982); no data are available for northern California. A comparison of the length–frequency distributions of emergent (animals visible without the need to turn rocks or use a light) abalone from Santa Rosa Island (Tegner *et al.*, 1989) with Mendocino and Sonoma Counties (Parker *et al.*, 1988) suggests that recruitment is much more regular in the south (Fig. 26.2). The northern size distribution is dominated by large, old animals, a majority of which are vulnerable to the fishery. These distributions are significantly different (Kolomogorov-Smirnov test, $p < 0.001$). The average size of 183 mm for northern red abalone contrasts sharply with the average of 75 mm for a central California population within the otter range, where animals of sport minimum legal size are rare despite protection from fishing (Hines & Pearse, 1982).

THE CALIFORNIA COMMERCIAL FISHERY FOR RED ABALONE

Landing summaries of commercial fisheries used in this chapter refer to south-central California, the area from San Diego through Santa Cruz Counties, and the San Francisco commercial area, San Mateo County to Point Lobos in San Francisco County and the Farallon Islands (Fig. 26.1). Landing statistics for 1973–88 were obtained from both published and unpublished compilations of California Department of Fish and Game (CDFG) market landing receipts (McAllister, 1975, 1976; Pinkas, 1977; Oliphant, 1979; Oliphant *et al.*, 1990). Red abalone landings for 1931–51 were taken from Cox (1962) and are not divided by subarea of California;

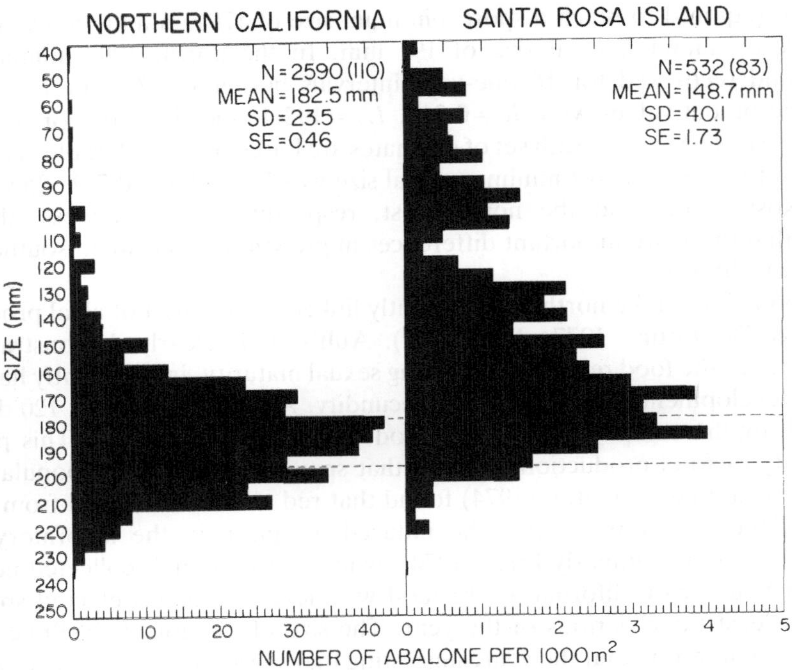

Fig. 26.2 Size–frequency distributions of emergent red abalone found in northern California (Mendocino and Sonoma Counties, Parker *et al.*, 1988) and southern California (Santa Rosa Island, Tegner *et al.*, 1989). The dashed lines indicate sport (178 mm) and commercial (197 mm) minimum legal sizes.

for 1951–72 abalone landings were not delineated by species in published catch bulletins and were obtained for south–central California from an unpublished report by Burge *et al.* (1975). Abalone landings reported in the San Francisco area were not delineated by species prior to 1973 and were assumed to be reds since black (*H. cracherodii*) abalone are rarely taken in this area.

Prior to 1916, red abalone stocks in California could be described as virginal, having accumulated in the recent absence of significant predation by either man (Cox, 1962) or sea otters (US Fish and Wildlife Service, 1987). From 1916 until the early 1960s, commercial harvest was focused in central California from Monterey to Point Arguello (Cox, 1962) (Fig. 26.1). The fishery slowly shifted southward after the opening of the mainland south of Point Conception and the Channel Islands in 1943. State-wide landings averaged over 900 metric tons wet weight per year from 1931 to 1968, with a peak landing of 1744 t in 1935 and a record low of 45 t in 1942 during World War II (Fig. 26.3). Management of the commercial fishery was based largely on a size limit. The CDFG assumed that an appropriate size limit, large enough to allow several spawnings but small enough that the size would be achieved within a reasonable number of years after settlement, could maintain the stocks (Burge *et al.*, 1975). Since 1968, commercial catches in south–central California have declined steadily to current levels of less than 136 t per year. Limited entry to the commercial fishery (and tighter limitations on sport

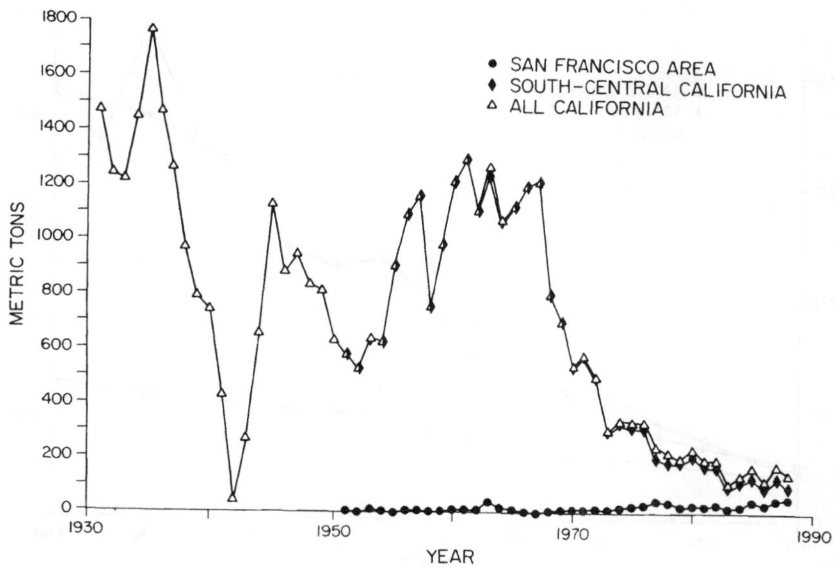

Fig. 26.3 Commercial red abalone landings from all California for 1931 to 1988 and broken down into south-central California and the San Francisco area for 1951 to 1988. The total and south-central California catches are similar because the latter comprises the vast majority of the area open to commercial fishing. See the text for further explanation.

fishermen) was instituted in 1977, but the downward trend has continued. Neither commercial catch quotas nor separate area allocations for commercial and sport fishermen have been used as management tools for red abalone in California.

The declining catch in south–central California which began in the late 1960s has been attributed to the combined effects of greatly increased commercial effort, increased natural predation as a growing sea otter population expanded southward into the commercially-productive Beckets Reef and Point Estero beds and eventually to Point Conception, effects of pollution, area closures and competition with a growing sport fishery (Bissell & Hubbard, 1968; Ebert, 1968; Miller *et al.*, 1974; Burge *et al.*, 1975; Cicin-Sain *et al.*, 1977; US Fish and Wildlife Service, 1987; Tegner, 1989; Tegner *et al.*, 1989). The decline was delayed by a major increase in fishing power. From 1965 to 1970 'dead boats' or single-diver boats using hookah without a tender replaced the less efficient hardhat divers (E. Ebert, pers. comm.). In 1968, the Radon Craft was introduced and became the vessel of choice for the commercial fishery. Boats like the Radon, using multi-hose hookah, are capable of high speed and withstanding rough seas; these allowed exploitation of the remaining virgin stocks on the Channel and Farallon Islands from the 1970s onward. Yield-per-recruit and egg-per-recruit analyses suggested that present sport and commercial legal minimum sizes allow for adequate egg production to maintain stock sizes in southern California (Tegner *et al.*, 1989; see also Chapter 20, this volume). Nevertheless, the sharp and continuing decline in landings indicates that management heavily based on minimum legal size has not been successful.

Commercial harvest of red abalone in the San Francisco area was not reported yearly prior to 1951, although Cox (1962) stated that 238 t were landed from San

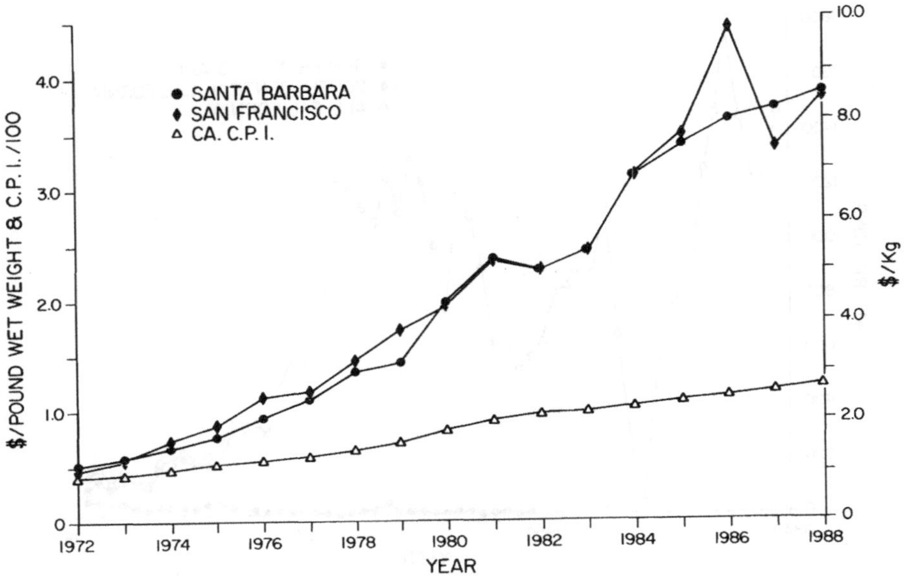

Fig. 26.4 Red abalone commercial ex-vessel price in Santa Barbara and the San Francisco area, and the California consumer price index for 1973 to 1988.

Francisco to Humboldt Bay in Northern California in the three years prior to the 1945 commercial closure north of Point Lobos (Fig. 26.1). Landings in the San Francisco area averaged only 9.6 t from 1951 through 1972. Since 1972, landings have increased to a peak of 52.7 t in 1988, which constitutes 34% of California's commercial red harvest (Fig. 26.3); most of the recent increase has been falsely reported as originating from the Farallon Islands. This increase cannot all be attributed to the small area of subtidal habitat around the Farallon Islands (3.8 km^2) considering its long history of commercial exploitation. CDFG Wildlife Protection staff suspect that increased illegal commercial harvest on the north coast, especially in association with the recent sea-urchin fishery, is the cause for some of the rise in landings.

Recent increases in the landed value of red abalone show an inverse relationship to decline in state-wide catches. The fisherman's price-per-pound in the shell has increased over twice the rate of inflation as shown in comparison with the California consumer price index (CPI) from 1973 through 1988 (Fig. 26.4). In both the south-central and San Francisco areas, the price paid to fishermen increased from about $0.50 in 1973 to over $3.80 per pound in 1988 ($1.10 to $8.38 per kilogram), or an 800% increase compared to 300% increase in the CPI during the same period.

THE NORTH-CENTRAL CALIFORNIA SPORT FISHERY FOR RED ABALONE

Sport fishery statistics, currently not available from southern California, were obtained for the north-central region of the state using a combined field 'creel' and randomized telephone survey initiated in 1985. The creel was focused in Sonoma, Mendocino, and Humboldt Counties, where the largest concentrations of sport fishermen were found. Prohibited from using scuba to take abalone north of Yankee Point in Monterey County, sport fishermen are restricted to snorkling (free diving) or 'shore picking', intertidal fishing without the use of fins. The open season is April through November with July and December through March closed, and there is a daily bag limit of four abalone for holders of sport fishing licences.

During each year of the 1985–8 creel, 1600–2800 divers or shore pickers were interviewed. Data collected included size of the abalone taken, take-per-picker-day (t/p), and county of fisherman origin. In addition, 9600–15 200 north-central California households from counties within 42 km of the ocean estuaries were randomly telephoned each year. Each phone contact was asked if a household member had participated in a sport abalone fishery; the fishermen were then asked the number of days spent abalone fishing in the preceding two months, county of the fishing trip, whether the fishermen were divers or shore pickers and the species targeted.

Expanded estimates of effort (Et) in red abalone fishing days and total weight in shell (Tw) were obtained separately for divers and shore pickers for each two month period by combining creel and telephone data. Expanded estimates of effort were obtained for each two month period using $Et = N \times (Rt/Rcc) \times (1/Re)$, where N is the census value for the number of occupied households in the telephoned counties, Rt the ratio of abalone fishing days to total contacted households, Rcc the creel ratio of fishermen from telephoned counties to all areas of origin and Re an adjustment for households that admitted abalone fishing effort prior to rejecting further questions. The expanded estimates of landed weight were obtained using $Wt = Et \times Wx \times t/p$, where Wx is the average total weight of red abalone obtained using measured lengths and the length–weight regression $W = aL^b$, with $a = 0.000161$ and $b = 3.02$ (Ault, 1982).

The 95% confidence intervals (CI) were computed for each landed weight estimate using 'bootstrapping', a method developed by Effron & Tibshirani (1986). Variances for each 2-month stratum were then combined to compute a yearly CI from the relationship $CI = \pm 1.96 \times$ (standard error). Deficiencies during the pilot year of the survey did not allow an April estimate or 95% CI computation for 1985 as a whole.

More than 93% of the sport fishing trips reported during the north-central California survey occurred in northern California, with 46% and 26% in Mendocino and Sonoma Counties, respectively. Effort in days spent abalone fishing differed seasonally between divers and shore pickers in this region. During the sport season from April through November, the 1985 to 1988 average diving effort was high for all months except November (Fig. 26.5). Shore picking effort was high only from April through May, declining in August, and negligible during September through November. Analysis of take-per-picker-day (t/p) showed a similar pattern of seasonal decline for shore pickers and not divers. The decrease in effort of shore pickers reflects in part fewer low tides later in the sport season; however, the concurrent decrease in t/p is also a factor. Ault & DeMartini (1987) found significant movement of red abalone subtidally, and suggested net movement onshore during winter months and offshore during the summer in this region. If

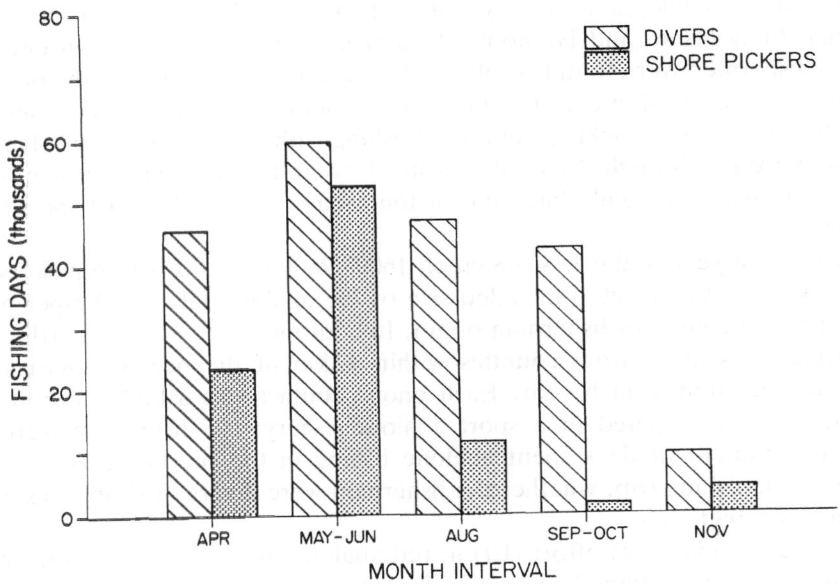

Fig. 26.5 Average effort by month interval for north-central California sport divers and shore pickers, 1985–8.

applicable to intertidal stocks, such movement could help explain seasonal differences in availability to shore pickers.

The catch for both modes of fishing combined ranged from a 1988 low of 521 ± 119 t to a high of 1550 ± 204 t in 1986. Because the data for 1985 did not include April landings, these results were not directly comparable to other years of the survey. The sport diver landings declined significantly in 1988 in comparison with both 1986 and 1987, and shore pickers consistently showed lower catches than divers in all four years of the survey (Fig. 26.6).

Parker *et al.* (1988) suggest that restricting harvesting to shore picking and free-diving protects a significant proportion of the north coast stock of fishable abalone, even in areas of heavy fishing. Nevertheless, timely management changes may be needed to reduce the level of harvest in this area, considering: (a) legal harvest by the sport fishery alone may already exceed sustainable levels for northern California with landings rivalling the commercial harvest in south-central California prior to the 1968 decline, (b) the suspected illegal commercial take, (c) lower growth rates and (d) lower levels of recruitment for Sonoma and Mendocino Counties compared with Santa Rosa Island in southern California.

CONCLUSIONS

With declining stocks in south-central California, commercial fishermen have pressured the CDFG for access to the north coast. Available population data indicate that this may be infeasible. For 11 of the 16 years of tagging data, L_∞ values were below commercial minimum legal size (J.D. DeMartini, unpublished data). The high proportion of emergent animals larger than sport minimum legal

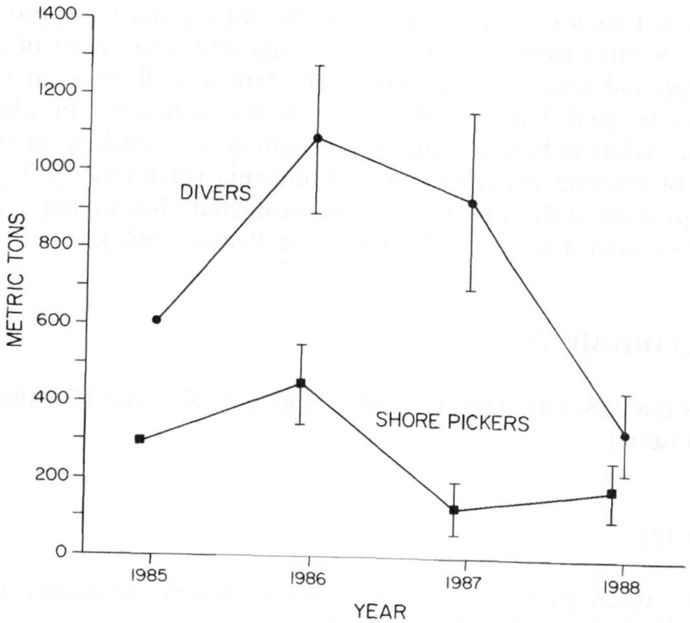

Fig. 26.6 Sport landing weight of red abalone by divers and shore pickers from 1985 to 1988 in north-central California, with 95% confidence intervals.

size (Fig. 26.2) indicates how severely this population would be impacted if the commercial minimum legal size were dropped or if sport fishermen were allowed to use scuba gear. The apparently low recruitment rates on the north coast may relate to oceanography as well as food supply; the region from Cape Blanco in Oregon to Point Conception is a region of maximum upwelling with very cool temperatures and strong offshore transport during most of the year (Parrish *et al.*, 1981). In contrast, southern California is characterized by warmer temperatures more suitable for abalone development and growth (Leighton, 1974), minimal offshore transport to disperse larvae and abundant food supplies. That we have not been able to stabilize stocks in southern California despite apparently better recruitment and growth rates underscores the need for a very conservative approach to management of remaining north coast stocks. In the absence of proven, cost-effective enhancement methods (Tegner, 1989; Tegner & Butler, 1989), the future of the fishery statewide depends upon better management of existing stocks.

We advocate

(1) immediate reduction in sport harvest through a reduced bag limit and/or seasonal closure coupled with continued monitoring,
(2) further reduction of commercial effort and establishment of mechanisms to effectively prevent illegal harvest on the North Coast, and
(3) research to refine models for stock management and to understand the ecological changes taking place in abalone habitat on the north coast caused by the sea-urchin fishery.

Emergent red abalone distributions and the fishery itself are probably recent phenomena resulting from the demise of the sea otter from most of California in the eighteenth and nineteenth centuries. The fate of both sport and commercial fisheries may be sealed if the otter range is not contained by effective zonal management. While debate continues as to whether overfishing or otters caused the decline of abalone stocks in central California (reviewed by Tegner, 1989), size–structure data make it clear that even protected abalone populations in this area cannot support a fishery today (Hines & Pearse, 1982).

ACKNOWLEDGEMENTS

The authors thank S. Farlinger, D. Parker and S.A. Shepherd for their comments on the manuscript.

REFERENCES

Ault J.S. (1982) *Aspects of laboratory reproduction and growth of the red abalone*, Haliotis rufescens *Swainson*. Ms thesis, Humboldt State University.

Ault J.S. (1985) Some quantitative aspects of reproduction and growth of the red abalone, *Haliotis rufescens* Swainson. *J. World Maricul. Soc.*, **16**, 398–425.

Ault J.S. & DeMartini J.D. (1987) Movement and dispersion of red abalone, *Haliotis rufescens*, in Northern California. *Calif. Fish Game*, **73**, 196–213.

Bissell H. & Hubbard F. (1968) *Report on the sea otter, abalone and kelp resources in San Luis Obispo and Monterey Counties and proposals for reducing the conflict between the commercial abalone industry and the sea otter*. California Department of Fish and Game, Sacramento. (Senate Concurrent Resolution, 74, 1967 Legislative Session.)

Burge R., Schultz S. & Odemar M. (1975) *Report on recent abalone research in California with recommendations for management*. Unpublished report to the California Fish and Game Commission, 17 January 1975.

Cicin-Sain B., Moore J.E. & Wyner A.J. (1977) Management approaches for marine fisheries: The case of the California abalone. *Univ. Calif. Sea Grant College Program, Sea Grant Publ.*, 54.

Cox K.W. (1962) California abalones, family Haliotidae. *Calif. Dept. Fish Game, Fish Bull.*, **118**, 1–133.

Deacon J. (1977) *Habitat selection and competition among abalone and sea urchins at the Sea Ranch, California*. Ms thesis, University of California, Davis.

Ebert E.E. (1968) A food habits study of the southern sea otter, *Enhydra lutris nereis*. *Calif. Fish Game*, **54**, 33–42.

Efron B. & Tibshirani R. (1986) Bootstrap methods for standard errors, confidence intervals, and other measures of statistical accuracy. *Stat. Sci.*, **1**(1), 54–77.

Gerard V.A. (1976) *Some aspects of material dynamics and energy flow in a kelp forest in Monterey Bay, California*. PhD dissertation, University of California, Santa Cruz.

Giorgi A.E. & DeMartini J.D. (1977) A study of the reproductive biology of the red abalone, *Haliotis rufescens* Swainson, near Mendocino, Calif. *Calif. Fish Game*, **63**, 80–94.

Haaker P.L., Parker D.O. & Henderson K.C. (1986) Red abalone size data from Johnsons Lee, Santa Rosa Island, collected from 1978 to 1984. *Calif. Dept. Fish Game, Mar. Res. Tech. Rpt*, No. 53.

Hines A.H. & Pearse J.S. (1982) Abalones, shells and sea otters: Dynamics of prey populations in Central California. *Ecology*, **63**, 1547–60.

Leighton D.L. (1966) Studies of food preference in algivorous invertebrates of Southern California kelp beds. *Pac. Sci.*, **XX**, 104–13.

Leighton D.L. (1974) The influence of temperature on larval and juvenile growth in three species of Southern California abalones. *Fish. Bull., US*, **72**, 1137–45.

McAllister R. (1975) California marine fish landings for 1973. *Calif. Dept Fish Game, Fish Bull.*, **163**, 1–53.

McAllister R. (1976) California marine fish landings for 1974. *Calif. Dept Fish Game, Fish Bull.*, **166**, 1–53.

Miller D.J., Geibel J. & Houk J. (1974) Results of the 1972 skin diving assessment survey, Pismo Beach to Oregon. *Calif. Dept. Fish Game, Mar. Res. Tech. Rpt*, **23**, 1–61.

North W.J. & Pearse J.S. (1970) Sea urchin population explosion in Southern California waters. *Science*, **167**, 209.

Oliphant M. (1979) California marine fish landings for 1976. *Calif. Dept Fish Game, Fish Bull.*, **170**, 1–56.

Oliphant M.S., Gregory P.A., Ingle B.J. & Madrid R. (1990) California marine fish landings for 1977–1986. *Calif. Dept. Fish Game, Fish. Bull.*, **173**, 1–52.

Parker D.O., Haaker P.L. & Henderson K.C. (1988) Densities and size compositions of red abalone, *Haliotis rufescens*, at five locations on the Mendocino and Sonoma County coasts, September 1986. *Calif. Dept Fish Game, Mar. Res. Div., Ad. Rpt*, No. 88–5.

Parrish R.H., Nelson C.S. & Bakun A. (1981) Transport mechanisms and reproductive success of fishes in the California Current. *Biol. Ocean.*, **1**, 175–203.

Pinkas L. (1977) California marine fish landings for 1975. *Calif. Dept. Fish Game, Fish Bull.*, **168**, 1–55.

Price P.S. (1974) *Aspects of the reproductive cycle of the red abalone*, Haliotis rufescens. Ms thesis, San Diego State University.

Smith S.V. (1972) Production of calcium carbonate on the mainland shelf of Southern California. *Limnol. Oceanogr.*, **17**, 28–41.

Tegner M.J. (1989) The California abalone fishery: Production, ecological interactions, and prospects for the future. In *Marine Invertebrate Fisheries: Their Assessment and Management* (Ed. by J.F. Caddy), pp. 401–20. Wiley, New York.

Tegner M.J., Breen P.A. & Lennert C.E. (1989) Population biology of red abalones, *Haliotis rufescens*, in Southern California and management of the red and pink, *H. corrugata*, abalone fisheries. *Fish. Bull. US*, **87**, 313–39.

Tegner M.J. & Butler R.A. (1985) The survival and mortality of seed and native red abalones, *Haliotis rufescens*, on the Palos Verdes Peninsula. *Calif. Fish Game*, **71**, 150–63.

Tegner M.J. & Butler R.A. (1989) Abalone seeding. In *Handbook of Culture of Abalones and Other Gastropods* (Ed. by K. Hahn), pp. 157–82. CRC Press, Boca Raton.

Tegner M.J. & Dayton P.K. (1991) Sea urchins, El Niños, and the long term stability of Southern California kelp forest communities. *Mar. Ecol. Prog. Ser.*, **77**, 49–63.

Tegner M.J. & Levin L.A. (1982) Do sea urchins and abalones compete in California kelp forest communities? In *International Echinoderms Conference, Tampa Bay 14–17 September, 1981* (Ed. by J.M. Lawrence), pp. 265–71. A.A. Balkema, Rotterdam.

US Fish and Wildlife Service (1987) *Final Environmental Impact Statement for Translocation of Southern Sea Otters*. USFWS, May 1987, Vol. I–III.

Chapter 27
Case histories for three species of California abalone, *Haliotis corrugata*, *H. fulgens* and *H. cracherodii*

DAVID O. PARKER, PETER L. HAAKER and HEIDI A. TOGSTAD
California Department of Fish and Game, Marine Resources Division, Long Beach, California 90802, USA

ABSTRACT: Pink, green and black abalone have been important components of commercial and recreational abalone fisheries in southern California. Fisheries have been intense, successively concentrating on the less valuable species. Harvests of pinks and greens have declined and are currently at 10% and blacks at 40% of long-term means. Management measures have been progressively more restrictive, with only limited effect. Speculation on the causes of the declines centres on overharvesting, effects of environmental variability and changes in community structure. Remedial actions must be taken to reduce the current fishery impacts on stocks of these species. More detailed information on fishery characteristics, population dynamics and community interactions will be required to understand better the history of these stocks and to manage for best resource use.

RESUMEN: Se presenta una revisión sobre el conocimiento actual de la biología y pesquería de tres abulones importantes de California, *Haliotis corrugata*, *H. fulgens* y *H. cracherodii*, con énfasis en los recursos del Sur de California. Se discute y analiza la literatura disponible, la investigación actual y la información no publicada, lográndose identificar las necesidades de investigación futura sobre la biología pesquera de estas especies. También se describen la historia y el estado actual de la pesquería, asi como la evolución de las medidas administrativas. En la actualidad sus rendimientos son aproximadamente del 10% de los valores máximos históricos y son consecuencia del esfuerzo pesquero pasado, de los cambios ecológicos y los efectos acumulativos de la población humana local, la cual sobrepasa los 12 millones. Cualquier rehabilitación futura de estos recursos dependerá de la administración, que deberá estar basada en una mayor comprensión de la biología y pesquería de estas especies.

INTRODUCTION

Three abalone, pink (*Haliotis corrugata*), green (*H. fulgens*) and black (*H. cracherodii*), have been important components of California's multi-species abalone fishery. These species, which predominate in southern California, became increasingly important after World War II as the central California red abalone fishery declined and recreational diving activities became ever more popular (Cox, 1962).

This chapter summarizes applicable biological knowledge, reviews fishery histories and management efforts, discusses possible factors involved in the declines and suggests future management information needs which will lead to more appropriate resource utilization and recovery.

BIOLOGICAL BACKGROUND

Pink abalone occur from Point Conception, California (35°N), to Punta Abreojos, Baja California (27°N), and inhabit the most widespread depth zone of the three species, from 6 to 24 m (Cox, 1962; Doi *et al.*, 1977). Green abalone range from Point Conception, California (35°N), to Bahia Magdalena, Baja California (24°N), and are found most often in depths of 1.5 to 8 m (Leighton *et al.*, 1981). Black abalone, a mainly intertidal species, are uncommon north of San Francisco, California, but are found from Coos Bay, Oregon (44°N), to Cabo San Lucas, Baja California (23°N) (Morris *et al.*, 1980). The depth and geographical distribution patterns of these abalone are determined by the temperature tolerances of larvae and juveniles (Leighton, 1974) and by exposure to surf and local food supply patterns for adults (Tutschulte, 1976).

Pink and green abalone reach a size of 254 mm in shell length, while black abalone are known to reach 203 mm (Cox, 1962). Growth patterns are similar to those of other haliotids, with rapid growth in early years followed by a slowing with age and size. Growth rates for these species have been estimated from several studies with estimates of the von Bertalanffy growth parameters K and L_∞, respectively, ranging from 0.09 to 0.47 and 165 to 202 mm for pinks, 0.09 to 0.38 and 167 to 205 mm for greens and 0.06 to 0.6 and 98 to 136 mm for blacks (Burge *et al.*, 1975; Wright, 1975; Doi *et al.*, 1977; Tutschulte & Connell, 1988a; Haaker, Shepherd *et al.* and Togstad *et al.*, unpublished data). The estimated ages of minimum commercial harvest sizes range from 8 to over 15 years or more. Significant differences in growth rates occur between locations and years, which probably reflect both short- and long-term differences in food availability and other habitat characteristics.

The potential reproductive capacity of these species is large. Adult female pink and green abalone can produce 3.5 million and 6.5 million eggs per year, respectively. Sexual maturity occurs at relatively small sizes, early in life: 40 mm for pinks, 60–70 mm for greens and 40 mm for blacks. All these species spawn during the spring to autumn period, with pinks and greens spawning several times during that period (Leighton & Boolootian, 1963; Tutschulte & Connell, 1981).

Mortality rates have been estimated in several cases, but are far from completely understood. Reported values of natural mortality (M) for pinks range from 0.35 to 0.5 and for greens a value of 0.9 has been estimated (Tutschulte, 1976; Doi *et al.*, 1977). The contribution of natural mortality (M) to total mortality (Z), the effects of fishery pressure and temporal and geographic variations in mortality rates are unknown.

A wide variety of organisms are known to prey on abalone. In southern California, several crabs, lobsters, gastropods, octopuses, fishes and a ray have all been identified as abalone predators (Cox, 1962; Tutschulte, 1976; Tegner & Levin, 1983; Tegner & Butler, 1985, 1989). Larger abalone appear to be protected to some degree by size from most predators. Two predators, the sea otter, *Enhydra lutris*, and the bat ray, *Myliobatis californica*, are known to prey selectively on larger abalone. The sea otter has been absent from southern California for over a century, but in central California its foraging has eliminated the red abalone fishery (Hardy *et al.*, 1982) and would probably have a similar impact in southern California.

Pink, green and black abalone are primarily drift algae feeders. The major portion of their diets consist of the kelp, *Macrocystis* spp. (Leighton, 1966; Tutschulte & Connell, 1988b). Other species of brown and red algae are also consumed and may be actively selected when available. Long-term disruptions of food supplies can severely affect abalone stocks, resulting in starvation and mortality in some cases (MacGinitie & MacGinitie, 1966).

Competitive interactions between abalone and other invertebrate herbivores for food and space have been the subject of speculation and some experimentation (Shepherd, 1973; Tegner & Levin, 1982). Three sea urchins, red, *Strongylocentrotus franciscanus*, purple, *S. purpurdatus*, and white, *Lytechinus* sp., are the most obvious competitors for food and space with pink and green abalone. Black abalone live in the intertidal zone, an area subject to intense competitive interactions, and must compete with numerous other species for food and space (Douros, 1987; Chapter 19, this volume).

Intraspecific competition for space and food has been hypothesized as a cause of stacking behaviour (Douros, 1987) and as a contributing cause of mass mortalities in black abalone (Chapters 17 and 19, this volume). Others interspecific interactions may occur at the depth–habitat zone boundaries of pink/green and green/black abalone.

Diseases or mass mortalities were unknown in these species until recently. Unexplained mortalities of black abalone (now known as withering syndrome) were first reported in 1987 from several southern California locations (Chapter 17, this volume; B.N. Tissot, pers. comm.). Mortalities have continued to the present, spreading slowly to unaffected areas. Black abalone populations have been reduced by nearly 100% in some areas. Similar mortalities in other species have not been reported. No definite cause has been identified, although a potential pathogen has been found (Chapter 16, this volume), and speculation also includes environmental stresses such as competition, food availability and water temperature as contributory causes (Chapter 19, this volume). There is no evidence that withering syndrome affects pink or green abalone.

FISHERY BACKGROUND

As with the better known California red abalone fishery (Cox, 1962; Tegner *et al.*, 1989; Chapter 26, this volume), commercial harvests of pink, green, and black abalone have declined dramatically from historical levels (Fig. 27.1). In addition, according to the limited data available, recreational take has declined (Fig. 27.2). Management measures primarily based on size limits have largely failed to protect abalone stocks. Harvests were controlled more by availability of abalone and fishery harvesting power than by management regulations. Numerous additional restrictions on both commercial and recreational effort have not stopped the decline in harvests of these three species of abalone.

The general pattern of the California abalone fishery has been successive exploitation of less valuable and accessible species such as green and black abalone as harvests of more desirable species such as reds and pinks have declined (Fig. 27.1). Commercial harvests of pink and green abalone are now only 10% of

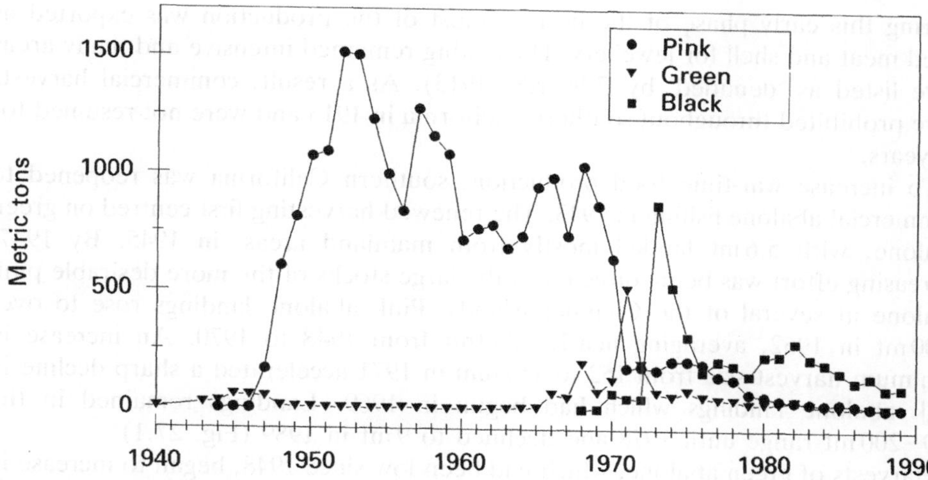

Fig. 27.1 Commercial harvests of pink, green, and black abalone in southern California. (From California Department of Fish and Game commercial fishery landing statistics.)

the mean levels for the 1960–70 and 1972–82 periods, respectively, while black abalone harvests have dropped to 40% of the 1975–85 mean.

Green and black abalone were the first species in southern California to come under commercial harvesting pressure in modern times. Chinese immigrants began harvesting the intertidal and shallow nearshore waters as early as 1850. Centred in San Diego, this fishery expanded to northern Baja California and several of the Channel Islands. Nearly 2000 metric tons (mt) were taken in 1879 (Cox, 1962). Local governments, concerned over depletion of stocks, banned harvesting from shallow nearshore areas. These restrictions shifted the fishery to Japanese divers with hard-hat gear who were able to exploit deeper stocks including pink abalone.

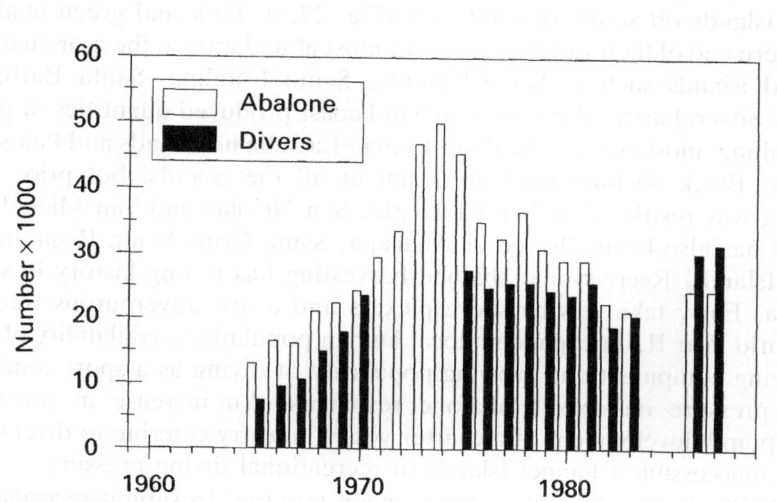

Fig. 27.2 Reported recreational take of pink, green and black abalone by southern California charter diving vessels. (From CDFG partyboat logbook records.)

During this early phase of the fishery, most of the production was exported as dried meat and shell for jewellery. Harvesting remained intensive and many areas were listed as 'denuded' by Edwards (1913). As a result, commercial harvests were prohibited throughout southern California in 1913 and were not resumed for 30 years.

To increase war-time food production, southern California was reopened to commercial abalone fishing in 1943. The renewed harvesting first centred on green abalone, with 5.6 mt landed mostly from mainland areas, in 1945. By 1947, increasing effort was being directed to the large stocks of the more desirable pink abalone at several of the Channel Islands. Pink abalone landings rose to over 1500 mt in 1952, averaging nearly 1000 mt from 1948 to 1970. An increase in minimum harvest size from 152 to 159 mm in 1971 accelerated a sharp decline in pink abalone landings which had begun in 1969. Landings remained in the 150–200 mt range until 1978 and declined to 9 mt in 1989 (Fig. 27.1).

Harvests of green abalone, which had been low since 1948, began to increase in 1966. Demand for green abalone increased as pinks became less available, and more fishing effort shifted to southern California as sea otters expanded their range in central California. Landings of greens rose to 500 mt in 1971, aided by a decrease in minimum harvest size from 184 to 178 mm, but returned to previous levels within two years. Since that time, landings of greens declined slowly to 7 mt in 1989 (Fig. 27.1).

Black abalone, considered unacceptable for domestic markets due to poor meat quality, were not harvested again until 1968 when laws prohibiting export of abalone were repealed. Initial small landings increased rapidly to 870 mt by 1973. Restrictive regulations on black abalone harvesting, including a 240 abalone/day per diver quota, closed areas and increased harvest size from 140 to 146 mm, were instituted in 1974, and landings were in the 150 to 200 mt range from 1976 to 1986. Landings for 1988 and 1989 were 92 and 95 mt, respectively, and may reflect the presence of withering syndrome.

The commercial fishery for these species has concentrated primarily at the Channel Islands off southern California (Fig. 27.3). Pink and green abalone (at the northern end of their range here) occur most abundantly at the more temperate-influenced islands such as San Clemente, Santa Catalina, Santa Barbara and Anacapa. Several areas along the mainland coast produced quantities of pink and green abalone, most notably the Point Loma–La Jolla headlands and Palos Verdes Peninsula. Black abalone were abundant at all the islands, but prior to 1984 harvesting was restricted to San Clemente, San Nicolas and San Miguel Islands and since has also been allowed at Anacapa, Santa Cruz, Santa Rosa and Santa Barbara Islands. Recreational abalone harvesting has a long history in southern California. Early take was by shore pickers and a few adventurous free-divers. After World War II, an expanding local human population, availability of reliable scuba diving equipment and growing popularity of diving as a sport combined to increase pressure on coastal abalone resources. An increase in private boat ownership and development of a charter vessel industry catering to divers opened the once inaccessible Channel Islands to recreational diving pressure.

Since 1958, charter diving boats have been required to submit records of their passengers' take (Fig. 27.2). Only about 2500 abalone were taken in 1958 and 1959 by 1700 divers (Young, 1961). Take rose steadily to nearly 14 000 by 1964

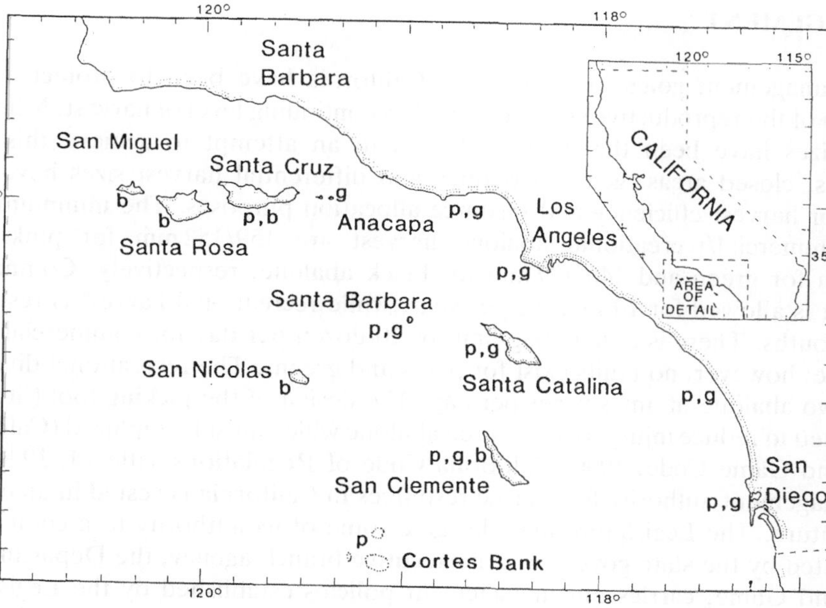

Fig. 27.3 Major harvest areas for pink (p), green (g) and black (b) abalone in southern California. (From CDFG commercial fishery landing statistics.)

(Wood, 1964, 1967) and had risen to nearly 18 000 abalone for 15 000 divers in 1968, a catch rate of 1.2 abalone/diver (Young, 1973). The peak years for the charter fleet were 1973–5, when the mean annual take was 36 000 pink and green abalone. The take of abalone dropped to about 24 000 in 1986 and 1987, but the number of divers increased to over 30 000 in 1986, a catch rate of 0.8 abalone/diver.

Divers using privately-owned boats and those operating from shore are two other segments of the recreational fishery. Estimates for take of pink, green and black abalone by private boat divers during 1975–81 averaged 9500 per year, decreasing from 13 500 in 1977 to 5200 in 1981 (Wine, 1978, 1979a,b, 1982). Unfortunately, data on this segment of the fishery have not been collected since 1982. Shore-based diving effort, once a major portion of the recreational fishery, now occurs in only a few areas, reflecting the low level of most mainland abalone stocks (D. Parker, pers. obs.).

Recreational take now occurs predominantly at the Channel Islands. Charter vessel records show that effort and take have shifted from the more accessible and/or protected islands to more remote or exposed locations over time. Pink abalone are the most heavily harvested recreational species. Their depth distribution coincides with the 'comfort zone' of most recreational divers, making them vulnerable to even casual abalone seekers. Green abalone occupy the high subtidal zone, where wave action and turbulence often make diving uncomfortable, and are thus less available to divers. The black abalone's reputation for tough, dark meat makes it a species shunned by most recreational divers. However, they are the only species available to non-diving shore pickers in Southern California and have been heavily harvested in many mainland areas.

MANAGEMENT

The management goals for abalone in California have been to protect a large portion of the reproductive stock to provide a continuing level of harvest. Minimum legal sizes have been the main tool used in an attempt to achieve this goal. Seasons, closed areas, gear restrictions and differential harvest sizes have been used for harvest efficiency and resource allocation purposes. The minimum sizes for commercial/recreational abalone harvest are 159/152 mm for pink, 178/152 mm for green and 146/127 mm for black abalone, respectively. Commercial harvest is allowed for 10 months per year, while recreational harvest is restricted to 7 months. There is a daily bag limit of 20 dozen per day for commercial black abalone; however, no limits exist for pinks and greens. The recreational diver can take two abalone of any species per day. The design of the picking tool ('iron') is regulated to reduce injury to undersized abalone which must be replaced (California Fish and Game Code, 1990; California Code of Regulations Title 14, 1990).

Management authority for marine resources in California is vested in an elected Legislature. The Legislature may delegate some of its authority to a commission appointed by the state governor. An executive branch agency, the Department of Fish and Game, carries out management policies established by the Legislature and Commission. Obviously, such a system is a political process and allows for much input from interested parties. In practice, most changes in management regulations are the product of compromise and concensus addressing the concerns of resource managers, user groups and others. Wyner *et al.* (1977) thoroughly analyzed how this process worked in the case of the abalone fishery.

Major changes took place in the fishery between the late 1960s and early 1970s; the commercial fishery became concentrated in southern California, commercial harvesting efficiency and power increased and the recreational fishery expanded throughout the Channel Islands. Landings of pink abalone declined sharply, green abalone landings rose and fell rapidly and a renewed fishery for black abalone developed during this period. In an effort to prevent further fishery collapse and resource depletion, several additional management measures were put into effect (Burge *et al.*, 1975; Wyner *et al.*, 1977). A limited entry system was established to reduce harvesting pressure and part-time, inexperienced divers in the commercial fishery. The commercial black abalone fishery was limited geographically and placed under a daily harvesting quota. The recreational daily bag limit was reduced from five to four and a 67 km section of mainland coast was closed to all abalone take.

These measures have been modified several times. Commercial permits were reduced from 499 in 1976, prior to limited entry, to 130 at present. A new permit ceiling of 70, daily harvest quotas, and a two-for-one permit transfer provision are now in place. The recreational season has been shortened to seven months and the daily bag limit has been reduced to two abalone per person.

Despite these measures, harvests of pink, green and black abalone have continued to decline, and abalone stocks appear to be declining also. To date, abalone management has been reactive, responding to changes in the fishery and stocks after the fact, rather than based on actual stock productivity. Such forward-looking management requires knowledge of stock size, recruitment, mortality and growth. Since these parameters are incompletely understood for pink, green and black abalone, the ineffectiveness of management to date is not surprising.

DISCUSSION

The rapid decline and continued decreasing production of California's abalone fishery have been attributed to a number of causes (Burge *et al.*, 1975; Cicin-Sain *et al.*, 1977; Henderson *et al.*, 1988). Excessive harvest pressure and resulting depletion of pre-fishery stocks, loss of productive central California red abalone stocks, illegal take, habitat degradation and changes in community structure have all played a role in the present state of pink and green abalone stocks in southern California. Despite 20 years of harvesting, black abalone remained abundant in many areas but are now at low levels over a wide area due to withering syndrome mortality (Chapter 17, this volume).

Recruitment overfishing has been suggested as the main cause of the fishery's decline and is one which can be most directly addressed through corrective management measures. Tegner *et al.* (1989) used yield analysis to conclude that simple recruitment overfishing was not the cause of the decline of pink abalone and that 20 years of apparent stability of pink abalone harvests were due, in part, to stock productivity. The initial stock size of pink abalone was very large, covering wider depth and geographic ranges than any other subtidal species in southern California. Stability of the pink abalone harvest may be illusory; the long-term harvest trend was a decreasing one from 1952 to 1970, despite increases in fishery size and harvesting power.

Green abalone are less widely distributed in southern California and occupy a narrow habitat zone. Thus, even with high abundance within that zone, the stock size of green abalone was much smaller than that of pinks. When the fishery depended upon green abalone as a substitute for the declining pinks, harvests quickly rose and fell to previous levels as available stocks were removed.

Stocks of abalone experiencing regular recruitment would seem likely to sustain higher harvest levels than are presently seen for pink and green abalone. These species can spawn for five or more years before reaching harvest size, which should leave a considerable spawning stock in place, given regular recruitment. Pink and green abalone in southern California are at the northern extreme of their ranges, and successful spawning and recruitment may depend on the occurrence of favourable environmental conditions which are often irregular and variable. During the 1982–4 El Niño event in southern California, recruitment of cold temperate species such as red abalone was depressed, while warm temperate species such as pink abalone showed increased recruitment (Tegner & Dayton, 1987). Under such conditions, growth rates for these species may be lower than elsewhere and are often nil at harvest size in southern California (Burge *et al.* 1975; Togstad *et al.*, unpublished data). Once stocks of harvestable size abalone have been removed, annual production for the fishery is low. The collapse of pink and green abalone fisheries in Baja California (Morales, 1983), where environmental conditions may be closer to optimum, indicates that such a scenario is incomplete and that other factors may be involved.

The recreational fishery, which operates under smaller harvesting sizes, has added a complicating factor which has become more significant as commercial stocks decline and recreational diving increases. Recreational and commercial effort takes place in many of the same areas, placing the two fisheries in direct competition. As a result of the differential harvest sizes, the recreational fishery intercepts a portion of the stock which would have become available for commercial harvest.

Black abalone stocks remained abundant in many areas subject to harvesting pressure. Size distributions indicate the removal of much of the harvestable stock by the fishery (Chapter 17, this volume). Growth rates of harvestable size abalone are also low and portend continued harvest declines if recruitment is low or infrequent. Mass mortality from withering syndrome has reduced stocks in several important fishery areas and will complicate future management.

Rehabilitation of depleted abalone stocks by seeding cultured juveniles or transplanting excess native abalone seemed an attractive way to reverse declines and develop new local spawning stocks. Unfortunately, such efforts in southern California have proven unsuccessful or impractical (Tegner & Butler, 1989; Chapter 34, this volume; pers. obs.). High mortality of seeded juveniles in the field, poor availability of donor stocks for transplantations and the costs and personnel required for such operations have reduced interest and research efforts into abalone enhancement techniques.

Southern California now supports a human population of over 12 million, most of whom reside within 50 km of the coastline. Development of this magnitude brings with it a number of impacts on the marine environment. Large ocean discharges of municipal and industrial wastes, coastal developments such as harbours, residential areas and increased coastal access points all result in degradation and loss of nearshore habitats. Even if direct individual use of marine resources is small, the sheer numbers of potential users can have a large cumulative effect. Unorganized, 'casual' illegal take, even from areas closed to take, in such a situation may have as much or more impact as regulated fisheries. In addition, fisheries for other nearshore resources such as lobsters, sea urchins and predatory fishes can have both direct and indirect effects on abalone stocks (Tegner, 1980). Even under the most conservative and enlightened management plan, the future of abalone stocks in southern California is uncertain.

In the absence of more detailed information on abalone population dynamics, community interactions and fishery characteristics, the causes of the declines will remain fertile ground for speculation. Long-term series of data are needed to make accurate assumptions about past and present conditions and to predict future changes in stock levels and responses to management. In the short term, the commercial fishery should be further curtailed in terms of the number of participants, season length, areas and total catch. Remaining pink and green abalone stocks at several mainland and Channel Island locations should be protected by closures. Black abalone stocks unaffected by withering syndrome should be protected as a reservoir for future stock recovery.

Specifically, programmes are needed to measure the temporal and geographic patterns of abalone recruitment throughout southern California. Fishery-independent stock assessment methods are required to institute management based on stock productivity and avoid the reactive system of the past. Closer monitoring of the fisheries, including creel censuses and fishing effort, needs to be established to improve understanding of the effects on stocks and the effectiveness of management measures. In light of the failure of past management, alternative management methods such as harvest refugia and individual transerable quotas need to be investigated for the applicability to the abalone resource. These programmes will be expensive and will have to compete with management programmes for other fisheries (many larger than the abalone fishery) for the limited funds available for marine resource management in California. Only continued interest by both resource managers and users will ensure that these species of abalone remain an important part of California's marine resources.

ACKNOWLEDGEMENTS

The authors thank Joyce Underhill, Rosa Madrid, Steve Torres and Calvin Chun for their assistance in accessing commercial and recreational fishery data. Hal Clemens reviewed early drafts and made valuable suggestions.

REFERENCES

Burge R.S., Schultz S. & Odemar M.O. (1975) *Draft report on recent abalone research in California with recommendations for management*. Calif. Dept Fish and Game.

California, State of (1990) *Fish and Game Code*, Sections 8300–10.

California, State of (1990) *California Code of Regulations*, Title 14, Sections 29.15, 100–2.

Cicin-Sain B., Moore J.E. & Wyner A.J. (1977) Management approaches for marine fisheries: The case of the California abalone. *Univ. Calif. Sea Grant Coll. Prog., Sea Grant Publ.*, **54**, 1–223.

Cox K.W. (1962) California abalone, family Haliotidae. *Calif. Dept Fish Game, Fish Bull.*, **118**.

Doi T., Guzmán del Próo S.A., Marín A.,V., Ortiz Q.,M., Camacho A.,J. & Muñoz L.,T. (1977) Análisis de la población diagnostico de la pesquería de abulón amarillo (*Haliotis corrugata*) en el area de Punta Abreojos e Isla Cedros, B.C. Dirección General del Institudo National de Pesca, México. *Serie Cientifica*, No. 18.

Douros W.J. (1987) Stacking behavior of an intertidal abalone: an adaptive response or a consequence of space limitation? *J. Exp. Mar. Biol. Ecol.*, **108**, 1–14.

Edwards C.L. (1913) The abalone industry in California. *Calif. Dept Fish Game Fish Bull.*, **1**.

Hardy R., Wendell F. & De Martini J.D. (1982) A status report on California shell fish fisheries. In *Social Science Perspectives on Managing Conflicts between Marine Mammals and Fisheries* (Ed. by B. Cicin-Sain, P.M. Grifman & J.B. Richards), pp. 328–40. Marine Policy Program, University of California, Santa Barbara.

Henderson K.C., Parker D.O. & Haaker P.L. (1988) The survival and growth of transplanted adult pink abalone, *Haliotis corrugata*, at Santa Catalina Island. *Calif. Fish Game*, **74**, 82–6.

Leighton D.L. (1966) Studies on food preference in algivorous invertebrates of southern California kelp beds. *Pac. Sci.*, **20**, 104–13.

Leighton D.L. (1974) The influence of temperature on larval and juvenile growth in three species of southern California abalones. *Fish. Bull.*, **72**, 1137–45.

Leighton D.L. & Boolootian R.A. (1963) Diet and growth in the black abalone, *Haliotis cracherodii*. *Ecology*, **44**(2), 227–38.

Leighton D.L., Byhower M.T., Kelly J.C., Hooker G.N. & Morse D.E. (1981) Acceleration of development and growth in young green abalone, *Haliotis fulgens*, using warmed effluent sea water. *J. World Maricul. Soc.*, **12**(1), 170–80.

MacGinitie M. & MacGinitie G.E. (1966) Starved abalones. *Veliger*, **8**, 313.

Morales J.J. (1983) El abulón: requiem por una pesquería. *Techna Pesquera, La Revista de la Pesca Mexicana*, **16**(186), 13–19.

Morris R.H., Abbott D.P. & Haderlie E.C. (1980) *Intertidal Invertebrates of California*. Stanford University Press, Stanford.

Shepherd S.A. (1973) Competition between sea urchins and abalone. *Aust. Fish.*, **32**(6), 4–7.

Tegner M.J. (1980) Multispecies considerations of resource management in southern California kelp beds. *Can. Tech. Rep. Fish. Aquat. Sci.*, **954**, 125–43.

Tegner M.J., Breen P.A. & Lennert C.E. (1989) Population biology of red abalones, *Haliotis rufescens*, in southern California and management of the red and pink, *H. corrugata*, abalone fisheries. *Fish. Bull.*, **87**(2), 313–39.

Tegner M.J. & Butler R.A. (1985) The survival and mortality of seeded and native red abalones, *Haliotis rufescens*, on the Palos Verdes Peninsula. *Calif. Fish Game*, **71**(3), 150–63.

Tegner M.J. & Butler R.A. (1989) Abalone seeding. In *Handbook of Culture of Abalone and other Marine Gastropods* (Ed. by K.O. Hahn), pp. 157–82. CRC Press, Boca Raton.

Tegner M.J. & Dayton P.K. (1987) El Niño effects on southern California kelp forest communities. *Adv. Ecol. Res.*, **17**, 243–79.

Tegner M.J. & Levin L.A. (1982) Do sea urchins and abalones compete in California kelp forest communities? In *International Echinoderms Conference, Tampa Bay* (Ed. by J.M. Lawrence). A.A. Balkema, Rotterdam.

Tegner M.J. & Levin L.A. (1983) Spiny lobsters and sea urchins: analysis of a predator–prey interaction. *J. Exp. Mar. Biol. Ecol.*, **73**, 125.

Tutschulte T.C. (1976) *The comparative ecology of three sympatric abalones*. PhD dissertation, University of California, San Diego.

Tutschulte T.C. & Connell J.H. (1981) Reproductive biology of three species of abalones (*Haliotis*) in southern California. *Veliger*, **23**, 195–206.

Tutschulte T.C. & Connell J.H. (1988a) Growth of three species of abalones (*Haliotis*) in southern California. *Veliger*, **31**, 204–13.

Tutschulte T.C. & Connell J.H. (1988b) Feeding behavior and algal food of three species of abalones (*Haliotis*) in southern California. *Mar. Ecol. Prog. Ser.*, **49**, 57–64.

Wine V.L. (1978) Southern California independent sport fishing survey annual report no. 2. *Calif. Dept Fish Game, Mar. Res. Admin. Rpt*, No. 78–2.

Wine V.L. (1979a) Southern California independent sport fishing survey annual report no. 3. *Calif. Dept Fish Game, Mar. Res. Admin. Rpt*, No. 79–3.

Wine V.L. (1979b) Southern California marine sportfishing: private-boat catch and effort, 1975–1976. *Calif. Dept Fish Game, Mar. Res. Admin. Rpt*, No. 79–11.

Wine V.L. (1982) Southern California marine sportfishing: private-boat catch and effort during 1981. *Calif. Dept Fish Game, Mar. Res. Admin. Rpt*, No. 82–7.

Wood W.F. (1964) Partyboat logs show how skin divers fared during 1960, 1961, and 1962. *Calif. Fish Game*, **50**(2), 114.

Wood W.F. (1967) Partyboat logs show how skin divers fared during 1963 and 1964. *Calif. Fish Game*, **53**(3), 192–6.

Wright M. (1975) Growth in the black abalone, *Haliotis cracherodii*. *Veliger*, **18**, 194–9.

Wyner A.J., Moore J.E. & Cicin-Sain B. (1977) Politics and management of California abalone fishery. *Mar. Pol.*, **1**(4), 326–39.

Young P.H. (1961) Partyboat logs show how skin divers fared during 1958 and 1959. *Calif. Fish Game*, **47**(3), 303–5.

Young P.H. (1973) Partyboat logs show how skin and scuba divers fared 1965 through 1970. *Calif. Fish Game*, **59**(1), 62–72.

Chapter 28
Fisheries management and biology of northern abalone, *Haliotis kamtschatkana*, in the northeast Pacific

S. FARLINGER *Department of Fisheries and Oceans, Fisheries Branch, 417 West 2nd Avenue, Prince Rupert, British Columbia, Canada, V8J 1G8*
A. CAMPBELL *Department of Fisheries and Oceans, Biological Sciences Branch, Pacific Biological Station, Nanaimo, British Columbia, Canada, V9R 5K6*

ABSTRACT: Northern abalone are harvested in commercial dive, native food and sports fisheries in the northeast Pacific. Commercial abalone landings in Alaska and British Columbia (BC) increased rapidly in the late 1970s, but have since declined to low levels. Catch per unit effort has generally declined in BC during 1977–86 and stabilized at low levels. For Washington State there are only recreational and native food fisheries for which few landing data are available. Fishery management in BC is conservative and includes vessel quotas, whereas in Alaska area quotas are used. The total annual quota has been reduced to the present 47 t for BC and 15–26 t for Alaska. The reductions in landings and quota in BC have decreased the number of vessels and increased the unit price ($/kg) at a greater rate than the consumer price index, making the fishery lucrative for both legal and illegal participants. Abalone stock assessment in BC involves bi- or triennial dive surveys to provide indices of abundance from which quotas can be adjusted. A 1989 resurvey of a major fishing area in BC suggested a further decline in both legal and pre-recruit abundance indices. Future research should focus on factors affecting recruitment and an evaluation of the survey methodology and size limit.

RESUMEN: La captura del abulón norteño en el noroeste del Pacífico es perseguida, tanto para buceo comercial, como para pesca deportiva o para consumo ‘nativo’. Las descargas comerciales de abulón en Alaska y Columbia Británica (CB), se incrementaron rápidamente al final de los 70s, declinada desde entonces a niveles bajos. La captura por unidad de esfuerzo en CB, ha permanecido relativamente estable desde la declinación. En Washington no se cuenta con datos de las descargas comerciales de la perquería de abulón, ya que ésta es primordilamente recreativa. La administración de la pesquería de abulón en CB es conservativa, considera cuotas por barco, mientras que en Alaska se imponen cuotas por área. Actualmente en CB la cuota total es de 47 t, y de 15 a 26 t en Alaska. En años recientes la reducción de la cuota en CB, ha producido la disminución del número de barcos, e incrementado su precio por unidad. Esto último, sin embargo, ha aumentado en proporción mayor al índice de precios al consumidor, haciendo a la pesquería de abulón más lucrativa, tanto para los participantes legales como para los ilegales. La evaluación de stocks de abulón en CB, incluye análisis de la CPUE de las bitácoras de los buzos y prospecciones, bianuales o trianuales, que proporcionan índices de abundacia a partir de los cuales se pueden estimar las cuotas. Los factores que confunden las estimaciones de la CPUE y de abundancia del stock son los cambios en la experiencia del buzo y el sistema de cuotas, que cambia la estrategia de pesca de éstos. El muestreo de las tallas de las descargas de abulón en CB, ha demostrado en los últimos años, que la talla promedio de captura está por arriba del tamaño mínimo legal de 100 mm. Las investigaciones futuras deben incluir los factores que afectan el reclutamiento, evaluación del tamaño mínimo de captura y técnica de prospección.

INTRODUCTION

The 'northern' or 'pinto' abalone, *Haliotis kamtschatkana*, occurs from Sitka Island, Alaska, to Baja California (Cox, 1962; McLean, 1966; Paul & Paul, 1981); this is the only abalone species occurring in Washington State, British Columbia (BC) and Alaska (Fig. 28.1). Abalone generally occur in patchy distribution on exposed and semi-exposed coasts (Sloan & Breen, 1988). Abalone are harvested

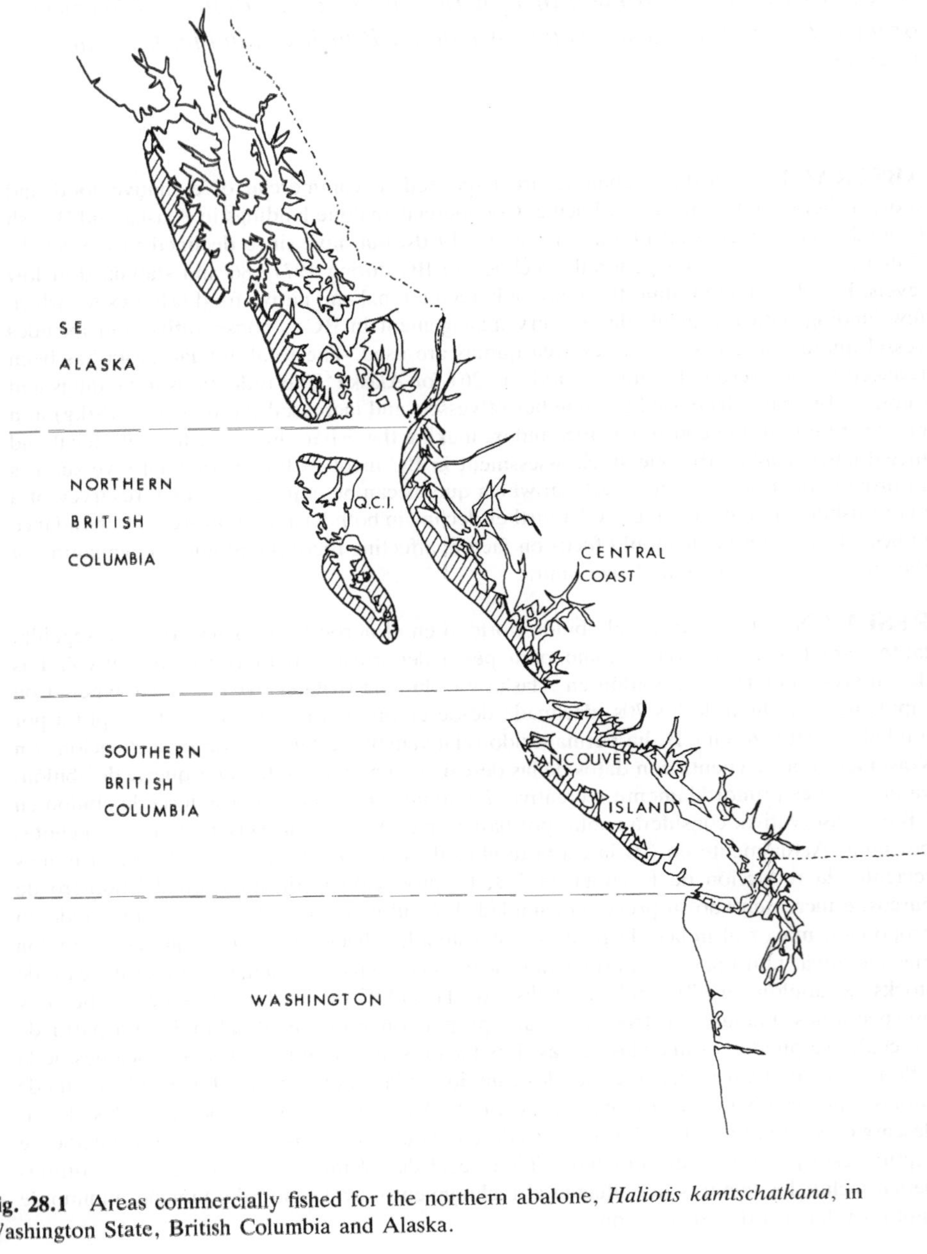

Fig. 28.1 Areas commercially fished for the northern abalone, *Haliotis kamtschatkana*, in Washington State, British Columbia and Alaska.

in BC and Alaska by native people, commercial and sport divers. In Washington State, there is no commercial fishery and there are few data available on the native food or recreational abalone harvests (Bargmann, 1984). The biology and commercial fisheries in BC are well documented (Federenko & Sprout, 1982; Bates, 1984, 1985; Farlinger & Bates, 1985; Jamieson & Francis, 1986; Sloan & Farlinger, 1987; Farlinger & Thomas, 1988; Sloan & Breen, 1988). The purpose of this chapter is to review recent developments in the fisheries and biology of northern abalone (taken to refer to *H. kamtschatkana*, unless otherwise stated), especially in BC, and to present recommendations for future research.

GENERAL BIOLOGY

Only a summary of relevant abalone biology is presented in this section since a comprehensive review has been published by Sloan & Breen (1988). Abalone occur in a wide variety of habitats from intertidal to subtidal depths of 100 m, although most adults are found at <10 m in BC (Sloan & Breen, 1988). Mature males and females are mass spawners broadcasting their gametes synchronously in shallow subtidal areas during summer (Mottet, 1978; Breen & Adkins, 1980a). As many as 2.3 million eggs may be produced by a female (S. Olsen, pers. comm., in Sloan & Breen, 1988). Fertilized eggs hatch in a few days into larvae which swim in the water column for about a week (Mottet, 1978; Olsen, 1984) being dispersed by local water currents (Quayle, 1971; Sloan & Breen, 1988). Juvenile abalone (<10 mm shell length, SL) are cryptic in habit and are sometimes found under rocks occurring mostly deeper than adults at 5–15 m depth (Breen & Adkins, 1979, 1982). Abalone are herbivorous and as the juveniles develop to maturity, they change their diet from benthic diatoms and microalgae, moving to shallower, more exposed areas to feed on macroalgae (Sloan & Breen, 1988). Abalone growth can vary considerably depending on the extent of exposure to wave action and availability and quality (i.e., algal spp.) of food (Breen & Adkins, 1979; Emmett & Jamieson, 1988; Sloan & Breen, 1988). Although aging of *H. kamtschatkana* is not possible by counting annual rings on the shell, an estimated minimum of six years is required for this species to reach the recruit (fishable) size of 100 mm SL in BC (Breen, 1986). Natural mortality of abalone is considered to be relatively low ($M = 0.15-0.20$) (Breen, 1986). Recruitment in recent years has been estimated to be poor in some areas of BC (Breen, 1986).

HARVESTING HISTORY

Prior to 1971 the commercial abalone fishery in BC was sporadic and local (Sloan & Breen, 1988), and landings remained lower than 20 t except for a slight rise during 1964 (Fig. 28.2). Shore picking was the principal harvest method, but after 1960 some scuba gear was used. Landings in the BC commercial fishery increased dramatically in the 1970s, peaking in 1977 at 474 t (Federenko & Sprout, 1982; Fig. 28.2). The fishery accelerated rapidly for several reasons, including improved scuba gear, reduced access and fishing time in the salmon and herring fisheries, acceptance of *H. kamtschatkana* into the Japanese market,

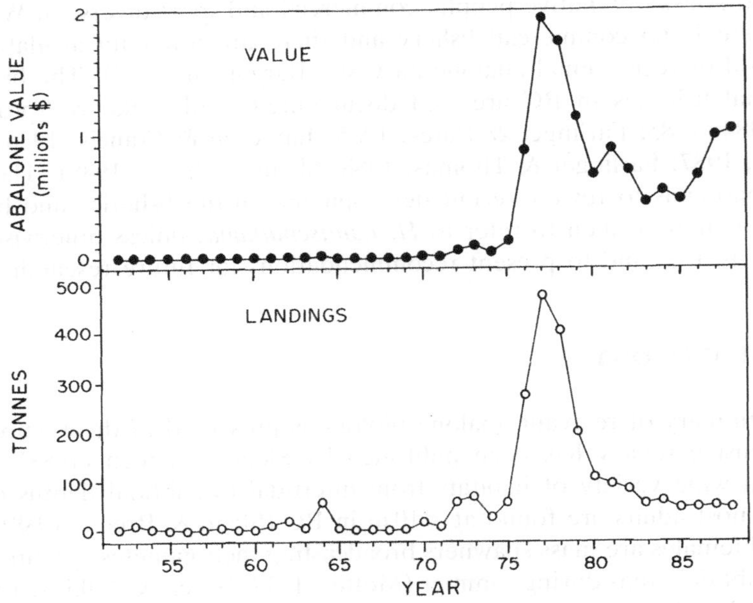

Fig. 28.2 Abalone landings and value (Can$) in the British Columbia fishery, 1952–88.

increasing product demand and the introduction of larger vessels with freezing capacity that allowed fishermen access to distant north-coast stocks (Federenko & Sprout, 1982; Farlinger & Thomas, 1989). Approximately 85% of commercial landings are now sold frozen and glazed. Fishing restrictions imposed in 1977, including a requirement to measure abalone underwater to reduce sublegal mortality and licence limitation, culminated in the introduction of vessel quotas in 1980. Landings have decreased since 1980 along with vessel quotas (Table 28.1), which were reduced as surveys indicated a continuing decline in the index of abalone abundance. The number of vessels participating in the fishery has decreased with the decline in vessel quotas to the point where the existing 26 licences were used by only 13 boats in 1988 (Fig. 28.3).

Native food fisheries have historically harvested or picked abalone from shore. Several north-coast Indian groups in BC have recently trained divers to participate in sea cucumber and sea urchin fisheries, consequently expanding the harvest techniques available to the native food fishery. No landing statistics are available for this fishery, although substantial areas of the coast are closed to commercial fishing for the purpose of supplying native food abalone. Recreational fishermen take abalone by both diving and shore picking. Diving charters to remote areas probably increased this catch in the mid-1980s, although no measure is available at present in BC.

The Alaska fishery followed a similar pattern to that of BC, with maximum landings of 168 t recorded in 1981, falling to 34 t in 1988–9 (Koeneman & Botelho, 1989). Commercial harvests were highly variable until 1976, when market conditions and relaxed harvest regulations encouraged the development of this

Table 28.1 Total quota, catch (metric tons) and effort data in the British Columbia abalone fishery, 1977–88; data are from harvest logs.

Year	Quota (t)	Catch (t)	Diver days	CPUE (kg/dd)	Diver hours	CPUE (kg/dh)
1977	none	474.8	2356	201.5	12680	37.4
1978	none	404.0	2169	186.3	11812	34.2
1979	227	197.2	1303	151.4	5816	33.9
1980	113	106.0	789	134.4	3282	32.3
1981	94	93.6	700	133.8	2741	34.2
1982	94	82.1	602	136.4	2587	31.7
1983	71	53.3	415	128.3	1584	33.6
1984	59	57.5	486	118.3	1903	30.2
1985	47	43.8	443	98.8	1367	32.0
1986	47	45.6	477	95.5	1851	24.6
1987	47	46.4	467	99.4	1726	25.8
1988	47	45.5	429	106.0	1750	26.0

dd = diver day, dh = diver hour, CPUE = catch per unit effort.

fishery (Koeneman & Botelho, 1989). High harvests were a result of increased effort (scuba) in the late 1970s. In the 1980s the quota held harvests down to a third of the previous average of 64 t, despite the participating number of vessels exceeding 40 in some years.

For Washington State a single creel census survey in 1983–4 suggested that the abalone harvest was probably about 12 t annually (R. Burge, pers. comm.; Bargmann, 1984).

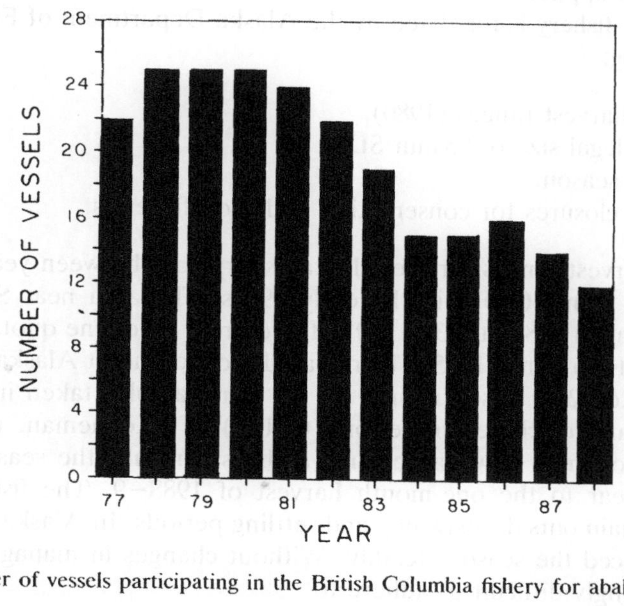

Fig. 28.3 Number of vessels participating in the British Columbia fishery for abalone.

MANAGEMENT STRATEGIES

Commercial fisheries

The federal Department of Fisheries and Oceans (DFO) manages marine species in BC. Evolution of the present management system in BC is well documented (Federenko & Sprout, 1982; Breen, 1984; Farlinger & Bates, 1985; Farlinger & Thomas, 1988; Sloan & Breen, 1988). The system currently used is elaborate and includes:

(1) a minimum (recruit) size limit of 100 mm SL (101.6 mm SL from 1908, changed to present level in 1981),
(2) vessel licence limitation to 26 (1977),
(3) vessel quotas (1980),
(4) seasonal restrictions (1988),
(5) numerous local permanent area closures to commercial fishing.

The recent fishing season closures (15 December to 15 January and 15 June to 15 September in the same year) were an attempt to curb illegal fishing by reducing the time vessels have abalone on board. The licence is a personal one and must name the vessel to be used; the fee has remained at Can$200 since 1976. Licence policy states that no more than four licences may be assigned to one vessel. The licence holder is required to notify the appropriate office of DFO prior to fishing and prior to leaving the fishing area. A validation form confirming weight is signed by a Fishery Officer upon landing. Harvest logs with charts and sales slips are required before the following year's licence is issued.

The quota declined steadily from 1980 to 1985 based on continued decline in survey results (Table 28.1), and remained at 1.8 t (4000 lb) per licence or 47.2 t for the fishery through 1988. For vessels which freeze abalone, a glaze allowance of 7% has been applied since 1981.

In Alaska the fishery is regulated by the Alaska Department of Fish and Game (ADF&G) using:

(1) guideline harvest ranges (1980),
(2) minimum legal size of 95 mm SL,
(3) restrictive season,
(4) local area closures for conservation and food fisheries.

Guideline harvest ranges for the Alaska fishery vary between years from 15 to 26 t, decreased from 46–57 t in the early 1980s. The area near Sitka was split from other areas in 1982 at 6.3 to 7.9 t; the current guideline quota for this area has remained at 3.6 t since 1985. The remainder of southeast Alaska has a varying quota of 11.3 to 22.6 t based mainly on the port samples taken in the previous year's fishery and catch-per-unit-effort (CPUE) (T. Koeneman, pers. comm.). The fishery is open to new participants and, as a result, the season has grown shorter each year to the one-month harvest of 1988–9. The fishery opens in October to remain outside spawning and settling periods. In Alaska the increasing effort has reduced the season steadily. Without changes in management, quotas will be increasingly difficult to adhere to.

Sport and native food fisheries

Local closures to commercial fishing, intended to provide opportunity for native food and sport fisheries, are numerous in BC; they are also used in Alaska. All groups are required to adhere to the size limit. There are bag limits for sport fishing in BC ranging from 4 to 12 per day, depending on area. The effect of the sport fishery in BC was considered minor by Breen (1986), although areas which are readily accessible may be depleted locally (S. Farlinger, pers. obs.). No data are available for these or the native food fishery, apart from a survey in the Strait of Georgia where abalone were not abundant and divers took less than two abalone per dive (McElderry & Richards, 1984). There have been some problems enforcing the size limit outside the commercial fishery.

In Alaska a bag limit of 50 per day applies to the native food or subsistence fishery. Regulations for the Washington State recreational dive fishery include a size limit of 89 mm and a daily bag limit of five abalone.

FISHERY ECONOMICS

In BC the decreasing quota by the early 1980s led to a fishery in which one quota did not provide enough income or fishing time to a fisherman per year. As a result, abalone fishermen leased their licences to others or continued abalone fishing as a sideline to other fisheries. In spite of a DFO policy that no more than four licences may be fished from one vessel, at least two operators have more than five licences under their control. This has reduced the number of vessels and fishermen participating (Fig. 28.3) to 13 in 1988 and 12 or less in 1989. With several licences fishing from one vessel, effort is probably focused in fewer locations in any one year; this may lead to local depletions of legal sized abalone. Although no formal rotational system is in place, fishermen move their effort between areas voluntarily. From late 1986 to mid-1988, the price of abalone rose from Can\$14.25/kg to Can\$22.37/kg, a rate greater than the consumer price index (Fig. 28.4). The mid-1988 income for a single licence was about Can\$40 000 for an

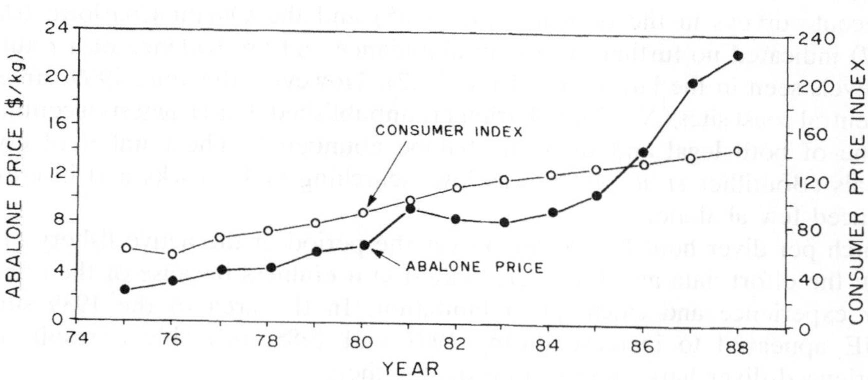

Fig. 28.4 Ex-vessel unit price (Can\$/kg) for abalone caught in British Columbia and Canadian consumer price index (% change from 1981).

average 16.5 diver days (estimated from harvest logs). The amount taken illegally may equal or exceed the quota according to estimates by commercial licence holders. The value of the fishery in 1977 when landings reached 474 t was estimated to be Can$1.9 million. In 1988, the landed value of 47 t was recorded at Can$1.05 million. When this value is compared to the consumer price index (Fig. 28.4), the landed return has only decreased to one-third of the peak value, while the landings were reduced by a factor of 10. At current prices, the fishery remains lucrative for participants even at low quota levels.

Koeneman & Botelho (1989) reported the price of abalone in Alaska at US$11.26/kg (approximately Can$13.40/kg). The combined effect of increased production and lack of limited entry has increased fishing intensity, making it difficult to remain within quotas; the fishing season was completed in one month in 1989.

SURVEYS AND COMMERCIAL FISHERY SAMPLING

Numerous dive surveys were conducted in the major abalone fishing areas of BC during 1977–89 (Adkins & Stefanson, 1977; Breen, 1980; Breen *et al.*, 1978, 1982; Adkins, 1978; Adkins & Stefanson, 1979; Breen & Adkins, 1979, 1980b, 1981, 1982; Boutillier *et al.*, 1984, 1985; Farlinger & Bates, 1986; Carolsfeld *et al.*, 1988; S. Farlinger, unpublished data). The surveys were intended to provide an index of abundance in commercially fished areas as well as in some closed areas. Some areas have been repeatedly resurveyed; for example, in the Queen Charlotte Islands, 58 sites originally surveyed in 1977 and 1978 were resurveyed in 1983, 1984 and 1987 (Boutillier *et al.*, 1985; Carolsfeld *et al.*, 1988); in the central coast, 25 sites were resurveyed in 1983, 1985 and 1989 (Boutillier *et al.*, 1984; Farlinger & Bates, 1986; S. Farlinger, unpublished data).

The sites were sampled using the method described by Breen & Adkins (1979), where 16 square metre quadrats were examined in a 7 m by 16 m area. Abalone on top of and under rocks were counted and measured in each quadrat. Numbers used for comparison are total and legal abalone per square metre, as well as numbers of pre-recruits (94–101 mm SL) and recruits (102–107 mm SL) per square metre.

Recent surveys in the central coast (1985) and the Queen Charlotte Islands (1987) indicated no further decline in abundance and no evidence of a return to the levels seen in the late 1970s (Table 28.2). However, the June 1989 survey of the central coast sites ($N = 25$, S. Farlinger, unpublished data) suggests a continuing decline of both legal and sublegal abalone abundance. The number of cryptic animals (Boutillier *et al.*, 1985) was low; searching under rocks and in crevices produced few abalone.

Catch per diver hour has declined over the period of the active fishery (Table 28.1); the effort data are, however, limited in usefulness because of the effect of diver experience and catch quota limitation. In the area of the 1989 survey, CPUE appeared to increase during 1987 and 1988, probably a result of an experienced diver harvesting several quotas there.

Port samples from the central coast suggested a relative abundance of individuals around 115 mm SL in 1987 (Table 28.3), which agreed with the slight rise in

Table 28.2 A summary of abalone densities (No./m^2) from surveys conducted in British Columbia, 1978–89.

	Central coast 25 sites		Queen Charlotte Is. 58 sites		
Year	Total	Legal	Total	Legal	Source
1978–9	–	–	2.82	0.36	Breen & Adkins (1979)
1979–80	2.74	0.65	–	–	Breen & Adkins (1980)
					Breen & Adkins (1981)
					Breen & Adkins (1982)
1983	1.91	0.31	–	–	Boutillier *et al.* (1984)
1984	–	–	0.54	0.09	Boutillier *et al.* (1985)
1985	1.53	0.36	–	–	Farlinger & Bates (1986)
1987	–	–	0.75	0.20	Carolsfeld *et al.* (1988)
1989	0.55	0.11	–	–	S. Farlinger, unpublished data

Table 28.3 Annual abalone port sampling summary, Prince Rupert, British Columbia, 1985–8.

Year	Mean shell length (mm)	*N*
1985	111 ± 7.8	425
1986	112 ± 8.0	829
1987	115 ± 7.9	557
1988	111 ± 7.7	723

Mean shell length ± 1 sd.
N = sample size.

recruits (102–107 mm SL) seen in the 1985 dive survey (Table 28.4). The commercial fishery showed landings of 10 t in 1985, and approximately 4 t per year since then from the central coast area. This would not account for the decline in pre-recruit abalone seen in the 1989 survey.

Breen & Adkins (1981) and Sloan & Breen (1988) suggested that the survey methodology was not inherently strong for use in comparing densities in specific

Table 28.4 Abalone densities (No./m^2) from 25 comparable sites surveyed in 1983, 1985 and 1989 in the central coast of British Columbia.

	Mean density (No./m^2)			Change		
	1983	1985	1989	1983–5	1983–9	1985–9
Total abalone	1.43	1.57	0.56	0.14	– 0.87	– 1.01
Legal abalone (100 mm SL)	0.22	0.34	0.11	0.12	– 0.11	– 0.23
Pre-recruit abalone (94–101 mm SL)	0.18	0.25	0.08	0.06	– 0.10	– 0.16
Recruit abalone (102–107 mm SL)	0.10	0.14	0.03	0.04	– 0.06	– 0.11

SL = shell length.

areas and recommended that new sampling designs be tested. A review of the survey data and their usefulness for comparative purposes should be attempted as surveys will continue to provide the basis for management, at least in the immediate future.

Samples are taken at the point of landing of the commercial catch and length frequencies obtained. Mean lengths recorded during 1985–9 in BC were 10–15 mm SL and in Alaska 5–10 mm SL above the size limit, suggesting that on average abalone experience at least a year's growth after reaching legal size before they are harvested. No stock assessment other than port sampling is available in Alaska at this time.

CONCLUSION

The cause of the recruitment decline of abalone in BC is difficult to determine (Breen, 1986; Sloan & Breen, 1988). Fishing mortality due to the combined commercial, native and sport fisheries (as well as unmeasured poaching including pre-recruits) has undoubtedly contributed to this decline; however, comparisons of indices of abundance in adjacent areas open and closed to the commercial fishery have not shown differences in abalone densities (Boutillier *et al.*, 1985; Carolsfeld *et al.*, 1988). Past and present abalone resource surveys may not be sufficiently robust to detect small (e.g. <30%) changes, although large changes (e.g. >50%) may be detectable; work is in progress to examine the reliability of this survey methodology.

Recent studies in Australia on *H. rubra* indicate that the short (3–7 days) larval pelagic period may allow completion of the larval phase near the parent, and that there is evidence for a relationship between local abundance of breeding stock and recruitment (McShane *et al.*, 1988; Prince *et al.*, 1988). Although *H. kamtschatkana* has a short (<1 week) larval period (Mottet, 1978; Olsen, 1984), a similar parent stock–recruitment relationship has yet to be shown. Biological factors other than exploitation by man, such as predation and disease, may have contributed to the decline in abalone abundance in BC. Recent survey divers reported unusual numbers of new (shiny) empty shells near some of the sites (G. Thomas, pers. comm.). Sea otters, *Enhydra lutris*, a known predator of abalone (Estes & Van Blaricom, 1985), may have contributed to abalone mortality within the central coast (recent sea otter sightings were reported by Province of BC biologists, pers. comm.). These recruitment declines might also have resulted from large-scale oceanographic or ecosystem fluctuations.

Should future surveys show that the decline in recruitment observed in the central coast is widespread throughout BC, then remedial management actions such as closing the fishery, reducing catches of all users and/or increasing the minimum size may be implemented. Even if the commercial fishery is restricted, data collection is required for the sport and native food fisheries for abalone in BC.

The suitability of the size limit was examined by Breen (1986) and discussed by Sloan & Breen (1988), who pointed out that results are dependent on assumptions of consistent growth, mortality and fecundity, which are unlikely considering the large variability of these parameters in wild abalone populations. They suggest that yield-per-recruit and egg-per-recruit analyses should be repeated when more is known about population dynamics in different habitats.

Future research is required on a number of subjects to improve understanding of abalone population dynamics:

(1) documentation of stock–recruitment relationships,
(2) estimates of juvenile growth rates and adult fecundity,
(3) evaluation of survey techniques.

The abalone fishery is, however, relatively small in BC and Alaska compared to salmon, herring and other shellfish fisheries which compete for research funding assistance and resources.

ACKNOWLEDGEMENTS

The authors thank Tim Koeneman, Alaska Department of Fish and Game, for his cooperation in providing data and comments on the Alaska abalone fishery, R. Burge and Greg Bargmann, the Washington Department of Fisheries, for providing data and information concerning the Washington abalone sport fishery and N. Bourne for reviewing an earlier draft of this paper.

REFERENCES

Adkins B.C. (1978) An examination of some commercially harvested abalone populations in the Moresby Island area. *Fish. Mar. Serv. MS Rpt*, **1455**.

Adkins B.C. & Stefanson A.P. (1977) An examination of harvested and unharvested abalone populations in the Moresby Island area. *Fish. Mar. Serv. MS Rpt*, **1435**.

Adkins B.C. & Stefanson A.P. (1979) North coast abalone survey in harvested areas, November 1978. *Fish. Mar. Serv. MS Rpt*, **1500**.

Bargmann G.G. (1984) Recreational diving in the State of Washington and the associated harvest of food fish and shellfish. *Wash. State Dept Fish. Tech. Rpt*, **82**.

Bates K.T. (1984) Review of the 1982 commercial abalone fishery in British Columbia. *Can. MS Rpt Fish. Aquat. Sci.*, **1749**.

Bates K.T. (1985) Review of the 1983 commercial abalone fishery in British Columbia. *Can. MS Rpt Fish Aquat. Sci.*, **1826**.

Boutillier J.A., Carolsfeld W., Breen P.A. & Bates K.T. (1984) Abalone survey in the Estevan group and Aristazabal Island, May 1983. *Can. MS Rpt Fish. Aquat. Sci.*, **1747**.

Boutillier J.A., Carolsfeld W., Breen P.A., Farlinger S. & Bates K.T. (1985) Abalone resurvey of the southeast Queen Charlotte Islands, July 1984. *Can. MS Rpt Fish. Aquat. Sci.*, **1818**.

Breen P.A. (1980) Measuring fishing intensity and annual production in the abalone fishery of British Columbia. *Can. Tech. Rept. Fish. Aquat. Sci.*, **947**.

Breen P.A. (1984) Recommendations for the 1983 abalone season. In *1982 Shellfish Management Advice, Pacific Region* (Ed. by G.S. Jamieson). *Can. MS Rpt Fish. Aquat. Sci.*, **1774**, 53–6.

Breen P.A. (1986) Management of the British Columbia fishery for northern abalone (*Haliotis kamtschatkana*). In *North Pacific Workshop on Stock Assessment and Management of Invertebrates* (Ed. by G.S. Jamieson & N. Bourne). *Can. Spec. Publ. Fish. Aquat. Sci.*, **92**, 300–12.

Breen P.A. & Adkins B.E. (1979) A survey of abalone populations on the east coast of the Queen Charlotte Islands, August 1978. *Fish. Mar. Serv. MS Rpt*, **1490**.

Breen P.A. & Adkins B.E. (1980a) Spawning in a British Columbia population of northern abalone, *Haliotis kamtschatkana*. *Veliger*, **23**, 177–9.

Breen P.A. & Adkins B.E. (1980b) Observations of abalone populations in Emily Carr Inlet and Lotbiniere Bay, April 1980. *Can. MS Rpt Fish. Aquat. Sci.*, **1576**.

Breen P.A. & Adkins B.E. (1981) Abalone surveys and tagging conducted during 1979. *Can. MS Rpt Fish. Aquat. Sci.*, **1623**.

Breen P.A. & Adkins B.E. (1982) Observations of abalone populations on the north coast of British Columbia, July 1980. *Can. MS Rpt Fish. Aquat. Sci.*, **1633**.

Breen P.A., Adkins B.E. & Heritage G.D. (1978) Observations of abalone and subtidal communities made during a survey of the Queen Charlotte Strait and upper Johnstone Strait areas July 13–20, 1977. *Fish. Mar. Serv. Tech. Rpt*, **789**.

Breen P.A., Adkins B.E. & Sprout P.E. (1982) Abalone populations on the west coast of Banks Island, June 1980. *Can. MS Rpt Fish. Aquat. Sci.*, **1640**.

Breen P.A., Stefanson A.P. & Adkins B.E. (1978) North Coast abalone surveys in harvested areas, spring 1978. *Fish. Mar. Serv. MS Rpt*, **1480**.

Carolsfeld W., Farlinger S., Kingzett B.C., Sloan N.A. & Thomas G. (1988) Abalone resurvey in the southeast Queen Charlotte Islands, June 1987. *Can. MS Rpt Fish. Aquat. Sci.*, **1966**.

Cox K.W. (1962) California abalones, family *Haliotidae*. *Calif. Fish. Game Fish. Bull.*, **118**, 1–133.

Emmett B. & Jamieson G.S. (1988) An experimental transplant of northern abalone, *Haliotis kamtschatkana*, in Barkley Sound, British Columbia. *Fish Bull.*, **87**, 95–104.

Estes J.A. & Van Blaricom G.R. (1985) Sea-otters and shellfisheries. In *Marine Mammals and Fisheries* (Ed. by J.R. Beddington, R.S.H. Beverton & D.M. Lavigne), pp. 187–235. Allen and Unwin, London.

Farlinger S. & Bates K.T. (1985) Review of shellfish fisheries in northern British Columbia to 1984. *Can. MS Rpt Fish. Aquat. Sci.*, **1841**.

Farlinger S. & Bates K.T. (1986) Abalone survey in the Estevan group and Aristazabal Island, June 1985. *Can. MS Rpt Fish. Aquat. Sci.*, **1896**.

Farlinger S. & Thomas G.A. (1988) Review of shellfish fisheries on northern British Columbia 1985 and 1986. *Can. MS Rpt Fish. Aquat. Sci.*, **1998**.

Federenko A.Y. & Sprout P.E. (1982) Abalone biology, fishery regulations, commercial catch (1952–1980), and current state of resource in British Columbia. *Can. MS Rpt Fish. Aquat. Sci.*, **1658**.

Jamieson G.S. & Francis K. (Eds) (1986) Invertebrate and marine plant fishery resources of British Columbia. *Can. Spec. Publ. Fish. Aquat. Sci.*, **91**.

Koeneman T. & Botelho C.A. (1989) Miscellaneous species. In: *Report to the Board of Fisheries, Southeast Alaska and Yakutat (Region 1). Regional Information Report No. 1J89–01*, pp. 6.1–6.28. Alaska Dept of Fish and Game, Division of Commerical Fisheries, Juneau, Alaska.

McElderry H. & Richards L. (1984) Recreational SCUBA diving in the Strait of Georgia: An analysis of the distribution of diving effort and the importance of collecting marine animals. *Can. MS Rpt Fish. Aquat. Sci.*, **1794**.

McLean J.H. (1966) *West American prosobranch Gastropoda: Superfamilies Patellacae, Pleurotomariacea, and Fissurellacea*. PhD dissertation, Stanford University.

McShane P.E., Black K.P. & Smith M.G. (1988) Recruitment processes in *Haliotis rubra* (Mollusca: Gastropoda) and regional hydrodynamics in southeastern Australia imply localized dispersal of larvae. *J. Exp. Mar. Biol. Ecol.*, **124**, 175–203.

Mottet M.G. (1978) A review of the fishery biology of abalones. *Wash. State Dept Fish. Tech. Rpt*, **37**.

Olsen S. (1984) *Shellfish Enhancement Report. NOAA, NMFS, Final Report for Project 1-144-R.* Washington State Department of Fisheries.

Paul A.J. & Paul J.M. (1981) Temperature and growth of maturing *Haliotis kamtschatkana* Jonas. *Veliger*, **23**, 321–4.

Prince J.D., Sellers T.L., Ford W.B. & Talbot S.R. (1988) Confirmation of a relationship between the localized abundance of breeding stock and recruitment for *Haliotis rubra* Leach (Mollusca: Gastropoda). *J. Exp. Mar. Biol. Ecol.*, **122**, 91–104.

Quayle D.B. (1971) Growth, morphometry and breeding in the British Columbia abalone (*Haliotis kamtschatkana* Jonas). *Fish. Res. Bd Can. Tech. Rpt*, **279**.

Sloan N.A. & Breen P.A. (1988) Northern abalone, *Haliotis kamtschatkana*, in British Columbia: fisheries and synopsis of life history information. *Can. Spec. Publ. Fish. Aquat. Sci.*, **103**.

Sloan N.A. & Farlinger S. (1987) Abalone. In *1986 Shellfish Management Advice, Pacific Region* (Ed. by G.S. Jamieson & R.M. Harbo). *Can. MS Rpt Fish. Aquat. Sci.*, **1756**, 63–7.

Chapter 29
Australian abalone fisheries and their management

J.D. PRINCE *PO Box 209, Leederville, Western Australia 6007*
S.A. SHEPHERD *Department of Fisheries, 135 Pirie Street, Adelaide, South Australia*

ABSTRACT: The Australian abalone industry is divided among five states and is managed independently by each. The fisheries and the evolution of their current management strategies are described. Most states began managing their abalone fisheries with open access and size limits. They then controlled expansion of their fisheries by the introduction of limited entry. Saleable licences were eventually introduced in each state to facilitate the turnover of divers and caused a substantial increase in fishing pressure. A form of individual transferable quotas (ITQ) was then introduced in every state to control the catch. Limited entry fulfilled a useful function in controlling the rate at which each industry expanded, but proved of limited use in controlling the level of catch. Licence transferability increased the debt burden of the industry and the incentive of divers to fish, but encouraged a turnover of participants in the industry. The ITQs are successful because they directly control the catch. ITQs, size limits and closures will continue to be used extensively in the abalone industry to manage the fisheries, although continued refinement will be necessary to correct existing weaknesses.

RESUMEN: La industria Australiana del abulón comprende cinco estados, y en cada estado su administración es independiente. En general, todos los estados maniobraron rápidamente para controlar el crecimiento inicial de la industria. La pesquería en cada estado se describe junto con la evolución de su actual estrategia administrativa. Por lo común, cada estado comenzó administrando la pesquería de abulón de forma abierta y con límites en la talla; siguieron hacia el acceso limitado, sin transferencia de licencias; después, por el acceso limitado y con transferencia del permiso; para finalmente adoptar la forma de ITQs (cuotas individuales de transferencia). El acceso restringido cumple una función útil ya que permite controlar la velocidad de expansión de cada industria pero no ha permitido el control de la captura a cualquier nivel particular. La introducción de la transferencia de permisos acentuó esta debilidad al aumentar el monto de la deuda de la industria e incrementar los incentivos a los buzos. Sin embargo, parece ser que ha sido necesario favorecer un cambio de posiciones de los participantes en la industria. Las ITQs son aparentemente exitosas, biológicamente hablando, debido a que permiten el control directo de las capturas; sin embargo dudamos que sean tan efectivas económicamente hablando, como pretenden los economistas. Sugerimos que dentro de la industria Australiana, los ITQs y el límite en la talla continuen úsandose extensivamente para el manejo biológico de la pesquería, sin embargo, es necesario un refinamiento continuo del sistema, con el fin de perfeccionar, desde los puntos de vista biológico, económico y social, la administración de las pesquerías.

INTRODUCTION

Since 1965 the Australian abalone fishery has grown from small beginnings to a major fishery now worth over $100 million annually. The growth of the fishery imposed increasing pressure on the stocks and management responded with increasingly complex methods of control. Management has been imperfect, and has proceeded at times with *ad hoc* measures and at times by trial and error. Seldom have managers published reasons for their decisions and there has been little attempt to develop a rigorous, experimental approach to management (see Walters, 1986). Often management has been considered an art (Waugh, 1984) or even non-science guided by vaguely stated wisdoms of witchcraft (Larkin, 1972). Yet an historical account of the abalone fishery is critical to understanding the present management schemes and provides an important body of information from which sound principles of management can be derived.

In this chapter we briefly review the biology of the three exploited species of abalone in Australia, and then the history of the abalone fisheries in each state. We provide essential information on the social milieu in which the fisheries developed because this, as much as biological imperatives, influenced management. Finally, we assess the strengths and weaknesses of the measures used and suggest future directions for the better management of these fisheries.

GENERAL BIOLOGY

Three species of abalone are fished commercially in Australia. The blacklip, *Haliotis rubra* (for synonymy of *H. conicopora* see Chapter 3, this volume), is taken predominantly in southeastern Australia to depths of 30 m, the greenlip, *H. laevigata*, predominantly in South Australia to similar depths and Roe's Abalone, *H. roei*, predominantly in Western Australia to a depth of about 3 m (Fig. 29.1).

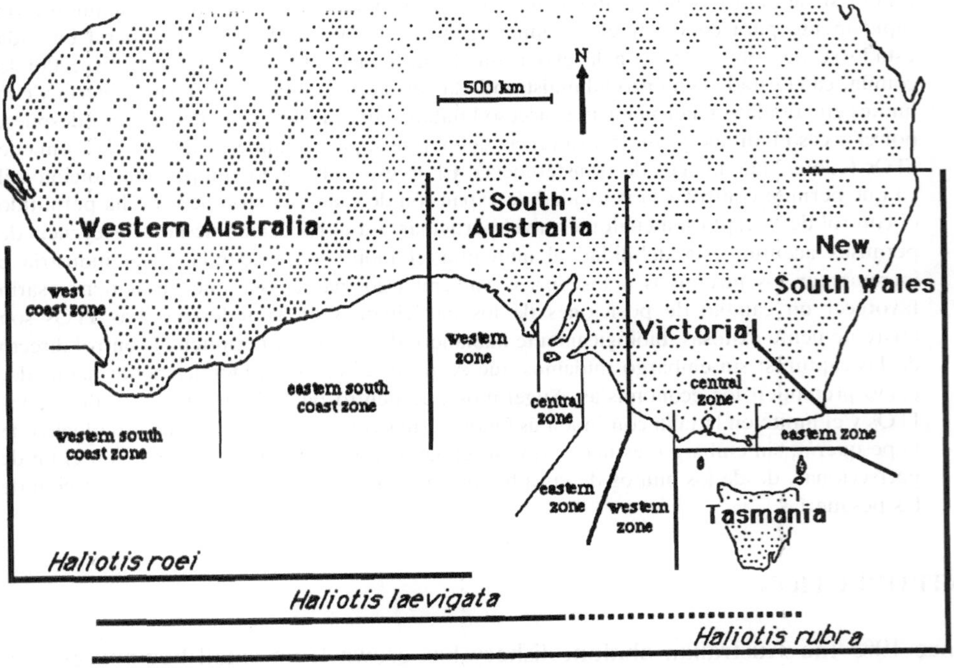

Fig. 29.1 The five southern Australian states with their abalone management zones, and the geographic range of the three commercially exploited species.

Abalone larvae settle preferentially on crustose coralline algae (reviewed in Chapter 11, this volume). All the Australian species are cryptic as juveniles, but, with increasing size, occupy sites which are more accessible to divers. As adults *H. roei* is the most cryptic and *H. laevigata* the least (Shepherd, 1973; Witherspoon, 1975). Until a size of about 10 mm is reached, juveniles feed principally on the bacterial and diatom epibiotas of crustose coralline algae, as well as on its epithelial tissue (Garland *et al.*, 1985; Shepherd & Cannon, 1988). Larger juveniles and adults feed selectively on foliose algae (Chapter 13, this volume). Growth of juvenile abalone is linear or sigmoidal, but the growth of adults can be described by using parameters of the von Bertalanffy growth equation. Growth studies of Australian species are reviewed in Chapter 12 (this volume).

Predators, parasites and harmful commensals and the natural mortality of abalone are reviewed in Chapter 21 (this volume). Reproduction has been described for all the commercial species (Shepherd & Laws, 1974; McShane *et al.*, 1986b; Prince *et al.*, 1988a; Wells & Keesing, 1989) and aggregative behaviour in relation to spawning by Shepherd (1986a). The spawning of *H. laevigata* is apparently restricted to summer. In the other two species it seems more flexible; spawning is possible all year but occurs mainly in the winter, spring or autumn, according to locality.

The work of Prince *et al.* (1987, 1988b), McShane *et al.* (1988) and the review by McShane (Chapter 11, this volume) suggest that in *H. rubra* larval dispersal is generally on a scale of metres to tens of metres. However, Shepherd & Lowe (in prep.) show that dispersal of larvae of *H. laevigata* is much greater and is of the order of hundreds of metres or even kilometres. Movement of abalone after settlement also appears to be limited. Juveniles have been recorded to move at rates of about 1 m per month (Shepherd & Godoy, 1989) and adults from a few metres up to about 200 m per month (Beinssen & Powell, 1979; Shepherd, 1986b; Prince, 1989). These studies suggest that the size of a unit stock of *H. rubra* and *H. laevigata* in Australia is in the order of hundreds of metres to kilometres, a scale which is consistent with the genetic studies of Brown & Murray (Chapter 3, this volume).

Stock assessments have been made in most states, and sustainable yields calculated using various models (Chapter 20, this volume). In Victoria, McShane *et al.* (1986a) published an atlas of Victorian reefs, Beinssen (1979a,b) and McShane & Smith (1989a) measured the fishing mortality and power of divers, and McShane (1987, 1988) assessed the stocks. In South Australia Sluczanowski (1984, 1986) used an egg-per-recruit model to examine different management strategies.

Despite this impressive accumulation of knowledge, some major gaps remain. The precise timing of spawning, or the cues that trigger spawning, are unknown. Prince *et al.* (1987) suggested that *H. rubra* spawns during periods of minimal water movement. The relationship between fecundity, aggregative behaviour of adults and fertilization success also remains unknown. However, the most serious gap is the lack of understanding of the process of recruitment to reefs. Nothing is known of larval behaviour, recruitment variability and the factors influencing it, or of the relationship with the size of the parent stock. A reason why progress in understanding the latter relationship cannot be expected is that a yet more basic problem exists – that of measuring abalone abundance.

The distribution of abalone is non-random and highly clumped. Moreover, this clumping or aggregation occurs at different spatial scales from metres to kilometres. This distribution pattern makes reliable and statistically sound measurements of absolute abundance difficult to obtain. Approximate methods have been developed and are in use (Shepherd, 1985; McShane & Smith, 1989a), but there has been little attempt to evaluate them rigorously. Greater effort must be made to develop techniques appropriate to each species and its habitat. Only then will it become possible to describe the relationship between stock and recruitment and measure its variability.

CASE STUDIES

Australia is a federation of six states, in which legislative powers are divided between federal and state governments. The Australian constitution confers specific powers on the federal government, and a residue of unspecified powers rests with the states. Subsequent constitutional agreements between federal and state governments gave to the states exclusive fisheries jurisdiction over territorial waters (up to 3 nautical miles from the coast) and over designated bays and gulfs. Abalone fisheries principally lie within state waters and so are managed independently by five different state governments.

Except for a period in the 1880s when some Chinese in Tasmania dried and exported abalone to China (Tenison-Woods, 1882), Australian abalone stocks were virtually unexploited before 1963. As recreational diving with scuba increased in popularity in the early 1960s, divers collected abalone during weekends and holidays from areas close to population centres. Australian processors either canned, dried or simply washed, salted and froze meats for export (Anon., 1964).

One of the first organized, full-time operations was mounted by an ex-serviceman from the USA with experience in the Californian fishery, who recognized the potential of the Australian stocks while on recreational leave from Vietnam. Together with a number of local divers he operated a mother-boat and harvested remote stocks on southern coasts. Between 1965 and 1968, as markets became more established, the total number of divers and the proportion of full-time participants increased rapidly. By the end of the decade a professional abalone fishery, now using surface compressors and hookah air hoses, was established in each of the five southern states (Harrison, 1969).

Legal minimum sizes, often intuitively determined, were imposed in each state to preserve parental stocks. Each state also introduced an annual licence, with a requirement to submit fishing returns, in order to control the infant fisheries better. This licensing system was used to encourage local participation in the industry and licences were issued to diver-operators only. Governments recognized early on the health hazards associated with diving (Shepherd & Gilligan, 1974) and determined that the industry should be owner-operated, so that the individual facing the risk should receive the financial reward.

Intertwined in the history is the development of fishing strategies and increased efficiency by divers. Better wetsuits, boats and equipment, increased incentive and improved knowledge of fishing grounds are all cited by Harrison (1986) as contributing to improved efficiency. In South Australia, 'shark cages' were de-

Fig. 29.2 A 'shark cage' used by South Australian divers. Buoyancy is adjusted by expelling water from flotation tanks. Abalone are collected in the bag hanging below the diver. An inquisitive white shark is on the left. By David Doubilet, © 1987 National Geographic Society.

veloped in 1975 (Anon., 1976) (Fig. 29.2) to give the diver propulsion, protection from shark attack and hot water for warmth. They are powered hydraulically or pneumatically from a motor on the diver's boat, and enable a diver to cover many kilometres of the sea floor. The cages proved to be especially suited to harvesting the widely dispersed greenlip stocks.

New South Wales (NSW)

G.D. Hamer, Division of Fisheries, NSW Agriculture and Fisheries, Sydney

The modern Australian abalone industry began and expanded rapidly in NSW, where abalone stocks, composed entirely of *H. rubra*, are readily accessible and close to large population centres. Management policies during the 1960s and 1970s were strictly non-interventionist, in the belief that market forces would operate to prevent the collapse of the stock. The early catch figures in this period are unreliable because the many small operators involved in the industry grossly understated their catch to minimize tax. Minor fishing began in 1959, but substantial catches were first landed in 1965, and the annual catch grew to 1200 t (all weights quoted are total weights using a $\frac{1}{3}$ conversion ratio for meat weight/total weight) in 1971 (see Waugh, 1984). The annual catch and CPUE then fluctuated, but have recently declined (Fig. 29.3).

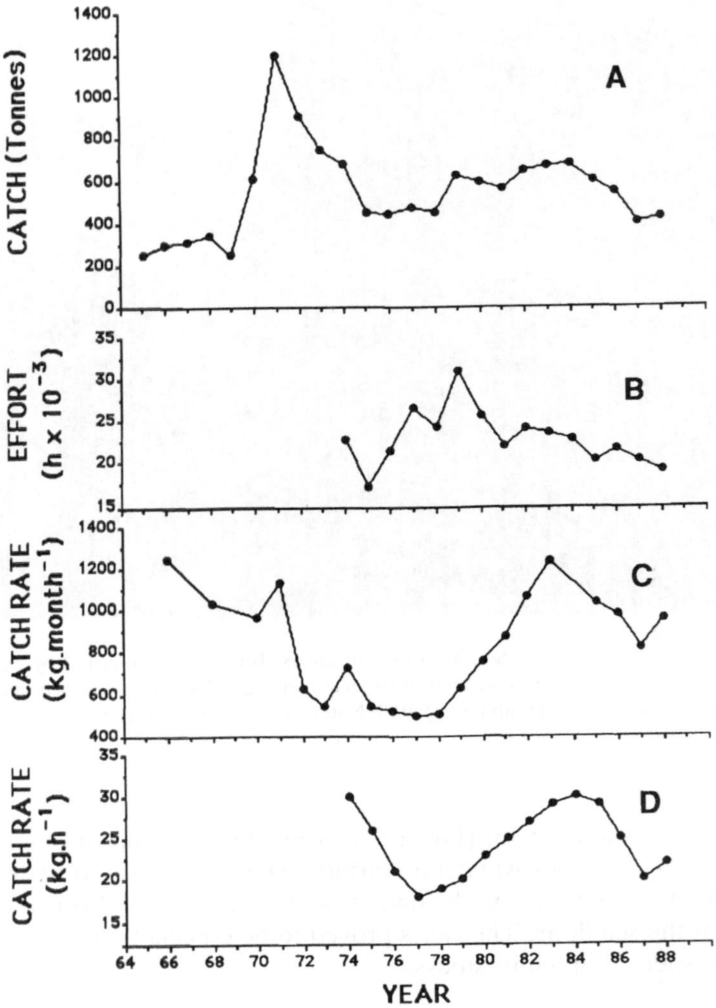

Fig. 29.3 Trends in the total catch (tonnes live weight) of *H. rubra* in New South Wales (A), effort in thousands of diving hours (B) and catch rates in kg diver month^{-1} (C) and kg h^{-1} (D).

It appears that fishing exhausted the accumulated virgin biomass by the early 1970s, causing the subsequent large decline in annual catches and CPUE. Since that time, fishing has apparently relied almost entirely upon annual recruitment (Hamer, 1983). The increase in catches and CPUE that occurred during the early 1980s could have resulted from improved recruitment and growth due to environmental factors, or reduced intra-specific competition, or from the increased size limit in 1980 (see below). Whatever the situation then, it seems likely that the stock is now in decline.

A size limit, an amateur bag limit and a requirement to land abalone in the shell were first imposed in 1973. In 1980 the Fisheries Department belatedly moved to stabilize the industry by limiting the number of licences to 59 and by increasing the size limit to 115 mm shell length (SL) (Francois & Gorman, 1980). Initially, licences were non-transferable (that is a diver leaving the industry relinquished his licence to the state) and a policy of not re-issuing licences when divers retired was followed to reduce the number of licences. In 1985 licences became transferable, allowing divers to sell their licence on the open market. The government also adopted a two-for-one policy, which required an entrant to buy two existing licences as a condition of entry, thus reducing the number of licences in the fishery.

These policies did not succeed in stabilizing the fishery and in 1989, in response to industry pressure, a system of individually transferable quotas (ITQs) was implemented. The total quota was fixed at about the then level of the catch. Subsequently, an industry-funded buy-back scheme was introduced to reduce further the catch by 50 t, and this measure, combined with the two-for-one policy, may be more effective in stabilizing the fishery at a lower level of about 290 t. However, the future of the fishery remains precarious because of the prevalence of illegal operators whose total catch is equal to the commercial catch (Prince & Hamer, unpublished data). Unless poaching is controlled, the recent conservation measures will do little to halt the decline of many overfished reefs.

Tasmania

J.D. Prince and W.J. Nash, Division of Sea Fisheries, Taroona, Tasmania

Tasmania has the largest abalone industry in Australia. Over 95% of the catch is *H. rubra*, with the rest comprising *H. laevigata*. A minimum size limit (127 mm SL) was introduced in 1962. The total catch rose rapidly from 1963 to 1967 (Fig. 29.4(A)) as the number of divers increased. The government introduced a licensing system in 1965, and in 1967 enforced the size limit by prohibiting shucking at sea (Anon., 1987).

After 1969 no new entrants were admitted to the fishery, and divers were required to earn their income primarily from the fishery; at the same time the annual licence fee was increased from $10 to $100 (Harrison, 1983). With these measures the number of divers decreased from over 250 to 120 and effort declined (Fig. 29.4(B)). Between 1969 and 1974 licences could not be transferred, but only re-allocated. Under this policy few divers left the industry. The declining CPUE after 1971 (Fig. 29.4(C)) may partly reflect the declining number of virgin stocks, but it more probably reflects the increasing wealth of divers and their declining incentive (Prince, 1989).

In 1974 licences became transferable. The initial period thereafter was marked by an influx, particularly in 1977, of experienced interstate divers, causing the value of licences to escalate from $10 000 to $40 000 (Harrison, 1983). The financial burden of purchasing a licence gave new divers a strong incentive to work hard. The real (inflation-adjusted) price of licences and of abalone rose steadily during the late 1970s, and further increased the financial commitments of divers entering the industry. Consequently catches, effort and CPUE all increased during this period. The rise in the price of abalone halted during 1978–9 and 1982 and this was mirrored by slight downturns in catch, effort, CPUE and the price of licences.

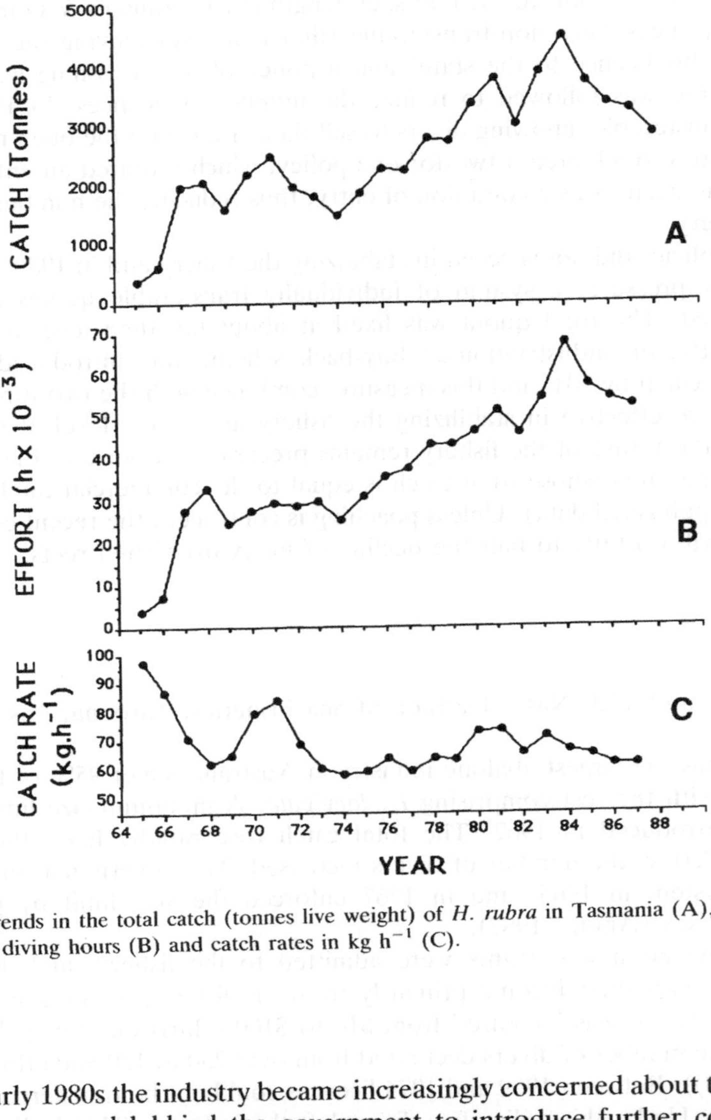

Fig. 29.4 Trends in the total catch (tonnes live weight) of *H. rubra* in Tasmania (A), effort in thousands of diving hours (B) and catch rates in kg h^{-1} (C).

In the early 1980s the industry became increasingly concerned about the growing fishing pressure and lobbied the government to introduce further conservation measures. During 1983 and 1984, in response to real price rises, divers increased their effort and catch by about 20–25%, demonstrating the then latent capacity in the industry for additional effort despite the policy of licence limitation. In 1969–70 and 1975–6, when the level of exploitation was lower, higher levels of effort were accompanied by marked and sustained increases in CPUE, but in 1983 they only resulted in a slight increase in CPUE, and in 1984 CPUE declined. This suggests that the industry had finally exhausted the virgin biomass and had begun to overfish the annual recruitment.

In 1985 a system of ITQs was introduced which reduced the catch by about 15%. Quotas were progressively reduced in the next four years, so that the annual catch declined from a record 36 tonnes per diver in 1984 to 17 tonnes per diver in 1989, or back to the level of catch taken during the mid-1970s.

Victoria

P.E. McShane, Marine Science Laboratory, Queenscliff, Victoria

Like the Tasmanian catch, the Victorian catch is almost entirely *H. rubra*. The government limited the number of licences in 1968 and did not re-issue licences when divers retired. This policy resulted in a rapid reduction of licensed divers from 164 to 100. The number of divers then stabilized because few chose to retire. In 1968 the coastline was divided into three zones (Fig. 29.1), and licences applied to one zone only in order to distribute fishing pressure more uniformly. Minimum size limits were introduced. These were: for *H. rubra* 110 mm SL for the central zone (100 mm in Port Phillip Bay), and 120 mm elsewhere, and for *H. laevigata* 130 mm.

Between 1969 and 1975 the reported catch and CPUE slowly declined (Fig. 29.5), apparently due to the increasing age, and decreasing incentive, of divers.

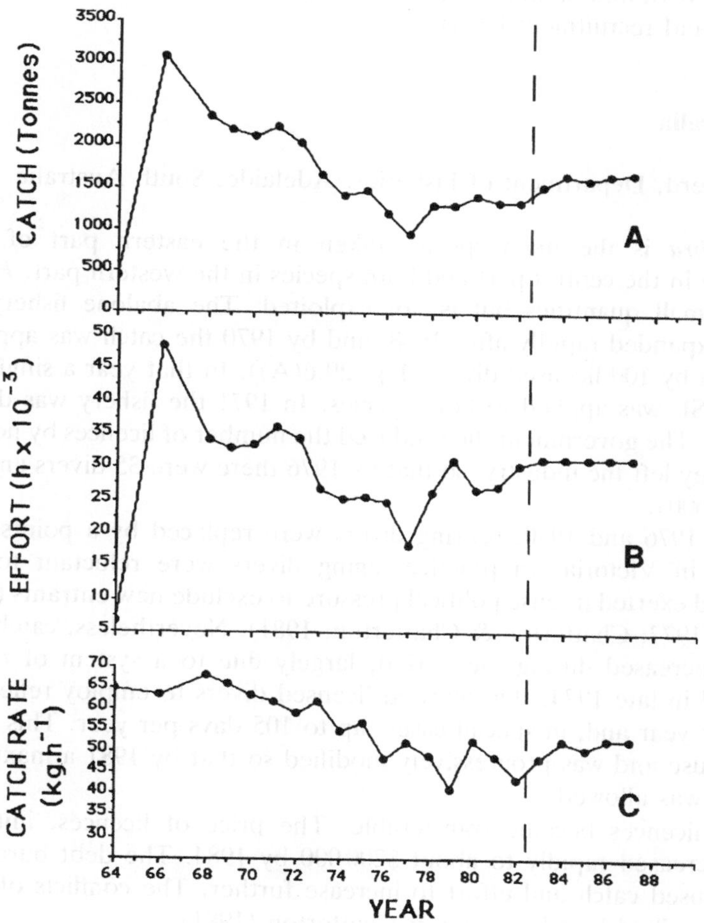

Fig. 29.5 Trends in the total catch (tonnes live weight) of *H. rubra* in Victoria (A), effort in thousands of diving hours (B), and catch rates in kg h^{-1} (C).

The government then created a mechanism to replace retiring divers. Applicants for licences were assessed according to their previous experience in the industry and other factors; where point scores were equal, a selection was made by ballot. The procedure was unpopular and opposed by divers. The problems and conflicts in managing the fishery until 1980 are well reviewed by Stanistreet (1982).

In 1984 the government, concerned at the slow decline of catches and catch rates, made licences transferable with a two-for-one provision (as in NSW). Despite the ensuing reduction in the number of divers in the central zone from 48 to 35, the effort and catch escalated as younger divers entered the central zone seeking to recover the cost of purchasing two licences. The official catch figures for 1987 (1700 t) are probably unreliable and the true catch was probably above 2000 t.

In 1988 the government introduced ITQs for each zone (initially set at 70% of the mean 10-year catch), and abandoned the two-for-one policy. The quota system stabilized the annual catch at about 1680 t and assessments around that time indicate that the Victorian stocks are stable (McShane, 1987, 1988). Catch quotas are reviewed annually, after taking into account indicators of stock abundance and recruitment levels.

South Australia

S.A. Shepherd, Department of Fisheries, Adelaide, South Australia

Haliotis rubra is the main species taken in the eastern part of the state, *H. laevigata* in the central part and both species in the western part. *Haliotis roei* occurs in small quantities but is not exploited. The abalone fishery in South Australia expanded rapidly after 1968, and by 1970 the catch was approximately 1600 t taken by 100 licensed divers (Fig. 29.6(A)). In that year a single size limit of 130 mm SL was applied to both species. In 1971 the fishery was divided into three zones. The government then reduced the number of licences by not replacing divers as they left the industry, so that by 1976 there were 35 divers and the catch was about 600 t.

Between 1976 and 1980, retiring divers were replaced by a points and ballot system, as in Victoria. In practice, aging divers were reluctant to leave the industry and exerted intense political pressure to exclude new entrants (Chatterton & Arnold, 1977; Chatterton & Chatterton, 1981). Nevertheless, catch and effort gradually increased during the period, largely due to a system of relief diving (introduced in late 1974) that allowed licensed divers to employ relief divers for 21 days per year and, in special cases, up to 105 days per year. This system was open to abuse and was progressively modified so that by 1981 a maximum of 21 relief days was allowed.

In 1980 licences became transferable. The price of licences, initially about $30 000, increased rapidly to about $200 000 by 1984. The debt burden on new entrants caused catch and effort to increase further. The conflicts of the period are well described by Chatterton & Chatterton (1981).

A review of the fishery (Lewis *et al.*, 1984) expressed concern for its long-term stability. As a result, short-term closures of overexploited areas were introduced

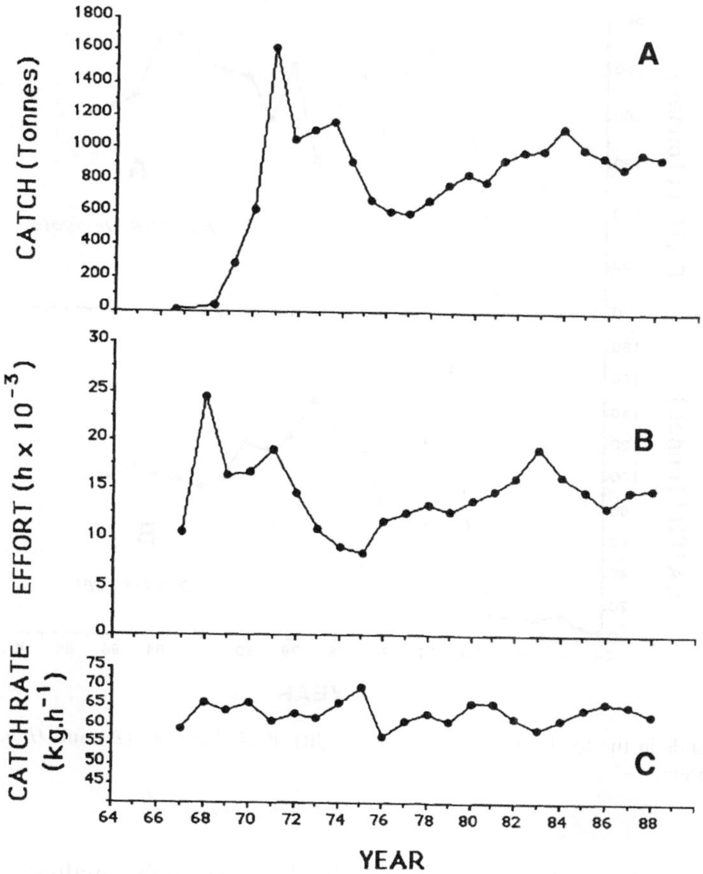

Fig. 29.6 Trends in the total catch (tonnes live weight) of *H. rubra* and *H. laevigata* (combined) in South Australia (A), effort in thousands of diving hours (B), and catch rates in kg h^{-1} (C).

and the size limit of greenlip was increased to 145 mm SL in the western zone to ensure that at least 40% of the egg production of the virgin stock was maintained in accordance with the egg production models of Sluczanowski (1984). An ITQ system was introduced into the western zone in 1985 and extended to the other zones by 1990 (see Muse & Schelle, 1989, for details). Quota levels were then 21 t per diver (western zone), 30.6 t per diver (central zone) and 18.5 t per diver (eastern zone). Since then, the quota of *H. laevigata* in the western zone has been reduced by one tonne. Quotas are assessed annually after taking into account abundance of recruits and adults from field surveys of specific sites, trends in effort and data, and the weight gradings of the catch.

Western Australia

L.M. Joll and J.W. Penn, WA Marine Research Laboratories, North Beach, Western Australia

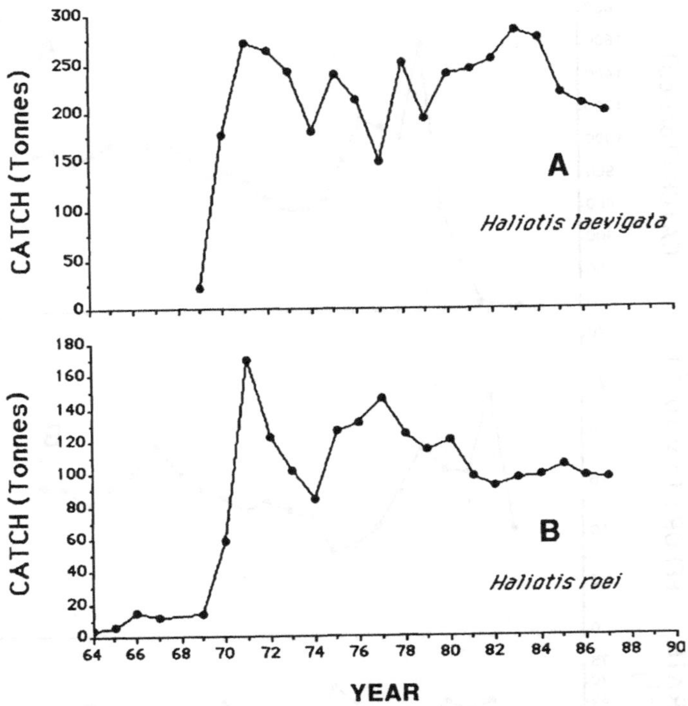

Fig. 29.7 Trends in the total catch (tonnes live weight) of *H. laevigata* (A) and *H. roei* (B) in Western Australia.

The abalone fishery of Western Australia is Australia's smallest, and mainly comprises *H. laevigata* and *H. roei*; *H. rubra* is also taken in small amounts (Penn, 1974). Because the major abalone beds are found in isolated parts of the state, the fishery began more slowly than elsewhere. Fishing began in 1964 but remained at a low level with part-time operators until 1969 (Fig. 29.7). In 1970 an influx of divers from other states caused a rapid expansion of the industry as all but the most remote areas were explored and fished.

To control the expansion of the industry and to prevent the possibility of overexploitation, licences were limited in 1971. Initially the 36 licences were non-transferable and were not re-issued when divers retired. With this policy the number of divers was reduced to 26 by 1975. In 1975 the Western Australian fishery was divided into three zones. Two of these zones are on the south coast of Western Australia (eastern and western), where *H. laevigata* is the main species harvested. The third zone is the west coast where only *H. roei* is taken. There are 6, 8 and 12 divers respectively in these zones.

Weight or size limits were also imposed for each species; these vary from 125 g to 300 g according to zone for *H. laevigata* and *H. rubra*, while the size limit of *H. roei* is 60 mm SL, except in the Perth area where it is 70 mm. To promote orderly fishing and reduce competition between divers an ITQ of 24 tonnes per diver was introduced in the eastern south coast zone in 1985 and in the western south coast zone in 1986. In 1988 the quota for the former zone was reduced to

18 t per diver because the divers were unable to catch the larger quota, and in 1989 the quota for the latter zone was reduced to 15 t per diver. Since 1985 the west coast divers who collect *H. roei*, now with a quota of 9 t per diver, have been given access to the stocks of this species along the south coast.

In addition to licence limitation and size limits, closed seasons, rolling closures and daily catch limits have been effectively used, particularly for the *H. roei* fishery, to maximize the catch rates and reduce operating costs. Because the fishery is small (<400 t annually), the government has allocated relatively few resources to research and management. In consequence, management issues have been resolved by consultation between government and divers.

ASSESSMENT OF MANAGEMENT MEASURES

Abalone fisheries developed during the period when the concepts of fisheries management world-wide were in flux. During the 1960s, thinking was dominated by ideas of static, population-based maximum sustainable yield (MSY). In the next decade, biologists, urged by economists, gradually replaced MSY with the concept of maximum economic yield (MEY). Finally, in the 1980s, the limitations of both became apparent and ideas of optimum yield (OY), which takes account of biological, economic and sociological objectives, developed (Larkin, 1977; Sinclair, 1978; Cunningham, 1981; Healey, 1984; Waugh, 1984). The management of abalone fisheries in general reflected these changing ideas, although policies were seldom enunciated. Initially all states espoused the principle of MSY, and introduced licence limitation and reduced the number of licences to prevent expansion beyond MSY (e.g. Harrison, 1983). In NSW an open access policy was pursued and defended (Francois & Gorman, 1980), and later abandoned only when the problems of overfishing caused by that policy became manifest (Waugh, 1984).

During the 1970s, government policies were increasingly influenced by economic considerations. In South Australia economic assessments motivated the policy to reduce the number of licences (Gleeson, 1976; Philipson, 1982). In Victoria management objectives were to obtain 90% of MSY and a high economic return on labour and capital (Anon., 1982). In Western Australia economic returns to divers were an important element in management measures (Ward, 1986). In all states social and political considerations played an important role, although they are best documented in South Australia (Chatterton & Chatterton, 1981, 1982; Olsen, 1982; Walkerden, 1982). Overall the states pursued similar policies and adopted similar measures, not always in step and with local variations, to achieve stability.

Size limits

Size limits were relied on in all states to protect the breeding stock and to improve biomass yield, rather than control the catch. In some states regional differences in growth rates were recognized with differential size limits, and sometimes these have been successful. However, limitations have become apparent. Prince (1989) and Nash (Chapter 23, this volume) have shown that *H. rubra* has great variability in growth rate between populations on a fine scale, and that sexual maturity is age-dependent rather than length-dependent. The implication is that a uniform size limit throughout a geographic area may 'lock up' slow-growing stocks and expose fast-growing stocks to increased fishing effort. These fast-growing, and usually most productive, stocks will then become the most vulnerable to recruitment overfishing. Thus size limits, correctly chosen, are optimal only where stocks grow at a uniform rate. Where the growth rate is heterogeneous, over- and/or under-fishing will occur, according to the extent of departure of populations from the selected model.

Overall, size limits have had mixed success in Australia. In Tasmania they have been too low in the south and too high in the north (Prince, 1989; Nash, Chapter 23, this volume). In Victoria they have been for the most part satisfactory, although slight increases are now proposed to improve stock protection (McShane, 1988). In South Australia and Western Australia they are mostly satisfactory, and in NSW they are only marginally suboptimal in a biological sense, but deliberately so due to the extent of poaching. In all states some stunted stocks (those in which few adults reach the legal size limit) are probably underexploited. In each state except NSW these are fished by temporarily reducing the size limit in restricted areas for short periods of time, thus ameliorating the disadvantages of a conservative size limit (see McShane & Smith, 1989b).

Despite the problems described, size limits can be used to protect the breeding stock. They should be set at a conservative level of egg production, after performing an egg-per-recruit analysis, (see Chapters 20 and 23, this volume). This may well be near the average maximum size (L_∞) of the population. Sluczanowski (1984, 1986), Prince (1989) and Nash (Chapter 23, this volume) show that modest increases in size limit can markedly increase egg production with little loss of yield per recruit.

In choosing the area over which the size limit is to operate, there is a trade-off between the size of the area and the administrative cost of enforcement. In addition, when an appropriate scale is chosen, a compromise may have to be made between a size limit that is high enough to conserve fast-growing stocks and the cost of lost production of slower-growing ones. In some places the effect of poaching also needs to be taken into account.

Closures

Seasonal, areal and rolling (temporal and areal) closures have been variously used in all states. Their success has depended on the purpose and the local conditions. In Western Australia rolling closures are used to maximize catch rates and reduce operating costs on the west coast (see Philipson, 1982); on the remote south coast of Western Australia the policy was discontinued due to enforcement problems, relating to poaching.

Closures have the effect of increasing egg production because they postpone fishing mortality or increase the period between visits in pulse fishing (Sluczanowski, 1984). In South Australia the fishing season for *H. laevigata* is manipulated to make a *de facto* closure. The opening of the fishing season occurs on 1 January when spawning is complete and divers generally take their quotas well before the next spawning season, so providing in effect a closure prior to and during spawning.

Closures have more often been used to control fishing mortality, but here their record of success is poor. Divers change their behaviour to accommodate the closure; they re-arrange holidays, increase their effort in areas that remain open or, with a 'goldrush mentality', intensively fish newly opened areas. The most promising use for closures, as now in operation in most states, is to facilitate exploitation of stunted stocks by closing simultaneously non-target stocks (see above) and to manipulate supplies to exporters at glut periods.

Licence limitation

The Australian experience with licence limitation parallels that in other fisheries (Crutchfield, 1979; Hancock, 1979; Waugh, 1984; Wilen, 1988). It controlled initial expansion and capitalization, but failed in the long term to curb either effort or catch. A diver's behaviour can change and he will, given sufficient incentive, substantially increase his effort and will in any event increase in efficiency (Harrison, 1983). Thus effort increased 50% in four years and catch by 100% in six years in the Tasmanian fishery (Harrison, 1986), and a similar increase in effort occurred in South Australia from 1980 to 1984 (Lewis *et al.*, 1984). The incentive in both cases was the introduction of transferable licences. Price increases, and the anticipated introduction of quotas, can also provide incentive for greater effort, as occurred in Tasmania in 1984. By the same token, stagnation has occurred in some fisheries. In Victoria there was a 40% decline in effort and a 40% decline in catch over the six years until 1975, and similar declines in South Australia, due in part to aging divers who had little incentive to fish hard. Thus licence limitation failed in the early years to maintain desired catch levels.

Transferable licences

The reluctance of divers to leave the industry, the inefficiency of the ballot systems used to replace them (Anon., 1982) and the political lobbying of divers (Chatterton & Chatterton, 1981) inevitably led to transferable licences. As summarized by Crutchfield (1979),

> 'efficiency in use, continuity of operation, and ease of administration are eloquent arguments for a relatively free and costless transferability of limited fishing rights. The price they will come to carry is both an economic barometer and an allocative device that no amount of administrative effort could match.'

But transferability was not without cost (Chatterton & Chatterton, 1981); it brought 'windfall profits' to the sellers, increased the capital investment in the fisheries and stimulated escalation of diver effort. The market value of the licence, usually priced at about three times the gross annual income of the licence, was worth about \$200000 in 1984, when the price of abalone to the diver was \$1.5 kg^{-1}, and in 1990 was worth about \$1–1.5 million in Tasmania, Victoria and South Australia, at an abalone value of \$20 kg^{-1}. In Western Australia the value of a licence is \$0.4–1 million according to zone, and in NSW it is about \$0.5 million. The cost of licences has induced structural changes in the industry. Owner-operators can seldom find the capital to buy a licence and are gradually being replaced by corporate ownership in which divers are employed under share-fishing or lease agreements. It is still too early to say what effects these changes will have.

Quotas

Since the early paper of Christy & Scott (1965), quotas have come to be seen by economists as an effective means of resolving the deficiencies of licence limitation. In particular, abalone fisheries which exploit long-lived species with stable and slowly changing stock levels may be best suited to management by quota (Copes, 1986). By creating quasi-property rights in fish, they offer much tighter control over each year's catch without the problems associated with controlling effort. They also reduce the competition between divers (vividly described by Black, 1976) and allow them to fish optimally while having regard to considerations of personal health and safety. As yet, various problems associated with quotas, such as 'quota busting' and data fouling (see Copes, 1986) have not been apparent, probably because divers have seen their interests best served by supporting the system. Yet quota busting is likely to be of increasing concern to managers due to the high price that abalone commands.

An expected consequence of quotas, flowing from different productivities and different exploitation costs of different substocks, is the spatial maldistribution of effort. Divers tend to overfish the most productive grounds close to port and underfish others. A second consequence is that the annual fixing of quotas requires much better knowledge of abalone stocks than was demanded by a simple licensing system. With the failure of CPUE data to reflect stock abundance reliably (see Chapter 20, this volume), a much better understanding of stock dynamics supplemented with appropriate techniques to monitor stock density and recruitment strength is now required to set quotas. Both consequences were predicted by Copes (1986).

Hitherto in Australia transfer of all or part of a quota allocation has been permitted only within a season. Now that supposed constitutional objections have been removed it is likely that full transferability will develop as in New Zealand (Chapter 30, this volume).

Licence fees

A number of factors have contributed to sharp increases in the cost of the management of Australian abalone fisheries in the 1980s. Greatly increased poaching, amounting to more than 30% of the stated catch in several states, has led to greater government effort to enforce the law. Second, the administration of quotas involves cost. These factors, and the need for better research information on the dynamics and abundance of stocks, have induced governments to seek to recover rent from conspicuously lucrative fisheries by way of increased licence fees. Most state governments have, by implication or explicitly, adopted the user-pay policy, namely that those who benefit from the effects of management should bear the associated costs. A recent, critical analysis of the policy is given by Haynes *et al.* (1986). In most states the annual licence fee is a fixed percentage (2–10%) of the annual value of the ITQ (i.e. $18000–$30000 per licence in Tasmania, Victoria and South Australia). Such fees now paid by divers have fulfilled the prediction of Geen & Nayar (1988) that divers as 'both the purchasers and beneficiaries of management can be increasingly expected to demand the types and levels of services that will offer them the greatest value for money'. In all states divers demand better enforcement, better fine-scale management and better research.

FUTURE MANAGEMENT

Management of abalone fisheries in Australia has attained a level of complexity such that no single management measure could possibly be effective. It is likely that ITQs will remain the cornerstone of control of fishing mortality. But more complex arrays of measures, such as differential size limits and closures, are likely to be used as increasingly fine-scale management evolves and problems like maldistribution of effort are addressed. Poaching is likely to be an endemic and increasing problem and, if not controlled, could cause the demise of a fishery whatever management scheme is in place. Better control of marketing outlets and stronger enforcement measures seem essential to reduce poaching.

What future directions might managers of abalone fisheries profitably pursue? One is to subdivide the present fishing zones into smaller units exploited by fewer divers, thus further reducing competition and encouraging divers to 'farm manage' the units optimally (see Wilen, 1988). Smaller units would better accord with present understanding of the geographic extent of substocks, and would facilitate reseeding reefs for stock enhancement. Another direction is to pursue the joint management of fisheries by government and industry. Kesteven (1988) argues that the transfer of greater responsibility for management to fishermen would foster efficient exploitation in both economic and conservational senses. Most states already have joint advisory committees with industry to consider problems of research and management, and the skill and experience of members of the industry have increased markedly. We foresee that management will become a partnership between government and industry, that draws on the diverse skills of both, and will increasingly focus on the fine-scale husbandry and enhancement of individual reefs.

ACKNOWLEDGEMENTS

The opinions in this chapter are those of the principal authors. The authors are grateful for the comments of A. Staniford, P.E. McShane, R.W. Day, W.J. Nash,

A.J. Harrison, M.J. Tegner, L.M. Joll, G.D. Hamer, Professor D. Cocheba, A.M. Olsen and A. Campbell on earlier drafts of this chapter.

REFERENCES

Anon. (1964) Abalone, where its found, how to process it. *Fisheries Newsletter*, June, 19.

Anon. (1976) Abalone divers continue work on undersea vehicle. *Aust. Fish.*, **36**(5), 15–16.

Anon. (1982) *Abalone Fishery Situation Report 10. SEFC.* CSIRO, Melbourne.

Anon. (1987) Abalone fishery. In *Fisheries Management History Series*. Department of Sea Fisheries, Tasmania.

Beinssen K. (1979a) The estimation of fishing mortality of abalone based on experiments by the fishing fleet. *Rapp. PV Reun. Cons. Int. Explor. Mer*, **175**, 27–9.

Beinssen K. (1979b) Fishing power of divers in the abalone fishery of Victoria, Australia. *Rapp. PV Reun. Cons. Int. Explor. Mer*, **175**, 20–2.

Beinssen K. & Powell D. (1979) Measurement of natural mortality in a population of blacklip abalone, *Notohaliotis ruber. Rapp. PV Reun. Cons. Int. Explor. Mer*, **175**, 23–6.

Black J.C. (1976) Abalone divers require special consideration. *Aust. Fish.*, **35**(1), 20–1.

Chatterton B. & Arnold L. (1977) Fisheries management and the new federalism. In *The Politics of New Federalism* (Ed. by D. Jaensch), pp. 175–81. Australian Political Studies Association, Adelaide.

Chatterton L. & Chatterton B. (1981) How much political compromise can fisheries management stand? Premiums and politics in closed coastal fisheries. *Marine Policy*, **5**, 114–34.

Chatterton L. & Chatterton B. (1982) The Chattertons reply. *Marine Policy*, **6**, 147–8.

Christy F.T. & Scott A. (1965) *The Common Wealth in Ocean Fisheries*. Johns Hopkins Press, Baltimore.

Copes P. (1986) A critical review of the individual quota as a device in fisheries management. *Land Economics*, **62**(2), 278–91.

Crutchfield J.A. (1979) Economic and social implications of the main policy alternatives for controlling fishing effort. *J. Fish. Res. Bd Can.*, **36**, 742–52.

Cunningham S. (1981) The evolution of the objectives of fisheries management during the 1970's. *Ocean Management*, **6**, 251–78.

Francois D.D. & Gorman T. (1980) Unique management methods in NSW abalone fishery. *Aust. Fish.*, **39**(9), 26–8.

Garland C.D., Cook S.L., Grant J.F. & McMeekin T.A. (1985) Ingestion of the bacteria on the cuticle of crustose (non-articulated) coralline algae by post-larval and juvenile abalone (*Haliotis ruber* Leach) from Tasmanian waters. *J. Exp. Mar. Biol. Ecol.*, **91**, 137–49.

Geen G. & Nayar M. (1988) Individual transferable quotas in the southern bluefin tuna fishery: an economic appraisal. *Marine Resource Economics*, **5**, 365–87.

Gleeson P.M. (1976) Estimation of the economic situation in the abalone fishery. *SAFIC*, **9**, 10–11.

Hamer G. (1983) NSW abalone stock assessment shows effort should be reduced. *Aust. Fish.*, **42**(8), 7–11.

Hancock D.A. (1979) Population dynamics and management of shellfish stocks. *Rapp. PV Reun. Cons. Int. Explor. Mer*, **175**, 8–19.

Harrison A.J. (1969) Australian abalone industry. *Aust. Fish.*, **28**(9), 3–13.

Harrison A.J. (1983) The Tasmanian abalone fishery. *Tasmanian Fish. Res.*, **26**, 1–43.

Harrison A.J. (1986) Gastropod fisheries of the Pacific with particular reference to Australian abalone. *Can. Spec. Publ. Fish. Aquat. Sci.*, **92**, 14–22.

Harrison A.J. & Grant J.F. (1971) Progress in abalone research. *Tasmanian Fish Res.*, **5**, 1–10.

Haynes J., Geen G. & Wilks L. (1986) Beneficiaries of fisheries management. *B.A.E. Discussion Paper 86.1*. Australian Government Publishing Service, Canberra.

Healey M.C. (1984) Multiattribute analysis and the concept of optimum yield. *Can. J. Fish. Aquat. Sci.*, **41**, 1393–1406.

Keesing J.K. & Wells F.E. (1989) Growth of the abalone *Haliotis roei* Gray *Aust. J. Mar. Freshwat. Res.*, **40**, 199–204.

Kesteven G.L. (1988) The conservation of fishery resources and management of their exploitation: the role of a licensing system. *Asian Fish. Sci.*, **1**, 123–33.

Larkin P.A. (1972) A confidential memorandum on fisheries science. In *World Fisheries Policy* (Ed. by B.J. Rothschild), pp. 189–97. University of Washington Press, Seattle.

Larkin P.A. (1977) An epitaph for the concept of maximum sustained yield. *Trans. Am. Fish. Soc.*, **106**, 1–11.

Lewis R.K., Shepherd S.A., Sluczanowski P. & Rohan G. (1984) An assessment of the South Australian abalone resource. *South Australian Department of Fisheries Report.*

McShane P.E. (1987) Victorian abalone stock assessment. Fourth review. *M.S.L. Program Review Series*, No. 83.

McShane P.E. (1988) Victorian abalone stock assessment. Fifth review. *M.S.L. Program Review Series*, No. 94.

McShane P.E., Beinssen K.H.H. & Foley S. (1986a) Abalone reefs in Victoria – a resource atlas. *Victorian Ministry of Conservation Forests and Lands*. Technical Report No. 47.

McShane P.E., Beinssen K.H.H., Smith M.G., O'Connor S. & Hickman J.F. (1986b) Reproductive biology of blacklip abalone *Haliotis ruber* Leach from four Victorian populations. *Mar. Sci. Lab. Tech. Rpt*, No. 55.

McShane P.E., Black K.P. & Smith M.G. (1988) Recruitment processes in *Haliotis rubra* (Mollusca: Gastropoda) and regional hydrodynamics in southeastern Australia imply localised dispersal of larvae. *J. Exp. Mar. Biol. Ecol.*, **124**, 175–203.

McShane P.E. & Smith M.G. (1989a) Direct measurement of fishing mortality in abalone (*Haliotis rubra*) off southeastern Australia. *Fish. Res.*, **8**, 93–102.

McShane P.E. & Smith M.G. (1989b) Victorian abalone fishery in good shape. *Aust. Fish.*, **48**(6), 20–5.

Muse B. & Schelle K. (1989) Individual fisherman's quotas: a preliminary review of some recent programs. *Alaska Commercial Fisheries Entry Commission*, No. 89–1.

Olsen A.M. (1982) Fisheries management: political inhibitions. *Marine Policy*, **6**, 140–6.

Penn J.W. (1974) West coast abalone. *Fins*, **7**(2), 26–8.

Philipson M. (1982) Economics of closures. *SAFIC*, **6**, 5–8.

Prince J.D. (1989) *The fisheries biology of the Tasmanian stocks of* Haliotis rubra. PhD thesis, University of Tasmania.

Prince J.D. & Ford W.B. (1985) Use of anaesthetic to standardise efficiency in sampling abalone populations (genus *Haliotis*; Mollusca: Gastropoda). *Aust. J. Mar. Freshwat. Res.*, **36**, 701–6.

Prince J.D., Sellers T.L., Ford W.B. & Talbot S.R. (1987) Experimental evidence for limited dispersal of haliotid larvae (*Haliotis*; Mollusca: Gastropoda). *J. Exp. Mar. Biol. Ecol.*, **106**, 243–63.

Prince J.D., Sellers T.L., Ford W.B. & Talbot S.R. (1988a) Recruitment, growth, mortality and population structure in a southern Australian population of *Haliotis rubra*. (Mollusca: Gastropoda). *Mar. Biol.*, **100**, 75–82.

Prince J.D., Sellers T.L., Ford W.B. & Talbot S.R. (1988b) Confirmation of a relationship between the localised abundance of breeding stock and recruitment for *Haliotis rubra* Leach (Mollusca: Gastropoda). *J. Exp. Mar. Biol. Ecol.*, **122**, 91–104.

Shepherd S.A. (1973) Studies of southern Australian abalone (genus *Haliotis*) I Ecology of five sympatric species. *Aust. J. Mar. Freshwat. Res.*, **24**, 217–57.

Shepherd S.A. (1985) Power and efficiency of a research diver, with a description of a rapid underwater measuring gauge: their use in measuring recruitment and density of an abalone population. In *Diving for Science 85* (Ed. by C.T. Mitchell), pp. 263–72. American Academy of Underwater Science, La Jolla.

Shepherd S.A. (1986a) Studies on southern Australian abalone (genus *Haliotis*). VII. Aggregative behaviour of *H. laevigata* in relation to spawning. *Mar. Biol.*, **90**, 231–6.

Shepherd S.A. (1986b) Movement of the South Australian abalone *Haliotis laevigata* in relation to crevice abundance. *Aust. J. Ecol.*, **11**, 295–302.

Shepherd S.A. & Cannon J. (1988) Studies on southern Australian abalone (genus *Haliotis*) X. Food and feeding of juveniles. *J. Malac. Soc. Aust.*, **9**, 21–6.

Shepherd S.A. & Gilligan J.E. (1974) Medical problems of diving. *Aust. Fish.*, **33**, 8–13.

Shepherd S.A. & Godoy C. (1989) Studies on southern Australian abalone (genus *Haliotis*). XI. Movement and natural mortality of juveniles. *J. Malac. Soc. Aust.*, **10**, 87–95.

Shepherd S.A. & Laws H.M. (1974) Studies on southern Australian abalone (genus *Haliotis*) II.

Reproduction of five species. *Aust. J. Mar. Freshwat. Res.*, **25**, 49–62.
Shepherd S.A. & Turner J.A. (1985) Studies on southern Australian abalone (genus *Haliotis*) VI. Habitat preference, abundance and predators of juveniles. *J. Exp. Mar. Biol. Ecol.*, **93**, 285–98.
Sinclair W.F. (1978) Management alternatives and strategic planning for Canada's fisheries. *J. Fish. Res. Bd Can.*, **35**, 1017–30.
Sluczanowski P.R. (1984) A management oriented model of an abalone fishery whose substocks are subject to pulse fishing. *Can. J. Fish. Aquat. Sci.*, **41**, 1008–14.
Sluczanowski P.R. (1986) A disaggregate model for sedentary stocks: the case of South Australian abalone. *Can. Spec. Publ. Fish. Aquat. Sci.*, **92**, 393–401.
Stanistreet K. (1982) Limited entry in the abalone fishery of Victoria. In *Policy and Practice in Fisheries Management* (ed. by N.H. Sturgess & Meany T.F.), pp. 139–52. Australian Government Publishing Service, Canberra.
Tenison-Woods J.E. (1882) *The Fish and Fisheries of New South Wales*. Government Printer, Sydney.
Walkerden R.D. (1982) Administrative aspects. *Marine Policy*, **6**, 148–9.
Walters C.J. (1986) *Adaptive Management of Renewable Resources*. Mcmillan, New York.
Ward T.J. (1986) Abalone biology and fisheries in Australia: a review of research. *Aust. Fish. Serv. Canberra. Aust. Fish. Pap.*, No. 86–12.
Waugh G. (1984) *Fisheries Management: Theoretical Developments and Contemporary Applications*. Westview Press, Boulder.
Wells F.E & Keesing J.K. (1989) Reproduction and feeding in the abalone *H. roei* Gray. *Aust. J. Mar. Freshwat. Res.*, **40**, 187–97.
Wilen J.E. (1988) Limited entry licensing: a retrospective assessment. *Mar. Res. Econ.*, **5**, 313–24.
Witherspoon N.B. (1975) Sizes in the Tasmanian commercial catch of blacklip abalone (*Haliotis ruber*) – possible effects of changing habitat during growth. *Tasmanian Fish. Res.*, **9**, 15–22.

Chapter 30
The paua (abalone) fishery of New Zealand

DAVID R. SCHIEL *Department of Zoology, University of Canterbury, Christchurch 1, New Zealand*

ABSTRACT: The modern commercial fishery for paua, *Haliotis iris*, in New Zealand dates from the mid-1940s, when shell was marketed and the meat was usually discarded. Both meat and shell were sold from the late 1950s. Since the 1986–7 fishing season, the fishery has been managed with an individual transferable quota (ITQ) system, with quota owned as a property right, and a total allowable catch (TAC) being established for each of 10 fishing regions. The current TAC for the entire country is 1147 metric tonnes (whole landed weight), an amount that is reviewed annually. The present annual value of the fishery, including exported and domestic meat product, the sale of shells and value-added products such as jewellery is estimated to be NZ$35 million (about US$21 million). The fishery may be dependent on accumulated biomass and recruitment overfishing may occur in some areas. Current research aims to investigate the status of populations (size/age structure and biomass) in the major fishing areas and to monitor the sizes and ages of paua landed in each region.

RESUMEN: La pesquería comercial moderna del abulón (*Haliotis iris* Martyn) de Nueva Zelandia se remonta a la mitad de la década de los cuarentas, cuando la concha era el producto comercial y la carne (callo) se desechaba. Después, a finales de los cincuentas, tanto el callo como las conchas se vendían. Desde la temporada de pesca de 1986–87 la pesquería se ha venido manejando mediante el sistema de Cuotas Individuales Transferibles (CIT), siendo estas consideradas con derecho de propiedad; y se ha establecido una Captura Permitida Total (CPT) en cada una de las diez regiones de pesca. La CPT actual anual, para todo el país, asciende a 1147 toneladas (peso bruto). El actual valor comercial de la pesquería; incluyendo las exportaciones y el consumo interno de carne, así como la venta de conchas; se estima que es de 35 millones de dólares NZ (unos 21 millones de dólares EUA). El enfoque utilizado en su manejo con las CTP's, es bastante cauto, por lo que son revisadas cada año. La pesquería puede estar dependiendo de la biomasa acumulada y reclutamiento; y al menos en algunas áreas, una sobre pesca puede ocurrir. La investigación actual persigue conocer la situación de las poblaciones en las principales áreas de pesca.

INTRODUCTION

Three species of abalone, locally called the Maori name 'paua', occur in shallow rocky habitats along the entire coastline of New Zealand. They are most abundant, however, in the cooler waters south of 41°S latitude. The major fishing areas are in the South Island, Stewart Island and the Chatham Islands (Fig. 30.1). Two species are fished commercially. The most important of these is the black-footed paua *Haliotis iris*, which grows to 180 mm shell length and is by far the most abundant species. The yellow-footed or queen paua, *H. australis*, grows to 110 mm, is much scarcer and is fished only occasionally. The third species, the white-footed or virgin paua, *H. virginea*, is smaller (70 mm maximum), consists of four regional subspecies (Powell, 1979) and is not fished commercially. This chapter concerns *H. iris* because it comprises virtually all of the commercial catch and there is relatively little information on *H. australis* and *H. virginea*.

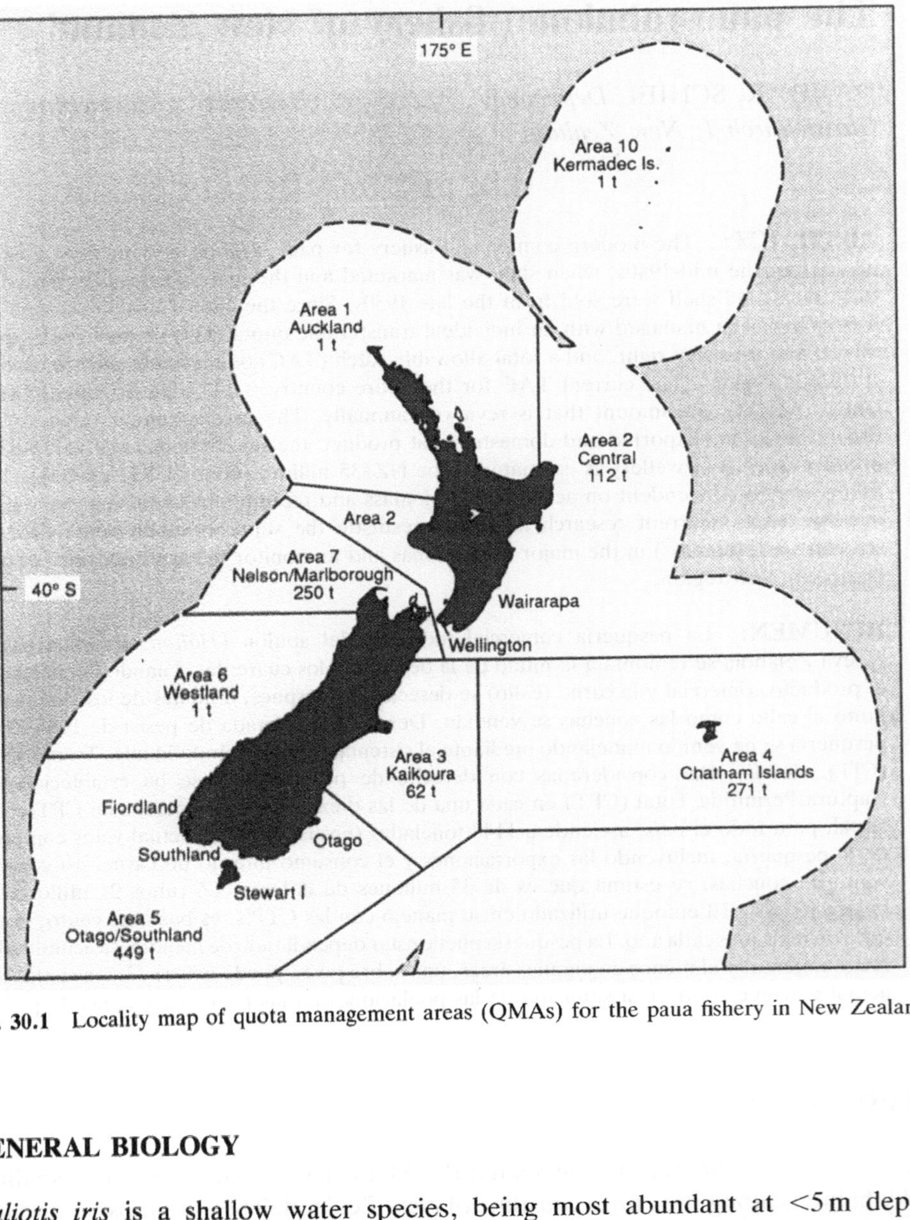

Fig. 30.1 Locality map of quota management areas (QMAs) for the paua fishery in New Zealand.

GENERAL BIOLOGY

Haliotis iris is a shallow water species, being most abundant at <5 m depth. Where rocky reef is available, however, its distribution may extend to 20 m (see Sainsbury, 1982a; Schiel, 1990). Settlement and recruitment of *H. iris* generally occur on coralline-encrusted rocks from the low tide mark to several metres in depth. Juveniles live and feed beneath rocks and boulders for three to five years, the length of this life stage varying among localities. Post-larvae feed on benthic diatoms such as *Navicula* spp. and *Nitzschia* spp. as well as red microalgae associated with coralline surfaces (P. Redfearn & L. Tong, Fisheries Research NZ, unpublished data) and may also consume bacteria, as do other species of haliotids (Garland *et al.*, 1985). At a juvenile stage of about 3 mm shell length, paua begin to graze macroalgae such as the green sea lettuce *Ulva lactuca*, the red seaweeds *Champia* sp. and *Pterocladia lucida* and several species of brown algae including *Macrocystis pyrifera*, *Lessonia variegata*, *Ecklonia radiata* and *Carpophyllum* spp. Diet is influenced by the local abundance and mixture of seaweeds (Poore, 1972a; Wilson, 1987).

The onset of sexual maturity occurs at 60–70 mm when paua are 3–5 years old. At this stage they begin to move from their cryptic habitats beneath boulders to more open rock surfaces, and may migrate into deeper water. As adults, they eat a wide range of seaweeds (Tunbridge, 1967), feeding predominantly on drift algae in most localities. They tend to be sedentary except when food is scarce (Poore, 1972b).

Reproduction in *H. iris* can occur throughout the cooler months (March–November) but spawning peaks occur in late summer–early autumn and sometimes in spring (Poore, 1973; Wilson, 1987). Large females can produce 11 million eggs and may spawn 7–8 million eggs. Successful spawnings may not be regular events, however, and entire populations fail to spawn in some years (Sainsbury, 1982a). Gonad maturation appeared to be normal in years of spawning failure, so some stimulus to spawn may have been absent. In some cases, temperature is correlated with spawning events (Poore, 1973; Wilson, 1987), but the exact stimuli necessary to induce spawning are unknown.

Settlement of *H. iris* has been reported both in the early autumn (around April) and in spring (around October) (Poore, 1972c; Wilson, 1987). This process is considerably variable among localities and years. It is estimated that paua may live to 60 years (T. Murray, Fisheries Research NZ, unpublished data). Furthermore, preliminary results from enzyme electrophoresis studies suggest that there are two genetic stocks of paua: one for the mainland and nearshore islands and one at the remote Chatham Islands (P. Smith, pers. comm.).

THE FISHERY

History of the fishery

Since pre-European times, paua meat has formed an important part of the traditional diet of the indigenous Maori people and the shells were used extensively in carvings, decorations and fishing lures (Phillipps, 1935; Best, 1977). The modern fishery began just after World War II and was centred on paua shells, which are valuable for making souvenirs and jewellery. The fishery was initially restricted to war veterans as a way to encourage their employment in remote areas of the country. The Customs Export Prohibition Order of 1946 banned the export of raw shell but allowed the export of manufactured items in an attempt to protect the local industry. Until the early 1960s, the meat was often discarded because of its low value and lack of markets and only the shell was sold (Brown, 1982).

The commercial fishery for paua meat began in the 1960s to service a growing domestic demand. Export of canned product was allowed and by the 1970s this led to considerable fishing pressure on the resource. The fishery was therefore closed nationwide for 4 months in 1972 to develop management measures for limiting harvests. This led to export restrictions being imposed in 1973. These aimed to control the catch of paua by limiting the total amount of meat for export. The quota between 1973 and 1984 was held at 71 400 dozen cans of processed product, or about 928 t (whole landed weight) of paua (Murray & Akroyd, 1984). In 1980, restrictions on the sale and export of shell were liberalized slightly and cabochons (small polished discs) were exported. The Paua Shell Exchange was established in 1982. All of the paua shell available in NZ is listed with the exchange so that local industry has an opportunity to buy it. If shell remains unsold after a period of 3 months (this period was 6 months until mid-1989), the exchange can arrange for an export licence to be issued. Legislation concerning the export of both meat and shell was liberalized in 1986 and again in late 1990. Restrictions were lifted so that live paua could be exported.

Individual quotas were established from the 1984–5 fishing year (October–September) and they became transferable from 1986–7. These individual transferable quotas (ITQs) were allocated on the basis of previous catch history, based on the average of the best two of the three years preceding the introduction of the quota system. Fishing histories were based on official fishery statistics supplied to the Ministry of Agriculture & Fisheries, but in the absence of fishing returns factory receipts could be used. Initial allocations were subject to an appeal process before finalizing the quotas and the total allowable catch (TAC). A quota appeal authority handles disputes and the TAC is set as the sum of the ITQs (1147 t when this was written, but with appeals was increased to 1287 t in 1991).

There are currently 201 quota holders, most of whom have only a few tonnes of quota (Fig. 30.2). Sixty-five percent have 5 t or less of quota, while only 11 have more than 20 t of allocation. Most quota holders belonged to the Abalone Divers of NZ Co-operative, which was set up in 1986. Prior to this, only a few processors were certified to handle paua for export, mostly in the form of canned, bleached meat. Quota holders may now sell their product, both meat and shell, in whatever

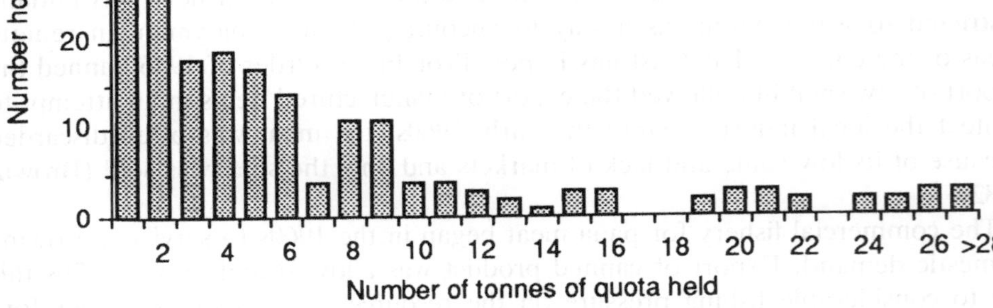

Fig. 30.2 Number of quota holders versus the amount of paua quota held in metric tonnes (whole landed weight).

form they wish to any market. The Abalone Divers Co-operative handled the marketing of about 75% of paua landed, giving divers more control over their product and the prices they received for it. The Co-operative folded in 1990.

The present fishery

By law, paua are fished exclusively by means of free diving (snorkelling). This method restriction allows access to most populations. A trowel or hooking device is used to dislodge animals from the substratum into hand-held nets. These are then either suspended from floating inner tubes or deposited into a small tender which is kept near the diver. Larger, fast runabouts capable of being transported overland by trailer are generally used to get to harvesting sites, but some shore diving also occurs. Paua are fished year-round, but the timing depends on the weather and sea conditions and on the current price of paua. For example, paua are now fished heavily from September to November to service a Christmas market, and fishing is discouraged in January and February when market demand is low. Unfortunately for the diver, the latter are summer months when weather and sea conditions tend to be mildest.

The minimum legal size for harvesting *H. iris* is 125 mm shell length. By the time a paua reaches this size, it should have spawned for several years. From yield-per-recruit analysis, Sainsbury (1982b) concluded that this minimum legal size was too large in some areas, but also pointed out that his analysis did not take account of reproductive potential and provided no indication of 'sustainability' of catches.

The fishery has been managed by means of a quota system since 1984 (see section above). The TAC is allocated among quota management areas, which encompass the major fishing regions (Fig. 30.1). The largest fisheries are in the Otago/Southland region (449 t), the Chatham Islands (271 t) and the Marlborough Sounds (250 t). In areas where there are only minor allocations (1 t each for Westland, the Auckland area and the Kermadec Islands), paua are either scarce or do not usually reach legal size. Under the ITQ system, an individual or company owns a quota allotment which they may sell, trade or lease in any amount. A resource rental, the amount of which is reviewed annually, is paid to the government on the amount of quota held. The resource rental may be used to compensate ITQ holders at market value should it become necessary to lower the TAC. To date, the TAC has not been lowered.

Fishery statistics

The total catch reached a peak in 1981 when 1961 t were landed (Fig. 30.3). There may have been a similar peak in catch in 1971, but catch data before 1974 are unreliable (Murray & Akroyd, 1984). Reported catches for the two years since ITQs were established were around 1100 t. The largest fishery has been in area 5 in southern NZ, but large tonnages have been removed from the five major fishing areas (Fig. 30.4). There were considerable fluctuations in landings in each region before the ITQ system was established. These were the result of many

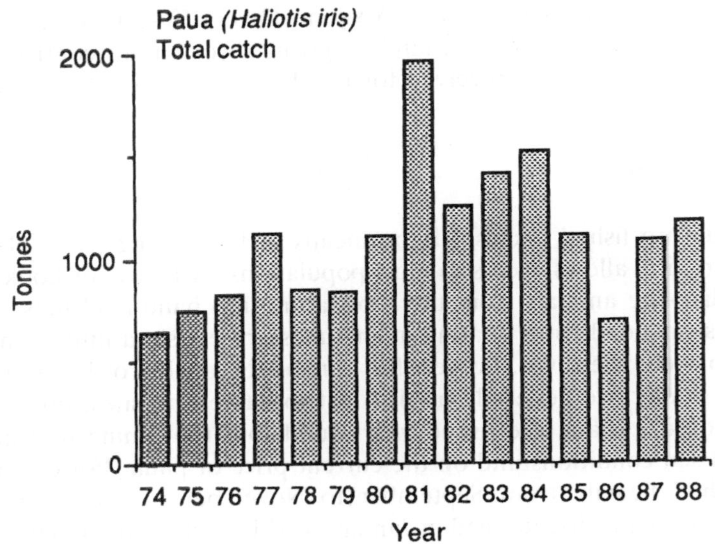

Fig. 30.3 Total annual catch of *Haliotis iris* in metric tonnes (whole landed weight) for all of New Zealand. Data are from Murray & Akroyd (1984) and Ministry of Agriculture & Fisheries Statistics Unit records.

influences, including the prices paid to divers and factors relating to employment in the more remote regions. Distance from markets and transport costs, particularly from the Chatham Islands (area 4), also affected landings in the past.

Catch per unit effort (CPUE) data were summarized by Cooper (1976), who detected a decline in CPUE. There was a 20% reduction in export quota in response to this trend (Murray & Akroyd, 1984). Murray (1982) reported that CPUE had stabilized between 1974 and 1981, although these data were highly variable (Fig. 30.5). The CPUE at the Chatham Islands was particularly variable. As for other abalone fisheries (Harrison, 1986; Chapter 22, this volume), patch fishing affects catch effort data. CPUE, therefore, is not necessarily a good gauge of the status of the paua fishery. Over four recent fishing years (Table 30.1), for example, CPUE has gone up and then down in area 2, stayed about the same in areas 3, 5 and 7 and abruptly increased in area 4. This latter increase has been explained by paua divers as the result of knowing where virtually all of the fishable stocks are located so that their effort is now more concentrated.

There have been few biomass surveys for paua populations (see Sainsbury, 1982a). In a large-scale survey, Petherick (1986, 1987) estimated the biomass along nine transects (250 m × 1 m) over 86.5 km of the Wairarapa coast. Extrapolating from these transects he estimated that there were 1030 t of paua in the survey area. This estimate, however, was not very reliable (coefficient of variation = 114%) because of the extremely patchy distribution of the paua, the logistic difficulties in sampling and the complexity of paua habitats on exposed coasts.

Patchiness in the distribution of paua is linked to both habitat heterogeneity and to biological effects associated with settlement and recruitment. Recruitment certainly appears to be highly variable both spatially and temporally. Sainsbury

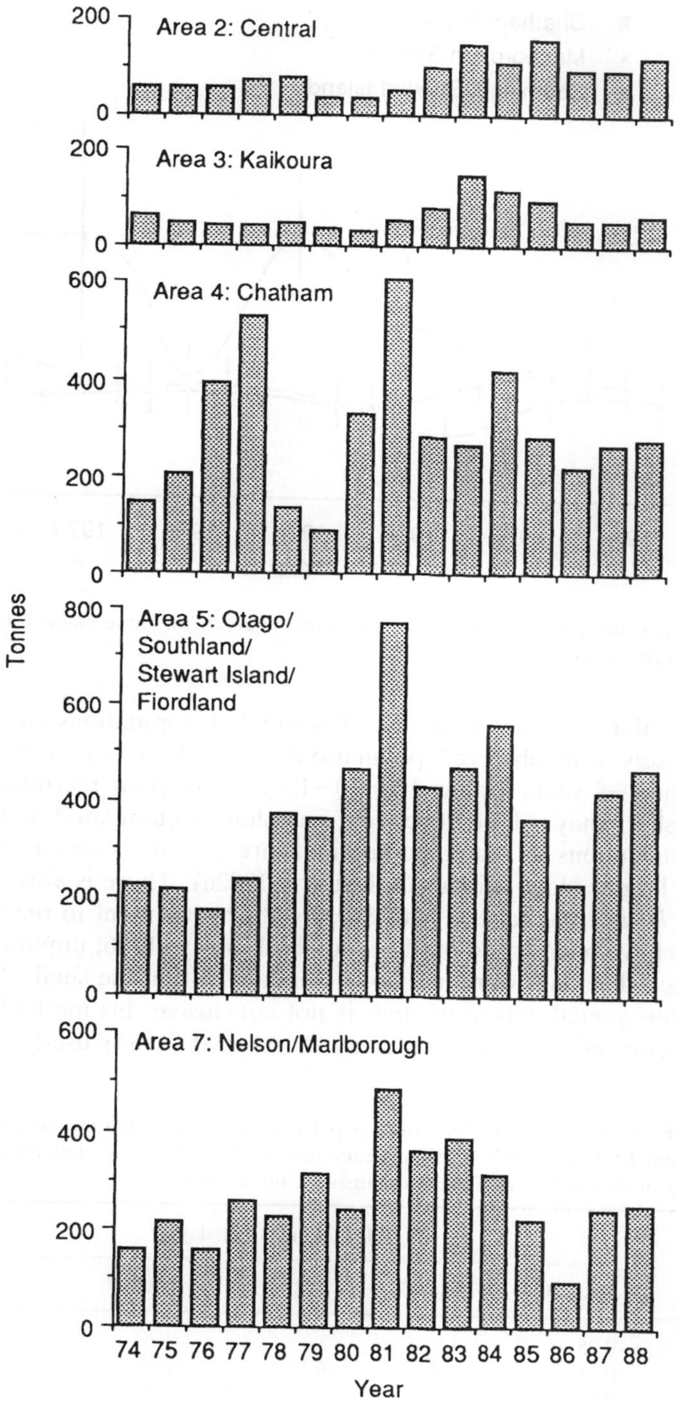

Fig. 30.4 Annual catch of *Haliotis iris* in metric tonnes (whole landed weight) for the major quota management areas.

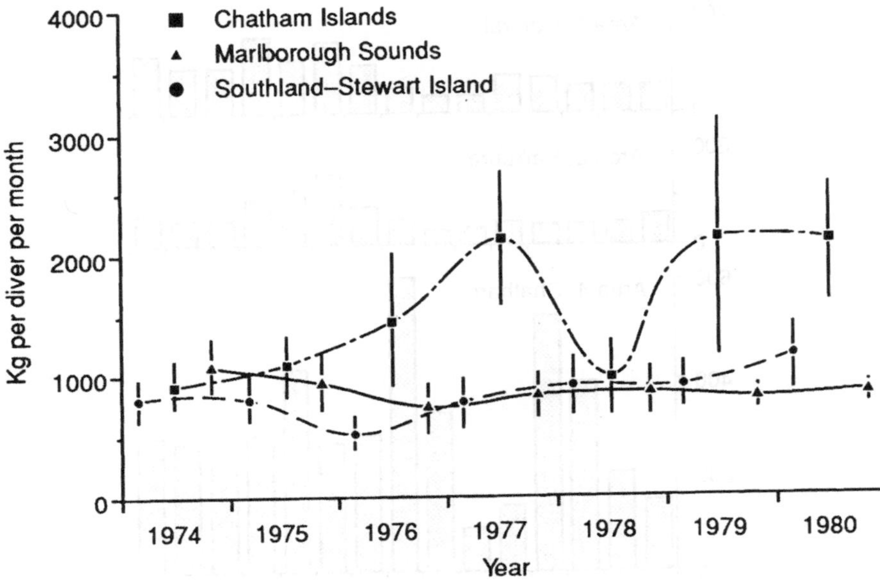

Fig. 30.5 Catch per unit effort (mean ±95% confidence interval) for three major fishing areas for *Haliotis iris* (Redrawn from Murray, 1982.)

(1982b) found that the size structure of simulated populations of *H. iris* corresponded closely with observed populations when there was a pulse of good recruitment for 4–8 years followed by 10–15 years of poor recruitment.

The paua fishery may still be dependent on fishing-accumulated biomass. Size–frequency distributions of paua populations are usually skewed, with a high proportion of larger, older animals (Sainsbury, 1982a). There is also considerable evidence that *H. iris* may take at least 8 years after settlement to reach legal size. Age estimates have been based on the study by Murray (1986; unpublished data), who determined that rings were laid down annually within the shell. Although his evidence for the annual nature of rings is not conclusive, his method of aging is the only one currently available. Murray & Akroyd (1984) used shell rings to

Table 30.1 Catch per unit effort (CPUE) (tonnes per day fished) for paua in quota management areas (QMAs) from 1983–4 to 1987–8 fishing years; dashes indicate that no data are available; data are from Ministry of Agriculture & Fisheries Statistics Unit records.

	CPUE (tonnes per day)				
QMA	1983–4	1984–5	1985–6	1986–7	1987–8
1	0.25	—	—	—	0.20
2	0.18	0.29	0.32	0.83	0.25
3	0.13	0.19	0.19	0.20	0.17
4	0.28	0.40	0.33	0.31	0.63
5	0.36	0.51	0.51	0.45	0.44
6	—	0.41	—	—	—
7	0.26	0.30	0.26	0.25	0.20

determine the ages of paua in several populations, finding considerable variation in length-at-age relationships among fishing regions. They also found that at different localities within a region there was considerable variability in age at entry to the fishery. In the Marlborough Sounds, for example, paua at one locality took 16 years to reach legal size, in another they took 26 years and in a third (unfished) locality they never reached legal size. This range of ages to reach legal size is typical among populations in all the fishing regions.

ECONOMICS OF THE FISHERY

The value of landed paua increased considerably over the period 1971–87 (Fig. 30.6). The most reliable data are export figures for paua meat, which had a value of NZ$14 million in 1987 ($1 NZ = $0.60 US in November 1989). Including domestic and export sales of all forms of meat, shell and value-added products the output from the fishery is currently valued at about NZ$35 million per annum.

The price paid to divers varies slightly throughout the year. Paua are landed whole and a figure for recovery rate of meat is calculated for each diver's landing at the processing plant. A good recovery rate is 40–45% of total weight. Divers are currently paid (June 1991) about NZ$31/kg for meat and about NZ$3/kg for shell.

The annual permit fee for paua divers is NZ$67.50. Resource rentals currently are NZ$126 per tonne of quota and there is a nominal fee paid if quota is transferred to new ownership.

The economics of the fishery and the status of stocks are undoubtedly influenced by recreational and illegal fishers. The harvest of paua by these groups has not been accounted for in setting the TACs, and the impact on the sustainability of the legal commercial harvest is unknown. There is no information on the harvest

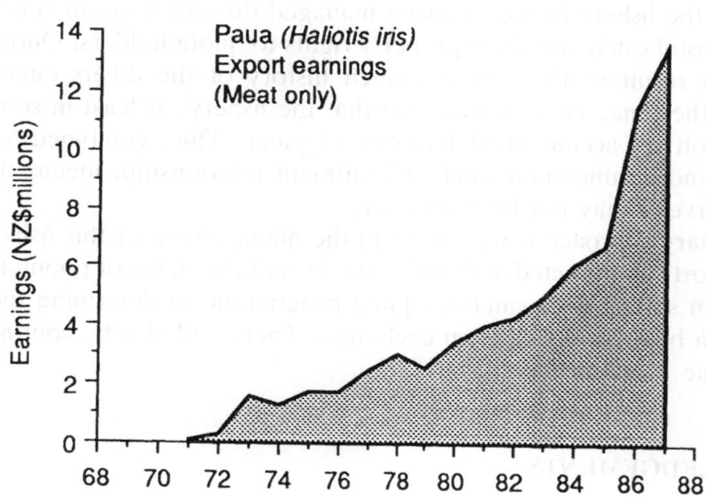

Fig. 30.6 Annual value of exported meat (fob) of *Haliotis iris*. This is processed product and does not include domestic sales and the value of shells. Data from the Fishing Industry Board of NZ.

by recreational divers. The effectiveness of the recreational catch limit of 10 paua per person per day, or 50 per party of five or more people, is also unknown. As New Zealand's coastline is largely accessible, paua occur in shallow water, and some 2.5% of the population engage in diving activities, there are likely to be at least localized effects of recreational fishing on the commercial sector.

There is also a large and organized illegal commercial fishery. For the 1987–8 fishing year the harvest was estimated to be at least 550 t. Considerable effort has been directed recently at intercepting and prosecuting illegal fishers as well as those marketing illegally-caught paua. The Fisheries Compliance Unit of the Ministry of Agriculture & Fisheries is responsible for ensuring that fishing regulations are complied with. This unit has mounted several major operations using field-based fisheries officers with sophisticated surveillance techniques, aided by police and traffic officers, to apprehend illegal fishers, distributors and product. The Abalone Divers Co-operative also assisted by advertising extensively and offering a $500 reward for information leading to the prosecution of those engaged in illegal fishing for paua. During August 1989 alone, 60 000 paua (about 10 t) were seized. These were export packed and stored frozen in a ship container, apparently bound eventually for markets in southeast Asia.

The ITQ system has provided some financial stability to the process of paua fishing by permitting divers to own the property rights to their quota. This has also contributed to an increase in price for paua. Since the system was established, many new divers have entered the fishery, incurring considerable financial commitments and repayments in purchasing quota and new equipment. For example, the current price of quota (June 1991) is about $50 000 per tonne, which takes at least three years to pay off for most divers.

CONCLUSION

Since 1986, the fishery in NZ has been managed through a quota system that has limited the total catch and given property rights to quota holders. Quotas were set largely as a result of the previous catch history of the divers engaged in the industry at the time. There is evidence that the fishery, at least in some areas, is dependent on an accumulated biomass of paua. This, combined with patchy settlement and an uncertain stock–recruitment relationship, means that current levels of harvests may not be sustainable.

A cautionary approach is warranted in the management of this fishery. Present research efforts are directed at defining stocks and the status of populations within them, and in setting up a catch-sampling programme to determine the sizes and ages of paua being removed from each area. There will also be continued efforts to reduce the illegal fishery.

ACKNOWLEDGEMENTS

The author thanks Talbot Murray for many useful comments and for access to his unpublished data, and Paul Breen and Warwick Nash for comments on this chapter.

REFERENCES

Best E. (1977) *Fishing Methods and Devices of the Maori* (reprinted from 1929 edn). NZ Government Printer, Wellington.

Brown C.H. (1982) Problems facing the commercial diver: past, present, and future. In *Proceedings of the Paua Fishery Workshop, Fisheries Research Division Occasional Publication No. 41* (Ed. by J.M. Akroyd, T.E. Murray & J.L. Taylor) pp. 28–32. (Obtainable from the Information Clerk, MAFFish Fisheries Research, Wellington.)

Cooper B. (1976) Paua fishery – past, present, and future. *Catch '76*, **3**(3), 3–7. (A Fisheries Research publication, obtainable as above.)

Garland C.D., Cooke S.L. & McMeekin T.A. (1985) Ingestion of the bacteria on the cuticle of crustose (non-articulated) coralline algae by post-larval and juvenile abalone (*Haliotis ruber* Leach) from Tasmanian waters. *J. Exp. Mar. Biol. Ecol.*, **91**, 137–49.

Harrison A.J. (1986) Gastropod fisheries of the Pacific with particular reference to Australian abalone. In *North Pacific Workshop on Stock Assessment and Management of Invertebrates* (Ed. by G.S. Jamieson & N. Bourne), pp. 14–22. Department of Fisheries and Oceans, Ottawa.

Murray T. (1982) Recent trends in the paua fishery. In *Proceedings of the Paua Fishery Workshop, Fisheries Research Division Occasional Publication No. 41* (Ed. by J.M. Akroyd, T.E. Murray & J.L. Taylor), pp. 9–13.

Murray T. (1986) Paua shell and its growth. *Shellfisheries Newsletter*, **30**, 10, in *Catch* **13**(3).

Murray T. & Akroyd J.M. (1984) The New Zealand paua fishery: an update of biological considerations to be reconciled with management goals. *Fisheries Research Centre Internal Report No. 5*.

Petherick C. (1986) Paua survey of the Wairarapa coast. *Shellfisheries Newsletter*, **31**, 8–9, in *Catch* **13**(6–7).

Petherick C. (1987) Preliminary survey of the Wairarapa coast. *Central Fishery Management Area Internal Report*, 87/2.

Phillipps W.J. (1935) The New Zealand paua-shell. *NZ J. Sci. Tech.*, **16**, 296–301.

Poore G.C.B. (1972a) Ecology of New Zealand abalones, *Haliotis* species (Mollusca: Gastropoda) 1. Feeding. *NZ J. Mar. Freshwat. Res.*, **6**, 11–22.

Poore G.C.B. (1972b) Ecology of New Zealand abalones, *Haliotis* species (Mollusca: Gastropoda) 2. Seasonal and diurnal movement. *NZ J. Mar. Freshwat. Res.*, **6**, 246–58.

Poore G.C.B. (1972c) Ecology of New Zealand abalones, *Haliotis* species (Mollusca: Gastropoda) 3. Growth. *NZ J. Mar. Freshwat. Res.*, **6**, 534–59.

Poore G.C.B. (1973) Ecology of New Zealand abalones, *Haliotis* species (Mollusca: Gastropoda) 4. Reproduction. *NZ J. Mar. Freshwat. Res.*, **7**, 67–84.

Powell A.W.B. (1979) *New Zealand Mollusca*. Collins, Auckland.

Sainsbury K.J. (1982a) Population dynamics and fishery management of the paua, *Haliotis iris*. 1. Population structure, growth, reproduction, and mortality. *NZ J. Mar. Freshwat. Res.*, **16**, 147–61.

Sainsbury K.J. (1982b) Population dynamics and fishery management of the paua, *Haliotis iris*. 2. Dynamics and management as examined using a size class population model. *NZ J. Mar. Freshwat. Res.*, **16**, 163–73.

Schiel D.R. (1990) Macroalgal assemblages in New Zealand: structure, interactions and demography. *Hydrobiologia*, **192**, 59–76.

Tunbridge B.R. (1967) Feeding habits of paua (*Haliotis iris* Martyn). *Fisheries Technical Report, NZ Marine Dept*, **20**.

Wilson N.H.F. (1987) *Reproduction and ecology of* Haliotis iris *and* H. australis *from Otago*. MSc thesis, University of Otago.

Chapter 31
The abalone fishery of South Africa

ROBERT J.Q. TARR *Sea Fisheries Research Institute, Private Bag X2, Rogge Bay 8012, Cape Town, Republic of South Africa*

ABSTRACT: The South African abalone fishery began in 1949 and is reliant on one species, *Haliotis midae*. Divers operate from the shore using small fibreglass dinghies or ski-boats. At the start of the fishery there were no controls on total effort by divers. Landings consequently fluctuated according to demand and then decreased rapidly as the availability of abalone declined. A production quota of 227 tonnes first limited total fishing effort in 1970. The production quota was changed to a whole mass quota in 1983. From 1986, total allowable catches (TACs) were set for each of seven fishing zones, and the sum of the TACs was 640 tonnes.

Research for management has recently been directed at obtaining data on habitat area for each fishing zone in order to facilitate annual TAC estimates for these grounds. Diving surveys have provided data on abalone densities and size composition in the fishing grounds and tagging studies will expand the knowledge on growth rates.

Since 1984, divers have been granted individual fishing rights to fixed percentages of the annual TAC, and are contractually obligated to deliver to specific factories. These 'entitlements' are fully heritable and saleable, but to date no new entrants to the fishery have resulted.

Catch statistics supplied monthly by the divers provide data on daily catch, diving time, place and area dived, and these data are presently used as indicators of fishing trends in specific fishing zones.

In the early days of the fishery a commercial abalone licence was available at a nominal £1, and is today a still nominal R60. Present export price for the canned or frozen product is estimated to be in excess of R110 per kg, with a total export value of some R16 million per annum.

Although expansion of the fishery is unlikely, the prospects for the existing fishery appear good, with catch rates improving.

RESUMEN: La pesquería de abulón de Sud Africa comenzó en 1949 y se basa en una sola especie: *Haliotis midae*. Los buceadores operan desde la costa utilizando botes de fibra de vidrio pequeños o medianos. Al comienzo de la pesquería no había controles sobre el esfuerzo total de los buceadores. Las capturas, en consecuencia, fluctuaban según la demanda y luego disminuyeron rápidamente en la medida en que la disponibilidad del abulón también se redujo. Una cuota de producción de 227 toneladas limitó en primera instancia el esfuerzo pesquero en 1970. La cuota de producción fue sustituida por una cuota de masa total en 1983. A partir de 1983 se establecieron Capturas Totales Permitidas (total allowable catches, TAC) en nueve zonas de pesca, y la suma de los tac fue 640 toneladas.

La investigación referente al manejo de la pesquería ha sido dirigida recientemente a la obtención de datos sobre las áreas habitat para cada zona de pesca con el objeto de facilitar las estimaciones anuales del tac para la zona. Prospecciones submarinas han aportado datos sobre densidades del abulón y composición por tallas en las zonas de pesca. Estudios de mercado permitirán un mejor conocimiento de las tasas de crecimiento.

Desde 1984 los buceadores recibían derechos pesqueros en forma individual sobre un porcentaje del tac anual, y por contrato estaban obligados a remitir a fábricas específicas. Estos derechos son hereditarios y vendibles pero hasta ahora no se ha registrado ninguna nueva entrada a la pesquería.

Las estadísticas de capturas entregadas mensualmente por los buceadores, proporcionan datos sobre captura diaria, tiempo de buceo, zona y área explorada, y estos datos son actualmente utilizados como indicadores de pautas pesqueras en zonas específicas.

En los comienzos de la pesquería una licencia para pesca comercial de abulón costabla 1 libra, y en la actualidad es 60 rands. El precio de exportación por el producto enlatado o congelado es 110 rand por kg, con un precio total de 16 milliones por año.

A pesar de que la expansión de la pesquería es probable, las perspectivas para la pesquería existente parecen buenas, con tasa de captura en plan de mejoría.

INTRODUCTION

This fishery relies on one species only, *Haliotis midae*, which is one of the largest of the Haliotids after the Californian *H. rufescens*. The adult fished population is mostly non-cryptic and readily visible, and most animals are to be found in depths shallower than 10 m in beds of the kelp *Ecklonia maxima*. The restricted habitat and largely non-cryptic behaviour make abalone vulnerable to divers, and the stocks are also fished by an ever-increasing number of recreational divers.

The growth, reproduction, movement and distribution of *H. midae* have been described by Newman (1964, 1966, 1967, 1968, 1969). Aspects of the assessment and management of the fishery comprise a chapter of Newman's PhD thesis (Newman, 1973). The diet was investigated by Barkai & Griffiths (1986), and the early development was described by Genade *et al.* (1988). A broad overview of the biology and fishery has been given by Tarr (1988).

THE FISHERY

The present fishery, based on subtidal stocks, began in Gans Bay in 1949, and soon expanded to cover the coastline from Cape Columbine to Quoin point near Cape Agulhas, which is still its present day extent (Fig. 31.1). This area incorporates a coastline totalling some 580 km, although not all of this is fished due to

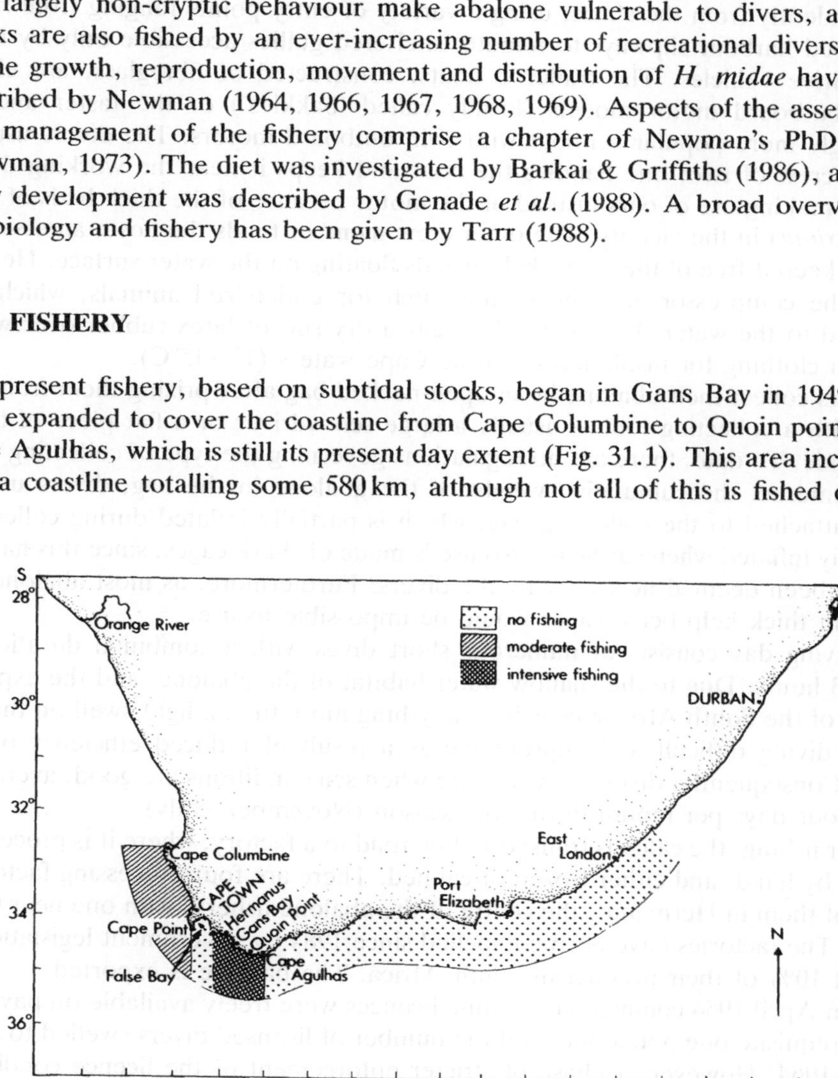

Fig. 31.1 The distribution of *Haliotis midae*, showing the commercially-fished area.

unsuitable (sandy) areas or closures. Newman (1964) described the fishery and research activities of the early 1960s.

A size limit of 10.16 cm shell breadth (12.4 cm shell length) was imposed in 1953, and increased to 11.43 cm (13.8 cm shell length) in 1954, which is still the standard today.

The fishery began with divers using traditional 'hard hat' helmet diving gear. With the dense kelp forests of *E. maxima* hindering progress they soon changed to the less cumbersome 'hookah' system which afforded divers greater mobility underwater. The hookah system is still used by today's divers so that, unlike many other commercial fisheries, today's abalone divers do not utilize advanced technology to increase their fishing efficiency. This has relevance to the use of catch-per-unit-effort indices in assessing the resource strength. The divers operate independently from the shore, using a variety of entry points ranging from fully-developed harbour slipways to remote, sheltered gullies accessible only by four-wheel-drive vehicles. The trailered boats, constructed of fibreglass, are either single outboard motor models (locally called 'bakkies') or the faster and increasingly more popular ski-boat with twin outboard motors. The divers usually have a crew of two; one handles the boat and keeps it near the working diver, usually making use of oars instead of the motor because of the thick beds of kelp (*E. maxima*) in the vicinity. The other crew member tends the hose, an essential task to keep it free of the dense kelp fronds floating on the water surface. He also tends the compressor and checks the catch for undersized animals, which are returned to the water. Divers usually wear a dry suit of latex rubber over warm woollen clothing for insulation from the Cape waters (12–15°C).

Divers collect their abalone in an open-necked bag after prizing them from the rock with a collecting tool – usually a large screwdriver or a flat piece of blunt iron blade. To date, there are no regulations governing the type of collecting tool. To counteract the cumulative weight of the abalone in the bag, divers use an airlift attached to the collecting bag, which is partially inflated during collection and fully inflated when surfacing. No use is made of shark cages, since this has not to date been deemed necessary by the divers. Furthermore, as most abalone are found in thick kelp beds, cages would be impossible to use.

A diving day consists of numerous short dives with a combined duration of about 3 hours. Due to the shallow water habitat of the abalone, and the exposed nature of the South African coastline, anything more than a light swell on the sea makes diving difficult and unprofitable as a result of reduced efficiency of the diver. Consequently, divers only operate when sea conditions are good, averaging some four days per month during the season (November–July).

After fishing, the catch is transported by road to a factory, where it is processed, mostly by hand, and either frozen or canned. There are four processing factories, three of them in Hermanus, the centre of the abalone fishery, with one near Cape Town. The factories have been compelled since 1984 by government legislation to market 10% of their product in South Africa. The balance is exported.

From April 1954 commercial abalone licences were freely available on payment of the requisite one pound fee, and the number of licensed divers swelled to some 104 in 1964. However, a phase of stricter enforcement of the licence conditions applicable to the divers forced many participants to leave, and the numbers were reduced to 68 by 1970 and to 47 by 1989.

The divers initially delivered freely according to their individual fishing effort to the various factories. From 1968, when the first production quota was set, each factory was granted the processing rights to a fixed percentage of the quota. In 1982, one of the processing factories was found guilty of major infringements of the permit conditions and its packing quota was cut permanently by 11 240 kg. This was granted to the newly formed Abalone Divers Association. Each diver therefore had an equal share of an export concession, which considerably improved their income.

In 1984, the divers became legally obligated according to their past affiliations to deliver to specific factories, ensuring the factories reliable deliveries and increasing the control of abalone transported overland. Furthermore, on the basis of individual past performance, the divers were each granted fishing rights to a fixed percentage of the existing quota, which was then called an 'entitlement'. This is fully transferable, heritable and saleable. With this development the divers were suddenly placed in a position of greater financial security, having a fixed 'investment' in the industry. A condition of sale, however, is that the members of the Abalone Divers' Association have first rights to any quota offered. This has resulted in the fishery remaining closed because any 'entitlements' which are offered for sale are bought by individual members or groups of divers, further increasing those members' share of the overall annual TAC. This has the effect of increasing the individual's total investment in the fishery, but at the same time has deprived the fishery of any 'new blood'. With the average age of the divers presently at about 45 years, however, the situation will soon change.

HISTORY OF THE FISHERY: PRODUCTION AND QUOTAS

Records of the quantity of flesh marketed after processing are only available from 1953 onwards. These are the best available indicators of historic levels of fishing and have been converted to whole mass to obtain an estimate of the annual abalone harvest. A 24% conversion rate, which is the approximate percentage meat weight (i.e. 'production') routinely recovered after processing (±23% for canning and 25% for freezing) was used up to 1975, after which actual catches are reported (Fig. 31.2).

In 1953, 770 tonnes of abalone were harvested, all of which were canned and exported to east Asia. The annual catch rose to 1500 tonnes in 1954, but then fluctuated at around 500 tonnes until 1959 because of decreased demand resulting from competition for the Oriental abalone market by other countries. Thereafter, the harvest increased considerably and peaked at 2800 tonnes in 1965. After this, despite a reasonable market, the catch declined annually as the original dense beds of abalone (accumulated stocks) became depleted.

Concern over the declining catch in the late 1960s resulted in the imposition in 1968 of a maximum production quota of 386 tonnes, which was allocated to six factories then in production on the basis of their past performance in the industry. However, this quota was not filled, and was reduced to 341 tonnes the following year, but again was not filled. Only in 1970 was the production quota of 227 tonnes filled, thereby limiting catches for the first time. It was, however,

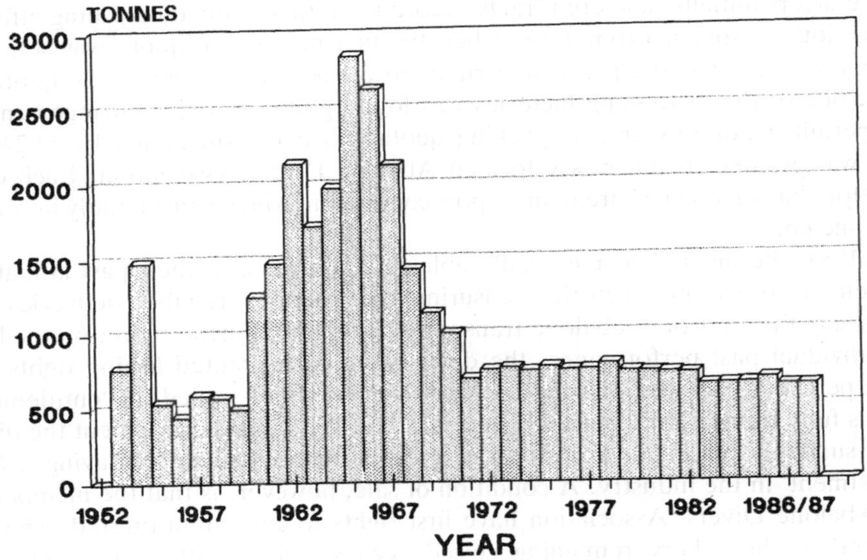

Fig. 31.2 Abalone landings over the period 1952 to 1988.

immediately further reduced to 181 tonnes in 1971, due to continued concern over the state of the resource.

From 1979 to 1982, the production quota was reduced by 10% to 163 tonnes because of concern that falling catch rates reflected declining stocks. In 1983, after certain irregularities in the implementation of the quota system had been discovered, the control system was changed from a production quota to a whole mass quota and monitored on the basis of whole mass delivered to the factories. This came about when an investigation (P. Goosen & A. Slater, unpublished report) showed that whole abalone were being cut up and marketed in pieces which, under the existing system, were not being declared against the quota. Using conversion factors arrived at during the investigation, the new quota was set at 660 tonnes whole mass, some 4% less than the whole mass delivery totals of the preceding years. The expectation that the new quota would enable production to be maintained was indicative of a substantial degree of wasteful processing in earlier years. The whole mass quota remained at 660 tonnes for a further two years.

In 1984, in response to complaints about the non-availability of abalone for local consumers, it was made compulsory to market 10% of production within the Republic of South Africa. This measure was also intended to reduce the level of black-market trade within the country. Up to 1985, the annual quota had been effective from January for each year, with the result that the quota was usually filled by November. With the advent of 10% local marketing it was suggested that the availability of fresh abalone over the holiday period was low due to the quota being filled by then. Accordingly, the season was opened two months earlier, in November 1985, and a three-month closed season was proclaimed from August to October. Since commercial quotas remained unchanged, the closed season only limited effort by recreational divers. There is no evidence that *H. midae* behaviour

Plate 31.1 Infra-red aerial photograph of kelp forests at extreme low tide.

changes during spawning periods, so a closed season was previously not considered necessary on biological grounds.

In 1986 the quota was allocated as the sum of total allowable catches (TACs) in each of seven specific fishing zones in order to balance effort between the various fishing grounds. The more accessible fishing grounds were showing signs of excessive fishing pressure, whereas remote grounds showed higher catch rates and mean sizes. (The most heavily fished ground in the Hermanus area had also been subjected to a voluntary annual limit of 90 tonnes since 1983.) The TAC estimates for the 1986 season were based mainly on trends in catch rates and commercial sampling data. Area closures resulted in an effective 3% drop in the TAC to 640 tonnes in 1986. However, almost 50 tonnes of abalone were condemned as unmarketable due to discoloration, and the divers made special application to recover this loss by additional fishing. This was granted, so landings for 1986 are almost 50 tonnes above the TAC for that year (Fig. 31.2).

In 1986–7, a project to map the area and depth contours of the kelp forests was completed (Tarr, in prep.). This provided estimates of abalone habitat areas and TACs were then determined using these data and size and density data from diving surveys as a guideline. The sum of the TACs remained at 640 tonnes.

RECENT RESEARCH FOR MANAGEMENT

Tagging of *H. midae* has been carried out in five of the seven fishing zones in order to determine whether growth rates differ between them (Tarr, in prep.). More than 6000 abalone have been tagged, with a recovery rate of about 18%.

Stock-assessment surveys have been carried out by research divers in all the major fishing zones. Numerous 600 m^2 belt transects were swum at 5 m depth intervals and data collected on density and size composition of the abalone populations (Tarr, in prep.).

More than 90% of the abalone population are found within kelp (*E. maxima*) stands visible from the sea surface. To obtain data on the habitat area of fishing zones, kelp forests were photographed from the air using infra-red film (Plate 31.1) and maps were prepared at a scale of 1:10 000 (Tarr, in prep.). These data are presently used in TAC estimates for the four main fishing zones, which together yield over 80% of the total annual TAC.

Because *H. midae* distribution is related to depth, depth contouring of the main fishing zones was performed using an inflatable boat and echo sounder. Position fixing for mapping purposes was carried out by shore stations using theodolites (Tarr, in prep.).

STATISTICS

From the inception of the industry in 1949 until 1972, divers were not required to submit catch returns. From 1960 to 1970, rough catch per unit effort (CPUE) analyses were derived from factory delivery and production records. The only measure of effort then obtainable was 'licensed diver months' (Newman, 1973). After this period, divers submitted monthly catch returns which provided data on

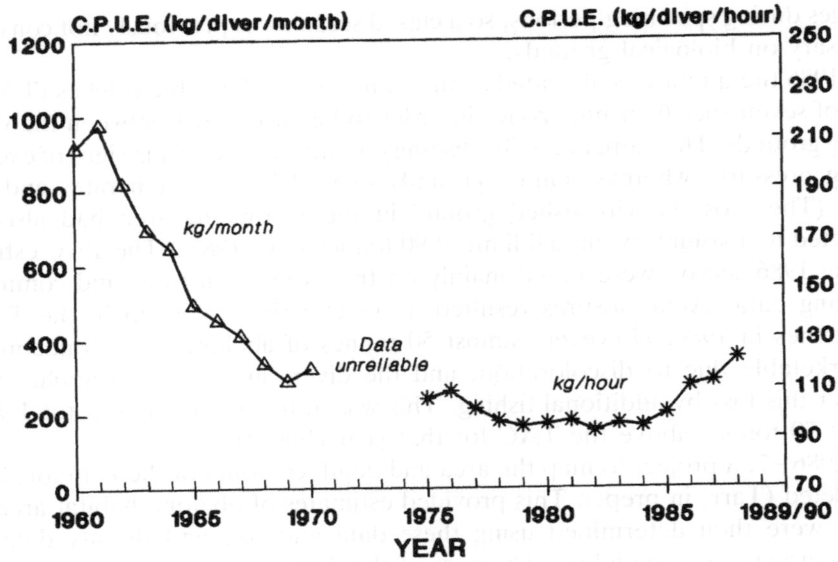

Fig. 31.3 Catch per unit effort (CPUE) trends for all areas combined using two different indices.

numbers and mass landed, diving place, total hours dived and the factory to which the catch was delivered. These data were used from 1975 for CPUE analyses, with CPUE being measured as kg whole mass per diver hour. The CPUE trends are shown in Fig. 31.3, using two different *Y*-axis scales. (Data from 1970 to 1975 in prep.). A clear declining trend is apparent up to 1970, which may have continued at a slower rate up to 1982. It appears, however, that an overall improvement in catch rates began in 1984. In the present fishery management, CPUE from the entire fishery is not used directly for assessment since combined figures may partly reflect changes in fishing patterns and not purely abalone abundance. Accordingly, CPUE per fishing zone is now used with other information in TAC estimations (Fig. 31.3).

From 1981, more detailed information was required regarding the locality dived in order to assess fishing effort more accurately on the various fishing grounds. Various checks were introduced to ensure greater accuracy of the catch and effort data.

When the whole mass quota system was changed for the 1985–6 season to a system of TACs for discrete fishing zones, the allocation of TACs was based mainly on fishery trends such as total catch per zone and CPUE assessed for each fishing zone. Thus, the statistics submitted by the divers contribute valuable data which are used directly in the management of the resource.

ECONOMICS

From 1953 to 1955, the total value of abalone exports was estimated to be between £70 000 and £85 000 per annum, which amounted to an export price of about £0.40 per kg (approximately R1.60/kg). A commercial abalone licence was then obtainable at £1, or the equivalent of about 2.5 kg exported product. The licence fee was raised to a still-nominal R40 in 1966 and to R60 in 1986–7. The current export price is kept secret among the various companies, but is conservatively estimated to be above R110/kg. Thus the cost of a licence for a diver has decreased from the equivalent of 2.5 kg export product to about 0.6 kg export product. The present value of the export industry is estimated to be R16 million, which is all marketed to Hong Kong.

In recent years the market for shells has improved, with most being exported to east Asia for use in various handicraft industries. Present export price estimates range from R500 to R800 per tonne. *Haliotis midae* does not, however, have a very good quality shell for craft use due to a high incidence of boring organisms.

Abalone divers in the past were considerably underpaid by the factories, for example in 1973 at R0.15 per kg whole mass. After the formation of the Abalone Divers Association in 1982 the negotiating power of the divers increased and their price improved. Divers are presently (1989) being paid R9/kg. This, combined with their recently acquired export concession, has meant that by virtue of their early entry into the fishery the present abalone divers have attained a position of considerable financial security for a relatively low capital investment.

CONCLUSIONS

With a highly accessible and relatively non-cryptic adult population and with no clear indication as to what minimum stock size is adequate to ensure continued recruitment, management strategies incorporate a number of safeguards. Present measures include the following:

(1) A relatively large size limit which ensures 3 to 4 years' protected spawning during which natural aggregations are not disturbed.
(2) Commercial fishing effort is monitored and regulated on an annual basis for each of the seven fishery zones. These TAC estimates are based on the habitat area of each zone as well as analysis of the normal fishery trends.
(3) Legislation ensures maximum availability of abalone products on the local market during peak holiday periods, thereby reducing black-market pressures.
(4) Recreational divers require a permit to catch the daily bag limit of five per person per day.
(5) A 3-month closed season limits the effort of the recreational diver.
(6) An abalone reserve (proclaimed in 1983) and a general marine reserve are situated within the most heavily fished coastline, and may provide an additional larval supply to adjacent fished grounds.

The policy of limited catches per area which was first implemented voluntarily by the divers for one area (near Hermanus) in 1983, and then formally for the whole fishery in 1985–6, appears to be working. The area near Hermanus which has been the most heavily fished area since the start of the fishery, due to its close proximity to the homes of most of the divers, serves as the best example. Figure 31.4 shows the CPUE trends for this zone since 1977, together with the annual percentage of the sampled commercial catch which occurred within the first 5 mm

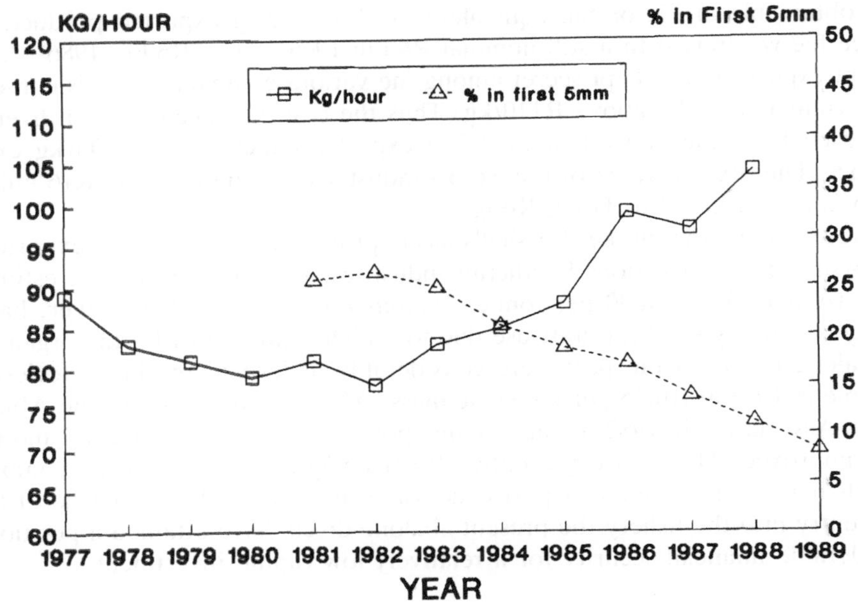

Fig. 31.4 Catch per unit effort (CPUE) and percentage recruit fishing in zone C.

size class above the size limit (i.e. the approximate percentage of the commercial catch reliant on growth recruitment). These indicate a trend of increasing catch per hour since 1982, which is mirrored by a steadily decreasing dependence of this fishing zone on growth recruitment.

The abalone fishery in the southwestern Cape appears to have progressed from a 'free-for-all' situation in the 1960s with few controls, when dense accumulated beds of large abalone were harvested at will, to a closely monitored and tightly managed fishery which is apparently showing signs of an improvement for the first time.

ACKNOWLEDGEMENTS

The author is grateful to Dr D.R. Schiel and Dr S.A. Shepherd for constructive comments on the manuscript.

REFERENCES

Barkai R. & Griffiths C.L. (1986) Diet of the South African abalone *Haliotis midae*. *S. Afr. J. Mar. Sci.*, **4**, 37–44.

Genade A.B., Hirst A.L. & Smit C.J. (1988) Observations of the spawning, development, and rearing of the South African abalone *Haliotis midae* Linn. *S. Afr. J. Mar. Sci.*, **6**, 3–12.

Newman G.G. (1964) Abalone (perlemoen) research and its application to an industry. *Commer. Ind.*, **23**(2), 76–8.

Newman G.G. (1966) Movements of the South African abalone *Haliotis midae*. *Invest. Rept Div. Sea Fish., Repub. S. Afr.*, **56**, 1–19.

Newman G.G. (1967) Reproduction of the South African abalone *Haliotis midae*. *Invest. Rept Div. Sea Fish., Repub. S. Afr.*, **64**, 1–24.

Newman G.G. (1968) Growth of the South African abalone. *Invest. Rept Div. Sea Fish., Repub. S. Afr.*, **67**, 1–24.

Newman G.G. (1969) Distribution of the abalone (*Haliotis midae*) and the effect of temperature on productivity. *Invest. Rept Div. Sea Fish., Repub. S. Afr.*, **74**, 1–7.

Newman G.G. (1973) *Assessment and management of some marine stocks in Southern Africa*. PhD thesis, University of Cape Town.

Tarr R.J.Q. (1988) Abalone. In *Oceans of Life off Southern Africa* (Ed. by A.I.I. Payne & R.J.M. Crawford), pp. 62–9. Vlaeberg, Cape Town.

Chapter 32
The Sultanate of Oman abalone fishery

D.W. JOHNSON, ALI AL-HARASSY and MOHM'D AL-HARTHY
Marine Science and Fisheries Centre, PO Box 467, Muscat, Oman

ABSTRACT: Harvest of the endemic abalone *Haliotis mariae* in the Sultanate of Oman is restricted to two areas of the North Arabian Sea where the kelp *Ecklonia radiata* community is maintained by cooler upwelling waters. About 1.75 million abalone are exported annually with a value of nearly 1.5 million Rial Omani (US$4 million). *Al sufailah*, the Omani abalone, has the highest economic yield per kilogramme of all Omani fisheries. They are almost exclusively dried and eventually reach the Chinese market. It has been an important source of income for several generations of Omani divers.

RESUMEN: Le pesca del abulón endémico (*Haliotis mariae*) es restringida en el Sultanato de Omán a dos áreas del norte del Mar de Arabia donde la comunidad del kelp (*Ecklonia radiata*) es mantenida por la surgencia de aguas frías. Cerca de 1.75 millones de abulones son exportados anualmente con un valor aproximado de 1.5 millones de Rial Omani (4 millones USD). De todas las pesquerías de Omán, el *Al sulfailah*, el abulón de Omán, tiene el mas grande rendimiento económico por kilogramo. Ellos son casi exclusivamente secados y eventualmente alcanzan el mercado de China. Por muchas generaciones éste ha sido un importante recurso de ingreso para los pescadores submarinos de Omán.

INTRODUCTION

Large abalone, such as the *Haliotis mariae* endemic to the Sultanate of Oman, are normally associated with cooler waters and their presence suggests that the current upwelling system has existed since Lower Miocene – for 15 million years (Taylor, 1985). Water temperatures of 25–27°C in May–June decrease to 19–21°C in August, before increasing again to 26–28° in November; a February decrease to 22–24°C is accompanied by a seasonal decrease of algal community production. Algal production is greatest from September through November, before water circulation and temperature decrease (Barratt *et al.*, 1986). A traditional harvest of abalone by unassisted divers during months when the sea permits (excluding the southwestern monsoon months of May to September) has existed at least since the 1970s. Previous to that, abalone were harvested only for their pearls (Anon., 1984).

There has been little effort allocated to the assessment or development of this resource. The Ministry of Agriculture and Fisheries sponsored a review of the status of the abalone fishery and a 3-day survey in 1981, which were followed by recommendations for increased development of the fishery (Anon., 1983). A United Nations FAO/UNDP preliminary stock assessment followed a visit during the last week of March in 1982 by Sanders (1982).

THE FISHERY

The abalone, *H. mariae*, is found associated with algal communities dominated by *Ecklonia radiata* and species of *Sargassopsis* and *Sargassum* from shallow inshore

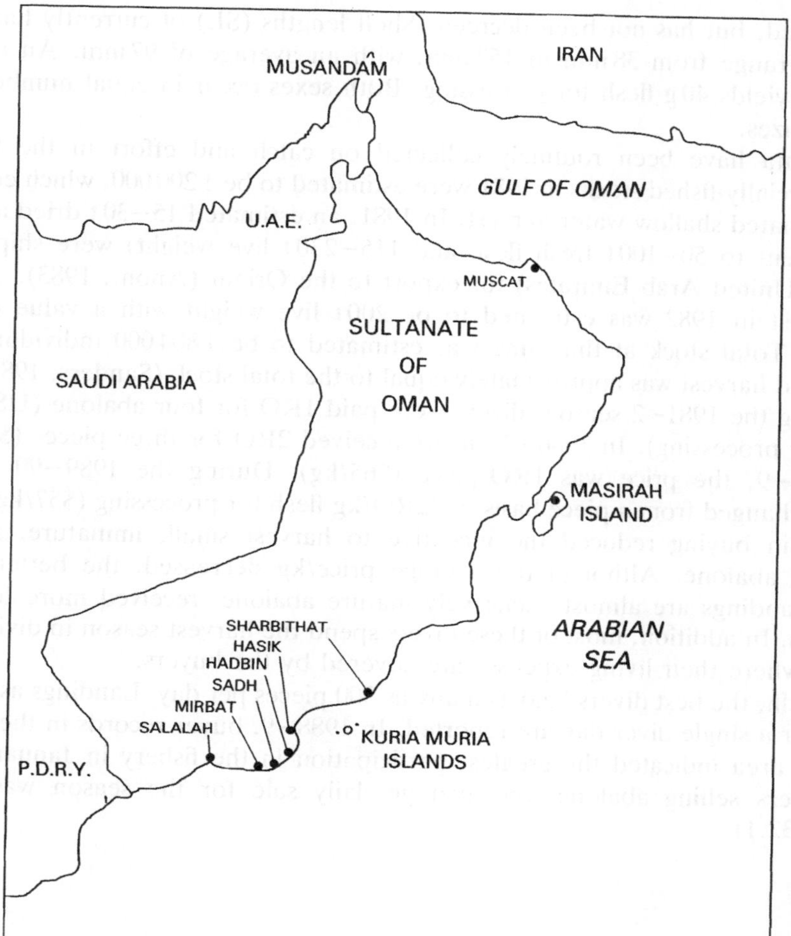

Fig. 32.1 Abalone landing localities in the Sultanate of Oman.

waters to 20 m depth. Adults are found in cracks and crevices of rocks, with the largest numbers in depths of 6–10 m. Juveniles (<25 mm) are found nearer to shore on the undersides of rocks (Anon., 1983).

Traditional divers harvest in waters less than 10 m depth using only a knife or pry tool; community consensus limits divers to the use of face masks. The foot is separated, boiled and sun dried at harvest areas prior to being marketed by buyers.

The centre of the abalone fishery is between Sadh and Hadbin, with an active fishery between Mirbat and Hasik, as well as an additional camp of divers at Sharbithat. Other populations have been fished in the past off the Kuria Muria Islands and Masirah Island (Fig. 32.1).

The only regulation on the fishery presently in force is an annual closed season from 1 April to 1 October, introduced in 1987. A minimum harvest size has been considered, but has not been decreed. Shell lengths (SL) of currently harvested abalone range from 38 mm to 152 mm, with an average of 97 mm. An average abalone yields 40 g flesh for processing. Both sexes occur in equal numbers and similar sizes.

No data have been routinely collected on catch and effort in the fishery. Commercially-fished stocks in 1981 were estimated to be 1 200 000, which equalled the estimated shallow water harvest. In 1981, an estimated 15–30 t dried abalone (equivalent to 50–100 t fresh flesh and 115–230 t live weight) were shipped to Dubai (United Arab Emirates) for export to the Orient (Anon., 1983).

Harvest in 1982 was estimated to be 200 t live weight with a value of $1.3 million. Total stock at that time was estimated to be 1 804 000 individuals; the estimated harvest was approximately equal to the total stock (Sanders, 1982).

During the 1981–2 season, divers were paid 1RO for four abalone (US$18/kg flesh for processing). In 1986–7, divers received 2RO for three pieces ($43/kg). In 1988–9, the price was 1RO/piece ($65/kg). During the 1989–90 season, buyers changed from a piece basis to 22RO/kg flesh for processing ($57/kg). That change in buying reduced the incentive to harvest small, immature, rapidly-growing abalone. Although the average price/kg decreased, the better divers whose landings are almost exclusively mature abalone, received more for their landings. In addition, most of these divers spend the harvest season in diving area camps where their living expenses are covered by the buyers.

In Sadh, the best divers land as many as 300 pieces per day. Landings as high as 19 kg for a single diver day are reported. In 1988–9, buyer records in the Sadh–Hadbin area indicated the greatest participation in the fishery in January, with 329 divers selling abalone; the average daily sale for the season was 2.2 kg (Table 32.1).

Table 32.1 Estimated 1988–9 abalone harvest in the Sultanate of Oman.

Date	No. of divers	Landings* average (kg)	Days dived	Earnings† average (RO)		Estimated landings (t) flesh	live
Estimated landings from the Sadh–Hadbin area (from Bouhlel, 1989)							
1988							
OCT	236	2.8	15	924		9.9	23.4
NOV	246	2.3	20	1012		11.3	26.6
DEC	281	2.0	16	704		9.0	21.2
1989							
JAN	329	1.1	20	484		7.2	17.0
FEB	266	2.9	14	893		10.1	23.8
MAR	109	1.6	17	598		3.0	7.1
Average/month	245	2.1	17	769	TOTAL	50.5	119.1
Estimate of total commercial season landings: Mirbat, Sadh, Hadbin, Hasik, and Sharbithat							
	350	200	100	4400		70	165

* Fresh meat weight.
† Based on meat weight value of 22RO/kg (1RO = US$2.60).

Table 32.2 Estimated shell lengths (in mm) for given ages based on data of Sanders (1982) compared with recent observations.

Ages (years)	0.5	1.0	1.5	2.0	2.5	3.0	4.0	5.0	6.0	Max. size obs.
Sanders (1982)	42	66	83	94	102	107	113	116	118	142
Unpublished data		50		75		100	125	138	150	152

FISHERY BIOLOGY

For the last decade there has been increasing concern about overharvest and protection of the reproductive potential of the fishery. The absence of a comprehensive research programme (distribution, feeding, growth rates, reproductive maturity, spawning activity and population dynamics) and development plan has been noted since 1981. In 1981, most abalone appeared to be harvested before reproductive maturity (at <76 mm SL) and a 90 mm SL minimum harvest size was recommended (Anon., 1983). First spawning in the third year (2+ cohort) was suggested by Sanders (1982).

Sanders (1982) assumed 85 mm SL to be the mean size at entry into the fishery. The 1982 survey obtained mean SLs of 96 mm for exploited stocks (<8 m depth) and 105 mm for unexploited (>8 m depth). Length frequencies examined (25 samples; 6330 abalone) during the 1987–8 and 1988–9 seasons suggest entry into the fishery at about 65 mm SL, with a mean SL of 97 mm, the same as the two samples measured by Sanders in 1982.

Growth rates and cohorts were assigned to size classes by Sanders (1982) after evaluation of the shell check marks on 42 shells; he assumed that annual growth checks could be identified and from them a growth rate derived. Recent observations (unpublished data) suggest that the growth rates and other population parameters derived by Sanders are incorrect. Estimated shell lengths for given ages based on Sanders (1982) and recent observations are given in Table 32.2. During two seasons of landings sampling (1987–9), 12% of the catch exceeded 115 mm SL and observed fishing pressure began during the first year of growth. It seems unlikely that many abalone remain in the fishery after their fifth year.

Change in the size composition of the catch appears to reflect increasing fishing pressure. The percentage of landed abalone less than 70 mm SL increased from 2.1% in 1984 to 3.2% in 1985 to 7.2% in 1987–8. The mean size and the proportion of small abalone in the catch also vary over time (Fig. 32.2(A)) and between localities (Fig. 32.2(B)). The localities in Fig. 32.2(B) are ranked in order of decreasing fishing intensity from Ras Janjairi (near Mirbat) to Sharbithat. The mean size at these localities increases from 82 mm to 105 mm SL, whereas the proportion less than 80 mm SL decreases from 47% to 4%.

CONCLUSIONS

Little is known of the biology of the Omani abalone *H. mariae*; consultants and advisors have consistently recommended research programmes in order to provide

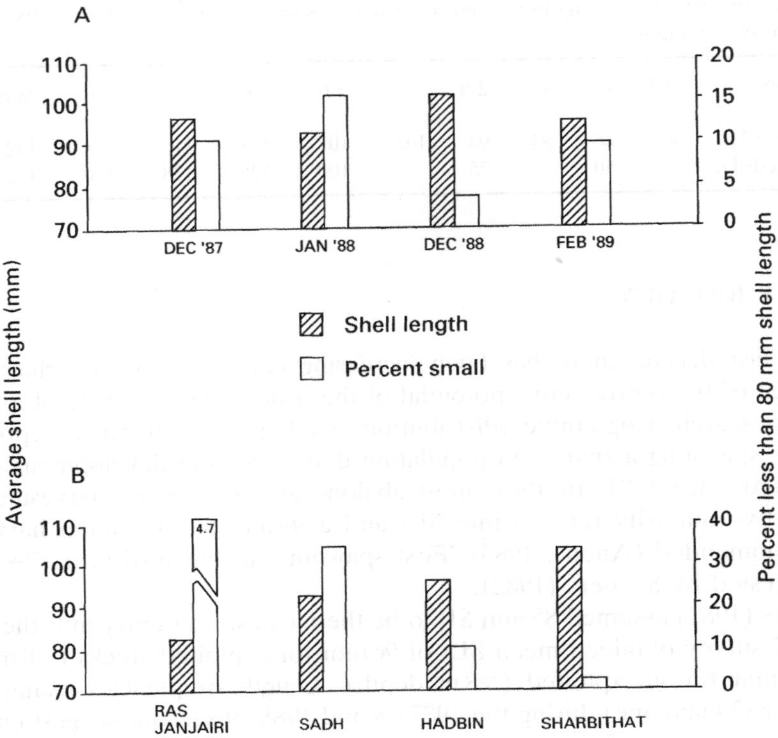

Fig. 32.2 Changes in average shell length (SL) and percentage less than 80 mm SL of landed abalone (*H. mariae*) at (A) Hadbin from 1987 to 1989; (B) four localities ranked in order of decreasing fishing pressure in February 1989.

a biological basis for wise management, full development and conservation of that fishery. A minimum size limit is essential to improve the biomass yield and maintain the breeding stocks. The switch by buyers from a piece to a weight basis and a minimum size regulation could be used by Ministry fisheries extension workers to increase production by discouraging the harvest of immature abalone.

The presence of unexploited populations below the present depth limit of divers has been considered a possible buffer against overfishing of the parental stock, but the presence or extent of such stocks has not been demonstrated. The use of new harvest methods (such as more advanced diving gear) should continue to be discouraged to prevent exploitation of whatever deeper stocks might exist.

Hatchery production for commercial culture and for enhancement of over-exploited stocks is the most promising means of increasing abalone production in the long term in the Sultanate of Oman.

ACKNOWLEDGEMENT

The authors thank Dr S.A. Shepherd for critical comments on the manuscript.

REFERENCES

Anon. (1983) *Report of abalone and lobster study in the Dhofar region*. Prepared in accord with the agreement between Waleed Associates and the Ministry of Agriculture and Fisheries. Development Application Corporation, Los Altos, California.

Anon. (1984) *Abalone research plan. RDA-FDP-WP2*, June 1984. Directorate of Fisheries, Ministry of Agriculture and Fisheries.

Barratt L., Ormond R.F.G & Wrathall T.J. (1986) *Ecological studies of southern Oman kelp communities. Part 1*. Tropical Marine Research Unit, Biology Department, University of York.

Bouhlel M. (1989) *Abalone stock in the Sultanate of Oman – its biological characteristics and exploitation*. December 1989. Directorate of Fisheries, Ministry of Agriculture and Fisheries.

Sanders M.J. (1982) Preliminary stock assessment for the abalone taken off the southeast coast of Oman – March 24 to April 2 1982. *FAO/UNDP FI: DP/RAB/80/015/3*. Food and Agriculture Organization, Rome.

Taylor J.D. (1985) *Marine Mollusca of Southern Oman*. British Museum (Natural History).

Chapter 33
The ormer (*Haliotis tuberculata*) fishery of France and the Channel Islands

JACQUES CLAVIER *ORSTOM, BP A5, Nouméa, New Caledonia*

ABSTRACT: *Haliotis tuberculata* is a medium-sized, white-fleshed abalone. It makes a much appreciated dish with a relatively high value for local market or export. *Haliotis tuberculata* exploitation is of minor economic importance in Europe. The fishing grounds are restricted to the intertidal zone where the populations are overexploited. The bulk of the stock is therefore distributed below low tide level and a potential small fishery exists. At present, biological knowledge is sufficient to define management options for the fishery and exploitation is likely to increase.

RESUMEN: De tamaño medio, *Haliotis tuberculata* tiene un músculo pedal de color blanco muy apreciado comercialmente. La pesquería de *H. tuberculata* presenta una importancia debil en Europa. Actualmente, este pesquería se localiza en la zona de entremareas donde se observan signos de sobrepesca. Sin embargo, las mayores concentraciones se encuentran bajo el nivel de la baja marea y potenciales zonas de pesca existen. Ahora, los conocimientos biológicos son suficientes para determinar las medidas de administración mas seguras dentro de la cuales puede manejarse la pesquería y el incremento en la explotación.

INTRODUCTION

Only one species of *Haliotis*, the ormer (*H. tuberculata* L.) is taken commercially in Europe. One other species (*H. lamellosa*) inhabits the Mediterranean sea, but is too small to be harvested. *Haliotis tuberculata* lives on the Atlantic coast from the south of the English Channel down to the northwest of Africa (Fig. 33.1). The species reaches its highest densities at the northernmost limit of its distribution.

BIOLOGY

The first biological studies on the ormer date from the end of the nineteenth century (Wegman, 1884; Boutan, 1898, 1899). Since then, an abundant literature has been devoted to this species. The anatomy and reproduction are described by Crofts (1929), with later more detailed reproductive studies performed in France (Girard, 1972; Koike, 1978; Cochard, 1980) and in the Channel Islands (Hayashi, 1980a). The growth of *H. tuberculata* is described by Forster (1967), Hayashi (1980b) and Cochard (1982), and juvenile growth by Koike (1978) and Clavier & Richard (1986a). Movement *in vitro* has been observed by Robert (1907), Crofts (1929), Lissman (1945) and Hayashi (1982), and in nature by Clavier & Richard (1984). Feeding of the ormer is described by Stephenson (1924) and feeding preferences by Culley & Peck (1981). Information on predation is provided by Stephenson (1924), Crofts (1929) and Forster (1962). The factors influencing spatial distribution are described by Clavier & Chardy (1989).

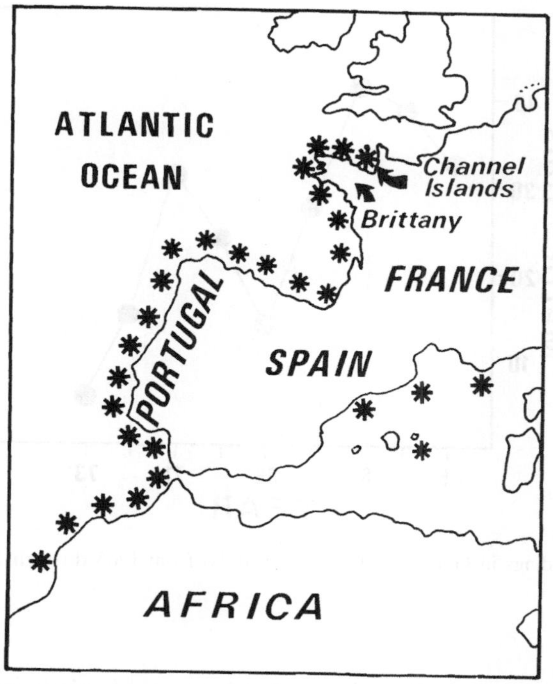

Fig. 33.1 Distribution of *Haliotis tuberculata*, after Gaillard (1958).

Abundance and biomass of ormer populations have been estimated in the vicinity of St Malo, in France (Clavier & Richard, 1986b), but no mathematical model for managing the fisheries has yet been published.

THE FISHERY

Reviews of the ormer exploitation in the Channel Islands and in France are given by Bossy & Culley (1976) and Berthou *et al.* (1985) and are summarized below.

The ormer is mostly gathered in the Channel Islands and in Britanny. The use of any diving gear is forbidden except in Guernsey, one of the Channel Islands, where the gathering of ormers using diving gear was permitted until 1973. Thus the fishing grounds are now restricted to the intertidal zone where the animal is levered off the rocks with a knife or metal bar.

Ormer gathering is a traditional activity for the coastal population. *Haliotis tuberculata* has always been regarded as a delicacy and a great number of fishermen search along the shore at each spring tide for it. For example, in March 1984, more than 2000 shore gatherers were counted in three days along a 5-mile 'ormer ground' in northern Brittany. This activity is mainly recreative and most of the ormers are taken for domestic use. The minimum legal size of *H. tuberculata* is 80 mm.

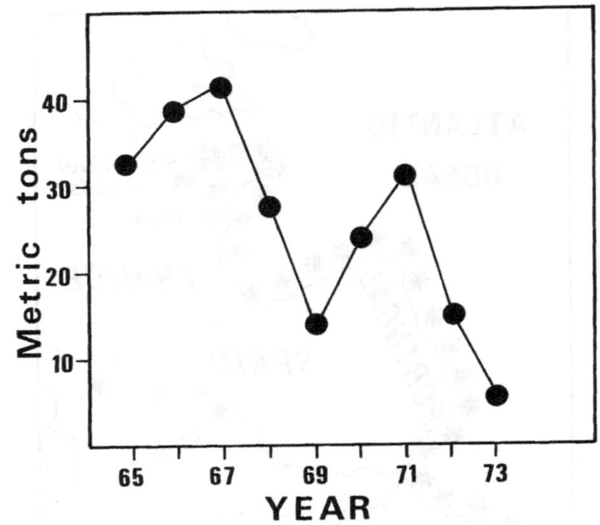

Fig. 33.2 Ormer landings in Guernsey (Channel Islands) from 1965 through 1973, after Bossy & Culley (1976).

Minor exploitation of *Haliotis* by scuba divers took place in Guernsey Island until 1973. An official ban on ormer fishing has been imposed since this date, because of the severe drop in the yields (Fig. 33.2).

Official statistics give a total annual catch of about 2 tonnes in France. This value is underestimated and the actual catch by shore gathering may be estimated to be about 20 tonnes. The illegal catch by divers must be added to this estimate, so that 40–50 tonnes of ormers are probably taken from the French coasts. About half of this production is marketed with a mean price to fishermen of 60–70 francs (about US$9–10) per kg.

ACKNOWLEDGEMENTS

The author wishes to thank Dr S.A. Shepherd for reading and commenting on a draft of this chapter.

REFERENCES

Berthou P., Beurier J.P., Buestel D. & Clavier J. (1985) Etat actuel des connaissances sur les stocks naturels d'ormeau: éléments halieutiques et juridiques. *La Pêche Marit.*, **1286**, 288–94.

Bossy S. & Culley M. (1976) Channel Islands 'ormer' fishery faces an uncertain future. *Fish. News Intern.*, Sept, 55–60.

Boutan L. (1898) Production artificielle des perles chez les *Haliotis*. *CR Acad. Sci. Paris*, **127**, 828–30.

Boutan L. (1899) La cause principale de l'asymétrie des mollusques gastéropodes. *Arch. Zool. Exp. Gén.*, **3**, 204–342.

Clavier J. & Chardy P. (1989) Investigation into the ecology of the ormer (*Haliotis tuberculata*, L.). Factors influencing spatial distribution. *Aquat. Living Resour.*, **2**, 191–7.

Clavier J. & Richard O. (1984) Etude expérimentale du déplacement de l'ormeau (*Haliotis tuberculata*) dans le milieu naturel. *Rev. Trav. Inst. Pêches Marit.*, **46**, 315–26.

Clavier J. & Richard O. (1986a) Growth of juvenile *Haliotis tuberculata* (Mollusca: Gastropoda) in their natural environment. *J. Mar. Biol. Assoc. UK*, **66**, 497–503.

Clavier J. & Richard O. (1986b) Estimation du stock naturel d'ormeaux (*Haliotis tuberculata*) dans la région de Saint-Malo. *Rev. Trav. Inst. Pêches Marit.*, **48**, 95–105.

Cochard J.C. (1980) *Recherches sur les facteurs déterminant la sexualité et la reproduction chez* Haliotis tuberculata *L*. Thèse 3e cycle, Univ. Bretagne occid.

Cochard J.C. (1982) La croissance de l'ormeau *Haliotis tuberculata* L. en rade de Brest. *Haliotis*, **12**, 61–9.

Crofts D.R. (1929) *Haliotis. LMBC Memoirs Typ. British Mar. Plants Anim.*, **29**, 1–174.

Culley M.B. & Peck L.S. (1981) The feeding preferences of the ormer, *Haliotis tuberculata* (L.). *Kieler Meeresforch.*, **5**, 570–2.

Forster G.R. (1962) Observations on the ormer population of Guernsey. *J. Mar. Biol. Ass. UK*, **42**, 493–8.

Forster G.R. (1967) The growth of *Haliotis tuberculata*: results of tagging experiments in Guernsey 1963–65. *J. Mar. Biol. Assoc. UK*, **47**, 287–300.

Gaillard J.M. (1958) *Haliotis tuberculata* Linne. Systématique et distribution. *Bull. Lab. Mar. Dinard*, **44**, 7–11.

Girard A. (1972) La reproduction de l'ormeau *Haliotis tuberculata* L. *Rev. Trav. Inst. Pêches Marit.*, **36**, 163–84.

Hayashi I. (1980a) The reproductive biology of the ormer, *Haliotis tuberculata*. *J. Mar. Biol. Assoc. UK*, **60**, 415–31.

Hayashi I. (1980b) Structure and growth of a shore population of the ormer *Haliotis tuberculata*. *J. Mar. Biol. Assoc. UK*, **60**, 431–9.

Hayashi I. (1982) Small scale laboratory culture of the ormer, *Haliotis tuberculata*. *J. Mar. Biol. Assoc. UK*, **62**, 835–44.

Koike Y. (1978) Biological studies on the propagation of the ormer, *Haliotis tuberculata* Linnaeus. 1 – Larval development and growth of juveniles. *La Mer*, **16**, 124–36.

Lissman H.W. (1945) The mechanism of locomotion in gastropod mollusks. I – Kinematics. *J. Exp. Biol.*, **21**, 58–69.

Robert A. (1907) Remarques sur la progression des Rhipidoglosses. *Bull. Soc. Zool. France*, **32**, 55–62.

Stephenson T.A. (1924) Note on *Haliotis tuberculata*. *J. Mar. Biol. Assoc. UK*, **13**, 480–95.

Wegman H. (1884) Contributions à l'histoire naturelle des Haliotides. *Arch. Zool. Exp. Gén.*, **2**, 289–378.

PART VIII
STOCK ENHANCEMENT: INTRODUCTIONS

Chapter 34
Brood-stock transplants as an approach to abalone stock enhancement

MIA J. TEGNER *Scripps Institution of Oceanography, La Jolla, California 92093–0201, USA*

ABSTRACT: When local abalone populations are eliminated by disturbance or over-exploitation, natural recovery will depend on recolonization by dispersing larvae. Recolonization by larvae depends upon the distance from brood stock, current patterns, and the length of the planktonic phase. Green abalone, *Haliotis fulgens*, populations on the Palos Verdes Peninsula in Southern California showed little evidence of recovery despite adequate food and a fishing closure. The short free-swimming period of the larvae and a drift-tube experiment suggested that dispersal between isolated populations is rare and stressed the importance of local brood stock. To provide local sources of larvae, 4453 sexually-mature green abalone were transplanted between November 1981 and September 1982 to two locations at Palos Verdes, where the drift-tube data indicated that the probability of larval retention on the peninsula was high. Abalone were tagged, measured and graded visually for gonadal condition before planting. Predator control efforts were associated with reductions in the short-term mortality associated with planting; total mortality observed by September 1983 was less than 10%. The abalone grew and gonadal changes suggested reproduction. Surveys at Palos Verdes in 1985 found a large number of animals within size categories expected from the brood-stock transplants; there was no comparable pulse of recruitment at sites distant from Palos Verdes. As in recent results from Australia, the recruitment apparently resulting from these transplants is local, but the scale observed ranges from hundreds of metres to a few kilometres. The results indicate that fishing closures will be an effective means of rehabilitating stocks only when there is an adequate source of larvae. Brood-stock transplants are expensive, but until the survival rates of cultured larval or seed releases improve, this may be the only approach for the recovery of badly depleted and isolated stocks.

RESUMEN: Cuando las poblaciones locales de abulón son elinimadas por disturbios o por sobre-explotación, la recuperación natural dependerá de la recolonización por larvas provenientes de otros lados. Esta recolonización depende de la distancia a donde se originaron las larvas, la duración de la fase planctónica y los patrones de corrientes. Las poblaciones de abulón verde *Haliotis fulgens*, en la Península de Palos Verdes, al sur de California, mostraron pocas evidencias de recuperación, a pesar de tener alimento adecuado y haberse vedado la pesquería. El período natatorio tan corto de la larva y un experimento con tubos a la deriva, sugieren que la dispersión entre poblaciones aisladas es rara, y resalta la importancia que las poblaciones locales tienen como stock reproductor. Para proporcionar fuentes locales de larvas se transplantaron 4453 abulones maduros entre Noviembre de 1981 y Septiembre de 1982 a dos localidades en la Península de Palos Verdes, donde los datos de deriva indicaron que la probabilidad de retención de larvas era alta. Antes del transplante éstos se marcaron, midieron, y se les valoró visualmente su condición gonadal. El control de depredadores estuvo asociado con una disminución a corto plazo de la mortalidad asociada al transplante; la mortalidad observada hasta Septiembre de 1983 fué menor al 10%. Los abulones crecieron y los cambios gonádicos sugieren que hubo reproducción. Evaluaciones realizadas en Palos Verdes en 1985 mostraron que un gran número de animales encontrados estuvieron dentro de los intervalos de tallas esperados del stock de reproductores provenientes del transplante; además, no hubo pulsos de reclutamiento comparables en sitios lejanos a Palos Verdes. Al igual que en resultados recientes de Australia, el reclutamiento aparentemente producido por los transplantes es local; sin embargo, la escala aquí observada va de los cientos de metros a algunos kilómetros. Los resultados indican que el cierre de la pesquería seriá una medida efectiva en la rehabilitación de poblaciones sólo cuando exista una fuente adecuada de larvas. Los transplantes de stocks de reproductores son caros, pero hasta que las tasas de sobrevivencia de larvas o de semillas liberadas no mejoren, este podría ser el único medio para la recuperación de poblaciones muy mermadas y aisladas.

INTRODUCTION

The California Legislature closed the coastline from Palos Verdes Point to Dana Point (Figs 34.1 and 34.2), an area adjacent to densely-populated urban areas of Los Angeles and Orange Counties, to all abalone fishing in 1977 to allow recovery of badly depleted stocks. The Palos Verdes Peninsula, which once supported a productive fishery for several species of abalone (unpublished California Department of Fish and Game (CDFG) block landing records), suffered major environmental degradation due to sewage pollution; there was little or no giant kelp (*Macrocystis pyrifera*), the major source of abalone food, for almost two decades.

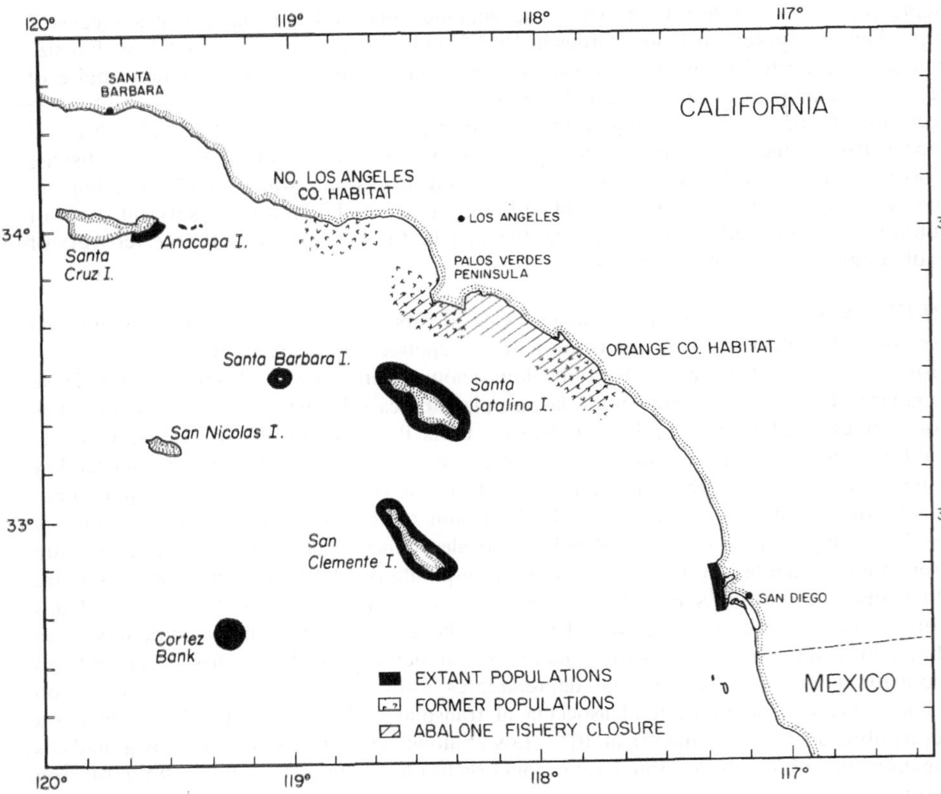

Fig. 34.1 Map illustrating green abalone (*Haliotis fulgens*) habitat in southern California. With the exception of Santa Cruz Island, green abalone populations remain relatively abundant at Channel Island locations but are depleted in former mainland habitats north of San Diego. The Palos Verdes Peninsula is isolated from other mainland kelp forest habitats by long stretches of soft bottom.

Pollution controls and kelp restoration efforts began to show results in 1974 and, by the early 1980s, the *Macrocystis* canopy was in the same size range as during the major years of the abalone fishery (Wilson *et al.*, 1977, unpublished data). Despite protection from fishing and abundant food, the green abalone (*Haliotis fulgens*) population of Palos Verdes showed no signs of recovery. The low number of recruits in apparently suitable habitat and the relatively short larval life of this species (Leighton *et al.*, 1981) suggested that recolonization of Palos Verdes by larvae originating elsewhere is a rare event. A drift-tube experiment (Tegner & Butler, 1985a) indicated that the probability of larval transport from offshore islands, where this species remains relatively abundant, to the depleted mainland within an appropriate time frame is low. However, a high proportion of the drift tubes deployed along the mainland were transported a few kilometres and recovered in suitable habitat within reasonable time frames. The results of the drift-tube experiment suggested that, in the virtual absence of local brood stock, the fishing closure would not promote recovery of mainland populations of green abalone in the near future. The objectives of this study were to test the hypothesis that local brood stock is required for the recovery of an isolated green abalone population and to consider the feasibility of adult transplants as a management tool.

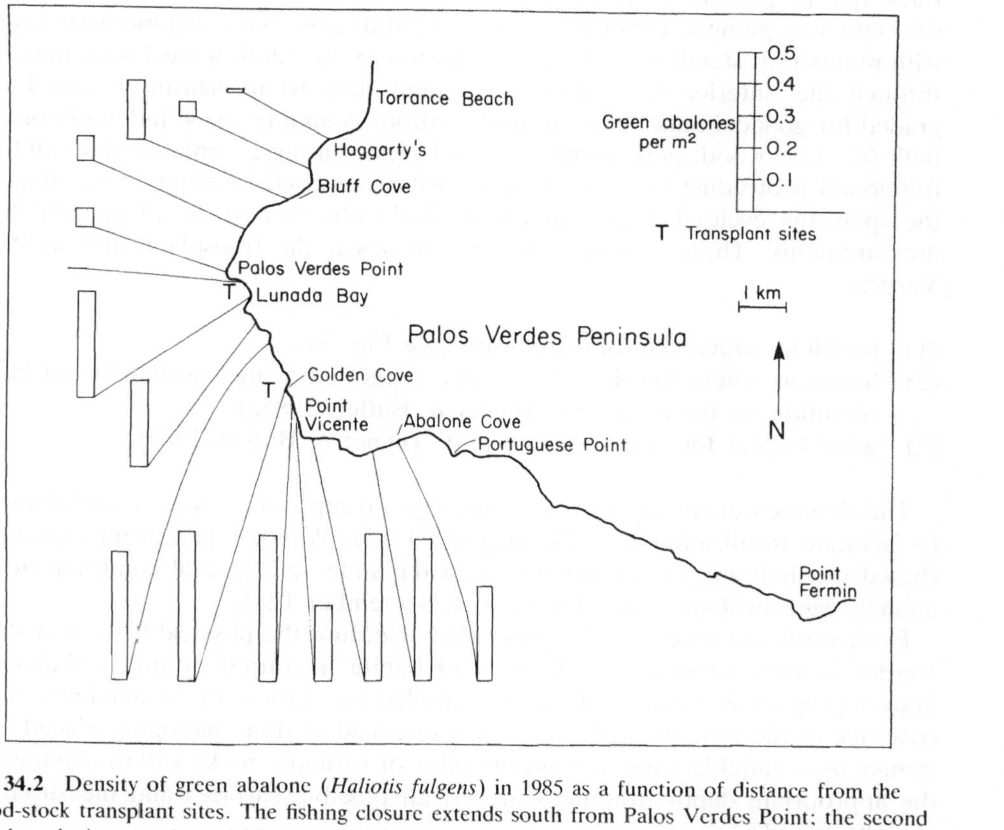

Fig. 34.2 Density of green abalone (*Haliotis fulgens*) in 1985 as a function of distance from the brood-stock transplant sites. The fishing closure extends south from Palos Verdes Point; the second brood-stock site was about 0.3 km east of this point. Torrance Beach and Portuguese Point represent the present boundaries of unsuitable habitat.

The Japanese began transplanting abalone as early as 1929 to improve growth rates (Saito, 1979), and moving slow-growing animals to better areas remains fairly common practice in that country (Mottet, 1978). Emmett & Jamieson (1989) concluded that transplantation of 50–100 mm *H. kamtschatkana* from exposed, food-poor habitat to sheltered, productive habitat was feasible to enhance growth, but that economic feasibility would be dependent on site-specific recovery rates and transplantation costs. Saito (1979) showed that spawning of transplants in a bay previously unoccupied by abalone led to susbstantial recruitment and development of fishable stocks. However, the rapid loss of a small brood-stock transplant at Santa Catalina Island (Henderson *et al.*, 1988) suggests that illegal take may be a critical problem in southern California.

MATERIALS AND METHODS

To provide local sources of larvae, the author, in conjunction with the CDFG, collected 4453 reproductively-mature (>100 mm, Tutschulte, 1976) green abalone from inside the shallow water (6 m) commercial fishing closure on the windward side of Santa Barbara Island between November 1981 and September 1982. The animals were transported in the flooded stern well, which had been lined with a thick net to prevent attachment directly to the deck, of the *RV Kelp Bass*; seawater was pumped continuously at a vigorous rate. The abalone were tagged with numbered, stainless-steel washers twisted on to stainless-steel wire threaded through the anterior respiratory pores, and they went measured, sexed and graded for gonadal bulk before transplantation. A simple visual index of gonadal bulk (G. Lockwood, pers. comm.), with 1 representing a depleted state and 6 a full gonad protruding from the mantle cavity, was used to estimate the stage of the spawning cycle. For consistency, a single observer made all gonadal bulk measurements. Three criteria were used to select the transplant sites at Palos Verdes:

(1) locations within the fishing closure (see Fig. 34.2),
(2) locations where the drift bottle data suggested a high probability of larval retention on the peninsula (Tegner & Butler, 1985a),
(3) good habitat for green abalone (e.g. Tegner & Butler, 1989).

The abalone were transplanted individually into appropriate habitat and clustered to facilitate fertilization (e.g. Pennington, 1985). Weather permitting, mortality (based on shell and tag collections), gonadal state and general condition of the animals were evaluated monthly through September 1983.

Evaluation was conducted in spring 1985. Because the physical habitat at Palos Verdes is quite variable and the type of habitat favoured by green abalone is known (Tegner & Butler, 1989), we sampled non-randomly to maximize areal coverage in the time available to us. Experienced abalone biologists placed 1 m^2 frames over suitable habitat, typically piles of turnable rocks within channels in the appropriate depth stratum, searched all possible surfaces and measured all abalone found.

RESULTS AND DISCUSSION

Survival and growth of the transplants

Short-term mortality associated with the stresses of transplantation was reduced considerably as we instituted and expanded predator and scavenger controls (Table 34.1). Live but stressed abalone attract attacks by scavengers, notably a whelk (*Kelletia kellettii*) and two asteroids of the genus *Pisaster*. No predators were collected during a small trial transplant to Golden Cove in November 1981, and the mortality rate by December 1982 was 42% based on the collection of tagged shells. We collected all predators in the immediate region of transplanted

Table 34.1 Short-term mortality experienced by adult green abalone (*Haliotis fulgens*) transplants as a function of increasing predator control; while the time periods for the mortality calculations differ, the results are comparable because most of the mortality took place within three months of transplantation (see text).

(a) Transplant record

Transplant date	Number transplanted	% Mortalities (through Dec. 1982)	Predator control?
Golden Cove			
Nov 1981	57	42	no
Feb 1982	679	21	yes
Apr 1982	1036	12	yes
Jun 1982	608	7	yes
Totals	2380	14	
Palos Verdes Point			
Aug 1982	1042	3	yes
Sep 1982	1031	4	yes
Totals	2073	4	

(b) Predator removal record; small numbers (4–10) of *Astrometis sertulifera*, *Loxorhynchus grandis*, *Panulirus interruptus* and *Scorpaenichthys marmoratus* were also removed.

Transplant date	*Octopus* spp.	*Kelletia kelleti*	*Pisaster* spp.
Golden Cove			
Nov 1981	No predators were removed		
Feb 1982	7	936	396
Apr 1982	16	3076	656
Jun 1982	7	1093	266
Totals	30	5105	1318
Palos Verdes Point			
Aug 1982	11	609	81
Sep 1982	4	605	44
Totals	15	1214	125

animals in a large-scale experiment in February 1982 and the mortality was reduced to 21%. A concerted effort was made to remove as many predators as possible in April, and mortality as of December 1982 was 12%. By the final Golden Cove transplant in June, reductions in the catch per unit effort indicated that our collections were clearly having an impact. Fewer predators were found and the mortality rate of these animals was 7%. The total mortality rate observed through December 1982 for the 2380 green abalone transplanted into Golden Cove was 14%.

Predator control was instituted from the beginning of the transplants to the Palos Verdes Point site. In August and September 1982 2073 animals were transplanted and the observed mortality through December 1982 was 4% (Table 34.1). Although this site did not have as large a scavenger population, intensive control combined with our greater experience in handling the abalone apparently resulted in many fewer deaths.

Most deaths probably occurred within the first few days after transplantation. Once abalone overcome the stress of handling and heal minor cuts or abrasions, they become immune to attacks by scavenging whelks and seastars. However, *Octopus* spp., batrays (*Myliobatis californica*), possibly spiny lobsters (*Panulirus interruptus*) and sheep crabs (*Loxorhynchus grandis*) prey on healthy adult abalone. We thus expected and observed a continuing low mortality rate of presumably adapted animals. The shell collections of the April transplant are typical. Fifty-seven shells (6% of the number planted and 45% of the mortalities as of December 1982) were found the week they were put down and are considered transplant mortalities. Sixteen shells which showed no growth were found in May, 14 in June, 1 in August, 3 in September, 4 in October, 3 in November and 4 in December. Some of these are doubtlessly transplant mortalities which were not found in April; others probably survived longer but did not grow. Starting in October, we began finding an equal number of shells which had increased in size since transplantation. A 152 mm animal grew 15 mm before its shell was found in December; a growth rate of almost 2 mm per month is outstanding for a large animal.

The extraordinary El Niño storms of 1982–3 (Tegner & Dayton, 1987) had dramatic effects in shallow green abalone habitat. *Macrocystis* was eliminated from both transplant sites; we assume that the animals had virtually no food from January until kelp recruited in late April. The large waves caused considerable sand movement at Golden Cove and rock rolling at Palos Verdes Point; both are potential sources of abalone mortality (Cox, 1962). Based on collections of shells or tags, 63 mortalities or 3.1% of the animals alive at the end of 1982 were found at Golden Cove and 22 mortalities or 1.1% of the remaining transplants were found at Palos Verdes Point. (Assuming that all abalone had adapted to their new habitat by December 1982, animals from the different transplant dates were subsequently lumped.) While some mortalities may have been buried or transported out of the area, extensive searches suggest that this number is low. The total mortality observed through September 1983, 9.1% of the 4453 animals transplanted, would appear to be an acceptable level for resource management considerations, especially considering the severity of the storms.

In September 1983, 356 animals were brought to the surface to be measured. Growth, which varied with abalone size, transplant date and location, ranged from 0 to 33 mm over the 12 to 23 months since transplantation. Forty per cent showed no growth or were actually slightly decreased in size due to shell abrasion. On a per-year basis, average growth of animals from the different transplant groups ranged from 1.8 to 2.9 mm/year. These low rates are not surprising given transplantation stress, low food availability during early 1983 and the size of the animals. The tagging data of Burge *et al.* (1975) indicate that green abalone of 156 mm, the average size of the transplants, are 8 to 9 years old and growing at a slow rate. In spring 1985, remeasurement of 206 animals yielded an average net growth of 7 mm at Golden Cove and 12 mm at Palos Verdes Point. Sea urchin grazing was having a major impact on kelp availability at Golden Cove in 1985.

Evidence for reproduction

While survival and growth are indications that the transplants have adapted to a new environment, an important criterion for the success of this experiment is evidence of reproduction. Leighton *et al.* (1981) report that the natural spawning period of *H. fulgens* extends from late spring to mid-summer and again from early to mid-autumn. The gonads of a haphazard sample of abalone from each transplant group were graded about monthly through September 1983. The initial and final gradings were done on board ship and the intervening assessments underwater. Some degree of spawning always took place on board ship before the gonads were graded prior to transplantation. The average gonad grade of the April group dropped one month after transplantation (Fig. 34.3). This may have been caused by additional spawning during planting (as was observed at Palos Verdes Point) or more likely because the animals mobilized energy reserves stored in the gonads to heal the inevitable abrasion or small cuts suffered during handling. By two months after planting, average gonad grades for each set of animals increased to well over their initial grade. The hypothesis that gonad grades dropped the first month after transplantation to deal with associated stress is supported by the February group's gonads developing in size before the April animals, which in turn developed before the June animals for an August spawning, whereas all three groups developed synchronously for a November spawning.

The evidence for a 1982 summer spawning of the Golden Cove animals was the increase in average gonad grade up to August followed by a sharp decline (Fig. 34.3). The average August grade for abalone transplanted in February and April was 3.5. We have observed spontaneous spawning when animals of grade 3.5 and above are handled, so we believe that a large proportion of the abalone spawned during the summer. The November 1982 peak was about 2.6, so fewer animals were ripe. Nevertheless, some reproduction probably did take place. Samples of both sexes from both transplant sites brought on board ship in November spawned spontaneously in flowing seawater aquaria.

While the winter of 1982–3 apparently did not have much effect on the survival of transplants, there was a striking effect on the gonad cycle. Gonad development was depressed and delayed in comparison with 1982 (Tegner & Dayton, 1987, Fig. 9). Uki & Kikuchi (1982) have shown in the laboratory that abalone given insufficient food show slower gonad growth and a lower rate of spawning success than satiated controls. As the transplanted animals were essentially without food

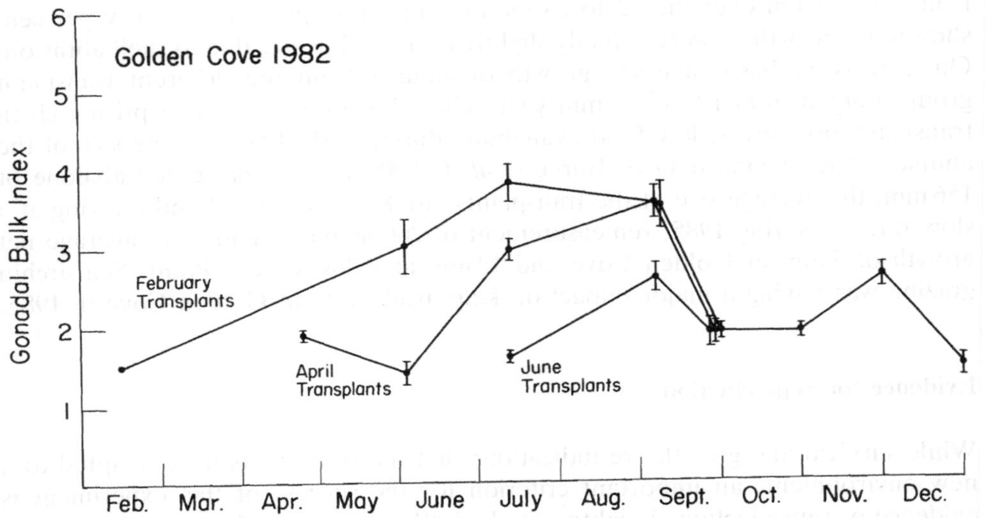

Fig. 34.3 Change in gonadal bulk index after transplantation to the Palos Verdes Peninsula in 1982. The scale varies from 1, a spawned-out condition, to 6; spontaneous spawning has been observed at ≥3.5. Variability is expressed as ± one standard error with the exception of February, which was estimated only. There was no difference between groups after September.

for three months, we assume that this was the explanation for the delayed onset and lower average gonad grade. Despite the low averages, some abalone at each transplant site were in spawning condition during summer 1983 (Fig. 34.4). Kelp

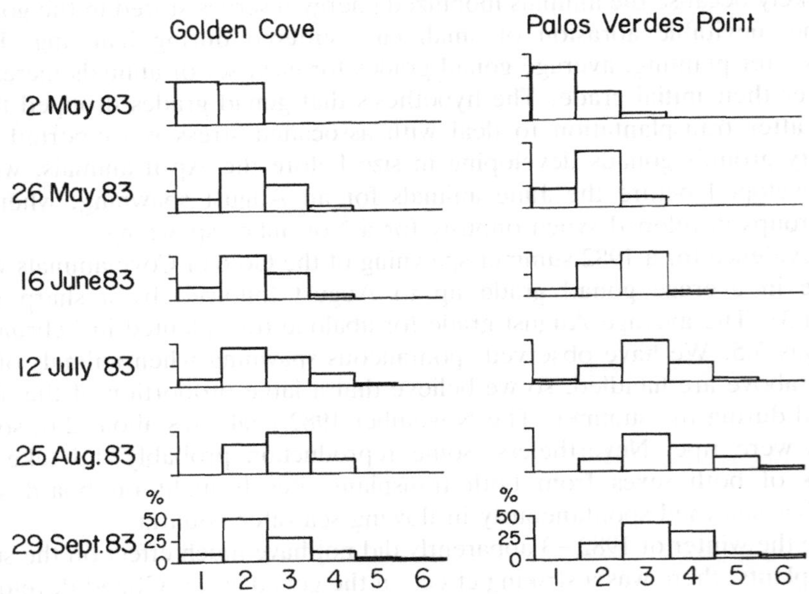

Fig. 34.4 Proportion of the sampled population in each gonad grade from May through September 1983 at the Golden Cove and Palos Verdes Point transplant sites.

canopies were luxuriant by midsummer and remained abundant through the rest of the study; there should have been much spawning from autumn 1983 onwards.

Evidence for recruitment resulting from the transplants

In spring 1985, Palos Verdes was surveyed for recruits to the green abalone population. On the basis of the drift-tube data (Tegner & Butler, 1985a), we believed that juveniles resulting from the brood-stock transplants were likely to be spread over several kilometres; sites examined are shown in Fig. 34.2. With insufficient resources to conduct proper density estimations in many locations, we chose to maximize the number of abalone sampled by selecting optimal habitat. The size–frequency distribution of the 348 green abalone found in 1985 is shown in Fig. 34.5. In the laboratory, green abalone may grow up to 39 mm in their first year (Leighton, 1974). Thus animals spawned in August 1982 could have been up to 100 mm, but most of the spawning undoubtedly took place after that. It is

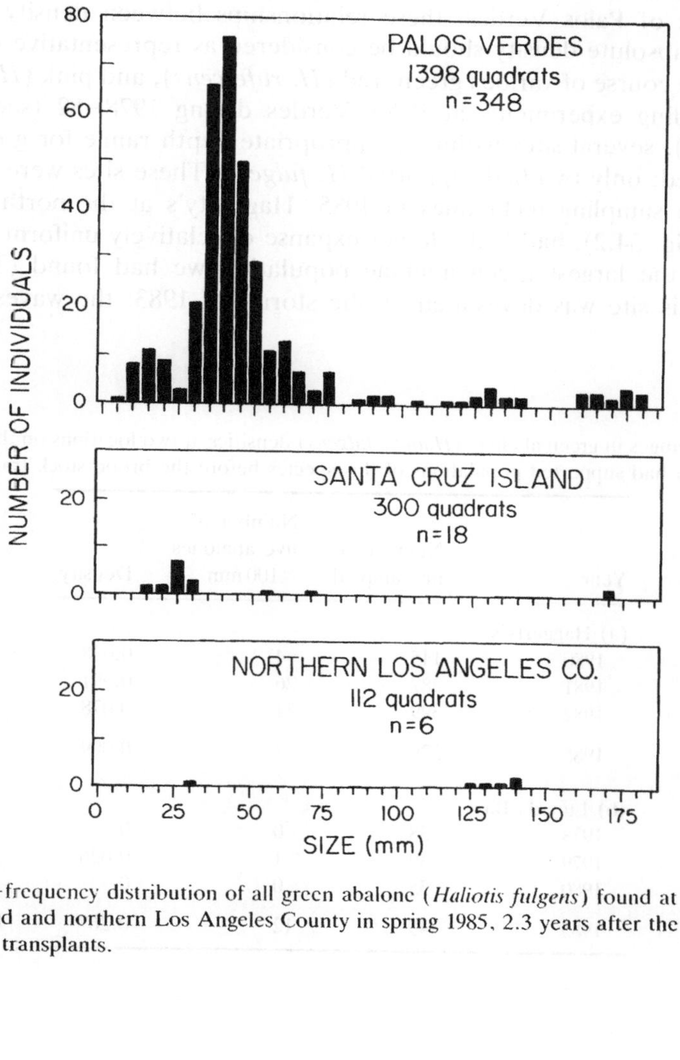

Fig. 34.5 Size–frequency distribution of all green abalone (*Haliotis fulgens*) found at Palos Verdes, Santa Cruz Island and northern Los Angeles County in spring 1985, 2.3 years after the completion of the brood-stock transplants.

apparent that virtually all of the abalone were in size categories which could have resulted from the brood-stock transplants. The abalone were spread around the suitable habitat of the peninsula (Fig. 34.4). Torrance Beach in the north marks the beginning of a long stretch of sandy coastline and landslides from Portugese Point east make that area unsuitable. The average size of the animals found from Point Vicente east was significantly larger than that of green abalone found in the central and northern areas of the peninsula (Kolmogorov-Smirnov test, $p < 0.05$). As food supplies appeared to be equally abundant, this observation suggests the possibility of variation among spawning seasons in the net current direction.

The relationship between the density in diver-selected quadrats and absolute density is strongly habitat dependent and thus site specific. The absolute density was determined at Lunada Bay with the use of randomly chosen quadrats on transects laid perpendicular to the shoreline. The density in the diver-selected quadrats was 32% higher than in the randomly chosen quadrats ($N = 68$ and 45, respectively). At a site between Golden Cove and Point Vicente, every third square metre was searched on three transects perpendicular to shore. Here the density in the diver-selected quadrats was almost twice the density in the regularly spaced quadrats ($N = 31$ and 75, respectively). Given the habitat variability characteristic of Palos Verdes, these relationships between density in selected habitat and absolute density should be considered as representative only.

During the course of various green, red (*H. rufescens*), and pink (*H. corrugata*) abalone seeding experiments at Palos Verdes during 1978–82 (see Tegner & Butler, 1989), several sites within an appropriate depth range for green abalone were surveyed; only two had supported *H. fulgens*. These sites were reexamined with random sampling techniques in 1985. Haggarty's at the north end of the peninsula (Fig. 34.2), had had a larger expanse of relatively uniform appropriate habitat and the largest green abalone population we had found (Table 34.2). However, this site was devastated by the storms of 1983; the waves apparently

Table 34.2 Changes in green abalone (*Haliotis fulgens*) densities at two locations on the Palos Verdes Peninsula which had supported populations of this species before the brood stock transplants.

Year	Number of m^2 sampled	Number of live abalones <100 mm	Density
(a) Haggarty's			
1980	145	4	0.028
1981	287	26	0.091
1982	399	31	0.078
1985	229	2	0.009
(b) Lunada Bay			
1978	38	0	0
1979	39	1	0.026
1980	35	0	0
1985	45	12	0.267

dispersed the boulder habitat, creating a sandy to silty environment with virtually no kelp. Not surprisingly, the abalone population had declined from previous observations. The results from the other site, a small kelp forest on an energetic point off Lunada Bay, suggested that there had been an increase in green abalone populations after the brood-stock transplants (Table 34.2). While the data in Fig. 34.4 are not absolute densities, a comparison with the 1980–82 data from Haggarty's, the densest green population observed prior to this experiment, supports order of magnitude or greater increases in density.

To control for potential large-scale changes in recruitment patterns which might have been caused by the 1982–4 El Niño (Tegner & Dayton, 1987), for example, similar searches were conducted in green abalone habitat on Santa Cruz Island and northern Los Angeles County, regions which the drift-bottle data suggested would not be influenced by the brood-stock transplants at Palos Verdes. Both regions had supported green abalone populations in the past (CDFG, unpublished block landing data); information about specific sites within each region had been synthesized for the drift-bottle study (Tegner & Butler, 1985a). Using the same method of sampling optimal habitat at a number of sites, one juvenile green was found at northern Los Angeles County and 16 at Santa Cruz Island (Table 34.3). We had no previous data for any of the northern Los Angeles County sites and eight of the nine Santa Cruz Island sites. No juvenile green abalone were found at Middle Anchorage on southeastern Santa Cruz Island in 1981 ($180\,m^2$, density of adults $>100\,mm = 0.15$) and none in 1985 ($49\,m^2$). While the size–frequency data at the two control sites suggest that 1985 may have represented increases over past recruitment, these sites clearly did not experience the scale of recruitment observed at Palos Verdes (Fig. 34.5).

Prince *et al.* (1987, 1988) observed a strong relationship between adult and recruit density for *H. rubra* in Australia, and suggested that recruitment is local, on a spatial scale of tens of metres. There was no relationship between adult and recruit density for *H. fulgens* at Palos Verdes; while there was evidence of reproduction at both sites, the Golden Cove adult transplant site had the highest and the Palos Verdes Point site the lowest recruit densities of the 15 sites surveyed (Fig. 34.4). The green abalone data support the hypothesis of primarily local recruitment, but on a scale of hundreds of metres to several kilometres. Thus it is likely that these larvae dispersed pelagically. Over longer time scales, pelagic dispersal would have been required to account for the distribution of this species among the islands of the Southern California Bight.

Table 34.3 Green abalone (*Haliotis fulgens*) selected quadrat data, 1985.

Site	Number of m^2 sampled	Number of live abalones $<100\,mm$	Density
Palos Verdes Peninsula	1398	330	0.236
Northern Los Angeles Co.	112	1	0.009
Santa Cruz Island	300	16	0.053

Management considerations

These results support the hypotheses generated by the drift-tube study that green abalone recruitment is relatively local and that, in the absence of local brood stock, a fishing closure will not support recovery of depleted stocks on short time scales. The size–frequency data and the apparent increases in density suggest that brood-stock transplants may be an effective approach to stock enhancement. Natural production of recruits by transplants eliminates the behavioural problems which apparently limit the survival of hatchery-reared juveniles, an approach to stock enhancement which has not been effective at Palos Verdes and in southern California generally (Tegner & Butler, 1985b, 1989; Schiel & Welden, 1987). While brood-stock transplants may be effective in depleted areas, the approach is expensive and limited by the availability of animals for transplantation. Nevertheless, until the survival rates of cultured larvae (Tong *et al.*, 1987) or seed (Tegner & Butler, 1985b, 1989) improve, this may be the only approach for the recovery of isolated stocks.

The effectiveness of the brood-stock approach is dependent upon the longevity of the transplants. Poaching appears to be a serious problem; Henderson *et al.* (1988) were able to locate fewer than 5% of the animals one year after a small transplant of pink abalone to Santa Catalina Island, a case where known mortalities were 18%. At Palos Verdes Point, migration of tagged animals into and out of a 400 m^2 marked square during the first year demonstrated that the green transplants were moving, but the extent of their dispersal over the period of the study is unknown. Using the number of abalone recovered per dive on the first day of a survey as an indication of abundance, the abundance declined by 63% at Golden Cove and by 36% at Palos Verdes Point between September 1983 and spring 1985. About 9% of the number originally planted in 1981–2 at each site were observed alive in 1985 in 27 dives at Golden Cove and 18 at Palos Verdes Point; these search efforts were too low to account for much dispersal. The natural mortality rate of sexually mature green abalone is not known, but it is likely to be low (Tutschulte, 1976). Despite less sport diver activity than at Santa Catalina Island, poaching was probably an issue at Palos Verdes as well. Future transplant efforts should involve increased public education and enforcement activity.

These results have general implications for fishery managers beyond the narrower issue of rehabilitation of depleted stocks. Local recruitment underscores the importance of maintaining local brood stocks at concentrations high enough to promote successful fertilization and an adequate larval supply as a basic management goal (see also Prince *et al.*, 1987). Fishery closures, a traditional management tool, will only work if adequate brood stock remains nearby, as isolated stocks cannot depend upon long-distance larval dispersal for replenishment. Thus managers are faced with the need for very fine-scale assessment of stocks, a difficult proposition in the financial climate of the 1990s in California.

ACKNOWLEDGEMENTS

The author thanks K. Barsky, R. Butler, P. Haaker, C. Lennert, D. Parker, K. Plummer, and K. Rudie for help in the field and with data analysis. The author is grateful to P. Breen, G. Davis and J. DeMartini for their constructive criticism of the manuscript. This work is the result of research sponsored by the California Department of Fish and Game, by the Los Angeles County Fish and Game Commission, by the National Marine Fisheries Service, by NOAA, National Sea Grant College Program and by the California State Resources Agency.

REFERENCES

Burge R., Schultz S. & Odemar M. (1975) *Recent abalone research in California with recommendations for management.* Draft report presented to the Calif. Fish Game Comm., San Diego, 17 January 1975.

Cox K.W. (1962) California abalones, family Haliotidae. *Calif. Fish Game Bull.*, **118** 1–133.

Emmett B. & Jamieson G.S. (1989) An experimental transplant of northern abalone, *Haliotis kamtschatkana*, in Barkley Sound, British Columbia. *Fish. Bull. US*, **87**, 95–104.

Henderson K.C., Parker D.O. & Haaker P.L. (1988) The survival and growth of transplanted adult pink abalone, *Haliotis corrugata*, at Santa Catalina Island. *Calif. Fish Game*, **74**, 82–6.

Leighton D.L. (1974) The influence of temperature on larval and juvenile growth in three species of Southern California abalones. *Fish. Bull. US*, **72**, 1137–45.

Leighton D.L., Byhower M.J., Kelly J.C., Hooker G.N. & Morse D.E. (1981) Acceleration of development and growth in young green abalone (*Haliotis fulgens*) using warmed effluent seawater. *J. World Maricul. Soc.*, **12**, 170–80.

Mottet M.G. (1978) A review of the fisheries biology of abalones. *State of Washington, Dept. Fish. Tech. Rep.*, **37**.

Pennington J.T. (1985) The ecology of fertilization of echinoid eggs: the consequences of sperm dilution, adult aggregation, and synchronous spawning. *Biol. Bull.*, **169**, 417–30.

Prince J.D., Sellers T.L., Ford W.B. & Talbot S.R. (1987) Experimental evidence for limited dispersal of haliotid larvae (genus *Haliotis*; Mollusca: Gastropoda). *J. Exp. Mar. Biol. Ecol.*, **106**, 243–63.

Prince J.D., Sellers T.L., Ford W.B. & Talbot S.R. (1988) Confirmation of a relationship between the localized abundance of breeding stock and recruitment for *Haliotis rubra* Leach (Mollusca: Gastropoda). *J. Exp. Mar. Biol. Ecol.*, **122**, 91–104.

Saito K. (1979) Studies on propagation of ezo abalone, *Haliotis discus hannai* Ino – I. Analysis of the relationship between transplantation and catch in Funka Bay Coast. *Bull. Jap. Soc. Sci. Fish.*, **45**, 695–704.

Schiel D.R. & Welden B.C. (1987) Responses to predators of cultured and wild red abalone, *Haliotis rufescens*, in laboratory experiments. *Aquaculture*, **60**, 173–88.

Tegner M.J. & Butler R.A. (1985a) Drift-tube study of the dispersal potential of green abalone (*Haliotis fulgens*) larvae in the Southern California Bight: implications for recovery of depleted populations. *Mar. Ecol.-Prog. Ser.*, **26**, 73–84.

Tegner M.J. & Butler R.A. (1985b) The survival and mortality of seeded and native red abalones. *Haliotis rufescens*, on the Palos Verdes Peninsula. *Calif. Fish Game*, **71**, 150–63.

Tegner M.J. & Butler R.A. (1989) Abalone seeding. In *Handbook of Culture of Abalones and Other Gastropods* (Ed. by K. Hahn), pp. 157–82. CRC Press, Boca Raton.

Tegner M.J. & Dayton P.K. (1987) El Niño effects on Southern California kelp forest communites. *Adv. Ecol. Res.*, **17**, 243–79.

Tong L.J., Moss G.A. & Illingworth J. (1987) Enhancement of a natural population of the abalone, *Haliotis iris*, using cultured larvae. *Aquaculture*, **62**, 67–72.

Tutschulte T.C. (1976) *The comparative ecology of three sympatric abalones.* PhD thesis, University of California, San Diego.

Uki N. & Kikuchi S. (1982) Influence of food levels on maturation and spawning of the abalone *Haliotis discus hannai* related to effective cumulative temperature. *Bull. Tohoku Regional Fish. Lab.*, **45**, 45–53.

Wilson K.C., Haaker P.L. & Hanan D.A. (1977) Kelp restoration in Southern California. In *The Marine Plant Biomass of the Pacific Northwest Coast* (Ed. by R.W. Krauss), pp. 183–202. Oregon State University Press, Corvallis.

Chapter 35
The enhancement of paua (*Haliotis iris* Martyn) populations in New Zealand

DAVID R. SCHIEL *Department of Zoology, University of Canterbury, Christchurch 1, New Zealand*

ABSTRACT: The release of hatchery-raised *Haliotis iris* into natural habitats has been considered as one option for the expansion of the paua industry in New Zealand. Small-scale (1 m^2), replicated experiments were used to investigate processes affecting survival and growth of juvenile paua within sites, and larger scale releases were carried out at several sites to mimic more realistic enhancement procedures. Larval releases were also done on a confined scale. Results from the small-scale experiments showed that predators had a minimal detectable effect on survival, whereas unstable substratum, particularly the storm-related movement of sand, gravel and small rocks, had major effects on survival. For the large-scale releases, 9–32% of juveniles survived the first year and survival after 3 years was as high as 24%. The release of larvae does not appear to be cost-effective, but the release of juveniles may be. Commercial scale seeding was started during late 1990, but results are not yet available.

RESUMEN: La liberación de *Haliotis iris* producido en granjas hacia el medio natural, se ha considerado como una de las formas de incrementar la industria del abulón (*Haliotis iris* Martyn) de Nueva Zelandia. En el presente capítulo se revisan los resultados de este programa experimental. Se montaron experimentos repetidos, a pequeña escala (1 m^2), para investigar los procesos que afectan, en algunos sitios, la sobrevivencia y el crecimiento de juveniles de paua, haciéndose en varios lugares liberaciones a mayor escala, tratando de simular los procedimientos de fomento poblacional con mayor realismo. También se efectuaron liberaciones de larvas, en una escala limitada. Los resultados a pequeña escala mostraron que la depredación tuvo efectos mínimos sobre la sobrevivencia, mientras que los cambios en el habitat, especialmente los movimientos de arena, grava y pequeñas rocas, tuvieron un mayor impacto en la sobrevivencia del abulón. La liberación de juveniles a mayor escala muestra una supervivencia anual que fluctuó del 9 al 32%, aunque después de tres años ésta no fue mayor del 24%. La liberación de larvas no parece ser costeable, aunque la de juveniles sí. Esto todavía requiere intentarse a una escala comercial mayor, con todo resulta, una promesa para la recuperación de las poblaciones naturales.

INTRODUCTION

The potential for enhancement of populations of commercially important invertebrates has received considerable attention both worldwide and in New Zealand (Yamakawa, 1983; Sloan & Breen, 1988; Tegner & Butler, 1989). The impetus for enhancement programmes is usually a severe depletion or the imminent threat of collapse of a natural fishery, coupled with the substantial market demand for abalone and the high prices of the product.

The particular life histories, habitats and ecological interactions of abalone species have rendered them much more difficult to enhance in inshore waters than bivalve species such as scallops, oysters and mussels. Bivalves share the characteristics of being filter-feeders, having life spans of only a few years, growing to adult sizes in a few years or less, residing in relatively simple and generally protected habitats inshore and being essentially sedentary once spat have settled. In contrast, abalone are slow-growing and long-lived, taking many years to reach adult size (Poore, 1972; Shepherd & Hearn, 1983). Furthermore, they are grazers which interact in complex ways with other organisms in heterogeneous rocky-reef environments.

The incentive for the paua enhancement programme in New Zealand has not resulted from an imminent threat to the fishery. The current quota system provides for annual reviews of the fishery in all areas of the country and adjustments to the total allowable commercial catch, if necessary (Chapter 30, this volume). The awareness that recruitment failures may occur (Sainsbury, 1982b) and that localized areas may become depleted of paua have been the background for programmes by the Ministry of Agriculture and Fisheries (MAF) Fisheries Research Centre and regional paua divers to investigate enhancement as one option for the long-term sustainability of the paua fishery.

This chapter reviews some of the experimental work done between 1986 and 1989 on the survival and growth of hatchery-reared paua that have been transplanted to coastal habitats. Much of this work was done on a relatively small spatial scale. The premise was that understanding processes at small scales would allow precision (*sensu* Andrew & Mapstone, 1987) in estimates of mortality, accurately identify sources of mortality and the scales over which they operate and be much less expensive than larger trials. If results were promising, trials could be scaled up to commercially realistic levels. The programme in New Zealand is currently in this latter stage of experimentation.

In this assessment of the feasibility of paua enhancement, many experiments were performed using juveniles and larvae. This chapter presents only the results of one specific transplantation experiment using juveniles to illustrate the techniques used and the variability of results. Larval experiments are discussed and the results of all experiments are used in composite form to discuss survival of transplanted paua and to present three scenarios which analyse the cost-effectiveness of transplanting hatchery-reared paua to natural habitats.

METHODS

A transplantation experiment using juveniles was carried out along the coast of Wellington (41°21′S, 174°45′E) from 16 December 1985 to 24 April 1986. An orthogonal experiment was used with the factors consisting of sites (×3), predators (two levels: access and exclusion), paua type (hatchery-reared and wild) and paua size (small and large juveniles). Eight enclosures made of plastic mesh (hole width = 10 mm) were constructed at each site. At a subtidal depth of approximately 3 m, enclosures were made by hammering lengths of steel bar into the rocky substratum to form the corners of a square with 1 m sides. The mesh was tied to the rods to a height of 40 cm and a 20 cm portion of the mesh was folded towards the outside of each enclosure along the substratum. Small rocks and boulders were placed on to this flap to help secure the enclosures, each of which was then filled with rocks from the surrounding reef. Enclosures were randomly arrayed at least 2 m apart over about 64 m^2. Four enclosures were randomly selected and mesh covers were installed over them to exclude predators. The roofs were not installed until paua were placed on to the reefs. The major predators to be excluded were the abundant inshore wrasses, *Notolabrus celidotus* and *N. fucicola*, and the blue cod, *Parapercis colias*. These fish are known to feed on gastropods inshore (Jones, 1988).

Hatchery-reared paua of two sizes were used: paua approximately 1 year old (mean shell length, 19.0 mm; standard error, 0.39 mm) and 2.5 years old (51.8, 0.62). These were grown at the MAF Mahanga Bay Hatchery using techniques described by Tong & Moss (Chapter 45, this volume). Growth and survival of these seeded paua were compared with those from natural habitats (sizes: 26.7, 0.26 and 44.6, 0.38). Because of problems of availability, the two sizes of wild paua used were only approximately the same as hatchery-reared paua. All paua were marked with a small amount of coloured epoxy. Wild paua were brought to the Mahanga Bay Hatchery for tagging and left overnight in tanks with running seawater.

On the day of transplantation, all paua were measured by a team of people and 20 paua in each of four categories (hatchery: small and large, wild: small and large) were put into bags made of plastic mesh. These were transported to the sites in ice chests (kept cool with freezer packs) and paua were placed on to the reefs by divers. Each bag of paua was randomly ascribed to a reef. There were 80 paua on each reef, a number which is well below maximum densities found at some sites along the Wellington coast (D.R. Schiel, unpublished data). Results were analysed with a four-way ANOVA.

A series of experiments was performed beginning in 1986, to assess mortality and growth from the larval stage through to settlement and recruitment in natural habitats. Larvae were reared at Mahanga Bay Hatchery. Larvae were used for seeding after they developed to the stage of having between eight and ten radular teeth and were competent to settle.

Larvae were sealed into plastic bags containing cool seawater (about 20 000 larvae to 1 litre). These bags were placed into ice chests for transport to the field. Reefs with rocks encrusted with coralline algae were previously marked and a 1 m^2 area was covered with a tent made of plankton mesh. Bags of larvae were opened into these, allowing larvae to escape onto the reefs. Tents were left in place for 24 hours, by which time larvae should have settled. Subsequent survival and growth was monitored (Schiel, in prep.).

RESULTS

Juvenile transplants

The initial experiment was assessed for survival on 24 January 1986, 5.5 weeks after transplantation. Of the 15 sources of variation in the experiment, only four were significant (Fig. 35.1). The main factor *site* was significant, with site 2 having lower survival overall than the others. No second or third order interactions involving *site* were significant. The second order interaction of *predators* × *paua type* was significant. Overall, there was no difference between exclusion treatments and those with predator access but a slightly higher average survival of wild paua (Fig. 35.1(B)) rendered the interaction significant. The second order interaction

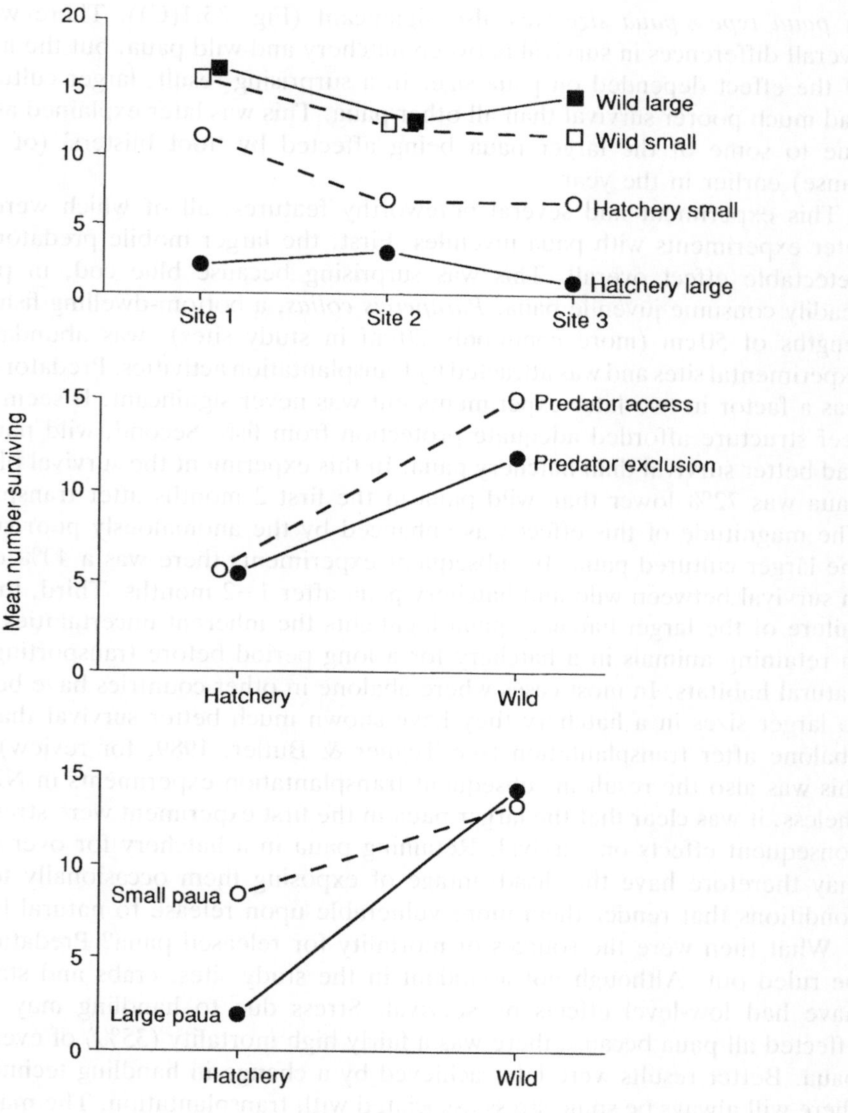

Factors	Significance (P)
Sites	.001,**
Predator exclusion	.21, n.s.
Wild *vs* Hatchery	.007,**
Small *vs* Large	.08, n.s.
Predator exclusion × (Wild/Hatchery)	.002,**
(Wild/Hatchery) × (Small/Large)	.05,*

Fig. 35.1 Results of the first juvenile transplantation experiment showing number surviving out of 20 in each paua category. Graphs are presented for interactions that were significant. Wild and hatchery refer to paua type, small and large refer to paua sizes. Significance was determined using 4-way ANOVA. SITE was treated as a random factor; all others were fixed. Cochran's test of variances was not significant for raw data.

of *paua type* × *paua size* was also significant (Fig. 35.1(C)). There were clear overall differences in survival between hatchery and wild paua, but the magnitude of the effect depended on paua size. In a surprising result, larger cultured paua had much poorer survival than all other paua. This was later explained as possibly due to some of the larger paua being affected by 'foot blisters' (of unknown cause) earlier in the year.

This experiment had several noteworthy features, all of which were seen in later experiments with paua juveniles. First, the larger mobile predators had no detectable effect overall. This was surprising because blue cod, in particular, readily consume juvenile paua. *Parapercis colias*, a bottom-dwelling fish reaching lengths of 50 cm (more commonly 20 cm in study sites), was abundant in the experimental sites and was attracted by transplantation activities. Predator exclusion was a factor in two later experiments but was never significant. It seems that the reef structure afforded adequate protection from fish. Second, wild paua always had better survival than hatchery paua. In this experiment the survival of hatchery paua was 72% lower than wild paua in the first 2 months after transplantation. The magnitude of this effect was enhanced by the anomalously poor survival of the larger cultured paua. In subsequent experiments there was a 40% difference in survival between wild and hatchery paua after 1–2 months. Third, the relative failure of the larger hatchery paua highlights the inherent uncertainties and risks in retaining animals in a hatchery for a long period before transporting them to natural habitats. In most cases where abalone in other countries have been raised to larger sizes in a hatchery they have shown much better survival than smaller abalone after transplantation (see Tegner & Butler, 1989, for review). Indeed, this was also the result in subsequent transplantation experiments in NZ. Nevertheless, it was clear that the larger paua in the first experiment were stressed, with consequent effects on survival. Retaining paua in a hatchery for over two years may therefore have the disadvantage of exposing them occasionally to adverse conditions that render them more vulnerable upon release to natural habitats.

What then were the sources of mortality for released paua? Predation cannot be ruled out. Although not abundant in the study sites, crabs and starfish may have had low-level effects on survival. Stress due to handling may also have affected all paua because there was a fairly high mortality (35%) of even the wild paua. Better results were later achieved by a change in handling techniques, but there will always be some stress associated with transplantation. The major source of mortality, however, was minor shifts in habitat. A curious feature of experiments with *H. iris* is the small number of shells recovered. This may be partially due to the small sizes of transplanted paua, because small shells are less likely to be recovered (Hines & Pearse, 1982). This may also point to removal by mobile predators. Even in the enclosures with roofs, however, shells were generally not found easily. When the reefs were finally dismantled after 5 months, many of the shells were discovered in the sediments that had accumulated at the bottom of each reef. During the experiment, some rocks had shifted, but the major source of mortality was the invasion of sand, silt and gravel. Replicates with the greatest mortality were always those worst affected by sedimentation. This process occurred over a small spatial scale, with proximal replicates often having quite different levels of sediment. The enclosures probably exacerbated this source of mortality. In times of normal water movement (i.e. in the absence of large storms), the enclosures tended to accumulate more sediments than areas around them, probably because the mesh walls reduced water movement. These experiments, therefore, demonstrate potential sources of mortality and the spatial and temporal scales over which they operate, rather than define the magnitude of effects at larger scales. Most reefs were destroyed by a large storm in April 1986. Although many paua survived, this was site dependent and the experiment was terminated.

Larval transplants

Relative to juveniles, larval paua are easy and cheap to produce. Their settlement and survival in natural habitats were examined to identify sources of mortality and to assess whether transplantation could be commercially viable. Results of specific experiments will be reported elsewhere (Schiel, in prep.) but are summarized here. Only about 10% of larvae enclosed over a reef actually settled on the experimental areas (Table 35.1). Although it was difficult to see clearly, it appeared that most larvae were no longer in the water column when the tents were removed. This indicated that they had escaped, which is unlikely, that they were active on the reef but had not settled or that they had died. In any case, there was no evidence of settlement on the rocks immediately outside the tents. After 4–5 months there were about 11 paua per reef remaining. The economic considerations are discussed below.

Economics of transplantation

Some economic scenarios based on data and plausible assumptions are presented here. They are somewhat simplistic and are not meant as comprehensive models, but rather serve to illustrate that transplantation is promising enough to proceed with larger, commercial-scale trials.

The transplantation of larvae is not economically viable at this stage (Table 35.1). Twenty thousand larvae from the hatchery cost \$4 (all values are in \$NZ; \$NZ1 approximates to \$US0.60), which is 3.6 times the cost of purchasing 11 juveniles at 5 mm. This does not take into account the cost of the somewhat meticulous work required to transplant larvae into natural habitats. There would also be considerable difficulties in scaling up larval transplants to commercially-useful levels.

Table 35.1 Survival and cost of released larvae of *H. iris*; numbers are means per m^2, with standard errors in brackets; recruitment is at 4–5 mm; monetary values are in \$NZ.

Number		
At release	Settlement	Recruitment
20 000 (1211.1)	1945 (541.5)	11.2 (4.85)
Cost of 20 000 larvae (at \$200/10^6)	= \$4	
Cost of seed (11 juveniles at 5 mm at \$0.02/mm)	= \$1.10	

The transplantation of juveniles, however, may be viable. Three cases are presented using different survival rates of adults and juveniles (Table 35.2). In each case, the number of transplanted juveniles is standardized to 1000, with an initial cost of \$250 from the hatchery (i.e., 1000 juveniles × 12.5 mm shell length × \$0.02/mm). The cost of transplanting paua to natural habitats is set at zero because it is assumed that eventual commercial transplants will be done by paua divers in coastal areas visited as part of normal commercial fishing. While this is a simplification of the social and monetary costs associated with transplantation, it is also realistic given the initial agreement with divers at the remote Chatham Islands (750 km east of mainland New Zealand) where the first commercial-scale seeding is being done. It is assumed that paua will be harvested after 9 years, by which time they should have achieved the minimum legal size of 125 mm (estimates from Murray & Akroyd, 1984).

The economic return is set at the current price for meat (\$28/kg) and shell (average of \$3.25/kg, although shell is graded and the price ranges from \$0.77 to \$6.60). The meat recovery per paua and the number of shells per kg are averages from unpublished fisheries statistics (the range of values varies among localities). The investment value (IV) of \$250 is set at 8% per year for 9 years (a conservative estimate based on government projections of 2–5% inflation during that period) and the internal rate of return (IRR) is calculated for the returned price of recovery of the paua after 9 years (if IRR is <8%, therefore, a loss is recorded). In cases where a loss is made, the price at which meat and shell would have to be sold to break even is calculated.

In the first case (Table 35.2(a)), the 3-year survival rate for released juveniles is set at 24.3%, giving an annual survival of 0.624 (instantaneous rate of natural mortality, $M = 0.472$). The figure was derived from a large transplant experiment where 2100 paua were put into each of three sites (Schiel, in prep.). The experiment was allowed to run for 3 years at one site, and the recovery was 24.3% (the other sites had considerable human interference after the first year). This survival rate is within the range of those reported elsewhere (see Tegner & Butler, 1985, for review). The adult mortality rate was set at $M = 0.2$, which is the upper limit calculated for *H. iris* by Sainsbury (1982a). The total recovery using these parameters is \$316.15, which represents a loss of \$146.59 (32% below the IV). To break even, prices would have to rise to \$41/kg for meat and \$4.75/kg for shell.

In the second case (Table 35.2(b)), juvenile survival remains the same, but adult mortality is set at $M = 0.1$, a figure considered probable by Sainsbury (1982a). The recovery is 133 paua with a value of \$576, which is a gain of \$113.26 over the IV.

In the third case (Table 35.2(c)), the annual survival of transplanted juveniles is set at 19.3% for the first year after release. This figure is the average survival of all transplanted paua at all three sites of coastal releases. Subsequent mortality of $M = 0.1$ is taken from Sainsbury (1982a) who considered this to be a reasonable figure for paua >30 mm, a size reached in the second year of growth. There is a recovery of 87 paua and a net loss of \$85.96 in this case, which is 18.5% below the IV of the original \$250. Meat and shell prices would need to rise only slightly to achieve the break-even figure.

These cases are useful in gauging the range of economic return associated with the release of hatchery-raised paua, but they are by no means definitive. Taxes

Table 35.2 Schedule of size, age, survival and cost of hatchery-raised *H. iris* released into natural habitats; the initial cost = \$250; the returns for meat and shell, at present values, are shown as is the internal rate of return (IRR) and the investment value of \$250 at 8% per year.

(a) 3-year survival = 24.3%; annual survival = 0.624, $M = 0.472$; for adults, $M = 0.2$.

Stage	Size	Age	Number left	Cost/paua
Release	10–15 mm	0.5 years	1000	\$ 0.25
Juvenile	70–80 mm	3 years	243	\$ 1.03
Adult	125 mm	9 years	73	\$ 3.43

Return:

Meat: at 145 g/paua × 73 paua × \$28/kg	=	\$296.38	
Shell: at 1 kg/12 shells × 73 shells × \$3.25/kg	=	19.77	
Total recovery	=	\$316.15	(\$4.33/paua)
IRR = 2.98%			
Investment value of \$250 after 9 years	=	\$462.74	(\$6.34/paua)
Net loss	=	**\$146.59**	

To break even, price/kg would have to be \$40.99 (meat) and \$4.75 (shell) (This assumes the price ratio of meat to shell remains constant).

(b) 3-year survival = 24.3%; annual survival = 0.624, $M = 0.472$; for adults, $M = 0.1$.

Stage	Size	Age	Number left	Cost/paua
Release	10–15 mm	0.5 years	1000	\$ 0.25
Juvenile	70–80 mm	3 years	243	\$ 1.03
Adult	125 mm	9 years	133	\$ 1.88

Return:

Meat: at 145 g/paua × 133 paua × \$28/kg	=	\$539.98	
Shell: at 1 kg/12 shells × 133 shells × \$3.25/kg	=	36.02	
Total recovery	=	\$576.00	(\$4.33/paua)
IRR = 11.0%			
Investment value of \$250 after 9 years	=	\$462.74	(\$3.48/paua)
Net gain	=	**\$113.26**	

To break even, price/kg would have to be \$22.51 (meat) and \$2.61 (shell)

(c) 1-year survival = 19.3%, $M = 1.645$; subsequent $M = 0.1$.

Stage	Size	Age	Number left	Cost/paua
Release	10–15 mm	0.5 years	1000	\$ 0.25
Juvenile	35–45 mm	1 year	193	\$ 1.29
Juvenile	70–80 mm	3 years	158	\$ 1.58
Adult	125 mm	9 years	87	\$ 2.87

Return:

Meat: at 145 g/paua × 87 paua × \$28/kg	=	\$353.22	
Shell: at 1 kg/12 shells × 87 shells × \$3.25/kg	=	23.56	
Total recovery	=	\$376.78	(\$4.33/paua)
IRR = 5.26%			
Investment value of \$250 after 9 years	=	\$462.74	(\$5.32/paua)
Net loss	=	**\$ 85.96**	

To break even, price/kg would have to be \$34.40 (meat) and \$3.99 (shell) and the labour associated with raising juveniles and transplanting them may increase costs in the future. Furthermore, survival and growth rates are variable and these cases clearly demonstrate that the return is sensitive to mortality rates. However, there may also be further resource returns through the natural fecundity of the seeded paua, which become mature at 70–100 mm (Poore, 1973; Wilson, 1987). An additional favourable factor is that paua may reach the minimum legal size in as little as 6 years at some sites (Schiel & Breen, in prep.).

DISCUSSION

Many variables affect the success of enhancement programmes. Predators, shifts in habitat, behavioural patterns of released abalone, handling techniques, the quality and sizes of seed and specific site characteristics all influence survival and growth after release. Perhaps more importantly, the magnitude of their effects varies spatially and temporally. This is evident from a localized scale within an area of coastline to a geographic scale in different countries. In Japan, for example, considerable effort has been directed at enhancement of abalone populations (Saito, 1984). Although results are variable, there have been several reported successes (reviewed by Tegner & Butler, 1989), which can be ascribed to several factors. The rearing programmes have been large, producing millions of seed abalone per year; transplantation to coastal habitats has been carried out for many years (Ino, 1966; Mottet, 1980); there appears to be a reasonable understanding of increasing survival with the sizes of seed abalone (Inoue, 1976) and predation of seed abalone is probably not as large as in other countries (Tegner & Butler, 1989). The Tōni Fisheries Cooperative Association in the Iwate Prefecture has been releasing seed abalone since 1977 at sites around Tōni Bay, reaching 600 000 seed in 1986. During 1984 to 1986, released abalone comprised about 32% of the total catch weight of abalone (figures provided by Iwate Prefecture Mariculture Centre). Annual survival rates of about 56% for 30 mm seed, with a recovery of about 18% after 3 years, were considered to be the break-even point financially.

These apparent successes of harvesting released abalone have not been reported elsewhere. Predation by a variety of mobile species is the major cause of mortality of released abalone in California (Tegner & Butler, 1985, 1989). Recovery of live seed after 1 year was 1.2–2.8% and shell recovery from dead abalone was 48% of original transplant numbers. Several species of fish, sea stars, crabs and especially octopus were responsible for the high mortality. The invertebrates, in particular, occupy the same portions of habitats as small abalone and it was not possible to remove them thoroughly enough to reduce their effects.

In NZ, the biology and ecology of *H. iris* present a different suite of conditions for enhancement endeavours. Predators undoubtedly consume paua, but their effects are not detectable against the background of overall mortality. The major source of mortality is shifts in habitat, particularly the movement of sand and gravel. During storms there is also considerable movement of the rocks to which paua attach, further reducing survival. As in California (Tegner & Butler, 1985; Henderson *et al.*, 1988), there also appears to be a relatively high proportion of total mortality in the weeks immediately after transplantation of seed paua. In

NZ, wild and hatchery-raised paua survived at the same rates after this initial period. California seed abalone are more vulnerable to predators than wild abalone in laboratory experiments (Schiel & Welden, 1987; Tegner & Butler, 1989). This was not seen in NZ field-based experiments, where it is more likely that stress and injuries due to handling contributed to differences in survival. The larger cultured paua in the first experiment were clearly stressed and consequently suffered a high mortality.

The larval seeding experiments showed that successful recruitment can result. However, the work associated with these transplants is fairly time-consuming and the process has proved to be neither reliable enough nor cost-effective for commercial use at this stage.

The economic cases presented for juveniles show that enhancement programmes may be financially viable in NZ. The smaller scale experimental results clearly highlighted the importance of site-specific effects on survival and growth. It is envisaged that the effects of sites will be averaged out by using several sites for each commercial transplantation endeavour. This will also hedge against the possibility of complete destruction of a site by storms.

Despite the encouragement for enhancement offered by experimental results, the programme is still in an early stage in NZ and awaits the results of commercial-scale trials. Of much more immediate importance to the NZ paua fishery is the understanding and management of wild stocks (Chapter 30, this volume). Because of the interest shown by paua quota holders, these programmes will proceed in tandem for at least the next few years.

ACKNOWLEDGEMENTS

The author thanks Geoff Baird, Rob Mattlin, John Zeldis, Scoresby Shepherd and A. Staniford for comments on the manuscript, and Paul Breen and John Schellevis for help with the economic projections.

REFERENCES

Andrew N.L. & Mapstone B.D. (1987) Sampling and the description of spatial pattern in marine ecology. *Oceanogr. Mar. Biol. Ann. Rev.*, **25**, 39–90.

Henderson K.C., Parker D.O. & Haaker P.L. (1988) The survival and growth of transplanted adult pink abalone, *Haliotis corrugata*, at Santa Catalina Island. *Cal. Fish Game*, **74**, 82–6.

Hines A.H. & Pearse J.S. (1982) Abalones, shells, and sea otters: dynamics of prey populations in central California. *Ecology*, **63**, 1547–60.

Ino T. (1966) (The abalone science and its propagation in Japan.) Nippin Suisan Shigen Hogo Kyokai, Suisan Zoyoshoku Soshu 11. (Translated from Japanese by *Fish. Res. Bd Can.*, Translation Series 1078, 1968.)

Inoue M. (1976) Awabi (abalone). In *Suisan Zoyoshoku Deeta Bukku*. (Translated by M.G. Mottet in *A Review of the Fisheries Biology of Abalones*, pp. 19–60. State of Washington, Dept. Fish, Olympia, 1978.)

Jones G.P. (1988) Ecology of rocky reef fish of north-eastern New Zealand. *NZ J. Mar. Freshwat. Res.*, **22**, 445–63.

Mottet M.G. (1980) Factors leading to the success of Japanese aquaculture with an emphasis on northern Japan. *Wash. State Dept Fish. Tech. Rep.*, **52**.

Murray T. & Akroyd J.M. (1984) The New Zealand paua fishery: an update of biological considerations to be reconciled with management goals. *Fisheries Research Centre Internal Report No. 5*.

Poore G.C.B. (1972) Ecology of New Zealand abalones *Haliotis* species (Mollusca: Gastropoda) 3. Growth. *NZ J. Mar. Freshwat. Res.*, **6**, 534–9.

Poore G.C.B. (1973) Ecology of New Zealand abalones, *Haliotis* species (Mollusca: Gastropoda) 4. Reproduction. *NZ J. Mar. Freshwat. Res.*, **7**, 67–84.

Sainsbury K.J. (1982a) Population dynamics and fishery management of the paua, *Haliotis iris*. 1. Population structure, growth, reproduction, and mortality. *NZ J. Mar. Freshwat. Res.*, **16**, 147–61.

Sainsbury K.J. (1982b) Population dynamics and fishery management of the paua, *Haliotis iris*. 2. Dynamics and management as examined using a size class population model. *NZ J. Mar. Freshwat. Res.*, **16**, 163–73.

Saito K. (1984) Ocean ranching of abalones and scallops in northern Japan. *Aquaculture*, **39**, 361–73.

Schiel D.R. & Welden B.C. (1987) Responses to predators of cultured and wild red abalone, *Haliotis rufescens*, in laboratory experiments. *Aquaculture*, **60**, 173–88.

Shepherd S.A. & Hearn W.S. (1983) Studies on southern Australian abalone (Genus *Haliotis*). IV. Growth of *H. laevigata* and *H. ruber*. *Aust. J. Mar. Freshwat. Res.*, **34**, 461–75.

Sloan N.A. & Breen P.A. (1988) Northern abalone, *Haliotis kamtschatkana* in British Columbia: fisheries and synopsis of life history information. *Canadian Spec. Pub. Fish. Aquat. Sci.*, **103**.

Tegner M.J. & Butler R.A. (1985) The survival and mortality of seeded and native red abalones, *Haliotis rufescens*, on the Palos Verdes Peninsula. *Calif. Fish Game*, **71**, 150–63.

Tegner M.J. & Butler R.A. (1989) Abalone seeding. In *Handbook of Culture of Abalone and other Marine Gastropods* (Ed. by K.O. Hahn), pp. 157–82. CRC Press, Boca Raton.

Wilson N.H.F. (1987) *Reproduction and ecology of* Haliotis iris *and* H. australis *from Otago*. MSc thesis, University of Otago.

Yamakawa H. (1983) Abalone (*Haliotis (Nordotis) discus*) culture. In *Modern Methods in Aquaculture* (Ed. by T. Kafuku & H. Ikenoue), pp. 172–82. Elsevier, Amsterdam.

Chapter 36
The introduction of abalone into Chile

CECILIA GODOY *Instituto de Fomento Pesquero, Universidad Austral de Chile, Casilla 401 Ancud, Chile*
GABRIEL JEREZ *Instituto de Fomento Pesquero, Avda. Pedro de Valdivia 2633, Casilla 1287, Santiago, Chile*
FRANCISCO PONCE *Subsecretaria de Pesca, Avda. Bellavista 168 p. 16, Valparaiso, Chile*
(Abridged translation by S.A. Shepherd)

ABSTRACT: Experiments on the red abalone, *Haliotis rufescens*, introduced into Chile for experimental purposes, are described briefly. The intensive culture of this abalone in a closed system has been achieved, but the high operating costs suggest that cage culture in the sea would be a preferred economic option. Because little is known of the effects of introducing an exotic abalone species on subtidal benthic communities, assessment of the impact is imperative. The considerations involved in such assessment are discussed.

RESUMEN: El estado de sobreexplotación de las principales pesquerías bentónicas chilenas, ha estimulado las investigaciones tendientes a aumentar la disponibilidad de materia prima y ha orientado los esfuerzos hacia la conservación de los recursos. Algunas alternativas propuestas son los cultivos intensivos, la repoblación de bancos sobreexplotados, el manejo de áreas de extracción y la introducción de especies no nativas de alto valor comercial. El abulón rojo (*Haliotis rufescens*) ha sido una de estas especies y cuya factibilidad de introducción ha sido motivo de estudio. Su cultivo intensivo en sistemas cerrados ha mostrado resultados satisfactorios, pero su elevado costo operativo apunta hacia otras alternativas como el cultivo en cautiverio en el medio natural. Sin embargo, se desconocen los posibles efectos de este molusco sobre las comunidades submareales nativas y por tanto, los estudios ecológicos son de primera prioridad. En este trabajo se resumen los resultados obtenidos de transplantes de abulón rojo a Chile, además de discutir los posibles efectos de la introducción y sus proyecciones futuras.

INTRODUCTION

The overexploitation of the main Chilean benthic fisheries, such as loco (*Concholepas concholepas*), the white urchin (*Loxechinus albus*) and the northern scallop (*Argopecten purpurata*), has encouraged biologists to examine the feasibility of introducing the Californian red abalone, *Haliotis rufescens*, into Chile. In this chapter we describe briefly the results of laboratory experiments on the survival and growth of red abalone produced from adults introduced into Chile, and make recommendations for the development of red abalone culture in Chile.

HISTORY

Viviani (1981) and Owen *et al.* (1984) have given preliminary information about the intensive culture of red abalone in Chile. Under an agreement between the Organización de los Estados Americanos, Fundación Chile and the Centro de Investigaciones Submarinas of the Universidad del Norte (in Coquimbo), in which avoidance of ecological harm was accorded special emphasis, 300 *H. rufescens* were transferred to a laboratory in Coquimbo, Chile, for experimentation.

The growth rate of 60 juveniles (mean length 16.6 mm) was 21.7 mm in 12 months, that of 40 subadults was 17.5 mm per year over 27 months, while that of 20 adults was 0–1 mm over 9 months. Various algal diets were used in the growth experiments on juveniles and *Lessonia flavicans* was found to promote the most rapid growth.

Later, adults were selected for reproduction, and from these 100 000 larvae were taken to the settlement stage (representing 18% survival). Subsequently 60% of the settling veligers survived to 9 months and grew to a mean length of 25 mm in a year. The juveniles were fed a diet of *L. nigrescens* and *L. flavicans*.

Other experiments on the behaviour of this abalone in the presence of various native Chilean species of predators and herbivores were conducted. We observed no interactions in the presence of common crabs of the intertidal and shallow subtidal shores of Chile, e.g. *Rhynchocinetes* sp., *Acanthocyclus* sp. and *Taliepus* sp.

The principal predators of adult abalone were judged to be crabs of the genera *Cancer* and *Homalaspis* and the sea star *Meyenaster* sp. Among possible competing herbivores, the sea urchins *Tetrapigus niger* and *Loxechinus albus* appeared to be the species whose behaviour and diet were most similar to those of *Haliotis*. However *T. niger* differs radically from *Haliotis* in that it grazes microscopic algae, whereas *T. albus*, like adult *Haliotis*, feeds preferentially on drift algae.

Possible sites (mostly in the north of Chile) to carry out seeding experiments were selected where the natural loco, sea urchin and oyster resources were almost exhausted. That initiative was subsequently abandoned because the authorities had no intention of allowing wild populations of abalone, and had considered only the cultivation of abalone in cages. Fundación Chile is developing a project to evaluate the economics and technology of cultivation of abalone in suspended cages and at the same time is carrying out an environmental impact study (EIS) to determine the effect of large-scale intensive culture of the red abalone on the environment.

ECOLOGICAL ASPECTS

The introduction of exotic species must be considered under the theoretical framework of a human disturbance, whose effect on the natural environment will depend on its intensity (e.g. the type of introduction that will occur), the initial number of individuals introduced and more generally on the particular biological characteristics of the species.

The abalone is a generalist herbivore that does not feed during its larval stage, grazes benthic diatoms in the juvenile stage and feeds on drift algae in the adult stage. Abalone are mobile with nocturnal feeding behaviour. Their natural predators are a wide range of vertebrates (sea otter, fishes) and invertebrates (sea stars, crabs). In Chilean ecosystems there are some species that could control abalone, but we believe that its principal predator will always be man.

REGULATORY ASPECTS

Chilean law requires that the importation and propagation of marine resources, whatever the state of their development, should only be allowed with the prior consent of the Under-Secretary of Fisheries. A requirement is that an import be accompanied by a certificate issued by the country of origin that the animals are free from disease. Regarding the propagation of marine resources, the Under-Secretary stipulates the conditions of importation. In 1989, fishery authorities established a requirement to provide sufficient biological and ecological data to allow the Under-Secretary to evaluate the social benefits of the proposed introduction. If the authorities lack sufficient information on the proposal they will require an EIS of the proponent, setting out an analysis of the behaviour of the species, in terms of the likelihood of introducing disease or causing harm to indigenous species or to the ecosystem.

When the EIS is completed, the Under-Secretary will evaluate it from a social point of view, determine the desirability of introducing and cultivating the exotic species and will rule on the sanitation and handling requirements.

DISCUSSION

It is clear that without effective control measures no regulatory system for the introduction of exotic species could give satisfactory results. Augsburger & Gallardo (1983) proposed the formulation of a regional code, taking as a guide those used in North America and Europe (e.g. EIFAC, 1988). That proposal implies the organization of work groups and the legislation of effective norms that transcend national boundaries. The fishery regulations now under discussion in Chile contain modifications which will speed up the procedures now in use.

The effect of an introduction will depend on the species' biology. Intensive culture in a closed system is the safest way to prevent any harmful effect on the environment. Direct seeding of natural reefs, on the other hand, is the least controlled method and the most likely to cause ecological changes in the natural system. The alternative of cultivation in captivity presupposes the control of variables such as biological interactions, food and harvesting, but cannot control the reproductive process. Abalone seed production in Chile in controlled systems is a good way to avoid introduction of parasites and epibionts. In respect of the last two, ecological studies should be designed in the long term to quantify the changes at a local and regional level and also, in the case of abalone, to evaluate its reproductive viability and determine the effect of larval settlement in natural communities. Our experience working in marine reserves, where people have restricted access, shows that manipulations at different trophic levels provokes different and complex changes to species' populations and the community as a whole (Moreno, 1984; Moreno *et al.*, 1986).

According to data from Ab Lab (California), the maximum annual production of one million abalone would need 754 t macroalgae (i.e. 14 t per week). If we take into account the total landings of the algae *Lessonia* and *Macrocystis* in 1988 as having a value equal to its total availability in the natural environment (69 186 t) we would need only 1.1% of the landings in a year. It is likely that concentrated foods at lower cost will be found, thus avoiding the use of natural resources for food.

It has been shown that the red abalone can be cultivated in Chile, at least in closed systems, and that proposals for cultivation in captivity are promising. Chile has definite economic advantages for the development of such projects, compared with other countries. In the final decision to introduce an exotic species, environmental and social considerations are an integral part of the assessment process. Such initiatives should be supported by legal measures that incorporate administrative procedures as well as methodologies to evaluate the environmental impact. In that way we will not only protect our unique marine ecosystem, but we will also promote the development of human enterprises involved in those ventures.

ACKNOWLEDGEMENTS

The authors thank Sr M Braun and Srta P Avendaño for assistance in preparing this chapter.

REFERENCES

Augsburger A. & Gallardo V. (1983) Necesidades de implementar un código de prácticas para reducir los riesgos de efectos originados por la introducción de especies exóticas. In *Proceedings Simposio Internacional Avances y Perspectivas de la Acuacultura en Chile* (Ed. by H. Fuentes, J. Castillo y L. DiSalvo), pp. 467–78. Universidad del Norte, Facultad de Ciencias del Mar, Coquimbo, Chile.

(EIFAC) European Inland Fisheries Advisory Commission (1988) Codes of practice and manual of procedures for consideration of introductions and transfers of marine and freshwater organisms. *EIFAC Occasional Paper*, 23. FAO.

Moreno C.A. (1986) Un resumen de las consecuencias ecológicas de la exclusión del hombre en la zona intermareal de Mehuin, Chile. *Estudios Oceanológicos*, **5**, 59–66.

Moreno C.A., Lunecke K. & Lepez I. (1986) The response of an intertidal *Concholepas concholepas* (Gastropoda) population to protection from man in southern Chile and effects on benthic sessile assemblages. *Oikos*, **46**, 359–64.

Moreno C.A., Sutherland J. & Jara H. (1984) Man as a predator in the intertidal zone of southern Chile. *Oikos*, **42**, 155–60.

Owen B., DiSalvo L., Ebert E. & Fonck E. (1984) Culture of the California red abalone *Haliotis rufescens* Swainson (1822) in Chile. *Veliger*, **27**(2), 101–5.

Viviani C.A. (1981) *Introducción y cultivo experimentales del abalón rojo de California* (Haliotis rufescens) *en Chile*. Informe Final Proyecto OEA-CIS, U. del Norte, Coquimbo, Chile.

PART IX
CULTURE

Chapter 37
Review of genetics and stock management of the Pacific abalone

K. FUJINO *School of Fisheries Sciences, Kitasato University, Sanriku, Iwate 022–01 Japan*

ABSTRACT: National policy on abalone aquaculture in Japan has been directed mainly toward ocean ranching for stock enhancement, with less emphasis on intensive culture in enclosures in the last decade. Results of the major effort in stock enhancement, however, have not become evident. For various reasons, interest in the intensive culture of the Pacific abalone is increasing. A series of genetic studies of the Pacific abalone, *Haliotis discus hannai*, was started in the latter half of the 1970s. These included genetic characterization of species and populations and development of technologies for genetic improvement.

This chapter reviews progress in genetic research and development of breeding technologies and discusses the possible impact of application of such technologies in abalone aquaculture and on the genetic resources of closely-related abalone species. Genetic characterization included identification of closely-related species and their hybrids by means of observations of chromosome number and zymogram and intraspecific differences in morphometric traits of shells between local populations, occurrence of physically-deficient animals in relation to inbreeding structure of population, differences in temperature tolerance associated with thermostability variations of isozymes and differential fertility and survival. Successful development of technologies for genome manipulation has made it possible to induce two kinds of triploid sib animals by inhibiting release of the first or second polar body. Comparison of performance between normal diploid and triploid sib animals showed some marked improvements in survival and temperature tolerance and some negative aspects of the application of genome manipulation. On the basis of these observations, possible strategies are discussed for improving productivity in abalone aquaculture and for the conservation of gene pools of wild abalone populations from the point of view of stock management under conditions of intensive culture and/or ocean ranching.

RESUMEN: En la última década, la política nacional sobre la acuacultura del abulón en Japón, ha estado dirigida principalmente hacia el sistema de 'ranching' en el océano con el propósito de mejorar las poblaciones 'stock', dándose un menor énfasis hacia el cultivo intensivo en encierros. A pesar de los grandes esfuerzos realizados para el mejoramiento de las poblaciones 'stock', los resultados no han sido evidentes. Debido a varias razones, los intereses hacia el cultivo intensivo del abulón del pacífico está en aumento. Por otro lado, una serie de estudios genéticos sobre el abulón del pacífico *Haliotis discus hannai* fué iniciada en la mitad de la década de 1970. Estos incluyen la caracterización genética de las especies y poblaciones y el desarrollo de las tecnologías para el mejoramiento genético.

El presente trabajo revisa el progreso en la investigación genética y en el desarrollo de las tecnologías de cultivo y discute además el posible impacto de la aplicación de tales tecnologías en la acuacultura del abulón, así como también hacia los recursos genéticos de las especies de abulón cercanamente relacionadas. La caracterización genética conducida incluye la identificación de las especies cercanamente relacionadas y sus híbridos a través de las observaciones del número cromosómico y del zimograma y las diferencias intra-específicas en los caracteres morfométricos de la concha entre las poblaciones locales, la ocurrencia de animales deficientes físicamente en relación a la estructura endogámica de la población, las diferencias en la tolerancia a la temperatura asociadas con las variaciones en la termoestabilidad de las isoenzimas así como también las diferencias en la fertilidad y la sobrevivencia. El desarrollo exitoso de las tecnologías para la manipulación del genoma ha hecho posible la inducción de las dos clases de animales triploides inhibiendo la liberación del primero o segundo cuerpos polares. La comparación del comportamiento entre los animales diploides normales y los triploides, mostró algunos mejoramientos significativos en la sobrevivencia y en la tolerancia a la temperatura así como también algunos aspectos negativos en la aplicación de la manipulación del genoma. Basándose en las observaciones, las posibles estrategias para el mejoramiento de la productividad en la acuacultura del abulón y para la conservación del 'pool' genético en sus poblaciones silvestres desde el punto de vista del manejo de poblaciones bajo las condiciones del cultivo intensivo y/o 'ranching' en el océano, son discutidas.

INTRODUCTION

Systematic mass production of seed of the Pacific abalone, *Haliotis discus hannai*, is one of the most advanced aquaculture technologies in Japan. Millions of abalone seed are now being produced and reared up to an average shell length of approximately 3 cm at hatcheries operated by prefectural governments and then by local fishermen's associations until the abalone are released into the sea for stock enhancement or reared further in enclosures up to harvest size. National policy on abalone aquaculture in Japan has been directed mainly toward stock enhancement, with less emphasis on intensive culture in enclosures. Results of the major effort in stock enhancement, however, have not become evident. The reasons for this have yet to be studied systematically.

Despite significant improvements in genetic engineering technologies and related genetic research, genetic principles and breeding technologies have not been applied adequately to large-scale abalone aquaculture in Japan. Progress in research and development in this subject has been reviewed, and possible applications of the results for increasing productivity of abalone aquaculture, as well as for conservation of genetic resources of abalone species, have been discussed (Fujino, 1988). This chapter updates the progress in genetic research and development of breeding technologies, discusses the possible impact of the application of such breeding technologies to productivity in abalone aquaculture and to genetic resources of the abalone species and proposes appropriate cautions for conservation of gene pools of wild populations of abalone species from the point of view of stock management.

GENETIC CHARACTERIZATION OF SPECIES AND POPULATIONS

Identification of closely-related species and their hybrids

Chromosome observations have made it possible to subdivide abalone species into at least three groups in terms of chromosome number: $2n = 36$ for *H. discus hannai*, *H. discus discus*, *H. gigantea* from Japan and *H. cracherodii* from California; $2n = 32$ for *H. diversicolor aquatilis*, *H. diversicolor*, *H. varia* and *H. planata* from Japan; and $2n = 28$ for *H. tuberculata* from Europe (Arai & Wilkins, 1986; Arai *et al.*, 1982, 1988). No difference in karyotype has been found between *H. discus hannai* and *H. discus*, called Ezo-awabi and Kuro-awabi, respectively, in Japan (Arai *et al.*, 1982). The former is classified as a northern form of the latter. Generally, *H. discus hannai* is found in cold-water habitats and *H. discus discus* in warm-water habitats.

Zymogram analyses for myogen and 13 enzymes from some 230 specimens collected from several localities on the northern and southern coasts of Japan indicated that the presence or absence of four loci, which code sorbitol dehydrogenase, adenylate kinase, malate dehydrogenase and phosphoglucose isomerase, among 26 loci examined, could be key characters for distinguishing between *H. discus hannai* and *H. discus discus* (Fujino *et al.*, 1980; Sasaki *et al.*, 1980). In addition, application of the key characters resulted in recognition of polymorphic hybrids between the two forms. More detailed analysis of this subject is now in progress.

Intraspecific genetic variations

Local populations

After comparing frequencies of alleles coding isozymes at 12 loci from some 460 specimens collected from nine localities along the Japan Sea and the Pacific coasts of the mainland of Japan and Hokkaido, Sasaki (1983) demonstrated the existence of local populations of the Pacific abalone and calculated genetic distances between them.

Marked geographical differences have been found in some morphometric and physiological traits. Comparison of shell:volume to shell:length ratios among the specimens collected from the wild populations of Okushiri Island, Hokkaido and Nakaminato, Ibaragi Prefecture, as well as those grown in Nakaminato after being transplanted from Okushiri Island, disclosed that the ratio is greatest in the transplanted Okushiri specimens; the next highest ratio is in Okushiri native specimens and the lowest is in Nakaminato specimens. The differences were significant, suggesting that the traits are under both genetic and environmental controls (Sasaki, 1983). Geographical differences in genotypic compositions of thermostability variations of enzymes and the association of these variations with differences in water temperature between the localities sampled have been shown (Wilkins *et al.*, 1980; Okumura *et al.*, 1981).

Inbreeding and deficient animals

After applying the Hardy-Weinberg law of genetic equilibrium to genotypic compositions at multiple loci of some hundred specimens collected from local waters of Iwate Prefecture, Fujino (1978a, 1979) observed an excessive occurrence of homozygous genotypes at multiple loci. He interpreted the observation, according to Wright (1921), as evidence of inbreeding structure in the wild population and estimated that the inbreeding coefficient for the population sampled ranged from $F = 0.25$ to 0.42.

The occurrence of deficient or malnourished abalone, locally called 'rho-gai' or 'yase-gai' and of low commercial value, has long been known to fishermen. A greater degree of homozygosity in these animals, distinguished from normal individuals by low weight of the soft parts relative to the total body weight, suggested that the occurrence of deficient animals is due to an inbreeding depression caused by a pair of deleterious recessive genes (Fujino, 1978b). An attempt was made to estimate the frequency of the deleterious recessive gene in relation to the frequencies of occurrence of deficient animals in the wild population (Fujino *et al.*, 1981). In addition, significantly greater proportions of genotypes homozygous for thermoresistant alleles were observed in deficient animals than in normal animals at the esterase (Es·M) and phosphoglucose isomerase (Pgi · S) loci. The observation suggested probable involvement of thermostability variations as a mechanism for development of deficient animals (Fujino *et al.*, 1984a).

Temperature tolerance

Wilkins *et al.* (1980) showed thermostability variations in enzymes for the first time by using phosphoglucomutase (Pgm) from the Pacific abalone and the following technique. Immediately after starch-gel electrophoresis of tissue extracts by normal procedures, the gel was sliced into two layers. One layer was placed in a refrigerator. The other was packed carefully in a thin plastic film and heated in a waterbath at 55° ± 1°C for 12 minutes. Then the unheated and heated gels were stained for Pgm under identical conditions. Thermoresistant or thermosensitive types were determined by comparing Pgm activities of the corresponding bands on the two layers of the gel.

Similar variations were found in three additional enzymes: phosphoglucose isomerase, malate dehydrogenase and esterase (Okumura *et al.*, 1981). Close association between the frequency of thermosensitive alleles and the lowest monthly average temperature of seawater was observed commonly at the aforementioned four loci when the allele frequencies were compared with the fluctuations of surface seawater among samples collected from five localities. This suggested that the four loci act as a polygenic system with various degrees of contribution to the temperature tolerance of individual animals.

Experimental comparisons of survival rates among various thermostability genotypes after exposure to high- or low-temperature stimuli provided further evidence of the association between the genetic traits and temperature tolerance in individual animals. In a test of tolerance to high-temperature (30°C) stress, significantly greater proportions of individuals homozygous for thermoresistant alleles (R · R) survived than did those of the other genotypes (R · S and S · S), both at the Pgm and the Mdh · F thermostability loci. In a test of tolerance to low-temperature (−1°C) stress, greater proportions of individuals homozygous for thermosensitive alleles (S · S) survived than did those of the other genotypes (R · S and R · R) at the Mdh · F thermostability locus. No marked differences in survival rates against low-temperature stress were found at the Pgm locus. Electrophysiological techniques showed an association between genotypes of thermostability variations and heart-rate responses to thermal stimuli (Fujino *et al.*, 1984b).

Differences in tolerance for high-temperature (30°C) stress were found among the normal diploid and the two kinds of triploid animals, induced through chromosome-set manipulation. In addition, a combined effect of the chromosome-set manipulation and selected genotypes of thermostability variations of enzymes was observed (Fujino *et al.*, 1987b).

Fertility and survival

Overdominance of heterozygous genotypes, on the basis of analyses of S. Wright's and Hardy-Weinberg's laws of genetic equilibrium, was observed at the esterase M (Es · M) and phosphogluconate dehydrogenase (Pgdh) loci in animals with shell lengths of approximately 8 cm or more that were collected from local waters in

Iwate Prefecture (Fujino, 1987a). A similar analysis of larger population data from Iwate and Akita Prefectures showed that the deviation in genotypic frequencies from equilibrium at the Es · M locus was associated with shell length. The ratios (O/E) between the observed (O) and expected (E) frequencies of heterozygotes are less than 1 in juveniles, increase gradually as shell length increases and exceed 1 at a shell length of approximately 8–9 cm, whereas those of homozygotes and the other genotypes decrease as shell length increases (Fujino & Sasaki, 1984). These results suggested that differential survival rates among the genotypes are associated with the age of the animals and are similar to results observed in similar studies of the serum transferring locus in the skipjack tuna *Katsuwonus pelamis* (Fujino & Kang, 1968). Whether the deviation observed in genotypic composition at the Es · M locus in juvenile animals is the result of differences in the fertility of genotypes or other factors is not yet known. No such association with age has been observed for the Pgdh locus.

Quantitative analyses of genotypic composition in parents and their triploid offspring populations, induced by genome manipulation, indicated overdominance of triploid heterozygous animals at the Pgdh locus. Whether the overdominance observed in both triploid and diploid (see the previous paragraph) animals at the Pgdh locus is caused by an identical mechanism is not yet known (Fujino *et al.*, 1987a).

Similar but more detailed analyses on genotypic relationships between parents and their diploid and triploid suboffsprings, induced by cold shock to inhibit release of first or second polar body, showed that differences in the maternal genotype for the locus that codes for variation in the thermostability of phosphoglucomutase are associated with the production of triploid offspring. Greater fitness was seen in oocytes of the homozygous genotype of the thermosensitive allele (S · S) than in the other two genotypes (R · S and R · R) (Fujino *et al.*, 1988a). Supporting evidence was provided by similar analyses of parentages in genotypic composition at multiple loci coding for phosphoglucose isomerase (Pgi · S) and esterase (Es · F) thermostability variations (Fujino *et al.*, 1988b).

Other traits

Despite significant variations within sib animals and aquacultural interest in several traits of commercial importance, no genetic study has been reported on the growth of the Pacific abalone.

TECHNOLOGIES FOR GENETIC IMPROVEMENTS

Development of genome manipulation technologies

After presentation of conventional technologies for genetic improvements, Fujino (1986) has discussed the significance and possible usefulness of chromosome engineering for genome manipulation and sex control in aquacultural species. Careful observation of various parameters involved in meiotic and early developmental processes is vital for successful development of these technologies. Figure 37.1 presents the relationships between ambient water temperature and time elapsed after fertilization until the release of the first or second polar body, or the

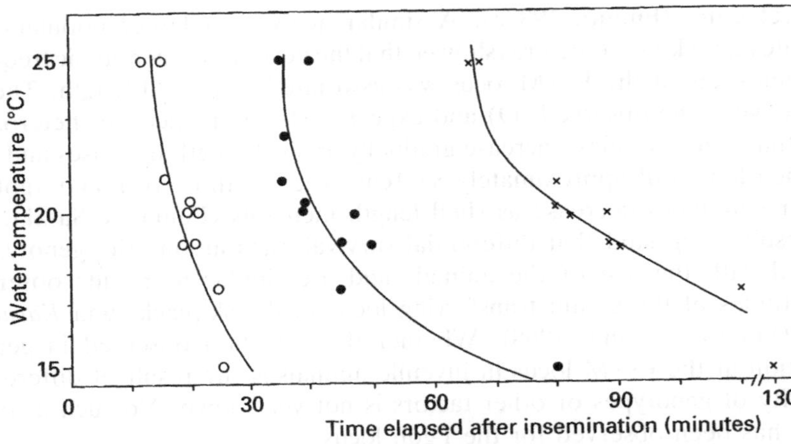

Fig. 37.1 Time lags in meiotic and early developmental processes in the Pacific abalone due to differences in ambient water temperature. The three curves represent the correlations between water temperature and elapsed time after fertilization until release of the first (open circles) or second (closed circles) polar body, or beginning of the first cleavage of the zygote (crosses) (from Arai *et al.*, 1986).

beginning of the first cleavage of the zygote in the Pacific abalone (Arai *et al.*, 1986).

Induction of triploid animals

Eggs and sperm were obtained by stimulating animals with sterilized and heated seawater according to the method of Kikuchi & Uki (1974). Eggs were fertilized by mixing them with a seawater suspension of sperm at 20°C. Cold-shock treatment of eggs (3°C for 15 minutes) 12 or 32 minutes after fertilization induced thousands of triploid animals by inhibiting release of the first or second polar body. Alternative treatments of eggs, heat shock (35°C for 3 minutes) or hydrostatic-pressure shock (200 kg/cm^2 for 5 minutes) 7 or 22 minutes after fertilization, also produced triploid animals. Ploidy status of individual animals was determined by observing chromosome numbers and/or genotypes of isozymes at multiple loci (Arai *et al.*, 1986). Quantitative analyses of genotypic composition of isozymes in parents and their offsprings showed, for the first time, that triploid animals retaining the first polar body and those retaining the second polar body were different from each other in genetic composition, especially in the degrees of heterozygosity/homozygosity (Fujino *et al.*, 1987a; Fujino, 1988).

Gynogenesis

Fertilization of eggs with spermatozoa irradiated by ultraviolet rays (42 erg/mm^2 for 30 to 60 s) successfully induced gynogenesis, indicating a high frequency of occurrence of haploid zygotes (Arai *et al.*, 1984). Treatment of the zygotes by cold shock under conditions similar to those used to inhibit release of the first or second polar body for triploidization has not been successful in inducing gynogenetic diploids. This indicated a very low rate of survival of the resulting animals. Further efforts for the improvement of the technique are now under way.

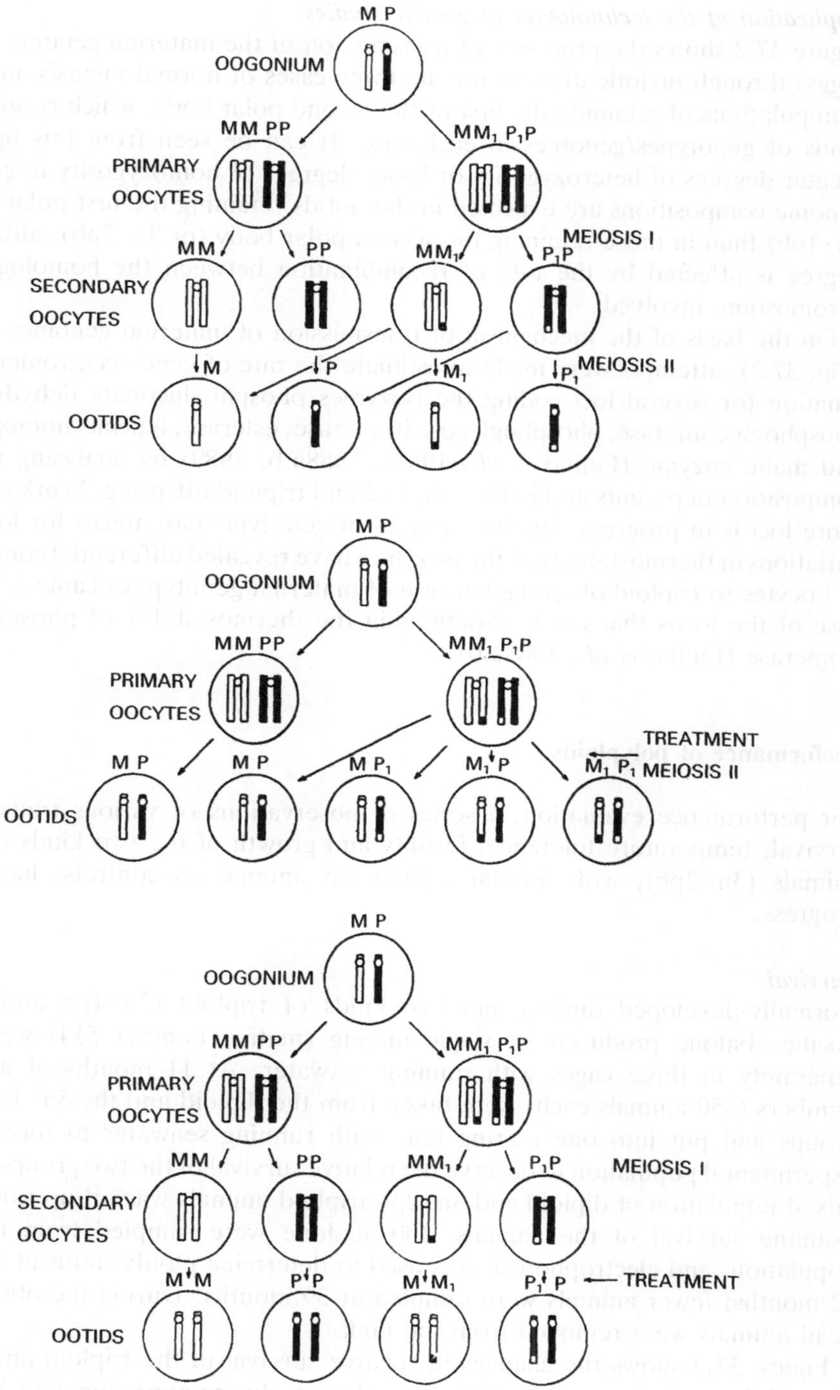

Fig. 37.2 Diagram of transmission of maternal genomes through normal meiotic processes (top) and manipulations for retaining first polar body (middle) and second polar body (bottom), demonstrating differences in the degrees of homozygosity/heterozygosity in genetic compositions of ootids produced (from Fujino *et al.*, 1987b).

Application of the technologies to genetic studies

Figure 37.2 shows the processes of transmission of the maternal genome to ootids (eggs) through meiotic division for the three cases of normal meiosis and for the manipulations of retaining the first or the second polar body, which results in four kinds of genotypes/genomes in each case. It can be seen from this figure that greater degrees of heterozygosity or lesser degrees of homozygosity in genotypic/genome compositions are expected in the ootids retaining the first polar body (or 3n · 1pb) than in those retaining the second polar body (or 3n · 2pb), although the degree is affected by the rate of recombination between the homologous loci/chromosome involved.

On the basis of the mechanism of transmission of maternal genomes to ootids (Fig. 37.2), attempts were made to estimate the rate of gene–centromere recombination for several loci coding the isozymes phosphogluconate dehydrogenase, phosphoglucomutase, phosphoglucose isomerase, esterase, leucine aminopeptidase and malic enzyme (Fujino *et al.*, 1987a, 1988a,b, 1989) by analysing genotypic composition in parents and in their diploid and triploid offspring. Work examining more loci is in progress. Similar analyses of genotypic parentages for loci coding variations in thermostability of the isozymes have revealed differential contributions of oocytes to triploid offspring because of maternal genotypes (Table 37.1) in the case of the locus that codes variations in the thermostability of phosphoglucose isomerase (Fujino *et al.*, 1988a,b).

Performance of polyploids

For performance evaluation, a series of observations of various traits, such as survival, temperature tolerance, fertility and growth of the two kinds of triploid animals (3n · 2pb), with normal diploid sib animals as controls, has been in progress.

Survival

Normally-developed diploid and two kinds of triploid (3n · 1pb and 3n · 2pb) Pacific abalone produced by single mating (mating number 514) were reared separately in three cages with running seawater. At 11 months of age, equal numbers (350 animals each) were taken from the diploid and the 3n · 1pb triploid groups and put into one rearing tank with running seawater to form a mixed experimental population to observe the relative survival of the two groups. Another mixed population of diploid and 3n · 2pb triploid animals was set up similarly. To examine survival of the animals, 100 abalone were sampled from the mixed population, and electrophoresis was used to determine ploidy status at 12, 16 and 22 months; fewer animals were sampled at 39 months. During the observations, dead animals were removed from the tank.

Figure 37.3 shows the changes in relative survival in the triploid animals with elapsed time, observed in the mixed populations during approximately 40 months. Triploid animals, both 3n · 1pb and 3n · 2pb, survived significantly longer than normal diploid animals in a mixed population, at least up to approximately 30 months of age. Continued observations on the survival of triploid animals are under way.

Table 37.1 Parent–F_1 offspring relationships of genotypes at Pgi · S R/S locus in mating No 514 of the Pacific abalone (from Fujino *et al*., 1988b).

(a) Parent–F_1 diploid relation

Parents	Females		Male
Genotype	$1^S \cdot 2^M$	$2^R \cdot 2^R$	$2^R \cdot 2^R$
Proportion	a	b	
Allele frequency	$p^S_M = \frac{a}{4}$		$p_P = 0$
	$q^R_M = \frac{a}{4} + \frac{b}{2}$		$q^R_P = 1$
where $a + b = 2$			
$F_1 \cdot 2n$ offspring			
Genotype	$1^S \cdot 2^R: \frac{a}{4}$		
Proportion	$2^R \cdot 2^R: \frac{a}{4} + \frac{b}{2}$		

(b) Parent–F_1 triploid relations

Parents	Females		Male
Genotype	$1^S \cdot 2^R$	$2^R \cdot 2^R$	$2^R \cdot 2^R$
Proportion	a	b	
Fitness	S	S_1	
where $a + b = aS + bS_1 = 2$			
Gametes mated for producing $3n \cdot 1pb$ Eggs			**Sperms**
Genotype/allele	$1^S \cdot 1^S: \frac{aS}{4} x$		$p_P = 0$
Proportion	$1^S \cdot 2^R: \frac{aS}{2}(1 - x)$		$q^R_P = 1$
	$2^R \cdot 2^R: \frac{aS}{4} x + \frac{bS_1}{2}$		
F_1 $3n \cdot 1pb$ offspring			
Genotype	$1^S \cdot 1^S \cdot 2^R: \frac{aS}{4} x$		
Proportion	$1^S \cdot 2^R \cdot 2^R: \frac{aS}{2}(1 - x)$		
	$2^R \cdot 2^R \cdot 2^R: \frac{aS}{4} x + \frac{bS_1}{2}$		
Gametes mated for producing $3n \cdot 2pb$ Eggs			**Sperms**
Genotype/allele	$1^S \cdot 1^S: \frac{aS}{4}(1 - 2x)$		$p_P = 0$
Proportion	$1^S \cdot 2^R: aSx$		$q^R_P = 1$
	$2^R \cdot 2^R: \frac{aS}{4}(1 - 2x) + \frac{bS_1}{2}$		
F_1 $3n \cdot 2pb$ offspring			
Genotype	$1^S \cdot 1^S \cdot 2^R: \frac{aS}{4}(1 - 2x)$		
Proportion	$1^S \cdot 2^R \cdot 2^R: aSx:$		
	$2^R \cdot 2^R \cdot 2^R: \frac{aS}{4}(1 - 2x) + \frac{bS_1}{2}$		

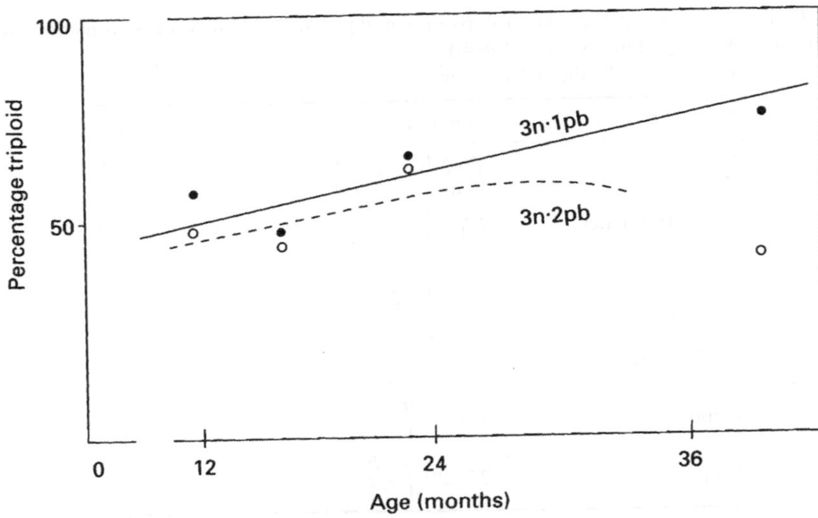

Fig. 37.3 Progressive increase of relative survival of triploid animals with age in comparison with normal diploid animals, observed in a mixed population of diploid and triploid F_1 sib offspring from mating number 514 of Pacific abalone reared in an enclosure. Solid line and solid circles represent those for 3n · 1pb animals and dotted line and open circles those for 3n · 2pb animals. The observations were started immediately after mixing 350 diploid and 350 triploid animals 11 months old. Ploidy status of individual animals was determined by examining genotypes at multiple loci.

Temperature tolerance

Fujino *et al.* (1987a) described differences in temperature tolerance between normal diploid and triploid animals. Diploid and two kinds of triploid (3n · 1pb and 3n · 2pb) abalone produced by a single mating group and reared separately in three cages were put into an experimental tank and exposed to high-temperature (30°C) stress. Significant differences in the survival rates were observed among the three sib groups. The highest rate was seen in the 3n · 1pb group; next, in order, were the 2n and 3n · 2pb groups (Fig. 37.4). In addition, a combined effect of chromosome manipulations and selected genotypes at loci for variations in thermostability of phosphoglucomutase was observed, indicating that the greatest survival rate was seen in the genotypes homozygous for the thermoresistant (R) allele (i.e. R · R · R) in the 3n · 1pb triploid.

Other traits

A series of observations of deviations in fertility and growth rate in triploid animals, in comparison with normal diploid sib animals, is under way.

STRATEGY FOR BREEDING PROGRAMMES AND FUTURE WORK

Elements influential in breeding programmes

After reviewing performance of polyploid animals induced by genome manipulation in finfish and shellfish, Fujino (1987a,b,c) mentioned some negative aspects in the

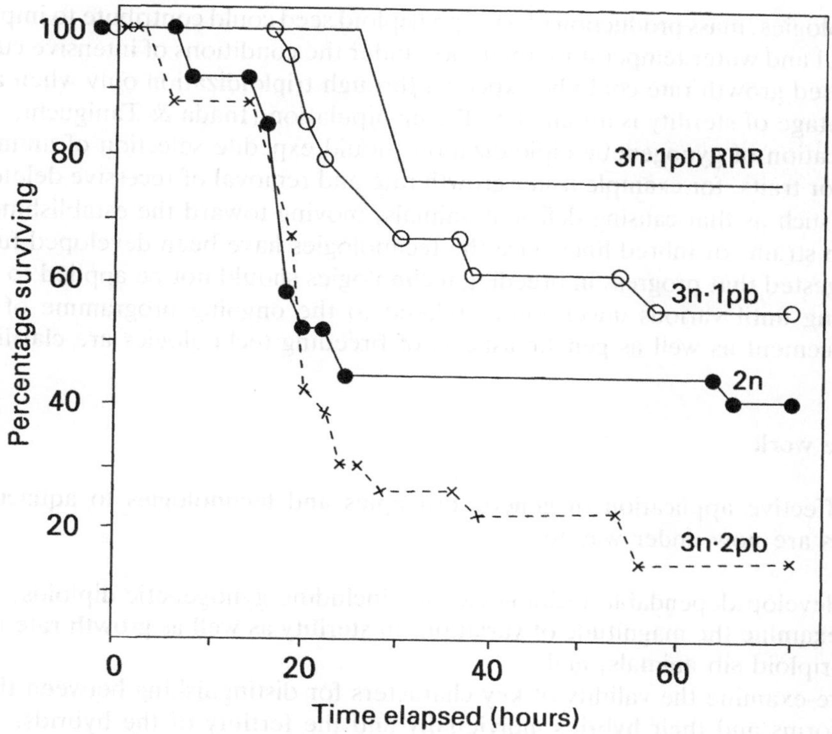

Fig. 37.4 Differences in tolerance to high-temperature (30°C) stress in normal diploid and triploid sib populations of the Pacific abalone. RRR indicates genotypes homozygous for the thermoresistant (R) allele (from Fujino *et al.*, 1987b, with some modification).

outcomes, such as inferior performance in triploid fish and 3n · 2pb triploid molluscs, mainly due to increased homozygosity of genomes. This was exemplified by the decreased temperature tolerance in 3n · 2pb triploid Pacific abalone (Fig. 37.4). He also mentioned that triploidization resulted in varying degrees of reduction in fertility or in complete sterility in several species examined. This, together with the observations on the Pacific abalone presented in the preceding paragraph, strongly suggests the necessity of various precautions in planning breeding programmes. Not only increasing productivity of aquaculture but also stock management or conservation of genetic resources of indigenous populations of closely-related species must be considered before existing aquacultural technologies or genome manipulation are applied. The most fundamental element to be considered should be whether breeding programmes are applied to intensive culture or to ocean ranching for stock enhancement in extensive areas.

Application of breeding technologies

For various reasons, interest in the intensive culture of the Pacific abalone in enclosures is increasing. On the basis of progress in genetic research and breeding technologies, mass production of 3n · 1pb triploid seed could contribute to improved survival and water temperature tolerance under the conditions of intensive culture. Increased growth rate could be expected through triploidization only when a high percentage of sterility is attained by the manipulation (Inada & Taniguchi, 1990). Application of gynogenetic diploidization should expedite selection of animals of superior traits, for example faster growth rate and removal of recessive deleterious genes such as that causing deficient animals, moving toward the establishment of certain strains or inbred lines once the technologies have been developed fully. It is suggested that progress in breeding technologies should not be applied to ocean ranching until various uncertainties related to the ongoing programme of stock enhancement as well as genetic aspects of breeding technologies are clarified.

Future work

For effective application of genetic principles and technologies to aquaculture, studies are now under way to

(1) develop dependable technologies for including gynogenetic diploids,
(2) examine the magnitude of variations in sterility as well as growth rate among triploid sib animals, and
(3) re-examine the validity of key characters for distinguishing between the two forms and their hybrids individually and the fertility of the hybrids.

REFERENCES

Arai K., Fujino K. & Kudo M. (1988) Karyotype and zymogram differences among three species of the abalones *Haliotis planata*, *H. varia*, and *H. diversicolor diversicolor*. *Nippon Suisan Gakkaishi*, **54**(12), 2055–64.

Arai K., Naito F. & Fujino K. (1986) Triploidization of the Pacific abalone with temperature and pressure treatments. *Nippon Suisan Gakkaishi*, **52**(3), 417–22.

Arai K., Naito F., Sasaki H. & Fujino K. (1984) Gynogenesis with ultraviolet ray irradiated sperm in the Pacific abalone. *Nippon Suisan Gakkaishi*, **50**(12), 2019–23.

Arai K., Tsubaki H., Ishitani Y. & Fujino K. (1982) Chromosomes of *Haliotis discus hannai* Ino and *H. discus* Reeve. *Nippon Suisan Gakkaishi*, **48**(12), 1689–91.

Arai K. & Wilkins N.P. (1986) Chromosomes of *Haliotis tuberculata* L. *Aquaculture*, **58**, 305–8.

Fujino K. (1978a) Genetic studies on the Pacific abalone. I. Inbreeding and overdominance as evidenced by biochemical polymorphism in a wild population. *Nippon Suisan Gakkaishi*, **44**(4), 357–61.

Fujino K. (1978b) Genetic studies on the Pacific abalone. II. Excessive homozygosity in deficient animals. *Nippon Suisan Gakkaishi*, **44**(7), 767–70.

Fujino K. (1979) Apparent physical deficiency associated with inbreeding structure suggested by biochemical polymorphism in the Pacific abalone. In *Proceedings XVIth International Conference Animal Blood Groups and Biochemical Polymorphism*, pp. 245–56. USSR Organizing Committee, Leningrad.

Fujino K. (1986) Impact of genetic factors on aquaculture and stock management. In *Realism in Aquaculture: Achievements, Constraints, Perspectives* (Ed. by M. Bilio, H. Rosenthal & C.J. Sindermann), pp. 421–48. European Aquaculture Society, Brendene.

Fujino K. (1987a) Japanese aquaculture program. In *Proceedings World Symposium on Selection, Hybridization, and Genetic Engineering in Aquaculture II* (Ed. by K. Tiews), pp. 462–73. Heinemann GmbH & Co. Berlin.

Fujino K. (1987b) Genome manipulation and performance evaluation of induced polyploids in fin-fish and shell-fish. In *Recent Progress in Breeding Science, 28*, pp. 75–86. Keigaku, Tokyo.

Fujino K. (1987c) Heterosis in aquacultural species of animals. In *Animal and Plant Breeding and Heterosis. VIIth Symposium on Fundamental Breeding Science* (Ed. by K. Nishikawa), pp. 29–39. Japan Society of Breeding, Univ. Gifu.

Fujino K. (1988) An attempt for analysing shifts of heterozygosity/homozygosity due to genome manipulations. *Fish Genet. Breed. Sci.*, **13**, 1–7.

Fujino K., Kamakura M., Mito R. & Arai K. (1987a) Overdominance of survivals at phosphogluconate dehydrogenase locus in triploid Pacific abalone populations. *Nippon Suisan Gakkaishi*, **53**(10), 1759–64.

Fujino K., Kaneko R., Ikeda T., Arai K. & Okumura S. (1989) Estimating rate of gene-centromere recombination at leucine-aminopeptidase, malic enzyme, and phosphoglucomutase loci in the Pacific abalone. *Nippon Suisan Gakkaishi*, **55**(11), 1875–83.

Fujino K. & Kang T. (1968) Transferrin groups of tunas. *Genetics*, **59**, 79–91.

Fujino K., Nakajima S. & Sawada H. (1988a) Differential contribution of oocytes to triploid offsprings due to maternal genotypes at phosphoglucomutase thermostability variation locus in the Pacific abalone. *Nippon Suisan Gakkaishi*, **54**(6), 953–8.

Fujino K., Nakagima S. & Takahashi T. (1988b) Differential contribution of oocytes to triploid offsprings due to maternal genotypes at multiple loci in the Pacific abalone. *Nippon Suisan Gakkaishi*, **54**(12), 2049–54.

Fujino K., Okumura S. & Inayoshi H. (1987b) Temperature tolerance differences among normal diploid and triploid Pacific abalone. *Nippon Suisan Gakkaishi*, **53**(1), 15–21.

Fujino K. & Sasaki K. (1984) Age association of genotypic proportions of isozymes in the Pacific abalone. *Nippon Suisan Gakkaishi*, **50**(1), 11–15.

Fujino K., Sasaki K., Okumura S. & Arai K. (1981) Progress of research on genetics and breeding science of the Pacific abalone. *Fish Genet. Breed. Sci.*, **6**, 19–29.

Fujino K., Sasaki K. & Okumura S. (1984a) Probable involvement of thermostability variations of enzymes in the mechanisms of occurrence of deficient abalone. *Nippon Suisan Gakkaishi*, **50**(4), 597–601.

Fujino K., Sasaki K. & Wilkins N.P. (1980) Genetic studies on the Pacific abalone. III. Differences in electrophoretic patterns between *Haliotis discus* Reeve and *H. discus hannai* Ino. *Nippon Suisan Gakkaishi*, **46**(5), 543–8.

Fujino K., Yamamori K. & Okumura S. (1984b) Heart-rate responses of the Pacific abalone against water temperature changes. *Nippon Suisan Gakkaishi*, **50**(10), 1671–5.

Inada Y. & Taniguchi N. (1990) Characteristics in induced triploid Ayu. *Fish Genet. Breed. Sci.*, **15**, 1–10.

Kikuchi S. & Uki H. (1974) Technical study on artificial spawning of abalone, genus *Haliotis*. II. Effect of irradiated sea water with ultraviolet ray on induction to spawn. *Bull. Tohoku Reg. Res. Lab.*, **33**, 79–86.

Okumura S., Sasaki K. & Fujino K. (1981) Thermostability variations at multiple loci in the Pacific abalone. *Nippon Suisan Gakkaishi*, **47**(12), 1727–30.

Sasaki K. (1983) *Genetic and breeding studies on the Pacific abalone*. Doctoral thesis, Kitasato University School of Fisheries Sciences.

Sasaki K., Kanazawa K. & Fujino K. (1980) Zymogram differences among five species of the abalones from the coasts of Japan. *Nippon Suisan Gakkaishi*, **46**(9), 1169–75.

Wilkins N.P., Fujino K. & Sasaki K. (1980) Genetic studies on the Pacific abalone. IV. Thermostability differences among phosphoglucomutase variants. *Nippon Suisan Gakkaishi*, **46**(5), 549–53.

Wright S. (1921) Systems of mating. *Genetics*, **6**, 111–78.

Chapter 38
Review of the nutritional requirements of abalone (*Haliotis* spp.) and development of more efficient artificial diets

NAGAHISA UKI *National Research Institute of Aquaculture, Nansei, Mie, 516–01, Japan*
TAKESHI WATANABE *Tokyo University of Fisheries, Konan, Minatoku, Tokyo, 108, Japan*

ABSTRACT: Quantitative requirements for major nutrients were studied as a basis for developing efficient artificial diets for abalone (*Haliotis* spp.). Semipurified experimental diets were developed which included casein and alginic acid as the protein and carbohydrate sources, respectively. This diet was used to determine the optimal level of each major nutrient on the basis of the maximal growth rate and feed conversion efficiency (FCE). The optimal levels of protein, lipid and minerals were 28%, 5% and 8%, respectively.

Feeding abalone on commercial white fishmeal diets resulted in a lower growth rate and FCE, 50% of that observed with the casein diet. Heat-treated protein sources gave inferior growth rates and lower FCE, especially those sources treated at higher temperatures and fed to larger sizes of abalone. These results suggest that inferior growth rates observed when white fishmeal is used are caused by heat treatment during the manufacturing process.

RESUMEN: Como base para el desarrollo de dietas artificiales eficientes se estudiaron las cantidades requeridas de los nutrientes principales. Se desarrollaron dietas experimentales semipurificadas, las cuales incluyeron caseína y ácido algínico como fuente de proteínas y carbohidratos respectivamente. Usando esta dieta se determinó el nivel óptimo de los principales nutrientes sobre la base de la máxima tasa de crecimiento y la eficiencia de conversión alimentaria (ECA). En la dieta los niveles óptimos de proteína, lípidos y minerales fueron de 28%, 5% y 8% respectivamente. La alimentación del abulón basada en harina blanca de pescado dió como resultado una tasa de crecimiento baja y una baja ECA, del 50% comparada con la dieta de caseína. Las proteínas tratadas con calor dieron tasas de crecimiento inferiores y bajos ECA, especialmente aquellas fuentes de proteínas tratadas a altas temperaturas y utilizadas para alimentar abulones de tamaño grande. Estos resultados sugieren tasas de crecimiento inferiores, usando harina de pescado tratada con calor durante el proceso de manufactura.

INTRODUCTION

Seedling production of abalone (genus *Haliotis*) is developing to the level of scheduled mass production in Japan since the technical establishment of maturation, spawning and seedling collection (Uki & Kikuchi, 1984). Benthic microalgae and macroalgae have been used as the diet for juveniles; however, artificial diets are gradually becoming popular and are fed together with algae. In the case of natural seaweeds, hatcheries often have limited access to seaweed sources, and even if it is possible to get natural seaweeds, availability is limited by season. An artificial diet will overcome these restrictions and make scheduled hatchery operation possible.

Many studies have evaluated algae as the diet for abalone (e.g. Uki *et al.*, 1986b). However, only a few studies (Ogino & Ohta, 1963; Ogino & Kato, 1964) have been carried out on the nutritional requirements of this mollusc. Therefore our studies (Uki *et al.*, 1985a,b, 1986a,c; Uki & Watanabe, 1986) on the quantitative requirements for major nutrients are reviewed here.

MATERIALS AND METHODS

Animals

The abalone (Ezo-awabi in Japanese) *Haliotis discus hannai* were collected near the Enoshima Islands in Onagawa, Miyagi Prefecture, for use as spawners. Juveniles were produced artificially from this stock after accelerated maturation by the conditioning method and spawning induced by ultraviolet irradiation (Uki & Kikuchi, 1984). Juveniles were reared with benthic microalgae and fed on the alga *Eisenia bicyclis* before the experiment.

Populations of one- to two-year-old juveniles, of 28–47 mm shell length and 2–13 g body weight, were used. Animals were arranged at the start of each rearing period into populations (here called lots) of equal number in a given trial consisting of 10 to 75 abalone of 100 g total lot weight.

Diet preparation

The diet was manufactured according to the method of Ogino & colleagues (Ogino & Ohta, 1963; Ogino & Kato, 1964) with some modifications. Sodium alginate (viscosity, 1000 cps) was added to the mixture of ingredients (Table 38.1). After the addition of water, 120–150% w/w (the amount varied among experiments), the mixture was formed into a paste and shaped into a cylinder 80 mm long and 10 mm in diameter. Cylinders were dipped in an aqueous solution of $CaCl_2$ (5% w/v) for one minute. By this treatment, sodium alginate was converted to an insoluble calcium alginate gel in which the nutrients were bound.

Each cylinder weighed 5 g on a dry weight basis. Diets were kept in a closed vessel and stored at −40°C. One cylinder was defrosted and used at each feeding. As determined in previous trials, moisture content had no effect on feeding and growth rates (Uki *et al.*, 1985a).

Culture system

Twenty-litre styrene resin tanks, 21 cm × 36 cm × 28 cm, were used for the experiments. The outside of each tank was painted black to prevent the growth of benthic algae. One roof-shaped plastic shelter was set on the bottom. Filtered seawater regulated at a temperature of about 20°C was supplied to each tank at 20 l/h. Gentle aeration was provided to facilitate the removal of faeces via the pipe set in the bottom corner. Five independent experiments were conducted over 40 or 80 (for the case of lipid) days.

Table 38.1 Composition of the experimental diets for abalone (values expressed as percentages).

Ingredient	Diet No. 1	2	3	4
Casein	30	30	0	
White fishmeal	0	0	47	
Dextrin	34	31	21	
Cellulose powder	5	5	5	
Mineral mixture[1]	4	4	4	
Vitamin mixture[2]	1.5	1.5	1.5	Live
Choline chloride	0.5	0.5	0.5	fronds
Lipid[3]	5	5	1	of
Algal powder[4]	0	3	0	*Eisenia*
Sodium alginate	20	20	20	*bicyclis*
	Nutrient content			
Crude protein	28.5	28.5	29.7	17.5
Crude lipid	5.0	5.0	5.3	1.2

[1] Composition of the mineral mixture is the same as those designed for carp and rainbow trout by Ogino (1980).
[2] Composition of the vitamin mixture is the same as those for carp designed by Ogino & Saito (1970); for details, see Uki *et al.* (1985b).
[3] Soybean oil:Pollock liver oil (3:2) containing 1% vitamin E.
[4] Powder of the dried 'Makonbu' *Laminaria japonica*.

Feeding method and determination of feeding amount

Feeding studies indicate that abalone have a clear nocturnal feeding periodicity from dusk to midnight (Uki, 1981) and that the amount fed should be regulated in terms of the speed and duration of feeding, which is related to temperature. In these trials, abalone were allowed to feed from evening to the next morning. The animals were fed daily, and uneaten feed was collected after 12 h and dried to determine consumption. The minimal amount needed to ensure maximal growth for each lot was between 2 and 3 g, on the basis of trials (Uki *et al.*, 1985a) (Fig. 38.1). The amount of diet supplied was 5 g in this series of experiments.

Feeding amount F(g) is calculated by the following equation

$$F = (GS/100) - R$$

where G is diet supplied (in g), S is percentage recovery of the diet (determined from controls without abalone), and R is diet remaining (in g) after the abalone have fed (all on a dry-weight basis). Factor S is almost constant under the same conditions of pellet form and content of the binder (i.e., 92%, 98% and 100% recovery rates for 20%, 30% and 40% sodium alginate).

Feed conversion efficiency (FCE), protein efficiency ratio (PER) and rate of net protein utilization (NPU) were calculated as follows:

$$\text{FCE} = W/F,\ \text{PER} = W/P \text{ and NPU} = (B - (B_K - I_K))/I$$

where W is weight gain (g wet weight), P is protein intake (in g), B is nitrogen content of the body fed on protein diet, B_K is nitrogen content of body fed on

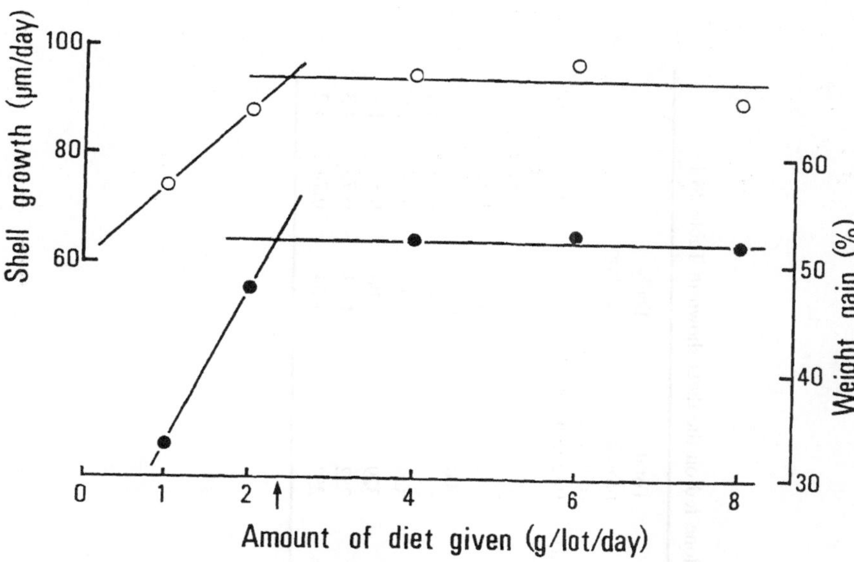

Fig. 38.1 Results of the preliminary experiment conducted to determine daily minimal amount of diet required for sustaining maximal growth rate in abalone. Each lot contained 44 abalone of 100 g initial total weight. Open circles and solid circles indicate average daily increase in shell length and weight gain, respectively, in 40 days. The arrow indicates the daily minimal amount of diet necessary.

non-protein diet, I is nitrogen intake from protein diet and I_K is nitrogen intake from non-protein diet (equals 0, as not detected).

Crude protein was determined by semimicro Kjeldahl analysis in which the nitrogen coefficient generally was assumed to be 6.25.

Composition of basal diet

Examples of the combinations of ingredients are shown in Table 38.1. The diet consists of nutrient sources of protein (casein or fishmeal), carbohydrate (sodium alginate, dextrin and cellulose), lipid (soybean and cod liver oil) and mixtures of minerals and vitamins. Depending on the experimental purpose, the nutrient sources and their content were changed by the use of dextrin.

RESULTS AND DISCUSSION

Optimal levels of fundamental nutrients

Casein was investigated as the protein source in the semipurified test diet and the optimal content of carbohydrate, mineral and lipid was evaluated on the basis of growth and FCE.

Casein diet

The dietary value of the casein diet was comparable or superior to that of the live alga *E. bicyclis*, which has a high dietary value (Uki *et al.*, 1986b) (Table 38.2).

Table 38.2 Growth, feeding, feed conversion efficiency and protein efficiency ratio results for abalone fed on the diets shown in Table 38.1.

Lot[1] No.	Diet No.	Initial size[2] SL (mean ± sd) (mm)	Initial size[2] BW (g)	Avg. growth gain SL (mm)	Avg. growth gain BW (g)	Daily increment in SL (μm)	BW gain (%)	Total diet consumed (g/individual)	Total protein consumed (mg/individual)	Daily feeding rate (%)	FCE[3]	PER[4]
1–1	1	28.8 ± 2.4	3.07	3.29	1.28	110	42	0.85	247	0.77	1.23	4.3
1–2	1	28.9 ± 2.7	3.07	2.80	1.22	93	40	0.82	238	0.74	1.21	4.3
2–1	2	28.5 ± 2.2	3.08	3.63	1.33	121	43	1.03	300	0.92	1.09	3.8
2–2	2	28.3 ± 2.5	3.07	3.51	1.14	117	37	0.95	275	0.87	0.99	3.5
3–1	3	29.3 ± 2.7	3.07	1.77	0.78	59	25	1.17	349	1.12	0.58	1.9
3–2	3	29.0 ± 2.1	3.08	1.91	0.73	64	24	1.10	329	1.06	0.57	1.9
4–1	4	28.9 ± 2.5	3.06	3.17	1.09	106	36	1.44	252	1.33	0.76	4.3
4–2	4	28.9 ± 2.1	3.06	3.18	1.08	106	35	1.45	253	1.34	0.74	4.2

[1] Each group contains 35 individuals.
[2] shell length and body weight are abbreviated to SL and BW, respectively.
[3] FCE: feed conversion efficiency (g gain/g feed).
[4] PER: protein efficiency ratio (g gain/g protein intake).

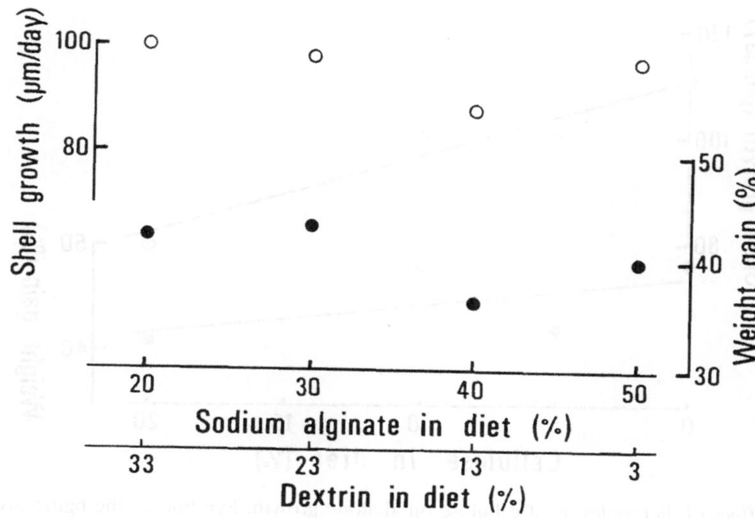

Fig. 38.2 Effect of dietary levels of sodium alginate on abalone growth. Symbols in the figure are the same as in Fig. 38.1.

This result indicates that the casein diet is suitable as a semipurified test diet for abalone. As there were no differences between the diets with and without algal powder, algal powder has no additional effect as a feeding attractant at this level of algal content. Feeding abalone on white fishmeal diets resulted in lower growth, FCE and PER, 50% to 60% of the values obtained with the casein diet, even though high feeding rates were observed. The difference in growth might be due mainly to differences in digestibility of the protein sources because of the lower FCE.

Alginic acid

No clear relationship was found between growth and dietary levels of sodium alginate (Fig. 38.2). It is well known that alginic acid is a major component of the polysaccharides found in the Phaeophyta, with a content of 20% to 30% (Tsuchiya, 1952). Abalone have alginase (enzyme code 3.2.1.16, poly-β-1.4-mannuronide glycanohydrolase) (Oshima, 1931a). The value of FCE and PER increased from 0.80 and 2.8 to 0.98 and 3.5, respectively, with an increase in the content of alginic acid. This suggests that abalone can use alginic acid more efficiently than dextrin as an energy source. The dextrin content also had no effect on abalone growth in the diet trials because the content of alginic acid was inversely related to dextrin. The optimal level of sodium alginate in the test diet is 20% to 30%, depending on the stability of the pellet.

Cellulose

Growth was slightly reduced with increasing dietary cellulose levels (Fig. 38.3). FCE and PER also were reduced, from 0.91 and 3.2 to 0.74 and 2.6, respectively, with the increase in content of cellulose. Cellulose is found in seaweeds at a level of 2% to 15% (Tsuchiya, 1952). Abalone have cellulase (enzyme code 3.2.1.4, β-1, 4-glucan 4-glucanohydrolase) (Oshima, 1931b). However, the FCE and

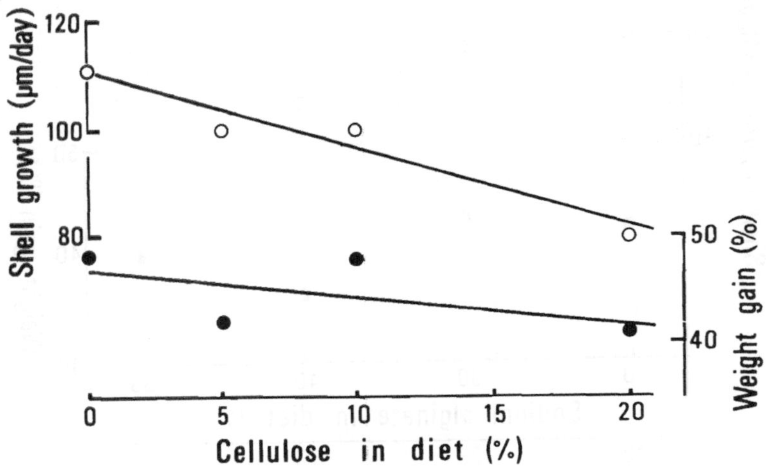

Fig. 38.3 Effect of dietary levels of cellulose on abalone growth. Symbols in the figure are the same as in Fig. 38.1.

PER results indicate that the assimilation efficiency of cellulose was not high in this herbivore. Addition of cellulose to the diet was not essential, because the effect on improvement of growth and FCE did not appear as in the case of some fishes (Buhler & Halver, 1961).

Minerals
The relationship between mineral content and abalone growth is shown in Fig. 38.4. The optimal level of mineral mixture is thought to be about 8%, on the basis of the improvement in growth that occurred with increasing mineral content up to

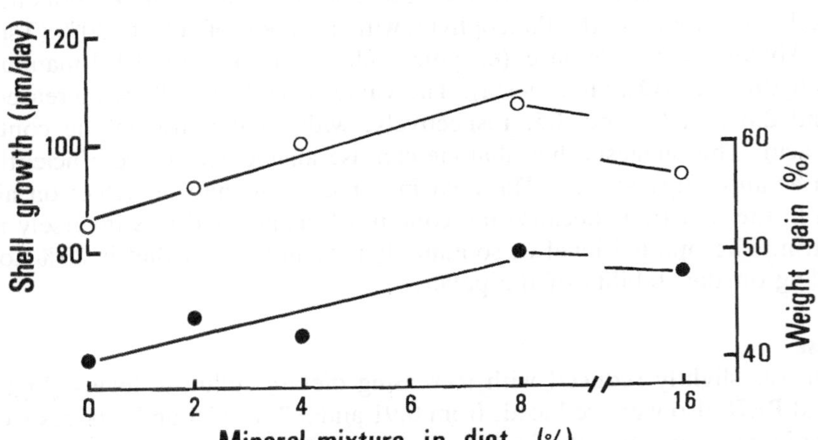

Fig. 38.4 Effect of dietary levels of mineral mixture on abalone growth. Symbols in the figure are the same as in Fig. 38.1.

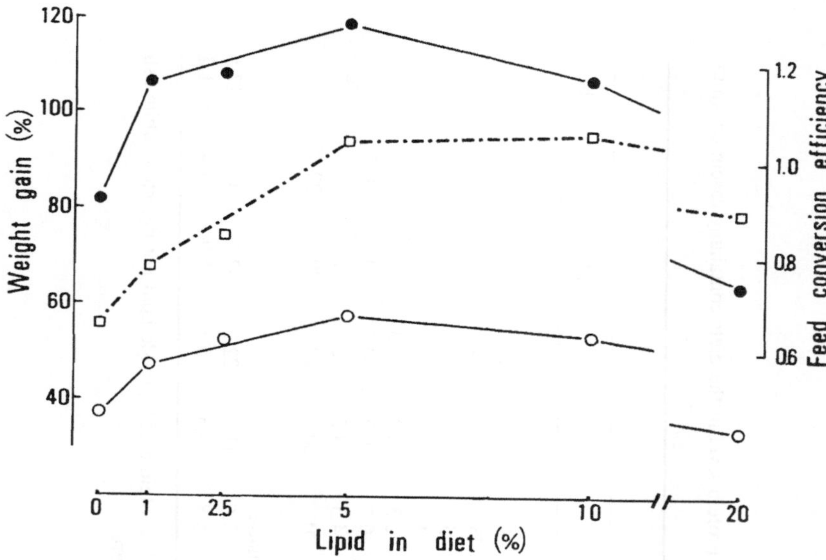

Fig. 38.5 A relationship between dietary lipid levels and abalone growth. Open and closed circles indicate weight gain (%) at the 40th and 80th day, respectively. Open squares show feed conversion efficiency in 80 days.

8%. Aquatic animals have a special formation mechanism for biomineralization of their hard tissues. The materials used depend on both environmental water and diet (Wada & Fujinuki, 1974). The optimal level of 8% in abalone is higher than that in fishes (Ogino, 1980) and lower than that in crustacea (Deshimaru & Kuroki, 1974). The mineral content in further test diets was set at 4% after taking into consideration the improvement of the solubility of the pellet, because the recovery rate of the diet, including the 8% mineral content, is lower than for those diets in which the mineral content is 4%.

Lipid

The relationship between lipid content and growth and FCE of abalone is shown in Fig. 38.5. Growth rate and FCE increased with an increase in lipid content. Superior growth at the 5% level of lipid caused a higher FCE. Seaweeds generally contain lower lipid levels (0.1% to 2% dry weight) (Tsuchiya, 1952). It is particularly noteworthy that abalone show superior growth when fed a diet that has such a high lipid content because this mollusc has no chance to feed on such levels in nature. Inferior growth of abalone fed on the non-lipid diet might be caused by the decrease of usable energy or the lack of an essential factor contained in the lipid. It was established recently that *H. discus hannai* requires n3 and n6 highly unsaturated fatty acids (HUFA) as essential fatty acids (Uki *et al.*, 1986c). The requirement of the abalone for n3 HUFA was estimated to be about 1% of the diets containing 5% lipid. Growth rates decreased at lipid levels higher than 10%. Growth rates on the 20% lipid diet were lower than those for non-lipid diets. These results suggest that lipase activity is not so high that a protein-sparing effect cannot be expected, as is the case with fishes (Ogino *et al.*, 1976). On the basis of these results, the optimal lipid level is about 5% in an artificial diet.

Table 38.3 Composition of the experimental abalone diets used for determination of the optimal protein level in the diets containing casein or white fishmeal as protein sources (values expressed as percentages).

	No of diet													
Ingredient	1	2	3	4	5	6	7	8	9	10	11	12	13	14
Casein	0	5	10	20	30	40	50	60						*Eisenia bicyclis*
White fish meal									8	16	32	47	63	
Dextrin	63	58	53	43	33	23	13	3	55.5	48.5	34	20	5	
Cellulose	5	5	5	5	5	5	5	5	5	5	5	5	5	
Mineral mixture*	4	4	4	4	4	4	4	4	4	4	4	4	4	
Vitamin mixture*	2.3	2.3	2.3	2.3	2.3	2.3	2.3	2.3	2.3	2.3	2.3	2.3	2.3	
Choline chloride	0.7	0.7	0.7	0.7	0.7	0.7	0.7	0.7	0.7	0.7	0.7	0.7	0.7	
Lipid*	5	5	5	5	5	5	5	5	4.5	3.5	2	1	1	
Sodium alginate	20	20	20	20	20	20	20	20	20	20	20	20	20	
	Nutrient content of dry matter													
Crude protein (%)	0	4.8	9.5	19.0	28.2	37.5	46.5	55.4	5.6	11.1	22.2	32.5	43.1	17.5
Crude lipid (%)	5.6	5.6	5.5	5.5	5.5	5.4	5.4	5.4	5.8	5.5	5.5	5.9	7.4	1.2

*Composition is the same as those of Table 38.1; nutrient content in a protein source: casein: crude protein 86.1%, crude lipid not detected; white fish meal: crude protein 63.2%, crude lipid 9.3%; each mixture of ingredients contains 8–10% moisture.

Optimal protein level

The abalone were allowed to feed on diets containing different amounts of casein or fishmeal (Table 38.3). The weight gain, protein gain and FCE were proportional to the dietary protein levels until a plateau was reached (Figs 38.6 and 38.7 (left)).

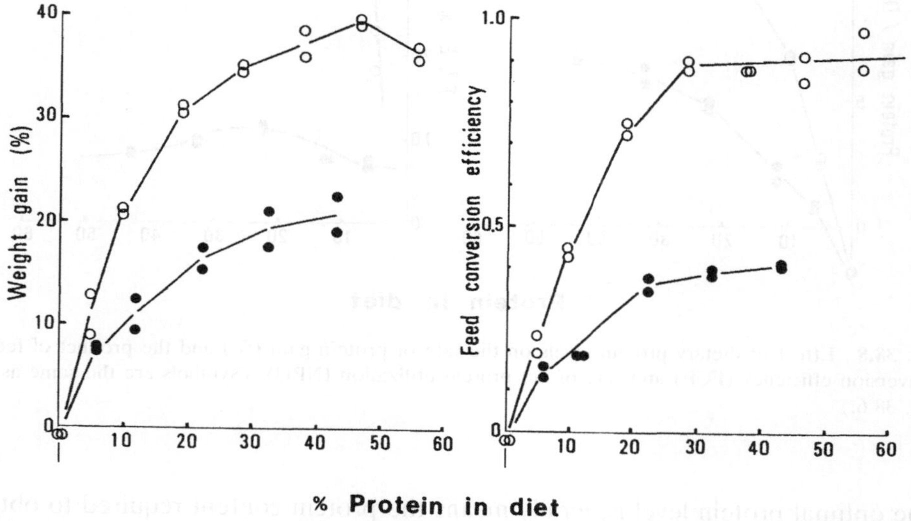

Fig. 38.6 Effect of dietary protein levels on gain in body weight (%) and feed-conversion efficiency. Open circles indicate non-protein and casein diets, and closed circles indicate white fishmeal diets.

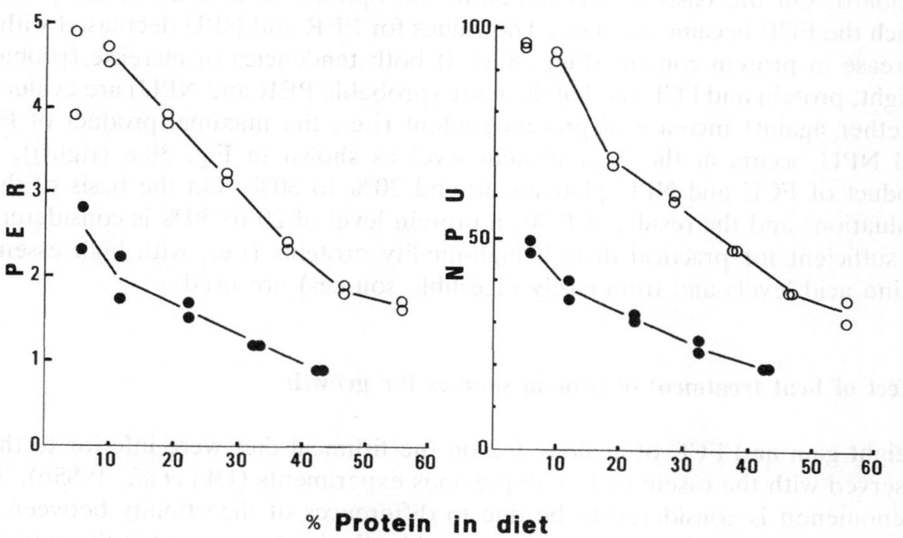

Fig. 38.7 The relationship between dietary protein levels and the values for protein efficiency ratio (PER) and rate of net protein utilization (NPU). Symbols are the same as in Fig. 38.6.

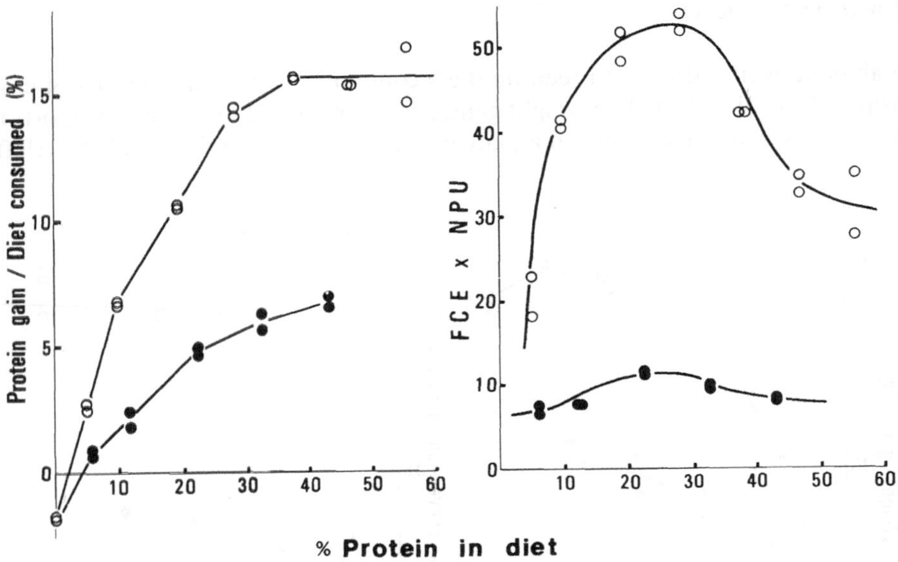

Fig. 38.8 Effect of dietary protein levels on the rate of protein gain (%) and the product of feed-conversion efficiency (FCE) and rate of net protein utilization (NPU). (Symbols are the same as in Fig. 38.6.)

The optimal protein level generally means the protein content required to obtain either maximal growth rate or maximal gain in protein. The optimal level is about 50% on the basis of maximal growth rate. However, as this experiment was conducted with *ad libitum* feeding, and some differences in feeding rate were seen among lots, weight gain per amount of diet consumed (i.e., FCE) is a better standard. On the basis of this standard, the optimal level is 28%, the point at which the FCE became constant. The values for PER and NPU decreased with an increase in protein content (Fig. 38.8). If both tendencies of increase (probable weight, protein and FCE) and of decrease (probable PER and NPU) are evaluated together against increase of protein content (i.e., the maximal product of FCE and NPU occurs at the 28% protein level as shown in Fig. 38.8 (right)), the product of FCE and NPU plateaus around 20% to 30%. On the basis of these evaluations and the results of FCE, a protein level of 20 to 30% is considered to be sufficient for practical diets if high-quality proteins (i.e., with high essential amino acid levels and from easily digestible sources) are used.

Effect of heat treatment of protein sources for growth

Weight gain and FCE of abalone fed on the fishmeal diet were inferior to those observed with the casein diet in all previous experiments (Uki *et al.*, 1985b). This phenomenon is considered to be due to differences of digestibility between the protein sources and is assumed to be caused by the heat treatment in the manufacturing procedure. Relations between heat treatment and growth of the mollusc were surveyed in populations of two different sizes of animals. All diets were

Table 38.4 Conditions for heat treatment of dietary protein sources.

Protein source	Heat treatment	
	Temperature (°C)	Time (h)
Casein	—	—
Heat-treated casein	130	2
Freeze-dried fishmeal	—	—
Heat-treated fishmeal*	110	2
Raw fresh meat		

Dashes indicate no treatment.
* Pollock meat was boiled (100°C for 20 minutes) in advance.

adjusted to almost the same content of crude protein and lipid. The conditions for heat treatment are described in Table 38.4.

No-heat treatment of casein, freeze-dried fishmeal and fish flesh showed superior growth rates compared with heat-treated protein sources (Fig. 38.9). No differences were seen among the FCE values (1.3) of these diets except for freeze-dried fishmeal (FCE 0.76), where a slight effect of drying on the quality might be suspected. Differences in growth rates are probably due to differences in food intake in these diets. However, growth of abalone and FCE was remarkably

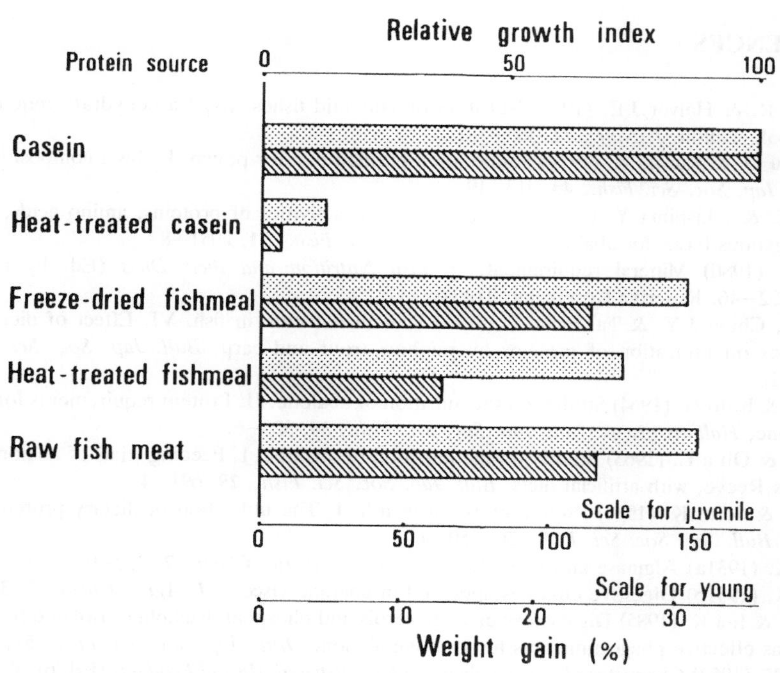

Fig. 38.9 Effect of heat treatment of protein sources on abalone growth. Upper and lower columns indicate the growth of juveniles (shell length 13 mm) and young adults (shell length 52 mm), respectively.

reduced by feeding diets containing heat-treated casein (FCE 0.17) and fishmeal (FCE 0.74). The influence of heat treatment on growth and FCE was more marked with stronger heat treatment and also in larger sizes of abalone.

These results suggest that stronger heat treatment affects the solubility of materials, resulting in lower digestibility, and that the activities of abalone proteases decrease and become more specialized with age (Fig. 38.9). Manufacturing processes of fishmeal include boiling, steaming, drying and grinding, with heat treatments several hours long. When fishmeal is chosen as a protein source for practical diets, the treatment used in the manufacturing process should be carefully controlled. Abalone have a low protease activity compared with fish and crustacea (Oshima, 1931b). Further detailed examination will be needed to clarify the relative relationships between physiological characteristics of digestibility and solubility of materials. In these studies, we have tried to define the fundamental and quantitative requirements for abalone nutrition. Also, abalone feeding stimulants have recently been elucidated (Sakata & Ina, 1985; Harada & Akishima, 1985). Application of those factors is expected to increase the feeding rate on practical diets.

ACKNOWLEDGEMENTS

The authors wish to express their gratitude to Dr Brian Davy, International Development Research Center, for his kindness in reviewing the English.

REFERENCES

Buhler D.R. & Halver J.E. (1961) Nutrition of salmonid fishes. IX. Carbohydrate requirements of chinook salmon. *J. Nutr.*, **74**, 307–18.

Deshimaru O. & Kuroki K. (1974) Studies on a purified diet for prawn. I. Basal composition of diet. *Bull. Jap. Soc. Sci. Fish.*, **40**, 413–19.

Harada K. & Akishima Y. (1985) Feeding attraction activities of proteins, amino acids, lipids and nitrogenous bases for abalone. *Bull. Jap. Soc. Sci. Fish.*, **51**, 2051–8.

Ogino C. (1980) Mineral requirement. In *Fish Nutrition and their Diets* (Ed. by C. Ogino), pp. 232–46. Koseisha-Koseikaku, Tokyo.

Ogino C., Chiou J.Y. & Takeuchi T. (1976) Protein nutrition in fish. VI. Effect of dietary energy sources on utilization of proteins by rainbow trout and carp. *Bull. Jap. Soc. Sci. Fish.*, **42**, 213–18.

Ogino C. & Kato N. (1964) Studies on the nutrition of abalone. II. Protein requirements for growth of abalone, *Haliotis discus*. *Bull. Jap. Soc. Sci. Fish.*, **30**, 523–6.

Ogino C. & Ohta E. (1963) Studies on the nutrition of abalone. I. Feeding trials of abalone, *Haliotis discus* Reeve, with artificial diets. *Bull. Jap. Soc. Sci. Fish.*, **29**, 691–4.

Ogino C. & Saito K. (1970) Protein nutrition in fish. I. The utilization of dietary protein by young carp. *Bull. Jap. Soc. Sci. Fish.*, **36**, 250–4.

Oshima K. (1931a) Alginase and their characteristics. *J. Agric. Chem.*, **7**, 332–9.

Oshima K. (1931b) Digestive enzymes appeared in abalone viscera. *J. Agric. Chem.*, **7**, 328–31.

Sakata K. & Ina K. (1985) Digalactosyldiacylglycerols and phosphatidylcholines isolated from a brown alga as effective phagostimulants for a young abalone. *Bull. Jap. Soc. Sci. Fish.*, **51**, 659–65.

Tsuchiya Y. (1952) Chemistry of marine algae. In *Chemistry of Marine Products* (Ed. by Y. Tsuchiya), pp. 373–432. Koseisha-Koseikaku, Tokyo.

Uki N. (1981) Feeding behavior of experimental populations of the abalone *Haliotis discus hannai*. *Bull. Tohoku Reg. Fish. Res. Lab.*, **43**, 53–8.

Uki N., Kemuyama A. & Watanabe T. (1985a) Development of semipurified test diets for abalone. *Bull. Jap. Soc. Sci. Fish.*, **51**, 1825–33.

Uki N., Kemuyama A. & Watanabe T. (1985b) Nutritional evaluation of several protein sources. *Bull. Jap. Soc. Sci. Fish.*, **51**, 1835–9.

Uki N., Kemuyama A. & Watanabe T. (1986a) Optimum protein level in diets for abalone. *Bull. Jap. Soc. Sci. Fish.*, **52**, 1005–12.

Uki N. & Kikuchi S. (1984) Regulation of maturation and spawning of an abalone, *Haliotis* (Gastropoda) by external environmental factors. *Aquaculture*, **39**, 247–61.

Uki N., Sugiura M. & Watanabe T. (1986b) Dietary value of seaweeds occurring on the Pacific coast of Tohoku for growth of the abalone *Haliotis discus hannai*. *Bull. Jap. Soc. Sci. Fish.*, **52**, 257–66.

Uki N., Sugiura M. & Watanabe T. (1986c) Requirements of essential fatty acids in the abalone *Haliotis discus hannai*. *Bull. Jap. Soc. Sci. Fish.*, **52**, 1013–23.

Uki N. & Watanabe T. (1986) Effect of heat treatment of dietary protein sources on their protein quality for abalone. *Bull. Jap. Soc. Sci. Fish.*, **52**, 1199–204.

Wada K. & Fujinuki T. (1974) Physiological regulation of shell formation in mollusks. I. Chemical composition of extrapallial fluid. *Bull. Natl Pearl Res. Lab.*, **18**, 2085–110.

Chapter 39
A study of polyspermy in abalone

JOSÉ LUIS STEPHANO *Escuela Superior de Ciencias, Universidad Autónoma de Baja California, A.P. 2921, Ensenada, BCN México*

ABSTRACT: The susceptibility to polyspermy of abalone (*Haliotis rufescens*) oocytes inseminated at various times after spawning was determined by counting sperm pronuclei in preparations stained with bisbenzimide and viewed by fluorescence microscopy. Oocytes were highly resistant to polyspermy at all times tested (up to $6\frac{1}{4}$ h after spawning), even when the sperm/oocyte ratio was increased to 66 000, almost two orders of magnitude above that required for successful monospermic fertilization (*ca* 1000). The biological significance of these results may be that abalone oocytes are normally exposed to high sperm densities during natural spawning. Intracellular microelectrode recordings showed that *H. kamtschatkana* oocytes respond to sperm with a rapid, positive-going shift in membrane potential, suggesting that abalone oocytes have a fast electrical polyspermy block. When the interval between spawning and insemination of *H. rufescens* oocytes was longer than 2 h, fertilizability declined, with no development occurring in oocytes inseminated after 5 h. However, when oocytes from late inseminations were stained and viewed by fluorescence microscopy, it was observed that many of them were fertilized but had failed to activate, indicating that one component of aging is the loss of the oocyte's ability to respond to sperm.

RESUMEN: Se investigó la susceptibilidad a la polispermia en el abulón *Haliotis rufescens* en función del tiempo transcurrido entre el desove y la fertilización. Los ovocitos se tiñeron con bisbenzimida y se registraron los que incorporaron el pronúcleo masculino, con ayuda de microscopía de fluorescencia. Los resultados mostraron que los ovocitos de abulón resistan grandemente a la polispermia, a todos los tiempos probados (hasta $6\frac{1}{4}$ horas después del desove), aun a proporciones de espermas/ovocitos de 66 000, casi dos órdenes de magnitud arriba de la requerida para una exitosa fertilización monospérmica (~1000). El significado biológico de estos resultados puede ser que, durante el desove natural, los ovocitos de abulón están expuestos a concentraciones muy altas de espermatozoides. Registros con un eléctrodo intracelular mostraban que los ovocitos de *H. kamtschatkana* reaccionan a los espermatozoides mediante un cambio rápido y positivo en el potencial de la membrana, lo cual sugiere que los ovocitos de abulón poseen un bloqueo eléctrico rápido a la polispermia. Cuando el intervalo entre desove e inseminación en *H. rufescens* fue más largo que 2 horas, disminuyó la fertilidad y no había desarrollo en ovocitos inseminados después de 5 horas. Sin embargo, cuando ovocitos inseminados tardiamente fueron teñidos y examinados por microscopía de fluorescencia, se observó que muchos fueron penetrados por espermatozoides aunque no activaron, indicando que un componente del envejecimiento de ovocitos es la pérdida de la capacidad para responder a la entrada del espermatozoide.

INTRODUCTION

In a study of polyspermy in oyster (*Crassostrea gigas*) oocytes, it was observed that oocytes recently removed from the ovary were highly susceptible to polyspermy, but acquired resistance if they were incubated for an hour in seawater before insemination (Stephano & Gould, 1988). Oocytes obtained by spawning were more resistant to polyspermy, but their resistance also increased somewhat when they were incubated in seawater for $1\frac{1}{2}$ h before insemination. Since polyspermy leads to abnormal development (Boveri, 1902; Jaffe & Gould, 1985; Stephano & Gould, 1988), these results indicate that the methods for obtaining oocytes, time of insemination and sperm/oocyte ratios are important considerations for optimizing the yield of normal larvae in experimental laboratories and hatcheries.

The above observations led to the question of whether susceptibility to polyspermy might vary with time of insemination after spawning in other marine invertebrates. Because of their commercial importance, abalone were chosen for further investigation. So far, there is no method for successful fertilization of abalone oocytes excised from the ovary. Therefore, oocytes were obtained by spawning and inseminated at intervals thereafter with different sperm/oocyte ratios to determine their susceptibility to polyspermy. It was found that abalone oocytes are very resistant to polyspermy at all times tested (13 m to 6 $\frac{1}{4}$ h) after spawning. Fertilization potentials were recorded during insemination of abalone oocytes, suggesting they possess a rapid electrical polyspermy block.

MATERIALS AND METHODS

Animals and gametes

Adult *Haliotis rufescens* were purchased from The Abalone Farm (Cayucos, CA) and maintained in the dark at 15–17°C in seawater tanks equipped with biological filters. They were fed *Macrocystis* and tanks were cleaned frequently by siphoning detritus from the bottom and replacing portions with fresh seawater. Spawning was induced by the Tris-peroxide method of Morse *et al.* (1978). Oocytes were washed in Millipore-filtered (0.45 μm pore) natural seawater and maintained at 15–17°C. Sperm suspensions were stored at 4°C, since motility was maintained longer at this temperature.

Sperm and oocyte concentrations were determined with samples fixed in an equal volume of 15% formaldehyde + 85% seawater. Sperm were counted with a haemocytometer and aliquots of oocyte suspensions were counted at 100×. Inseminations were performed in 2.5–5 ml filtered natural seawater in 10 ml beakers. Oocyte concentrations ranged from 120 to 2400/ml and sperm/oocyte ratios were varied from 60 to 66 000, with final sperm concentrations of 34 000 to 83.4×10^6/ml. Cultures were maintained at 15–17°C.

Assay for polyspermy

Five minutes after insemination, oocytes were collected with a hand centrifuge and resuspended in sperm lysin to dissolve the vitelline layers and remove supernumerary sperm embedded in these layers. Sperm lysin was the 12 000 × *g* supernate from a concentrated sperm suspension that had been frozen and thawed twice in seawater (see Lewis *et al.*, 1982, for a description of abalone sperm lysin). Lysin was added dropwise to the oocyte pellet until the vitelline layers disappeared, as monitored by light microscopy. Usually 3 to 5 min were required. Oocytes were then washed twice in seawater. Gentle centrifugations were required since oocytes without vitelline layers are fragile. After washing, oocytes were resuspended in seawater and aliquots were fixed by adding one-quarter volume 15% formaldehyde in seawater (final concentration 3.7%). Another portion of each culture was incubated without removing the vitelline layers for observations of subsequent development.

Fixation was at 15 to 20 min after insemination since the best time to observe male pronuclei is around the time of first polar body formation (see Stephano & Gould, 1988; after second polar body formation, pronuclei are too decondensed to score reliably). Chromosomes and pronuclei were stained by removing the supernate from the fixed eggs and adding several drops of a solution containing 25 μg/ml bisbenzimide (H33258, Calbiochem, La Jolla, CA) in seawater, 10 mM TAPS (Sigma), pH 8. After about an hour, preparations were squashed and examined at 600× with an Olympus BH2 fluorescence microscope. More than 50 oocytes were scored for each sample. Photographs were taken with Kodak T Max 100 or Ektachrome 100 film.

Electrophysiology

Haliotis kamtschatkana gametes were obtained from spontaneous spawning and kept at 4°C until used. Electrophysiological methods are described by Jaffe *et al.* (1979). Oocytes with intact jelly and vitelline layers were placed in small plastic petri dishes on a water-cooled stage at about 12°C (ambient seawater temperature) and observed with a stereomicroscope. Electrodes of 30–60 MΩ were filled with 3 M KCl. A single electrode was used for recording voltage and passing current. The IR drop across the electrode was electronically balanced by a bridge circuit in the amplifier (Bio-Dyne Electronics Laboratory, Santa Monica, CA). Membrane potentials and applied currents were recorded on a Gould-Brush (Cleveland, OH) chart recorder.

RESULTS

Figure 39.1 shows the appearance of fixed and stained eggs in the assay for polyspermy. Nuclei of penetrated sperm are more decondensed than nuclei of external sperm (Fig. 39.1(a)); nevertheless, it is necessary to remove external sperm (see Methods) for a clear view of the egg cytoplasm. Monospermic and polyspermic eggs are shown in Figs 39.1(b) and (c), respectively.

Abalone oocytes were highly resistant to polyspermy, whether they were inseminated within 1 or 2 h of spawning or up to $6\frac{1}{4}$ h later. The data in Fig. 39.2 show that oocytes fertilized less than 1 h (13–53 min) after spawning were no more polyspermic than oocytes fertilized later (1 h 19 min to 2 h). Oocytes inseminated more than 2 h after spawning were also highly resistant to polyspermy. For example, when inseminated $5\frac{1}{2}$– $6\frac{1}{4}$ h after spawning with 35 000 sperm/oocyte, all 74 of the oocytes penetrated by sperm were monospermic. Of a total of 1765 fertilized oocytes inseminated with sperm/oocyte ratios up to 66 000 from 13 min to $6\frac{1}{4}$ h after spawning, only 14 were polyspermic (11 dispermic and 3 trispermic).

As the interval between spawning and insemination was prolonged for more

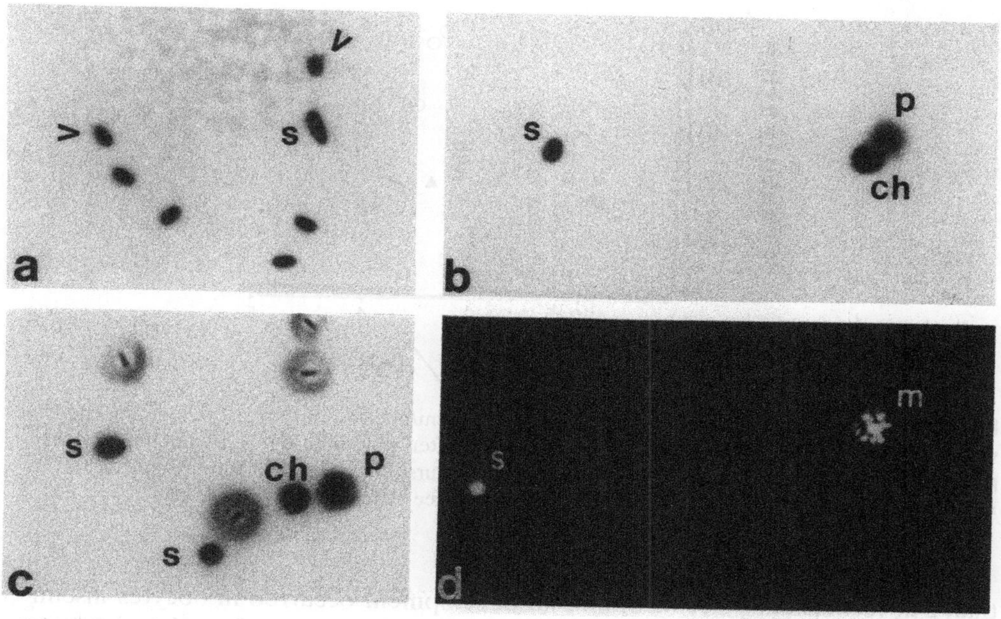

Fig. 39.1 Fertilized *Haliotis rufescens* eggs stained with bisbenzimide, squashed and viewed by fluorescence microscopy. (a) Monospermic egg showing the distinction between the decondensed pronucleus of a penetrated sperm (s) and the condensed nuclei of six externally-bound sperm (arrows). (b) Typical monospermic egg with decondensed sperm pronucleus (s), first polar body (p) and female chromosomes prior to second polar body formation (ch). (c) Polyspermic egg with two sperm pronuclei(s), first polar body (p) and female chromosomes (ch). Four external sperm nuclei (with halos) are visible out of focus. (d) Oocyte inseminated $6\frac{1}{4}$ hours after spawning. A sperm(s) has penetrated, but the oocyte failed to activate; female chromosomes are still in metaphase of the first meiotic division (m). (a)–(c) were printed from colour slides and (d) from a black-and-white negative. ×1725.

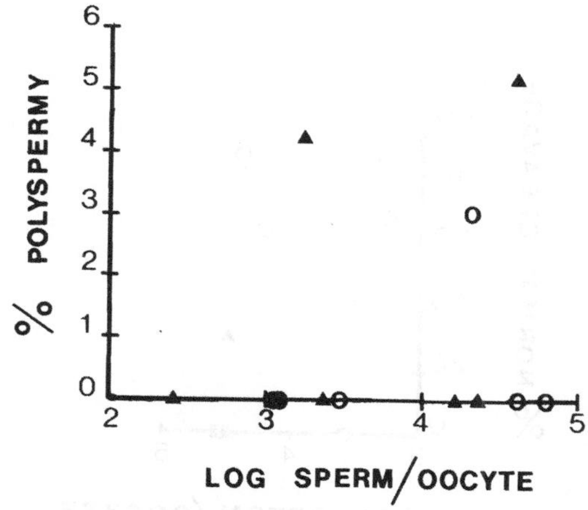

Fig. 39.2 Percentage of polyspermic eggs inseminated at various sperm/oocyte ratios within 1 hour of spawning (○) or between 1 and 2 hours after spawning (▲). No difference in susceptibility to polyspermy was observed between the two groups. Data are from four different female/male pairs.

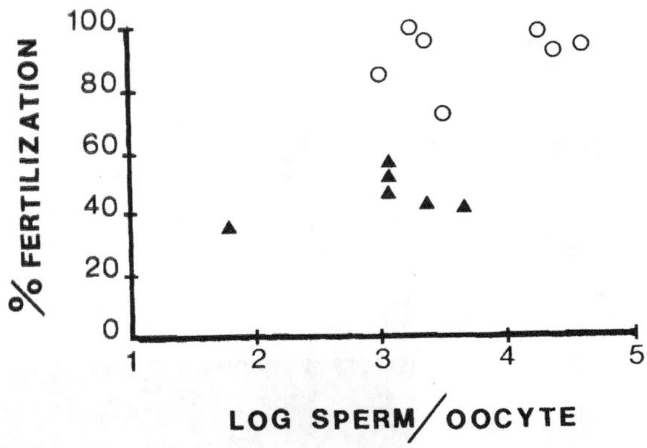

Fig. 39.3 Percentage of fertilization as a function of sperm/oocyte ratio for oocytes inseminated within 2 hours of spawning (○) or between 2 and 3 hours after spawning (▲). Eggs containing a sperm pronucleus were scored as fertilized. During the first 2 hours after spawning, fertilization success remained high, but declined thereafter. Data are from three different female/male pairs.

than 2 h, fertilizability declined, and no development occurred in oocytes inseminated after 5 h (Figs 39.3 and 39.4). However, when oocytes from late inseminations were examined with fluorescence microscopy, it was observed that many of them were fertilized, but had failed to activate (Fig. 39.1(d)). For example, in one experiment when oocytes were inseminated (35 000 sperm/oocyte) at 5 h 40 min after spawning, 75% contained a sperm pronucleus, but only 16% had resumed meiosis and the rest remained in metaphase I. Unactivated oocytes containing a sperm pronucleus were only occasionally observed when the interval between spawning and insemination was 2 h or less, but became increasingly frequent after this time.

The absence of polyspermy in abalone oocytes despite high sperm/oocyte ratios suggested these oocytes might have a fast electrical block to polyspermy (see Discussion). Support for this hypothesis comes from observations that abalone

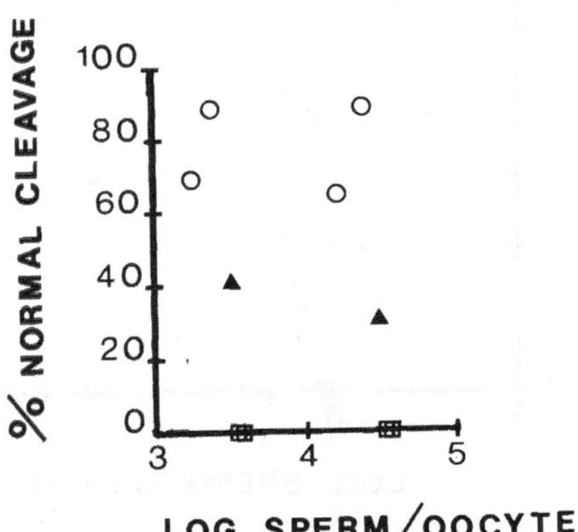

Fig. 39.4 Percentage of eggs with normal cleavage (2–4-cell stages) when insemination was at various times after spawning: ○ before 2 hours; ▲2–3 hours; □5–6 ¼hours. Data from 1 pair.

Fig. 39.5 Intracellular microelectrode recording from an oocyte of *Haliotis kamtschatkana* during insemination showing a fertilization potential. 0 mV is indicated on the left, positive potentials are above this line and negative potentials below. The resting potential was −45 mV; $1\frac{3}{4}$ minutes after sperm addition the fertilization potential rose to a maximum of +35 mV.

oocytes respond to insemination with fertilization potentials. Figure 39.5 shows the membrane potential during insemination of an *H. kamtschatkana* oocyte $1\frac{3}{4}$ min after sperm were introduced nearby. The potential rose rapidly (rise time <1 s) to +35 mV (resting potential was −45 mV) and remained positive for 10 min 25 s, then returned to −50 mV. In another recording from a different oocyte, the potential jumped from −42 to +40 mV 24 s after sperm were added, and remained positive for 4 min. Presumably the delay between sperm addition and the onset of the fertilization potential was the time required for sperm to swim to the oocyte and penetrate the jelly and vitelline layers, since in other species fertilization potentials have been observed to occur within 1 s of sperm contact with the oocyte plasma membrane (see Jaffe & Gould, 1985).

DISCUSSION

The data presented here show that abalone oocytes are highly resistant to polyspermy, whether they are inseminated immediately (13 min) or $6\frac{1}{4}$ h after spawning. High percentages of fertilization were obtained with about 1000 sperm/oocyte (Fig. 39.3), and this ratio was increased to 66 000 without increasing polyspermy. Although percentages of fertilization began to decline when insemination was delayed for more than 2 h after spawning, those oocytes that were fertilized were monospermic.

Sperm/oocyte ratios used for successful fertilization have been published in previous studies, but numbers of sperm penetrating oocytes were not counted. For example, Kikuchi & Uki (1974a,b) reported that a minimum sperm/oocyte ratio of about 1000 was necessary for a high percentage of fertilization in *H. discus*. Sperm/oocyte ratios of 20 000 to 50 000 have also been used with this species (Kikuchi & Uki, 1974a,b; Arai *et al.*, 1984). Ebert & Hamilton (1983) obtained 100% normal morulae from *H. rufescens* oocytes inseminated with 20 000 sperm/oocyte. Mill & McCormick (1991) reported high percentages of normal cleavage with sperm/oocyte ratios of 5000 to 30 000 in *H. rufescens* and *H. corrugata* and 1500 to 30 000 in *H. fulgens*. Corresponding sperm concentrations in these studies were 34 000 to 83.4×10^6/ml (present study), 100 000 to 1.9×10^6/ml (Kikuchi & Uki, 1974a,b), 1.9×10^6/ml (Arai *et al.*, 1984), 400 000/ml (Ebert & Hamilton, 1983) and 15 000 to 300 000/ml (Mill & McCormick, 1991). Egg concentrations were 120 to 2400/ml in the present study and about 10 to 100/ml in the other studies cited. Leighton & Lewis (1982) reported that a higher concentration of sperm, 10^6/ml, was necessary for optimal fertilization in *H. rufescens*, but did not specify the oocyte concentration used, so comparison with other studies is difficult. In another study that failed to mention oocyte concentration, 45 000 sperm/ml were recommended as optimal for *H. diversicolor*, and cultures with more than 100 000 sperm/ml had reduced percentages of hatching larvae (Chen, 1989). The reduced yield was atttributed to polyspermy, but there were no data to support this claim (Chen, 1989).

Thus, it can be concluded that abalone will tolerate a wide range of sperm/oocyte ratios since they are highly resistant to polyspermy. Nevertheless, very high concentrations of sperm should be avoided since extensive washing is then required to prevent fouling by degenerating supernumerary sperm (see Chen, 1989; perhaps fouling was the reason for his reduced yields with high sperm concentrations). Furthermore, with extremely high sperm concentrations, supernumerary sperm will release sufficient acrosomal lysin to dissolve the oocyte vitelline layers (Kikuchi & Uki, 1974b; Leighton & Lewis, 1982); without this protective coat, eggs and embryos are quite fragile and prone to lysis (see Methods above).

A decline in successful fertilizations with increased intervals between spawning and insemination was previously described (Kikuchi & Uki, 1974c; Leighton & Lewis, 1982; Ebert & Hamilton, 1983; Chen, 1989) and shown to be temperature-dependent (Kikuchi & Uki, 1974c; Ebert & Hamilton, 1983). With *H. rufescens* gametes at 15°C, fertilization success remained high for 3 h and declined thereafter (Ebert & Hamilton, 1983). When gametes were held at 15°C, sperm viability declined faster than oocyte viability, since 72% of 6-h-old oocytes developed when inseminated with fresh sperm, but no development occurred when they were inseminated with 6-h-old sperm (Ebert & Hamilton, 1983). Ebert and Hamilton's (1983) criterion for successful fertilization with aged gametes was normal cleavage to the morula stage. In the present study, sperm nuclear penetration was also examined and revealed that substantial numbers of aged oocytes were in fact penetrated by sperm, although they failed to activate and cleave. Thus, one component of oocyte aging is the loss of the activation response to fusing sperm. It would be interesting to determine what mechanisms are involved.

How abalone oocytes prevent polyspermy is not yet known. The presence of fertilization potentials suggests they have an electrical polyspermy block. In sea urchins, starfish, the marine worm *Urechis* and frogs the first known response of eggs to the fertilizing sperm is a rapid, positive-going shift in the egg membrane potential, which prevents additional sperm entries during the first minute or more; after the fertilization potential is over, the egg is protected by the more slowly established permanent polyspermy block (reviewed by Jaffe & Gould, 1985). The present study shows that abalone oocytes have fertilization potentials. To establish that they have an electrical polyspermy block will require additional experiments showing that holding the oocyte membrane potential positive prevents sperm entry, while negative potentials cause polyspermy.

In sea urchins, the permanent polyspermy block is produced by exocytosis of egg cortical granules whose contents modify the vitelline layer, making it impenetrable by sperm (see Jaffe & Gould, 1985). In *Urechis* and some other marine invertebrates including various molluscs, cortical granules are present but do not undergo exocytosis at fertilization; the permanent polyspermy block is presumably due to a change in the egg plasma membrane that prevents supernumerary sperm from fusing with it (see Jaffe & Gould, 1985). Whether abalone oocytes even have cortical granules is not clear. These structures were not mentioned in a fine structural study which included observations of mature ovarian oocytes (Martin *et al.*, 1983). In the present study, many sperm were observed within the perivitelline space of heavily inseminated but normally developing eggs, supporting the existence of a polyspermy block at the egg plasma membrane. Lewis *et al.* (1982) reported that vitelline layers of 2-cell *H. rufescens* embryos retained the ability to bind sperm and be dissolved by sperm lysin. Studies are underway to characterize further the permanent polyspermy block in abalone eggs.

The fact that abalone oocytes resist polyspermy even at very high sperm/oocyte ratios suggests they may normally encounter high sperm densities in nature. Shepherd (1986) observed that sexually mature *H. laevigata* show a tendency to aggregate during the spawning season. (Breen & Adkins 1980) observed many *H. kamtschatkana* piled one on top of the other during spawning in the field. In addition, spawners had congregated on the highest elevation available in their immediate area (Breen & Adkins, 1980). During a spontaneous spawn in the laboratory, individuals of the same species were also observed to have climbed to the top of the tanks (Quayle, 1971). These observations suggest abalone oocytes are exposed to high sperm densities during natural spawning and that this is the adaptive explanation for their high resistance to polyspermy.

ACKNOWLEDGEMENTS

Lourdes A. Solano-Estrada, J.M. Ornelas Orozco and M. de R. Aldaz Velarde are thanked for their help in the laboratory, and M. Gould is thanked for help with the English translation. The electrophysiological recordings were performed at the Friday Harbor Marine Laboratories, Friday Harbor, WA; the author thanks Drs L.A. Jaffe, R. Kado and D. Kline for the use of their equipment. This work was supported by grants from CONACYT, UABC and Friday Harbor Marine Laboratories.

REFERENCES

Arai K., Naito F., Sasaki H. & Fujino K. (1984) Gynogenesis with ultraviolet ray irradiated sperm in the Pacific abalone. *Bull. Jap. Soc. Sci. Fish.*, **50**, 2019–23.

Boveri T. (1902) On multipolar mitosis as a means of analysis of the cell nucleus. In, *Foundations of Experimental Embryology*, 2nd edn. 1964 (Ed. by B.H. Willier & J.M. Oppenheimer), pp. 75–97. Hafner Press, New York.

Breen P.A. & Adkins B.E. (1980) Spawning in a British Columbia population of northern abalone, *Haliotis kamtschatkana*. *Veliger*, **23**, 177–9.

Chen H.-C. (1989) Farming the small abalone, *Haliotis diversicolor supertexta*, in Taiwan. In *Handbook of Culture of Abalone and Other Marine Gastropods* (Ed. by K.O. Hahn), pp. 265–83. CRC Press, Boca Raton.

Ebert E.E. & Hamilton R.M. (1983) Ova fertility relative to temperature and to the time of gamete mixing in the red abalone. *Haliotis rufescens*. *Calif. Fish Game*, **69**, 115–20.

Jaffe L.A. & Gould M. (1985) Polyspermy-preventing mechanisms. In *Biology of Fertilization*, Vol. 3 (Ed. by C.B. Metz & A. Monroy), pp. 223–50. Academic Press, New York.

Jaffe L.A., Gould-Somero M. & Holland L.Z. (1979) Ionic mechanism of the fertilization potential of the marine worm, *Urechis caupo* (Echiura). *J. Gen. Physiol.*, **73**, 469–92.

Kikuchi S. & Uki N. (1974a) Technical study on artificial spawning of abalone, genus Haliotis II. Effect of irradiated sea water with ultraviolet rays on inducing to spawn. *Bull. Tohoku Reg. Fish. Res. Lab.*, **33**, 29–86.

Kikuchi S. & Uki N. (1974b) Technical study on artificial spawning of abalone genus Haliotis III. Reasonable sperm density for fertilization. *Bull. Tohoku Reg. Fish. Res. Lab.*, **34**, 67–72.

Kikuchi S. & Uki N. (1974c) Technical study on artificial spawning of abalone, genus Haliotis IV. Duration of fertility related to temperature. *Bull. Tohoku Reg. Fish. Res. Lab.*, **34**, 73–5.

Leighton D.L. & Lewis C.A. (1982) Experimental hybridization in abalones. *Int. J. Invert. Reprod.*, **5**, 273–82.

Lewis C.A., Talbot C.F. & Vacquier V.D. (1982) A protein from abalone sperm dissolves the egg vitelline layer by a non-enzymatic mechanism. *Develop. Biol.*, **92**, 227–39.

Martin G.G., Romero K. & Miller-Walker C. (1983) Fine structure of the ovary in the red abalone *Haliotis rufescens* (Mollusca:Gastropoda). *Zoomorphology*, **103**, 89–102.

Mill T.S. & McCormick T.B. (1991) Optimum sperm density for fertilization in three species of (*Haliotis*) abalone. In *Abalone of the world: biology, fisheries and culture* (Supplementary papers). (Eds S.A. Guzmán de Próo, M.J. Tegner & S.A. Shepherd) *Res. Pap. Dept. Fish.* (S. Aust) (in press).

Morse D.E., Hooker N. & Morse A. (1978) Chemical control of reproduction in bivalve and gastropod molluscs, III. An inexpensive technique for mariculture of many species. *World Maricul. Soc. Proc.*, **9**, 543–7.

Quayle D.B. (1971) Growth, morphometry and breeding in the British Columbia abalone (*Haliotis kamtschatkana* Jonas). *Fish Res. Bd Can. Tech. Rept*, No. 279.

Shepherd S.A. (1986) Studies on southern Australian abalone (genus *Haliotis*) VII. Aggregative behavior of *H. laevigata* in relation to spawning. *Mar. Biol.*, **90**, 231–6.

Stephano J.L. & Gould M. (1988) Avoiding polyspermy in the oyster (*Crassostrea gigas*). *Aquaculture*, **73**, 295–307.

Chapter 40

Growth models and food conversion of cultured juvenile red abalone (*Haliotis rufescens*)

JENNIFER L. GREENIER *149 Salinas Drive, Vacaville, California 95688 USA.*

JOHN Y. TAKEKAWA *US Fish & Wildlife Service, Northern Prairie Wildlife Research Center, 6924 Tremont Road, Dixon, California 95620, USA*

ABSTRACT: Logistic, Gompertz, and von Bertalanffy growth models were fitted to juvenile red abalone data. Juvenile abalone (33 weeks) averaged 11.23 mm in length at the beginning of the experiment. Growth measurements were taken on abalone subjected to three density and three light treatments. Abalone grew an average of 8.37 mm in length, 5.36 mm in width and 0.90 g in weight during the 6-month study. Non-linear regression techniques were used to fit models to the data. Gompertz and von Bertalanffy models fit the data equally well, but the von Bertalanffy was selected as the best model because it had a smaller residual error (0.19). Feeding rate (FR) and feed conversion (FC) were also examined. Feeding rate values indicate that the abalone consumed between 3.3 and 10.9% per day of their body weight in kelp. Feed conversion ratios indicate that 3.6–8.9 g kelp per day are required to produce a 1 g increase in abalone weight. No clear relationship between FR, FC and abalone growth could be discerned.

RESUMEN: Se ajustaron datos de juveniles de abulón rojo a los modelos de crecimiento logístico de Gompertz y von Bertalanffy. El abulón juvenil (33 semanas) logró un promedio de longitud de 11.23 mm al inicio del experimento. Se tomaron medidas de crecimiento de abulones sometidos a 3 tratamientos de densidad y de luz. Durante los 6 meses de estudio, el abulón creció un promedio de 8.37 mm en lo largo, 5.36 mm en lo ancho y 0.90 g en el peso. Se usaron técnicas de regresión no lineal para ajustar los datos a los modelos. Los modelos de Gompertz y von Bertalanffy ajustan a los datos igualmente bien pero el de von Bertalanffy fue seleccionado como el mejor modelo porque tenía un menor error residual (0.19). También fueron examinadas la tasa de alimentación y la conversion de alimentos. Los valores de la tasa de alimentación indican que el abulón consumió entre 3.3–10.9% del peso de su cuerpo en sargazo gigante por día. Las relaciones de la conversion de alimento indican que se requieren 3.6–8.9 g de sargazo por día para producir un incremento de 1 g en el peso del abulón. No pudo ser discernida una relación clara entre la tasa de alimentación, conversión de alimento y el crecimiento del abulón.

INTRODUCTION

Interest in the artificial propagation of the red abalone (*Haliotis rufescens*) has increased in recent years (Ebert & Houk, 1984). Depletion of wild populations has created a greater market for commercially-produced animals. In addition, development of improved cultivation techniques has made commercial production of the red abalone more economically viable (Hooker & Morse, 1984).

Maximization of juvenile abalone growth is a critical component of successful commercial production. Many mathematical models have been used to describe growth (Clow & Urquhart, 1974; Ricker, 1975). Growth equations can be used to predict 'age to length' or time necessary to reach a certain size. Hatchery applications include projections of age for weaning (the period during which abalone can forage kelp but strongly prefer diatoms, when they are 3–5 mm in length, Ebert & Houk, 1984), grow-out (stage at which abalone have completely switched to kelp as their main food source, when ⩾12 mm in length, Ebert & Houk, 1984) or age at marketable size (51–102 mm, Hahn, 1989; Rosenberry, 1989).

A variety of factors are known to influence growth in young abalone (Leighton, 1972; Chen, 1984), including rearing density (Shibui, 1972; Koike *et al.*, 1979) and light (Shibui, 1972; Uki, 1981; Olsen, 1982). The primary goal of this study is to develop a model to describe juvenile red abalone growth under a variety of conditions for broad application to commercial production.

Food indices are used to predict intake, assimilation and production efficiencies. A feed conversion (FC) ratio is the amount of food needed to produce a unit of animal weight gain (Parker, 1987) and can be used to calculate feed costs. Feeding rate (FR) is a measure of the amount of food an animal consumes relative to its body weight (Ebert & Houk, 1984) and provides another way of estimating food requirements. Feeding rate and FC are also examined to describe relationships between these variables and juvenile abalone growth.

MATERIALS AND METHODS

Location and experimental design

Research was conducted at Granite Canyon Marine Resources Laboratory, a California Department of Fish and Game (CDFG) facility in Monterey, California, between November 1988 and April 1989. Juvenile red abalone (33 weeks) were spawned from laboratory broodstock in March 1988. Abalone were grown in cylindrical plastic containers (14 cm inside diameter × 12 cm height) on a rectangular water table (2.2 m × 0.9 m × 0.3 m). Filtered seawater was distributed to containers with a PVC pipe (2.5 cm) manifold. Water (15°C ± 0.5°C) was supplied at 1.5 ± 0.5 l/min and containers were aerated with air stones.

Abalone were stocked at three densities: low (189/m^2), medium (568/m^2) and high (1420/m^2) or 10, 30 and 75 abalone per container. Light treatments were natural (sunlight filtered through fibreglass panels), artificial (fluorescent light simulating the natural photoperiod) and *no* light (24-h darkness). Black plastic partitions were used to separate light treatments.

The experimental unit was defined as a single container subjected to a treatment combination. Abalone were treated as repeated measures (Hurlburt, 1984) and growth measurements were averaged for each container. Treatments (replicated three times) were randomly assigned to containers, and densities were randomly positioned within each light section.

Growth measurements

Growth measurements (shell length, shell width and wet body weight) were taken every 3 weeks. Length and width were measured with vernier calipers to 0.05 mm. Weight was measured with a Mettler® digital balance to 0.01 g.

Correlation of growth variables

Pearson correlation analysis (SAS, 1988) was used to determine relationships among growth parameters (length, width and weight). Samples from each 3-week measurement period were not independent, precluding use of correlation analysis across sample periods (Underwood, 1981). Therefore, correlation was performed only on observations from the last measurement period. A linear regression was used to relate length and weight. Length measurements were used for analyses because shell length is usually used to determine time for weaning, grow-out and harvest.

Growth models

Three mathematical growth models were fit to the length data:

(1) logistic (Clow & Urquhart, 1974)

$$L = L_{min} \frac{1 + b}{1 + be^{-kt}}$$

where L is shell length (mm) at time t; L_{min} is the shell length at the beginning of the experiment; and k and are b positive growth constants;

(2) Gompertz (Clow & Urquhart, 1974)

$$L = L_{max}\, e^{-ce^{-kt}}$$

where L_{max} is the asymptotic or maximum shell length and c and k are positive growth constants; and

(3) von Bertalanffy (Clow & Urquhart, 1974)

$$L = L_{max}\,(1 - be^{-kt})$$

where L_{max} is the maximum shell length and b and k are positive growth constants.

In the equations, t is equal to the age of the abalone in weeks minus 33.

Iterative least squares non-linear regressions were used to fit each model to the data and estimate model parameters (Marquardt, 1963; SAS, 1988). L_{max} was defined as 'the length that an average animal would achieve if it continued to live and grow indefinitely' (Ricker, 1975). Wild red abalone have been reported to exceed 280 mm in length (Howorth, 1988) and Hooker & Morse (1984) stated that red abalone may grow to a length of 300 mm. Based on this information, we set L_{max} at 300 mm. The model with the smallest residual error was selected as the model that best fitted the data (Fabens, 1965).

Food consumption

Abalone were maintained on an *ad libitum* diet of kelp (*Macrocystis* spp.). Containers were cleaned, the remaining food was collected and weighed and fresh

Table 40.1 Correlation of growth variables measured for juvenile red abalone at 57 weeks of age; values for the Pearson correlation coefficient, *r*, are presented relating length, width and weight ($N = 957$).

Variable	Length	Width	Weight
Length	1	—	—
Width	0.9872	1	—
Weight	0.9533	0.9529	1

kelp was added each week. Estimates of FC were calculated following Parker (1987):

$$\text{FC} = \frac{\text{weight of feed ingested (g)}}{\text{weight gain by abalone (g)}}$$

Feeding rate (Ebert & Houk, 1984) was determined by the equation

$$\text{FR (\%)} = \frac{F_1 - F_2}{t \times w} \times 100$$

where F_1 = initial food weight, F_2 = final food weight, t = number of days and w = mean abalone weight. Analysis of variance (ANOVA) procedures were used to test the significance of FC and FR among treatments. ANOVA tests were deemed significant at $P \leqslant 0.05$.

RESULTS

Correlation of variables

Growth variables were highly correlated (Table 40.1). The variables length and width were most closely related, with a Pearson correlation coefficient of 0.9872. A linear regression relating length and weight resulted in the following equation (where $r^2 = 0.9814$, $N = 8$, $P = 0.0001$):

$$\text{Weight} = -1.0338 + 0.1049\ (\text{length})$$

Growth of abalone

Abalone averaged 11.23 mm (range 7.25–18.00 mm) in length and 0.19 g (range 0.05–0.70 g) in weight when the experiment began. They grew an average of 8.37 mm in length and 0.90 g in weight over the 6-month study (Table 40.2; Fig. 40.1).

Growth models

All models fit the data well with similar residuals (Fig. 40.1, Table 40.3). An ANOVA test of residual values for all treatment combinations revealed that

Table 40.2 Mean growth of juvenile red abalone over a 6-month period; *N* is the number of containers represented, with 10–75 abalone per container; one container of abalone was lost at week 51 due to cessation of water flow to that container.

Age (weeks)	*N*	Length (mm)	Width (mm)	Weight (g)
33	27	11.23	7.55	0.19
36	27	11.74	7.88	0.22
39	27	12.26	8.18	0.27
42	27	13.45	8.98	0.35
45	27	14.76	9.87	0.46
48	27	16.04	10.68	0.59
51	27	17.54	11.65	0.78
54	26	18.37	12.15	0.91
57	26	19.60	12.91	1.09
Mean increase		8.37	5.36	0.90

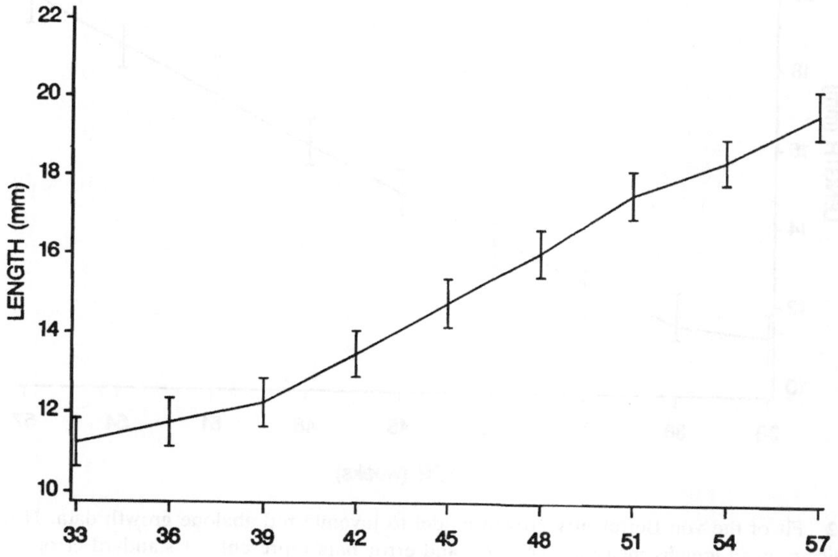

Fig. 40.1 Growth in length (mm) of juvenile red abalone reared under various light and density conditions for 6 months. Each value is the mean of 27 containers of abalone and error bars represent ±1 standard error.

Gompertz and von Bertalanffy models fit the data significantly better than the logistic model (Table 40.3). Although there was no significant difference between Gompertz and von Bertalanffy models, the von Bertalanffy model had the smallest mean residual error (Table 40.3) and was selected as the model that best fits the growth data (Fig. 40.2).

Table 40.3 Comparison of growth models applied to juvenile red abalone growth data collected every 3 weeks over a 6-month period; abalone were 33 weeks of age and averaged 11.23 mm in length at the beginning of the experiment; values represent the residual errors of each model after application of non-linear regression procedures.

	Residual		
Model	Total	Mean	se
Logistic	14.21	0.30*	0.0320
Gompertz	9.70	0.20	0.0202
von Bertalanffy	9.27	0.19	0.0188

* Significantly different from the other two models.

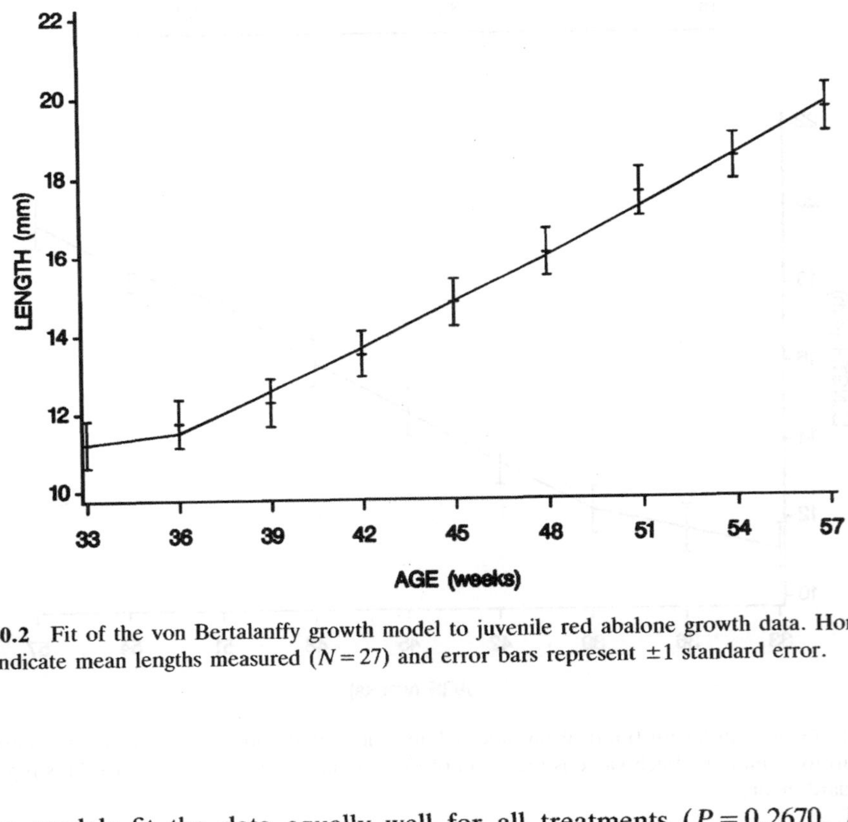

Fig. 40.2 Fit of the von Bertalanffy growth model to juvenile red abalone growth data. Horizontal bars indicate mean lengths measured ($N = 27$) and error bars represent ±1 standard error.

The models fit the data equally well for all treatments ($P = 0.2670$, $N = 6$, $F = 1.27$). Differences observed among the light and density treatments with regard to their effect on growth will be presented elsewhere. Growth curves were generated for all treatments from length observations using the von Bertalanffy growth equation (Figs 40.3 and 40.4). All treatments resulted in a similar pattern of growth. Estimates of von Bertalanffy growth parameters b and k were calculated (Table 40.4). Values for b ranged from 0.6661 to 0.6752, while k ranged from 0.00446 to 0.00653.

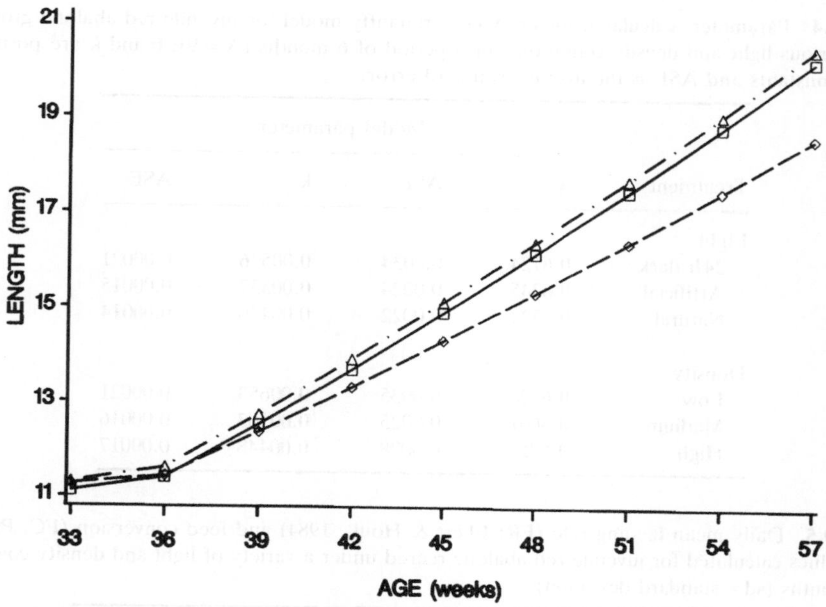

Fig. 40.3 Comparison of growth curves for juvenile red abalone subjected to three different light conditions over a 6-month period. Triangles represent mean lengths for abalone reared in containers ($N = 9$) kept in 24-h darkness. Squares represent mean lengths for abalone ($N = 9$) subjected to artificial light. Diamonds signify mean lengths for abalone ($N = 9$) maintained in sunlight.

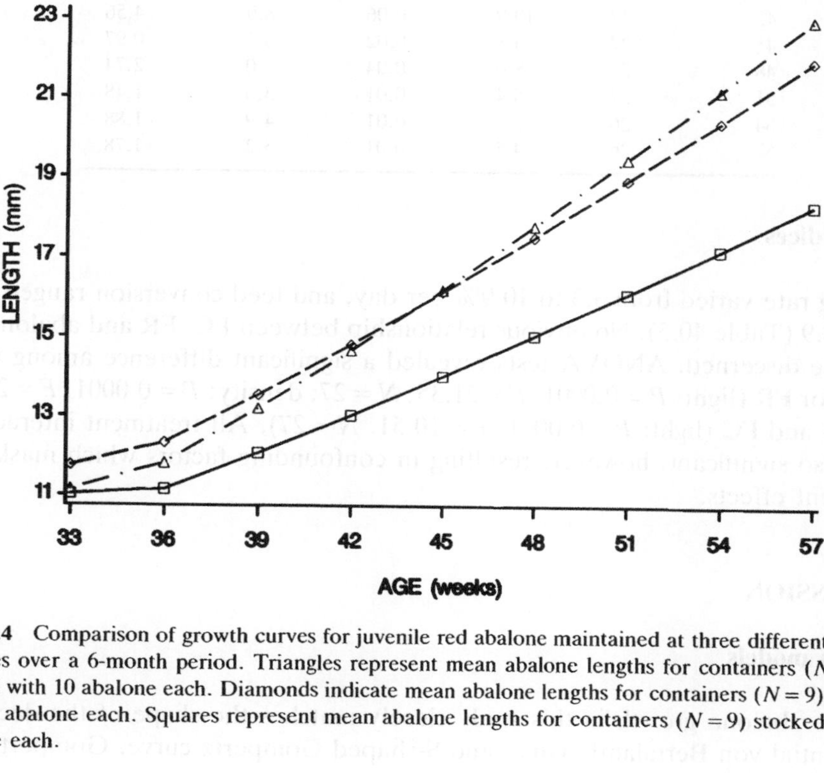

Fig. 40.4 Comparison of growth curves for juvenile red abalone maintained at three different rearing densities over a 6-month period. Triangles represent mean abalone lengths for containers ($N = 9$) stocked with 10 abalone each. Diamonds indicate mean abalone lengths for containers ($N = 9$) stocked with 30 abalone each. Squares represent mean abalone lengths for containers ($N = 9$) stocked with 75 abalone each.

Table 40.4 Parameters calculated for the von Bertalanffy model for juvenile red abalone grown under various light and density conditions for a period of 6 months ($N=9$); b and k are positive growth constants and ASE is the average standard error.

	Model parameters			
Treatment	b	ASE	k	ASE
Light				
24 h dark	0.6724	0.0034	0.00536	0.00021
Artificial	0.6745	0.0024	0.00537	0.00015
Natural	0.6724	0.0022	0.00446	0.00014
Density				
Low	0.6728	0.0035	0.00653	0.00021
Medium	0.6661	0.0025	0.00563	0.00016
High	0.6752	0.0028	0.00448	0.00017

Table 40.5 Daily mean feeding rate (FR: Ebert & Houk, 1984) and feed conversion (FC: Parker, 1987) values calculated for juvenile red abalone reared under a variety of light and density conditions for 6 months (sd = standard deviation).

		FR(%)		FC	
Age (weeks)	*N*	Mean	sd	Mean	sd
36	27	3.9	0.01	4.4	1.27
39	27	4.9	0.02	4.9	2.17
42	27	10.9	0.06	8.9	4.56
45	27	5.8	0.02	4.1	0.97
48	27	8.0	0.04	7.0	2.74
51	27	5.4	0.01	3.6	1.48
54	26	3.3	0.01	4.9	1.88
57	26	4.5	0.01	5.2	1.78

Food indices

Feeding rate varied from 3.3 to 10.9% per day, and feed conversion ranged from 3.6 to 8.9 (Table 40.5). No obvious relationship between FC, FR and abalone age could be discerned. ANOVA tests revealed a significant difference among treatments for FR (light: $P=0.0001$, $F=21.35$, $N=27$; density: $P=0.0001$, $F=23.07$, $N=27$) and FC (light: $P=0.0001$, $F=10.51$, $N=27$). All treatment interactions were also significant, however, resulting in confounding factors which mask true treatment effects.

DISCUSSION

Growth models

Juvenile abalone growth in this study closely matches the shape of the decaying exponential von Bertalanffy curve and S-shaped Gompertz curve. Gompertz and von Bertalanffy models fit the data similarly, and may be equally appropriate for describing observed growth data (Ricker, 1975). The Gompertz model, however, is most often used to describe 'weight at age' relationships (Ricker, 1975). The upper portion of the S-shaped Gompertz curve is similar in shape to the von Bertalanffy curve. Only the portion of the curve beyond the inflection point usually fits the data when applied to length data (Ricker, 1975).

The von Bertalanffy model has been shown to characterize growth for abalone and other aquatic animals (Beverton & Holt, 1957; Sainsbury, 1982; Shepherd & Hearn, 1983; Tutschulte & Connell, 1988). This model is probably a better choice than the Gompertz model for describing growth of juvenile abalone, since growth is most often expressed in terms of length.

The von Bertalanffy growth equation does not have an inflection point demarcating maximum growth rate. Therefore, the Gompertz curve should be applied in cases where an inflection point is indicated. An inflection point may be useful in determining age at which abalone should be harvested for market, since growth rate drops off beyond the inflection point on the Gompertz curve.

Application of the second derivative of the Gompertz equation to these data indicates a peak growth rate at 2.7 years of age. Caution must be exercised, however, in using a growth model to make predictions beyond the range of the data. This estimate is only valid if L_{max} is set correctly and growth is asymptotic in older abalone. 'Age to length' projections beyond the size range examined in this study may not be accurate since the pattern of abalone growth varies with time (Leighton, 1972; Ebert & Houk, 1984).

Tegner *et al.* (1989) use an L_∞ value of 201 mm for wild red abalone. However, this value is an estimate of maximum mean length for abalone from a population subjected to predation and exploitation. This value is just above the minimum allowable commercial harvest size of 197 mm.

We assume L_{max} to be 300 mm, which approximates the maximum attainable length of red abalone (Hooker & Morse, 1984). Setting L_{max} on the basis of maximum length is the approach used by Tutschulte & Connell (1988). Sensitivity analyses (Lehmann, 1975), however, indicate predicted values differ less than 1% when using 201 mm (Tegner *et al.*, 1989) versus 300 mm for L_{max}.

The von Bertalanffy growth equation may be applied to these data to predict the age at which transfer for weaning (5–10 mm) or grow out (12 mm) should occur since these sizes are within the range of lengths observed in this experiment. This model predicts a grow-out age of 38 weeks.

Ideally, growth models should be constructed from data gathered during the entire commercial life cycle of the abalone. This information could be applied by hatchery managers who wish to make 'age to length' projections for stages of abalone growth to facilitate management decisions.

Food indices

Although no distinct pattern emerged relating FR to FC to individual treatments, these results are not unusual as abalone are known to be erratic feeders, sensitive to a number of environmental and physiological influences (Shibui, 1972; Koike *et al.*, 1979; Uki, 1981). In this experiment, FR values indicate that the abalone consumed between 3.3 and 10.9% of their body weight in kelp per day. Feed conversion ratios indicate that 3.6–8.9 g kelp per day are required to produce a 1 g increase in abalone weight.

Food use indices are based on the weight of the animal. However, if size in length is known, FR and FC values can be related to abalone length by using the regression equation developed in this study for converting length to weight. Feed conversion and FR may be used to estimate how much kelp juvenile abalone 10–30 mm in size will consume and convert to weight gain.

ACKNOWLEDGEMENTS

This research was supported by grants from CDFG, Earl and Ethel Meyers Foundation, David Packard Foundation and the Conchologists Society of America. The authors would like to thank E.E. Ebert (CDFG) and J. Nybakken (Moss Landing Marine Laboratories) for their assistance and P. Haaker, K. Norman-Boudreau and M.J. Tegner for reviewing the manuscript.

REFERENCES

Beverton R.H. & Holt S.J. (1957) On the dynamics of exploited fish populations. *GB Min. Agric. Fish., Fish. Invest., Ser. 2*, **19**, 1–533.

Chen H. (1984) Recent innovations in cultivation of edible molluscs in Taiwan, with special reference to the small abalone *Haliotis diversicolor* and the hard clam *Meretrix lusoria*. *Aquaculture*, **39**, 11–27.

Clow D.J. & Urquhart N.S. (1974) *Mathematics in Biology: Calculus and Related Topics*. W.W. Horton, New York.

Ebert E.E. & Houk J.L. (1984) Elements and innovations in the cultivation of red abalone *Haliotis rufescens*. *Aquaculture*, **39**, 375–92.

Fabens A.J. (1965) Properties and fitting of the von Bertalanffy growth curve. *Growth*, **29**, 265–89.

Hahn K.O. (1989) *Handbook of Culture of Abalone and other Marine Gastropods*. CRC Press, Boca Raton.

Hooker N. & Morse D.E. (1984) Abalone: the emerging development of commercial cultivation in the United States. In *Recent Innovations in the Cultivation of Pacific Molluscs* (Ed. by D.E. Morse, K. Chew & R. Mann), pp. 365–413. Elsevier, New York.

Howorth P.C. (1988) *The Abalone Book*. Naturegraph Publishers, Happy Camp, California.

Hurlburt S.H. (1984) Pseudoreplication and the design of ecological field experiments. *Ecol. Monogr.*, **54**, 187–211.

Koike Y., Flassch J. & Mazurier J. (1979) Biological and ecological studies on the propagation of the ormer, *Haliotis tuberculata* Linnaeus. II. Influence of food and density on the growth of juveniles. *La Mer*, **17**, 43–52.

Lehmann E.L. (1975) *Nonparametrics: Statistical Methods based on Ranks*. Holden-Day, San Francisco.

Leighton D.L. (1972) Laboratory observations on the early growth of the abalone, *Haliotis sorenseni*, and the effect of temperature on larval development and settling success. *Fish. Bull.*, **70**(2), 373–81.

Marquardt D.W. (1963) An algorithm for least-squares estimation of nonlinear parameters. *J. Soc. Indust. Appl. Math.*, **11**, 431–41.

Olsen S. (1982) Abalone and scallop culture in Puget Sound. In *Abstracts*, NSA West Coast Section Meeting, Olympia, WA.

Parker N.C. (1987) Feed conversion indices: controversy or convention? *Prog. Fish Cult.*, **49**, 161–6.

Ricker W.E. (1975) Computation and interpretation of biological statistics of fish populations. *Fish. Res. Bd Can. Bull.*, **91**.

Rosenberry B. (1989) Abalone unlimited, Inc. *Mollusk Farming USA*, December.
Sainsbury K.J. (1982) Population dynamics and fishery management of the paua, *Haliotis iris* I. Population structure, growth, reproduction and mortality. *NZ J. Mar. Freshwat. Res.*, **16**, 147–61.
SAS Institute (1988) *SAS/STAT user' guide, release 6.03 edition*. SAS Institute, Cary, NC.
Shepherd S.A. & Hearn W.S. (1983) Studies on southern Australian abalone (genus *Haliotis*). IV. Growth of *H. laevigata* and *H. ruber*. *Aust. J. Mar. Freshwat. Res.*, **34**, 461–75.
Shibui T. (1972) On the normal development of the eggs of Japanese abalone, *Haliotis discus hannai* Ino, and ecological and physiological studies of its larvae and youngs. *Bull. Iwate Pref. Fish. Exp. Stat.*, **2**, 1–60.
Tegner M.J., Breen P.A. & Lenner C.E. (1989) Population biology of red abalones, *Haliotis rufescens*, in southern California and management of the red and pink, *H. corrugata*, abalone fisheries. *Fish Bull.* **87**(2), 313–39.
Tutschulte T.C. & Connell J.H. (1988) Growth of three species of abalones (*Haliotis*) in southern California. *Veliger*, **31**, 204–13.
Uki N. (1981) Feeding behavior of experimental populations of the abalone, *Haliotis discus hannai*. *Bull. Tohoku Reg. Fish. Res. Lab.*, **43**, 53–8.
Underwood A.J. (1981) Techniques of analysis of variance in experimental marine biology and ecology. *Oceanogr. Mar. Biol. Ann. Rev.*, **19**, 513–605.

Chapter 41
Desarrollo y estado actual del cultivo de abulón en México
(Development and present status of abalone culture in Mexico)

ALFREDO E. SALAS GARZA and RICARDO SEARCY BERNAL
Instituto de Investigaciones Oceanológicas, Universidad Autónoma de Baja California, Apdo. Postal 453, Ensenada, BC, México

ABSTRACT: Abalone culture is considered an alternative for the drastic reduction of natural abalone populations. In the 1970s the activities in this field took place in the Estación de Biología Pesquera, El Sauzal, BC, where the first abalone spawnings and larval cultures were done. A seed production programme was developed for reseeding purposes, involving more than 10000 seed in two years. Another programme of artificial spawning in commercial boats and liberation of fertilized eggs in the sea bottom was also promoted.

A second seed production programme with reseeding goals was started in 1983 between the cooperative organizations and the Secretaría de Pesca, in two regional laboratories. Since then, they have been producing around 5000 juveniles annually. Another cooperative laboratory started operations in 1985, producing 25000 seed each year.

The Instituto de Investigaciones Oceanológicas (IIO) of the University of Baja California started studies on abalone juvenile ecology and reseeding experiments, with recovery rates of 1 to 6% in eight months. It also developed successfully the first experimental culture in captivity, and gave technological advice to a cooperative, that currently performs the first pilot scale abalone culture on floating devices in Baja California.

Currently all the abalone seed production will be used primarily for culture in captivity, and two more cooperatives' laboratories will start operations this year. The problem of an inefficient seed production technology, still remains the main obstacle for the development of this activity in México. The IIO, is performing research dealing with this problem.

RESUMEN: El cultivo de abulón surge como una alternativa al drástico agotamiento de las poblaciones naturales. En la década de los 1970s las actividades en este campo son desarrolladas por la Estación de Biología Pesquera del Sauzal, BC, donde se realizaron los primeros desoves y cultivos larvarios de abulón. Se realizó el primer programa de producción de semilla y repoblamiento por dos años, que culminó con la siembra de más de 10000 juveniles. Se realizaron desoves a bordo de embarcaciones comerciales y se liberaron posteriormente los huevos fertilizados en el fondo marino.

Es hasta 1983 cuando se inicia un segundo programa de producción de semilla con fines de repoblamiento, entre el Sector Cooperativo Pesquero y la Secretaría de Pesca en dos laboratorios de la región. Se producen desde entonces alrededor de 5000 semillas por año. Otro laboratorio de ese sector que opera desde 1985, ha producido alrededor de 25000 semillas anualmente.

La Universidad Autónoma de Baja California, a través del Instituto de Investigaciones Oceanológicas (IIO) en 1984 realizó estudios sobre la ecología de juveniles y siembras experimentales con recuperaciones del 1 al 6% en ocho meses. Se realizó también el primer cultivo experimental en cautiverio con éxito y se dió asesoría a una cooperativa que desarrolla el primer cultivo piloto de abulón en artes suspendidas en Baja California.

Actualmente la semilla producida en México se destinará fundamentalmente al cultivo en cautiverio, estando por iniciar operaciones otros dos laboratorios del sector cooperativo en este año. Sin embargo, prevalece el problema de la deficiente producción de semilla, como el principal obstáculo para el desarrollo de esta actividad en el país. El IIO, realiza investigaciones encaminadas a dar alternativas a este problema.

INTRODUCCION

En México el abulón es un recurso que se ha explotado comercialmente desde hace más de 120 años, llegando a ser el principal productor del mundo hace más de una década, cuando aportaba el 34% de la captura mundial con 2000 y 3000 toneladas anuales (FAO, 1975). A partir de 1976, a pesar de las medidas regulatorias impuestas por la Secretaría de Pesca para protejer el recurso, tales como épocas de veda y talla mínima de captura, la producción declinó de manera constante hasta llegar a menos de 300 toneladas en 1986.

En este capítulo se pretende hacer una reseña histórica de los hechos más sobresalientes del cultivo de abulón en México, para lo cual fue necesario rescatar trabajos que, a pesar de ser una contribución importante, fueron de escasa circulación. Se recurrió además, a entrevistar a personas directamente relacionadas con los sectores oficial y productivo, para enriquecer este documento sobre acontecimientos que por su naturaleza no figuran en ningún trabajo publicado. Los autores piden disculpas por la omisión no intencional de algún hecho relevante en esta revisión. Finalmente, se refiere también al estado actual del cultivo en nuestro país, haciéndose un llamado a la comunidad a colaborar en el desarrollo de esta empresa.

DESARROLLO HISTORICO

En la década de los 1960s se manifestaron algunas inquietudes por cultivar el abulón, efectuándose un transplante de más de 2000 abulones adultos que fueron extraídos en Isla de Cedros, BC, y sembrados en varios sitios de Bahía de los Angeles, BC. Al parecer los abulones murieron ya que el único registro, efectuado ocho meses después de la siembra, fue de 52 abulones vivos en la Isla del Piojo (Ortiz Quintanilla, 1966). El intento de introducir la especie a otra localidad geográfica, fué el inicio de una serie de acciones que después le seguirían con la intención de incrementar el recurso abulonero del país. Pocos años después, la alternativa de la acuacultura tomaría una mayor fuerza, ante la situación tan crítica de esta pesquería. Algunos científicos mexicanos fueron a Japón para familiarizarse con las técnicas de ese país, que son de las más avanzadas en este campo. En el Japón los esfuerzos acuaculturales se han encaminado principalmente a el repoblamiento de bancos naturales y en menor grado al cultivo en cautiverio.

Es en 1970 cuando en la Estación de Biología Pesquera del Sauzal, BC (hoy Centro Regional de Investigaciones Pesqueras) dependiente del Departamento de Pesca, (Fig. 41.1, Tabla 41.1) se efectúan los primeros desoves de abulón rojo, *Haliotis rufescens* y abulón azul *H. fulgens*, utilizando como inductores el shock térmico, la estimulación mecánica y el estímulo con esperma en el medio. El trabajo de Félix Cota (1970) describe el desarrollo embrionario y larvario previo al asentamiento.

En ese mismo Centro, en 1972 se inició el primer programa de cultivo de abulón del país, que consistió en la producción de semilla en el laboratorio para repoblar zonas sobreexplotadas (Félix Cota, 1972). Este proyecto fue financiado por un fondo constituído por cooperativas e industrias y manejado por un patronato. En este Centro se trabajó con un sistema de acuarios de 24 m^3 de agua de mar

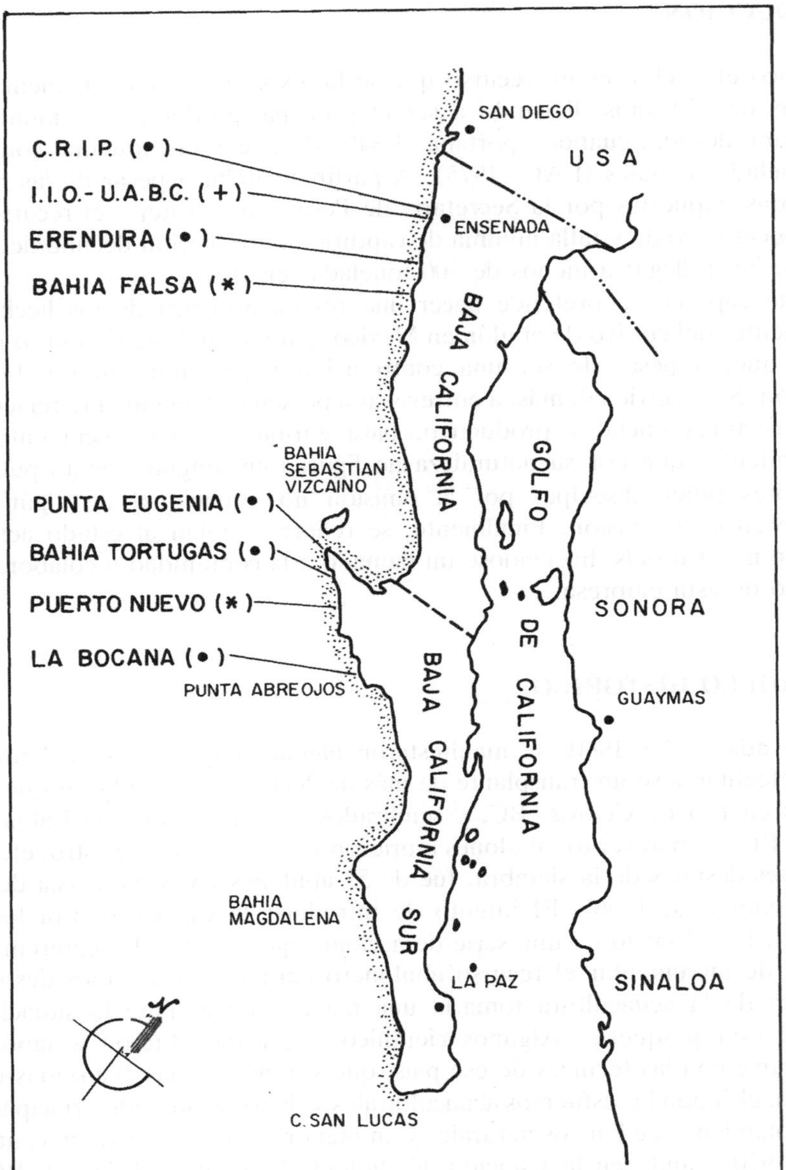

Fig. 41.1 Localización de laboratorios de producción (•) de semilla de abulón o investigación (+) y granjas (*) de engorda en la península de Baja California.

filtrada, en los cuales se pretendía producir anualmente 100 000 juveniles de 3.0 cm de longitud, mediante la aplicación de las técnicas japonesas, además de investigar costos y sistemas de producción. Félix Cota (1973) registró sobrevivencias del 1% del total de la larva fijada hasta los 50 días de edad, destacando una mayor resistencia y un mejor crecimiento de las postlarvas de *H. rufescens* sobre *H. fulgens*. Este programa culminó con la liberación en el océano de más de

Table 41.1 Relación de laboratorios y granjas (*) dedicados al cultivo de abulón en México.

Laboratorio o granja(*)	En operación	Actividad	Producción total aproximada	Destino
CRIP El Sauzal, BC	1970–6	Producción de semilla	>10 000	Repoblamiento
Eréndira, BC	1981–2 1984–8	Producción de semilla	>15 000	Repoblamiento
Bahía Tortugas, BCS	1983–9	Producción de semilla	25 000	Repoblamiento
La Bocana, BCS	1986–9	Producción de semilla	55 000	Repoblamiento
Bahía Falsa(*), BC	1985–9	Engorda	1 ton. c/concha	
IIO–UABC BC	1987–9	Investigación	–	–
Punta Eugenia, BCS	1989	Producción de semilla y engorda	–	Cultivo
Puerto Nuevo(*) BCS	1989	Engorda	–	Cultivo

10 000 juveniles de *H. rufescens* mayores de 2.0 cm de longitud. La siembra la realizaron pescadores en las zonas de influencia de las Cooperativas Pesqueras; Punta Abreojos al Sur de Bahía Tortugas; Leyes de Reforma contigua a Bahía Asunción; Bahía Tortugas en la bahía del mismo nombre y la Cooperativa Ensenada que abarca aproximadamente tres cuartas partes del Estado de Baja California.

En 1973, durante el desarrollo del programa antes mencionado, el Biólogo Martín Ortiz Quintanilla (Ortiz Quintanilla, 1974, 1980) buscó otras alternativas para incrementar las poblaciones naturales de abulón. Empezó dando asesoría a diversas cooperativas para que los equipos, durante la extracción del abulón, realizaran lo que se denominó 'mareas de cría'. Estas consisten en exponer machos y hembras de abulón al sol por una hora y hacerlos desovar, para después fertilizar los óvulos en una cubeta y colocarlos en una bolsa de plástico, que el buzo baja al fondo para liberarlos. Esta actividad se realizó por algunos equipos en una o dos temporadas de pesca, pretendiendo ayudar a la recuperación del recurso natural.

Las 'mareas de cría' fueron en ese tiempo muy cuestionadas por algunos científicos mexicanos y extranjeros, sin embargo, actualmente (13 años después) se parte de ese principio básico para actividades de repoblamiento que aún están en evaluación. Estas consisten en el cultivo de larvas en el laboratorio y una vez que están próximas al asentamiento, se colocan en una bolsa de plástico para transportarlas al lugar donde un buzo las liberará en el fondo. Estas actividades se están realizando en países como Japón, (Mizuguchi, 1983) Nueva Zelanda (Tong *et al.*, 1987) y Estados Unidos (M.J. Tegner, com. personal).

En 1975, a través de un intercambio de cooperación técnica entre los gobiernos de México y Japón, en la Estación de Biología Pesquera del Sauzal, BC, dos expertos japoneses en cultivo de abulón, Shogo Kikuchi y Nagoshisma Uki, enseñaron a científicos y técnicos mexicanos el uso de las técnicas de cultivo japonesa. El entrenamiento comprendió principalmente: el desove de sementales expuestos al agua de mar irradiada con luz ultravioleta (UV), el cultivo puro de diatomeas bentónicas para inducir las larvas al asentamiento, la colecta de crías mediante láminas transparentes colonizadas con diatomeas y el transplante y mantenimiento de las crías en tanques exteriores de concreto (Kikuchi y Uki, 1976). La falta de infraestructura no permitió el cumplimiento total de los objetivos. Los científicos japoneses determinaron que las condiciones naturales de México, son mucho más propicias que en su país para el cultivo natural de crías, por lo que habría de esperar resultados positivos de las inversiones que se efectuaron. También consideraron urgente el desarrollo de técnicas de siembra y el entrenamiento de técnicos para su aplicación, ya que de lo contrario la semilla sembrada sólo serviría de alimento a depredadores sin beneficiar a las poblaciones naturales.

En 1978, la Secretaría de Pesca presentó un estudio de factibilidad de cultivo de abulón (Ortiz Quintanilla y Medina, 1978) para producir un millón de juveniles anualmente en el laboratorio del Ejido Eréndira, BC. Se proponía vender la semilla a cooperativas pesqueras para repoblar bancos naturales y que fuera recuperada tres o cuatro años después. Las Industrias Pesqueras Paraestatales en 1979, solicitaron los servicios de Living Marine Resources Inc. (LMR) en California, EU para estudiar el recurso abulonero y rehabilitarlo en regiones costeras selectas. En este proyecto participaron tres científicos norteamericanos y dos mexicanos. Se evaluó el estado de poblaciones de abulón, utilizando filmaciones con cámara submarina en transectos en Bahía Tortugas BCS, Isla de Cedros e Islas San Benito BC. Además en octubre de 1979 se realizaron 'mareas de cría' con *H. corrugata* en Isla de Cedros BC y se fracasó en el intento de sembrar postlarva, debido a la baja eclosión de las gametas fertilizadas para ese efecto. (Swint *et al.*, 1979).

En 1981 y 1982 en el laboratorio de Eréndira, BC (Fig. 41.1, Tabla 41.1) se realizaron desoves y se produjeron algunos cientos de semillas de abulón por técnicos de la Secretaría de Pesca. Sin embargo, es hasta 1983 cuando, mediante un convenio, la Federación de Sociedades Cooperativas de Producción Pesquera y la Secretaría de Pesca, iniciaron lo que fué el segundo programa de producción de semilla de abulón con fines de repoblamiento en el país. Las metas fueron la producción de 100 000 juveniles de 2.0 cm por año y, además, la capacitación de personal en la biotecnia, incluyendo personal técnico por parte del sector cooperativo (Javier Amador Buenrostro, com. personal). Para el efecto, se trabajó en el laboratorio de Eréndira, BC y en otro de reciente construcción en Bahía Tortugas, BCS (Fig. 41.1, Tabla 41.1).

En este programa que duró más de dos años, se aplicaron principalmente las técnicas japonesas. La inducción al desove se hizo utilizando los métodos (combinados en ocasiones) de desecación, choque térmico y luz UV. El cultivo de larvas se realizó en cubetas con cambios diarios de agua y el asentamiento en tanques conteniendo placas de plástico cubiertas con diatomeas. Los juveniles se mantenían así hasta tener varios milímetros para luego ser liberados en tanques para crecimiento. La producción anual por laboratorio fluctuó entre 2000 y 5000 semillas, debido parcialmente a la falta de infraestructura adecuada. La mayor parte de la semilla producida fue liberada por los buzos de las cooperativas en zonas adyacentes a Bahía Tortugas y en algunas islas cercanas (Mario Andrade, com. personal).

Hasta ese momento las instituciones oficiales y el sector productivo habían encauzado la mayor parte de sus esfuerzos a tratar de producir grandes cantidades de semilla y muy poco se había hecho de investigación, lo cual pudo haber contribuido al lento avance de la adecuación de las técnicas extranjeras a las condiciones particulares del país.

El Instituto de Investigaciones Oceanológicas (IIO) de la Universidad Autónoma de Baja California (UABC), (Fig. 41.1, Tabla 41.1) luego de realizar un diagnóstico de la problemática del cultivo de abulón, inició en 1984 un programa que brindara alternativas de uso de la semilla de abulón que se produjera en los laboratorios mexicanos para complementar los esfuerzos de los sectores mencionados contándose con el apoyo de la Secretaría de Educación Pública y del Consejo Nacional de Ciencia y Tecnológía. Este programa empezó con el estudio de la ecología de juveniles en el medio natural (Searcy Bernal y Salas Garza, 1990). Posteriormente en el Rosario y Punta Banda, BC se realizaron con buceo autónomo tres siembras experimentales de semilla de *H. rufescens* de longitudes de 14 a 33 mm, con un total de 1720 juveniles. Se utilizaron tres técnicas diferentes: manual, sustratos artificiales y jaulas (Searcy Bernal y Salas Garza, 1987a). Los sustratos artificiales consistieron en losas de concreto de 30 × 40 cm con relieve inferior, que proveía de refugio a 30 ó 50 juveniles que ahí se colocaban.

Debido a las bajas recuperaciones en las siembras (1–7% en ocho meses), en 1985 se experimentó como alternativa el cultivo en cautiverio, el cual consistió en un arte de cultivo flotante con barriles de plástico perforados (con capacidad de 100 litros) para permitir la circulación de agua. El experimento se realizó en la Bahía de Todos Santos, BC, con semilla de *H. rufescens*, utilizando organismos con dos tallas promedio, 20 y 31 mm. Se alimentaron cada cinco o siete días con dietas diferentes (*Macrocystis pyrifera* y *M. pyrifera* con *Egregia menzienzii*); después de 18 meses los abulones crecieron un promedio anual de 27 mm y sobrevivieron más del 80%. Esto significa que un juvenil alcanza la talla comercial de 7–8 cm (que corresponde al abulón cultivado) en un período no mayor de tres años (Salas Garza y Searcy Bernal, 1987b). Su tasa de crecimiento es similar a la reportada en California EU por lo que debe ser factible realizar cultivos en cautiverio, considerando los altos costos de mano de obra de ese país en relación con México. Por lo anterior, fue considerado como una mejor opción para la utilización de la semilla. En este mismo año la Cooperativa Bahía Falsa (Fig. 41.1, Tabla 41.1) solicitó asesoría al IIO para iniciar el primer cultivo piloto comercial en el país utilizando *H. cracherodii* y actualmente con asesoría de la Secretaría de Pesca continua el cultivo con *H. rufescens*, *H. fulgens* y *H. corrugata* (con semilla producida en California, EU) en la Isla San Martín, BC.

Con estos antecedentes, la Secretaría de Pesca a través de la Dirección General de Administración de Pesquerías, realizó en 1987 cambios en su estrategia proponiendo 'destinar las semillas y juveniles de abulón producidos en los centros de acuacultura, al cultivo y no para repoblación' (Secretaría de Pesca, 1987).

La Cooperativa Pesquera Progreso construyó un laboratorio de producción de semilla de abulón en La Bocana, BCS (Fig. 41.1, Tabla 41.1) que entró en operaciones en 1986. Con un cuarto para desoves, una área de fijación de larva de 63 m^2 y cinco pilas de concreto de 7 × 1.5 × 1 m para la cría de juveniles se pretende producir 50 000 semillas anualmente, las cuales se destinarían al repoblamiento. Los sementales de *H. fulgens* son extraídos en la localidad, induciéndolos al desove con choque térmico y luz UV. El asentamiento de larvas se realiza en placas de plástico colonizadas de diatomeas bentónicas, donde se mantienen los juveniles hasta los 8–10 mm para luego ser transferidos a tanques de concreto, donde se crían hasta los 20 o 30 mm. La sobrevivencia de larvas a juveniles se calcula que es entre el 1 y 2%, produciendo en los tres años de operación 25 000, 15 000 y 25 000 juveniles, colocándose este laboratorio a la vanguardia en esta actividad en el país. El 95% de esta producción se ha liberado en el mar y parte de esta en un arrecife natural para hacer una evaluación de la recuperación en su tamaño comercial. El 5% restante se mantiene en cautiverio, y se realizan estimaciones periódicas de su crecimiento. (Daniel Aguilar, com. personal).

ESTADO ACTUAL

En este año iniciará operaciones el laboratorio de cultivo de abulón de Punta Eugenia BCS (Tabla 41.1), el cual es financiado por siete Cooperativas pertenecientes a la Federación de Sociedades Cooperativas de Producción Pesquera, 'Baja California'. Sus metas de producción en los próximos años son de 50 000, 100 000 y 200 000 juveniles, los cuales serán engordados en pilas exteriores hasta que alcancen los 7 u 8 cm. Cuando se tengan excedentes de larva, se pretende realizar las denominadas 'mareas de cría' en lugares previamente seleccionados para repoblamiento. Se encuentra en construcción una granja en Puerto Nuevo, BCS (Fig. 41.1, Tabla 41.1), que financía la Cooperativa Emancipación, para la engorda de abulón en estanques de concreto exteriores, con posibilidades en un futuro de producir también su propia semilla (Raúl Celis, com. personal).

El desarrollo del cultivo de abulón en México ha sido lento, debido en parte a la falta de una biotecnia adecuada para producción de semilla. Ante esto, la alternativa más inmediata es su compra en California EU, donde existen varios laboratorios que producen comercialmente semilla de 15 mm de longitud a un precio de $0.35 a $0.50 dólares. Sin embargo, es indispensable continuar los esfuerzos para que en el país se produzca semilla de abulón a nivel masivo y de esta manera iniciar cultivos comerciales autosuficientes, que puedan ser planeados a mediano y largo plazo.

En 1986, el IIO inició algunos experimentos relacionados con la producción de semilla en el laboratorio, tales como desoves, cultivos de larvas, cultivos de diatomeas, etc. Sin embargo, es hasta 1987 cuando se inicia un programa enfocado prioritariamente a esta fase del cultivo. Se realizaron con éxito desoves de *H. rufescens* y *H. fulgens* por el método de peróxido de hidrógeno y UV. Se realizaron experimentos utilizando como inductores al asentamiento: diatomeas bentónicas, mucus de abulón, cloruro de potasio (KCl) y ácido gama aminobutírico (GABA), obteniéndose una metamorfosis del 73% y 79% de la larva con las dos últimas. Se continuaron los experimentos con GABA, encontrándose que al alimentar la postlarva un día después del asentamiento, la sobrevivencia de *H. corrugata* 22 días después, fue del 70%. En otro experimento similar con *H. rufescens*, aplicando GABA para inducir al asentamiento y alimentando cuatro días después, la sobrevivencia en 22 días fue del 34% (Flores Aguilar, 1989; Searcy Bernal, *et al.* 1989). Estos experimentos se realizaron a nivel piloto comercial, por lo que será necesario probarlos en laboratorios comerciales. No obstante, constituyen una inovación importante que puede incrementar significativamente la producción de semilla.

El IIO ha planteado ante los sectores productivo y oficial (Salas Garza y Searcy Bernal 1987b), la necesidad de un mayor impulso a esta actividad, a través una organización entre estos sectores, incluyendo el académico, para realizar un diagnóstico del cultivo de abulón y en conjunto, redoblar esfuerzos para dar solución a los problemas inmediatos y aquéllos que surgieran a mediano y largo plazo. De esta manera, el cultivo de abulón podrá ser una actividad comercial en el país en unos pocos años.

BIBLIOGRAFIA

FAO (1975) *Year Book of Fisheries Statistics*, Vol. 40. United Nations Food and Agricultural Organization, Rome.

Félix Cota I. (1970) *Fecundación artificial y desarrollo embrionario de* Haliotis fulgens *Phillipi, 1845 y* Haliotis rufescens *Swainson, 1822 en condiciones de Acuario*. Tesis Profesional Escuela Superior de Ciencias Marinas, Universidad Autónoma de Baja California.

Félix Cota I. (1972) Programa Recurso Abulonero. *Boletín Informativo. Centro de Promoción Pesquera de El Sauzal*, **II**, 1–6.

Félix Cota I. (1973) Primeros resultados del cultivo de abulón en Baja California. *Boletín Informativo. Centro de Promoción Pesquera de El Sauzal*, **X**, 1–8.

Flores Aguilar R.A. (1989) *Estudio sobre la inducción del asentamiento y metamorfosis de larvas de dos especies de abulón* (Haliotis rufescens *y* H. corrugata). Tesis de Licenciatura, Facultad de Ciencias Marinas, UABC, Ensenada.

Kikuchi, S. y Uki N. (1976) Informe técnico sobre al cultivo y pesquería de abulón en Mexico. *Informe Técnico INP/SI Mexico D.F. No. 96.*

Mizuguchi K. (1983) Relationship between development of coastal fisheries and aquaculture. *Bull. Jap. Soc. Fish. Oceanogr.*, No. 43, 72–4.

Ortiz Quintanilla M. (1966) Informe preliminar de las investigaciones sobre la Biología y pesca del abulón comercial de las Islas de Cedros, Benito y Guadalupe, BC SIC *Dirección General de Pesca e Industrial conexas: Instituto Nacional de Investigaciones Biológicas Pesqueras. Serie de Trabajos de Divulgación*, IX No. 109.

Ortiz Quintanilla M. (1974) Algo nuevo por el abulón. *Técnica Pesquera*, No. 20, 20–6.

Ortiz Quintanilla M. (1980) Un Sistema para inducir el desove de abulón es los campos pesqueros de Baja California, México. *Memorias del 2do. Simposium Latinoamericano de Acuacultura. Vol. I. México*, pp. 871–81. Departamento de Pesca.

Ortiz Quintanilla M. y Medina I. (1978) *Proyecto para la Construcción de Laboratorio Productor de Juveniles de Abulón en BCS*. Dirección General de Acuacultura, Departmento de Pesca.

Salas Garza A.E. y Searcy Bernal R. (1987a) Perspectivas de cultivo de abulón en Baja California. In *Informe Final*, pp. 40–6. Proyecto sobre Ecología y Transplante de Juveniles de Abulón (*Haliotis* spp.) en Baja California 1984–1986. IIO – Universidad Autónoma de Baja California.

Salas Garza A.E. y Searcy Bernal R. (1987b) Alternativas de cultivo de abulón en Baja California. II Siembra en Artes Flotantes. In *Informe Final*, p. 71–93. Proyecto sobre Ecología y Transplante de Juveniles de Abulón (*Haliotis* spp.) en Baja California. 1984–1986. IIO Universidad Autónoma de Baja California.

Searcy Bernal R. y Salas Garza A.E. (1987) Alternativas de cultivo de abulón en Baja California. I. Siembra en el medio natural. In *Informe Final*, p. 49–70. Proyecto sobre Ecología y Transplante de Juveniles de Abulón (*Haliotis* spp.) en Baja California. 1984–1986. IIO Universidad Autónoma de Baja California.

Searcy Bernal R. y Salas Garza A.E. (1990) Investigaciones sobre cultivo de abulón en la Universidad Autónoma de Baja California. *Serie Científica (UABCS, México*, **1**, 44–5.

Secretaría de Pesca (1987) *Esquema de Regulación propuesto para la Administración de la Pesquería de Abulón*. Dirección General de Administración de Pesquerías.

Swint W., Evans T., Candy M., Vergue P., Ortiz Quintanilla M. y Endo Meguro R. (1979) Baja California abalone survey and spawning experiments. *Progress Report Number 2. Cedros Island and San Benito Islands.* Living Marine Resources Inc., San Diego.

Tong L.J., Moss G.A. & Illingworth J. (1987) Enhancement of a natural population of the abalone, *Haliotis iris*, using cultured larvae. *Aquaculture*, **62**, 67–72.

Chapter 42
Investigaciones en México sobre la etapa critica de la producción de semilla de abulón (*Haliotis* spp.)

(Research in Mexico on the critical stage of abalone (*Haliotis* spp.) seed production)

RICARDO SEARCY BERNAL, ALFREDO E. SALAS GARZA and ROBERTO A. FLORES AGUILAR *Instituto de Investigaciones Oceanológicas, Universidad Autónoma de Baja California, Apartado Postal 453, Ensenada, BC, México*

ABSTRACT: Two experiments are described on metamorphosis induction and early postlarval survival, aspects that in the world practice of abalone seed production are considered as the critical stage (with mortalities from 90 to 99%).

In the first experiment, with *H. corrugata*, metamorphosis was induced only with gamma-aminobutyric acid (GABA) without antibiotics, varying the time of initial addition of food (benthic diatoms), as a strategy to reduce bacterial interference. Metamorphosis after 24 h was 100% in all treatments, and the best survival was obtained when food was added one day after induction (69.5%, 22 days after settlement).

In the second experiment, with *H. rufescens*, different inductors were compared: benthic diatoms, mucus of adult abalone and GABA (with and without antibiotics, and different times of initial feeding). Metamorphosis, 24 h after induction, was estimated at 11.8, 60.2 and 84.2% respectively. There were no significant differences among the various GABA treatments. The worst postlarval survival, 22 days after settlement, was obtained using diatoms as inductor (12.1%), and the best, using GABA and feeding four days after induction (34.5%).

It is suggested that delaying the time of feeding, could improve GABA effectiveness as a metamorphosis inductor, and increase initial postlarval survival in abalone culture practice.

RESUMEN: Se describen dos experimentos sobre la inducción de la metamorfosis larval y la sobrevivencia postlarval inicial, que en la práctica mundial de la producción de semilla de abulón son señalados como la etapa crítica (con mortalidades del 90 al 99%).

En el primer experimento, con *H. corrugata*, se indujo la metamorfosis sólo con ácido gamma-aminobutírico (GABA) sin antibióticos, variando el tiempo de adición inicial de alimento (diatomeas bentónicas), como una estrategia para reducir la interferencia bacteriana. La metamorfosis a las 24 horas fue de un 100% en todos los tratamientos y la mejor sobrevivencia se obtuvo al alimentar un día después de la inducción (69.5% a los 22 días del asentamiento).

En el segundo experimento, con *H. rufescens*, se comparó el efecto inductor de diatomeas bentónicas, mucus de abulones adultos y GABA (con y sin antibióticos y distintos tiempos de primera alimentación). La inducción de la metamorfosis a las 24 horas se estimó en 11.8, 60.2 y 84.2% respectivamente. No se detectaron diferencias significativas entre los distintos tratamientos con GABA. La sobrevivencia postlarval mínima a los 22 días del asentamiento se obtuvo usando diatomeas como inductor (12.1%) y la máxima, con GABA alimentando cuatro días después de la inducción (34.5%).

Se sugiere que el retrasar el tiempo de alimentación puede mejorar la efectividad del GABA como inductor de la metamorfosis e incrementar la sobrevivencia postlarval inicial en la práctica del cultivo de abulón.

INTRODUCCION

En México, al igual que en otros países, se ha venido presentando un agotamiento de las existencias naturales de abulón (*Haliotis* spp.) y se ha planteado la acuacultura como una de las estrategias para solucionar esta problemática (Salas Garza y Searcy Bernal, 1990).

El Instituto de Investigaciones Oceanológicas (IIO) de la Universidad Autónoma de Baja California (UABC), desarrolla un proyecto de investigación sobre la acuacultura del abulón, para apoyar los esfuerzos de los sectores oficial y cooperativo. En una primera etapa (1984–6), se realizaron experimentos sobre siembra de semilla en el fondo marino, así como su cultivo en condiciones de cautiverio; mientras que en la etapa actual se abordan fundamentalmente aspectos relacionados con la obtención de semilla en laboratorio (Searcy Bernal y Salas Garza, 1990).

La producción de semilla de abulón en un proceso que requiere de 10 a 12 meses generalmente, y las etapas que comprende son descritas detalladamente por Seki (1980) y Ebert y Houk (1984), entre otros. En diversos laboratorios de investigación y producción del mundo, se ha coincidido en señalar que el período de asentamiento y metamorfosis, así como los primeros uno a tres meses de vida postlarval, constituyen la etapa crítica ya que la mortalidad asciende usualmente a un 90 ó 99% (Tabla 42.1).

Se han propuesto distintas hipótesis para explicar esa elevada mortalidad inicial, incluyendo problemas nutricionales y genéticos (Morse *et al.*, 1979b). Recientemente se ha descubierto que las larvas de abulón requieren un estímulo externo para poder realizar la metamorfosis, que es proporcionado de manera muy ineficiente en los sistemas convencionales, a través de los sustratos colonizados previamente con diatomeas, bacterias y otros microorganismos, que se utilizan en

Tabla 42.1 Algunos registros de mortalidad inicial de postlarvas de haliotidos, en laboratorios de distintos países.
(Some records of initial mortality of haliotid postlarvae, in laboratories of different countries.)

Especie	País	Mortalidad promedio (%)	Periodo	Referencia
H. discus	Japón	99.8	2.5 meses	Sagara, 1975[1]
	Japón	86.0	2–3 meses	Toole, 1987
	Corea	92.1	6 días	Pyen *et al.*, 1981
	Corea	94.3[2]	1 mes	Rho *et al.*, 1985
H. rufescens	EU	90.0	1 mes	Ebert y Houk, 1984
		92.5	3 meses	Ebert y Houk, 1984
	EU	99.0	1 mes	Morse, 1984
	Chile	99.6	2 meses	Owen *et al.*, 1984
	México	90.0	8 días	Flores Aguilar, 1989
		91.3	22 días	Flores Aguilar, 1989
H. fulgens	México	99.0	1–2 meses	Felix Cota, 1973
	México	99.2	10 días	Searcy Bernal *et al.*, 1988
		99.5	41 días	Searcy Bernal *et al.*, 1988

[1] Citado por Mottet (1978).
[2] Cifra calculada en base a tasas de asentamiento y sobrevivencia.

esta etapa del proceso. De esta manera, se puede propiciar un cultivo anormalmente retardado, que es fácil presa de bacterias patógenas y protozarios, desencadenándose así un episodio de mortalidad masiva (Morse *et al.*, 1979a,b; Hooker y Morse, 1985).

En Japón, se ha observado que el mucus secretado por abulones juveniles o adultos, induce efectivamente el asentamiento y metamorfosis larval (Seki y Kan-no, 1981a), y este método ha sustituído al tradicional en algunos laboratorios de ese país (Seki, 1980; Toole, 1987) y Corea (Y.J. Chang, 1987, com. pers.); sin embargo, no se dispone de cifras sobre el incremento en la eficiencia a nivel comercial, como resultado de la adopción de esa metodología. Se considera que el mucus puede contener substancias inductoras y servir adicionalmente como alimento de las postlarvas (Seki y Kan-no, 1981a); aunque también se ha propuesto que podría ser un medio de cultivo de bacterias patógenas (D.E. Morse, com. pers. citado por Slattery, 1987).

Por otro lado, Morse *et al.* (1979a) descubrieron que las principales substancias que inducen la metamorfosis de estos gasterópodos en la naturaleza, se encuentran en las algas incrustantes coralinas. Una de ellas, el ácido gamma-aminobutírico (GABA), puede ser usado en laboratorio a muy bajo costo, lográndose virtualmente un 100% de metamorfosis, de una manera rápida y sincrónica.

Ese descubrimiento, producto de experimentos a pequeña escala y utilizando antibióticos, se trató de aplicar a escala piloto y comercial en distintos laboratorios del mundo con resultados poco satisfactorios, ya que la mortalidad en la etapa crítica continuó siendo tan elevada como con la tecnología convencional (Akashige *et al.*, 1981; Slattery, 1987; Searcy Bernal y Salas Garza (1990); Flores Aguilar, 1989; Hahn, 1989).

Akashige *et al.* (1981), mencionan incluso que el GABA no indujo la metamorfosis de *H. discus*. Sin embargo, Morse (1984) considera que los resultados negativos de estos autores, se debieron a un control inadecuado de la calidad del agua y en particular de las bacterias, que pueden interferir significativamente en el proceso de inducción con GABA (Morse *et al.*, 1979b). En general, parece existir el consenso de que el uso efectivo de este inductor requiere la presencia de antibióticos (Slattery, 1987; D.E. Morse 1988, com. pers), lo cual podría repercutir negativamente en su aplicación comercial.

Una de las hipótesis centrales en la línea de investigación actual del IIO, es que el retrasar la alimentación inicial de las postlarvas de abulón (como una estrategia para reducir la interferencia bacteriana en el proceso de la inducción y aumentar la sobrevivencia postlarval inicial), puede ser una alternativa para utilizar efectivamente el GABA a escala masiva, sin tener que recurrir al uso de antibióticos. Muestreos y observaciones preliminares, sugieren que en los sistemas de cultivo de postlarvas, más del 75% de las bacterias heterótrofas se encuentran asociadas a la película de diatomeas bentónicas (Hinojosa-Rivera, en preparación) y, por otro lado, se ha documentado que las larvas de abulón poseen reservas suficientes para la metamorfosis y que inician su alimentación después de ese proceso (Seki y Kan-no, 1981b; Hooker y Morse, 1985; Norman-Boudreau *et al.*, 1986).

En esta contribución se describen dos experimentos enmarcados en el contexto anterior. En el primero, con abulón amarillo (*H. corrugata*), se prueba la efectividad del GABA variando los tiempos de adición inicial de alimento. En el segundo, con abulón rojo (*H. rufescens*), además de cubrir el aspecto anterior, se comparan otros inductores (diatomeas y mucus) y se evalúa la efectividad del GABA, aplicando antibióticos en las larvas (después de la eclosión) y en el alimento inoculado inicialmente, como una medida de control en estos dos aspectos del cultivo, que se consideran fuentes potenciales de contaminación bacteriana.

MATERIALES Y METODOS

Descripción general de experimentos

Ambos experimentos se realizaron en recipientes de plástico de 18 litros, colocados aleatoriamente en mesas de cultivo con iluminación continua.

El primer experimento se realizó entre octubre y noviembre de 1987, con larvas de *H. corrugata* producida en un laboratorio de California. EU, con métodos convencionales (Ebert y Houk, 1984) y transportada al IIO a los cinco días de la fertilización. Se colocaron alrededor de 2000 larvas veliger avanzadas (competentes para la metamorfosis) en cada uno de seis recicipientes correspondientes a tres tratamientos con duplicado: adición de alimento un día antes, un día después y dos días después de la inducción con GABA 1 μM (Sigma).

Las diatomeas bentónicas y microorganismos asociados, que se utilizaron como alimento, se captaron en recipientes de un sistema anexo en lo que se permitía un flujo lento de agua de mar filtrada a 5 y 1 μM, bajo iluminación continua. Después de una a dos semanas, se colectaban las diatomeas con una brocha suave y se filtraban con una bolsa de 'fieltro' de 1 μM (Ebert y Houk, 1984). Con este método se obtienen diatomeas menores de 10 μM que son adecuadas para la alimentación inicial de postlarvas (Hahn, 1989). Los filtrados de distintos recipientes de captación se mezclaron para propiciar una composición similar del alimento inicial, y se inocularon los recipientes experimentales con una densidad de 1.6×10^4 células/cm^2, que equivale a la utilizada por Ebert y Houk (1984).

El segundo experimento se desarrolló entre febrero y marzo de 1989, con larvas de *H. rufescens* producidas en el IIO. Después de la eclosión, se trató la mitad de la larva trocófora durante tres horas con una mezcla de penicilina sódica y sulfato de estreptomicina (100–130 mg/l c/u), con el objetivo de eliminar bacterias que pudieran haber quedado en el sistema, asociadas a membranas de huevos o detritus larval, que no siempre se logran separar por completo de las larvas eclosionadas con los procedimientos convencionales.

El alimento utilizado fue captado mediante el método señalado anteriormente y una parte de éste, fue tratado con la mezcla de antibióticos mencionada durante 24 h antes de la inoculación. La densidad de inóculo en este experimento fue de 7.2×10^4 células/cm^2.

La asignación de los distintos tratamientos en el segundo experimento, se señala en la Tabla 42.2. En el tratamiento con mucus, se permitió el crecimiento de diatomeas a partir del inóculo durante dos días y posteriormente se colocaron dos abulones adultos (*H. rufescens*, 7–8 cm) por recipiente durante las 24 h previas a la introducción de las larvas. En el noveno tratamiento se prolongó el tiempo de primera alimentación lo más posible hasta detectar, mediante observaciones microscópicas diarias, indicios de un menor crecimiento o mayor mortalidad de postlarvas, en relación a los recipientes con alimento.

Tabla 42.2 Descripción de los tratamientos y número de réplicas en el experimento con *H. rufescens*. Cada replica corresponde a un recipiente con 15 litros de agua de mar esterilizada (UV) a la que se agregaron 2000 larvas pedivelíger approximadamente.
(Description of treatments and number of replicates in the *H. rufescens* experiment; each replicate was a 15 l container with UV sterilized seawater. About 2000 pediveliger larvae were added to each container).

Inductor	No. de replicas	Tiempo de primera alimentacion	Antibioticos[1] en	
			Larva	Alimento
1. Diatomeas	3	1 dia antes	Si	No
2. Mucus	3	3 dias antes	Si	No
3. GABA	3	1 dia antes	Si	No
4. GABA	2	1 dia antes	No	Si
5. GABA	2	1 dia antes	Si	Si
6. GABA	1	1 dia antes	No	No
7. GABA	3	1 dia después	Si	No
8. GABA	2	1 dia después	No	No
9. GABA	2	Indefinido	Si	No

[1] Se utilizó una mezcla de penicilina sódica y sulfato de estreptomicina (100–130 mg/l c/u); la larva se trató durante 3 horas después de la eclosión y el alimento durante 24 horas antes de la inoculación.

En ambos experimentos se mantuvieron inicialmente los recipientes con cambios diarios de agua UV y después del décimo día, en el primer experimento, y del cuarto, en el segundo, se abrió un flujo lento (~300 ml/min) de agua de mar filtrada con cartuchos de 10 y 5 μM. No se proporcionó aereación y la temperatura fluctuó entre 18 y 22°C en el primer experimento, y entre 14 y 17°C en el segundo.

Criterios de evaluación

La competencia o capacidad de las larvas para efectuar la metamorfosis, se determinó por la presencia de cuatro ramificaciones en los tentáculos cefálicos (Ebert y Houk, 1984).

Se consideró que se indujo la metamorfosis efectivamente cuando se observó la pérdida del velum (Seki y Kan-no, 1981b), mediante el examen microscópico. Se practicó una evaluación a las 24 h de la inducción, en un mínimo de 20 postlarvas por recipiente, colectadas aleatoriamente de un tamiz (durante los cambios de agua) y del fondo (mediante un pincel o agua a presión). Se calculó un promedio ponderado de estas dos muestras mediante un índice de asentamiento estimado cualitativamente antes del muestreo. Este índice tomó valores de 0, 1, 2, 3 y 4, cuando se consideró un asentamiento del 5, 25, 50, 75 y 95% respectivamente y éstos últimos valores fueron los factores de ponderación.

A los cuatro días de la inducción en el segundo experimento, se colectaron 924 larvas que aún permanecían nadando en los tres tratamientos con diatomeas y se separaron en cuatro cristalizadores. En dos de ellos se agregó GABA (1 μM) y los otros dos sirvieron como control. A las 24 h se evaluó el porcentaje de organismos metamorfizados.

La sobrevivencia final, a los 22 días de la inducción, se estimó mediante el conteo directo de postlarvas en cada recipiente, por tres observadores. Se calculó la sobrevivencia en tiempos intermedios, mediante la colecta y conteo de postlarvas muertas. Las conchas de estas se clasificaron de acuerdo a los estadios ilustrados por Leighton (1974): sin concha periostomal (SP), con periostomo inicial (CP), midasimétrica (MD) y circular (CR).

Los porcentajes de metamorfosis y sobrevivencia se compararon entre tratamientos, mediante análisis de varianza de una vía con datos transformados (arco seno de la raíz cuadrada), confirmándose previamente la homogeneidad de varianzas con pruebas de Bartlett (Sokal y Rohlf, 1981). En caso de significación estadística ($P < 0.05$), se realizó un análisis de comparaciones multiples a posteriori (diferencia mínima significativa, LSD).

Otras variables registradas que no se consideran en esta contribución por motivos de espacio, son el crecimiento de postlarvas en ambos experimentos, y los niveles de bacterias heterótrofas totales (medio Zobell) y tipo *Vibrio* (medio TCBS), en el segundo experimento.

RESULTADOS

Primer experimento (*H. corrugata*)

A las 24 horas de la inducción, se observó que todas las postlarvas estaban asentadas y en las muestras de todos los recipientes colectados en el tamiz ($5 < n < 10$), se comprobó que el 100% de las postlarvas vivas estaban metamorfizadas y se consideró innecesario colectar muestras del fondo.

Las curvas de sobrevivencia por tratamiento se presentan en la Figura 42.1 y es evidente la mayor sobrevivencia de los abulones alimentados un día después en relación a los alimentados un día antes y dos días después de la inducción con GABA. Esta diferencia sólo fue significativa en las dos últimas fechas (ANOVA y LSD).

En la Figura 42.2 se muestra la distribución de estadios de las conchas colectadas en cada tratamiento acumuladas hasta el día 14 después de la inducción. En los dos tratamientos que existió mayor mortalidad, ésta ocurrió en el estadio midasimétrico; sin embargo, en los recipientes que se alimentaron dos días después de la inducción las conchas estaban menos desarrolladas.

Segundo Experimento

En el tratamiento nueve (Tabla 42.2), se inoculó alimento al cuarto día después de la inducción, ya que en muestras colectadas el día anterior se registró una mortalidad del 6.2% en esos recipientes, que aún no recibían alimento, contra un 0% en recipientes con alimento.

No se detectaron diferencias significativas en el porcentaje de larvas metamorfizadas a las 24 horas de la inducción con GABA, entre los tratamientos con y sin antibióticos alimentados un día antes (claves 3, 4, 5 y 6 de la Tabla 42.2), ni entre los que aún no recibían alimento (claves 7, 8 y 9), por lo que se decidió agruparlos [illegible] análisis posterior.

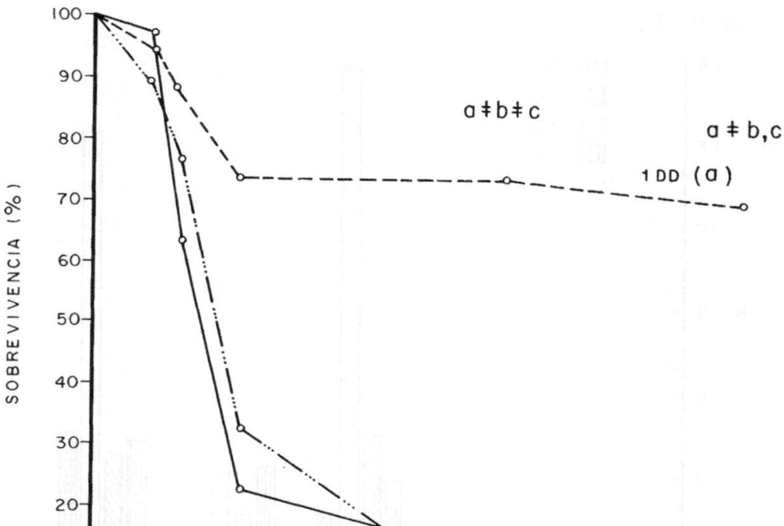

Fig. 42.1 Sobrevivencia de *H. corrugata* despues de la inducción de la metamorfosis con GABA, variando el tiempo de alimentacion inicial (1DA = un dia antes, 1DD = un dia despues y 2DD = dos dias despues de la inducción). Cada punto es el promedio de dos replicas. Los resultados de las comparaciones múltiples significativas se muestran arriba.

Fig. 42.1 Survival of *H. corrugata* after metamorphosis induction with GABA, varying initial feeding schedule (1DA = one day before, 1DD = one day after and 2DD = two days after induction). Each point is an average of two replicates. Results of significant multiple comparisons are shown above.

Tabla 42.3 Indice de asentamiento (IA) y porcentaje de metamorfosis (%M) de *H. rufescens* a las 24 horas de la inducción con distintos estímulos; la última columna (%Mp) es un promedio ponderado de las muestras de tamiz y fondo, usando IA como factor de ponderación.
(Settlement index (IA) and percentage of metamorphosis (%M) of *H. rufescens*, 24 h after induction with different stimuli. The last column (%Mp) is a weighted average of sieve (tamiz) and bottom (fondo) samples, using IA as the weighting factor.)

			Tamiz		Fondo		
Inductor	No. réplicas	IA[1]	N	%M	N	%M	%Mp
Diatomeas	3	0.3	146	7.5	14	61.1	11.8
Mucus	3	2.3	145	10.1	60	95.1	60.2
GABA (S/alimento)	7	3.6	254	33.1	236	93.5	85.5
GABA (C/alimento)	7	3.9	143	50.7	292	85.0	82.9

[1] Variable estimada cualitativamente en cada réplica (rango de 0 a 4).

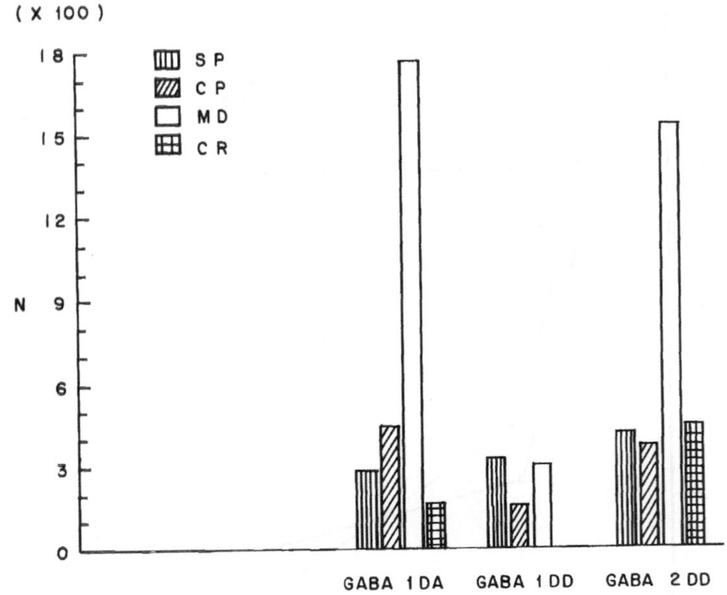

Fig. 42.2 Distribución de estadios de conchas de postlarvas muertas de *H. corrugata* acumuladas hasta el dia 14 después de la inducción con GABA y distintos tiempos de alimentación inicial (ver Fig. 42.1) (SP = sin concha peristomial, CP = concha peristomial inicial, MD = midasimétrico, CR = circular).

Fig. 42.2 Distribution of stages of *H. corrugata* dead postlarvae shells, accumulated until day 14 after GABA induction with different initial feeding schedules (see Fig. 42.1) (SP = without peristomial shell, CP = with initial peristomial shell, MD = midasymmetric, CR = circular).

En la Tabla 42.3, se resume la evaluación de la inducción de la metamorfosis a las 24 horas del inicio del experimento. Se registraron diferencias significativas entre tratamientos en el índice de asentamiento y la prueba LSD separó tres grupos (diatomeas, mucus y los dos tratamientos con GABA), así como en el porcentaje de metamorfosis, donde la prueba LSD sólo diferenció al tratamiento con diatomeas del resto (mucus y GABA).

El resultado del experimento sobre inducción de la metamorfosis, en larvas que aún permanecían nadando al cuarto día en el tratamiento con diatomeas, se presenta en la Figura 42.3. En los recipientes control (sin GABA), sólo un 28% de las larvas metamorfizaron, contra el 100% registrado en los critalizadores con GABA.

No resultaron significativas las comparaciones de sobrevivencia entre tratamientos con y sin antibióticos, tanto en los que se alimentó un día antes (claves 3, 4, 5 y 6) como en los inoculados un día después (claves 7 y 8), por lo que éstos grupos de tratamientos se combinaron.

En la Figura 42.4, se presentan las curvas de sobrevivencia para los diferentes tratamientos. Los recipientes inducidos con GABA y alimentados cuatro días después, alcanzan el valor final mayor (34.5%) y los correspondientes a diatomeas el menor (12.1%). La sobrevivencia final en mucus, GABA 1da y GABA 1dd fue de 27.6, 30.6 y 28.9, respectivamente. Las diferencias sólo fueron significativas al día 3, cuando la prueba LSD agrupó a los tratamientos con diatomeas y mucus en

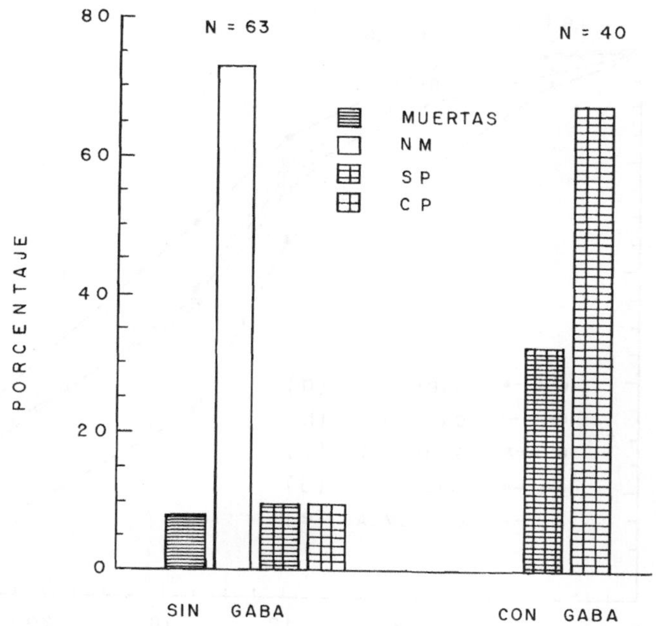

Fig. 42.3 Inducción de la metamorfosis después de 24 h de tratar con y sin GABA, larvas, de *H. rufescens* que aún permanecian nadando después de cuatro dias en el tratamiento de diatomeas (NM = no metamorfizadas; SP y CP = metamorfizadas sin y con concha peristomial respectivamente).

Fig. 42.3 Induction of metamorphosis after 24 h of treatment with and without GABA, larvae of *H. rufescens* that were still swimming after four days in the diatom treatment (muertas = dead; NM = not metamorphized; SP and CP = metamorphized without and with peristomial shell respectively).

un subgrupo, y a todos en los que se utilizó GABA, en otro; así como al día 7, en que la prueba LSD sólo diferenció la sobrevivencia menor en diatomeas, de todos los demás tratamientos (homogéneos entre sí).

En la Figura 42.5, se presenta la distribución de estadios de conchas de organismos muertos, acumulados hasta el día 22 después del asentamiento. En los tratamientos con mucus y GABA, la mayor mortalidad correspondió a estadíos midasimétrico y circular; mientras que con diatomeas, ésta ocurrió en organismos que aún no formaban concha peristomial.

DISCUSION

Es conveniente separar en el análisis la inducción del asentamiento y metamorfosis, de la sobrevivencia postlarval inicial.

Inducción del asentamiento y metamorfosis

Efectividad del GABA

En ambos experimentos se comprobó la alta efectividad del GABA para inducir

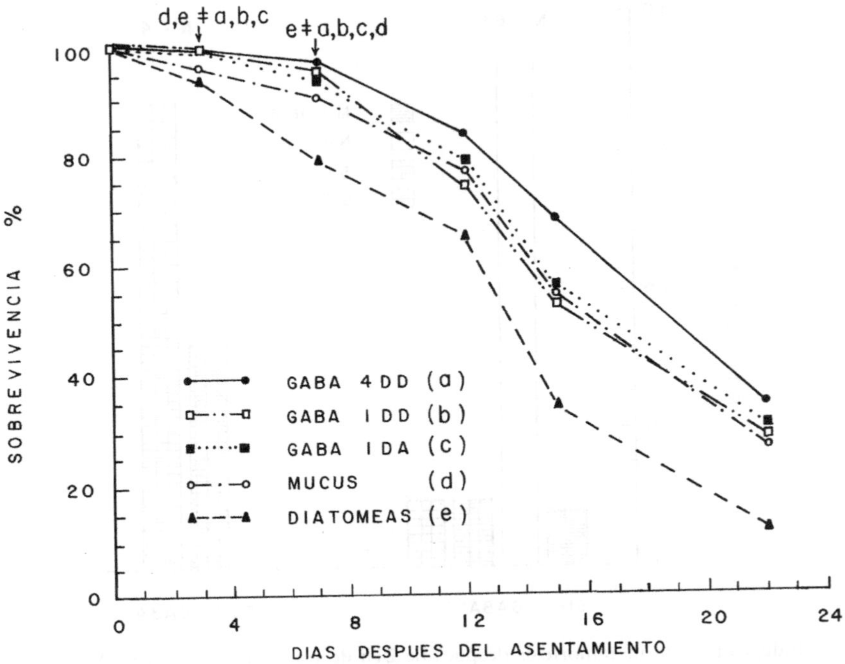

Fig. 42.4 Sobrevivencia de postlarvas de *H. rufescens* asentadas con distintos métodos. Los tres tratamientos con GABA difieren en el tiempo de alimentación inicial: un dia antes (1DA), un dia después (1DD) y cuatro dias después (4DD) de la inducción. Los resultados de las comparaciones múltiples significativas se muestran arriba.

Fig. 42.4 Survival of *H. rufescens* postlarvae settled with different methods. GABA treatments differ in the time of initial feeding: one day before (1DA), one day after (1DD) and four days after (4DD) induction. Results of significant multiple comparisons are shown above.

el asentamiento y metamorfosis de larvas de abulón, incluso en presencia de diatomeas y bacterias sin usar antibióticos. Este es el primer reporte en ese sentido.

El hecho de no encontrar diferencias significativas en el porcentaje de larvas metamorfizadas, entre tratamientos con y sin alimento presente en el momento de la inducción con GABA, sugiere que en ambos experimentos no existió una interferencia importante de bacterias en el proceso de inducción. Esto implica que no hubo un consumo microheterotrófico importante del aminoácido, ni algún tipo de acción directa (patógena, p. ej.) que inhibiera la detección del estímulo y el desencadenamiento de la metamorfosis. El resultado anterior contrasta con el de Morse *et al.* (1979b) y Akashige *et al.* (1981), quienes reportan que el GABA no induce la metamorfósis en ausencia de antibióticos; aunque es posible que en esos experimentos haya existido un nivel mayor de contaminación bacteriana.

El uso de antibióticos después de la eclosión y en los inóculos de alimento en el segundo experimento, no mejoró el efecto inductivo del GABA, sugiriendo un aporte no crítico de bacterias por ambas fuentes. Sin embargo, en este experimento se detectaron niveles iniciales de bacterias muy bajos (datos no publicados) y es posible que se trate de un caso poco representativo de la práctica común.

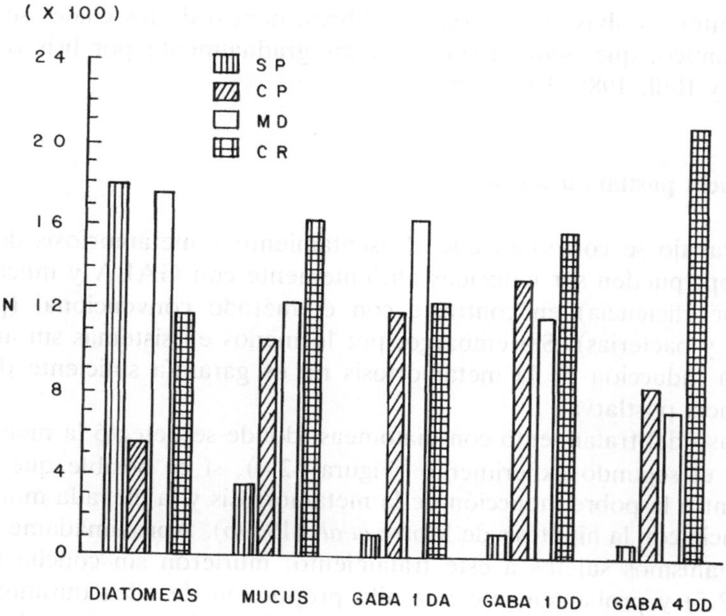

Fig. 42.5 Distribución de estadios de conchas de postlarvas muertas de *H. rufescens* acumuladas hasta el dia 22 despues de la induccion con diferentes tratamientos (simbologia como en la figura 42.2).

Fig. 42.5 Distribution of stages of *H. rufescens* dead postlarvae shells, accumulated until day 22 after induction with different treatments (symbols as in Fig. 42.2).

Efectividad del mucus y diatomeas

Después del GABA, el mucus figura como el segundo estímulo en eficiencia para inducir el asentamiento y metamorfosis en el experimento con *H. rufescens*, aunque su acción es aparentemente más lenta (Tabla 42.3, Figura 42.5). Este resultado positivo con el mucus, es consistente con los reportes de Seki y Kan-no (1981a) y Slattery (1987). Se considera que el mucus debe contener algún tipo de substancias inductoras, y que puede servir como una fuente alimenticia inicial para las postlarvas (Seki y Kan-no, 1981a).

El inductor menos eficiente de la metamorfosis, fue la película de diatomeas y bacterias usada tradicionalmente (Tabla 42.3). Es interesante observar, sin embargo, un alto porcentaje (70.1%) de organismos metamorfizados (vivas más muertas con concha postlarval) al final del experimento (Figura 42.3). Esto implica un mecanismo gradual (aunque ineficiente) de inducción que podría descansar en la acumulación de substancias inductoras.

Morse *et al*. (1979a,b) sugieren que no son las diatomeas, sino las bacterias asociadas, las que pueden provocar la liberación de GABA o ácido glutámico (precursor del GABA con propiedades inductoras menos eficientes), a través de su acción sobre el detritus larval u otras fuentes de materia orgánica. Dado que las evaluaciones de esos autores fueron a las 5 y 17 horas, no se puede descartar la contribución directa de las diatomeas a más largo plazo. Estas poseen concentraciones internas altas de aminoácidos libres, dentro de los cuales se destaca el ácido glutámico, que podrían ser liberados gradualmente por lisis o exudación (Fuhrman y Bell, 1985; Bjornsen, 1988).

Sobrevivencia postlarval inicial

En este trabajo se corrobora que el asentamiento y metamorfosis de larvas de *Haliotis* spp. pueden ser inducidas eficientemente con GABA y mucus (aunque con menor eficiencia) en contraste con el método convencional (película de diatomeas y bacterias). Sin embargo, por lo menos en sistemas sin antibióticos, una buena inducción de la metamorfosis no es garantía suficiente de una alta sobrevivencia postlarval.

En el caso del tratamiento con diatomeas, donde se detectó la menor sobrevivencia en el segundo experimento (Figura 42.4), sí es posible que exista una relación entre la pobre inducción de la metamorfosis y la elevada mortalidad, en concordancia con la hipótesis de Morse *et al.* (1979b). Aproximadamente un 30% de los organismos sujetos a este tratamiento, murieron sin concha peristomial (Figura 42.5) y probablemente una alta proporción de ellos durante la primer semana del experimento (Figura 42.4), sin haber iniciado su metamorfosis. Ebert y Houk (1984) señalan que en su laboratorio, donde se utiliza sólo este inductor, aproximadamente la mitad de la mortalidad inicial consiste en larvas no metamorfizadas.

En el tratamiento con mucus se duplicó la sobrevivencia final de *H. rufescens* (Figura 42.4) sugiriéndose que los beneficios proporcionados por éste inductor (Seki, 1980; Seki y Kan-no, 1981a), exceden los posibles perjuicios como medio de cultivo bacteriano (D.E. Morse, com. pers. citado por Slattery, 1987); sin embargo, como se mencionó anteriormente, en este experimento se detectaron niveles bajos de bacterias. Slattery (1987) reporta para la misma especie una sobrevivencia del 50% en 11 semanas y no se cuenta con elementos para tratar de explicar esta diferencia, ya que los sistemas de cultivo fueron distintos.

Tanto Slattery (1987) como algunos laboratorios comerciales de California (Flores Aguilar, 1989), han utilizado GABA sin antibióticos con resultados negativos en términos de sobrevivencia inicial. Un aspecto común en esos casos, es la adición del inductor en presencia de sustratos colonizados por diatomeas y bacterias. Una posible explicación de lo anterior, podría resultar de la combinación de varios factores:

(a) un rápido y casi total asentamiento larval;
(b) posible interferencia bacteriana (consumo de GABA o acción patógena) en la inducción;
(c) mortalidad inicial de algunas postlarvas debido a la degradación local del ambiente como producto de una alta densidad de postlarvas y bacterias;
(d) aumento progresivo de la concentración de bacterias patógenas que utilizan las larvas o postlarvas muertas como fuente de energía y
(e) una especial sensibilidad de las larvas durante la metamorfosis y primeros estadios postlarvales, a la infección bacteriana y otras condiciones ecológicas adversas.

Los resultados de este trabajo, sugieren que el uso del GABA aunado a un retraso óptimo de la adición de alimento, puede producir un incremento importante (Figura 42.1) o moderado (Figura 42.4) en la sobrevivencia inicial de postlarvas de haliótidos, comparado con su administración en la presencia de la película de diatomeas bacterias.

El tiempo óptimo de retraso puede ser variable y se debe realizar más investigación sobre las causas que lo determinan. En el caso de *H. corrugata*, se logró al primer día despues de la inducción, y al parecer al segundo día las postlarvas ya habían agotado sus reservas y murieron como consecuencia de la falta de alimento (Flores Aguilar, 1989). En el experimento con *H. rufescens* fue posible retrasar la alimentación hasta cuatro días sin observar efectos secundarios adversos; de hecho, en este tratamiento se obtuvo la mejor sobrevivencia (Figura 42.4). Se considera que la temperatura menor en este segundo caso, podría contribuir a explicar esa diferencia.

El efecto distinto del retraso de la primera alimentación en la sobrevivencia de postlarvas inducidas con GABA, en los experimentos con *H. corrugata* (diferencia significativa) y *H. rufescens* (diferencia no significativa), sugiere que esta estrategia puede rendir beneficios variables. Algunos factores que podrían incidir en este aspecto son: cantidad y calidad de bacterias y diatomeas, temperatura, densidad de postlarvas, especie y condiciones generales de cultivo (tipo de recipientes y sustratos, circulación de agua, aireación, calidad del agua, etc.). Por ejemplo, el hecho de haber mantenido agua UV durante 10 días en el primer experimento, contra sólo cuatro en el segundo, pudo haber tenido una influencia decisiva en la protección inicial de la postlarvas. Actualmente, se realiza investigación en el IIO, en este sentido.

AGRADECIMIENTOS

Los autores agradecen al Dr Earl E. Ebert, del Departamento de Caza y Pesca en Monterey, California, la donación de las larvas de *H. corrugata* utilizadas en este trabajo; así como el apoyo financiero de la Secretaría de Educación Pública, México, DF (convenios C85010155 y C86010281).

BIBLIOGRAFIA

Akashige S., Seki T., Kan-no H. & Nomura T. (1981) Effects of gamma-aminobutyric acid and certain neurotransmitters on the settlement of the larvae of *Haliotis discus hannai* Ino (Gastropoda). *Bull. Tohoku Reg. Fish. Res. Lab.*, **43**, 37–45.

Bjornsen P.K. (1988) Phytoplankton exudation of organic matter: Why do healthy cells do it? *Limnol. Oceanogr.*, **33**, 151–4.

Ebert E.E. & Houk J.L. (1984) Elements and innovations in the cultivation of the red abalone *Haliotis rufescens*. *Aquaculture*, **39**, 375–92.

Félix Cota I. (1973) Primeros resultados del cultivo de abulón en Baja California. *Boletín Informativo INP*, **10**, 4–5.

Flores Aguilar R.A. (1989) *Estudio sobre la inducción del asentamiento y metamorfosis de larvas de dos especies de abulón* (Haliotis rufescens *y* H. corrugata). Tesis de licenciatura, Facultad de Ciencias Marinas, UABC, Ensenada.

Fuhrman J.A. & Bell T.M. (1985) Biological considerations in the measurement of dissolved free aminoacids in seawater and implications for chemical and microbiological studies. *Mar. Ecol. Prog. Ser.*, **25**, 13–21.

Hahn K.O. (1989) *Handbook of Culture of Abalone and other Marine Gastropods*. CRC Press, Boca Raton, Florida.

Hinojosa Rivera P.R. (en preparación) *Cuantificación de bacterias heterótrofas y tipo* Vibrio *en el desarrollo de un cultivo de larvas y postlarvas de abulón y sus efectos en la mortalidad*. Tesis de licenciatura, Facultad de Ciencias Marinas, UABC, Ensenada.

Hooker N. & Morse D.E. (1985) Abalone: the emerging development of commercial cultivation in the United States. In *Crustacean and Mollusc Aquaculture in the United States* (Ed. J.V. Huner & E.E. Brown) pp. 365–413. AVI Publishers, Westport.

Leighton D.L. (1974) The influence of temperature on larval and juvenile growth in three species Southern California abalones. *Calif. Dept Fish Game Bull.*, **72**, 1137–45.

Morse D.E. (1984) Biochemical and genetic engineering for improved production of abalones and other valuable molluscs. *Aquaculture*, **39**, 263–82.

Morse D.E., Hooker N., Duncan H. & Jensen L. (1979a) Gamma-aminobutyric acid, a neurotransmitter, induces planktonic larvae to settle and begin metamorphosis. *Science*, **204**, 407–10.

Morse D.E., Hooker N., Jensen L. & Duncan H. (1979b) Induction of larval abalone settling and metamorphosis by gamma-aminobutyric acid and its congeners from crustose red algae. II: Aplications to cultivation, seed-production and bioassays; principal causes of mortality and interference. *Proc. World. Marical. Soc.*, **10**, 81–91.

Mottet M.G. (1978) A review of the fishery biology of abalones. *Dept. Fish. Tech. Rep.*, **37**, 1–81.

Norman-Boudreau K., Burns D., Cook C.A. & Austin A. (1986) A simple technique for detection of feeding in newly metamorphosed abalone. *Aquaculture*, **51**, 313–17.

Owen B., DiSalvo L.H., Ebert E.E. & Fonck E. (1984) Culture of the California red abalone *Haliotis rufescens* Swainson (1822) in Chile. *Veliger*, **27**, 101–5.

Pyen C.K., Jo J.Y., Jang K.N. & Yang K.H. (1981) The induced spawning and the rearing of the early stage of the abalone, *Haliotis discus hannai*. *Bull. Fish. Res. Dev. Agency*, **26**, 37–49.

Rho S., Kho C.S. & Kim S.H. (1985) Study on the propagation of abalone 4. Seedling collection in simplified outdoor canvas. *Bull. Fish. Res. Dev. Agency*, **36**, 69–77.

Sagara J. (1975) Abalone culture. In *Culture of Marine Life*, pp. 51–72. Japan International Cooperation Agency, Tokyo.

Salas Garza A.E. y Searcy Bernal R. (1990) Problemas y alternativas del cultivo de abulón en México. *Serie Científica (UABCS, México)*, **1**, 51–8.

Searcy Bernal R. y Salas Garza A.E. (1990) Investigaciones sobre el cultivo de abulón en la Universidad Autónoma de Baja California. *Serie Científica (UABCS, Mexico)*, **1**, 44–50.

Searcy Bernal R., Salas Garza A.E. y Flores Aguilar R.A. (1988) Crecimiento de postlarvas y juveniles de abulón azul (*Haliotis fulgens*) en un laboratorio mexicano. *Ciencias Marinas*, **14**, 57–72.

Seki T. (1980) An advanced biological engineering system for abalone seed production. In *Proceedings of the International Symposium on Coastal Pacific Marine Life*, pp. 45–54. Bellingham, Washington.

Seki T. & Kan-no H. (1981a) Induced settlement of the Japanese abalone *Haliotis discus hannai* veliger by the mucous trails of the juvenile and adult abalone. *Bull. Tohoku Reg. Fish. Res. Lab.*, **43**, 29–34.

Seki T. & Kan-no H. (1981b) Observations on the settlement and metamorphosis of the veliger of the Japanese abalone, *Haliotis discus hannai* Ino. Haliotidae, Gastropoda. *Bull. Tohoku Reg. Fish. Lab.*, **42**, 31–42.

Slattery M. (1987) *Settlement and metamorphosis of red abalone* (Haliotis rufescens) *larvae. A critical examination of mucus, diatoms and gamma aminobutyric acid (GABA) as inductive substrates.* Master of Arts thesis, San Jose State University.

Sokal R.R. & Rohlf R.J. (1981) *Biometry. The Principles and Practice of Statistics in Biological Research*. W.H. Freeman, San Francisco.

Toole C. (1987) *Report of a study trip to Japan. 1. Molluscan aquaculture and fisheries*. University of California, Eureka.

Chapter 43
Producción de semilla de abulón en el laboratorio: el método de Bahía Tortugas BCS, México

(Abalone seed production in the laboratory)

JOSÈ MANUEL MAZON-SUASTEGUI *Centro de Investigaciones Biológicas de Baja California Sur, AC Ocampo 312, AP 128, La Paz BCS 23000, México*

LUIS BAZUA-SICRE, GUSTAVO LUCERO-MARTINEZ Y RUBEN RODRIGUEZ-RAMOS *Secretaría de Pesca, Delegación Federal en Baja California Sur, Jalisco y Madero, La Paz, BCS 23000, México*

ABSTRACT: The Centro de Acuacultura in Bahía Tortugas, BCS, dependent on Mexico Fisheries Department is the first hatchery facility intended for applied research, technology development and seed production of abalone in Mexico. The technical staff has proved the feasibility of abalone culture and, at present, fishermen's cooperatives are involved in commercial projects. Adult spawning is induced by air exposure and thermal changes in ultraviolet-light-irradiated seawater. After 4–6 days in static culture, 1–2% of the larvae succesfully settle on clear PVC sheets with a diatom film. At least 20% of the postlarvae survive to 5 mm size, growing 40–50 microns/day on a diatom diet. The 5 mm seed begin to eat young macroalgae and rearing can be continued outdoors with negligible mortality and a growth rate of 75–150 microns/day, depending on the macroalgae/diatom diet. Grow out in suspended trays averaged 106 microns/day and nearly 100% survival in a 1-year period. The medallion size (5–7 cm) can be attained in less than 2 years.

RESUMEN: El Centro de Acuacultura en Bahía Tortugas, BCS, dependiente de la Secretaría de Pesca, es el primer laboratorio mexicano proyectado para la investigación aplicada, desarrollo tecnológico y producción de semilla de abulón. Sus técnicos han demostrado la factibilidad y despertado el interés en el cultivo de abulón y actualmente, las Cooperativas Pesqueras operan proyectos comerciales. El desove se induce mediante la exposición al aire y estimulación térmica de los adultos en agua irradiada con luz ultravioleta. Después de 4–6 días de cultivo estático, 1–2% de la larva se fija exitosamente en placas de PVC claro fitocolonizadas y al menos un 20% sobrevive a la talla de 5 mm, creciendo 40–50 micras/día y alimentándose de diatomeas. La semilla de 5 mm ya consume macroalgas tiernas y su preengorda puede continuarse con mínima mortalidad fuera del laboratorio, donde crece 75–150 micras/día en función de la dieta (macroalgas/diatomeas). Durante un año de engorda en suspensión, se ha registrado un crecimiento promedio de 106 micras/día y supervivencia cercana al 100%, por lo que la talla medallón (5–7 cm) puede alcanzarse en menos de dos años.

INTRODUCCION

El cultivo de abulón en México se originó como una alternativa a la sobreexplotación pesquera del recurso. El Centro de Acuacultura en Bahía Tortugas, BCS, dependiente de la Secretaría de Pesca, inició su operación a mediados de 1984, teniendo como objetivos la investigación aplicada, adecuación e innovación tecnológica y la producción masiva de semilla de abulón azul (*Haliotis fulgens*) y amarillo

(*H. corrugata*) con fines de repoblación y cultivo intensivo a la talla medallón (5–7 cm), promoviendo en esta forma una nueva actividad económica para el País.

El esquema de producción y la estrategia operativa del laboratorio, fueron diseñados por Bazúa Sicre (1983), a partir de los trabajos de Ino (1952), Kikuchi y Uki (1974a,b,c) y Shepherd (1976) entre otros, incorporando nuevos procedimientos y experiencias adquiridas en varios años de trabajo. En este aspecto, el proceso de diseño e innovación tecnológica continúa.

MATERIAL Y METODOS

Producción de alimento

El sistema de producción se basa en el policultivo, dentro y fuera del laboratorio, de microalgas y diatomeas bentónicas silvestres menores de 10 micras, en tanques de fibra de vidrio que albergan cada uno 26 m^2 de placas de PVC translúcido, ofreciendo una superficie fitocolonizable de 32 m^2. La diversidad específica varía estacionalmente, con abundancia de algunas diatomeas como *Navicula* sp. y *Nitzschia* sp.

Para la obtención de la cepa inicial, se deja correr agua marina filtrada a 10 micras en un tanque de fibra de vidrio, lo cual propicia la fitocolonización natural de su superficie. Después de 20–30 días, el tanque se drena y se desprenden las diatomeas débiles ó semibentónicas, rociando agua marina filtrada. El perifiton remanente se desprende tallando suavemente con un trozo de hule espuma humedecida y se transfiere a cubetas de plástico con agua marina filtrada a 10 micras. Después de 3–4 horas, el inóculo se obtiene por decantación, eliminando así las diatomeas mayores.

A fín de propiciar el asentamiento del inóculo, la primera fase del cultivo (34–36 h) es estática y se proveen los macronutrientes y silicatos del medio f/2 de Guillard (1972). Posteriormente se proporciona aireación y flujo contínuo de agua marina filtrada a 10 micras. Dentro del laboratorio se provee iluminación fluorescente y al exterior se aprovecha la luz solar difusa.

Acondicionamiento y desove de reproductores

Los reproductores se obtienen mediante buceo en los bancos abuloneros regionales y, a manera de acondicionamiento previo al desove, se mantienen durante al menos una semana en tanques de fibra de vidrio con falso fondo de malla sintética y flujo contínuo de agua marina filtrada (10 micras, 2–3 l/min) a temperatura estable. Cada tanque aloja 30 individuos, que son alimentados '*at libitum*' con sargazo (*Macrocystis pyrifera*), el cual se retira dos días antes del desove a fín de depurar el tracto digestivo, reducir la producción de heces y obtener gametos limpios.

Para la inducción al desove se aplican tres diferentes métodos de estimulación. Inicialmente los reproductores son expuestos al aire y al sol hasta que la concha muestra los primeros signos de desecación, después de lo cual se transfieren a la sombra, donde continúa la exposición al aire hasta completar 60–75 minutos.

Enseguida, una vez separados por sexo, los organismos se colocan en tanques semicilíndricos de fibra de vidrio con una superficie utilizable de 5 m^2, con agua marina filtrada a 10 micras e irradiada con luz ultravioleta, siguiendo el método de Kikuchi y Uki (1974a,b).

Finalmente, se provoca una variación de 2°C en la temperatura del agua, a razón de 1°C/h. Si los reproductores han sido acondicionados a 16–19°C, la temperatura se incrementa y si han sido acondicionados a más de 20°C, la temperatura se disminuye. Después de aplicar esta variación térmica, la temperatura en los tanques de inducción al desove se mantiene estable durante una hora, para retornar gradualmente a su nivel inicial a razón de 1°C/h, lo que hace un tiempo total de cinco horas.

Fertilización y cultivo de huevos y larvas

Los óvulos normales se sedimentan en el fondo de los tanques de desove, de donde son extraídos por medio de un sifón y sometidos a limpieza y tamizado selectivo en agua, con malla de 212, 150 y 75 micras. El esperma es obtenido en la parte superior de la columna de agua y tamizado con malla de 75 micras.

La fertilización se realiza de preferencia antes de que transcurran 30 minutos del desove. Los óvulos (retenidos en un tamíz de 100 micras), se sumergen (1–3 min) en una suspensión de esperma (100 000–125 000 cel/ml) obtenida de varios machos y una vez fecundados se lavan vigorosoamente con agua marina filtrada a 5 micras. Después de 10 minutos de reposo los huevos normales se han sedimentado y mediante decantación cuidadosa se elimina el esperma excedente y óvulos inviables flotantes, repitiendo el procedimiento unas 10 veces.

La etapa de incubación se inicia de 60 a 90 minutos después de la fecundación y cuando ya empieza la división celular. Los huevos se colocan en cubetas de plástico (19 l) con agua marina filtrada a 5 micras, a tal densidad que formen una capa uniforme y sin sobreposición en el fondo. No se proporciona alimento, flujo de agua ó aíreación y se evitan variaciones mayores de 1°C en la temperatura del agua.

El cultivo larval se continúa por espacio de 4–6 días en las mismas cubetas, a densidad de 10 larvas/ml y temperatura estable, evitando variaciones mayores de 1.5°C. Cada 12 horas se realiza un cambio total de agua y tamizado selectivo con malla de 100 y 150 micras, para larvas trocófora y véliger respectivamente. Estas se lavan en agua dentro del tamiz receptor, a chorro suave de manguera para evitar el desprendimiento de la concha larval, hasta eliminar el tejido blando en las larvas muertas. Los recipientes y accesorios de cultivo se lavan con agua dulce clorada, se enjuagan abundantemente y de ser necesario se exponen al sol durante algunas horas.

Fijación

La fijación se realiza sobre placas onduladas de PVC translúcido de 32 × 40 cm, fitocolonizadas por microalgas y diatomeas menores de 10 micras y en las paredes de los tanques de cultivo donde éstos se encuentran instalados. El Centro dispone de 358 m^2 de colectores y destina 148 m^2 de estanquería semicilíndrica de fibra de vidrio para el cultivo de fijaciones y crías.

Cuando se detecta la presencia de al menos un 20% de larvas en fase pedivéliger avanzada en las cubetas de cultivo, se drenan los tanques de fijación, se eliminan las diatomeas débiles rociando las placas colectoras y las paredes de los tanques con agua marina filtrada a 5 micras, dejando sólo una película delgada de color ligeramente amarillo. A continuación se llenan los tanques al 50% de su capacidad con agua marina filtrada a 5 micras, se agrega larva pedivéliger en cantidad tal que se establezca una densidad de 3–5/cm^2 y se proporciona un pequeño flujo de agua, calculando que transcurridas 12–15 horas los tanques estén a punto de derramarse. Lo anterior, a fín de ampliar la franja de fijación e inducir cambios en la tensión superficial que favorecen el proceso (Matsunaga *et al.*, 1987).

Al concluir la fijación los colectores se disponen en posición vertical y se proporciona aireación y flujo contínuo de agua marina filtrada (5 micras, 2–3 l/min/tanque). El agua se distribuye a lo largo del tanque por medio de un tubo de PVC con múltiples perforaciones. Los desechos acumulados en el fondo, se eliminan periódicamente por medio de sifoneo, tamizando en agua el producto obtenido para evitar pérdida de postlarvas.

Preengorda

Esta etapa comprende la crianza de postlarvas hasta obtener semilla. La alimentación de organismos menores de 5 mm se basa exclusivamente en las microalgas y diatomeas disponibles en las placas colectoras y en las paredes de los tanques de fijación/preengorda ubicados dentro del laboratorio. Estas superficies de pastoreo se inoculan ó resiembran periodicamente. Se evita toda manipulación innecesaria de las postlarvas y se proporciona aereación suave y flujo contínuo de agua marina filtrada (5 micras, 3–4 l/min).

Una vez alcanzada la talla de 5 mm, la pre-semilla se transfiere a tanques de preengorda ubicados a cielo abierto fuera del laboratorio, donde reciben aireación y flujo de agua marina filtrada (20 micras, 4–6 l/min/tanque) y una dieta mixta constituída por microalgas y diatomeas bentónicas y brotes tiernos de macroalgas. Cada semana se realiza un drenaje total para limpieza de los tanques y reposición de macroalgas.

Cultivo intensivo en suspensión

La preengorda puede continuarse hasta la talla de 25 mm ó transferir la semilla al medio natural y cultivarla en artes suspendidas. Para organismos de 15 mm de longitud se han utilizado canastas ostrícolas y para tallas menores, algunos prototipos fabricados localmente con malla de acero y plástico. La preengorda en suspensión se da como alternativa a la limitación de espacio en el laboratorio y representa además, el inicio del cultivo intensivo ó engorda a una talla comercial propuesta para acuacultura (5–7 cm), que es menor a la actualmente establecida para el producto de la pesquería. La limpieza de las artes de cultivo y reposición de alimento (*Macrocystis pyrifera*), se realiza cada 5–7 días.

RESULTADOS

Producción de alimento

El sistema de cultivo de microalgas y diatomeas bentónicas silvestres descrito ha asegurado hasta hoy la disponibilidad de alimento natural en la cantidad, calidad, variedad y talla requeridas para la fijación larval y preengorda de semilla de abulón en el laboratorio. Este procedimiento resulta bastante versátil ya que, modificando el tiempo de reposo en las cubetas-inóculo, pueden separarse diatomeas de diferente tamaño.

El método de selección de cepas e inoculación, puede combinarse con un filtrado selectivo (Ebert y Houk, 1984) al momento de la cosecha a fín de eliminar las formas mayores. A diferencia de los cultivos realizados con iluminación artificial fluorecente dentro del laboratorio, en aquellos que reciben luz solar y se ubican a la intemperie se obtienen por regla general diatomeas más grandes; al parecer, la diferencia se debe a la fuente luminosa.

Acondicionamiento y desove de reproductores

A la fecha no existen problemas para la obtención de reproductores maduros silvestres; durante los meses de septiembre a diciembre puede colectarse abulón azul (*H. fulgens*) y de diciembre a abril abulón amarillo (*H. corrugata*). Se seleccionan ejemplares jóvenes y sanos, de 4–5 años de edad y 10–14 cm de longitud de concha.

El estímulo combinado de la exposición al aire, variación térmica y agua irradiada con luz ultravioleta, produce una respuesta favorable superior al 90% de los reproductores sexualmente maduros. La expulsión de gametos se produce a través de los orificios respiratorios, mediante movimientos de elevación y descenso de la concha con un breve movimiento de rotación que ejerce presión y masaje sobre la gónada. En condiciones normales, el esperma se mantiene en suspensión mientras que los óvulos tienden a sedimentarse; en caso contrario, los gametos tienen escasa ó nula viabilidad.

Fertilización y cultivo de huevos y larvas

La concentración de esperma se establece en función de su motilidad y de la atracción ejercida por el óvulo, ya que ambos factores son mínimos al existir una máxima proporción de gametos anormales y viceversa. Es necesario evaluar al microscopio el porcentaje de fecundación para evitar sobreexposición, ya que el tiempo de inmersión (1–3 min) de los óvulos en la suspensión de esperma, varía de acuerdo a la calidad y viabilidad de cada lote de gametos. Los óvulos son bastante resistentes al manejo, lo cual permite una limpieza enérgica post-fecundación que elimina adecuadamente el exceso de esperma, reduciendo la mortalidad y malformaciones larvales asociadas a la polispermia.

A una temperatura de 20–24°C, la eclosión se presenta regularmente después de 8–12 horas de incubación, mientras que a 16–20°C las primeras larvas trocófora libres se pueden observar hasta 12–15 horas después de la fecundación. Siguiendo la metodología descrita, es posible obtener un 60–100% de larvas trocófora normales.

Fijación

Durante la etapa de fijación y desarrollo postlarval temprano, se obtiene un rendimiento de 1–2% a partir de larva pedivéliger. Al respecto y para el abulón azul (*H. fulgens*), Searcy Bernal *et al.* (1988) reportan una supervivencia de 0.8% y 0.5% a los 10 y 41 días de cultivo post-fijación y Leighton *et al.* (1981), un rendimiento menor del 1% de larva pedivéliger a pre-semilla de 2–3 mm. Para el abulón rojo, puede esperarse una supervivencia de al menos 7.5% al tercer mes de la fijación (Ebert y Houk, 1984).

El asentamiento larval se presenta con mayor intensidad cerca de la superficie del agua y en un período de 10–12 horas. Para obtener los mejores resultados, la fijación debe iniciarse cuando en las cubetas de cultivo larval se detecta la presencia de al menos un 20% de larvas en fase pedivéliger avanzada. Para entonces se ha desarrollado el pie, los tentáculos cefálicos y la mancha ocular y las larvas alternan su natación activa con un comportamiento reptante.

Preengorda

A partir de la fijación, el crecimiento promedio registrado dentro del laboratorio es del orden de 40–50 micras/día y la supervivencia de 20–30%, obteniendo pre-semilla de 5 mm en 3–4 meses de preengorda. Partiendo de pre-semilla de 5 mm, el crecimiento promedio registrado fuera del laboratorio es de 75 micras/día en grupos alimentados exclusivamente con *Macrocystis pyrifera* y de 116–150 micras/día al utilizar dietas mixtas constituídas por microalgas, diatomeas y macroalgas.

Un cambio drástico en el tipo de alimento retrasa el crecimiento, mientras que al utilizar dietas mixtas se propicia una adaptación progresiva y gradual del pastoreo de perifiton típico del juvenil temprano al aprovechamiento de macroalgas, que caracteriza a la fase adulta. Mediante este procedimiento se puede obtener semilla de 15 mm en un período de 2 meses de preengorda, con una supervivencia de 99% a partir de pre-semilla de 5 mm.

Cultivo en suspensión

En relación al cultivo intensivo en suspensión, se ha registrado una supervivencia del 100% y un crecimiento promedio de 106.51 micras/día durante un ciclo anual, para semilla de abulón azul de 15 mm producida en el Centro de Acuacultura en Bahía Tortugas y cultivada en canastas tipo Nestier con una dieta mixta de las macroalgas *Macrocystis pyrifera*, *Eisenia arborea*, *Ulva* sp., *Laminaria* sp., *Enteromorpha* sp. y *Corallina* sp. (Franco Santiago, 1986).

DISCUSION

Acondicionamiento y desove de reproductores

A la fecha no existe limitante para la obtención de reproductores silvestres maduros, de manera que el acondicionamiento gonádico se reduce a una etapa de adaptación a las condiciones del laboratorio. *Haliotis corrugata* parece ajustarse bién al modelo de inducción individual en recipientes pequeños; en cambio, *H. fulgens* responde mejor al ser estimulado en grupo en los tanques de desove.

Generalmente la rutina de inducción se inicia al mediodía, y el desove suele presentarse al atardecer. En caso contrario, se suspende la adición de agua irradiada y se agregan pequeñas dosis de acido acetil salicílico para evitar un desove nocturno incontrolado. Al día siguiente se repite el procedimiento y en caso de respuesta negativa se elige otro grupo de reproductores ó se amplía el tiempo de acondicionamiento.

Fertilización y cultivo de huevos y larvas

Al ser liberados, los óvulos miden alrededor de 210 micras y son fácilmente observables a simple vista por su coloración verde obscuro que contrasta con el color blanco de los tanques de desove. Desde que se inicia la inmersión de los óvulos en la suspensión de esperma, se observa al microscopio el proceso de fecundación, detectando la cantidad de espermas que hacen contacto con los óvulos, la aparición de los cuerpos polares y la membrana de fertilización. Esto es indispensable para evitar sobreexposición, ya que normalmente, en un período de 1–3 minutos, al menos un 80% de los óvulos han sido fecundados.

La eclosión se presenta a nivel de trocófora tardía, que al romper la membrana y debido a su fototropismo positivo y conducta gregaria, se desplaza activamente cerca de la superficie del agua, formando columnas de alta densidad. Las cápsulas, óvulos no fecundados y huevos inviables, se depositan en el fondo de las cubetas y de esta manera pueden ser fácilmente eliminados mediante sifoneo. Las larvas véliger se caracterizan por su concha larval y el vélum que extienden hacia arriba durante la natación, que alternan con periodos durante los cuales detienen su actividad velar y tienden a sedimentarse por efecto de la gravedad. Durante ésta etapa de cultivo su manipulación debe ser cuidadosa para evitar desprendimiento de la concha.

Producción de alimento y fijación larval

La etapa de fijación es altamente crítica en la producción de semilla, siendo de la mayor importancia mantener una adecuada cantidad de alimento y excelente calidad de agua. A fín de optimizar la fijación y metamorfosis larval deben utilizarse colectores con una capa delgada de microalgas/diatomeas, que en el caso del Centro de Acuacultura se obtienen a los 7 días de cultivo. Una capa muy gruesa se desprende con mayor facilidad y ocasiona pérdida de postlarvas y pre-semilla.

Una vez que las larvas se han asentado, e independientemente de la dimensión del tanque de fijación ó preengorda inicial, debe proveerse un flujo de agua suficiente para obtener un recambio diario de 100–200% del agua (Matsunaga *et al.*, 1987), con lo cual se asegura una eficiente remoción de desechos orgánicos.

Preengorda

Una vez que se ha desarrollado el primer orificio respiratorio (45–60 días) la pre-semilla puede transferirse a otro substrato, desprendiéndola con un pincel de cerda suave, espátula de plástico, mediante shock térmico breve y moderado ó utilizando anestésicos (Sagara, 1985; Matsunaga *et al.*, 1987). Sin embargo, lo más recomendable es reducir al máximo la manipulación.

Durante la etapa de preengorda al exterior, con pre-semilla mayor de 5 mm, y sin importar el tamaño del tanque, el flujo debe proveer un recambio diario de 300–500% (Matsunaga *et al.*, 1987). Las macroalgas pueden suministrarse con la periodicidad y en la cantidad que demande su consumo. La duración del fotoperiodo debe modificarse en función de la velocidad con que los juveniles pastorean la capa de perifiton y la rapidez con que la misma se recupera.

El rendimiento observado al utilizar dietas mixtas en los tanques exteriores, ha sido notablemente superior al registrado con una dieta de diatomeas dentro del laboratorio. La productividad y diversidad de especies que colonizan los colectores en los tanques exteriores es considerablemente mayor y contribuye a acentuar diferencias en el crecimiento de la semilla. De acuerdo con Matsunaga *et al.* (1987), las postlarvas pueden transferirse al exterior 10 días después de la fijación, lo cual, de ser aplicable en la localidad, posiblemente elevaría eficiencia y productividad, al permitir un mejor aprovechamiento del area dentro del laboratorio.

Cultivo intensivo

La engorda en canastas suspendidas ofrece algunas facilidades para el manejo de la semilla en campo, permitiendo destinar la estanquería del laboratorio a la fijación y desarrollo temprano. Se estima que esta variante de preengorda-engorda puede llegar a tener buenos resultados en un corto plazo ya que, incluso, se ha observado que las canastas pueden actuar como colectores de semilla silvestre.

CONCLUSION

Después de 5 años de trabajo, el Centro de Acuacultura en Bahía Tortugas, BCS, ha desarrollado una tecnología de ciclo completo para el cultivo de abulón, la cual ha sido descrita a lo largo de este trabajo. La investigación continúa y los procedimientos actualmente utilizados seguramente serán modificados en el futuro a fín de optimizar su eficiencia y productividad.

El objetivo principal del Centro de Acuacultura ha sido propiciar un interés creciente del sector pesquero mexicano en el cultivo de abulón, actividad que en otros países ha demostrado eficacia y rentabilidad. La creación de dos laboratorios-granja operados actualmente por el Sector Cooperativista Pesquero en Baja California Sur, son una muestra de ello.

BIBLIOGRAFIA

Bazua Sierc L.A. 1983. *Cultivo de Abulón en Baja California Sur*. SEPESCA, Depto. Acuacultura BCS, México.

Ebert E.E. y Houk J.L. (1984) Elements and innovations in the cultivation of red abalone *Haliotis rufescens*. *Aquaculture*, **39**, 375–92.

Franco Santiago J.R. (1986) Informe Preliminar sobre el crecimiento de juveniles de abulón azul (*Haliotis fulgens*), mantenidos en canastas Nestier en el campo pesquero 'El Rincón', BCS. *Fed. Reg. Soc. Coop. Prod. Pesq. 'Baja California' F.C.L.*, México.

Guillard R.L. (1972) Culture of phytoplankton for feeding marine invertebrate animals. In *Culture of Marine Invertebrate Animals* (Ed. by W.L. Smith & M.H. Chanley), pp. 29–60. Plenum Press, New York.

Ino T. (1952) Biological studies on the propagation of Japanese abalone (genus *Haliotis*). *Bull. Tokai Fish. Res. Lab.*, **5**, 1–108.

Kikuchi S. y Uki N. (1974a) Technical study on artificial spawning of abalone, genus *Haliotis* II. Effect of irradiated seawater with ultraviolet rays on inducing to spawn. *Bull. Tohoku Reg. Fish. Res. Lab.*, **33**, 79–85.

Kikuchi S. y Uki N. (1974b) Technical study on artificial spawning of abalone, genus *Haliotis* III. Reasonable sperm density for fertilization. *Bull. Tohoku Reg. Fish. Res. Lab.*, **34**, 67–71.

Kikuchi S. y Uki N. (1974c) Technical study on artificial spawning of abalone, genus *Haliotis* IV. Duration of fertility related to temperature. *Bull. Tohoku Reg. Fish. Res. Lab.*, **34**, 73–6.

Leighton D.L., Byhower M.J., Kelly J.C., Hooker G.N. y Morse D.E. (1981) Acceleration of development and growth in young green abalone (*Haliotis fulgens*) using warmed effluent water. *J. World Maricul. Soc.* **12**(1), 170–80.

Matsunaga N., Yoshida M.K. y Atienzo Cota J.A. (1987) Técnicas de producción de larvas de abulón. En *Introducción al Conocimiento del Medio Acuático. Serie de Educación Pesquera, Manual de Prácticas No. 4* (Ed. by SEP/JICA), pp. 1–21, Secretaría de Educación Publica (México) and Japanese International Co-operative Agency, Mexico City.

Sagara J. (1985) Cultivo de abulón. En *Generalidades de Acuacultura. Serie de Textos Didácticos en Ciencia y Tecnología del Mar* (Ed. by SEP/JICA), p. 41–54. Secretaría de Educación Publica (México) and Japanese International Co-operative Agency, Mexico City.

Searcy Bernal R., Salas Garza A. y Flores Aguilar R.A. (1988) Crecimiento de postlarvas y juveniles de abulón azul (*Haliotis fulgens*) en un laboratorio mexicano. *Ciencias Marinas*, **14**(4), 57–72.

Shepherd S.A. (1976) Breeding, larval development and culture of abalone. *Aust. Fish.*, **35**(4), 7–10.

Chapter 44
Abalone aquaculture: a North America regional review

EARL E. EBERT *California Department of Fish and Game, Marine Resources Division, 2201 Garden Road, Monterey, California 93940, USA*

ABSTRACT: Remarkable progress has been made in recent years in the cultivation of abalone species endemic to the west coast of North America. The fledgling abalone aquaculture industry that began in the mid-1960s in central California is finally coming of age. Two California-based hatcheries are now operating at full capacity, with a profit margin and in either the planning or development stages for further expansion. At least four other hatcheries are in various stages of production, while a number of companies are in the research and development phase.

Three abalone species are under cultivation and being marketed: the red, *Haliotis rufescens*, green, *H. fulgens*, and pink, *H. corrugata*, abalone. *Haliotis rufescens* constitutes the bulk of production, exceeding 95%. Much of the progress of the expanding abalone aquaculture industry in the western region is attributable to determined private entrepreneurship, comprising of individuals possessing a true pioneering spirit. University researchers, principally Dr Dan Morse and his colleagues at the University of California Santa Barbara, have made significant contributions to abalone culture. Noteworthy is their technique for spawning induction, now in wide use, and their procedure to induce larval settlement.

Another contributor to the overall development of technological as well as biological procedures for abalone cultivation is the California Department of Fish and Game's Marine Resources Laboratory (MRL), which has been conducting research on abalone cultivation methods since 1970.

Cultivation practices may vary widely between hatcheries, for instance larvae may be settled directly into nursery tanks or an intermediate settlement container may be used, but a 'common thread' is also apparent. Abalone are grown-out in either land-based tanks or in-the-sea barrels that are suspended from a raft or wharf. Animals in the barrel culture system are harvested at about 6–7 cm shell length. Those produced in the land-based tanks are harvested at about 8 cm shell length. The industry has long recognized the need for an artificial diet, particularly for the early juvenile stages. Major giant kelp, *Macrocystis* spp., beds prevail along much of the west coast, from central California to central Baja California, and the adjacent islands. Giant kelp is used almost exclusively for grow-out of abalone to harvest size. This kelp resource appears sufficient to supply the present as well as the anticipated future needs of the abalone aquaculture industry. However, severe El Niño events could result in short-term kelp supply problems.

RESUMEN: En los últimos años, se ha efectuado un avance notable, en el cultivo de las especies endémicas del abulón de la Costa Occidental de Norteamerica. La recientemente establecida industria del cultivo de abulón, que arrancó a mediados de los sesentas en la parte contral de California; finalmente se acerca a su madurez. Dos de los principales criaderos de California se encuentran ahora operando a toda su capacidad, con márgenes de ganancia, asi como en etapas de planeación o desarrollo para un futuro crecimiento. Por lo menos otros tres criaderos se encuentran en varias etapas de producción, mientras otras compañías están en las fases de investigación y desarrollo.

Tres son las especies de abulón que se cultivan y son comercializadas; el abulón rojo, *Haliotis rufescens*; el verde, *H. fulgens* y el rosado, *H. corrugata*; constituyendo el abulón rojo, mas de 95% de la producción. Gran parte del avance experimentado por la industria abulonera de la región Occidental, puede atribuirse a la determinación pionera de empresarios privados. Investigadores Universitarios, principalmente el Dr Morse y sus colegas de la Universidad de Santa Barbara, California, han contribuido significativamente a la ciencia del cultivo de abulón; desarrollando técnicas para la inducción del desove de los abulones-de amplio uso en la actualidad-asi como los procedimientos para obtener la fijación de las larvas.

Otro contribuyente al desarrollo de los métodos tecnológicos y biológicos para su cultivo, es el Laboratorio de Recursos Marinos del Departamento de Caza y Pesca de California; de Monterey, CA, que ha efectuado investigaciones al respecto desde 1970.

Las prácticas de cultivo varían con cierta amplitud entre los criaderos, aunque cierta tendencias aparentemente son comunes entre los mismos, sobre todo entre los elementos de cultivos, por ejemplo: el que las larvas de abulón se fijen directamente en los tanques de cría o en recipientes de fijación intermedia; que los abulones jóvenes son 'engordados' en tanques sobre el suelo o en barriles en el mar; suspendidos de balsas o de un muelle. Los abulones de barril son cosechados con concha de 5 cm y los de tanques sobre suelo, a longitudes de 8 cm. La investigación en los criaderos artificiales, se ha dirigido a aspectos nutricionales. La industria ha reconocido desde hace largo tiempo la necesidad de contar con dietas artificiales para el abulón, especialmente para las etapas juveniles. Los principales mantos de algas gigantes, de *Macrocystis* spp., predominan en la mayor parte de la Costa Occidental, del contro de California al centro de la Península de Baja California, asi como en las Islas adyacentes. Estas algas son empleadas casi exclusivamente para crecer el abulón hasta la talla comercial. El recurso algal parece bastar por el momento, asi como para los requerimientos futuros de la industria abulonera. Sin embargo, casos severos como el evento de 'el Niño' pueden provocar crisis a corto plazo en la disponibilidad del alga.

INTRODUCTION

Remarkable progress has been made in recent years in the cultivation of abalone species endemic to the west coast of North America. The fledgling abalone aquaculture industry that began in the mid-1960s in central California is finally coming of age. Much of the progress is attributable to determined private entrepreneurship, individuals possessing a true pioneering spirit. But progress has not come easily.

Prospective abalone farmers are confronted with all of the regulatory constraints frequently encountered by aquaculturists. In California there are several agencies that require permits. The permitting process is lengthy, often exceeding a year, and is expensive. Bowden (1981) reported on institutions that constrained the development of coastal aquaculture. Despite these obstacles, interest in cultivation of abalone in California remains high, prompted in part by a sharply declining commercial fishery, a high market demand for the farmed product and a good price to growers.

The California commercial abalone fishery consistently yielded more than 1814 tons annually through 1968 but then began a precipitous decline that has not abated (Table 44.1). This fishery decline is even more pronounced, since black abalone, *Haliotis cracherodii*, were not fished commercially until 1968. By 1972 the black and red, *H. rufescens*, abalone became the principal species landed. However, a recent malady of unknown aetiology has resulted in extensive mortality of black abalone at certain of the southern California Channel Islands (Chapter 17, this volume).

California's private sector principally grows three abalone species: the red, pink, *H. corrugata*; and the green, *H. fulgens*. The latter two are warm-water

Table 44.1 California commercial fishery abalone landings, 1968–89 (values in kg in shell).

	Species					
Year	Red	Pink	Black	Green	White	Total*
1968	805618	1029721	318	193748	383	2029788
1969	709523	863212	2173	71334	12705	1659304
1970	541956	639086	6952	122563	5086	1315809
1971	582226	175154	48263	510581	19684	1335996
1972	500037	182983	460355	192519	65236	1402400
1973	301154	168445	867714	71126	37700	1448417
1974	340681	206535	519552	55141	51604	1177089
1975	336920	169646	311817	77532	322882	1222054
1976	335492	195566	161822	54654	37153	786006
1977	241654	144473	208738	44190	6281	645344
1978	221714	130403	190530	42226	1653	586527
1979	197832	70361	147325	27164	227	442909
1980	223677	60202	230312	27732	281	542205
1981	194021	42725	231493	28999	73	497311
1982	195185	38845	287150	40223	409	561813
1983	104137	30244	219176	25626	217	379446
1984	196838	25812	196838	14242	—	372795
1985	167280	31127	163221	10865	750	373703
1986	119434	23510	121316	11680	397	278979
1987	177484	14306	140492	13138	—	345421
1988	147254	8620	91448	10659	—	258015
1989	212923	10192	99107	8946	10	331178
1990**	168631	10512	41162	12215	8	232626

* Includes unspecified or miscellaneous abalone species.
** Preliminary.

species and are cultivated at several southern-California-based hatcheries. The red abalone is the mainstay of the industry and accounts for more than 95% of total production.

California aquaculturists either grow out abalone in land-based tanks or employ an in-ocean barrel-culture system. With the latter system the barrels are suspended from either a dock or a floating raft. Abalone in the barrels are harvested at about 6 to 7 cm shell length. Ocean ranching of abalone on state-leased seabeds has been experimented with, but has not proven feasible.

ABALONE AQUACULTURISTS

Baja California abalone aquaculture operations are not included in this review, but are covered elsewhere in this volume (see Chapters 41 and 43).

North American abalone aquaculture operations not only began in California, but with one exception (Hawaii), abalone are not produced elsewhere in the western region. An abalone hatchery was established on Vancouver Island, British Columbia, Canada (Pacific Trident Mariculture, Ltd) to grow the pinto abalone,

H. kamtschatkana. However, they are not presently producing abalone. Experimental culture of the red abalone was tested in Washington State (Puget Sound), using a barrel-culture-raft system. However, this research was recently terminated, although the culture system, and abalone growth rate, appeared to be satisfactory. The Washington State Department of Fisheries (WSDF) expressed concern over cultivation of a non-native species. This concern was exacerbated following the black abalone mortality problem in California. However, WSDF is receptive to the aquaculture of the pinto abalone, an endemic species (R. Burge, pers. comm.).

In 1984 Monterey Abalone Farms moved their hatchery operations from the central California coast to Keahole Point, on the island of Hawaii. Their company has since been renamed Ocean Farms of Hawaii, and they continue to cultivate the red abalone. Cool seawater is pumped from a considerable depth to grow abalone in land-based tanks.

There are 16 registered abalone aquaculturists in California (Table 44.2). However, only three companies have full production operations and three companies have pilot-production facilities with full production facilities under construction. Several companies have pilot-production operations and are looking for sites to develop full production facilities. The majority of the registered companies are in an experimental base. Data on some representative abalone aquaculture firms are presented below:

Abalone Farm, Inc.

Located in central California, this was the first company to develop a full production hatchery and grow-out system in California. The company has undergone two financial reorganizations since it was first established as California Marine Associates in 1967. The Abalone Farm employs a land-based, concrete tank, grow-out system and grows only red abalone (Fig. 44.1). They use hydrogen peroxide to induce spawning (Morse *et al.*, 1977); static larval and postlarval growing containers; circular, 300 l tanks with through-flowing seawater for cultivation of juvenile stages up to about 10 mm lengths; and hanging 'baskets' for juvenile stages from 10 to 25 mm long. Following this, the abalone are transferred to the concrete grow-out tanks. They are nurtured on giant kelp, *Macrocystis* spp., from the juvenile (basket) stage to harvest. They are in the concrete tanks for about 20 months when they achieve market size, about 8 cm shell length.

About half of the product is sold alive while the remainder is processed. The Abalone Farm sells live abalone in Japan, Hawaii, New York, Chicago and San Francisco. Processed steaks (medallions) are for domestic sales. This company has a staff of approximately 30 and is presently undergoing expansion. They harvested 335 000 abalone in 1990 (Table 44.3), and project a 1000 ton harvest (3 500 000 abalone) by 1993.

Ab Lab

This was founded in 1972, but began intensive research on abalone cultivation methods in 1974 at Port Hueneme (southern California) (McMullen & Thompson,

Table 44.2 Registered abalone aquaculturists in California.

Company	Operations	Species
Abalone International PO Box 1640 Crescent City, 95531	Raft-barrel-culture systems in Crescent City harbour (northern CA), produces 6–7 cm abalone for domestic market	red, *H. rufescens*
Pacific Abalone Farms 1312 Lincoln Ave Pacific Grove, 93950	R&D (northern CA)	red, *H. rufescens*
Pacific Mariculture PO Box 692 Carmel Valley, 93924	Land-based pilot-production; full production hatchery under development (central CA)	red, *H. rufescens*
David LaPota 6678 Hemingway Drive San Diego, 92120	R&D (southern CA)	red, *H. rufescens* green, *H. fulgens* pink, *H. corrugata*
Maritech PO Box 90190 San Diego, 92109	R&D (southern CA)	red, *H. rufescens* green, *H. fulgens* pink, *H. corrugata* white, *H. sorenseni*
Pacific Mariculture Unlimited PO Box 4606 Lancaster, 93534	Cultivation in seafloor habitats; Avila Beach Harbor (central CA)	red, *H. rufescens*
The Abalone Farm, Inc. PO Box 136 Cayucos, 93430	Land-based cultivation of 8 cm abalone. Domestic and foreign markets (central CA)	red, *H. rufescens*
Abalone Unlimited, Inc. PO Box 730 Guadalupe, 93434	Pilot production, barrel-culture system in Morro Bay; major hatchery, land-based operation, under development near Santa Maria (central CA)	red, *H. rufescens*
Sea Ventures Enterprises, Inc. PO Box 4171 Santa Barbara, 93140	Pilot production; raft culture (southern CA)	*Haliotis* spp.
Bodega Farms- California Fish Growers, Inc. 625 Martin Ave Rohnert Park, 94928	Land-based pilot-production; full production facility in construction phase (northern CA)	red, *H. rufescens* green, *H. fulgens*
South Coast Shellfish Co. 4724 Avalon Avenue Santa Barbara, 93111	Experimental culture (southern CA)	—
The Cultured Abalone 336 Coronado Drive Goleta, 93117	R&D, Goleta Pt (southern CA)	red, *H. rufescens*
The Ab Lab c/o NCEL Port Hueneme, 93043	Production of seed abalone; barrel culture of 6 to 7 cm abalone for domestic and foreign markets; land-based grow-out facility in planning stages; joint venture in Chile (southern CA)	principally red, *H. rufescens*

Table 44.2 *Continued*

Company	Operation	Species
McCormick & Associates 1211 Spruce Street Berkeley, 94709	Pilot production of abalone seed and 8 cm abalone for domestic and foreign markets (southern CA)	red, *H. rufescens* pink, *H. corrugata* green, *H. fulgens*
U.S. Abalone Shaffer Road Santa Cruz, 95060	Pilot production of seed and 8 cm abalone for domestic and foreign markets (central CA)	red, *H. rufescens*
Gregorio Aquatech 1073-17th Street San Pedro, 90731	Production of abalone: facility expansion on Terminal Island in 1991	green, *H. fulgens*

Table 44.3 Production of red abalone (8 cm long) at the Abalone Farm, Cayucos, California, 1985–90.

Year	No of abalone	Metric tonnes
1985	120 000	
1986	179 450	
1987	201 000	
1988	216 000	19.5
1989	325 000	28.0
1990	335 000	30.0

1989). The Ab Lab is the largest seed abalone producer in the state. They principally grow *H. rufescens*, but have also cultivated green abalone. In their growing system, six-day-old larvae are settled directly into circular, 300 l rearing tanks. About 700 tanks are used, arranged in four tiers (Fig. 44.2). The abalone are harvested directly from these tanks after about 5 to 6 months, when about 8–10 mm long, and either sold to other growers or stocked in their in-ocean, barrel-culture system that is suspended from a pier at Port Hueneme.

This company was the first to develop and market a 5 cm 'cocktail' size abalone. The Ab Lab operates with a staff of three. Production data are presented in Table 44.4. The company is currently going through the permitting process for development of a large, land-based grow-out facility near Port Hueneme. Also, Ab Lab has entered into partnership with a Chilean firm to produce red abalone in southern Chile, using the barrel-culture system. The culture system will accommodate 600 000 seed abalone in the first year and double that production during the second year. A hatchery for production of red abalone seed is planned for southern Chile.

Abalone International

Abalone International, located in Crescent City harbour (northern California), has been in operation since 1986. The company employs a raft, barrel-culture

Fig. 44.1 The Abalone Farm, Cayucos, CA. (a) Partitioned concrete tanks used for grow-out of red abalone, *H. rufescens*; (b) red abalone being harvested from the growing tanks; (c) baskets, arranged in concrete tanks, used for growing juvenile red abalone; (d) processing red abalone for the domestic market.

Fig. 44.1 *Continued.*

Fig. 44.2 The Ab Lab, Port Hueneme, CA. (a) The four-tier-high culture system used to grow seed abalone; (b) larval abalone are settled directly into these 3001 tanks.

Table 44.4 Production of red abalone at the Ab Lab, Port Hueneme, California, 1985–9.

Year	Seed sales (No of abalone)	Restaurant sales (No of abalone)
1985	15 333	16 000
1986	100 000	8 400
1987	76 667	10 963
1988	273 333	9 467
1989*	280 000	15 000

system (Fig. 44.3). Red abalone seed (8 mm long) are purchased from the Ab Lab to stock a total of 100 plastic barrels, suspended from rafts. The barrels are stocked with 2000 animals each and bull kelp, *Nereocystis luetkeana*, is supplied weekly. The abalone are in the barrels from 20 to 28 months and mortality is about 8%. The abalone are harvested at 5–6 cm shell length and are presently sold for $2.00 each. The company began marketing in May 1990. Abalone International requires a staff of two or three to operate. Their annual production goal, for the 1.1 acre farm, is 250 000–300 000 abalone.

RESEARCH

Although a considerable amount of research on abalone aquaculture has been carried out by the private sector, particularly from a technological standpoint, university and state government scientists have made significant contributions.

The California Sea Grant programme has supported abalone aquaculture research since 1971 and has provided funding for some very important research projects. The initial significant contributions were made by David Leighton (Leighton, 1972, 1974, 1977). Especially significant has been the research by Dan Morse and his colleagues at the University of California, Santa Barbara. Major contributions include their research on spawning induction in *Haliotis* using hydrogen peroxide (Morse *et al.*, 1977) and the induction of larval settling using γ-aminobutyric acid (Morse *et al.*, 1979).

Abalone culture investigations at the California Department of Fish and Game's Marine Resources Laboratory (MRL), located at Granite Canyon in central California, commenced in 1970. The objectives were to investigate procedures for mass cultivation, to develop these procedures and to disseminate information (Ebert & Houk, 1984, 1989). In 1987 the broad mission of the MRL was redirected to fin-fish culture investigations. Limited abalone aquaculture research will continue, at least for the near term, and include primarily research that was in progress prior to 1987, such as a selective strain study (Fig. 44.4).

The selective strain study at the MRL is an extension of brood stock management and has been in progress for about 10 years. The objective is to determine if a faster growing strain can be developed without sacrificing meat yield. The assumption is that growth rate is an inherited character and can be isolated by selection of faster growers amongst siblings of successive generations. Abalone

Fig. 44.3 Abalone International, Crescent City, CA. (a) Raft used for hanging the culture barrels, and the 3.7 × 6.1 m maintenance shed; (b) the plastic culture barrel; (c) a red abalone nearing the 6–7 cm harvest size.

from three geographic regions of California (north, central and southern) are being used in the experiment. Breeding is being conducted using animals within and between these geographic regions. The data was analyzed and prepared for publication in 1991.

The MRL staff was augmented with a shellfish pathologist in 1990. A significant time allocation is being devoted to research on the black abalone wasting disease.

Fig. 44.4 California Department of Fish and Game, Marine Resources Laboratory, Granite Canyon. (b) Culture containers used for the selective strain study; (a) variable growth of siblings cultured under identical conditions. These particular specimens represent the only known hybrids of the black, *H. cracherodii* × pink, *H. corrugata* abalone. They were hatched at the Ab Lab and reared at Granite Canyon.

FUTURE PROSPECTS

Coastal land sites suitable for abalone aquaculture in California are becoming limited and expensive. Ocean ranching may prove feasible, but will require considerable research. The declining commercial fishery may make ocean ranching a more attractive option.

Giant kelp beds are extensive along the central and southern California coastal regions and offshore islands (North & Hubbs, 1968; Foster & Schiel, 1985). The supply could be limiting, on a short-term basis, for abalone aquaculturists during an El Niño event (Dayton & Tegner, 1984). Kelp harvest is regulated by the California Department of Fish and Game. Kelp beds are delineated geographically and most of them may be bid for on exclusive harvest rights; however, some beds always remain 'open' for multiple users. A kelp harvester's permit is required annually ($65 fee) and there is a harvest cost of $1.90 per wet ton (907 kg).

While abalone aquaculture will most probably expand several-fold during the next decade in the western region, much of this expansion will be in California and possibly Baja California. The prevailing attitude of restricting the introduction of the red abalone into Washington state and British Columbia and the more limited, patchy, kelp distributions in the Pacific northwest will most probably inhibit the growth of abalone aquaculture in that region.

REFERENCES

Bowden G. (1981) *Coastal Aquaculture Law and Policy: A Case Study of California*. Westview Press, Boulder.

Dayton P.K. & Tegner M.J. (1984) Catastrophic storms, El Niño, and patch stability in a southern California kelp community. *Science*, **224**, 283–5.

Ebert E.E. & Houk J.L. (1984) Elements and innovations in the cultivation of red abalone *Haliotis rufescens*. *Aquaculture*, **39**, 375–92.

Ebert E.E. & Houk J.L. (1989) Abalone cultivation methods used at the California Department of Fish and Game's Marine Resources Laboratory. In *Handbook of Culture of Abalone and other Marine Gastropods* (Ed. by K.O. Hahn), pp. 239–54. CRC Press, Boca Raton.

Foster M.S. & Schiel D.R. (1985) The ecology of kelp forests in California: a community profile. *US Fish Wildl. Serv. Biol. Rept*, **85**(7.2).

Leighton D.L. (1972) Laboratory observations on the early growth of the abalone, *Haliotis sorenseni*, and the effect of temperature on larval development and settling success. *Fish. Bull.*, **70**(2), 373–81.

Leighton D.L. (1974) The influence of temperature on larval and juvenile growth in three species of southern California abalones. *Fish. Bull.*, **72**(4), 1137–45.

Leighton D.L. (1977) Some problems and advances in culture of North American abalones (*Haliotis*). Paper presented to the First Symposium in the Latin American Aquaculture Association, Maracay, Venezuela, 5–12 November 1977.

McMullen J. and Thompson T. (1989) Abalone culture in an urban environment. In *Handbook of Culture of Abalone and other Marine Gastropods* (Ed. by K.O. Hahn), pp. 227–38. CRC Press, Boca Raton.

Morse D.E., Duncan H., Hooker N. & Morse A. (1977) Hydrogen peroxide induces spawning in molluscs, with activation of prostaglandin endoperoxide synthetase. *Science*, **196**, 298–300.

Morse D.E., Hooker N., Jensen L. & Duncan H. (1979) Induction of larval abalone settling and metamorphosis by γ-aminobutyric acid and its congeners from crustose red algae. II: Applications to cultivation, seed-production and bioassay; principal causes of mortality and interference. *Proc. World Maricul. Soc.*, **10**, 81–91.

North W.J. & Hubbs C.L. (1968) Utilization of kelp bed resources in southern California. *Calif. Dept Fish Game, Fish Bull.*, 139.

Chapter 45
The New Zealand culture system for abalone

L.J. TONG and G.A. MOSS *Fisheries Research Centre, PO Box 297, Wellington, New Zealand*

ABSTRACT: The hatchery production of larvae and juveniles of the New Zealand abalone (*Haliotis iris*) is reported. Techniques are described for broodstock management, controlled spawning using hydrogen peroxide and for the intensive rearing of larvae. Competence to settle is measured using the number of rows of chitinized teeth on the larval radula. Larvae with 10–12 rows settle faster and have a higher post-settlement survival than those with fewer rows.

Settlement and juvenile rearing to 10–15 mm takes place in a V-shaped tank system which provides a high surface area to volume ratio and requires a low labour input. Surface texture and adequate preparation of the surface with bacteria and diatoms are essential for ensuring high settlement rates and post-settlement survival.

Juveniles are fed on a variety of brown and red seaweeds. The best growth rates have been achieved with cultured *Gracilaria* sp.

Techniques for transporting larvae and juveniles are described.

RESUMEN: Se presenta la producción de larvas y juveniles en criaderos de abulón de Nueva Zelandia (*Haliotis iris*). Se describen las técnicas para el manejo de reproductores, la inducción al desove con peróxido de hidrógeno, y la cría intensiva de larvas. Se midió la competencia en la fijación mediante el número de hileras de dientes queratinizados en la radula de las larvas. Las larvas de 10–12 hileras se fijan mas rápido y presentan mayor sobrevivencia después de la fijación que las larvas con menor número de hileras. La fijación y crecimiento de juveniles, hasta los 10–15 mm tiene lugar en un sistema de tanques en form de V, el cual presenta una gran superficie del tanque, asi como su adecuada preparación con crecimento de bacterias y diatomeas, son esenciales para asegurar altas tasas de fijación y de la sobrevivencia posterior. Los juveniles fueron alimentados con varias especies de algas marinas, cafés y rojas. Las mejores tasas de crecimiento se lograron con *Gracilaria* sp. Se describen técnicas para el transporte de larvas y juveniles.

INTRODUCTION

Studies on the wild stocks of the New Zealand abalone, *Haliotis iris* Martyn, locally-named paua, show that recruitment to the fishery is generally low and, in some years, non-existent (Poore, 1973; Sainsbury, 1982; Murray & Ackroyd, 1984). In 1980 a project was started by the New Zealand Fisheries Research Division to determine whether paua could be produced for reseeding and enhancement trials. Studies on breeding, larval rearing, settlement and juvenile production of the paua were carried out at the Mahanga Bay Hatchery, Wellington.

BROODSTOCK MANAGEMENT

Paua spawn in late summer and early autumn (Poore, 1973), but the exact time may vary considerably from year to year and locality to locality. Within a wild population of paua some animals can be spawned at almost any time of the year. For example, from 1981–9, paua sampled from the Wellington south coast were

Table 45.1 The spawning success of broodstock held in spa-pool tanks.

	Females	Males
Number stimulated	80	79
Number failing to spawn	48 (60%)	41 (52%)
Number spawned	32 (40%)	38 (48%)

spawned from March to December inclusive. However, this source of broodstock is not reliable because in some years spawnings will only occur over a two- or three-month period.

To hold paua in breeding condition the broodstock are normally collected in early autumn prior to natural spawning. On capture they are checked for damage, sexed and tagged so that the history of individual animals can be recorded. Up to 150 gravid animals are held at ambient temperature in each of two 1000 litre tanks. The tanks have a whirlpool, water-jet circulation system ('spa pool'). The water is recirculated through a reservoir tank to which fresh seawater is added at a rate that changes the water in the system five times a day. The spa pool has a false bottom of 7 mm mesh. The paua stay attached to the sides of the tank and rarely move on to the mesh, through which faecal matter and fragmented seaweed pass. Once a week the tank is cleaned by opening the bottom valve and flushing out waste products.

The broodstock are kept at low light levels on a 12/12 photoperiod and are fed liberally on locally-abundant brown seaweeds such as *Macrocystis pyrifera*, *Lessonia variegata* and *Durvillaea antarctica*. The seaweed is held adjacent to the sides of the tank with an inner sleeve of 70 mm mesh. The spa pool maintains the seaweed in good condition and at the same time continuously exposes the paua to flowing, well-oxygenated water.

Gravid animals from the wild have been held in this system for up to 18 months before being successfully spawned, but the normal procedure is to replace the stock each year. A 40–50% spawning success has been obtained with paua held in this manner between March and November (Table 45.1).

SPAWNING AND LARVAL REARING

Spawning

To ensure an adequate supply of fertilized eggs, a minimum of eight female and six male paua are selected for stimulation at each spawning. The method used for the induction of spawning is similar to that described by Morse and co-workers (Tong, 1982). Pairs of paua of the same sex are placed in 12-litre tanks (Fig. 45.1). These sit in a water bath which allows the temperature to be controlled. Normally all spawning and larval rearing is carried out at 15°C using 1 micron-filtered water.

Sodium hydroxide (13 ml of a 1 M solution) and hydrogen peroxide (40 ml of a freshly made 6% solution) are added to each tank of seawater and the paua left in this solution for 3 hours. The tanks are then flushed with clean seawater for about

Fig. 45.1 Spawning tanks (foreground) and washing unit with sieves.

15 minutes before being covered and left undisturbed. Between 1 and 2 hours after flushing, spawning should begin, although some animals start later. Spawning may continue for up to 4 hours. Failure to spawn can often result from a poor or old batch of hydrogen peroxide (30% solution) and our procedure is to obtain freshly manufactured chemical every three months and to keep it at 4°C.

Fertilization

The eggs are siphoned off the bottom of the spawning tanks, through a 450-micron mesh sieve (to remove faecal matter), into an 80-micron mesh sieve sitting inside a 20-litre bucket. Up to 400 000 eggs (estimated) are collected into each bucket sieve and washed for 15 minutes.

Sperm from all spawning males is collected and checked for motility. A measured volume is added to the eggs to give a final concentration of about 100 000 sperm per ml.

The eggs and sperm are left to fertilize for 15 minutes, with occasional stirring. The sieves are gently lifted from the buckets and transferred to a washing unit for a further 15 minutes. To facilitate the handling of a large number of eggs a spray unit sits over each sieve, which in turn sits in a shallow trough of running water (Fig. 45.1). Up to 10 sieves can be handled at one time.

The gametes are fertilized as soon as possible after spawning but some delay can be tolerated. Studies on *H. virginea* showed that the viability of sperm and eggs was 4 hours and 6 hours respectively, but optimal fertilization occurred within 2 hours of spawning (McCardle, 1984).

The fertilized eggs are evenly spread over the bottom of the hatching tank (Fig. 45.2) in a layer one egg thick. A gentle flow of seawater, that does not

Transport of larvae

Intensive rearing of larvae has enabled trials on larval reseeding to be implemented (Tong *et al.*, 1987) and commercial farms to be supplied from the hatchery.

The system developed for transporting paua larvae in New Zealand is similar to that used overseas for oyster larvae (Burnell, 1985). A batch of 250 000 larvae, which are ready to settle, are sieved on to a square of 80-micron nylon mesh. The mesh is then folded and wrapped in moist paper towels. These are cooled to 4°C before being packed into polystyrene boxes for shipment. Trials have shown that after 24 hours out of water, 86% of larvae recover on being re-suspended.

SETTLEMENT

Criteria for settlement

Seki & Kan-no (1981) determined the competence of abalone larvae to settle by observing larval behaviour and by counting the number of buds on the cephalic tentacles. Four buds were discovered to be a good indicator of the larvae's readiness to settle.

In New Zealand we use the degree of development of the radula as the criterion for competence to settle. Between 30 and 40 larvae are decalcified with hydrochloric acid, dehydrated, cleared with potassium hydroxide and then squashed on to a microscope slide with a coverslip. Any radulae with more than three rows of chitinized teeth are easily seen. At 7 days, the radula has five or six rows of chitinized teeth which equate to four buds on the cephalic tentacle.

Larvae with three or four rows of chitinized teeth can settle, and most with five or six rows will settle. We have found that the initial rate of settlement and post-settlement survival is increased by retarding settlement until the larvae have 10 to 12 rows (Table 45.2).

Settlement tanks

Settlement and on-growing trials were first carried out in 800-litre, circular, flat-bottomed tanks with a water flow of 4 litres per minute. A central column aeration system was used to increase water movement. This system achieved between 1% and 7% survival of animals from pre-settlement to produce 1000–4000, 10-mm juveniles per tank (Illingworth, 1986). The disadvantages of these tanks were the low settlement surface to volume ratio and the flat bottom on which few postlarvae survived.

The circular tank has been superseded by the 1.2 m long, 900-litre capacity, V-shaped tank (Fig. 45.3). This has the advantage of no horizontal bottom and a large surface area, which is increased by the addition of vertical plates. Aeration, to create water movement, is provided through a series of holes in a pipe running the length of the tank at the base of the V. Water flows through the tank at 5 litres per minute. A floating collar of bubble plastic, attached to the tank at the water surface, prevents postlarvae and juveniles from crawling out.

Table 45.2 Settlement and post-settlement survival, on a *Lithothamnion* surface, for larvae of different ages (indicated by the number of rows of chitinized teeth in the radula).

Number of rows of chitinized teeth in the radula	Percentage settled at 2 h	Percentage settled at 6 h	Percentage surviving at 5 days
1–2	NT	15	1
3–4	66	77	29
5–6	65	82	39
7–8	71	81	36
9–10	89	93	48
11–12	85	96	65

NT = not tested.

V-shaped tanks tested to date have produced an average of 10000, 10–15 mm animals in 6–8 months. These tanks all had acrylic plates. Recent experiments using coated fibre-cement plates, with different surface textures, have resulted in an increased production of juveniles.

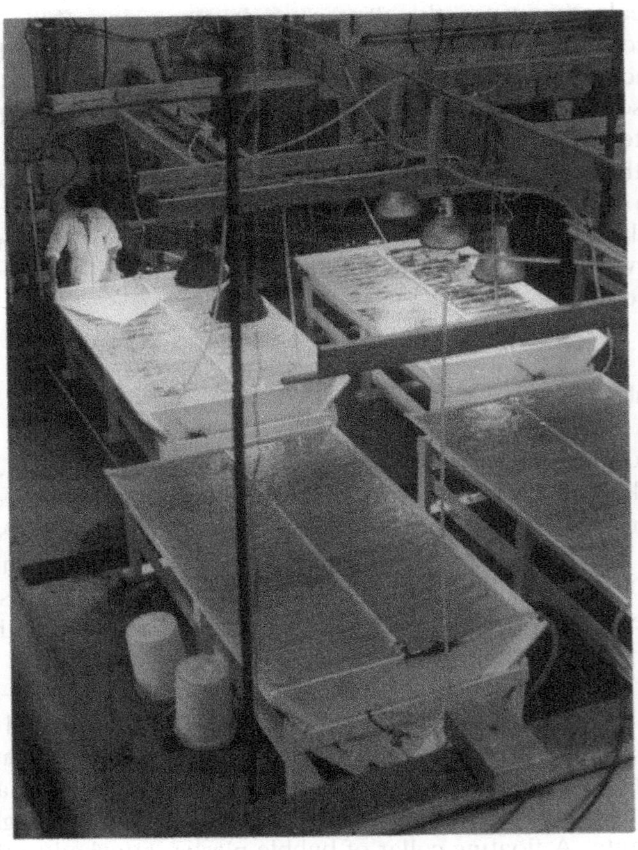

Fig. 45.3 Prototype V-shaped tanks.

Settlement procedure (for V-shaped tanks)

The V-shaped tank is cleaned with fresh water prior to being filled with 5-micron filtered seawater. A flow of water through the tank for 3 days allows a primary bacterial layer to build up on the plates and tank walls, which assists in inducing settlement. In experiments where this stage has been omitted or shortened the larvae are more reluctant to settle, and may remain swimming for 2–3 days after being introduced to the tank.

Suitable benthic diatom cultures, such as *Navicula* and *Nitzschia*, are cultured in the hatchery. A concentration of diatom cells is resuspended in a bucket and then added to the tanks to give the plates and walls a light 'dusting'. The plates can be placed on an angle to enhance diatom coverage.

After a further 24 hours, 200 000 paua larvae are added to the tank and aeration ensures that they are evenly distributed. The plates are left on an angle to improve settlement. The water flow is maintained and an 80-micron banjo sieve over the outlet prevents the larvae from escaping. Settlement usually occurs within 24 hours.

Abalone larvae can be induced to settle on almost any surface using gamma-aminobutyric acid (GABA), as described by Morse *et al.* (1979). Our experience has shown that a properly prepared surface with suitable food will result in a consistent settlement with or without the use of GABA. Most larvae will die within 24 hours if the surface is unsuitable.

ON GROWING

Post-settlement

To maintain optimum densities of benthic diatoms, light levels must be controlled. Outside tanks occasionally require shading to slow diatom growth and prevent smothering of the postlarvae. Conversely, tanks held inside require artificial lighting to promote sufficient diatom growth. The manipulation of light levels is essential for maintaining the right balance between postlarval growth and survival and available food. Food levels can also be enhanced by further dosing of the tanks with cultured diatoms. If heavy settlement of paua has occurred in a tank the plates can be shifted with ease to another tank to reduce the numbers of paua per tank to a manageable density.

Juveniles

On reaching 3–5 mm the juvenile paua are weaned on to macroalgae in the V-shaped tank. Common seaweeds used are the brown algae *L. variegata* and *M. pyrifera*, and a range of red algae including *Gracilaria sordida* and *Pterocladia lucida*. Juvenile paua fed cultured *Gracilaria* can grow up to 126 microns/day (Tong, 1983), but growth rate is dependent on temperature and the growth history of the seaweed (Pickering, 1986).

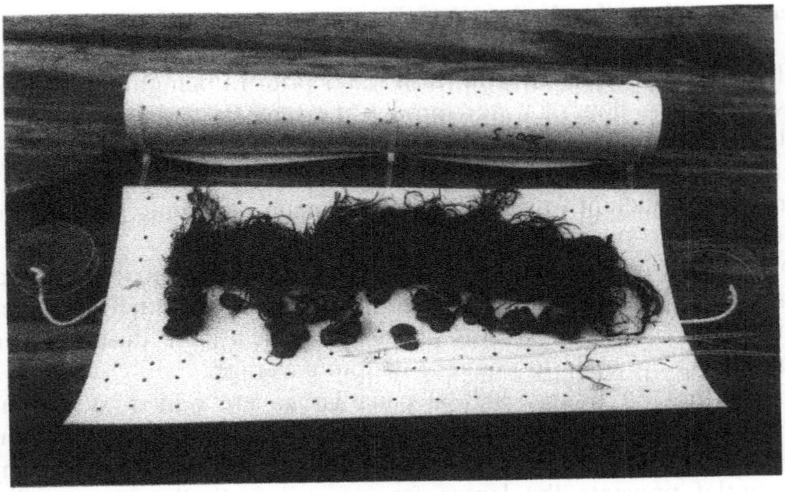

Fig. 45.4 Juvenile transport tubes.

Juveniles are best removed from the V-shaped tank and plates at 10–15 mm size (6–9 months old) using benzocaine, a water jet or with warm water (5–10°C warmer than ambient).

Transport of juveniles

Juveniles are transported from the hatchery to farms and reseeding sites in tubes (350 mm long and 75 mm in diameter) made from perforated, 1.6-mm styrene sheeting (Fig. 45.4). These tubes are secured using electrical ties and the ends are sealed with plastic petri dish lids. The juveniles are placed in the tubes together with moist seaweed.

The paua can then be held in running seawater both prior to shipping and at the point of destination. For packing, the tubes are drained and placed in polystyrene boxes with seaweed and ice packs. By this technique the paua are not crushed or smothered and are easily removed at their destination; the plastic ties are snipped, the tube unrolled and the animals simply brushed off the smooth sheet. The flat sheet with attached petri dish lids can then be reused.

Farming

Two commercial land-based paua farms have been established in New Zealand during 1987 and 1988. Their juvenile production is based on the V-shaped tank system (Henriques *et al.*, 1989). On growing to a marketable size (50–75 mm) is being carried out in a variety of experimental systems using drift and harvested seaweed as the food supply. Other paua farms are being planned, some in conjunction with *Gracilaria* farms.

Considerable interest has been shown by other commercial groups in a sea-based, barrel-culture system. Early trials indicated the feasibility of farming paua by this method (Hollings, 1988) and a new pump-action barrel system has been developed and awaits commercial trials.

ACKNOWLEDGEMENTS

The authors are grateful to their colleagues at the Mahanga Bay Shellfish Hatchery for their contribution to the ideas and techniques described and to Mr Bob Hickman for refereeing the chapter.

REFERENCES

Burnell G. (1985) Eyed larvae. *Aquaculture. Ireland*, **19**, 9–10.

Henriques P.R., Brown R.E. & Brown P.D. (1989) Trial commercial aquaculture of black-footed paua (*Haliotis iris*). In *Proceedings of AQUANZ '88, a National Conference on Aquaculture* (Ed. by M.F. Beardsell). *NZ Fish. Occ. Pub.*, No. 4, 53–5.

Hollings T. (1988) Paua barrel culture. *NZ FIB*, file number 10/06/2.

Illingworth J. (1986) Paua culture: systems and design. *Catch 86*, **13**(6–7), 4–5.

McCardle I.A. (1984) *Gametes from New Zealand Haliotidae and consideration of some factors affecting their viability*. MSc thesis, Victoria University of Wellington.

Morse D.E., Hooker N., Duncan H. & Jensen L. (1979) Gamma-aminobutyric acid, a neurotransmitter, induces planktonic abalone larvae to settle and begin metamorphosis. *Science*, **204**, 407.

Murray T.E. & Ackroyd J. (1984) The New Zealand paua fishery. *Unpub. NZ Fish. Res. Div. Int. Rept*, No. 5.

Pickering T.D. (1986) *Gracilaria* – a food for cultured paua? *Catch 86*, **13**(6–7), 7.

Poore G.C.B. (1973) Ecology of New Zealand abalones, *Haliotis* species (Mollusca:Gastropoda). IV. Reproduction. *NZ J. Mar. Freshwat. Res.*, **7**, 67–84.

Sainsbury K.J. (1982) Population and fishery management of the paua *Haliotis iris*. I. Population structure, growth, reproduction and mortality. *NZ J. Mar. Freshwat. Res.*, **16**, 147–61.

Seki T. & Kan-no H. (1981) Observations of the settlement and metamorphosis of the veliger of the Japanese abalone, *Haliotis discus hannai* Ino, Haliotidae, Gastropoda. *Bull. Tohoku Reg. Fish. Res. Lab.*, **42**, 31–9.

Tong L.J. (1982) The potential for aquaculture of paua in New Zealand. In *Proceedings of the Paua Fisheries Workshop* (Ed. by J.M. Ackroyd, T.E. Murray & J.L. Taylor). *NZ Fish. Res. Div. Occ. Publ.*, No. 41, 36–40.

Tong L.J. (1983) Paua research shows progress. *Catch 83*, **10**(8), 18.

Tong L.J., Moss G.A. & Illingworth J. (1987) Enhancement of a natural population of the abalone, *Haliotis iris*, using cultured larvae. *Aquaculture*, **62**, 67–72.

Wilson N. (1988) *Ecology and reproductive biology of* Haliotis iris *and* H. australis *in the Otago region*. MSc thesis, University of Otago.

Chapter 46
A review of abalone culture in China

ZONG QING NIE *Fuzhou Fisheries Research Institute, Gutian Road, Fuzhou, PR of China*

ABSTRACT: Seven species of abalone occur in China, but all except two are very small. The two largest species, *Haliotis discus hannai* and *H. diversicolor*, support a small fishery with an annual production of about 60 tonnes. The range, natural habitat and feeding of the two species are reviewed.

Abalone culture is developing in the north of China, but progress is hampered by the adverse environment. There are low winter sea temperatures in the north, wide intertidal zones and unsuitable sandy or muddy habitats in the central coasts and typhoons in the south. Stock enhancement and selective breeding are developing but are still experimental.

RESUMEN: Siete especies de abulón existen en China, pero todas, con excepción de dos, son muy pequeñas. Las dos especies más grandes, *Haliotis discus hannai* y *H. diversicolor*, sostienen una pequeña pesquería con una producción anual alrededor de 60 t. Se proporciona información sobre la distribución, habitat y alimentación. El cultivo del abulón se desarrolla en el norte y sur de China, pero el medio ambiente adverso impide su progreso. Hay bajas temperaturas invernales marinas en el norte, amplias zonas entre mareas e inapropiado habitat arenoso o fangoso en las costas centrales y tifones en el sur. El incremento de los stocks y la cría selectiva están siendo desarrolladas pero todavía son experimentales.

INTRODUCTION

China was one of the earliest countries in human history to use abalone. After capture by skindivers, the abalone were shelled, salted, cooked and dried in the sun, and then sold. Abalone meat has traditionally been a luxury food. The shell has been used in Chinese traditional medicine and is called 'Shi-Jue-Ming', which is thought to be beneficial for the eyesight and for the liver. The shell is also used for ornaments and works of art.

The production of abalone in mainland China is not great. The highest annual yield was around 100 tonnes (whole weight) in the 1950s, but has now declined to about 60 tonnes (excluding culture) due to overharvesting. The average size of capture has also declined.

Studies on the biology and culture of abalone in China are quite recent. In the late 1950s Zhang & Nie (1982) studied the ecology and reproduction of *H. discus hannai* in the islands of Changdao County (Fig. 46.1). The taxonomy of the Haliotidae was reviewed by Zhang & Qi (1962) and revised by Lu (1978). In the early 1970s, seed of *H. discus hannai* and *H. diversicolor* were first reared successfully in northern and southern China respectively. Seed of *H. discus hannai* were successfully transported to southern China and reared there (Chen *et al.*, 1977; Yang & You, 1975; Liu *et al.*, 1982, 1984). In the 1980s, cage culture (the long-line method) and culture in indoor tanks developed in Liaoning and Shangdong Provinces, and the techniques used for abalone culture rapidly improved (Du & Guo, 1981; Liu *et al.*, 1982, 1985; Nie, 1982; Nie & Chen, 1984; Nie *et al.*, 1984).

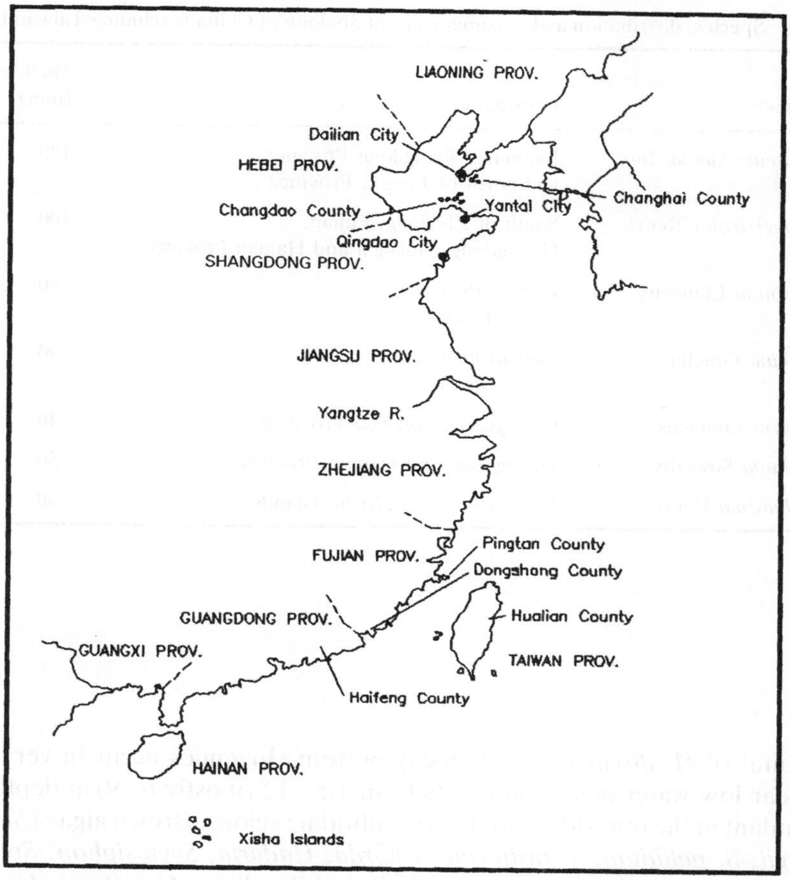

Fig. 46.1 Map of coastal provinces of China.

DISTRIBUTION AND BIOLOGY

Seven species of abalone (excluding *H. diversicolor supertexta* in Taiwan) are found in China (Table 46.1). They are all small in size, but the two largest, *H. discus hannai* in the north and *H. diversicolor* in the south, have the highest production.

Geographically, *H. discus hannai* occurs in Dailian City and its Changhai County (an archipelago county) in Liaoning Provice, and in Changdao County (an archipelago county) and Qingdao City and its counties in Shangdong Province. The major areas of *H. diversicolor* are Zhelang, Haifeng County and Naozhou Island, Zhangjiang City in Guangdong Province and Dongshang County (an island county) in Fujian Province (Zhang & Qi, 1962; Lu, 1978) (Fig. 46.1). Temperature and salinity ranges for the two largest species are given in Table 46.2.

Table 46.1 Species, distribution and maximum size of abalone in China (excluding Taiwan Province).

Species	Distribution	Shell length (mm)
H. discuss hannai Ino	Liaoning, Shangdong Province and north of Jiangsu Province	130
H. diversicolor Reeve	South of Zhejiang, Fujian, Guangdong, Guangxi and Hainan Provinces	100
H. asinina Linnaeus	Hainan Province Xusga Islands	80
H. ovina Gmelin	Hainan Province Xisha Islands	90
H. varia Linnaeus	Guangdong, Guangxi Province	40
H. planta Sowerby	Guangdong and Hainan Province	50
H. clathrata Reeve	Hainan Province, Xisha Islands	30

Habitat

The habitat of *H. discus hannai* is rocky bottom. Juveniles occur in very shallow water near low water mark and adults from 1.5–12 (mostly 6–9) m depth. Algae are abundant in the intertidal and shallow subtidal regions. Brown algae (*Sargassum thunbergii*, *S. pallidum*, *S. fusiforme*, *Chorda*, *Undaria*, *Scytosiphon*, *Sphacelaria* and *Punctaria*) are dominant, but red algae (*Corallina*, *Gracilaria*, *Gloiopeltis*, *Porphyra*, *Ceramium*, *Gelidium*, *Polysiphonia* and *Gymnogongrus*) and green algae (*Ulva*, *Codium*, *Monostroma* and *Enteromorpha*) are also common.

After April, the abalone migrate to shallow water where *S. thunbergii* and *S. pallidum* occur, and return to deeper water in winter (Zhang & Nie, 1982).

Food

The food eaten by *H. discus hannai* is known mainly from Huangchang Islands and Changdao County (Zhang & Nie, 1982). *Sargassum thunbergii* is the dominant food item, except in winter. Minor components are *Sphacelaria*, *Cladophora*, *Polysiphonia* and *Corallina*. Benthic diatoms also occur frequently in the gut, partly because diatoms are common epiphytes on preferred algae and partly because diatoms are the only food available when algae are scarce or absent.

Haliotis discus hannai were fed in cages under a long-line raft in the sea near Qingdao (Nie & Yan, 1985). Feeding peaks occurred in late spring to early summer and again in late autumn for adult abalone (Fig. 46.2). Declines in the amount of food eaten coincided with sea temperature maxima and minima and the spawning season (mid-June to early July). Feeding ceased altogether if the temperature fell below 3°C, and mortality increased.

Table 46.2 Data on salinity and temperature in habitats of commercial species (Nie, 1989)

		Month means (S = ‰)			Month mean (°C)		
Species	Site	Highest (month)	Lowest (month)	Year mean	Highest (month)	Lowest (month)	Year mean
H. discus hannai	Xiao Changshang Island, Changhai County, Liaoning Prov.	31.77 (March)	29.44 (June)	30.71	22.6 (Aug.)	0.0 (Feb.)	11.5
	Taiger Beach, Dalian City	31.86 (March)	30.07 (Aug.)	31.32	21.0 (Aug.)	1.5 (Feb.)	11.0
	Xiaomaidao, Qingdao City	31.74 (Jan.)	30.53 (Sep.)	31.15	24.5 (Aug.)	2.8 (Feb.)	13.6
H. diversicolor	Dongshang County, Fujian Prov.	32.96 (July)	31.08 (Jan.)	32.04	26.7 (Sep.)	13.9 (Feb.)	21.0
	Pingtan County, Fujian Prov.	34.03 (July)	30.31 (Dec.)	31.56	26.4 (Sep.)	11.8 (Feb.)	19.8
	Zhelang, Haifen County, Guangdong Prov.	33.81 (Apr.)	30.02 (July)	32.61	27.9 (Sep.)	15.4 (Feb.)	22.3
H. diversicolor supertexta	Hualian County Taiwan Prov.	34.50 (Feb.)	33.60 (Aug.)		27.8 (Aug.)	22.7 (Feb.)	

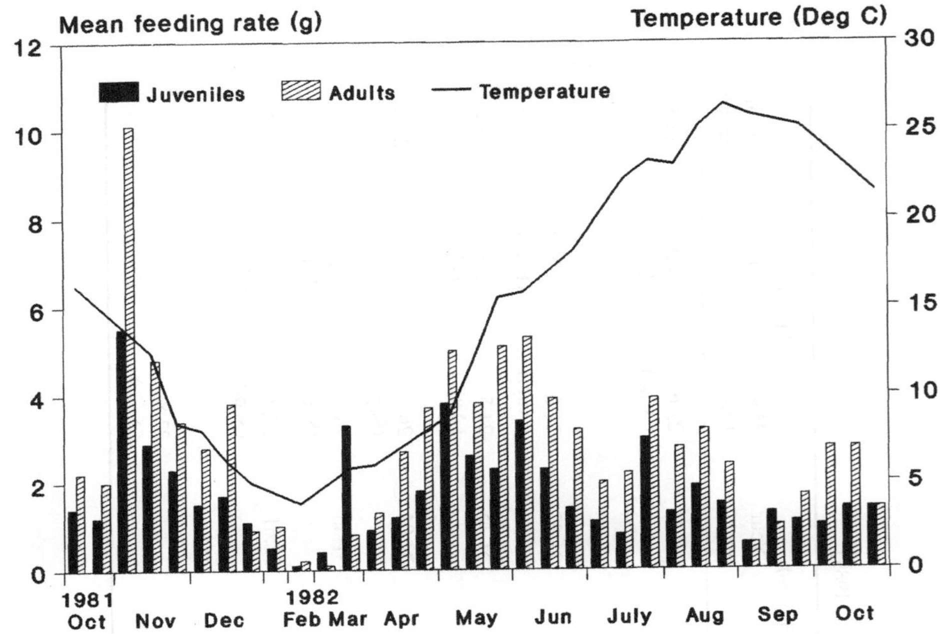

Fig. 46.2 Mean food consumption of juvenile and adult *H. discus hannai*, and seawater temperature per 10-day intervals from 20 October 1981 to 20 October 1982. Diet: *Laminaria japonica*, with *Ulva* sp. from October to December 1981. Juveniles 66–85 mm SL; adults 91–110 mm SL.

Preference trials by Nie & Yan (1985) showed that *Macrocystis pyrifera* (introduced from Mexico in 1978), *Undaria pinnatifida* and *Laminaria japonica*, followed by *Monostroma* and *Ulva*, were preferred, while *S. thunbergii*, *S. pallidum* and *Gelidium amansii* were least readily eaten. In general, the foliose and tender species were preferred. Food consumption was not constant on a daily basis, but fluctuated in a more or less regular way (Nie & Yan, 1985). Studies in Taiwan on *H. diversicolor supertexta* have given similar results (Jiang & Chun, 1974). Species of *Scytosiphonia* and *Monostroma* were preferred, followed by *Ulva* and *Enteromorpha*, with *Sargassum duplicatum*, *S. incanum* and *Gelidium japonicum* the least readily eaten. Again, tender algae seemed to be preferred.

Reproduction

The median size of sexual maturity (when 50% are mature) for *H. discus hannai* is from 46 to 50 mm shell length (SL) for males and at about 50 mm for females; at 60 mm SL all are mature. Zhang & Nie (1982) found that the sex ratio (males to females) was 1:1.15 ($N = 740$).

Both species have a brief annual spawning season (Table 46.3), which varies between provinces. Presumably the effective accumulated temperature (EAT) is reached earlier in some regions than in others (Nie, 1980; Liu *et al.*, 1985). During the spawning season, *H. discus hannai* aggregates on the highest parts of rocky reefs.

CULTURE

Originally, parental stock were taken from the sea during the spawning season and induced to spawn in the laboratory. However, the shortness of the season, the frequent poor weather and the difficulty in finding enough mature adults for large-scale production led to maturation of adults in culture using the technique of Kikuchi & Uki (1974a).

Parent stock are maintained in polyethylene cages at a density of 3–5 kg m^{-2} in aquaria which are aerated continuously with daily turnover of the seawater. Water temperatures are allowed to increase to 20°C and then kept at that level. Because *H. discus hannai* is adapted to lower winter temperatures in China than in Japan, it is believed that the maximum temperature and EAT for gonad maturation are also lower. Experience in China has shown that when the EAT reaches 800°C days for males and 1000°C days for females (and assuming 7.6°C is the biological zero at which no maturation occurs), animals are ready to spawn. In practice, temperatures begin to increase in February and spawning is induced in early May to June.

The present practice for spawning induction is to use seawater irradiated with ultra violet light (Kikuchi & Uki, 1974b). The usual dose is 500–800 mWh per litre. Animals are first exposed to the air for an hour and then immersed in irradiated seawater at a slightly elevated temperature.

After the larvae settle on corrugated plastic plates, their density is controlled at 0.10–0.12 cm^{-2}. At 3–5 mm SL they are anaesthetized in 3% alcohol seawater solution, removed from the transparent plates and transferred to dark plates in rearing tanks where they receive artificial food. By late November or early December they are 10–15 mm long. Production rates are 3000–7000 individuals per square metre of rearing tank (Nie, 1989).

Low seawater temperatures are a serious problem. In the north of China, sea temperature minima are usually <1°C, at which most juveniles <10 mm long die. For this reason animals must be reared to over 15 mm SL before the onset of winter, otherwise survival is low or there are increased costs for heating the water (Nie, 1989). Hence, it is important to use artificial food to increase the growth rate and survival (Wagatsuma & Nakamura, 1977; Nie *et al.*, 1986).

At 1 mm SL, creeping larvae are fed powdered algae such as *L. japonica*, *U. pinnatifida*, *S. thunbergii* and *M. pyrifera*, because benthic diatoms growing on the corrugated plates cannot meet their requirements. With these supplements the mean growth rate is 0.06–0.08 mm per day. Thereafter, powdered formulated foods are fed to juveniles of 3–7 mm SL, and granules to larger ones (Nie *et al.*, 1984).

Artificial foods are more efficient than some natural foods and are now widely used. For example, the mean growth rate of experimental animals 13–16 mm SL was 0.14 mm per day on a diet of formulated food, 1.78 times the growth rate of controls fed on *L. japonica* (Nie *et al.*, 1986). In other experiments in cages suspended from long lines, animals 23–24 mm SL grew 0.18 mm per day, which was 2.33 times faster than controls fed on *L. japonica* (Nie *et al.*, 1986).

The juveniles are transferred after a year to cages on rafts, and grown in them for 2 to 3 years (Fig. 46.3). The long-line raft is the most popular method in China

CULTURE

Originally, parental stock were taken from the sea during the spawning season and induced to spawn in the laboratory. However, the shortness of the season, the frequent poor weather and the difficulty in finding enough mature adults for large-scale production led to maturation of adults in culture using the technique of Kikuchi & Uki (1974a).

Parent stock are maintained in polyethylene cages at a density of 3–5 kg m^{-3} in aquaria which are aerated continuously with daily turnover of the seawater. Water temperatures are allowed to increase to 20°C and then kept at that level. Because *H. discus hannai* is adapted to lower winter temperatures in China than in Japan, it is believed that the maximum temperature and EAT for gonad maturation are also lower. Experience in China has shown that when the EAT reaches 800°C days for males and 1000°C days for females (and assuming 7.6°C is the biological zero at which no maturation occurs), animals are ready to spawn. In practice, temperatures begin to increase in February and spawning is induced in early May to June.

The present practice for spawning induction is to use seawater irradiated with ultra violet light (Kikuchi & Uki, 1974b). The usual dose is 500–800 mWh per litre. Animals are first exposed to the air for an hour and then immersed in irradiated seawater at a slightly elevated temperature.

After the larvae settle on corrugated plastic plates, their density is controlled at 0.10–0.12 cm^{-2}. At 3–5 mm SL they are anaesthetized in 3% alcohol seawater solution, removed from the transparent plates and transferred to dark plates in rearing tanks where they receive artificial food. By late November or early December they are 10–15 mm long. Production rates are 3000–7000 individuals per square metre of rearing tank (Nie, 1989).

Low seawater temperatures are a serious problem. In the north of China, sea temperature minima are usually <1°C, at which most juveniles <10 mm long die. For this reason animals must be reared to over 15 mm SL before the onset of winter, otherwise survival is low or there are increased costs for heating the water (Nie, 1989). Hence, it is important to use artificial food to increase the growth rate and survival (Wagatsuma & Nakamura, 1977; Nie *et al.*, 1986).

At 1 mm SL, creeping larvae are fed powdered algae such as *L. japonica*, *U. pinnatifida*, *S. thunbergii* and *M. pyrifera*, because benthic diatoms growing on the corrugated plates cannot meet their requirements. With these supplements the mean growth rate is 0.06–0.08 mm per day. Thereafter, powdered formulated foods are fed to juveniles of 3–7 mm SL, and granules to larger ones (Nie *et al.*, 1984).

Artificial foods are more efficient than some natural foods and are now widely used. For example, the mean growth rate of experimental animals 13–16 mm SL was 0.14 mm per day on a diet of formulated food; 1.78 times the growth rate of controls fed on *L. japonica* (Nie *et al.*, 1986). In other experiments in cages suspended from long lines, animals 23–24 mm SL grew 0.18 mm per day, which was 2.39 times faster than controls fed on *L. japonica* (Nie *et al.*, 1986).

The juveniles are transferred after a year to cages on rafts, and grown in them for 2 to 3 years (Fig. 46.3). The long-line raft is the most popular method in China

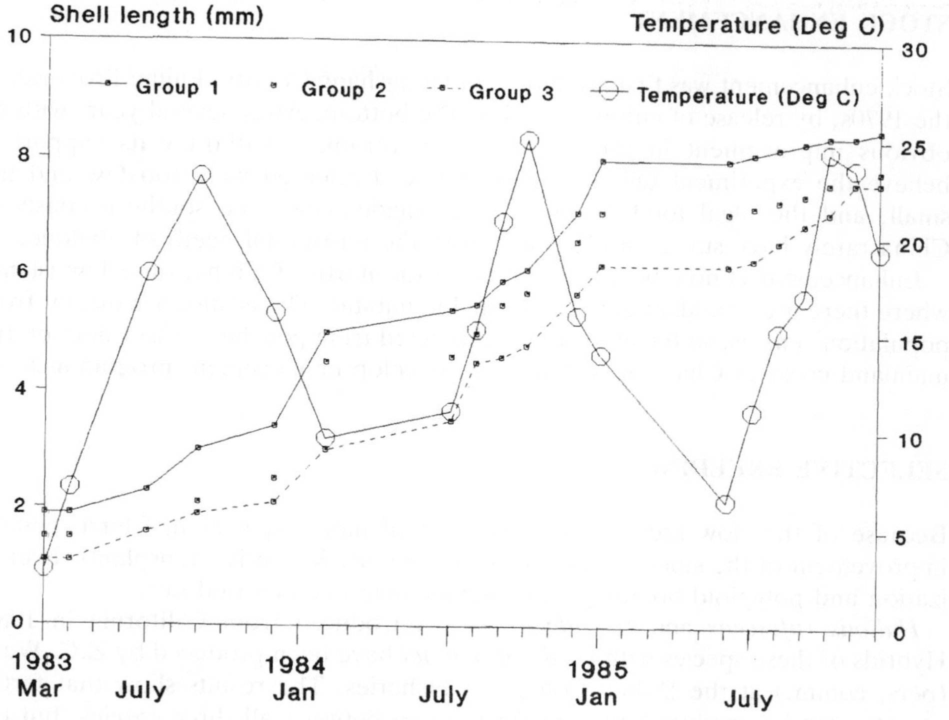

Fig. 46.3 Mean growth curves for three groups of juvenile *H. discus hannai* of different sizes in Dailian Sea from 1 March 1983 to 10 October 1985. Sea-surface temperatures are shown for the same period.

for seaweed and mollusc culture. A standard raft consists of a rope 60 m in length and 40 floats 25 cm in diameter. Two types of cage are used. One is a polyethylene tube 60 cm in length and 25 cm in diameter, with the ends covered by a mesh whose diameter varies according to the size of the abalone. The other is a vertical cylinder made of net 40 cm in diameter, with four tiers 80 cm apart. On the floor of each tier is a dark PVC plate. Survival over the whole period is 80–85%. Each long-line raft with 144 tubes yields about 450 kg (11 500 individuals). One worker can manage four rafts to produce 1800 kg abalone (Nie, 1989). The marketed size is 5 cm SL (to Hong Kong) to 7.5 cm SL (to Japan).

The main food used is the laminarian, *L. japonica*, which is cultured extensively in the north and south of China and is readily available. The efficiency of conversion (total weight fed divided by the increase in weight of the abalone) is 20 or more (Nie, 1989).

The limiting factor for production in the north of China is the long period of low sea temperatures in winter. The temperatures are <7°C for 4 months from mid-December to late April. In this period, abalone eat little food and their weight decreases (Nie, 1989). They stop feeding altogether at temperatures below 3 or 4°C.

STOCK ENHANCEMENT

Stock enhancement was first attempted in Dongshang County, Fujian Province, in the 1970s, by release of cultured seed on the bottom. After several years with no obvious improvement in the stocks, the government withdrew its support. I believe the experiment failed because the seed released were too few and too small, and the algal food resources were inadequate. The southern coasts of China rarely have sufficient algae to satisfy the nutritional needs of abalone.

Enhancement is now being attempted in the north of China, on a few islands where there are abundant algae and suitable habitat. The islands are distant from population centres, so the abalone are protected from poachers. Elsewhere on the mainland coast of China, it is difficult to develop enhancement programmes.

SELECTIVE BREEDING

Because of the slow growth and small size of native species in China, genetic improvement of the stocks would be advantageous. Recently, transplants, hybridization and polyploid breeding experiments have been carried out.

Haliotis rufescens and *H. fulgens* were introduced from California in 1985. Hybrids of these species with *H. discus hannai* have been produced by Z.C. Wang (pers. comm.) at the Dalian College of Fisheries. The results show that fertile gametes can be produced by cross-fertilization between all three species, but the highest fertilization rate is between *H. rufescens* and *H. discus hannai*. The first cross of male *H. rufescens* and female *H. discus hannai* showed the fastest growth rate, and proved better adapted to the cold water than the other crosses. These experiments established the feasibility of hybrid introductions on the north coast (Nie, 1989).

The Japanese black abalone *H. discus* was transported from Nawasaki to Fuzhou in late 1986 (unpublished data), and showed faster growth than in its native habitat. The transplants matured and spawned in the autumn after transplantation.

Z.C. Wang (pers. comm.) has carried out polyploid breeding and produced a few triploid juveniles of *H. discus hannai* by the application of cytochalasin B. However, the survival rate was low. Selective breeding continues but progress is slow because of the relatively long life cycle of abalone.

CONCLUSIONS

In recent decades the demand for abalone has exceeded the supply, the price has increased rapidly and greater attention has been given to abalone culture. The future of this industry will be assured once the various problems of culturing abalone are overcome. These are the low winter sea temperatures in northern China, the wide intertidal, sandy or muddy bottoms on the central coasts of China (Jiangsu and Zhejiang Provinces) and typhoons in the south. The culture of abalone in indoor tanks or coastal pools in the south, however, shows much promise.

ACKNOWLEDGEMENTS

The author thanks Dr S.A. Shepherd, for his English translation of the chapter, and Drs R.W. Day and P. Hone for criticism of the manuscript.

REFERENCES (in Chinese unless stated otherwise)

Chen H.-C. & Yang H.-H. (1979) Artificial reproduction of abalone, *Haliotis diversicolor supertexta*. *China Fisheries Monthly* (The Province of Taiwan), **314**, 3–9.

Chen M., Lu H.-K., Chen S.-J., Xie K.-E., Chen B.-N. & Laboratory of Abalone and Pearl-Oyster Culture, Dongshan, Fujian (1977) A preliminary study on the artificial rearing of the larvae and juveniles of *Haliotis discus hannai* Ino. *Acta Zoologica Sinica*, **23**(1), 35–45 (English abstr.).

Du E.-G. & Guo D.-L. (1981) Preliminary study on the artificial breeding of *Haliotis discus hannai* Ino. *J. Mar. Sci.*, **3**, 25–7.

Fujian (1976) (The abalone experiment group of the Institute of Oceanology, Academia Sinica and the Dongshang Laboratory of Abalone and Pearl-Oyster culture, Fujian Province). A preliminary study on rearing the seed of the abalone, *Haliotis diversicolor* Reeve. *Studia Marina Sinica*, **11**, 185–200.

Jiang Y.-M. & Chun F.-L. (1974) Studies on the preference for feed of abalone. *Haliotis diversicolor supertexta* Lischke. *China Fisheries Monthly* (The Province of Taiwan), **284**, 6–7.

Kikuchi S. & Uki N. (1974a) Technical study on artificial spawning of abalone genus Haliotis I. Relation between water temperature and advancing sexual maturity of *Haliotis discus hannai* Ino. *Bull. Tohoku Reg. Fish. Res. Lab.*, **33**, 69–78 (in Japanese).

Kikuchi S. & Uki N. (1974b) Technical study on artificial spawning of abalone genus Haliotis. II. Effect of irradiated sea water with ultraviolet rays on inducing spawning. *Bull. Tohoku Reg. Fish. Res. Lab.*, **33**, 79–86 (in Japanese).

Liang X.-Y. (1959) Anatomy of abalone, *Haliotis discus hannai* Ino. *Bull. Biol.*, **2**, 62–7.

Liu Y.-X., Liu Y.-F. & Gao X.-S. (1982) Initial research on promoted gonadal maturity of abalone (*Haliotis discus hannai* Ino) in indoor tank. *Fish. Sci.*, **1**, 10–13.

Liu Y.-F., Liu Y.-X. & Gao X.-S. (1984) Initial research on the artificial prepared feed for juvenile abalone *Haliotis discus hannai* Ino. *Fish. Sci.*, **3**, 14–19.

Liu Y.-F., Liu Y.-X., Sui X.-L. & Gao X.-S. (1985) A study on reproduction of disk abalone (*Haliotis discus hannai* Ino) in Dailian Coast. *J. Fish. China*, **9**(4), 311–20 (English abstr.).

Lu D.-H. (1978) A study on the Haliotidae from the coast of China. *Studia Marina Sinica*, **14**, 89–100.

Nie Z.-Q. (1980) Studies on abalone biology. *Foreign Fisheries*, **3**, 7–11.

Nie Z.-Q. (1982) Advances in studies on abalone culture and enhancement. *Bull. Oceanol. Limnol.*, **3**, 67–75.

Nie Z.-Q. (1989) *Abalone Culture and Enhancement*. China Agriculture Press, Beijing.

Nie Z.-Q. & Chen W.-H. (1984) Studies on rearing conditions of abalone, *Haliotis discus hannai* Ino. II. The effects of salinity and inorganic nutrients on the development of fertilized eggs and living of larvae. *Mar. Fish. Res.*, **6**, 41–8 (English abstr.)

Nie Z.-Q., Chen W.-H. & Ji M.-F. (1984) Studies on rearing conditions of abalone, *Haliotis discus hannai* Ino. I. The effects of temperature and food on the growth of larvae and young. *Mar. Fish. Res.*, **6**, 35–40 (English abstr.).

Nie Z.-Q., Wang P., Wang Z.-Q. & Yan J.-P. (1986) Experiments on preparing of formulated feed and feeding efficiency of young abalone, *Haliotis discus hannai* Ino. *Mar. Fish. Res.*, **7**, 53–64 (English abstr.).

Nie Z.-Q. & Yan J.-P. (1985) Studies on food habit of adult abalone, *Haliotis discus hannai* Ino. *J. Fish. China*, **9**(1), 19–27 (English abstr.).

Wagatsuma T. & Nakamura H. (1977) Production of abalone seed and artificial food. *Culture*, **9**, 69–77 (in Japanese).

Yang R.-Q. & You J.-H. (1975) Preliminary study on artificial breeding of *Haliotis diversicolor* Reeve. *Chinese J. Zool.*, **1**, 9–12.

Yang R.-Q., You Y.-H. & Cai T.-L. (1985) Preliminary study on the multiplication and cultivation of abalone, *Haliotis diversicolor* Reeve. *J. Fish. China*, **9**(1), 79–85.
Zhang L.-Y. & Nie Z.-Q. (1982) Preliminary studies on biological characteristic of abalone, *Haliotis discus hannai* Ino. *Bull. Mar. Fish. Res.*, **28**, 61–8.
Zhang X. & Qi Z.-Y. (1962) China Economic Fauna Mollusc. *China Science Press*, 6–12.

Index

Books published by **Fishing News Books**

Free catalogue available on request from Fishing News Books, Blackwell Science, Osney Mead, Oxford OX2 0EL, England

Abalone farming
Abalone of the world
Advances in fish science and technology
Aquaculture and the environment
Aquaculture & water resource management
Aquaculture development – progress and prospects
Aquaculture: principles and practices
Aquaculture in Taiwan
Aquaculture systems
Aquaculture training manual
Aquatic ecology
Aquatic microbiology
Aquatic weed control
Atlantic salmon: its future
The Atlantic salmon: natural history etc.
Bacterial diseases of fish
Better angling with simple science
Bioeconomic analysis of fisheries
British freshwater fishes
Broodstock management and egg and larval quality
Business management in fisheries and aquaculture
Cage aquaculture
Calculations for fishing gear designs
Carp farming
Carp and pond fish culture
Catch effort sampling strategies
Commercial fishing methods
Common fisheries policy
Control of fish quality
Crab and lobster fishing
The crayfish
Crustacean farming
Culture of bivalve molluscs
Design of small fishing vessels
Developments in electric fishing
Developments in fisheries research in Scotland
Dynamics of marine ecosystems
Ecology of fresh waters
The economics of salmon aquaculture
The edible crab and its fishery in British waters
Eel culture
Engineering, economics and fisheries management
The European fishing handbook 1993–94
FAO catalogue of fishing gear designs
FAO catalogue of small scale fishing gear
Fibre ropes for fishing gear
Fish catching methods of the world
Fisheries biology, assessment and management
Fisheries oceanography and ecology
Fisheries of Australia
Fisheries sonar
Fishermen's handbook
Fisherman's workbook
Fishery development experiences
Fishery products and processing
Fishing and stock fluctuations
Fishing boats and their equipment
Fishing boats of the world 1
Fishing boats of the world 2
Fishing boats of the world 3
Fishing ports and markets
Fishing with electricity
Fishing with light
Freshwater fisheries management
Fundamentals of aquatic ecology
Glossary of UK fishing gear terms
Handbook of trout and salmon diseases
A history of marine fish culture in Europe and North America
How to make and set nets
The Icelandic fisheries
Inland aquaculture development handbook
Intensive fish farming
Introduction to fishery by-products
The law of aquaculture: the law relating to the farming of fish and shellfish in Great Britain
A living from lobsters
Longlining
Making and managing a trout lake
Managerial effectiveness in fisheries and aquaculture
Marine climate, weather and fisheries
Marine fish behaviour in capture and abundance estimation
Marine ecosystems behaviour & management
Marketing: a practical guide for fish farmers
Marketing in fisheries and aquaculture
Mending of fishing nets
Modern deep sea trawling gear
More Scottish fishing craft and their work
Multilingual dictionary of fish and fish products
Multilingual dictionary of fishing vessels/safety on board
Multilingual dictionary of fishing gear
Multilingual illustrated dictionary of aquatic animals & plants
Multilingual illustrated guide to the world's commercial coldwater fish
Multilingual illustrated guide to the world's commercial warmwater fish
Navigation primer for fishermen
Netting materials for fishing gear
Net work exercises
Ocean forum
Pair trawling and pair seining
Pelagic and semi-pelagic trawling gear
Pelagic fish: the resource and its exploitation
Penaeid shrimps — their biology and management
Planning of aquaculture development
Pollution and freshwater fish
Purse seining manual
Recent advances in aquaculture IV
Recent advances in aquaculture V
Refrigeration on fishing vessels
Rehabilitation of freshwater fisheries
The rivers handbook, volume 1
The rivers handbook, volume 2
Salmon and trout farming in Norway
Salmon aquaculture
Salmon farming handbook
Salmon in the sea/new enhancement strategies
Scallop and queen fisheries in the British Isles
Scallop farming
Seafood science and technology
Seine fishing
Shrimp capture and culture fisheries of the US
Spiny lobster management
Squid jigging from small boats
Stability and trim of fishing vessels and other small ships
The state of the marine environment
Stock assessment in inland fisheries
Study of the sea
Sublethal and chronic toxic effects of pollution on freshwater fish
Textbook of fish culture
Trends in fish utilization
Trends in ichthyology
Trout farming handbook
Tuna fishing with pole and line

Printed and bound in the UK by
CPI Antony Rowe, Eastbourne

Printed and bound by CPI Group (UK) Ltd, Croydon, CR0 4YY
16/09/2026
14963385-0007